CONTROL THEORY

IEEE Press
445 Hoes Lane, P.O. Box 1331
Piscataway, NJ 08855-1331

IEEE Control Systems Society, *Sponsor*
CSS Liaison to IEEE Press, Bruce M. Krogh

Cover design: William T. Donnelly, *WT Design*

CONTROL THEORY

Twenty-Five Seminal Papers (1932–1981)

Edited by

Tamer Başar
University of Illinois at Urbana-Champaign

IEEE Control Systems Society, *Sponsor*

A Selected Reprint Volume

IEEE PRESS

The Institute of Electrical and Electronics Engineers, Inc., New York

This book and other books may be purchased at a discount
from the publisher when ordered in bulk quantities. Contact:

IEEE Press Marketing
Attn: Special Sales
445 Hoes Lane, P.O. Box 1331
Piscataway, NJ 08855-1331
Fax: +1 732 981 9334

For more information on IEEE Press products, visit the
IEEE Online Catalog and Store: http://www.ieee.org/store.

Printed in the United States of America.

10 9 8 7 6 5 4 3 2 1

ISBN 0-7803-6021-4

IEEE Order No. PC5870

Library of Congress Cataloging-in-Publication Data

Control theory : twenty-five seminal papers (1931–1981) / edited by Tamer Başar.
p. cm.
"IEEE Control Systems Society, sponsor."
"A selected reprint volume."
ISBN 0-7803-6021-4
1. Automatic control. 2. Control theory. I. Başar, Tamer. II. IEEE Control Systems Society.

TJ213.7.C66 2000
629.8—dc21

00-058171
CIP

Contents

Preface

CONTROL is in one sense a fairly young discipline. Even though it would be possible to push its historical origins back by about two millennia to the days of the Babylonians, in modern terms the real creation of the field has been in the twentieth century. It is in this century that CONTROL became a scientific discipline, with an intellectual core shaped by revolutionary ideas, novel concepts, and a wealth of analytical and computational tools. As a young and intellectually stimulating discipline, it attracted some of the brightest minds to its ranks and, with its theory driven by real applications, it provided versatile tools for generations of practicing engineers.

Prior to the twentieth century, there were of course also several key contributions. Perhaps the first time the study of control systems attracted serious scientific attention was toward the end of the eighteenth century, following James Watts's invention of the governor in 1788, which was designed to regulate the speed of the rotary steam engine. A related work on governors by Huygens[1] actually predated that of Watts by about a century; Huygens invented the centrifugal governor as a means of regulating a clock, which was adapted for windmills and water wheels in the Netherlands as early as the seventeenth century. *Feedback* played an important role in all these inventions, and soon it was realized and widely acknowledged that it is a concept that lies at the foundation of any successful control design. However, to make effective use of feedback, there was a need for a careful mathematical study of its impact on control design. James Clerk Maxwell was the first to realize this need and to respond to it by developing in his now famous paper,[2] considered by many to be the starting point of the scientific approach to control research, mathematical models for various governor mechanisms based on linear differential equations. He worked out in his paper a complete theory of stability for constant coefficient linear differential equations up to fourth order, and obtained some conditions for stability of fifth-order systems. Around the same time, and independently of Maxwell, a Russian engineer, Vyshnegradskii, had recognized the importance of control in industrial applications, and the need for developing a sound theory.[3] Where Maxwell left off was then picked up by Edward John Routh,[4] and independently by Adolf Hurwitz,[5] who came up with what is known today as the *Routh-Hurwitz stability criteria*, solving completely the problem of stability of constant coefficient linear differential equations of any finite order. At about the same time, and as the nineteenth century was coming to a close, another trend-setting development took place, again in the area of stability, but this time for nonlinear dynamical systems. Motivated by problems that arise in astronomy in connection with the motion of the planets, a topic studied earlier by Henry Poincaré,[6] among others, Aleksandr Mikhailovich Lyapunov developed in his doctoral thesis in Russia a new approach for testing the stability of the equilibrium of a system described by nonlinear ordinary differential equations, known today as the *Second Method of Lyapunov*.

Hence, there was quite a bit of accumulated activity in control at the beginning of the twentieth century. But what this century, and particularly its second half, delivered was something different in terms of both content and sheer volume of diverse contributions. The incessant growth caused by an explosion of new fresh ideas, and driven by numerous applications from different domains, brought this activity to unprecedented levels. As we are coming to the close of this century, we thought that it would be useful to reflect back and ask the questions: What have been the major results of this century in control? What have been the greatest hits in control? How has control theory evolved since the times of Maxwell, Routh, Hurwitz, and Lyapunov (among others)? There is of course no unique way of answering all these questions, but one possible way is to collect under one cover,

[1] C. Huygens, "Horologii oscillatorii," Part 5, Paris, 1673.

[2] J. C. Maxwell, "On governors," *Proc. Royal Soc. London*, **16**, 1868, pp. 270–283.

[3] J. Vyshnegradskii, "Sur la théorie générale des régulateurs (On the general theory of control)," *Compt. Rend. Acad. Sci. Paris*, **83**, 1876, pp. 318–321.

[4] E. J. Routh, *A Treatise on the Stability of a Given State of Motion*, Macmillan (London), 1877.

[5] A. Hurwitz, "Über the Bedingungen unter welschen eine Gleichung nur Wurzeln mit negativen reelen Teilen besitzt (On the conditions under which an equation has only roots with negative real parts)," *Mathematische Annalen*, **46**, 1895, pp. 273–284.

[6] H. Poincaré, *Les Méthodes Nouvelles de la Méchanique Céléste* (The New Methods of the Cellestial Mechanics), Vol. 1, Gauthier-Villars (Paris), 1892.

and as a representative of the major research developments and accomplishments in control in this century, some key papers that have made major impact on the field, along with some introductory material for each.

These considerations have led to the present volume, which contains twenty-five carefully selected seminal papers covering the period 1932–1981, and begins with Harry Nyquist's famous "Regeneration Theory" paper, which introduced the so-called Nyquist criterion (still a versatile tool for control engineers) and laid the foundations of a frequency-domain approach to stability analysis of linear control systems. The volume ends with the 1981 paper by George Zames, which marks the beginning of the "robustness" era in control—an era that we are leaving to the coming generations to evaluate (perhaps by the middle of the twenty-first century), along with other exciting developments the control field has experienced for the past two decades, and will undoubtedly continue to do so in the next century.

This volume was prepared under the auspices of the IEEE Control Systems Society, by an Editorial Board consisting of twelve members, namely:

Brian D.O. Anderson, Karl J. Åström, John Baillieul, Tamer Başar (Chair), Bruce A. Francis, Alberto Isidori, Petar V. Kokotović, Huibert Kwakernaak, William J. Levine, Lennart Ljung, David Q. Mayne, and *Jan C. Willems.*

Based on nominations received in response to solicitations that appeared in the IEEE Control Systems Magazine, and in the E-LETTER on Systems, Control, and Signal Processing (based in the Netherlands), and nominations generated by the Board members, and after several rounds of voting, the Board unanimously agreed on the selection of the twenty-five papers included in this volume. The preambles to each paper were written by the Board members (in some cases jointly), with the primary author(s) in each case identified by his (their) initials.

The twenty-five papers included in this volume cover a broad spectrum of major developments in control theory in the twentieth century, but still the volume should not be viewed as providing an exhaustive coverage of all topical areas of control, as this has not been a criterion set by the Editorial Board in their selection of the papers. The focus here has been the **papers** selected rather than the areas of research in control. Still, we believe that the selected papers clearly outline the *path* which the control discipline has followed during its rapid growth from the 1930s through the 1980s. To help the reader in this *journey*, in the preambles to individual papers we have discussed developments in areas neighboring the topic of a particular paper, so as to place its contributions and impact in proper perspective, and to maintain continuity in the flow of ideas from one topic to another.

What is the *path* we can trace from the chronologically ordered papers in the volume? First comes the basic feedback theory, as represented by the first three papers, by Nyquist, Black, and Bode, which was the outcome of research conducted at the Bell Laboratories in the 1930s driven by the need to develop electronic feedback amplifiers for long telephone lines. To this was added later, as a practical tool, the root locus method of Evans. In the 1940s, Wiener's work on prediction, filtering, and smoothing for time series, impacted control in many ways, fostering further developments not only in filtering theory but also in control design. The "Sputnik effect" and the ensuing space research propelled the development of a mathematically advanced optimal control theory in the late 1950s and early 1960s, with dynamic programming, maximum principle, and the LQ regulator design (with its associated concepts of controllability and observability) as its centerpieces, as represented in the works of Bellman, Pontryagin, and Kalman included in this volume. Also developed during this period were computational techniques to make dynamic programming and maximum principle practicable, as represented by the paper of Bryson and Denham. During the same period, efforts were intensified to develop an applicable stability theory of nonlinear feedback systems in the absolute stability framework, represented by the papers of Popov, Yakubovich, and Zames and, as extensions and refinements of Lyapunov concepts, in the papers by LaSalle and Willems. The need to operate in the presence of noise and other disturbances was also recognized in the early 1960s, as shown in papers by Kalman and Feldbaum. Then came the establishment of the precise relationship between input-output descriptions and state-space representations of linear systems, with its multifold ramifications, as presented in the 1963 paper by Kalman, and the establishment of a geometric theory for linear systems, in the paper by Wonham and Morse, with the introduction of the novel concept of controlled invariance, which found applications in much broader domains (than linear systems) later. We see in the 1970s the emergence of a nonlinear system theory, with associated richer concepts of controllability and observability, as shown in the papers by Brockett, Sussmann and Jurdjevic, and Hermann and Krener. Adaptive control is another area where a comprehensive theory started emerging in the 1970s, where the important notion of "self-tuning" was introduced (in the paper by Åström and Wittenmark), and methodologies were developed for establishing the convergence of adaptive control algorithms, as well as recursive stochastic algorithms, as presented in the papers by Goodwin, Ramadge and Caines, and Ljung. Robust control is yet another topic that gained steam in the late 1970s, with the 1981 paper by Zames (the last paper in this volume) marking the beginning of some intense activity in this domain. Of course, there were many other accomplishments in control during this period, than those represented by the twenty-five papers selected, and it is hoped that the preambles to the papers will convey to the reader this richness in the field.

I hope that the reader will enjoy this *journey*, and will develop a real sense of the evolution of the control field during the fifty-year period, 1932–1981, through the milestone accomplishments embodied in these twenty-five seminal papers and further discussed in the introductory material provided by our Editorial Board. It is our hope that the volume will be a valuable resource in the twenty-first century (and beyond) especially for young control researchers and engineers, and will be instrumental in the realization of an even more explosive century for control theory and its applications.

Tamer Başar
University of Illinois at Urbana-Champaign

Regeneration Theory

H. NYQUIST

By 1932 feedback systems were used extensively in applications such as power generation and transmission, steering of ships, autopilots for aircrafts, and process control; see [2] and [3]. Theoretical investigations typically consisted of stability analysis. This was done by linearizing the equations describing the systems and investigating the roots of the characteristic equation by the Routh-Hurwitz method. Nice examples of this approach are found in the early textbook on turbine control by Tolle [11] and in the analysis of an autopilot for ship steering by Minorsky [9]. A severe drawback with the approach was that the analysis gave no guidelines for modifying an unstable system to make it stable.

This paper by Nyquist, and the closely related papers by Black and Bode, which are the second and third papers in this volume, represent a paradigm shift because they approached the problem of analyzing a feedback system in a totally different way. Black invented the electronic feedback amplifier, and Bode and Nyquist developed original and powerful theories for its analysis and design. Even though the work was strongly focused on feedback amplifiers, it became apparent several years later that the result could actually be applied to all control systems.

Black, Bode, and Nyquist all worked at Western Electric, a forerunner of Bell Laboratories. They were faced with a major challenge, namely to develop electronic feedback amplifiers for long telephone lines. Several telephone conversations were frequency multiplexed and sent over one line. The application required amplifiers that were linear with constant gain. The problem was aggravated because there were many amplifiers on the cables. The following quote from Bode in a paper [4] from 1960 illustrates the difficulties:

> "Most of you with hi-fi systems are no doubt proud of the quality of your audio amplifiers, but I doubt whether many of you would care to listen to the sound after the signal had gone in succession through several dozen or several hundred of your fine amplifiers."

Black's idea to introduce feedback made it feasible to build electronic amplifiers for the demanding telephone applications. Use of feedback did, however, create instabilities, which were called *singing*, because of the acoustic effects. The problem of understanding the mechanisms that created the instabilities and to avoid them was a key issue which was solved in Nyquist's paper. The problem of stability is intimately connected with feedback. Instabilities are frequently encountered whenever feedback is used. The problem of understanding the stability problem had earlier been a very strong motivation both for Maxwell [8], in connection with governors, and for Stodola [10], in connection with water turbines. Before the publication of Nyquist's paper, stability was investigated by studying the roots of the characteristic equation of the linearized equations of motion.

Nyquist departed drastically from previous work by developing frequency domain criteria for stability. Instead of looking at the roots of a characteristic equation, Nyquist introduced the quantity $AJ(j\omega)$ as follows:

> "Let the complex quantity $AJ(j\omega)$ represent the ratio by which the amplifier and feedback circuit modify the current in one round trip."

The quantity $-AJ(s)$ is what is today called the *loop transfer function* of a feedback system and $AJ(j\omega)$ represents the steady-state transmissions of sinusoids around the loop. Because of the negative sign, the critical point in Nyquist's paper is $+1$ instead of -1 as is commonly used today. It seemed intuitively reasonable that an oscillation could be maintained at a frequency ω_0 if $AJ(j\omega_0) = 1$. Intuitively one would also expect that the system would be stable if $|AJ(j\omega_0)| < 1$ for the frequency ω_0 where the argument of $AJ(j\omega_0)$ is zero. This means that in steady state the signal is attenuated after one trip around the feedback loop. It had been observed experimentally, however, that the system could be stable even if $|AJ(j\omega_0)| > 1$. A related observation is that some unstable amplifiers could be stabilized by increasing the feedback gain. This puzzling phenomenon was called *conditional stability*. Nyquist solved the problem and showed that a stability criterion could be developed by analyzing the graph of the function $AJ(j\omega)$ and in particular the encirclements of the critical point.

Nyquist's original analysis was fairly complicated, but his main result had a lasting value. It was later pointed out by MacColl [6] that the result can in fact be easily obtained from a theorem of Cauchy or from the *Argument Principle* of complex analysis [1]. This principle says that if N is the number of poles and P the number of zeros in a region D of a function $f(z)$, then

$$N - P = \frac{1}{2\pi i} \int_D \frac{f'(z)}{f(z)} dz = \frac{1}{2\pi} \Delta_R \arg f(z)$$

The immediate implication of Nyquist's theorem is very well described by his colleague Bode [4] in the following passage:

> "Although the broad implications of the negative feedback invention were rather quickly apparent, Black's original ideas required support and confirmation in a number of directions before engineering exploitation of the field could go forward expeditiously and with confidence. For example, the invention relied basically upon amplifiers with loop gains much in excess of unity. The fact that such structures could exist, without instability, seemed doubtful in 1927 to many people. Obviously, in order to go further, we needed a practical criterion which would tell the design engineer what properties he should seek to ensure a stable system. Fortunately, for the application of the negative feedback amplifier, the required solution was quickly forthcoming in a classic analysis by Nyquist."

The long-range impact of Nyquist's paper on the emerging field of automatic control was even more profound. Previous methods for stability analysis based on a study of the roots of the characteristic equation provided no guidance on how to modify an unstable system to make it stable. Nyquist's results changed the situation completely. Since the basic quantity *AJ* is the product of the transfer function of the plant and the controller, it is straightforward to see how the controller should be modified to avoid the critical point. Nyquist thus provided a stability criterion that could be the basis of a design method. The key idea was to shape the Nyquist curve to obtain the desired properties. Nyquist's results were thus an inspiration for many of the design methods that are called today *loop shaping*. Nyquist also realized the importance of the quantity *AJ* which is called today the *loop transfer function*.

Nyquist's theorem is an important ingredient in all introductory courses in automatic control and circuit theory. The theorem has been generalized to multivariable systems in [7]. There have also been many extensions to nonlinear systems. The small gain theorem, the passivity theorem, and the circle criterion in the 1966 paper by Zames in this collection is one set of generalizations, and the describing function ([12], [5]) is yet another.

The works of Nyquist, Bode, and Black are excellent examples of good control engineering. They were driven by a major engineering problem: to realize telephone conversations over long distances. A significant advantage was that the process and the controller were designed concurrently; they are inherently combined in the electronic amplifier. The research group at Bell Laboratories had a nice mix of inventors, engineers, and theoreticians, and there were ample resources at their disposal to make it possible for them to take a long-range view of the problems.

References

[1] L. V. Ahlfors, *Complex Analysis*, McGraw-Hill (New York), 1966.

[2] B. Bennet, *A History of Control Engineering 1800–1930*, Peter Peregrinus (London), 1979.

[3] B. Bennet, *A History of Control Engineering 1930–1955*, Peter Peregrinus (London), 1993.

[4] H. W. Bode, "Feedback—The history of an idea," In *Proceedings of the Symposium on Active Networks and Feedback Systems*, Polytechnic Institute of Brooklyn, New York, 1960.

[5] R. Kochenburger, "Frequency response method for analysis of a relay servomechanism," *Trans. AIEE*, **69**:270–283, 1950.

[6] L. A. MacColl, *Fundamental theory of servomechanisms*, Van Nostrand (Princeton, NJ), 1945.

[7] A. G. J. MacFarlane and I. Postlethwaite, "The generalized Nyquist stability criterion and multivariable root loci," *Internat. J. Contr.*, **25**:81–127, 1977.

[8] J. C. Maxwell, "On governors," *Proc. Royal Society of London*, **16**:270–283, 1868. Also published in "Mathematical Trends in Control Theory" edited by R. Bellman and R. Kalaba, Dover Publications, New York, 1964, pp. 3–17.

[9] N. Minorsky, "Directional stability of automatically steered bodies," *J. Amer. Soc. of Naval Engineers*, **34**(2):280–309, 1922.

[10] A. B. Stodola, "Über die Regulierung von Turbinen," *Schweitzer Bauzeitung*, **22**:113–117, 121–122, 126–128, 134–135, 1893.

[11] M. Tolle, *Die Regulung der Kraftmaschinen*, Springer (Berlin), 1905.

[12] A. Tustin, "The effects of backlash and of speed-dependent friction on the stability of closed-cycle control systems," *J. Inst. Elec. Engrs. Pt. IIA*, **94**:143–151, 1947.

K.J.Å

Regeneration Theory

By H. NYQUIST

Regeneration or feed-back is of considerable importance in many applications of vacuum tubes. The most obvious example is that of vacuum tube oscillators, where the feed-back is carried beyond the singing point. Another application is the 21-circuit test of balance, in which the current due to the unbalance between two impedances is fed back, the gain being increased until singing occurs. Still other applications are cases where portions of the output current of amplifiers are fed back to the input either unintentionally or by design. For the purpose of investigating the stability of such devices they may be looked on as amplifiers whose output is connected to the input through a transducer. This paper deals with the theory of stability of such systems.

Preliminary Discussion

WHEN the output of an amplifier is connected to the input through a transducer the resulting combination may be either stable or unstable. The circuit will be said to be stable when an impressed small disturbance, which itself dies out, results in a response which dies out. It will be said to be unstable when such a disturbance results in a response which goes on indefinitely, either staying at a relatively small value or increasing until it is limited by the non-linearity of the amplifier. When thus limited, the disturbance does not grow further. The net gain of the round trip circuit is then zero. Otherwise stated, the more the response increases the more does the non-linearity decrease the gain until at the point of operation the gain of the amplifier is just equal to the loss in the feed-back admittance. An oscillator under these conditions would ordinarily be called stable but it will simplify the present paper to use the definitions above and call it unstable. Now, this fact as to equality of gain and loss appears to be an accident connected with the non-linearity of the circuit and far from throwing light on the conditions for stability actually diverts attention from the essential facts. In the present discussion this difficulty will be avoided by the use of a strictly linear amplifier, which implies an amplifier of unlimited power carrying capacity. The attention will then be centered on whether an initial impulse dies out or results in a runaway condition. If a runaway condition takes place in such an amplifier, it follows that a non-linear amplifier having the same gain for small current and decreasing gain with increasing current will be unstable as well.

Reprinted with permission from *Bell System Technical Journal*, H. Nyquist, "Regeneration Theory," Vol. 11, 1932, pp. 126–147.

Steady-State Theories and Experience

First, a discussion will be made of certain steady-state theories; and reasons why they are unsatisfactory will be pointed out. The most obvious method may be referred to as the series treatment. Let the complex quantity $AJ(i\omega)$ represent the ratio by which the amplifier and feed-back circuit modify the current in one round trip, that is, let the magnitude of AJ represent the ratio numerically and let the angle of AJ represent the phase shift. It will be convenient to refer to AJ as an admittance, although it does not have the dimensions of the quantity usually so called. Let the current

$$I_0 = \cos \omega t = \text{real part of } e^{i\omega t} \tag{a}$$

be impressed on the circuit. The first round trip is then represented by

$$I_1 = \text{real part of } AJe^{i\omega t} \tag{b}$$

and the nth by

$$I_n = \text{real part of } A^n J^n e^{i\omega t}. \tag{c}$$

The total current of the original impressed current and the first n round trips is

$$I_n = \text{real part of } (1 + AJ + A^2J^2 + \cdots A^nJ^n)e^{i\omega t}. \tag{d}$$

If the expression in parentheses converges as n increases indefinitely, the conclusion is that the total current equals the limit of (d) as n increases indefinitely. Now

$$1 + AJ + \cdots A^nJ^n = \frac{1 - A^{n+1}J^{n+1}}{1 - AJ}. \tag{e}$$

If $|AJ| < 1$ this converges to $1/(1 - AJ)$ which leads to an answer which accords with experiment. When $|AJ| > 1$ an examination of the numerator in (e) shows that the expression does not converge but can be made as great as desired by taking n sufficiently large. The most obvious conclusion is that when $|AJ| > 1$ for some frequency there is a runaway condition. This disagrees with experiment, for instance, in the case where AJ is a negative quantity numerically greater than one. The next suggestion is to assume that somehow the expression $1/(1 - AJ)$ may be used instead of the limit of (e). This, however, in addition to being arbitrary, disagrees with experimental results in the case where AJ is positive and greater than 1, where the expression $1/(1 - AJ)$ leads to a finite current but where experiment indicates an unstable condition.

The fundamental difficulty with this method can be made apparent by considering the nature of the current expressed by (*a*) above. Does the expression cos ωt indicate a current which has been going on for all time or was the current zero up to a certain time and cos ωt thereafter? In the former case we introduce infinities into our expressions and make the equations invalid; in the latter case there will be transients or building-up processes whose importance may increase as n increases but which are tacitly neglected in equations (*b*) – (*e*). Briefly then, the difficulty with this method is that it neglects the building-up processes.

Another method is as follows: Let the voltage (or current) at any point be made up of two components

$$V = V_1 + V_2, \tag{f}$$

where V is the total voltage, V_1 is the part due directly to the impressed voltage, that is to say, without the feed-back, and V_2 is the component due to feed-back alone. We have

$$V_2 = AJV. \tag{g}$$

Eliminating V_2 between (*f*) and (*g*)

$$V = V_1/(1 - AJ). \tag{h}$$

This result agrees with experiment when $|AJ| < 1$ but does not generally agree when AJ is positive and greater than unity. The difficulty with this method is that it does not investigate whether or not a steady state exists. It simply assumes tacitly that a steady state exists and if so it gives the correct value. When a steady state does not exist this method yields no information, nor does it give any information as to whether or not a steady state exists, which is the important point.

The experimental facts do not appear to have been formulated precisely but appear to be well known to those working with these circuits. They may be stated loosely as follows: There is an unstable condition whenever there is at least one frequency for which AJ is positive and greater than unity. On the other hand, when AJ is negative it may be very much greater than unity and the condition is nevertheless stable. There are instances of $|AJ|$ being about 100 without the conditions being unstable. This, as will appear, accords closely with the rule deduced below.

Notation and Restrictions

The following notation will be used in connection with integrals:

$$\int_I \phi(z)dz = \lim_{M\to\infty} \int_{-iM}^{+iM} \phi(z)dz, \tag{1}$$

the path of integration being along the imaginary axis (see equation 9), i.e., the straight line joining $-iM$ and $+iM$;

$$\int_{s+} \phi(z)dz = \lim_{M\to\infty} \int_{-iM}^{iM} \phi(z)dz, \tag{2}$$

the path of integration being along a semicircle [1] having the origin for center and passing through the points $-iM$, M, iM;

$$\int_C \phi(z)dz = \lim_{M\to\infty} \int_{-iM}^{-iM} \phi(z)dz, \tag{3}$$

the path of integration being first along the semicircle referred to and then along a straight line from iM to $-iM$. Referring to Fig. 1 it

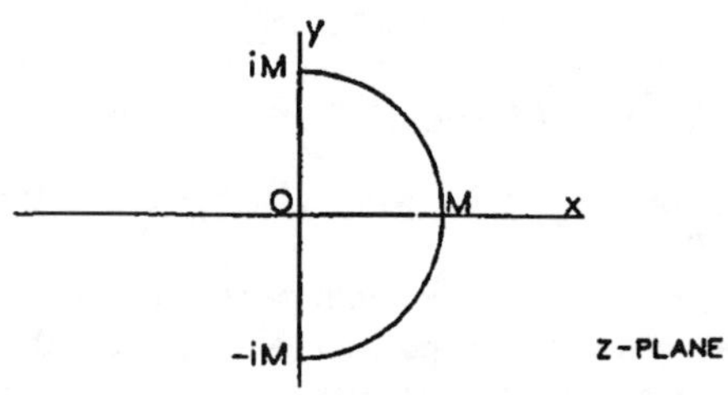

Fig. 1—Paths of integration in the z-plane.

will be seen that

$$\int_{s+} - \int_I = \int_C . \tag{4}$$

The total feed-back circuit is made up of an amplifier in tandem with a network. The amplifier is characterized by the amplifying ratio A which is independent of frequency. The network is characterized by the ratio $J(i\omega)$ which is a function of frequency but does not depend on the gain. The total effect of the amplifier and the network is to multiply the wave by the ratio $AJ(i\omega)$. An alternative way of characterizing the amplifier and network is to say that the amplifier is

[1] For physical interpretation of paths of integration for which $x > 0$ reference is made to a paper by J. R. Carson, "Notes on the Heaviside Operational Calculus," *B. S. T. J.*, Jan. 1930. For purposes of the present discussion the semicircle is preferable to the path there discussed.

characterized by the amplifying factor A which is independent of time, and the network by the real function $G(t)$ which is the response caused by a unit impulse applied at time $t = 0$. The combined effect of the amplifier and network is to convert a unit impulse to the function $AG(t)$. Both these characterizations will be used.

The restrictions which are imposed on the functions in order that the subsequent reasoning may be valid will now be stated. There is no restriction on A other than that it should be real and independent of time and frequency. In stating the restrictions on the network it is convenient to begin with the expression G. They are

$$G(t) \text{ has bounded variation}, \quad -\infty < t < \infty. \tag{AI}$$

$$G(t) = 0, \quad -\infty < t < 0. \tag{AII}$$

$$\int_{-\infty}^{\infty} |G(t)|\,dt \text{ exists.} \tag{AIII}$$

It may be shown [2] that under these conditions $G(t)$ may be expressed by the equation

$$G(t) = \frac{1}{2\pi i}\int_I J(i\omega)e^{i\omega t}d(i\omega), \tag{5}$$

where

$$J(i\omega) = \int_{-\infty}^{\infty} G(t)e^{-i\omega t}dt. \tag{6}$$

These expressions may be taken to define J. The function may, however, be obtained directly from computations or measurements; in the latter case the function is not defined for negative values of ω. It must be defined as follows to be consistent with the definition in (6):

$$J(-i\omega) = \text{complex conjugate of } J(i\omega). \tag{7}$$

While the final results will be expressed in terms of $AJ(i\omega)$ it will be convenient for the purpose of the intervening mathematics to define an auxiliary and closely related function

$$w(z) = \frac{1}{2\pi i}\int_I \frac{AJ(i\omega)}{i\omega - z}d(i\omega), \qquad 0 < x < \infty, \tag{8}$$

where

$$z = x + iy \tag{9}$$

and where x and y are real. Further, we shall define

$$w(iy) = \lim_{x \to 0} w(z). \tag{10}$$

[2] See Appendix II for fuller discussion.

The function will not be defined for $x < 0$ nor for $|z| = \infty$. As defined it is analytic[3] for $0 < x < \infty$ and at least continuous for $x = 0$.

The following restrictions on the network may be deduced:

$$\lim_{y \to \infty} y|J(iy)| \text{ exists.} \tag{BI}$$

$$J(iy) \text{ is continuous.} \tag{BII}$$

$$w(iy) = AJ(iy). \tag{BIII}$$

Equation (5) may now be written

$$AG(t) = \frac{1}{2\pi i}\int_I w(z)e^{zt}dz = \frac{1}{2\pi i}\int_{s+} w(z)e^{zt}dz. \tag{11}$$

From a physical standpoint these restrictions are not of consequence. Any network made up of positive resistances, conductances, inductances, and capacitances meets them. Restriction (AII) says that the response must not precede the cause and is obviously fulfilled physically. Restriction (AIII) is fulfilled if the response dies out at least exponentially, which is also assured. Restriction (AI) says that the transmission must fall off with frequency. Physically there are always enough distributed constants present to insure this. This effect will be illustrated in example 8 below. Every physical network falls off in transmission sooner or later and it is ample for our purposes if it begins to fall off, say, at optical frequencies. We may say then that the reasoning applies to all linear networks which occur in nature. It also applies to other linear networks which are not physically producible but which may be specified mathematically. See example 7 below.

A temporary wave $f_0(t)$ is to be introduced into the system and an investigation will be made of whether the resultant disturbance in the system dies out. It has associated with it a function $F(z)$ defined by

$$f_0(t) = \frac{1}{2\pi i}\int_I F(z)e^{zt}dz = \frac{1}{2\pi i}\int_{s+} F(z)e^{zt}dz. \tag{12}$$

$F(z)$ and $f_0(t)$ are to be made subject to the same restrictions as $w(z)$ and $G(t)$ respectively.

Derivation of a Series for the Total Current

Let the amplifier be linear and of infinite power-carrying capacity. Let the output be connected to the input in such a way that the

[3] W. F. Osgood, "Lehrbuch der Funktionentheorie," 5th ed., Kap. 7, § 1, Hauptsatz. For definition of "analytic" see Kap. 6, § 5.

amplification ratio for one round trip is equal to the complex quantity AJ, where A is a function of the gain only and J is a function of ω only, being defined for all values of frequency from 0 to ∞.

Let the disturbing wave $f_0(t)$ be applied anywhere in the circuit. We have

$$f_0(t) = \frac{1}{2\pi}\int_{-\infty}^{+\infty} F(i\omega)e^{i\omega t}d\omega \tag{13}$$

or

$$f_0(t) = \frac{1}{2\pi i}\int_{I+} F(z)e^{zt}dz. \tag{13'}$$

The wave traverses the circuit and on completing the first trip it becomes

$$f_1(t) = \frac{1}{2\pi}\int_{-\infty}^{\infty} w(i\omega)F(i\omega)e^{i\omega t}d\omega \tag{14}$$

$$= \frac{1}{2\pi i}\int_{I+} w(z)F(z)e^{zt}dz. \tag{14'}$$

After traversing the circuit a second time it becomes

$$f_2(t) = \frac{1}{2\pi i}\int_{I+} Fw^2e^{zt}dz, \tag{15}$$

and after traversing the circuit n times

$$f_n(t) = \frac{1}{2\pi i}\int_{I+} Fw^ne^{zt}dz. \tag{16}$$

Adding the voltage of the original impulse and the first n round trips we have a total of

$$s_n(t) = \sum_{k=0}^{n} f_k(t) = \frac{1}{2\pi i}\int_{I+} F(1 + w + \cdots w^n)e^{zt}dz. \tag{17}$$

The total voltage at the point in question at the time t is given by the limiting value which (17) approaches as n is increased indefinitely [4]

$$s(t) = \sum_{k=0}^{\infty} f_k(t) = \lim_{n\to\infty}\frac{1}{2\pi i}\int_{I+} S_n(z)e^{zt}dz, \tag{18}$$

where

$$S_n = F + Fw + Fw^2 + \cdots Fw^n = \frac{F(1 - w^{n+1})}{1 - w}. \tag{19}$$

[4] Mr. Carson has called my attention to the fact that this series can also be derived from Theorem IX, p. 49, of his Electric Circuit Theory. Whereas the present derivation is analogous to the theory expressed in equations (a)–(e) above, the alternative derivation would be analogous to that in equations (f)–(h).

Convergence of Series

We shall next prove that the limit $s(t)$ exists for all finite values of t. It may be stated as of incidental interest that the limit

$$\int_{I+} S_\infty(z)e^{izt}dz \tag{20}$$

does not necessarily exist although the limit $s(t)$ does. Choose M_0 and N such that

$$|f_0(\lambda)| \leq M_0. \qquad 0 \leq \lambda \leq t. \tag{21}$$

$$|G(t-\lambda)| \leq N. \quad 0 \leq \lambda \leq t. \tag{22}$$

We may write [5]

$$f_1(t) = \int_{-\infty}^{\infty} G(t-\lambda)f_0(\lambda)d\lambda. \tag{23}$$

$$|f_1(t)| \leq \int_0^t M_0 N d\lambda = M_0 N t. \tag{24}$$

$$f_2(t) = \int_{-\infty}^{\infty} G(t-\lambda)f_1(\lambda)d\lambda. \tag{25}$$

$$|f_2(t)| \leq \int_0^t M_0 N^2 t dt = M_0 N^2 t^2/2! \tag{26}$$

Similarly

$$|f_n(t)| \leq M_0 N^n t^n/n! \tag{27}$$

$$|s_n(t)| \leq M_0(1 + Nt + \cdots N^n t^n/n!). \tag{28}$$

It is shown in almost any text [6] dealing with the convergence of series that the series in parentheses converges to e^{Nt} as n increases indefinitely. Consequently, $s_n(t)$ converges absolutely as n increases indefinitely.

Relation Between $s(t)$ and w

Next consider what happens to $s(t)$ as t increases. As t increases indefinitely $s(t)$ may converge to zero, indicating a condition of stability, or it may go beyond any value however large, indicating a runaway condition. The question which presents itself is: *Referring to (18) and (19), what properties of $w(z)$ and further what properties of $AJ(i\omega)$ determine whether $s(t)$ converges to zero or diverges as t increases*

[5] G. A. Campbell, "Fourier Integral," *B. S. T. J.*, Oct. 1928, Pair 202.
[6] E.g., Whittaker and Watson, "Modern Analysis," 2d ed., p. 531.

indefinitely? From (18) and (19)

$$s(t) = \lim_{n \to \infty} \frac{1}{2\pi i} \int_{\rho+} F\left(\frac{1}{1-w} - \frac{w^{n+1}}{1-w}\right) e^{zt} dz. \tag{29}$$

We may write

$$s(t) = \frac{1}{2\pi i} \int_{\rho+} [F/(1-w)] e^{zt} dz - \lim_{n \to \infty} \frac{1}{2\pi i} \int_{\rho+} [Fw^{n+1}/(1-w)] e^{zt} dz \tag{30}$$

provided these functions exist. Let them be called $g_0(t)$ and $\lim_{n \to \infty} g_n(t)$ respectively. Then

$$g_n(t) = \int_{-\infty}^{\infty} g_0(t-\lambda) \phi(\lambda) d\lambda. \tag{31}$$

where

$$\phi(\lambda) = \frac{1}{2\pi i} \int_{\rho+} w^{n+1} e^{z\lambda} dz. \tag{32}$$

By the methods used under the discussion of convergence above it can then be shown that this expression exists and approaches zero as n increases indefinitely provided $g_0(t)$ exists and is equal to zero for $t < 0$. Equation (29) may therefore be written, subject to these conditions

$$s(t) = \frac{1}{2\pi i} \int_{\rho+} [F/(1-w)] e^{zt} dz. \tag{33}$$

In the first place the integral is zero for negative values of t because the integrand approaches zero faster than the path of integration increases. Moreover,

$$\int_I [F/(1-w)] e^{zt} dz \tag{34}$$

exists for all values of t and approaches zero for large values of t if $1 - w$ does not equal zero on the imaginary axis. Moreover, the integral

$$\int_C [F/(1-w)] e^{zt} dz \tag{35}$$

exists because

1. Since F and w are both analytic within the curve the integrand does not have any essential singularity there,
2. The poles, if any, lie within a finite distance of the origin because $w \to 0$ as $|z|$ increases, and
3. These two statements insure that the total number of poles is finite.

We shall next evaluate the integral for a very large value of t. It will suffice to take the C integral since the I integral approaches zero. Assume originally that $1 - w$ does not have a root on the imaginary axis and that $F(z)$ has the special value $w'(z)$. The integral may be written

$$\frac{1}{2\pi i}\int_C [w'/(1 - w)]e^{zt}dz. \tag{36}$$

Changing variables it becomes

$$\frac{1}{2\pi i}\int_D [1/(1 - w)]e^{zt}dw, \tag{37}$$

where z is a function of w and D is the curve in the w plane which corresponds to the curve C in the z plane. More specifically the imaginary axis becomes the locus $x = 0$ and the semicircle becomes a small curve which spirals around the origin. See Fig. 2. The function

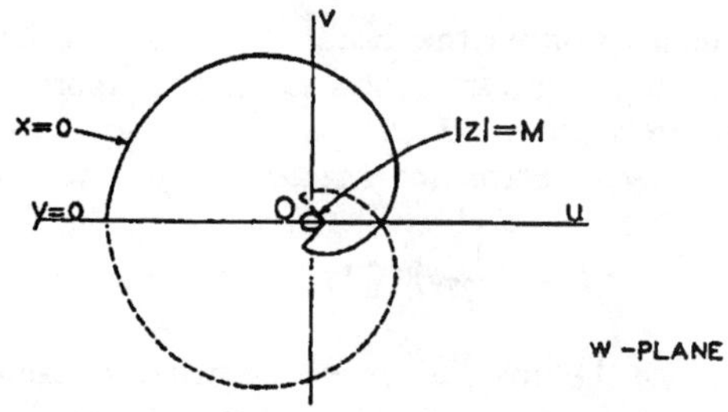

Fig. 2—Representative paths of integration in the w-plane corresponding to paths in Fig. 1.

z and, therefore, the integrand is, in general, multivalued and the curve of integration must be considered as carried out over the appropriate Riemann surface.[7]

Now let the path of integration shrink, taking care that it does not shrink across the pole at $w = 1$ and initially that it does not shrink across such branch points as interfere with its passage, if any. This shrinking does not alter the integral [8] because the integrand is analytic at all other points. At branch points which interfere with the passage of the path the branches stopped may be severed, transposed and connected in such a way that the shrinking may be continued past the branch point. This can be done without altering the value of the integral. Thus the curve can be shrunk until it becomes one or more very small circles surrounding the pole. The value of the total integral

[7] Osgood, loc. cit., Kap. 8.
[8] Osgood, loc. cit., Kap. 7, § 3, Satz 1.

(for very large values of t) is by the method of residues [9]

$$\sum_{j=1}^{n} r_j e^{z_j t}, \tag{38}$$

where z_j $(j = 1, 2 \cdots n)$ is a root of $1 - w = 0$ and r_j is its order. The real part of z_j is positive because the curve in Fig. 1 encloses points with $x > 0$ only. The system is therefore stable or unstable according to whether

$$\sum_{j=1}^{n} r_j$$

is equal to zero or not. But the latter expression is seen from the procedure just gone through to equal the number of times that the locus $x = 0$ encircles the point $w = 1$.

If F does not equal w' the calculation is somewhat longer but not essentially different. The integral then equals

$$\sum_{j=1}^{n} \frac{F(z_j)}{w(z_j)} e^{z_j t} \tag{39}$$

if all the roots of $1 - w = 0$ are distinct. If the roots are not distinct the expression becomes

$$\sum_{j=1}^{n} \sum_{k=1}^{r_j} A_{jk} t^{k-1} e^{z_j t}, \tag{40}$$

where A_{jr_j}, at least, is finite and different from zero for general values of F. It appears then that unless F is specially chosen the result is essentially the same as for $F = w'$. The circuit is stable if the point lies wholly outside the locus $x = 0$. It is unstable if the point is within the curve. It can also be shown that if the point is on the curve conditions are unstable. We may now enunciate the following

Rule: Plot plus and minus the imaginary part of $AJ(i\omega)$ against the real part for all frequencies from 0 to ∞. If the point $1 + i0$ lies completely outside this curve the system is stable; if not it is unstable.

In case of doubt as to whether a point is inside or outside the curve the following criterion may be used: Draw a line from the point $(u = 1, v = 0)$ to the point $z = -i\infty$. Keep one end of the line fixed at $(u = 1, v = 0)$ and let the other end describe the curve from $z = -i\infty$ to $z = i\infty$, these two points being the same in the w plane. If the net angle through which the line turns is zero the point $(u = 1, v = 0)$ is on the outside, otherwise it is on the inside.

If AJ be written $|AJ|(\cos\theta + i\sin\theta)$ and if the angle always

[9] Osgood, loc. cit., Kap. 7, § 11, Satz 1.

changes in the same direction with increasing ω, where ω is real, the rule can be stated as follows: The system is stable or unstable according to whether or not a real frequency exists for which the feed-back ratio is real and equal to or greater than unity.

In case $d\theta/d\omega$ changes sign we may have the case illustrated in Figs. 3 and 4. In these cases there are frequencies for which w is real and

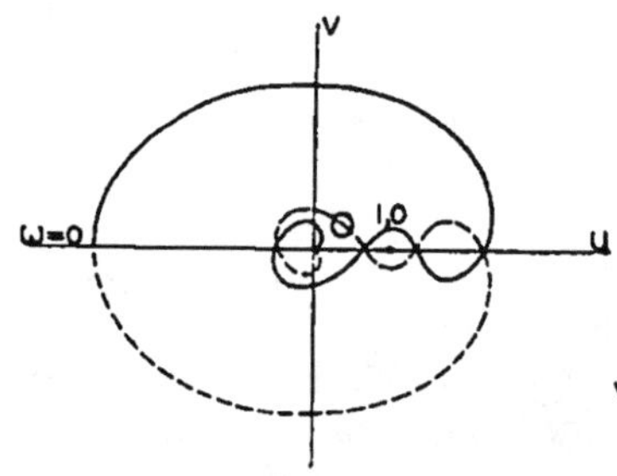

Fig. 3—Illustrating case where amplifying ratio is real and greater than unity for two frequencies, but where nevertheless the path of integration does not include the point 1, 0.

greater than 1. On the other hand, the point (1, 0) is outside of the locus $x = 0$ and, therefore, according to the rule there is a stable condition.

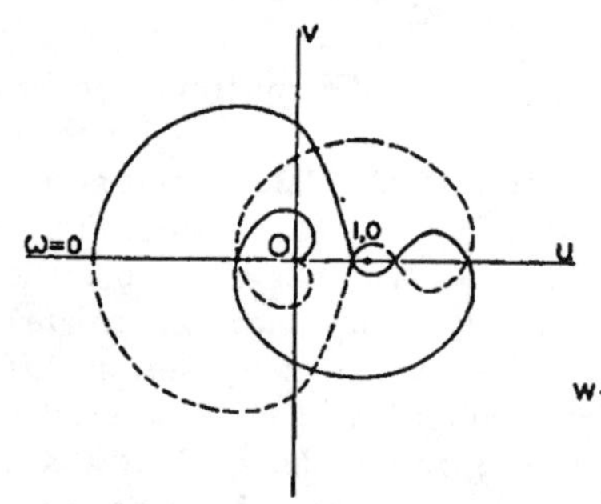

Fig. 4—Illustrating case where amplifying ratio is real and greater than unity for two frequencies, but where nevertheless the path of integration does not include the point 1, 0.

If networks of this type were used we should have the following interesting sequence of events: For low values of A the system is in a stable condition. Then as the gain is increased gradually, the system becomes unstable. Then as the gain is increased gradually still further, the system again becomes stable. As the gain is still further increased the system may again become unstable.

EXAMPLES

The following examples are intended to give a more detailed picture of certain rather simple special cases. They serve to illustrate the previous discussion. In all the cases F is taken equal to AJ so that f_0 is equal to AG. This simplifies the discussion but does not detract from the illustrative value.

1. Let the network be pure resistance except for the distortionless amplifier and a single bridged condenser, and let the amplifier be such that there is no reversal. We have

$$AJ(i\omega) = \frac{B}{\alpha + i\omega}, \tag{41}$$

where A and α are real positive constants. In (18) [10]

$$\begin{aligned} f_n &= \frac{1}{2\pi i}\int_I A^{n+1}J^{n+1}(i\omega)e^{i\omega t}di\omega \\ &= Be^{-\alpha t}(B^n t^n/n!). \end{aligned} \tag{42}$$

$$s(t) = Be^{-\alpha t}(1 + Bt + B^2t^2/2! + \cdots). \tag{43}$$

The successive terms f_0, f_1, etc., represent the impressed wave and the successive round trips. The whole series is the total current.

It is suggested that the reader should sketch the first few terms graphically for $B = \alpha$, and sketch the admittance diagrams for $B < \alpha$, and $B > \alpha$.

The expression in parentheses equals e^{Bt} and

$$s(t) = Be^{(B-\alpha)t}. \tag{44}$$

This expression will be seen to converge to 0 as t increases or fail to do so according to whether $B < \alpha$ or $B \geq \alpha$. This will be found to check the rule as applied to the admittance diagram.

2. Let the network be as in 1 except that the amplifier is so arranged that there is a reversal. Then

$$AJ(i\omega) = \frac{-B}{\alpha + i\omega}. \tag{45}$$

$$f_n = (-1)^{n+1}Be^{-\alpha t}(B^n t^n/n!). \tag{46}$$

The solution is the same as in 1 except that every other term in the series has its sign reversed:

$$\begin{aligned} s(t) &= -Be^{-\alpha t}(1 - Bt + B^2t^2/2! + \cdots) \\ &= -Be^{(-\alpha-B)t}. \end{aligned} \tag{47}$$

[10] Campbell. loc. cit. Pair 105.

This converges to 0 as t increases regardless of how great B may be taken. If the admittance diagram is drawn this is again found to check the rule.

3. Let the network be as in 1 except that there are two separated condensers bridged across resistance circuits. Then

$$AJ(i\omega) = \frac{B^2}{(\alpha + i\omega)^2} \cdot \tag{48}$$

The solution for $s(t)$ is obtained most simply by taking every other term in the series obtained in 1.

$$\begin{aligned} s(t) &= Be^{-\alpha t}(Bt + B^3t^3/3! + \cdots) \\ &= Be^{-\alpha t} \sinh Bt. \end{aligned} \tag{49}$$

4. Let the network be as in 3 except that there is a reversal. Then

$$AJ(i\omega) = \frac{-B^2}{(\alpha + i\omega)^2} \cdot \tag{50}$$

The solution is obtained most directly by reversing the sign of every other term in the series obtained in 3.

$$\begin{aligned} s(t) &= -Be^{-\alpha t}(Bt - B^3t^3/3! + \cdots) \\ &= -Be^{-\alpha t} \sin Bt. \end{aligned} \tag{51}$$

This is a most instructive example. An approximate diagram has been made in Fig. 5, which shows that as the gain is increased the

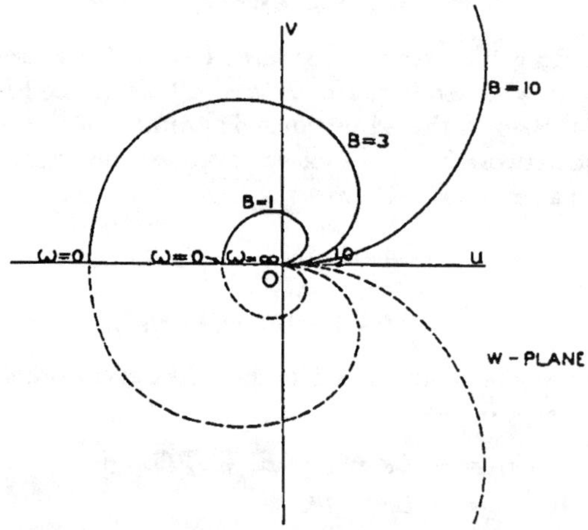

Fig. 5—Illustrating Example 4, with three values for B.

feed-back ratio may be made arbitrarily great and the angle arbitrarily small without the condition being unstable. This agrees with the expression just obtained, which shows that the only effect of increasing the gain is to increase the frequency of the resulting transient.

5. Let the conditions be as in 1 and 3 except for the fact that four separated condensers are used. Then

$$AJ(i\omega) = \frac{B^4}{(\alpha + i\omega)^4}. \tag{52}$$

The solution is most readily obtained by selecting every fourth term in the series obtained in 1.

$$\begin{aligned} s(t) &= Be^{-\alpha t}(B^3t^3/3! + B^7t^7/7! + \cdots) \\ &= \tfrac{1}{2}Be^{-\alpha t}(\sinh Bt - \sin Bt). \end{aligned} \tag{53}$$

This indicates a condition of instability when $B \geqq \alpha$, agreeing with the result deducible from the admittance diagram.

6. Let the conditions be as in 5 except that there is a reversal. Then

$$Y = \frac{-B^4}{(\alpha + i\omega)^4}. \tag{54}$$

The solution is most readily obtained by changing the sign of every other term in the series obtained in 5.

$$s(t) = Be^{-\alpha t}(-B^3t^3/3! + B^7t^7/7! - \cdots). \tag{55}$$

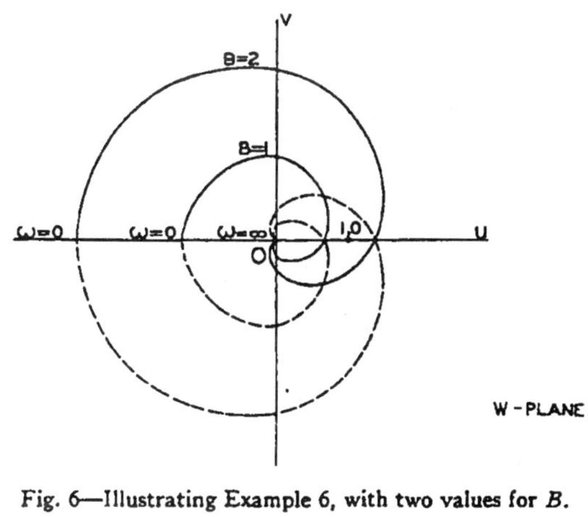

Fig. 6—Illustrating Example 6, with two values for B.

For large values of t this approaches

$$s(t) = -\tfrac{1}{2}Be^{(B/\sqrt{2}-\alpha)t}\sin(Bt/\sqrt{2} - \pi/4). \tag{56}$$

This example is interesting because it shows a case of instability although there is a reversal. Fig. 6 shows the admittance diagram for $B\sqrt{2} - \alpha < 0$ and for $B\sqrt{2} - \alpha > 0$.

7. Let

$$AG(t) = f_0(t) = A(1 - t), \qquad 0 \leq t \leq 1. \tag{57}$$

$$AG(t) = f_0(t) = 0, \qquad -\infty < t < 0, \qquad 1 < t < \infty. \tag{57'}$$

We have

$$\begin{aligned} AJ(i\omega) &= A\int_0^1 (1 - t)e^{-i\omega t}dt \\ &= A\left(\frac{1 - e^{-i\omega}}{\omega^2} + \frac{1}{i\omega}\right). \end{aligned} \tag{58}$$

Fig. 7 is a plot of this case for $A = 1$.

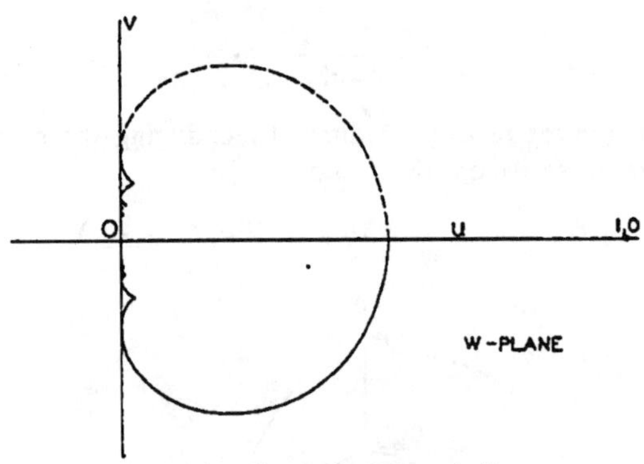

Fig. 7—Illustrating Example 7.

8. Let

$$AJ(i\omega) = \frac{A(1 + i\omega)}{(1 + i2\omega)}. \tag{59}$$

This is plotted on Fig. 8 for $A = 3$. It will be seen that the point 1 lies outside of the locus and for that reason we should expect that the system would be stable. We should expect from inspecting the diagram that the system would be stable for $A < 1$ and $A > 2$ and that it would be unstable for $1 \leq A \leq 2$. We have overlooked one fact, however; the expression for $AJ(i\omega)$ does not approach zero as ω

increases indefinitely. Therefore, it does not come within restriction (BI) and consequently the reasoning leading up to the rule does not apply.

The admittance in question can be made up by bridging a capacity in series with a resistance across a resistance line. This admittance

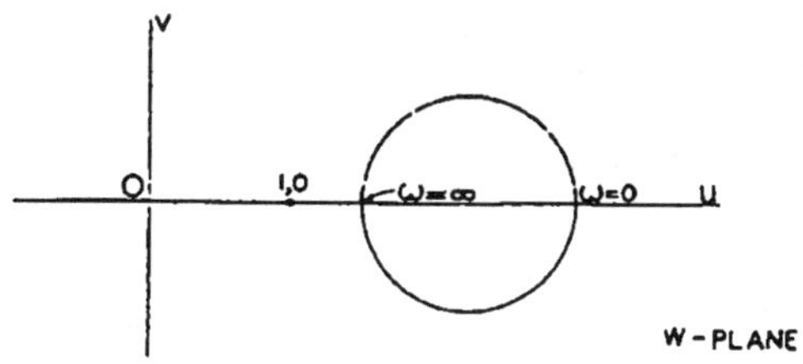

Fig. 8—Illustrating Example 8, without distributed constants.

obviously does not approach zero as the frequency increases. In any actual network there would, however, be a small amount of distributed capacity which, as the frequency is increased indefinitely, would cause the transmission through the network to approach zero. This is shown graphically in Fig. 9. The effect of the distributed capacity is

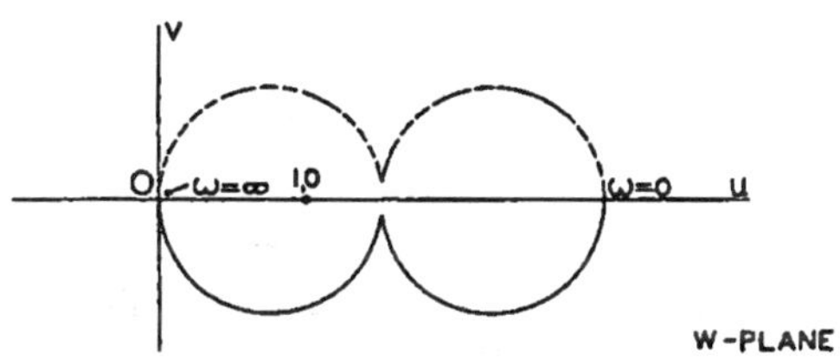

Fig. 9—Illustrating Example 8, with distributed constants.

essentially to cut a corridor from the circle in Fig. 8 to the origin, which insures that the point lies inside the locus.

Appendix I

Alternative Procedure

In some cases $AJ(i\omega)$ may be given as an analytic expression in $(i\omega)$. In that case the analytic expression may be used to define w for all values of z for which it exists. If the value for $AJ(i\omega)$ satisfies all the restrictions the value thus defined equals the w defined above for $0 \leq x < \infty$ only. For $-\infty < x < 0$ it equals the analytic continuation of the function w defined above. If there are no essential singularities anywhere including at ∞, the integral in (33) may be evaluated by the theory of residues by completing the path of integration so that all the poles of the integrand are included. We then have

$$s(t) = \sum_{j=1}^{j=n} \sum_{k=1}^{r_j} A_{jk} t^{k-1} e^{z_j t}. \tag{60}$$

If the network is made up of a finite number of lumped constants there is no essential singularity and the preceding expression converges because it has only a finite number of terms. In other cases there is an infinite number of terms, but the expression may still be expected to converge, at least, in the usual case. Then the system is stable if all the roots of $1 - w = 0$ have $x < 0$. If some of the roots have $x \geqq 0$ the system is unstable.

The calculation then divides into three parts:

1. The recognition that the impedance function is $1 - w$.[11]

2. The determination of whether the impedance function has zeros for which $x \geqq 0$.[12]

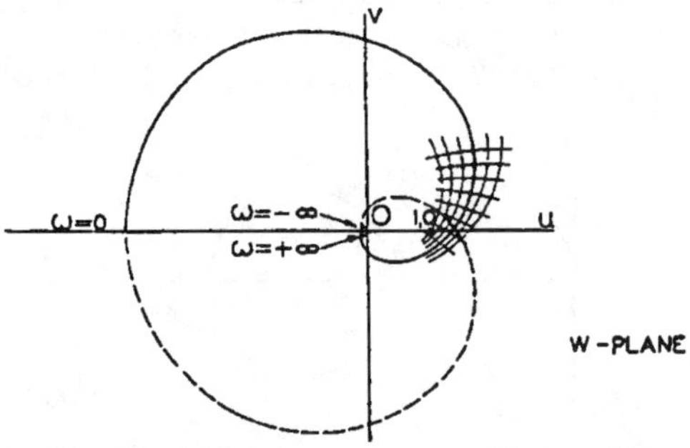

Fig. 10—Network of loci x = const., and y = const.

3. A deduction of a rule for determining whether there are roots for which $x \geqq 0$. The actual solution of the equation is usually too laborious.

To proceed with the third step, plot the locus $x = 0$ in the w plane, i.e., plot the imaginary part of w against the real part for all the values of y, $-\infty < y < \infty$. See Fig. 10. Other loci representing

$$x = \text{const.} \tag{61}$$

and

$$y = \text{const.} \tag{62}$$

[11] Cf. H. W. Nichols, *Phys. Rev.*, vol. 10, pp. 171–193, 1917.
[12] Cf. Thompson and Tait, "Natural Philosophy," vol. I, § 344.

may be considered and are indicated by the network shown in the figure in fine lines. On one side of the curve x is positive and on the other it is negative. Consider the equation

$$w(z) - 1 = 0$$

and what happens to it as A increases from a very small to a very large value. At first the locus $x = 0$ lies wholly to the left of the point. For this case the roots must have $x < 0$. As A increases there may come a time when the curve or successive convolutions of it will sweep over the point $w = 1$. For every such crossing at least one of the roots changes the sign of its x. We conclude that if the point $w = 1$ lies inside the curve the system is unstable. It is now possible to enunciate the rule as given in the main part of the paper but there deduced with what appears to be a more general method.

Appendix II

Discussion of Restrictions

The purpose of this appendix is to discuss more fully the restrictions which are placed on the functions defining the network. A full discussion in the main text would have interrupted the main argument too much.

Define an additional function

$$n(z) = \frac{1}{2\pi i}\int_I \frac{AJ(i\lambda)}{i\lambda - z}d(i\lambda), \qquad -\infty < x < 0. \tag{63}$$

$$n(iy) = \lim_{x \to 0} n(z).$$

This definition is similar to that for $w(z)$ given previously. It is shown in the theorem [13] referred to that these functions are analytic for $x \neq 0$ if $AJ(i\omega)$ is continuous. We have not proved, as yet, that the restrictions placed on $G(t)$ necessarily imply that $J(i\omega)$ is continuous. For the time being we shall assume that $J(i\omega)$ may have finite discontinuities. The theorem need not be restricted to the case where $J(i\omega)$ is continuous. From an examination of the second proof it will be seen to be sufficient that $\int_I J(i\omega)d(i\omega)$ exist. Moreover, that proof can be slightly modified to include all cases where conditions (AI)–(AIII) are satisfied.

[13] Osgood, loc. cit.

For, from the equation at top of page 298 [13]

$$\left|\frac{w(z_0 - \Delta z) - w(z_0)}{\Delta z} - \frac{1}{2\pi i}\int_I \frac{AJ(i\lambda)}{(i\lambda - z_0)^2}d(i\lambda)\right|$$

$$\leq |\Delta z| \left|\frac{1}{2\pi i}\int_I \frac{AJ(i\lambda)d(i\lambda)}{(i\lambda - z_0 - \Delta z)(i\lambda - z_0)^2}\right|, \qquad x_0 > 0. \quad (64)$$

It is required to show that the integral exists. Now

$$\int_I \frac{AJ(i\lambda)d(i\lambda)}{(i\lambda - z_0 - \Delta z)(i\lambda - z_0)^2}$$

$$= \int_I \frac{AJ(i\lambda)d(i\lambda)}{(i\lambda - z_0)^3}\left(1 + \frac{\Delta z}{i\lambda - z_0} + \frac{\Delta z^2}{i\lambda - z_0} + \text{etc.}\right) \quad (65)$$

if Δz is taken small enough so the series converges. It will be sufficient to confine attention to the first term. Divide the path of integration into three parts,

$$-\infty < \lambda < -|z_0| - 1, \qquad -|z_0| - 1 < \lambda < |z_0| + 1, \qquad |z_0| + 1 < \lambda < \infty.$$

In the middle part the integral exists because both the integrand and the range of integration are finite. In the other ranges the integral exists if the integrand falls off sufficiently rapidly with increasing λ. It is sufficient for this purpose that condition (BI) be satisfied. The same proof applies to $n(z)$.

Next, consider $\lim_{x \to 0} w(z) = w(iy)$. If iy is a point where $J(iy)$ is continuous, a straightforward calculation yields

$$w(iy) = AJ(iy)/2 + P(iy). \quad (66a)$$

Likewise,

$$n(iy) = -AJ(iy)/2 + P(iy) \quad (66b)$$

where $P(iy)$ is the principal value [14] of the integral

$$\frac{1}{2\pi i}\int_I \frac{AJ(i\lambda)}{i\lambda - iy}d(i\lambda).$$

Subtracting

$$w(iy) - n(iy) = AJ(iy) \quad (67)$$

If (iy) is a point of discontinuity of $J(iy)$

$$|w| \text{ and } |n| \text{ increase indefinitely as } x \to 0. \quad (68)$$

Next, evaluate the integral

$$\frac{1}{2\pi i}\int_{x+I} w(z)e^{zt}dz,$$

[14] E. W. Hobson, "Functions of a Real Variable," vol. I, 3d edition, § 352.

where the path of integration is from $x - i\infty$ to $x + i\infty$ along the line $x =$ const. On account of the analytic nature of the integrand this integral is independent of x (for $x > 0$). It may be written then

$$\lim_{x\to 0} \frac{1}{2\pi i}\int_{x+I} w(z)e^{zt}dz = \lim_{x\to 0}\frac{1}{2\pi i}\int_{x+I}\frac{1}{2\pi i}\int_{I}\frac{AJ(i\lambda)}{i\lambda - z}e^{zt}d(i\lambda)dz$$

$$= \lim_{x\to 0}\frac{1}{2\pi i}\int_{x+I}\frac{1}{2\pi i}\lim_{M\to\infty}\left[\int_{-iM}^{iy-i\delta} + \int_{iy-i\delta}^{iy+i\delta} + \int_{iy+i\delta}^{iM}\right]\frac{AJ(i\lambda)}{i\lambda - z}e^{zt}d(i\lambda)dz$$

$$= \lim_{x\to 0}\left[\frac{1}{2\pi i}\int_{x+I}\frac{1}{2\pi i}\int_{iy-i\delta}^{iy+i\delta}\frac{AJ(i\lambda)}{i\lambda - z}e^{zt}d(i\lambda)dz + Q(t,\delta)\right],\ x>0, \qquad (69)$$

where δ is real and positive. The function Q defined by this equation exists for all values of t and for all values of δ. Similarly,

$$\lim_{x\to 0}\frac{1}{2\pi i}\int_{x+I} n(z)e^{zt}dz$$

$$= \left[\lim_{x\to 0}\frac{1}{2\pi i}\int_{x+I}\frac{1}{2\pi i}\int_{iy+i\delta}^{iy+i\delta}\frac{AJ(i\lambda)}{i\lambda - z}e^{zt}d(i\lambda)dz + Q(t,\delta)\right],\ x<0, \qquad (70)$$

Subtracting and dropping the limit designations

$$\frac{1}{2\pi i}\int_{x+I} w(z)e^{zt}dz - \frac{1}{2\pi i}\int_{x+I} n(z)e^{zt}dz = \frac{1}{2\pi i}\int_{I} AJ(i\lambda)e^{i\lambda t}d(i\lambda). \qquad (71)$$

The first integral is zero for $t < 0$ as can be seen by taking x sufficiently large. Likewise, the second is equal to zero for $t > 0$. Therefore,

$$\frac{1}{2\pi i}\int_{x+I} w(z)e^{zt}dz = \frac{1}{2\pi i}\int_{I} AJ(i\omega)e^{i\omega t}d(i\omega) = AG(t), \quad 0 < t < \infty \qquad (72)$$

$$-\frac{1}{2\pi i}\int_{x+I} n(z)e^{zt}dz$$

$$= \frac{1}{2\pi i}\int_{I} AJ(i\omega)e^{i\omega t}d(i\omega) = AG(t) \quad -\infty < t < 0. \qquad (73)$$

We may now conclude that

$$\int_{I} n(iy)e^{iyt}d(iy) = 0, \qquad -\infty < t < \infty \qquad (74)$$

provided

$$G(t) = 0, \qquad -\infty < t < 0. \qquad \text{(AII)}$$

But (74) is equivalent to

$$n(z) = 0, \qquad (74')$$

which taken with (67) gives

$$w(iy) = AJ(iy). \qquad \text{(BIII)}$$

(BIII) is, therefore, a necessary consequence of (AII). (74′) taken with (68) shows that

$$J(iy) \text{ is continuous.} \qquad \text{(BII)}$$

It may be shown [15] that (BI) is a consequence of (AI). Consequently all the B conditions are deducible from the A conditions.

Conversely, it may be inquired whether the A conditions are deducible from the B conditions. This is of interest if $AJ(i\omega)$ is given and is known to satisfy the B conditions, whereas nothing is known about G.

Condition AII is a consequence of BIII as may be seen from (67) and (74). On the other hand AI and AIII cannot be inferred from the B conditions. It can be shown by examining (5), however, that if the slightly more severe condition

$$\lim_{y\to\infty} y^{\gamma} J(iy) \text{ exists,} \qquad (\gamma > 1), \qquad \text{(BI}a\text{)}$$

is satisfied then

$$G(t) \text{ exists,} \qquad -\infty < t < \infty, \qquad \text{(AI}a\text{)}$$

which, together with AII, insures the validity of the reasoning.

It remains to show that the measured value of $J(i\omega)$ is equal to that defined by (6). The measurement consists essentially in applying a sinusoidal wave and determining the response after a long period. Let the impressed wave be

$$E = \text{real part of } e^{i\omega t}, \qquad t \geq 0. \qquad (75)$$

$$E = 0, \qquad t < 0. \qquad (75')$$

The response is

$$\text{real part of} \int_0^t AG(\lambda)e^{i\omega(t-\lambda)}d\lambda$$

$$= \text{real part of } Ae^{i\omega t}\int_0^t G(\lambda)e^{-i\omega\lambda}d\lambda. \qquad (76)$$

For large values of t this approaches

$$\text{real part of } Ae^{i\omega t}J(i\omega). \qquad (77)$$

Consequently, the measurements yield the value $AJ(i\omega)$.

[15] See Hobson, loc. cit., vol. II, 2d edition, § 335. It will be apparent that K depends on the total variation but is independent of the limits of integration.

Stabilized Feedback Amplifiers

H. S. BLACK

THIS paper by Black describes the principle of feedback as it applies to electronic amplifiers based on vacuum tubes. It grew out of the intensive work at Bell Laboratories to develop electronic amplifiers that could be used in telephone communication over long distances. The work was carried out in a very stimulating and conducive research environment with participants such as Bode and Nyquist, whose seminal papers from this period also appear in this volume. The background that led to this ground-breaking work is described more fully in the preamble to Nyquist's paper.

Black invented the feedback amplifier in 1927, but this paper was published 7 years after the invention and two years after the publication of Nyquist's paper. Another paper [1] written by Black 50 years after the invention gives his personal reflections, including some noteworthy remarks on his struggle with the problem. A good perspective is also given by Bode in his 1960 paper [2].

Today it is easy to underestimate the significance of *feedback* and the fact that its use often leads to revolutionary consequences. Feedback had certainly been used for a long time before Black's invention, but not for electronic amplifiers. Even so it took nine years before the patent was granted, because the Patent Office did not believe that the amplifier would work. It is also easy not to realize the broad scope of automatic control. At the time of Black's work, the relations between the feedback amplifier and other earlier uses of feedback were not apparent. It also took a long time before the theory developed for the feedback amplifier was applied to other feedback systems. This reflects the fact that the concept of feedback is far from being trivial even if we now tend to take it for granted.

To devise an amplifier that could be used for telephone transmission, Black had to overcome two major problems: substantial variations in the gains of amplifiers and the interference between the channels caused by nonlinearities in the amplifiers. A major contribution by Black was to introduce feedback in amplifiers, or to quote from the paper:

> "... by building an amplifier whose gain is deliberately made say 40 decibels higher than necessary (10 000 fold excess on energy basis) and then feeding the output back on the input in such a way as to throw away excess gain, it has been found possible to effect extraordinarily improvement in constancy of amplification and freedom from nonlinearity."

In today's terminology, we could say that Black used feedback to design an amplifier that was robust to variations in the gain of the process. As a side-effect, he also obtained a closed-loop system that was extremely linear.

The invention of feedback is undoubtedly the key contribution of the paper. Another contribution is the initial attempt to develop methods for analysis and design of the feedback amplifier. This effort was completed in the paper by Bode, which is the third paper in this volume.

Black used the $j\omega$ method, developed by Steinmetz, to describe the dynamics of the system.[1] The notation of the paper is different from what is standard today: the process or the plant is denoted by μ and the controller by β. The gain is defined in such a way that β is positive for positive feedback. This is the same convention used by Nyquist. Black derived equations by static analysis, assuming that μ and β are constants. Dynamics was then introduced by regarding μ and β as functions of the frequency ω. The first step in the analysis is Equation (2) of the paper, which describes the amplifier output as a function of input, noise and distortion. Many conclusions are then drawn from this equation. The role of the sensitivity function

$$S = \frac{1}{1 - \mu\beta}$$

is emphasized both to express rejection of disturbances and sensitivity to parameter variations. Black classifies feedback as positive if $|S| > 1$ and negative if $|S| < 1$. Stability is investigated using the Nyquist criterion.

Several diagrams are introduced in the paper to guide analysis and design. Figure 4 is a plot of the phase of $\mu(\omega)\beta(\omega)$ as a function of its magnitude. The plot also had loci for constant magnitude of the sensitivity function, and is similar to the

[1] The Heaviside operator calculus [5], which was available at the time, was not used. Use of Laplace transforms to describe linear systems appeared later see, e.g., [3], [4], [7].

Nichols plot. The only difference is that the axes are reversed and that logarithmic scale is used for the magnitude of the loop transfer function in the Nichols plot. The Nyquist curves are also plotted with logarithmic scale for the magnitude of the loop transfer function. Diagrams that showed the gain of the amplifier with and without feedback were also introduced.

The invention of the electronic feedback amplifier has had an immense impact. It solved the immediate problems that the telecommunications industry was faced with, but it also had much broader implications. It was possible to build amplifiers where the gain only changed by 0.01 dB when the plate voltage of the tubes changed from 240 V to 260 V and the intermodulation was 75 dB below the output signal at full load. Black was awarded the IEEE Lamme Medal in 1957 "for his many outstanding contributions to telecommunication and allied electronic arts, especially the invention of the negative feedback amplifier and the successful development and application of the negative feedback amplification principle." On that occasion Mervin B. Kelley of Bell Laboratories articulated the impact of Black's work as follows [1]:

> "Although many of Harold's inventions have made great impact, that of the negative feedback amplifier is indeed the most outstanding. It easily ranks coordinate with De Forest's invention of the audion as one of the two inventions of broadest scope and significance in electronics and communications of the past 50 years. . . . It is no exaggeration to say that without Black's invention, the present long-distance telephone and television networks which cover our entire country and the transoceanic telephone cables would not exist. . . . Thus, the entire explosive extension of the area of control, both electrical and mechanical, grew out of an understanding of the feedback principle."

Apart from the invention as such, there are ideas in the paper which have also had strong impact. The equations derived by Black that relate the error to measurement noise, disturbances, and process uncertainties are still the starting point for analysis of control systems. They have been generalized in many ways, to multi-variable systems and to nonlinear systems. The idea to use graphical methods for the analysis of feedback systems became an essential part of the servomechanism practice.

In earlier work on ship steering and process control, feedback had been used primarily for stabilization and regulation. Black, however, used feedback to obtain insensitivity to parameter variations. The combined work of Black, Bode, and Nyquist was an important basis for the field of automatic control as it emerged in the 1940s; see [6].

References

[1] H.S. Black, "Inventing the negative feedback amplifier," *IEEE Spectrum*, pp. 55–60, December 1977.

[2] H.W. Bode, "Feedback—The history of an idea," In *Proc. Symp. Active Networks and Feedback Systems*, Polytechnic Institute of Brooklyn, New York, 1960.

[3] M.A. Gardner and J.L. Barnes, *Transients in Linear Systems*, John Wiley (New York), 1942.

[4] A.C. Hall, "Application of circuit theory to the design of servomechanisms," *Journal of the Franklin Institute*, **242**:279–307, 1946.

[5] O. Heaviside, *Electrical Papers*, MacMillan (New York), 1892.

[6] H.M. James, N.B. Nichols and R.S. Phillips, *Theory of Servomechanisms*, McGraw-Hill (New York), 1947.

[7] L.A. MacColl, *Fundamental Theory of Servomechanisms*, Van Nostrand (Princeton, NJ), 1945.

K.J.Å

The Bell System Technical Journal

January, 1934

Stabilized Feedback Amplifiers*

By H. S. BLACK

This paper describes and explains the theory of the feedback principle and then demonstrates how stability of amplification and reduction of modulation products, as well as certain other advantages, follow when stabilized feedback is applied to an amplifier. The underlying principle of design by means of which singing is avoided is next set forth. The paper concludes with some examples of results obtained on amplifiers which have been built employing this new principle.

The carrier-in-cable system dealt with in a companion paper [1] involves many amplifiers in tandem with many telephone channels passing through each amplifier and constitutes, therefore, an ideal field for application of this feedback principle. A field trial of this system was made at Morristown, New Jersey, in which seventy of these amplifiers were operated in tandem. The results of this trial were highly satisfactory and demonstrated conclusively the correctness of the theory and the practicability of its commercial application.

Introduction

DUE TO advances in vacuum tube development and amplifier technique, it is now possible to secure any desired amplification of the electrical waves used in the communication field. When many amplifiers are worked in tandem, however, it becomes difficult to keep the overall circuit efficiency constant, variations in battery potentials and currents, small when considered individually, adding up to produce serious transmission changes for the overall circuit. Furthermore, although it has remarkably linear properties, when the modern vacuum tube amplifier is used to handle a number of carrier telephone channels, extraneous frequencies are generated which cause interference between the channels. To keep this interference within proper bounds involves serious sacrifice of effective amplifier capacity or the use of a push-pull arrangement which, while giving some increase in capacity, adds to maintenance difficulty.

However, by building an amplifier whose gain is deliberately made, say 40 decibels higher than necessary (10,000 fold excess on energy basis), and then feeding the output back on the input in such a way

* Presented at Winter Convention of A. I. E. E., New York City, Jan. 23-26, 1934. Published in *Electrical Engineering*, January, 1934.

[1] "Carrier in Cable" by A. B. Clark and B. W. Kendall, presented at the A. I. E. E. Summer Convention, Chicago, Ill., June, 1933; published in *Electrical Engineering*, July, 1933, and in *Bell Sys. Tech. Jour.*, July, 1933.

as to throw away the excess gain, it has been found possible to effect extraordinary improvement in constancy of amplification and freedom from non-linearity. By employing this feedback principle, amplifiers have been built and used whose gain varied less than 0.01 db with a change in plate voltage from 240 to 260 volts and whose modulation products were 75 db below the signal output at full load. For an amplifier of conventional design and comparable size this change in plate voltage would have produced about 0.7 db variation while the modulation products would have been only 35 db down; in other words, 40 db reduction in modulation products was effected. (On an energy basis the reduction was 10,000 fold.)

Stabilized feedback possesses other advantages including reduced delay and delay distortion, reduced noise disturbance from the power supply circuits and various other features best appreciated by practical designers of amplifiers.

It is far from a simple proposition to employ feedback in this way because of the very special control required of phase shifts in the amplifier and feedback circuits, not only throughout the useful frequency band but also for a wide range of frequencies above and below this band. Unless these relations are maintained, singing will occur, usually at frequencies outside the useful range. Once having achieved a design, however, in which proper phase relations are secured, experience has demonstrated that the performance obtained is perfectly reliable.

Circuit Arrangement

In the amplifier of Fig. 1, a portion of the output is returned to the input to produce feedback action. The upper branch, called the μ-circuit, is represented as containing active elements such as an amplifier while the lower branch, called the β-circuit, is shown as a passive network. The way a voltage is modified after once traversing each circuit is denoted μ and β respectively and the product, $\mu\beta$, represents how a voltage is modified after making a single journey around amplifier and feedback circuits. Both μ and β are complex quantities, functions of frequency, and in the generalized concept either or both may be greater or less in absolute value than unity.[2]

Figure 2 shows an arrangement convenient for some purposes where, by using balanced bridges in input and output circuits, interaction between input and output is avoided and feedback action and amplifier impedances are made independent of the properties of circuits connected to the amplifier.

[2] μ is not used in the sense that it is sometimes used, namely, to denote the amplification constant of a particular tube, but as the complex ratio of the output to the input voltage of the amplifier circuit.

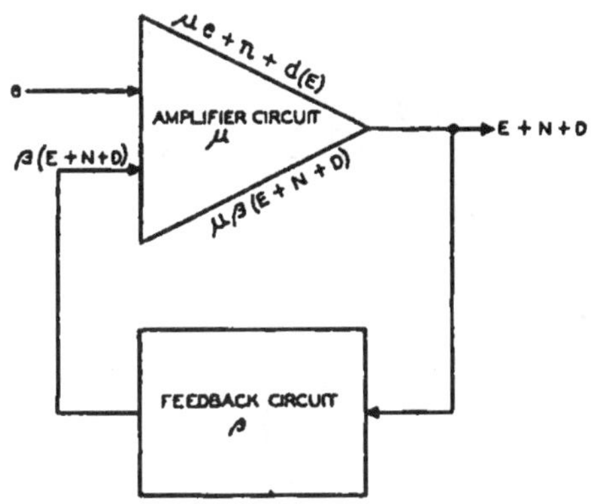

Fig. 1—Amplifier system with feedback.

e—Signal input voltage:
μ—Propagation of amplifier circuit.
μe—Signal output voltage without feedback.
n—Noise output voltage without feedback.
$d(E)$—Distortion output voltage without feedback.
β—Propagation of feedback circuit.
E—Signal output voltage with feedback.
N—Noise output voltage with feedback.
D—Distortion output voltage with feedback.

The output voltage with feedback is $E + N + D$ and is the sum of $\mu e + n + d(E)$, the value without feedback plus $\mu\beta[E + N + D]$ due to feedback.

$$E + N + D = \mu e + n + d(E) + \mu\beta[E + N + D]$$

$$[E + N + D](1 - \mu\beta) = \mu e + n + d(E)$$

$$E + N + D = \frac{\mu e}{1 - \mu\beta} + \frac{n}{1 - \mu\beta} + \frac{d(E)}{1 - \mu\beta}$$

If $|\mu\beta| \gg 1$, $E \doteq -\frac{e}{\beta}$. Under this condition the amplification is independent of μ but does depend upon β. Consequently the over-all characteristic will be controlled by the feedback circuit which may include equalizers or other corrective networks.

General Equation

In Fig. 1, β is zero without feedback and a signal voltage, e_0, applied to the input of the μ-circuit produces an output voltage. This is made up of what is wanted, the amplified signal, E_0, and components that are not wanted, namely, noise and distortion designated N_0 and D_0 and assumed to be generated within the amplifier. It is further assumed that the noise is independent of the signal and the distortion generator or modulation a function *only of the signal output.* Using the notation of Fig. 1, the output without feedback may be written as:

$$E_0 + N_0 + D_0 = \mu e_0 + n + d(E_0), \tag{1}$$

where zero subscripts refer to conditions without feedback.

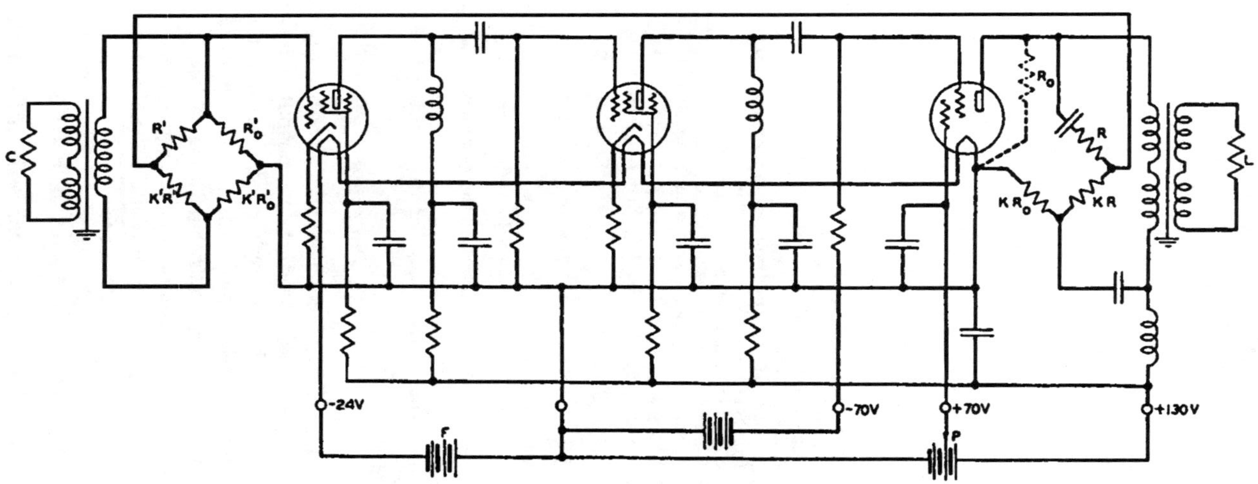

Fig. 2—Circuit of a negative feedback amplifier.

With feedback, β is not zero and the input to the μ-circuit becomes $e_0 + \beta(E + N + D)$. The output is $E + N + D$ and is equal to $\mu[e_0 + \beta(E + N + D)] + n + d(E)$ or:

$$E + N + D = \frac{\mu e_0}{1 - \mu\beta} + \frac{n}{1 - \mu\beta} + \frac{d(E)}{1 - \mu\beta}. \tag{2}$$

In the output, signal, noise and modulation are divided by $(1 - \mu\beta)$, and assuming $|1 - \mu\beta| > 1$, all are reduced.

Change in Gain Due to Feedback

From equation (2), the amplification with feedback equals the amplification without feedback divided by $(1 - \mu\beta)$. The effect of adding feedback, therefore, usually is to change the gain of the amplifier and this change will be expressed as:

$$G_{CF} = 20 \log_{10} \left| \frac{1}{1 - \mu\beta} \right|, \tag{3}$$

where G_{CF} is *db change in gain due to feedback*. $1/(1 - \mu\beta)$ will be used as a quantitative measure of the effect of feedback and the feedback referred to as *positive feedback* or *negative feedback* according as the absolute value of $1/(1 - \mu\beta)$ is greater or less than unity. Positive feedback increases the gain of the amplifier; negative feedback reduces it. The term *feedback* is not limited merely to those cases where the absolute value of $1/(1 - \mu\beta)$ is other than unity.

From $\mu\beta = |\mu\beta| \underline{|\Phi}$ and (3), it may be shown that:

$$10^{-G_{CF}/10} = 1 - 2|\mu\beta| \cos\Phi + |\mu\beta|^2, \tag{4}$$

which is the equation for a family of concentric circles of radii $10^{-G_{CF}/20}$ about the point 1, 0. Figure 3 is a polar diagram of the vector field of $\mu\beta = |\mu\beta| \underline{|\Phi}$. Using rectangular instead of polar coordinates, Fig. 4 corresponds to Fig. 3 and may be regarded as a diagram of the field of $\mu\beta$ where the parameter is db change in gain due to feedback. From these diagrams all of the essential properties of feedback action can be obtained such as change in amplification, effect on linearity, change in stability due to variations in various parts of the system, reduction of noise, etc. Certain significant boundaries have been designated similarly on both figures.

For example, boundary A is the locus of zero change in gain due to feedback. Along this parametric contour line where the absolute magnitude of amplification is not changed by feedback action, values of $|\mu\beta|$ range from zero to 2 and the phase shift, Φ, around the amplifier and feedback circuits equals $\cos^{-1} |\mu\beta|/2$ and, therefore, lies between $-90°$ and $+90°$ For all conditions inside or above this boundary, the gain with feedback is increased; outside or below, the gain is *decreased.*

STABILITY

From equation (2), $\mu e_0/(1-\mu\beta)$ is the amplified signal with feedback and, therefore, $\mu/(1-\mu\beta)$ is an index of the amplification. It is of course a complex ratio. It will be designated A_F and referred to as the amplification with feedback.

To consider the effect of feedback upon stability of amplification, the stability will be viewed as the ratio of a change, δA_F, to A_F where δA_F is due to a change in either μ or β and the effects may be derived by assuming the variations are small.

$$A_F \doteq \frac{\mu}{1-\mu\beta}, \tag{5}$$

$$\left[\frac{\delta A_F}{A_F}\right]_\mu \doteq \frac{\left[\frac{\delta\mu}{\mu}\right]}{1-\mu\beta}, \tag{6}$$

$$\left[\frac{\delta A_F}{A_F}\right]_\beta = \frac{\mu\beta}{1-\mu\beta}\left[\frac{\delta\beta}{\beta}\right]. \tag{7}$$

If $\mu\beta \gg 1$, it is seen that μ or the μ-circuit is stabilized by an amount corresponding to the reduction in amplification and the effect of introducing a gain or loss in the μ-circuit is to produce no material change in the overall amplification of the system; the stability of amplification as affected by β or the β-circuit is neither appreciably improved nor degraded since increasing the loss in the β-circuit raises the gain of the amplifier by an amount almost corresponding to the loss introduced and vice-versa. If μ and β are both varied and the variations sufficiently small, the effect is the same as if each were changed separately and the two results then combined.

In certain practical applications of amplifiers it is the change in gain or ammeter or voltmeter reading at the output that is a measure of the stability rather than the complex ratio previously treated. The conditions surrounding gain stability may be examined by considering the absolute value of A_F. This is shown as follows: Let (db) represent the gain in decibels corresponding to A_F. Then

$$(db) = 20 \log_{10} |A_F|,$$

$$\delta(db) \doteq 8.686\left[\frac{\delta|A_F|}{|A_F|}\right]. \tag{8}$$

To get the absolute value of the amplification: Let

$$\mu\beta = |\mu\beta| \underline{|\Phi}, \tag{9}$$

$$|A_F| = \frac{|\mu|}{\sqrt{1 - 2|\mu\beta| \cos \Phi + |\mu\beta|^2}}. \tag{10}$$

The stability of amplification which is proportional to the gain stability is given by:

$$\left[\frac{\delta|A_F|}{|A_F|}\right]_{|\mu|} \doteq \frac{1 - |\mu\beta| \cos \Phi}{|1 - \mu\beta|^2} \left[\frac{\delta|\mu|}{|\mu|}\right], \tag{11}$$

$$\left[\frac{\delta|A_F|}{|A_F|}\right]_{|\beta|} \doteq \left|\frac{\mu\beta}{1 - \mu\beta}\right| \left[\frac{\cos \Phi - |\mu\beta|}{|1 - \mu\beta|}\right] \left[\frac{\delta|\beta|}{|\beta|}\right], \tag{12}$$

$$\left[\frac{\delta|A_F|}{|A_F|}\right]_{\Phi} \doteq -\left|\frac{\mu\beta}{1 - \mu\beta}\right| \left[\frac{\sin \Phi}{|1 - \mu\beta|}\right] [\delta\Phi]. \tag{13}$$

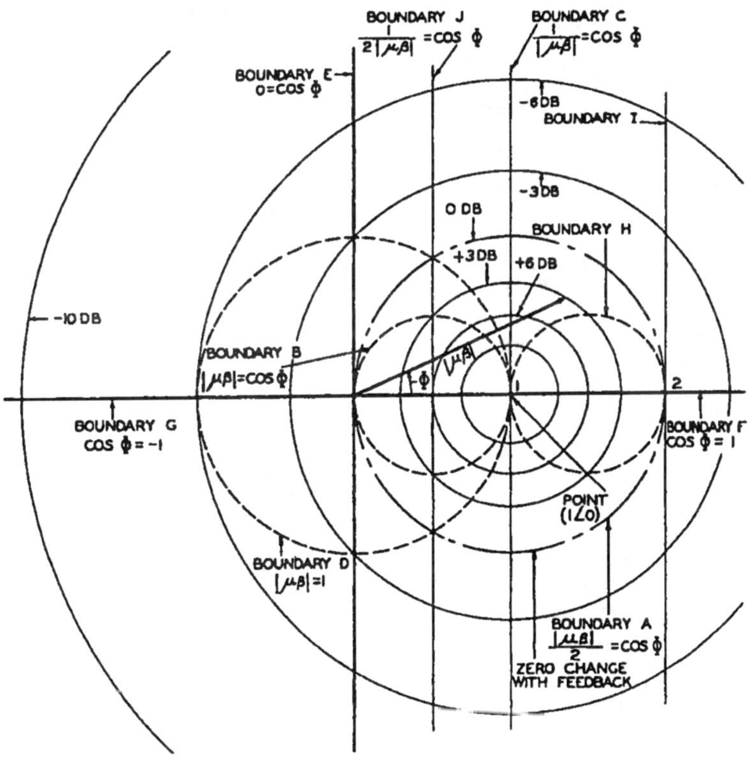

Fig. 3—The vector field of $\mu\beta$. See caption for Fig. 4.

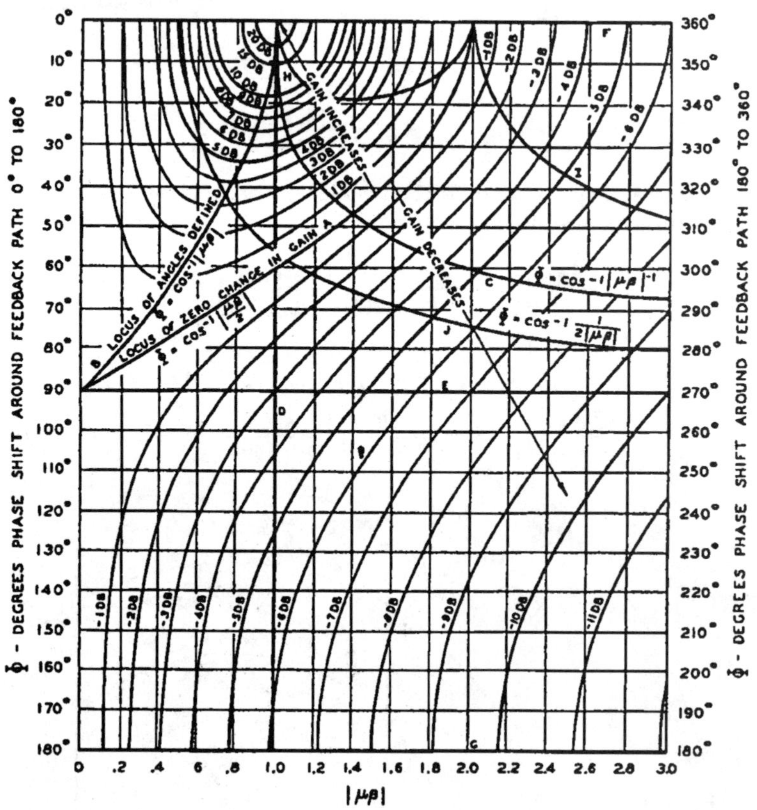

Fig. 4—Phase shift around the feedback path plotted as a function of $|\mu\beta|$, the absolute value of $\mu\beta$.

$\mu\beta$ is a complex quantity which represents the ratio by which the amplifier and feedback (or more generally μ and β) modify a voltage in a single trip around the closed path.

First, there is a set of boundary curves indicated as *A, B, C, D, E, F, G, H, I*, and *J* which gives either limiting or significent values of $|\mu\beta|$ and Φ.

Secondly, there is a family of curves in which db change in gain due to feedback is the parameter.

Boundaries

A. Conditions in which gain and modulation are unaffected by feedback.

B. Constant amplification ratio against small variations in $|\beta|$.

Constant change in gain, $\frac{1}{|1-\mu\beta|}$, against variations in $|\mu|$ and $|\beta|$.

Stable phase shift through the amplifier against variations in Φ_β.

The boundary on which the stability of amplification is unaffected by feedback.

C. Constant amplification ratio against small variations in $|\mu|$.

Constant phase shift through the amplifier against variations in $\Phi\mu$.

The absolute magnitude of the voltage fed back $\frac{|\mu\beta|}{|1-\mu\beta|}$ is constant against variations in $|\mu|$ and $|\beta|$.

A curious fact to be noted from (11) is that it is possible to choose a value of $\mu\beta$ (namely, $|\mu\beta| = \sec\Phi$) so that the numerator of the right hand side vanishes. This means that the gain stability is perfect, assuming differential variations in $|\mu|$. Referring to Figs. 3 and 4, contour C is the locus of $|\mu\beta| = \sec\Phi$ and it includes all amplifiers whose gain is unaffected by small variations in $|\mu|$. In this way it is even possible to stabilize an amplifier whose feedback is positive, i.e., feedback may be utilized to raise the gain of an amplifier and, at the same time, the gain stability with feedback *need* not be degraded but on the contrary improved. If a similar procedure is followed with an amplifier whose feedback is negative, the gain stability will be theoretically perfect and independent of the reductions in gain due to feedback. Over too wide a frequency band practical difficulties will limit the improvements possible by these methods.

With negative feedback, gain stability is always improved by an amount at least as great as corresponds to the reduction in gain and generally more; with positive feedback, gain stability is never degraded by more than would correspond to the increase in gain and under appropriate conditions, assuming the variations are not too great, is as good as or much better than without feedback. With positive feedback, the variations in μ or β must not be permitted to become sufficiently great to cause the amplifier to sing or give rise to instability as defined in a following section on "Avoiding Singing."

MODULATION

To determine the effect of feedback action upon modulation produced in the amplifier circuit, it is convenient to assume that the output of undistorted signal is made the same with and without feedback and that a comparison is then made of the difference in modulation with and without feedback. Therefore, with feedback, the input is changed to $e = e_0(1 - \mu\beta)$ and, referring to equation (2), the output voltage is μe_0, and the generated modulation, $d(E)$, assumes its value without feedback, $d(E_0)$, and $d(E)/(1-\mu\beta)$ becomes $d(E_0)/(1-\mu\beta)$ which is $D_0/(1 - \mu\beta)$. This relationship is approximate because the

---

D. $|\mu\beta| = 1$.
E. $\Phi = 90°$. Improvement in gain stability corresponds to twice db reduction in gain.
F and *G.* Constant amplification ratio against variations in Φ.
Constant phase shift through the amplifier against variations in $|\mu|$ and $|\beta|$.
H. Same properties as *B.*
I. Same properties as *E.*
J. Conditions in which $\frac{|\mu|}{|1 - \mu\beta|} = \frac{-1}{|\beta|}$ the overall gain is the exact negative inverse of the transmission through the β-circuit.

voltage at the input without feedback is free from distortion and with feedback it is not and, hence, the assumption that the generated modulation is a function *only of the signal output* used in deriving equation (2) is not necessarily justified.

From the relationship $D = D_0/(1 - \mu\beta)$, it is to be concluded that modulation with feedback will be reduced db for db as the effect of feedback action causes an arbitrary db reduction in the gain of the amplifier, i.e., when the feedback is negative. With positive feedback the opposite is true, the modulation being increased by an amount corresponding to the increase in amplification.

If modulation in the β-circuit is a factor, it can be shown that usually in its effect on the output, the modulation level at the output due to non-linearity of the β-circuit is approximately $\mu\beta/(1 - \mu\beta)$ multiplied by the modulation generated in the β-circuit acting alone and without feedback.

Additional Effects

Noise

A criterion of the worth of a reduction in noise is the reduction in signal-to-noise ratio at the output of an amplifier. Assuming that the amount of noise introduced is the same in two systems, for example with and without feedback respectively, and that the signal outputs are the same, a comparison of the signal-to-noise ratios will be affected by the amplification between the place at which the noise enters and the output. Denoting this amplification by a and a_0 respectively, it can be shown that the relation between the two noise ratios is $(a_0/a)(1 - \mu\beta)$. This is called the *noise index.*

If noise is introduced in the power supply circuits of the last tube, $a_0/a = 1$ and the noise index is $(1 - \mu\beta)$. As a result of this relation less expensive power supply filters are possible in the last stage.

Phase Shift, Envelope Delay, Delay Distortion

In the expression $A_F = [\mu/(1 - \mu\beta)]\underline{|\theta}$, θ is the overall phase shift with feedback, and it can be shown that the *phase shift through the amplifier with feedback may be made to approach the phase shift through the β-circuit plus 180 degrees.* The effect of phase shift in the β-circuit is not correspondingly reduced. It will be recalled that in reducing the change in phase shift with frequency, envelope delay, which is the slope of the phase shift with respect to the angular velocity, $\omega = 2\pi f$, also is reduced. The *delay* distortion likewise is reduced because a measure of delay distortion at a particular frequency is the difference between the envelope delay at that frequency and the least envelope delay in the band.

β-Circuit Equalization

Referring to equation (2), the output voltage, E, approaches $-e_0/\beta$ as $1 - \mu\beta \doteq -\mu\beta$ and equals it in absolute value if $\cos\Phi = \dfrac{1}{2|\mu\beta|}$ where $\mu\beta = |\mu\beta| \underline{|\Phi}$. Under these circumstances increasing the loss in the β-circuit one db raises the gain of the amplifier one db and vice-versa, thus giving any gain-frequency characteristic for which a like loss-frequency characteristic can be inserted in the β-circuit. This procedure has been termed β-circuit equalization. It possesses other advantages which cannot be dwelt upon here.

Avoid Singing

Having considered the theory up to this point, experimental evidence was readily acquired to demonstrate that $\mu\beta$ might assume large values,

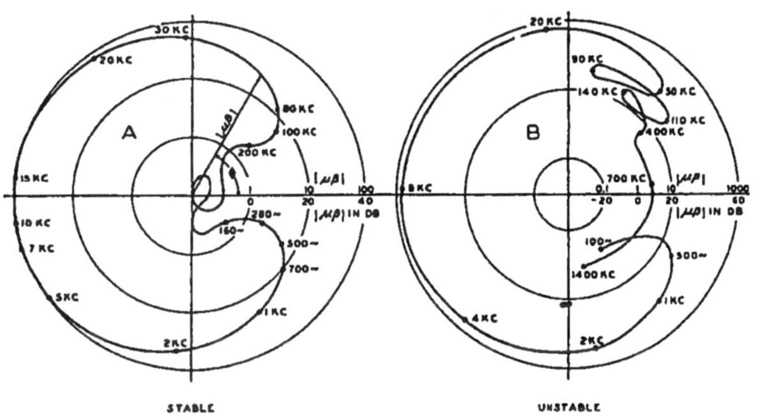

Fig. 5—Measured $\mu\beta$ characteristics of two amplifiers.

for example 10 or 10,000, provided Φ was not at the same time zero. However, one noticeable feature about the field of $\mu\beta$ (Figs. 3 and 4) is that it implies that even though the phase shift is zero and the absolute value of $\mu\beta$ exceeds unity, self-oscillations or singing will not result. This may or may not be true. When the author first thought about this matter he suspected that owing to practical non-linearity, singing would result whenever the gain around the closed loop equalled or exceeded the loss and simultaneously the phase shift was zero, i.e., $\mu\beta = |\mu\beta| + jo \geqq 1$. Results of experiments, however, seemed to indicate something more was involved and these matters were described to Mr. H. Nyquist, who developed a more general criterion for freedom from instability [3] applicable to an amplifier having linear positive constants.

To use this criterion, plot $\mu\beta$ (the modulus and argument vary with frequency) and its complex conjugate in polar coordinates for all values of frequency from 0 to $+\infty$. If the resulting loop or loops do not enclose the point (1, 0) the system will be stable, otherwise not.[3] The envelope of the transient response of a stable amplifier

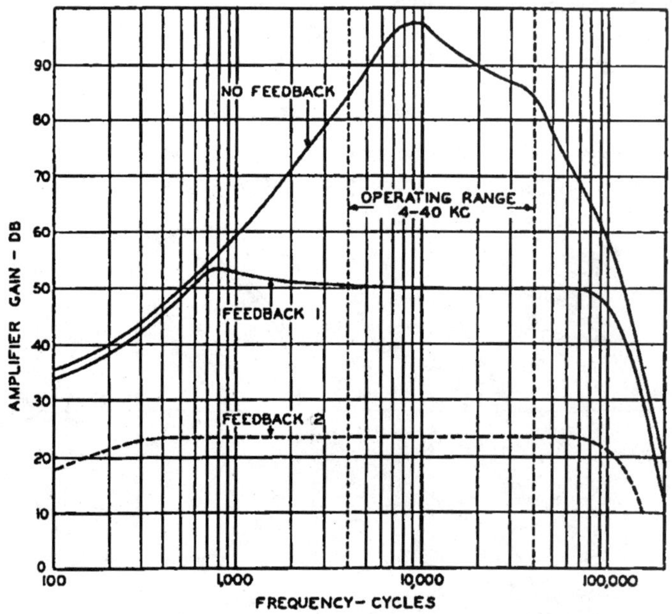

Fig. 6—Gain frequency characteristics with and without feedback of amplifier of Fig. 2.

always dies away exponentially with time; that of an unstable amplifier in all physically realizable cases increases with time. Characteristics A and B in Fig. 5 are results of measurements on two different amplifiers; the amplifier having $\mu\beta$-characteristic denoted A was stable; the other unstable.

The number of stages of amplification that can be used in a single amplifier is not significant except insofar as it affects the question of avoiding singing. Amplifiers with considerable negative feedback

[3] For a complete description of the criterion for stability and instability and exactly what is meant by enclosing the point (1, 0), reference should be made to "Regeneration Theory"—H. Nyquist, *Bell System Technical Journal,* Vol. XI, pp. 126–147, July, 1932.

have been tested where the number of stages ranged from one to five inclusive. In every case the feedback path was from the output of the last tube to the input of the first tube.

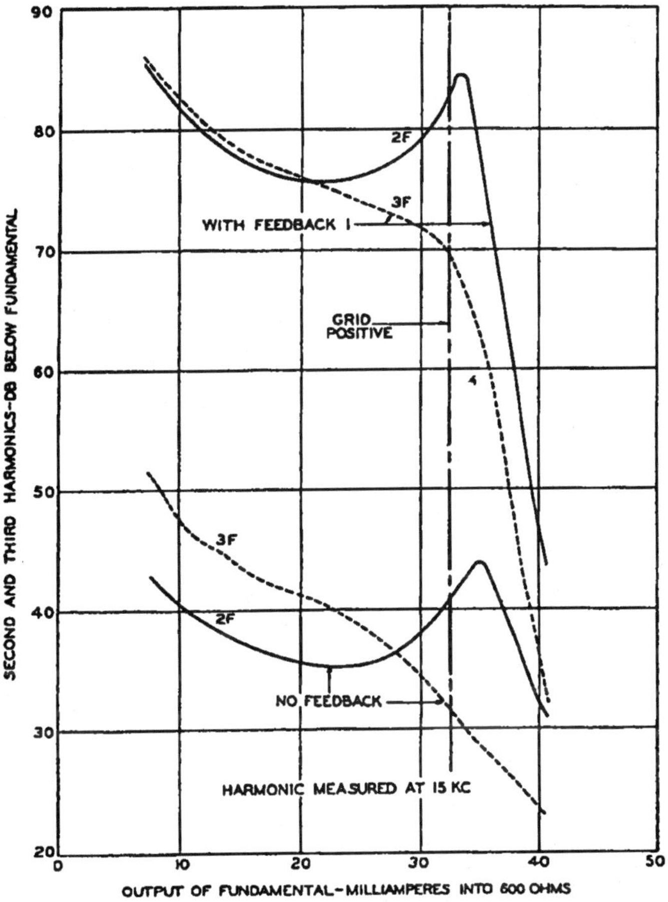

Fig. 7—Modulation characteristics with and without feedback for the amplifier of Fig. 2.

Experimental Results

Figures 6 and 7 show how the gain-frequency and modulation characteristics of the three-stage impedance coupled amplifier of Fig. 2 are improved by negative feedback. In Fig. 7, the improvement in harmonics is not exactly equal to the db reduction in gain. Figure 8 shows measurements on a different amplifier in which harmonics are reduced as negative feedback is increased, db for db over a 65 db range.

That the gain with frequency is practically independent of small variations in $|\mu|$ is shown by Fig. 9. This is a characteristic of the Morristown amplifier described in the paper by Messrs. Clark and Kendall [1] which meets the severe requirements imposed upon a repeater amplifier for use in cable carrier systems. Designed to amplify frequencies from 4 kc

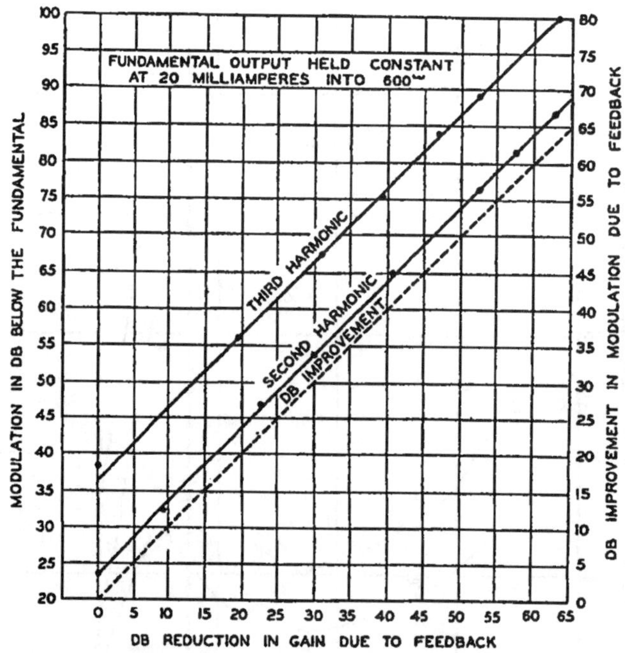

Fig. 8—Improvement of harmonics with feedback. One example of another amplifier in which with 60 db feedback, harmonic currents in the output are only one-thousandth and their energy one-millionth of the values without feedback.

to 40 kc the maximum change in gain due to variations in plate voltage does not exceed 7/10000 db per volt and at 20 kc the change is only 1/20000 db per volt. This illustrates that for small changes in $|\mu|$, the ratio of the stability without feedback to the stability with feedback, called the *stability index*, approaches $|1 - \mu\beta|^2/(1 - |\mu\beta| \cos \Phi)$ and gain stability is improved at least as much as the gain is reduced and usually more and is theoretically perfect if $\cos \Phi = 1/|\mu\beta|$.

[1] Loc. cit.

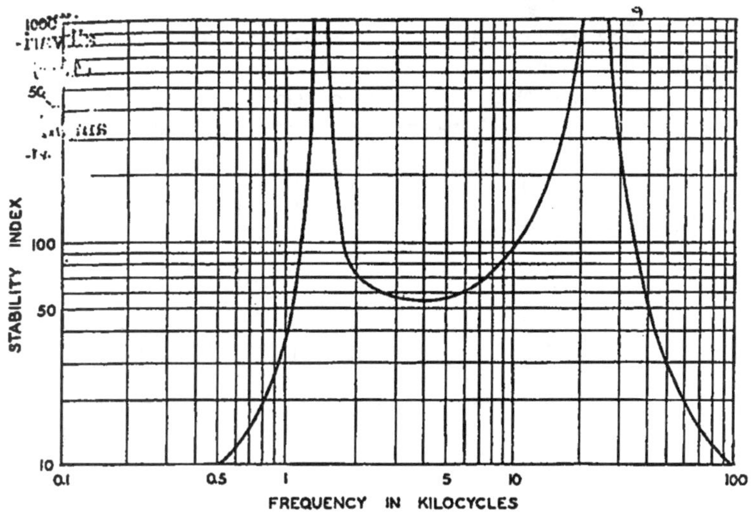

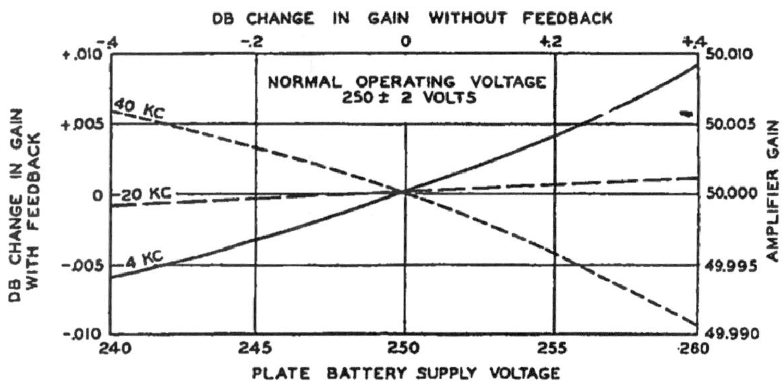

Fig. 9—Representative gain stability of a single amplifier as determined by measuring 69 feedback amplifiers in tandem at Morristown, N. J.

The upper figure shows the absolute value of the stability index. It can be seen that between 20 and 25 kc the improvement in stability is more than 1000 to 1 yet the reduction in gain was less than 35 db.

The lower figure shows change in gain of the feedback amplifier with changes in the plate battery voltage and the corresponding changes in gain without feedback. At some frequencies the change in gain is of the same sign as without feedback and at others it is of opposite sign and it can be seen that near 23 kc the stability must be perfect.

Figure 10 indicates the effectiveness with which the gain of a feedback amplifier can be made independent of variations in input amplitude up to practically the overload point of the amplifier. These measurements were made on a three-stage amplifier designed to work from 3.3 kc to 50 kc.

Figure 11 shows that negative feedback may be used to improve phase shift and reduce delay and delay distortion. These measurements

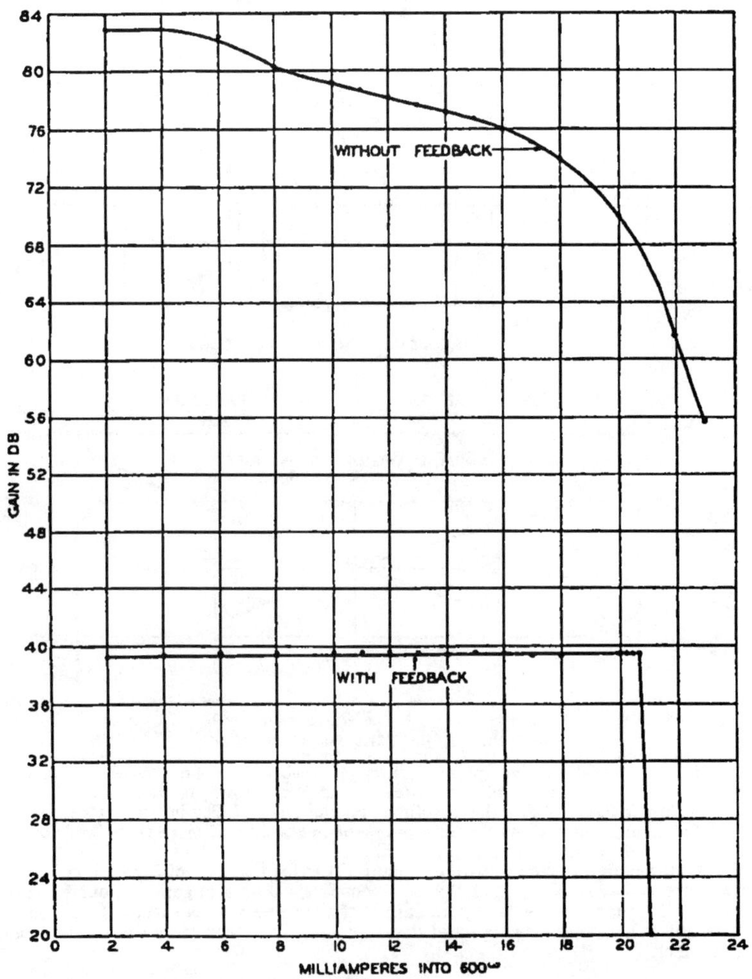

Fig. 10—Gain-load characteristic with and without feedback for a low level amplifier designed to amplify frequencies from 3.5 to 50 kc.

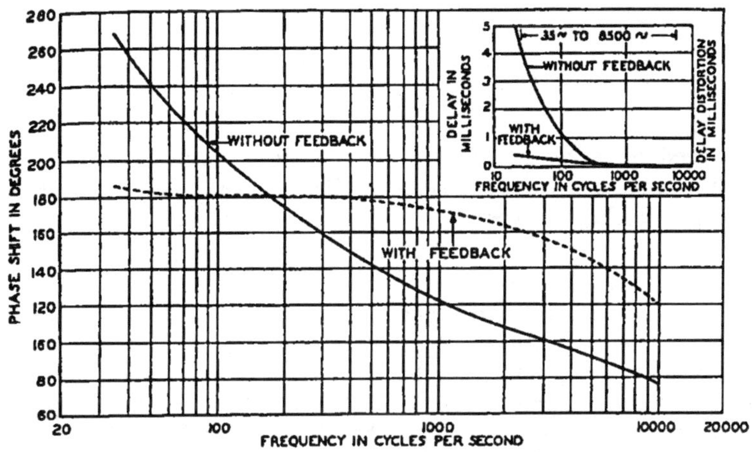

Fig. 11—Phase shift, delay, and delay distortion with and without feedback for a single tube voice frequency amplifier.

were made on an experimental one-tube amplifier, 35–8500 cycles, feeding back around the low side windings of the input and output transformers.

Figure 12 gives the gain-frequency characteristic of an amplifier with and without feedback when in the β-circuit there was an equalizer

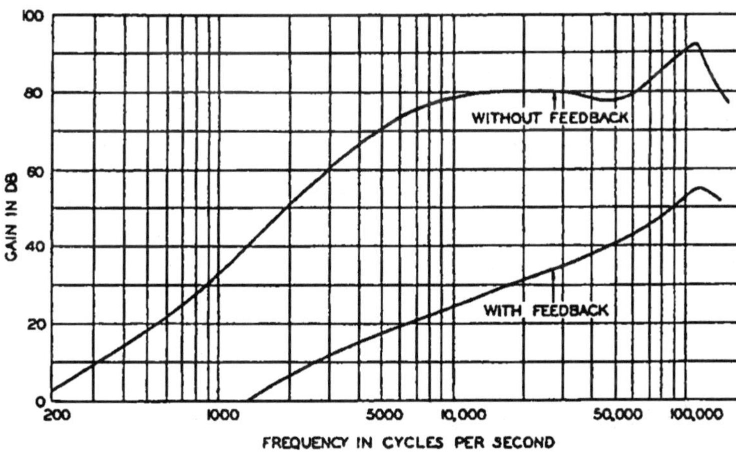

Fig. 12—Gain-frequency characteristic of an amplifier with an equalizer in the β-circuit. This was designed to have a gain frequency characteristic with feedback of the same shape as the loss frequency characteristic of a non-loaded telephone cable.

designed to make the gain-frequency characteristic of the amplifier with feedback of the same shape as the loss-frequency characteristic of a non-loaded telephone cable.

Conclusion

The feedback amplifier dealt with in this paper was developed primarily with requirements in mind for a cable carrier telephone system, involving many amplifiers in tandem with many telephone channels passing through each amplifier. Most of the examples of feedback amplifier performance have naturally been drawn from amplifiers designed for this field of operation. In this field, vacuum tube amplifiers normally possessing good characteristics with respect to stability and freedom from distortion are made to possess superlatively good characteristics by application of the feedback principle.

However, certain types of amplifiers in which economy has been secured by sacrificing performance characteristics, particularly as regards distortion, can be made to possess improved characteristics by the application of feedback. Discussion of these amplifiers is beyond the scope of this paper.

Relations Between Attenuation and Phase in Feedback Amplifier Design

H. W. BODE

LIKE the first two papers in this volume, Bode's paper also grew out of the research at Bell Laboratories aimed at developing the feedback amplifier. As indicated earlier in the preambles to the first two papers, Black invented the feedback amplifier, Nyquist developed a novel method of analyzing its stability, and Bode developed concepts and methods for its effective design. Bode's paper appeared eight years after Nyquist's, and six years after Black's, and this historical precedence is in fact reflected in the paper. Even though both Black and Nyquist hinted at design issues, it was Bode's paper which first gave a very mature view of design of feedback amplifiers.

A key result of the paper, and one that is central to feedback design, is the relationship between the magnitude and the phase of a transfer function. The paper contains, however, several other important results. It starts with a broad discussion of fundamental limitations, their role in engineering, and in particular in the design of feedback amplifiers. This discussion, written in a very lucid style, is still very readable and can be recommended to all engineers interested in design. Active bandpass filters implemented by feedback amplifiers are used to multiplex several phone conversations on one cable. The filters must have a certain bandwidth to accommodate a conversation with reasonable fidelity and the gain of the filters should drop off quickly outside the frequency band to allow many conversations on the same cable. Bode's analysis showed that (p. 422)

> "...the loop gain cannot be reduced too abruptly outside the frequency range which is to be transmitted if we wish to secure an unconditionally stable amplifier. It is necessary to allow at least a certain minimum interval before the loop gain can be reduced to zero."

Designers of feedback amplifiers were in the enviable position that they could simultaneously design the process and the controller. Bode's paper is thus a very early example of integrated process and controller design.

The key result of the paper alluded to earlier is embodied in Equation (2) of the paper, which gives the relationship between the magnitude and the phase of a transfer function. To formulate his results, Bode introduced the important concepts of *minimum phase* and *non-minimum phase*. He associated non-minimum phase with all-pass networks and transmission lines. There are no formal derivations in the paper, but Bode states in footnote 3 that the results can be derived by contour integration in the complex plane. Complete results can be found in Bode's book [1], which appeared later in 1945, and which gives a very complete presentation of theory for the analysis and design of feedback amplifiers. It contains, for example, the following interesting result on fundamental limitations, which is closely related to the results of this paper. The sensitivity function of a feedback system has the property

$$\int_0^\infty \log |S(i\omega)| d\omega = \pi \sum p_k$$

where p_k are the poles of the loop transfer function in the right half plane. This shows that if the sensitivity is reduced at some frequencies, then it must necessarily increase at other frequencies.

Another interesting result given in the paper is captured by its Equation (1), which states that for a minimum phase system the logarithm of the total gain is proportional to the area under the phase curve. This became known as *Bode's phase-area formula*.

The paper develops design procedures for feedback amplifiers. Because of the technology available at the time, much of the design was based on graphical constructions. A key observation is that the magnitude curve in the Bode plots could be approximated by straight line segments. This made it possible to design systems of very high order. An interesting perspective on this is provided by Bode in [2]:

> "As a matter of idle curiosity, I once counted to find what the order would have been, if I had worked with the differential equations directly. It turned out to be 55."

Bode noticed that for design purposes it is useful to have concepts that express the degree of stability. For this purpose he introduced the notions of *phase margin* and *amplitude margin*.

A key problem in the design of a feedback amplifier is to come up with a feedback loop so that the performance of the closed loop is invariant to changes in the amplifier gain. Bode presented a very elegant solution to this robust design problem, which he called the *ideal cutoff characteristic*. In today's terminology we could call it the *ideal loop transfer function*. Bode's idea was that a system whose Nyquist plot is a straight line through the origin as in Figure 18 of the paper has a phase margin that is invariant to gain changes. In practice, the Nyquist curve will only be linear in a limited frequency range. Bode showed how large the frequency range should be to accommodate a given gain variation. He thus came up with the first robust design method. He also rotated Nyquist's original diagram to put the critical point at -1. A more comprehensive treatment of this material can again be found in Bode's book [1].

Bode's paper had a tremendous impact. It contributed to a paradigm shift toward frequency response, which led to the emergence of servomechanism theory [8], [7]. The notions of minimum phase, non-minimum phase, gain margin, phase margin, Bode plots, Bode relations—all introduced in this paper—became key elements of the field of automatic control that emerged in the 1940s [7]. Bode's results are still essential elements of automatic control, which appear in practically all text books. His results on fundamental limitations were pursued further by Westcott [11] in 1952, but after that the problem lay dormant for many years. This is quite surprising particularly since results on fundamental limitations played such an important role in the neighboring field of communications [10]. There was, however, a revival of interest in the problem starting with the paper by Freudenberg and Looze [4], and lately there has been a substantial interest in fundamental limitations, see, e.g., [9]. An interesting problem that arises here is whether the limitations imposed by Bode's relations can be avoided by using nonlinear systems. A case in point is the Clegg integrator [3], which is a nonlinear system where the magnitude curve has the slope -1 and the phase lag is only $38°$.

Bode's ideal loop transfer function is not much discussed in the literature anymore, although it can be considered a natural beginning of robust control, as evidenced, for example, in the qualitative control theory of Horowitz [5], [6].

References

[1] H.W. Bode, *Network Analysis and Feedback Amplifier Design*, Van Nostrand (New York), 1945.

[2] H.W. Bode, "Feedback—The history of an idea," in *Proc. Symp. Active Networks and Feedback Systems*, Polytechnic Institute of Brooklyn, 1960.

[3] J.C. Clegg, "A nonlinear integrator for servomechanisms," *Trans. AIEE Part II*, **77**:41–42, 1958.

[4] J.S. Freudenberg and D.P. Looze, "Right half plane pole and zeros and design tradeoffs in feedback systems," *IEEE Trans. Automat. Contr.*, **AC-30**:555–565, 1985.

[5] I.M. Horowitz, *Synthesis of Feedback Systems*, Academic Press (New York), 1963.

[6] I.M. Horowitz, *Quantitative Feedback Design Theory (QFT)*, QFT Publications (Boulder, Col.), 1993.

[7] H.M. James, N.B. Nichols and R.S. Phillips, *Theory of Servomechanisms*, McGraw-Hill (New York), 1947.

[8] L.A. MacColl, *Fundamental theory of servomechanisms*. Van Nostrand (Princeton, NJ), 1945.

[9] M.M. Seron, J.H. Braslavsky and G.C. Goodwin, *Fundamental Limitations in Filtering and Control*, Springer (Berlin), 1997.

[10] C.E. Shannon, "A mathematical theory of communication," *Bell System Technical J.*, **7**:379–423 & 623–656, July–October 1948.

[11] J. Westcott, "The development of relationships from properties of gain frequency characteristics," in *Automatic and Manual Control*, A. Tustin, editor, Butterworth (London), 1952.

K.J.Å. & W.S.L.

Relations Between Attenuation and Phase in Feedback Amplifier Design

By H. W. BODE

INTRODUCTION

THE engineer who embarks upon the design of a feedback amplifier must be a creature of mixed emotions. On the one hand, he can rejoice in the improvements in the characteristics of the structure which feedback promises to secure him.[1] On the other hand, he knows that unless he can finally adjust the phase and attenuation characteristics around the feedback loop so the amplifier will not spontaneously burst into uncontrollable singing, none of these advantages can actually be realized. The emotional situation is much like that of an impecunious young man who has impetuously invited the lady of his heart to see a play, unmindful, for the moment, of the limitations of the $2.65 in his pockets. The rapturous comments of the girl on the way to the theater would be very pleasant if they were not shadowed by his private speculation about the cost of the tickets.

In many designs, particularly those requiring only moderate amounts of feedback, the bogy of instability turns out not to be serious after all. In others, however, the situation is like that of the young man who has just arrived at the box office and finds that his worst fears are realized. But the young man at least knows where he stands. The engineer's experience is more tantalizing. In typical designs the loop characteristic is always satisfactory—except for one little point. When the engineer changes the circuit to correct that point, however, difficulties appear somewhere else, and so on ad infinitum. The solution is always just around the corner.

Although the engineer absorbed in chasing this rainbow may not realize it, such an experience is almost as strong an indication of the existence of some fundamental physical limitation as the census which the young man takes of his pockets. It reminds one of the experience of the inventor of a perpetual motion machine. The perpetual motion machine, likewise, always works—except for one little factor. Evidently, this sort of frustration and lost motion is inevitable in

[1] A general acquaintance with feedback circuits and the uses of feedback is assumed in this paper. As a broad reference, see H. S. Black, "Stabilized Feedback Amplifiers," *B. S. T. J.*, January, 1934.

feedback amplifier design as long as the problem is attacked blindly. To avoid it, we must have some way of determining in advance when we are either attempting something which is beyond our resources, like the young man on the way to the theater, or something which is literally impossible, like the perpetual motion enthusiast.

This paper is written to call attention to several simple relations between the gain around an amplifier loop, and the phase change around the loop, which impose limits to what can and cannot be done in a feedback design. The relations are mathematical laws, which in their sphere have the same inviolable character as the physical law which forbids the building of a perpetual motion machine. They show that the attempt to build amplifiers with certain types of loop characteristics *must* fail. They permit other types of characteristic, but only at the cost of certain consequences which can be calculated. In particular, they show that the loop gain cannot be reduced too abruptly outside the frequency range which is to be transmitted if we wish to secure an unconditionally stable amplifier. It is necessary to allow at least a certain minimum interval before the loop gain can be reduced to zero.

The question of the rate at which the loop gain is reduced is an important one, because it measures the actual magnitude of the problem confronting both the designer and the manufacturer of the feedback structure. Until the loop gain is zero, the amplifier will sing unless the loop phase shift is of a prescribed type. The cutoff interval as well as the useful transmission band is therefore a region in which the characteristics of the apparatus must be controlled. The interval represents, in engineering terms, the price of the ticket.

The price turns out to be surprisingly high. It can be minimized by accepting an amplifier which is only conditionally stable.[3] For the customary absolutely stable amplifier, with ordinary margins against singing, however, the price in terms of cutoff interval is roughly one octave for each ten db of feedback in the useful band. In practice, an additional allowance of an octave or so, which can perhaps be regarded as the tip to the hat check girl, must be made to insure that the amplifier, having once cut off, will stay put. Thus in an amplifier with 30 db feedback, the frequency interval over which effective control of the loop transmission characteristics is necessary is at least four octaves, or sixteen times, broader than the useful band. If we raise the feedback to 60 db, the effective range must be more than a hundred times the useful range. If the useful band is itself large these factors

[3] Definitions of conditionally and unconditionally stable amplifiers are given on page 432.

may lead to enormous effective ranges. For example, in a 4 megacycle amplifier they indicate an effective range of about 60 megacycles for 30 db feedback, or of more than 400 megacycles if the feedback is 60 db.

The general engineering implications of this result are obvious. It evidently places a burden upon the designer far in excess of that which one might anticipate from a consideration of the useful band alone. In fact, if the required total range exceeds the band over which effective control of the amplifier loop characteristics is physically possible, because of parasitic effects, he is helpless. Like the young man, he simply can't pay for his ticket. The manufacturer, who must construct and test the apparatus to realize a prescribed characteristic over such wide bands, has perhaps a still more difficult problem. Unfortunately, the situation appears to be an inevitable one. The mathematical laws are inexorable.

Aside from sounding this warning, the relations between loop gain and loop phase can also be used to establish a definite method of design. The method depends upon the development of overall loop characteristics which give the optimum result, in a certain sense, consistent with the general laws. This reduces actual design procedure to the simulation of these characteristics by processes which are essentially equivalent to routine equalizer design. The laws may also be used to show how the characteristics should be modified when the cutoff interval approaches the limiting band width established by the parasitic elements of the circuit, and to determine how the maximum realizable feedback in any given situation can be calculated. These methods are developed at some length in the writer's U. S. Patent No. 2,123,178 and are explained in somewhat briefer terms here.

Relations Between Attenuation and Phase in Physical Networks[3]

The amplifier design theory advanced here depends upon a study of the transmission around the feedback loop in terms of a number of general laws relating the attenuation and phase characteristics of physical networks. In attacking this problem an immediate difficulty presents itself. It is apparent that no entirely definite and universal

[3] Network literature includes a long list of relations between attenuation and phase discovered by a variety of authors. They are derived typically from a Fourier analysis of the transient response of assumed structures and are frequently ambiguous, because of failure to recognize the minimum phase shift condition. No attempt is made to review this work here, although special mention should be made of Y. W. Lee's paper in the *Journal for Mathematics and Physics* for June, 1932. The proof of the relations given in the present paper depends upon a contour integration in the complex frequency plane and can be understood from the disclosure in the patent referred to previously.

relation between the attenuation and the phase shift of a physical structure can exist. For example, we can always change the phase shift of a circuit without affecting its loss by adding either an ideal transmission line or an all-pass section. Any attenuation characteristic can thus correspond to a vast variety of phase characteristics.

For the purposes of amplifier design this ambiguity is fortunately unimportant. While no unique relation between attenuation and phase can be stated for a general circuit, a unique relation does exist between any given loss characteristic and the *minimum* phase shift which must be associated with it. In other words, we can always add a line or all-pass network to the circuit but we can never subtract such a structure, unless, of course, it happens to be part of the circuit originally. If the circuit includes no surplus lines or all-pass sections, it will have at every frequency the least phase shift (algebraically) which can be obtained from any physical structure having the given attenuation characteristic. The least condition, since it is the most favorable one, is, of course, of particular interest in feedback amplifier design.

For the sake of precision it may be desirable to restate the situations in which this minimum condition fails to occur. The first situation is found when the circuit includes an all-pass network either as an individual structure or as a portion of a network which can be replaced by an all-pass section in combination with some other physical structure.[4] The second situation is found when the circuit includes a transmission line. The third situation occurs when the frequency is so high that the tubes, network elements and wiring cannot be considered to obey a lumped constant analysis. This situation may be found, for example, at frequencies for which the transit time of the tubes is important or for which the distance around the feedback loop is an appreciable part of a wave-length. The third situation is, in many respects, substantially the same as the second, but it is mentioned separately here as a matter of emphasis. Since the effective band of a feedback amplifier is much greater than its useful band, as the introduction pointed out, the considerations it reflects may be worth taking into account even when they would be trivial in the useful band alone.

It will be assumed here that none of these exceptional situations is found. For the minimum phase condition, then, it is possible to derive

[4] Analytically this condition can be stated as follows: Let it be supposed that the transmission takes place between mesh 1 and mesh 2. The circuit will include an all-pass network, explicit or concealed, if any of the roots of the minor Δ_{12} of the principal circuit determinant lie below the real axis in the complex frequency plane. This can happen in bridge configurations, but not in series-shunt configurations, so that all ladder networks are automatically of minimum phase type.

a large number of relations between the attenuation and phase characteristics of a physical network. One of the simplest is

$$\int_{-\infty}^{\infty} B du = \frac{\pi}{2}(A_\infty - A_0), \tag{1}$$

where u represents $\log f/f_0$, f_0 being an arbitrary reference frequency, B is the phase shift in radians, and A_0 and A_∞ are the attenuations in nepers at zero and infinite frequency, respectively. The theorem states, in effect, that the total area under the phase characteristic plotted on a logarithmic frequency scale depends only upon the difference between the attenuations at zero and infinite frequency, and not upon the course of the attenuation between these limits. Nor does it depend upon the physical configuration of the network unless a non-minimum phase structure is chosen, in which case the area is necessarily increased. The equality of phase areas for attenuation characteristics of different types is illustrated by the sketches of Fig. 1.

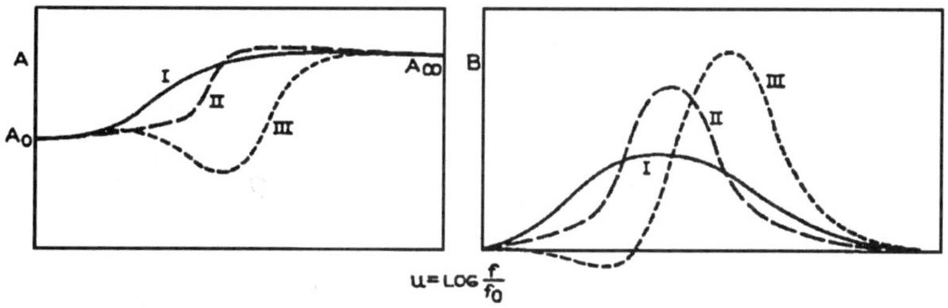

Fig. 1—Diagram to illustrate relation between phase area and change in attenuation.

The significance of the phase area relation for feedback amplifier design can be understood by supposing that the practical transmission range of the amplifier extends from zero to some given finite frequency. The quantity $A_0 - A_\infty$ can then be identified with the change in gain around the feedback loop required to secure a cut-off. Associated with it must be a certain definite phase area. If we suppose that the maximum phase shift at any frequency is limited to some rather low value the total area must be spread out over a proportionately broad interval on the frequency scale. This must correspond roughly to the cut-off region, although the possibility that some of the area may be found above or below the cut-off range prevents us from determining the necessary interval with precision.

A more detailed statement of the relationship between phase shift and change in attenuation can be obtained by turning to a second theorem. It reads as follows:

$$B(f_c) = \frac{1}{\pi} \int_{-\infty}^{\infty} \frac{dA}{du} \log \coth \frac{|u|}{2} du, \tag{2}$$

where $B(f_c)$ represents the phase shift at any arbitrarily chosen frequency f_c and $u = \log f/f_c$. This equation, like (1), holds only for the minimum phase shift case.

Although equation (2) is somewhat more complicated than its predecessor, it lends itself to an equally simple physical interpretation. It is clear, to begin with, that the equation implies broadly that the

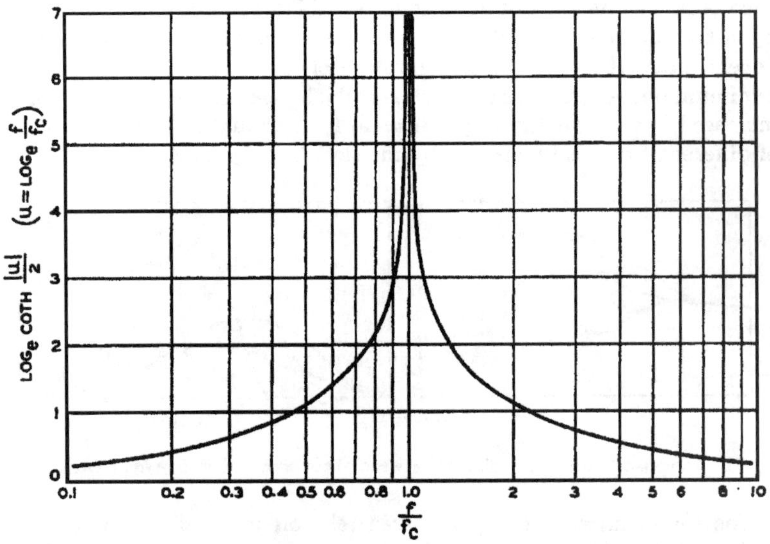

Fig. 2—Weighting function in loss-phase formula.

phase shift at any frequency is proportional to the derivative of the attenuation on a logarithmic frequency scale. For example, if dA/du is doubled B will also be doubled. The phase shift at any particular frequency, however, does not depend upon the derivative of attenuation at that frequency alone, but upon the derivative at all frequencies, since it involves a summing up, or integration, of contributions from the complete frequency spectrum. Finally, we notice that the contributions to the total phase shift from the various portions of the frequency spectrum do not add up equally, but rather in accordance with the function log coth $|u|/2$. This quantity, therefore, acts as a weighting function. It is plotted in Fig. 2. As we might expect physically it is much larger near the point $u = 0$ than it is in other regions. We can, therefore, conclude that while the derivative of attenuation at all frequencies enters into the phase shift at any particular frequency $f = f_c$ the derivative in the neighborhood of f_c is relatively much more important than the derivative in remote parts of the spectrum.

As an illustration of (2), let it be supposed that $A = ku$, which corresponds to an attenuation having a constant slope of 6 k db per octave. The associated phase shift is easily evaluated. It turns out, as we might expect, to be constant, and is equal numerically to $k\pi/2$ radians. This is illustrated by Fig. 3. As a second example, we may consider

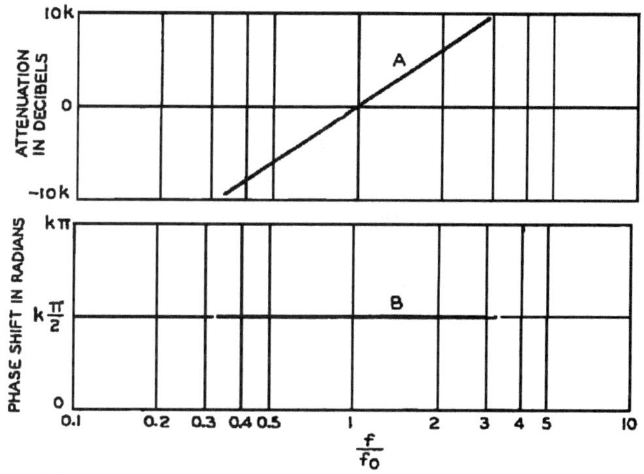

Fig. 3—Phase characteristic corresponding to a constant slope attenuation.

a discontinuous attenuation characteristic such as that shown in Fig. 4. The associated phase characteristic, also shown in Fig. 4, is proportional to the weighting function of Fig. 2.

The final example is shown by Fig. 5. It consists of an attenuation characteristic which is constant below a specified frequency f_b and has a constant slope of 6 k db per octave above f_b. The associated phase characteristic is symmetrical about the transition point between the two ranges. At sufficiently high frequencies, the phase shift approaches the limiting $k\pi/2$ radians which would be realized if the constant slope were maintained over the complete spectrum. At low frequencies the phase shift is substantially proportional to frequency and is given by the equation

$$B = \frac{2k}{\pi} \frac{f}{f_b}. \tag{3}$$

Solutions developed in this way can be added together, since it is apparent from the general relation upon which they are based that the phase characteristic corresponding to the sum of two attenuation

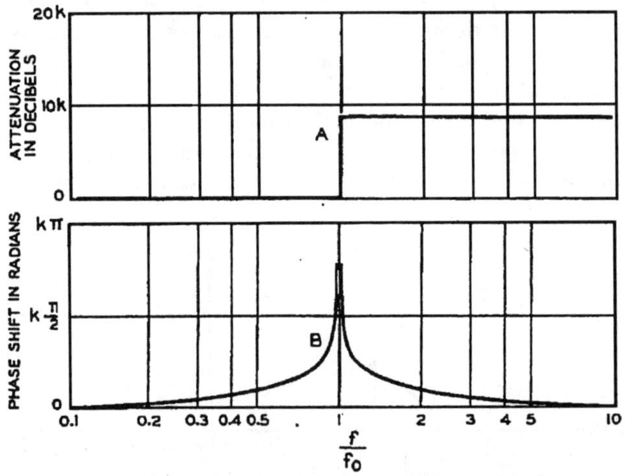

Fig. 4—Phase characteristic corresponding to a discontinuity in attenuation.

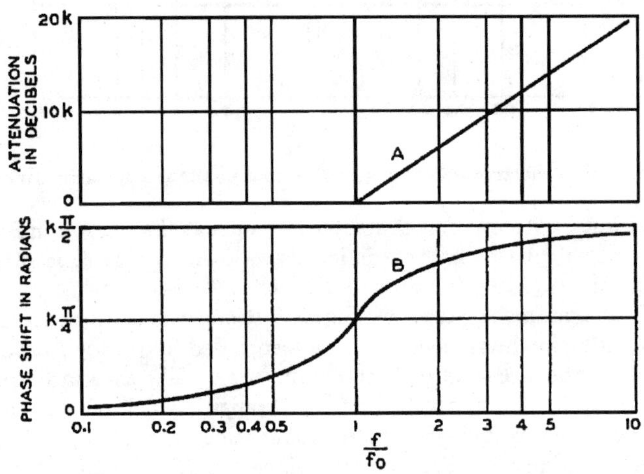

Fig. 5—Phase characteristic corresponding to an attenuation which is constant below a prescribed frequency and has a constant slope above it.

characteristics will be equal to the sum of the phase characteristics corresponding to the two attenuation characteristics separately. We can therefore combine elementary solutions to secure more complicated characteristics. An example is furnished by Fig. 6, which is built up from three solutions of the type shown by Fig. 5. By proceeding sufficiently far in this way, an approximate computation of the phase characteristic associated with almost any attenuation characteristic can be made, without the labor of actually performing the integration in (2).

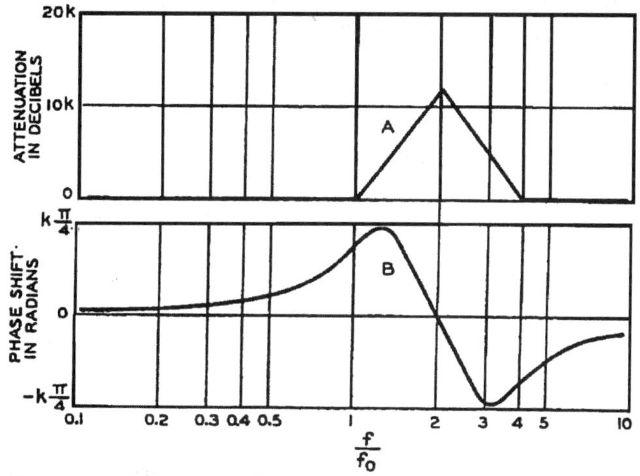

Fig. 6—Diagram to illustrate addition of elementary attenuation and phase characteristics to produce more elaborate solutions of the loss-phase formula.

Equations (1) and (2) are the most satisfactory expressions to use in studying the relation between loss and phase in a broad physical sense. The mechanics of constructing detailed loop cut-off characteristics, however, are simplified by the inclusion of one other, somewhat more complicated, formula. It appears as

$$\int_0^{f_0} \frac{A\,df}{\sqrt{f_0^2 - f^2}\,(f^2 - f_c^2)} + \int_{f_0}^{\infty} \frac{B\,df}{\sqrt{f^2 - f_0^2}\,(f^2 - f_c^2)}$$

$$= \frac{\pi}{2} \frac{B(f_c)}{f_c\sqrt{f_0^2 - f_c^2}}, \qquad f_c < f_0$$

$$= -\frac{\pi}{2} \frac{A(f_c)}{f_c\sqrt{f_c^2 - f_0^2}}, \qquad f_c > f_0, \quad (4)$$

where f_0 is some arbitrarily chosen frequency and the other symbols have their previous significance.

The meaning of (4) can be understood if it is recalled that (2) implies that the minimum phase shift at any frequency can be computed if the attenuation is prescribed at all frequencies. In the same way (4) shows how the complete attenuation and phase characteristics can be determined if we begin by prescribing the attenuation below f_0 and the phase shift above f_0. Since f_0 can be chosen arbitrarily large or small this is evidently a more general formula than either (1) or (2), while it can itself be generalized, by the introduction of additional irrational factors, to provide for more elaborate patterns of bands in which A and B are specified alternately.

As an example of this formula, let it be assumed that $A = K$ for $f < f_0$ and that $B = k\pi/2$ for $f > f_0$. These are shown by the solid lines in Fig. 7. Substitution in (4) gives the A and B characteristics in the rest of the spectrum as

$$B = k \sin^{-1}\frac{f}{f_0}, \qquad f < f_0$$

$$A = K + k \log\left[\sqrt{\frac{f^2}{f_0^2} - 1} + \frac{f}{f_0}\right], \qquad f > f_0. \tag{5}$$

These are indicated by broken lines in Fig. 7. In this particularly

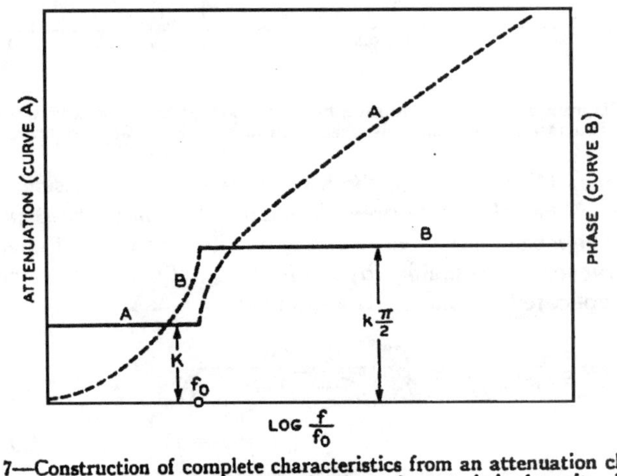

Fig. 7—Construction of complete characteristics from an attenuation characteristic specified below a certain frequency and a phase characteristic above it. The solid lines represent the specified attenuation and phase characteristics, and the broken lines their computed extensions to the rest of the spectrum.

simple case all four fragments can be combined into the single analytic formula

$$A + iB = K + k \log\left[\sqrt{1 - \frac{f^2}{f_0^2}} + i\frac{f}{f_0}\right]. \tag{6}$$

This expression will be used as the fundamental formula for the loop cut-off characteristic in the next section.

Overall Feedback Loop Characteristics

The survey just concluded shows what combinations of attenuation and phase characteristics are physically possible. We have next to determine which of the available combinations is to be regarded as representing the transmission around the overall feedback loop. The choice will naturally depend somewhat upon exactly what we assume that the amplifier ought to do, but with any given set of assumptions it is possible, at least in theory, to determine what combination is most appropriate.

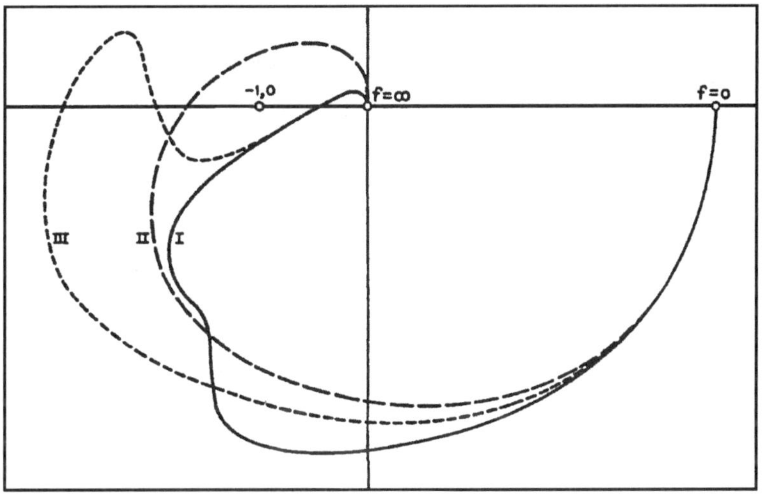

Fig. 8—Nyquist stability diagrams for various amplifiers. Curve I represents "absolute" stability, Curve II instability, and Curve III "conditional" stability. In accordance with the convention used in this paper the diagram is rotated through 180° from its normal position so that the critical point occurs at − 1, 0 rather than + 1, 0.

The situation is conveniently investigated by means of the Nyquist stability diagram [5] illustrated by Fig. 8. The diagram gives the path

[5] *Bell System Technical Journal*, July, 1932. See also Peterson, Kreer, and Ware, *Bell System Technical Journal*, October, 1934. The Nyquist diagrams in the present paper are rotated through 180° from the positions in which they are usually drawn, turning the diagrams in reality into plots of $-\mu\beta$. In a normal amplifier there is one net phase reversal due to the tubes in addition to any phase shifts chargeable directly to the passive networks in the circuit. The rotation of the diagram allows this phase reversal to be ignored, so that the phase shifts actually shown are the same as those which are directly of design interest.

traced by the vector representing the transmission around the feedback loop as the frequency is assigned all possible real values. In accordance with Nyquist's results a path such as II, which encircles the point $-1, 0$, indicates an unstable circuit and must be avoided. A stable amplifier is obtained if the path resembles either I or III, neither of which encircles $-1, 0$. The stability represented by Curve III, however, is only "Nyquist" or "conditional." The path will enclose the critical point if it is merely reduced in scale, which may correspond physically to a reduction in tube gain. Thus the circuit may sing when the tubes begin to lose their gain because of age, and it may also sing, instead of behaving as it should, when the tube gain increases from zero as power is first applied to the circuit. Because of these possibilities conditional stability is usually regarded as undesirable and the present discussion will consequently be restricted to "absolutely" or "unconditionally" stable amplifiers having Nyquist diagrams of the type resembling Curve I.

The condition that the amplifier be absolutely stable is evidently that the loop phase shift should not exceed 180° until the gain around the loop has been reduced to zero or less. A theoretical characteristic which just met this requirement, however, would be unsatisfactory, since it is inevitable that the limiting phase would be exceeded in fact by minor deviations introduced either in the detailed design of the amplifier or in its construction. It will therefore be assumed that the limiting phase is taken as 180° less some definite margin. This is illustrated by Fig. 9, the phase margin being indicated as $y\pi$ radians. At frequencies remote from the band it is physically impossible, in most circuits, to restrict the phase within these limits. As a supplement, therefore, it will be assumed that larger phase shifts are permissible if the loop gain is x db below zero. This is illustrated by the broken circular arc in Fig. 9. A theoretical loop characteristic meeting both requirements will be developed for an amplifier transmitting between zero and some prescribed limiting frequency with a constant feedback, and cutting off thereafter as rapidly as possible. This basic characteristic can be adapted to amplifiers with varying feedback in the useful range or with useful ranges lying in other parts of the spectrum by comparatively simple modifications which are described at a later point. It is, of course, contemplated that the gain and phase margins x and y will be chosen arbitrarily in advance. If we choose large values we can permit correspondingly large tolerances in the detailed design and construction of the apparatus without risk of instability. It turns out, however, that with a prescribed width of cutoff interval the amount of feedback which can be realized in the useful range is decreased as the assumed margins are increased, so that it is generally desirable to choose as small margins as is safe.

The essential feature in this situation is the requirement that the diminution of the loop gain in the cutoff region should not be accompanied by a phase shift exceeding some prescribed amount. In view of the close connection between phase shift and the slope of the attenuation characteristic evidenced by (2) this evidently demands that the amplifier should cut off, on the whole, at a well defined rate which is not too fast. As a first approximation, in fact, we can choose the cutoff

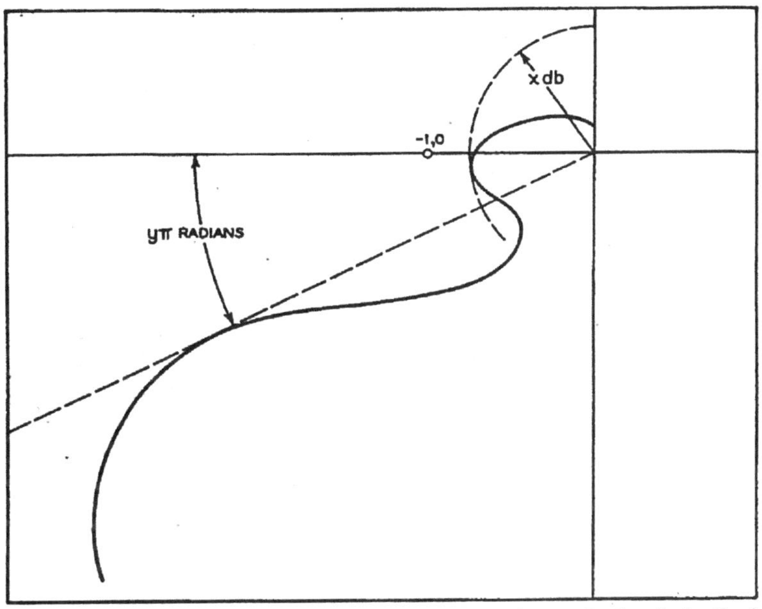

Fig. 9—Diagram to illustrate definitions of phase and gain margins for the feedback loop.

characteristic as an exactly constant slope from the edge of the useful band outward. Such a characteristic has already been illustrated by Fig. 5 and is shown, replotted,[6] by the broken lines in Fig. 10. If we choose the parameter corresponding to k in Fig. 5 as 2 the cutoff rate is 12 db per octave and the phase shift is substantially 180° at high frequencies. This choice thus leads to zero phase margin. By choosing a somewhat smaller k on the other hand, we can provide a definite

[6] To prevent confusion it should be noticed that the general attenuation-phase diagrams are plotted in terms of relative loss while loop cutoff characteristics, here and at later points, are plotted in terms of relative gain.

margin against singing, at the cost of a less rapid cutoff. For example, if we choose $k = 1.5$ the limiting phase shift in the $\mu\beta$ loop becomes 135°, which provides a margin of 45° against instability, while the rate of cutoff is reduced to 9 db per octave. The value $k = 1.67$, which corresponds to a cutoff rate of 10 db per octave and a phase margin of 30°, has been chosen for illustrative purposes in preparing Fig. 10. The loss margin depends upon considerations which will appear at a later point.

Although characteristics of the type shown by Fig. 5 are reasonably satisfactory as amplifier cutoffs they evidently provide a greater phase

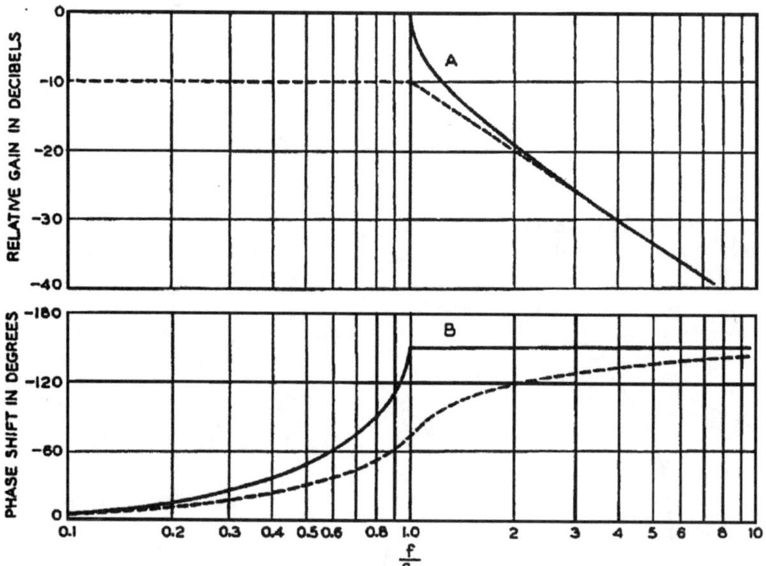

Fig. 10—Ideal loop cutoff characteristics. Drawn for a 30° phase margin.

margin against instability in the region just beyond the useful band than they do at high frequencies. In virtue of the phase area law this must be inefficient if, as is supposed here, the optimum characteristic is one which would provide a constant margin throughout the cutoff interval. The relation between the phase and the slope of the attenuation suggests that a constant phase margin can be obtained by increasing the slope of the cutoff characteristic near the edge of the band, leaving its slope at more remote frequencies unchanged, as shown by the solid lines in Fig. 10. The exact expression for the required curve can be found from (6), where the problem of determining such a characteristic appeared as an example of the use of the general formula (4).

At high frequencies the new phase and attenuation characteristics merge with those obtained from the preceding straight line cutoff, as Fig. 10 indicates. In this region the relation between phase margin and cutoff slope is fixed by the k in the equation (6) in the manner already described for the more elementary cutoff. At low frequencies, however, the increased slope near the edge of the band permits 6 k db more feedback.

It is worth while to pause here to consider what may be said, on the basis of these characteristics, concerning the breadth of cutoff interval required for a given feedback, or the "price of the ticket," as it was expressed in the introduction. If we adopt the straight line cutoff and assume the k used in Fig. 10 the interval between the edge of the useful band and the intersection of the characteristic with the zero gain axis is evidently exactly 1 octave for each 10 db of low frequency feedback. The increased efficiency of the solid line characteristic saves one octave of this total if the feedback is reasonably large to begin with. This apparently leads to a net interval one or two octaves narrower than the estimates made in the introduction. The additional interval is required to bridge the gap between a purely mathematical formula such as (6), which implies that the loop characteristics follow a prescribed law up to indefinitely high frequencies, and a physical amplifier, whose ultimate loop characteristics vary in some uncontrollable way. This will be discussed later. It is evident, of course, that the cutoff interval will depend slightly upon the margins assumed. For example, if the phase margin is allowed to vanish the cutoff rate can be increased from 10 to 12 db per octave. This, however, is not sufficient to affect the order of magnitude of the result. Since the diminished margin is accompanied by a corresponding increase in the precision with which the apparatus must be manufactured such an economy is, in fact, a Pyrrhic victory unless it is dictated by some such compelling consideration as that described in the next section.

Maximum Obtainable Feedback

A particularly interesting consequence of the relation between feedback and cutoff interval is the fact that it shows why we cannot obtain unconditionally stable amplifiers with as much feedback as we please. So far as the purely theoretical construction of curves such as those in Fig. 10 is concerned, there is clearly no limit to the feedback which can be postulated. As the feedback is increased, however, the cutoff interval extends to higher and higher frequencies. The process reaches a physical limit when the frequency becomes so high that parasitic effects in the circuit are controlling and do not permit the prescribed cutoff characteristic to be simulated with sufficient precision. For example, we are obviously in physical difficulties if the cutoff characteristic specifies a net gain around the loop at a frequency so high that the tubes themselves working into their own parasitic capacitances do not give a gain.

This limitation is studied most easily if the effects of the parasitic elements are lumped together by representing them in terms of the asymptotic characteristic of the loop as a whole at extremely high frequencies. An example is shown by Fig. 11. The structure is a

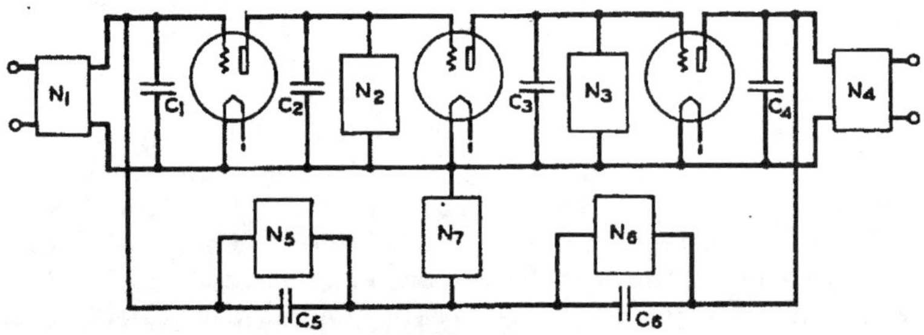

Fig. 11—Elements which determine the asymptotic loop transmission characteristic in a typical amplifier.

shunt feedback amplifier. The β circuit is represented by the T composed of networks N_5, N_6 and N_7. The input and output circuits are represented by N_1 and N_4 and the interstage impedances by N_2 and N_3. The C's are parasitic capacitances with the exception of C_5 and C_6, which may be regarded as design elements added deliberately to N_5 and N_6 to obtain an efficient high frequency transmission path from output to input. At sufficiently high frequencies the loop transmission will depend only upon these various capacitances, without regard to the N's. Thus, if the transconductances of the tubes are represented by G_1, G_2, and G_3 the asymptotic gains of the first two tubes are $G_1/\omega C_1$ and $G_2/\omega C_3$. The rest of the loop includes the third tube and the potentiometer formed by the capacitances C_1, C_4, C_5 and C_6. Its asymptotic transmission can be written as $G_3/\omega C$, where

$$C = C_1 + C_4 + \frac{C_1 C_4}{C_5 C_6}(C_5 + C_6).$$

Each of these terms diminishes at a rate of 6 db per octave. The complete asymptote is $G_1 G_2 G_3/\omega^3 C C_2 C_3$. It appears as a straight line with a slope of 18 db per octave when plotted on logarithmic frequency paper.

A similar analysis can evidently be made for any amplifier. In the particular circuit shown by Fig. 11 the slope of the asymptote, in units of 6 db per octave, is the same as the number of tubes in the circuit. The slope can evidently not be less than the number of tubes but it may be greater in some circuits. For example if C_5 and C_6 were omitted in Fig. 11 and N_5 and N_6 were regarded as degenerating into resistances the asymptote would have a slope of 24 db per octave and would lie below the present asymptote at any reasonably high frequency. In any event the asymptote will depend only upon the parasitic elements of the circuit and perhaps a few of the most significant design elements. It can thus be determined from a skeletonized version of the final structure. If waste of time in false starts is to be avoided such a determination should be made as early as possible, and certainly in advance of any detailed design.

The effect of the asymptote on the overall feedback characteristic is illustrated by Fig. 12. The curve $ABEF$ is a reproduction of the ideal

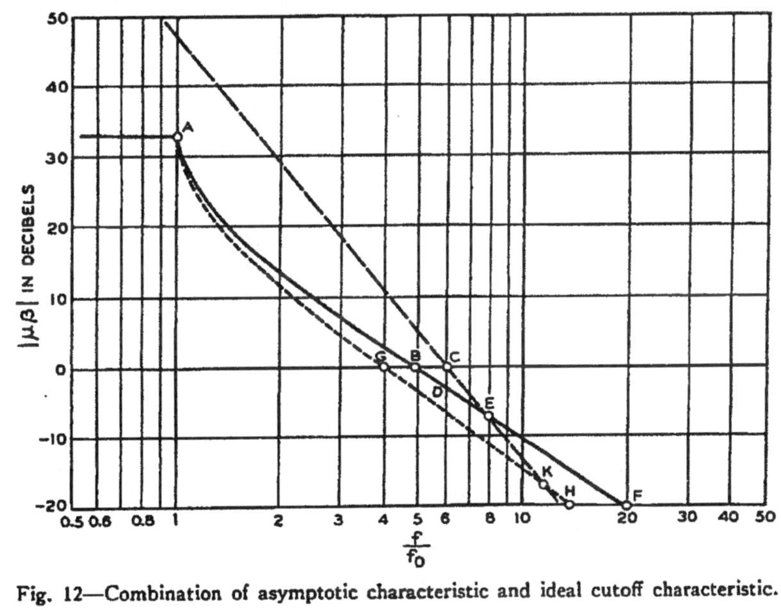

Fig. 12—Combination of asymptotic characteristic and ideal cutoff characteristic.

cutoff characteristic originally given by the solid lines in Fig. 10. It will be recalled that the curve was drawn for the choice $k = 5/3$, which corresponds to a phase margin of 30° and an almost constant slope, for the portion DEF of the characteristic, of about 10 db per octave. The straight line CEK represents an asymptote of the type just described, with a slope of 18 db per octave. Since the asymptote may be assumed to represent the practical upper limit of gain in the high-frequency region, the effect of the parasitic elements can be obtained by replacing the theoretical cutoff by the broken line characteristic $ABDEK$. In an actual circuit the corner at E would, of course, be rounded off, but this is of negligible quantitative importance. Since EF and EK diverge by 8 db per octave the effect can be studied by adding curves of the type shown by Fig. 5 to the original cutoff characteristic.

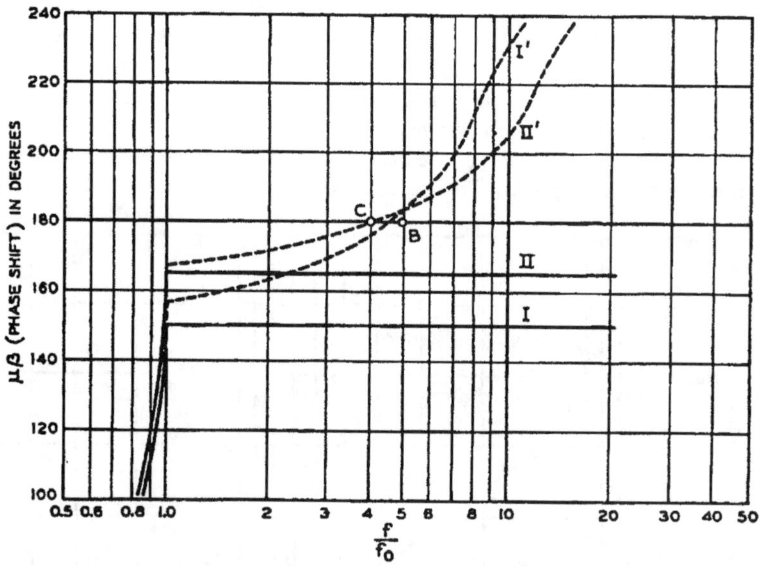

Fig. 13—Phase characteristics corresponding to gain characteristics of Fig. 12.

The phase shift in the ideal case is shown by Curve I of Fig. 13. The addition of the phase corresponding to the extra slope of 8 db per octave at high frequencies produces the total phase characteristic shown by Curve I′. At the point B where $|\mu\beta| = 1$, the additional phase shift amounts to 35 degrees. Since this is greater than the original phase margin of 30 degrees the amplifier is unstable when parasitic elements are considered. In the present instance stability can be regained by increasing the coefficient k to 1-5/6, which leads to the broken line characteristic $AGKH$ in Fig. 12. This reduces the nominal phase margin to 15 degrees, but the frequency interval between G and K is so much greater than that between B and E that the added phase is reduced still more and is just less than 15° at the new cross over point G. This is illustrated by II and II′ in Fig. 13. On the other hand, if the zero gain intercept of the asymptote CEK had occurred at a slightly lower frequency, no change in k alone would have been sufficient. It would have been necessary to reduce the amount of feedback in the transmitted range in order to secure stability.

The final characteristic in Fig. 13 reaches the limiting phase shift of 180° only at the crossover point. It is evident that a somewhat more efficient solution for the extreme case is obtained if the limiting 180° is approximated throughout the cutoff interval. This result is attained by the cutoff characteristic shown in Fig. 14. The characteristic con-

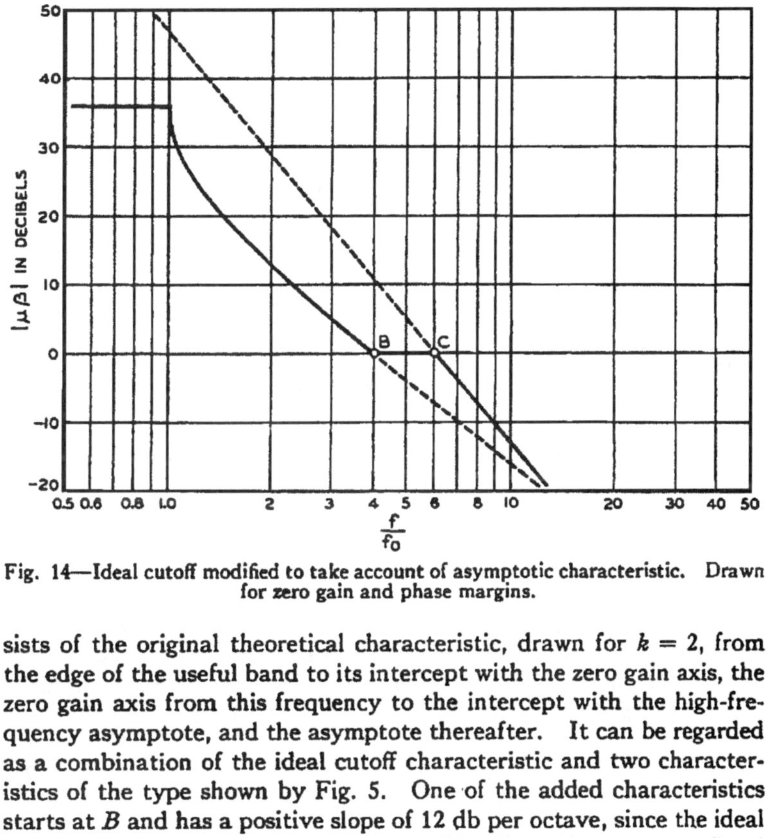

Fig. 14—Ideal cutoff modified to take account of asymptotic characteristic. Drawn for zero gain and phase margins.

sists of the original theoretical characteristic, drawn for $k = 2$, from the edge of the useful band to its intercept with the zero gain axis, the zero gain axis from this frequency to the intercept with the high-frequency asymptote, and the asymptote thereafter. It can be regarded as a combination of the ideal cutoff characteristic and two characteristics of the type shown by Fig. 5. One of the added characteristics starts at B and has a positive slope of 12 db per octave, since the ideal cutoff was drawn for the limiting value of k. The other starts at C and has the negative slope, − 18 db per octave, of the asymptote itself. As (3) shows, the added slopes correspond at lower frequencies to approximately linear phase characteristics of opposite sign. If the frequencies B and C at which the slopes begin are in the same ratio, 12 : 18, as the slopes themselves the contributions of the added slopes will substantially cancel each other and the net phase shift throughout the cutoff interval will be almost the same as that of the ideal curve alone. The exact phase characteristic is shown by Fig. 15. It dips

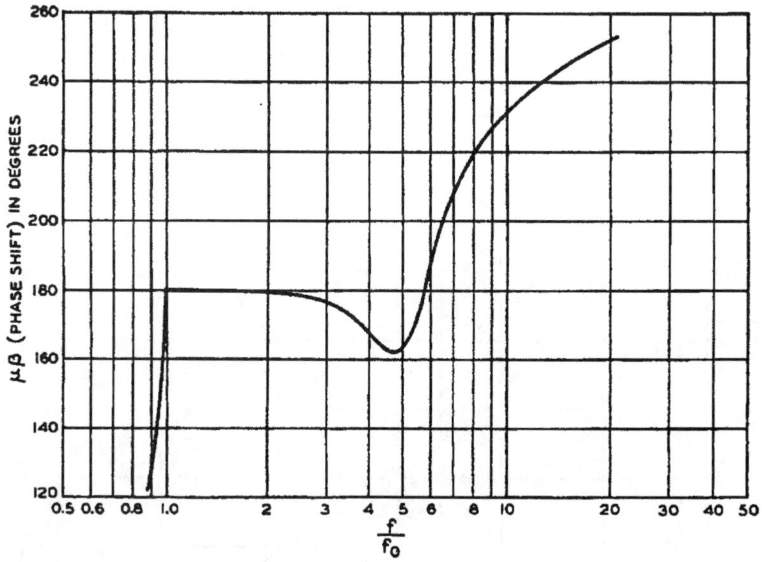

Fig. 15—Phase characteristic corresponding to gain characteristic of Fig. 14.

slightly below 180° at the point at which the characteristic reaches the zero gain axis, so that the circuit is in fact stable.

The same analysis can evidently be applied to asymptotes of any other slope. This makes it easy to compute the maximum feedback obtainable under any asymptotic conditions. If f_0 and f_a are respectively the edge of the useful band and the intercept (C in Figs. 12 and 14) of the asymptote with the zero gain axis, and n is the asymptotic slope, in units of 6 db per octave, the result appears as

$$A_m = 40 \log_{10} \frac{4f_a}{f_0}, \tag{7}$$

where A_m is the maximum feedback in db.[7]

[7] The formulæ for maximum feedback given here and in the later equation (8) are slightly conservative. It follows from the phase area law that more feedback should be obtained if the phase shift were exactly 180° below the crossover and rose

For the sake of generality it is convenient to extend this formula to include also situations in which there exists some further linear phase characteristic in addition to those already taken into account. In exceptional circuits, the final asymptotic characteristic may not be completely established by the time the curve reaches the zero gain axis and the additional phase characteristic may be used to represent the effect of subsequent changes in the asymptotic slope. Such a situation might occur in the circuit of Fig. 11, for example, if C_5 or C_6 were made extremely small. The additional term may also be used to represent departures from a lumped constant analysis in high-frequency amplifiers, as discussed earlier. If we specify the added phase characteristic, from whatever source, by means of the frequency f_d at which it would equal $2n/\pi$ radians, if extrapolated, the general formula corresponding to (7) becomes

$$A_m = 40 \log_{10} \frac{4}{nf_0} \frac{f_a f_d}{f_a + f_d}. \tag{8}$$

It is interesting to notice that equations (7) and (8) take no explicit account of the final external gain of the amplifier. Naturally, if the external gain is too high the available μ circuit gain may not be sufficient to provide it and also the feedback which these formulæ promise. This, however, is an elementary question which requires no further discussion. In other circumstances, the external gain may enter the situation indirectly, by affecting the asymptotic characteristics of the β path, but in a well chosen β circuit this is usually a minor consideration. The external gain does, however, affect the parts of the circuit upon which reliance must be placed in controlling the overall loop characteristic. For example, if the external gain is high the μ circuit will ordinarily be sharply tuned and will drop off rapidly in gain beyond the useful band. The β circuit must therefore provide a decreasing loss to bring the overall cutoff rate within the required limit. Since the β circuit must have initially a high loss to correspond to the high final gain of the complete amplifier, this is possible. Conversely, if the gain of the amplifier is low the μ circuit will be relatively flexible and the β circuit relatively inflexible.

rapidly to its ultimate value thereafter. These possibilities can be exploited approximately by various slight changes in the slope of the cutoff characteristic in the neighborhood of the crossover region, or a theoretical solution can be obtained by introducing a prescribed phase shift of this type in the general formula (4). The theoretical solution gives a Nyquist path which, after dropping below the critical point with a phase shift slightly less than 180°, rises again with a phase shift slightly greater than 180° and continues for some time with a large amplitude and increasing phase before it finally approaches the origin. These possibilities are not considered seriously here because they lead to only a few db increase in feedback, at least for moderate n's, and the degree of design control which they envisage is scarcely feasible in a frequency region where, by definition, parasitic effects are almost controlling.

In setting up (7) and (8) it has been assumed that the amplifier will, if necessary, be built with zero margins against singing. Any surplus which the equations indicate over the actual feedback required can, of course, be used to provide a cutoff characteristic having definite phase and gain margins. For example, if we begin with a lower feedback in the useful band the derivative of the attenuation between this region and the crossover can be proportionately reduced, with a corresponding decrease in phase shift. We can also carry the flat portion of the characteristic below the zero gain axis, thus providing a gain margin when the phase characteristic crosses 180°. In reproportioning the characteristic to suit these conditions, use may be made of the approximate formula

$$A_m - A = (A_m + 17.4)y + \frac{n-2}{n}x + \frac{2}{n}xy, \tag{9}$$

where A_m is the maximum obtainable feedback (in db), A is the actual feedback, and x and y are the gain and phase margins in the notation of Fig. 9. Once the available margin has been divided between the x and

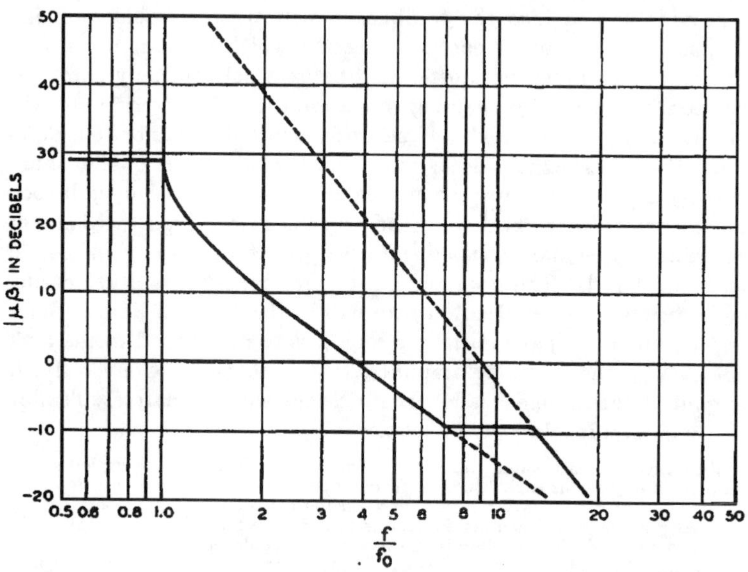

Fig. 16—Modified cutoff permitting 30° phase margin and 9 db gain margin.

y components by means of this formula, the cutoff characteristic is, of course, readily drawn in. An example is furnished by Figs. 16 and 17, where it is assumed that $A_m = 43$ db, $A = 29$ db, $x = 9$ db, $n = 3$ and $y = 1/6$. The Nyquist diagram for the structure is shown by Fig. 18. It evidently coincides almost exactly with the diagram postulated originally in Fig. 9.

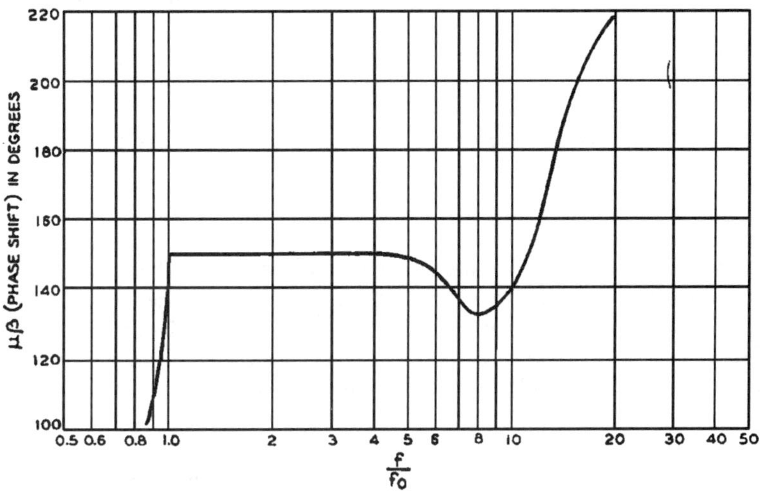

Fig. 17—Phase characteristic corresponding to gain characteristic of Fig. 16.

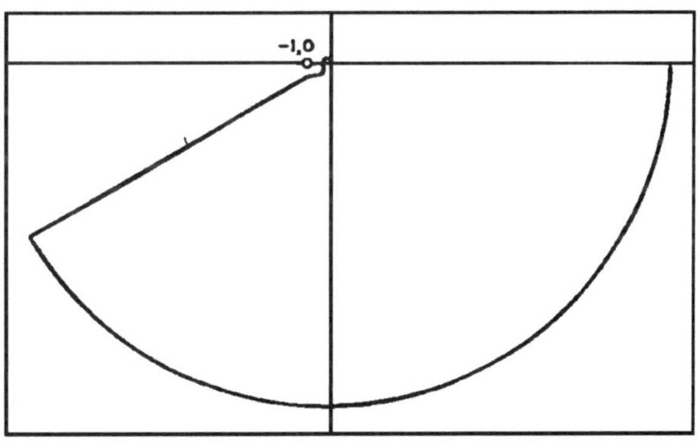

Fig. 18—Nyquist diagram corresponding to gain and phase characteristics of Figs. 16 and 17. As in Fig. 8 the diagram is rotated to place the critical point at $-1, 0$ rather than $+1, 0$.

With the characteristic of Fig. 16 at hand, we can return once more to the calculation of the total design range corresponding to any given feedback. From the useful band to the intersection of the cutoff characteristic with the zero gain axis the calculation is the same as that made previously in connection with Fig. 10. From the zero gain intercept to the junction with the asymptote, where we can say that design control is finally relaxed, there is, however, an additional interval of nearly two octaves. Although Fig. 16 is fairly typical, the exact breadth of the additional interval will depend somewhat on circumstances. It is increased by an increase in the asymptotic slope and reduced by decreasing the gain margin.

Relative Importance of Tubes and Circuit in Limiting Feedback [8]

The discussion just finished leads to the general conclusion that the feedback which can be obtained in any given amplifier depends ultimately upon the high-frequency asymptote of the feedback loop. It is a matter of some importance, then, to determine what fixes the asymptote and how it can be improved. Evidently, the asymptote is finally restricted by the gains of the tubes alone. We can scarcely improve upon the result secured by connecting the output plate directly to the input grid. Within this limit, however, the actual asymptotic characteristic will depend upon the configuration and type of feedback employed, since a given distribution of parasitic elements may evidently affect one arrangement more than another. The salient circuit problem is therefore that of choosing a general configuration for the feedback circuit which will allow the maximum efficiency of transmission at high frequencies.

The relative importance of tube limitations and circuit limitations is most easily studied if we replace (7) by

$$A_m = 40 \log_{10} \frac{4f_t}{nf_0} - \frac{2A_t}{n}, \tag{10}$$

where f_t is the frequency at which the tubes themselves working into their own parasitic capacitances have zero gain [9] and A_t is the asymptotic loss of the complete feedback loop in db at $f = f_t$. The first term

[8] The material of this section was largely inspired by comments due to Messrs. G. H. Stevenson and J. M. West.

[9] I.e., $f_t = \frac{G_m}{2\pi C}$, where G_m and C are respectively the transconductance and capacitance of a representative tube. The ratio $\frac{G_m}{C}$ is the so-called "figure of merit" of the tube. The analysis assumes that the interstage network is a simple shunt impedance, so that the parasitic capacitance does correctly represent its asymptotic behavior. More complicated four-terminal interstage networks, such as transformer coupling circuits and the like, are generally inadmissible in a feedback amplifier because of the high asymptotic losses and consequent high-phase shifts which they introduce.

of (10) shows how the feedback depends upon the intrinsic band width of the available tubes. In low-power tubes especially designed for the purpose f_t may be 50 mc or more, but if f_0 is small the first term will be substantial even if tubes with much lower values of f_t are selected. The second term gives the loss in feedback which can be ascribed to the rest of the circuit. It is evidently not possible to provide input and output circuits and a β-path without making some contribution to the asymptotic loss, so that A_t cannot be zero. In an amplifier designed with particular attention to this question, however, it is frequently possible to assign A_t a comparatively low value, of the order of 20 to 30 db or less. Without such special attention, on the other hand, A_t is likely to be very much larger, with a consequent diminution in available feedback.

In addition to f_t and A_t, (10) includes the quantity n, which represents the final asymptotic slope in multiples of 6 db per octave. Since the tubes make no contribution to the asymptotic loss at $f = f_t$ we can vary n without affecting A_t by changing the number of tubes in the circuit. This makes it possible to compute the optimum number of tubes which should be used in any given situation in order to provide the maximum possible feedback. If A_t is small the first term of (10) will be the dominant one and it is evidently desirable to have a small number of stages. The limit may be taken as $n = 2$ since with only one stage the feedback is restricted by the available forward gain, which is not taken into account in this analysis. On the other hand since the second term varies more rapidly than the first with n, the optimum number of stages will increase as A_t is increased. It is given generally by

$$n = \frac{A_t}{8.68} \tag{11}$$

or in other words the optimum n is equal to the asymptotic loss at the tube crossover in nepers.

This relation is of particular interest for high-power circuits, such as radio transmitters, where circuit limitations are usually severe but the cost of additional tubes, at least in low-power stages, is relatively unimportant. As an extreme example, we may consider the problem of providing envelope feedback around a transmitter. With the relatively sharp tuning ordinarily used in the high-frequency circuits of a transmitter the asymptotic characteristics of the feedback path will be comparatively unfavorable. For illustrative purposes we may assume that $f_a = 40$ kc. and $n = 6$. In accordance with (7) this would provide a maximum available feedback over a 10 kc. voice band of 17 db. It will also be assumed that the additional tubes for the low-power portions of the circuit have an f_t of 10 mc.[10] The corresponding A_t is 33 nepers [11] so that equation (11) would say that the feedback would be increased by the addition of as many as 27 tubes to the circuit. Naturally in such an extreme case this result can be looked upon only as a qualitative indication of the direction in which to proceed. If we add only 4 tubes, however, the available feedback becomes 46 db while if we add 10 tubes it reaches 60 db. It is to be observed that only a small part of the available gain of the added tubes is used in directly increasing the feedback. The remainder is consumed in compensating for the unfortunate phase shifts introduced by the rest of the circuit.

Amplifiers of Other Types

The amplifier considered thus far is of a rather special type. It has a useful band extending from zero up to some prescribed frequency f_0, constant feedback in the useful band, and it is absolutely stable. Departures from absolute stability are rather unusual in practical amplifiers and will not be considered here. It is apparent from the phase area relation that a conditionally stable amplifier may be expected to have a greater feedback for a cut-off interval of given breadth than a structure which is unconditionally stable, but a detailed discussion of the problem is beyond the scope of this paper.

Departures from the other assumptions are easily treated. For example, if a varying feedback in the useful band is desired, as it may be in occasional amplifiers, an appropriate cut-off characteristic can be constructed by returning to the general formula (4), performing the integrations graphically, if necessary. If the phase requirement in the cut-off region is left unchanged only the first integral need be modified. The most important question, for ordinary purposes, is that of determining how high the varying feedback can be, in comparison with a corresponding constant feedback characteristic, for any given asymptote. This can be answered by observing the form to which the first integral in (4) reduces when f_c is made very large. It is easily seen that the asymptotic conditions will remain the same provided the

[10] In tubes operating at a high-power level f_t may, of course, be quite low. It is evident, however, that only the tubes added to the circuit are significant in interpreting (11). The additional tubes may be inserted directly in the feedback path if they are made substantially linear in the voice range by subsidiary feedback of their own. This will not affect the essential result of the present analysis.

[11] It is, of course, not to be expected that the actual asymptotic slope will be constant from 40 kc. to 10 mc. Since only the region extending a few octaves above 40 kc. is of interest in the final design, however, the apparent A_t can be obtained by extrapolating the slope in this region.

feedback in the useful band satisfies a relation of the form

$$\int_0^{\pi/2} A\,d\phi = \text{constant}, \tag{12}$$

where $\phi = \sin^{-1} f/f_0$. Thus the area under the varying characteristic, when plotted against ϕ, should be the same as that under a corresponding constant characteristic having the same phase and gain margins and the same final asymptote. This is exemplified by Fig. 19, the

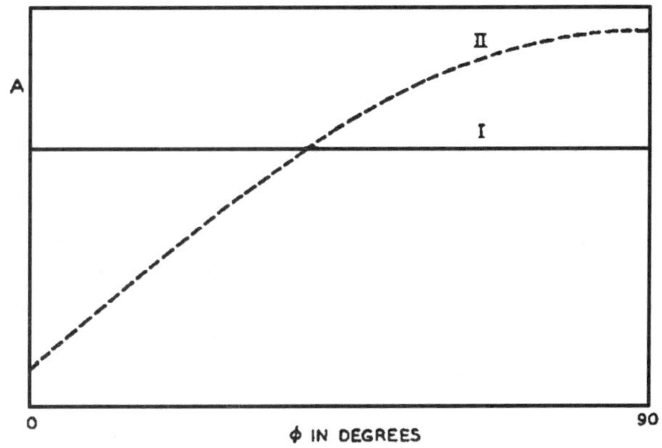

Fig. 19—Diagram to illustrate the computation of available feedback when the required feedback in the useful band is not the same at all frequencies.

varying characteristic being chosen for illustrative purposes as a straight line on an arithmetic frequency scale.

The most important question has to do with the assumption that the useful transmission band extends down to zero frequency. In most amplifiers, of course, this is not true. It is consequently necessary to provide a cut-off characteristic on the lower as well as the upper side of the band. The requisite characteristics are easily obtained from the ones which have been described by means of frequency transformations of a type familiar in filter theory. Thus if the cut-off characteristics studied thus far are regarded as being of the "low-pass" type the characteristics obtained from them by replacing f/f_0 by its reciprocal may be regarded as being of the "high-pass" type. If the band width of the amplifier is relatively broad it is usually simplest to treat the upper and lower cut-offs as independent characteristics of low-pass and high-pass types. In this event, the asymptote for the lower cut-off is furnished by such elements as blocking condensers and choke coils in the plate supply leads. The low-frequency asymptote is usually not so serious a problem as the high-frequency asymptote since it can be placed as far from the band as we need by using large enough elements in the power supply circuits. The superposition of a low-frequency cutoff on the idealized loop gain and phase characteristics of a "low-pass" circuit is illustrated by the broken lines in Fig. 20.

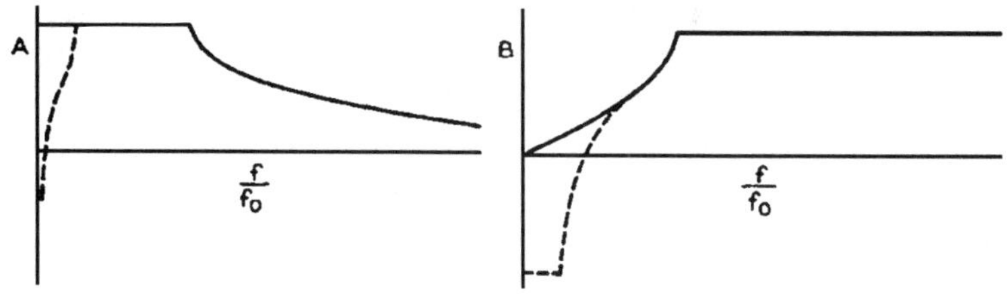

Fig. 20—Modification of loop characteristics to provide a lower cutoff in a broad-band amplifier.

If the band width is relatively narrow it is more efficient to use the transformation in filter theory which relates a low-pass to a symmetrical band-pass structure. The transformation is obtained by replacing f/f_0 in the low-pass case by $(f^2 - f_1 f_2/f(f_2 - f_1))$, where f_1 and f_2 are the edges of the prescribed band. It substitutes resonant and anti-resonant circuits tuned to the center of the band for the coils and condensers in the low-pass circuit. In particular each parasitic inductance is tuned by the addition of a series condenser and each parasitic capacity is tuned by a shunt coil. The parameters of the transformation must, of course, be so chosen that the parasitic elements have the correct values for use in the new branches.

This leads to a simple but important result. If the inductance of a series resonant circuit is fixed, the interval represented by $f_b - f_a$ in Fig. 21, between the frequencies at which the absolute value of the reactance reaches some prescribed limit X_0, is always constant and equal to the frequency at which the untuned inductance would exhibit the reactance X_0, whatever the tuning frequency may be. The same relation holds for the capacity in an anti-resonant circuit. Thus the frequency range over which the branches containing parasitic elements exhibit comparable impedance variations is the same in the band-pass structure and in the prototype low-pass structure. But since the transformation does not affect the relative impedance levels of the various branches in the circuit, this result can be extended to the complete $\mu\beta$ characteristic. We can therefore conclude that *the feedback which is obtainable in an amplifier of given general configuration and with given parasitic elements and given margins depends only upon the breadth*

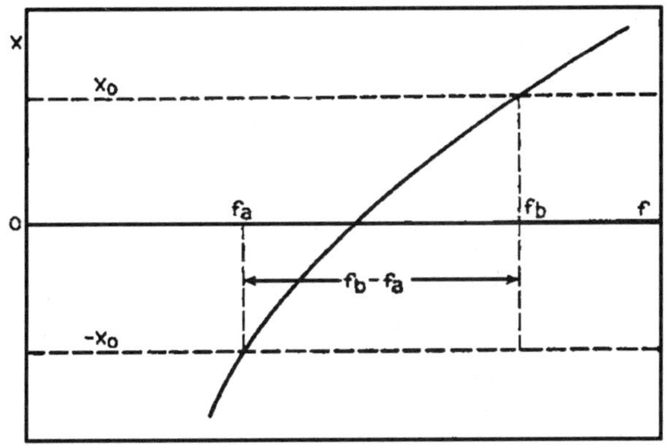

Fig. 21—Frequency interval between prescribed reactances of opposite sign in a resonant circuit with fixed inductance.

of the band in cycles and is independent of the location of the band in the frequency spectrum.

These relations are exemplified by the plots of a low-pass cutoff characteristic and the equivalent band-pass characteristic shown by Fig. 22. The equality of corresponding frequency intervals is indicated by the horizontal lines *A*, *B* and *C*.

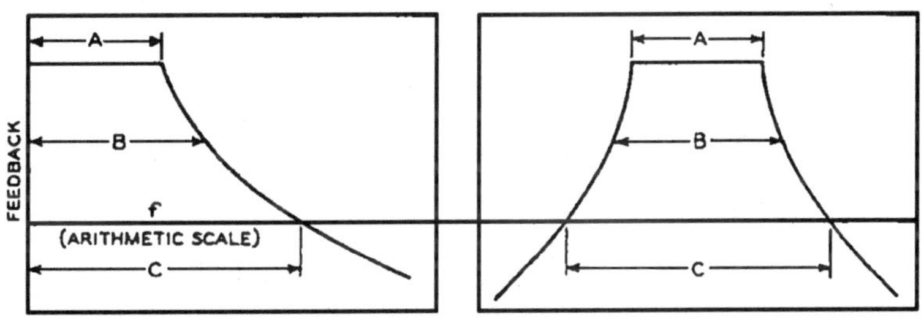

Fig. 22—Diagram to illustrate the conservation of band width in the low-pass to band-pass transformation with fixed parasitic elements. *A*, *B* and *C* represent typical corresponding intervals of equal breadth.

Example

An example showing the application of the method in an actual design is furnished by Fig. 23. The structure is a feedback amplifier intended to serve as a repeater in a 72-ohm coaxial line.[12] The useful frequency range extends from 60 to 2,000 kc. Coupling to the line is

[12] The author's personal contact with this amplifier was limited to the evolution of a paper schematic for the high frequency design. The other aspects of the problem are the work of Messrs. K. C. Black, J. M. West and C. H. Elmendorf.

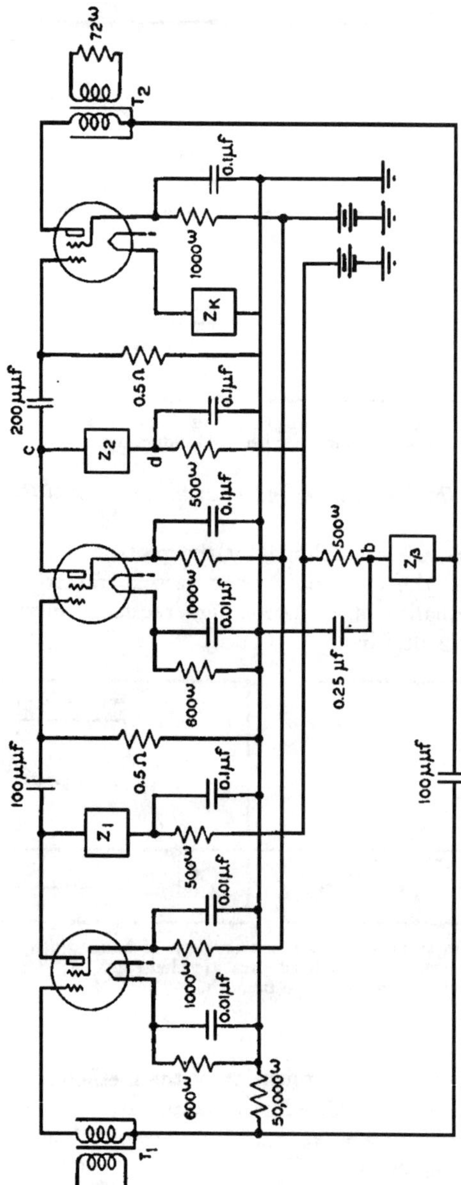

Fig. 23—General schematic of a 2 mc. feedback amplifier.

obtained through the shielded input and output transformers T_1 and T_2. The three stages in the μ circuit are represented in Fig. 23 as single tubes. Physically, however, each stage employes two tubes in parallel, the transconductances of the individual tubes being about 2000 micromhos. The principal feedback is obtained through the impedance Z_β. There is in addition a subsidiary local feedback on the power stage through the impedance Z_K. This is advantageous in producing a further reduction in the effects of modulation in this stage but it does not materially affect the feedback available around the principal loop.

The elements shown explicitly include resistance-capacitance filters in the power supply leads to the plates and screens, cathode resistances and by-pass condensers to provide grid bias potentials, and blocking condenser-grid-leak combinations for the several tubes. In addition to serving these functions, the various resistance-capacitance combinations are also used to provide the cutoff characteristic below the useful band. The low-frequency asymptote is established by the grid leak resistances and the associated coupling condensers and the approach of the feedback characteristic to the asymptote is controlled mainly by the cathode impedances and the resistance-capacitance filters in the power supply leads to the plates. The principal parts of the circuit entering into the $\mu\beta$ characteristic at high frequencies are the interstage impedances Z_1 and Z_2, the feedback impedance Z_β,[13] the cathode impedance Z_K, and the two transformers. The four network designs are shown in detail in Figs. 24, 25, 26, and 27.

The joint transconductance, 4000 micromhos, of two tubes in parallel operating into an average interstage capacity of 14 mmf, as indicated by Figs. 24 and 25, gives an f_t of about 50 mc. The parasitic capacities (chiefly transformer high side and ground capacities) in the other parts of the feedback loop provide a net loss, A_t, of about 18 db at this frequency. Since the asymptotic slope is 18 db per octave the intercept of the complete asymptote with the zero gain axis occurs about one octave lower, at slightly less than 25 mc. This is a relatively high intercept and may be attributed in part to the high gain of the vacuum tubes. The care used in minimizing parasitic capacities in the construction of the amplifier and the general circuit arrangement, including in particular the use of single shunt impedances for the coupling and feedback networks, are also helpful.

[13] The relative complexity of this network is explained by the fact that it actually serves as a regulator to compensate for the effects of changes in the line temperature. (See H. W. Bode, "Variable Equalizers," *Bell System Technical Journal*, April, 1938.) The present discussion assumes that the controlling element is at its normal setting. For this setting the network is approximately equal to a resistance in series with a small inductance. The fact that the amplifier must remain stable over a regulation range may serve to explain why the design includes such large stability margins.

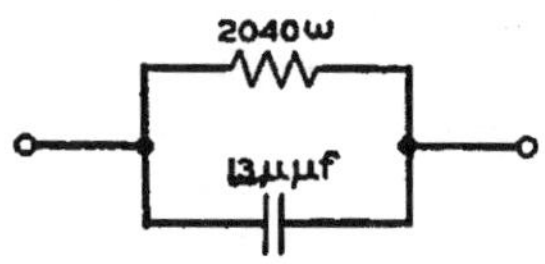

Fig. 24—First interstage for the amplifier of Fig. 23.

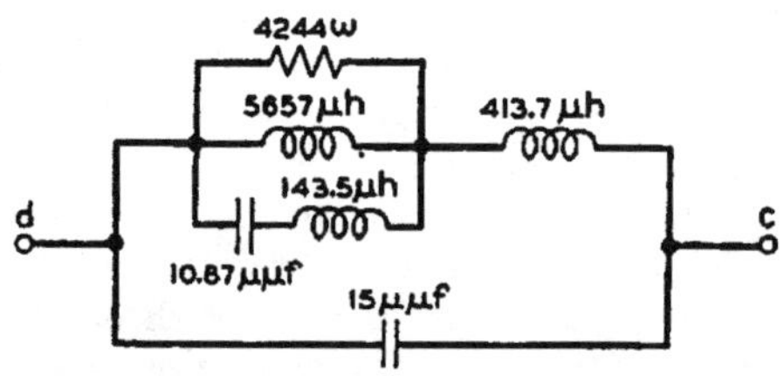

Fig. 25—Second interstage for the amplifier of Fig. 23.

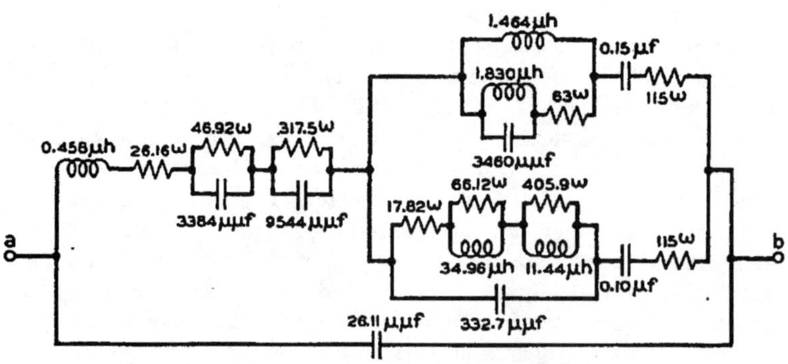

Fig. 26—β circuit impedance for the amplifier of Fig. 23.

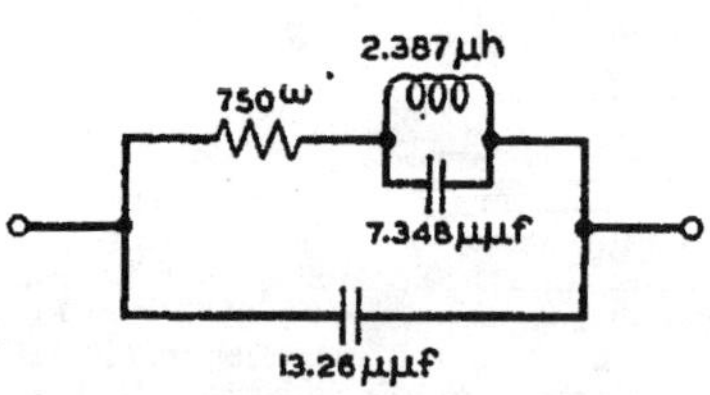

Fig. 27—Cathode impedance for the amplifier of Fig. 23.

In accordance with (7) the maximum available feedback A_m is 48 db. For design purposes, however, x and y in (9) were chosen as 15 db and 1/5 respectively. This reduces the actual feedback A to about 28 db. The theoretical cutoff characteristic corresponding to

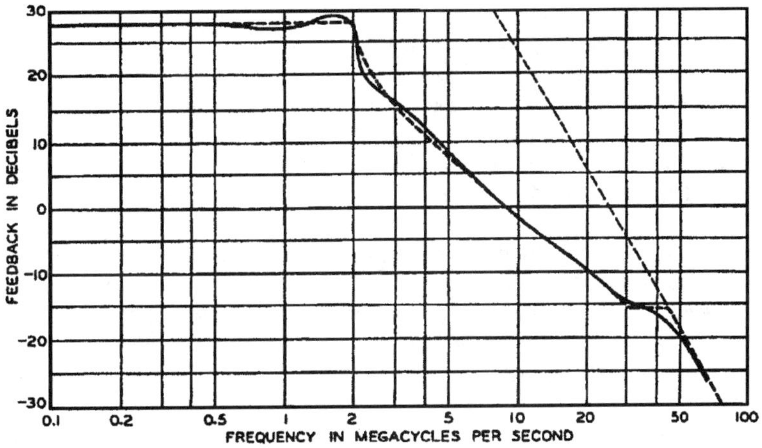

Fig. 28—Loop gain characteristic for the amplifier of Fig. 23.

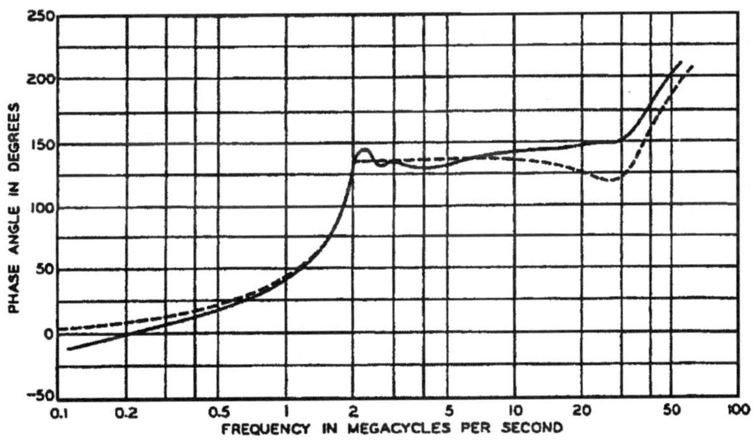

Fig. 29—Loop phase characteristic for the amplifier of Fig. 23.

these parameters is shown by the broken lines in Figs. 28 and 29, and the actual design characteristic by the solid lines. Since this is a structure in which the required forward gain is considerably less than the maximum available gain, the general course of the cutoff characteristic is controlled, in accordance with the procedure outlined previously, by the elements in the μ circuit. The sharp slope just beyond the edge of the useful band is obtained from a transformer anti-resonance. The relatively flat portion of the characteristic near its intersection with the asymptote is due partly to an anti-resonance of the β circuit with its distributed capacitance and partly to an increase in the gain of the third tube because of the filter-like action of the elements of Z_K in cutting out the local feedback on the tube in this region.

The large margins in the design made it possible to secure a substantial increase in feedback without instability. For example, with a loss margin as great as 15 db the feedback can be increased by adjusting the screen and plate voltages to increase the tube gains. A higher feedback can also be obtained by adjusting the resistance in the first interstage. As this interstage was designed, an increase in the resistance results in an increased amplifier gain and a correspondingly increased feedback which follows a new theoretical characteristic with a somewhat reduced phase margin. The adjustment, in effect, produces a change in the value of the constant k in equation (6). With this adjustment the feedback can be increased to about 40 db before the amplifier sings.

The Linear Filter for a Single Time Series

NORBERT WIENER

NORBERT Wiener is most widely known as the father of cybernetics [12], which he defined [11, p. 269] as "the theory of communication and control in the machine and in the living organism." Wiener's article in this volume treats that part of cybernetics known as filtering.

It is interesting that Wiener's work on prediction, filtering, and smoothing began with a control problem. During World War II, many scientists were engaged in research relevant to the war effort. For its obvious importance in anti-aircraft gunnery (fire control), Wiener and the engineer Julian Bigelow tackled the problem of predicting an airplane's future position from past observations. They thought [11, p. 244] to "develop the best predictor that mathematics allowed us to." As to what "best" should mean [11, pp. 244, 245], "the most natural quantity to choose at the start, if we should be guided by easy computation, if not military significance, was the mean square error of prediction." Thus the prediction problem posed by Wiener relates to a setup where the signal and noise are modeled as continuous-time zero-mean jointly stationary random processes with known covariances and cross-covariance; the observed signal, the input to the predictor, is signal plus noise; and the predictor is to produce an estimate of the signal advanced into the future by a specified amount of time. The problem is to obtain the causal linear time-invariant predictor that minimizes the rms error. Calculus of variations for this problem leads to an integral equation (the so-called Wiener-Hopf equation) of the type Wiener and Hopf had solved in 1931 via spectral factorization in the Fourier transform domain.

Wiener subsequently extended his work on prediction (or "extrapolation") to filtering and smoothing. His book [10], whose chapter III is reprinted here, first appeared in February, 1942, in limited circulation as a classified document. Because it was bound in yellow paper and the mathematics was challenging to the engineers reading it, it came to be known as the "yellow peril." To make the material in this document more accessible to engineers, Levinson developed Wiener's main ideas on prediction and filtering in a simpler mathematical form, and presented them in 1947 in the two papers [5], [6], which were subsequently reprinted as two appendices to Wiener's book [10]. Wiener intended this book to be a fusing of the subjects of time series and communication engineering, and in so doing he founded the statistical communication theory.

At the time Wiener studied the prediction problem, he was unaware of important earlier work by A. N. Kolmogorov beginning with [4]. Kolmogorov had given a treatment of the discrete-time prediction problem, though without an explicit formula for the optimal predictor. He is also credited for introducing the notion of *innovation sequence* in discrete-time filtering, as a tool for orthogonalizing the measurement sequence. This notion has later played an important role in recursive filtering, as discussed in the preamble to Kalman's 1960 filtering paper [3] included in this volume.

Wiener's theory of prediction, filtering, and smoothing (also referred to as Wiener-Kolmogorov theory in the discrete time) triggered a flurry of subsequent activity in that area, from making the theory accessible to control and communication engineers [1], to extending it to the finite-memory case [15]. These efforts, partly driven by the need to make the theory applicable to practical engineering problems, culminated in the late 1950s with the development of the Kalman filter [3], which provided a recursive scheme for the Wiener-Kolmogorov filter for the case when the signal and measurement are generated by finite-dimensional linear systems with white Gaussian inputs (see the preamble to [3]). For the earlier history and for the developments on filtering from Wiener until 1973, one may consult Kailath's comprehensive survey [2].

Shortly after their work on prediction, Wiener and Bigelow turned to anti-aircraft gunnery itself. As the war progressed and as radar became perfected, automatic feedback control of the gun became possible. They built an apparatus to track a moving spot of light on the ceiling. Their work had a profound influence at MIT in the Servomechanism Laboratory and the Radiation Laboratory. In particular, R. S. Phillips wrote a Radiation Laboratory report in May 1943, entitled "Servomechanisms," subsequently revised as [8].

Prior to Wiener's influence, servomechanisms were designed with regard to considerations of stability, of suitable damping, of the transient and steady-state error for a test input like a step or ramp, of the disturbance attenuation capability, and so on—no consideration had been given to the error response for actual inputs. Following Wiener, Phillips postulated that typical reference inputs can be described statistically. Then the basis of the rms-error approach is this [8, p. 308]: "To weight the error transfer function at each frequency according to its probability of occurrence and then to minimize the total weighted error." The components of Phillips's design procedure are to model the exogenous inputs (reference, disturbance, sensor noise) as stationary random processes; to specify the controller up to certain adjustable parameters; and then to optimize the parameters to minimize the rms tracking error. One drawback is that feedback stability is not handled directly—it is assumed that the optimal controller parameters are stabilizing. This approach to design, called the *rms-error approach*, *statistical design*, or the *Wiener-Hopf method*, became part of mainstream control engineering and appeared in many popular textbooks by the 1950s (e.g., [9] and [7]).

In minimizing the rms error, the controller should be constrained to stabilize the feedback system. This constrained optimization problem is intractable as it stands, being non-convex. A breakthrough was obtained in the mid-1970s by Youla, Bongiorno, and Jabr [13], [14]. They derived a parametrization of all stabilizing controllers for which the rms error is a convex function of the parameter. Their influential work can be considered the completion for multivariable linear time-invariant systems of the path initiated by Wiener.

The work from Wiener to Youla, Bongiorno, and Jabr was based on transfer function models of systems and covariance matrices of random signals. There was a parallel path from Kalman filtering to LQG theory based on state-space models. For linear time-invariant systems and stationary Gaussian signals, the two approaches are equivalent. The current name for this design method is $\mathcal{H}_2$ optimization and it remains a very popular design technique.

Wiener was a professor of mathematics at MIT from 1932 until 1960, and consequently was influenced by and influenced many MIT engineers of the day. One engineering student influenced by Wiener was George Zames, two of whose papers (one in a two-part form) are included in this volume.

References

[1] H.W. Bode and C.E. Shannon, "A simplified derivation of linear least-squares smoothing and prediction theory," *Proc. IRE*, **38**:417–425, 1950.

[2] T. Kailath, "A view of three decades of linear filtering theory," *IEEE Trans. Inf. Theory*, **20**:146–181, 1974.

[3] R.E. Kalman, "A new approach to linear filtering and prediction problems," *Trans. ASME, J. Basic Engineering*, **82D**(1):35–45, March 1960.

[4] A. Kolmogoroff, "Sur l'interpolation et extrapolation des suites stationnaires," *Comptes Rendus de l'Acad. Sci.*, **208**:2043–2045, 1939.

[5] N. Levinson, "The Wiener RMS (root mean square) error criterion in filter design and prediction," *J. Mathematics and Physics*, **XXV**(4):261–278, 1947.

[6] N. Levinson, "A heuristic exposition of Wiener's mathematical theory of prediction and filtering," *J. Mathematics and Physics*, **XXVI**(2):110–119, 1947.

[7] G.C. Newton, Jr., L.A. Gould, and J.F. Kaiser, *Analytic Design of Linear Feedback Controls*, Wiley (New York), 1957.

[8] R.S. Phillips, "RMS-error criterion in servomechanism design," Chapter 7 of *Theory of Servomechanisms*, H.M. James, N.B. Nichols, and R.S. Phillips, edts., McGraw-Hill (New York), 1947.

[9] J.G. Truxal, *Automatic Feedback Control System Synthesis*, McGraw-Hill (New York), 1955.

[10] N. Wiener, *Extrapolation, Interpolation, and Smoothing of Stationary Time Series*, MIT Press (Cambridge, MA), 1949.

[11] N. Wiener, *I am a Mathematician*, Doubleday, 1956.

[12] N. Wiener, *Cybernetics*, MIT Press (Cambridge, MA), 1961.

[13] D.C. Youla, J.J. Bongiorno, Jr., and H.A. Jabr, "Modern Wiener-Hopf design of optimal controllers, part I: the single-input-output case," *IEEE Trans. Auto. Contr.*, **AC-21**:3–13, 1976.

[14] D.C. Youla, H.A. Jabr and J.J. Bongiorno, Jr., "Modern Wiener-Hopf design of optimal controllers, part II: the multivariable case," *IEEE Trans. Auto. Contr.*, **AC-21**:319–338, 1976.

[15] L.A. Zadeh and J.R. Ragazzini, "An extension of Wiener's theory of prediction," *J. Applied Physics*, **21**(7):645–655, 1950.

B.A.F.

CHAPTER III

THE LINEAR FILTER FOR A SINGLE TIME SERIES

3.0 Formulation of the General Filter Problem

Let $f(t)$ and $g(t)$ be two complex time series. Let $f(t)$ represent a message and $g(t)$ a disturbance. We wish to determine that linear operator which, when applied to $f(t) + g(t)$, will give us the best approximation to $f(t + \alpha)$. In order to do this, we need statistical information concerning $f(t)$ and $g(t)$, which will be given by their auto-correlation and cross-correlation coefficients. We shall suppose that sufficient auxiliary conditions are satisfied to permit the free shifting of the time origin of the series as justified by the lemmas of the last chapter. Of course, the results of this chapter will include those of the last in the particular case in which $g(t)$ vanishes. In order to avoid as far as possible the duplication of arguments already given, we shall use a slightly different approach to the solution of the integral equation both of this chapter and of the last, although the methods of both chapters are applicable to either case.

Formally, the problem of this chapter exhibits certain differences according as the sign of α is positive or negative. If it is positive, we again have a prediction problem before us, albeit a prediction problem from perturbed data. If we start from correlation coefficients with rational Fourier transforms, the prediction operator on the frequency scale will likewise be rational. On the other hand, if α is negative, we arrive in the first instance at an operator which will be realizable, in that it will be a function of a complex variable with the desired behavior in one half-plane, but which will not be rational and will require another stage of approximation before it can be realized in an electric or mechanical network of a finite number of meshes. Offsetting this disadvantage, it will perform its filtering function as between the message $f(t)$ and the noise or disturbance $g(t)$ more perfectly and with increasing perfection as $\alpha \to -\infty$. In general the spectra of $f(t)$ and $g(t)$ will overlap, and when this is the case, not even an infinite lag in filtering will give perfect discrimination. The degree by which this discrimination fails to be perfect will determine the lag which may advantageously be designed into the filter and indirectly establish the number of network meshes justified.

The significance of lag in a filter may be seen from the following consideration: we shall find that, even though we approach the problem

Reprinted with permission from *Extrapolation, Interpolation, and Smoothing of Stationary Time Series* (Cambridge, MA: MIT Press, 1949), Norbert Wiener, "The Linear Filter for a Single Time Series (Chapter III)" pp. 81–103.

from the standpoint of time characteristic, we eventually arrive at a filter which is essentially an instrument for the separation of different frequency ranges. However, to determine the spectrum of an impulse with perfect sharpness, we must know the history of that impulse over all time. This we can never do, but the longer we wait before operating with the impulse, the more of that impulse has had an opportunity to come through and the more completely is its spectrum determined.

While our treatment of the filter, like all treatments, ultimately involves the spectra of the message and the noise or perturbations which it is supposed to separate, our criterion of performance is the mean square distortion of the message over the time and involves transient behavior as well as steady-state behavior. It is quite as critical of distortion of phase as of distortion of amplitude. In this it differs from methods more familiar in communication engineering theory, which have been devised for the most part with reference to voice-transmission and ultimately with reference to the human ear—an organ of unusually fine frequency discrimination, fair amplitude discrimination, and very bad phase discrimination. However, in television work and in other related techniques, the final receiving instrument is highly sensitive to phase, and existing filter techniques have proved distinctly inadequate. This has also been experienced in the case of mechanical filters, designed to smooth out irregularities in the input of accurate servo-mechanisms. It is hoped that the methods of this chapter may prove useful in these fields.

3.1 Minimization Problem for Filters

Let us consider the expression

$$\lim_{T\to\infty} \frac{1}{2T} \int_{-T}^{T} \left| f(t+\alpha) - \int_0^\infty [f(t-\tau) + g(t-\tau)]\, dK(\tau) \right|^2 dt. \tag{3.10}$$

Under very general conditions, this expression may be written

$$\varphi_{11}(0) - 2\mathrm{R}\left\{ \int_0^\infty [\varphi_{11}(\alpha+\tau) + \varphi_{12}(\alpha+\tau)]\, d\overline{K(\tau)} \right\} + \int_0^\infty dK(\sigma) \int_0^\infty d\overline{K(\tau)} \{\varphi_{11}(\tau-\sigma) + \varphi_{12}(\tau-\sigma) + \overline{\varphi_{12}(\sigma-\tau)} + \varphi_{22}(\tau-\sigma)\}, \tag{3.105}$$

where

$$\varphi_{11}(\tau) = \lim_{T\to\infty} \frac{1}{2T} \int_{-T}^{T} f(t+\tau)\overline{f(t)}\, dt, \tag{3.11}$$

$$\varphi_{12}(\tau) = \lim_{T\to\infty} \frac{1}{2T}\int_{-T}^{T} f(t+\tau)\overline{g(t)}\,dt, \tag{3.111}$$

$$\varphi_{22}(\tau) = \lim_{T\to\infty} \frac{1}{2T}\int_{-T}^{T} g(t+\tau)\overline{g(t)}\,dt. \tag{3.112}$$

Let us put

$$\varphi_{11}(\tau) + \varphi_{12}(\tau) + \overline{\varphi_{12}(-\tau)} + \varphi_{22}(\tau) = \varphi(\tau). \tag{3.12}$$

Then clearly

$$\varphi(\tau) = \overline{\varphi(-\tau)}, \tag{3.125}$$

and the expression which we wish to minimize will be

$$\varphi_{11}(0) - 2\mathrm{R}\left\{\int_0^\infty [\varphi_{11}(\alpha+\tau) + \varphi_{12}(\alpha+\tau)]\,d\overline{K(\tau)}\right\} + \int_0^\infty dK(\sigma)\int_0^\infty d\overline{K(\tau)}\varphi(\tau-\sigma). \tag{3.13}$$

If we write

$$\varphi_{11}(\alpha+\tau) + \varphi_{12}(\alpha+\tau) = h(\tau), \tag{3.135}$$

this minimization is reduced to that of the general real expression

$$-2\mathrm{R}\left\{\int_0^\infty h(\tau)\,d\overline{K(\tau)}\right\} + \int_0^\infty dK(\sigma)\int_0^\infty d\overline{K(\tau)}\varphi(\tau-\sigma). \tag{3.14}$$

Let us suppose that

$$h(\tau) = \int_0^\infty \varphi(\tau-\sigma)\,dQ(\sigma). \tag{3.145}$$

Then (3.14) becomes

$$\begin{aligned}
&-\int_0^\infty dQ(\sigma)\int_0^\infty \varphi(\tau-\sigma)\,d\overline{K(\tau)} \\
&-\int_0^\infty d\overline{Q(\sigma)}\int_0^\infty \varphi(\sigma-\tau)\,dK(\tau) \\
&+\int_0^\infty dK(\sigma)\int_0^\infty d\overline{K(\tau)}\varphi(\tau-\sigma) \\
&= -\int_0^\infty dQ(\sigma)\int_0^\infty \varphi(\tau-\sigma)\,d\overline{Q(\tau)} \\
&+\int_0^\infty d[K(\sigma)-Q(\sigma)]\int_0^\infty d[\overline{K(\tau)}-\overline{Q(\tau)}]\varphi(\tau-\sigma).
\end{aligned}$$

Let us now suppose that

$$\int_0^\infty e^{-i\omega t}\, dK(t) = k(\omega); \tag{3.15}$$

$$\int_0^\infty e^{-i\omega t}\, dQ(t) = q(\omega); \tag{3.16}$$

$$\varphi(t) = \frac{1}{2\pi}\int_{-\infty}^\infty \Phi(\omega)e^{i\omega t}\, d\omega. \tag{3.165}$$

Then the expression to be minimized becomes

$$-\frac{1}{2\pi}\int_{-\infty}^\infty |\, q(\omega)\,|^2\,\Phi(\omega)\, d\omega + \frac{1}{2\pi}\int_{-\infty}^\infty |\, k(\omega) - q(\omega)\,|^2\,\Phi(\omega)\, d\omega, \tag{3.17}$$

which will attain its minimum when

$$q(\omega) = k(\omega), \tag{3.175}$$

which minimum will be

$$-\frac{1}{2\pi}\int_{-\infty}^\infty \Phi(\omega)\,|\, q(\omega)\,|^2\, d\omega. \tag{3.18}$$

3.2 The Factorization of the Spectrum

In general, let us consider the integral equation

$$h(t) = \int_0^\infty \varphi(t-\tau)\, dK(\tau) \quad (t > 0), \tag{3.20}$$

where

$$h(t) = \frac{1}{2\pi}\int_{-\infty}^\infty H(\omega)e^{i\omega t}\, d\omega, \tag{3.205}$$

$$\varphi(t) = \frac{1}{2\pi}\int_{-\infty}^\infty \Phi(\omega)e^{i\omega t}\, d\omega, \tag{3.21}$$

and $\varphi(t) = \overline{\varphi(-t)}$, so that $\Phi(\omega)$ is real. We shall suppose that

$$\int_0^\infty e^{-i\omega t}\, dK(t) = k(\omega). \tag{3.215}$$

For the present, let us confine our attention to the case in which $\Phi(\omega)$ is rational. Here we shall limit ourselves to the case in which $\Phi(\omega)$ has no real zeros. We may then write

$$\Phi(\omega) = |\, \Psi(\omega)\,|^2, \tag{3.22}$$

where $\Psi(\omega)$ is a rational function free from zeros and poles in the lower half-plane. If we put

$$\psi(t) = \frac{1}{2\pi}\int_{-\infty}^{\infty} \Psi(\omega)e^{i\omega t}\,d\omega = 0 \quad \text{if} \quad t < 0, \tag{3.225}$$

we shall have

$$\varphi(\tau) = \begin{cases} \displaystyle\int_0^{\infty} \psi(t+\tau)\overline{\psi(t)}\,dt & (\tau > 0); \\ \displaystyle\int_{-\tau}^{\infty} \psi(t+\tau)\overline{\psi(t)}\,dt & (\tau < 0). \end{cases} \tag{3.23}$$

Then (3.20) becomes

$$\begin{aligned} h(t) &= \int_0^t dK(\tau)\int_0^{\infty} \psi(t-\tau+\sigma)\overline{\psi(\sigma)}\,d\sigma \\ &\quad + \int_t^{\infty} dK(\tau)\int_{\tau-t}^{\infty} \psi(t-\tau+\sigma)\overline{\psi(\sigma)}\,d\sigma \\ &= \int_0^{\infty} \overline{\psi(\sigma)}\,d\sigma \int_0^{\sigma+t} \psi(t-\tau+\sigma)\,dK(\tau) \quad (t > 0). \end{aligned} \tag{3.24}$$

Now let us put

$$h(t) = \int_0^{\infty} \overline{\psi(\sigma)}\,L(\sigma+t)\,d\sigma \quad (-\infty < t < \infty). \tag{3.245}$$

We see that, if we extend φ and dK to cover negative arguments, for which dK is to vanish,

$$\int_{-\infty}^{\infty} \psi(t-\tau)\,dK(\tau) = \begin{cases} L(t) & (t > 0); \\ 0 & (t < 0). \end{cases} \tag{3.25}$$

If

$$l(\omega) = \int_{-\infty}^{\infty} L(t)e^{-i\omega t}\,dt, \tag{3.255}$$

equation (3.245) becomes

$$H(\omega) = l(\omega)\cdot\overline{\Psi(\omega)}, \tag{3.26}$$

[where $H(\omega)$ is the Fourier transform of $h(t)$] which leads to

$$l(\omega) = \frac{H(\omega)}{\overline{\Psi(\omega)}}, \tag{3.265}$$

or

$$L(t) = \frac{1}{2\pi}\int_{-\infty}^{\infty} \frac{H(\omega)}{\overline{\Psi(\omega)}}\,e^{i\omega t}\,d\omega. \tag{3.27}$$

Then (3.25) becomes

$$\Psi(\omega)k(\omega) = \int_0^\infty L(t)e^{-i\omega t}\,dt, \tag{3.275}$$

which again leads to

$$k(\omega) = \frac{1}{2\pi\Psi(\omega)}\int_0^\infty e^{-i\omega t}\,dt \int_{-\infty}^\infty \frac{H(u)}{\overline{\Psi(u)}}e^{iut}\,du. \tag{3.28}$$

The denominator of this expression is a rational function free from zeros in the lower half-plane, while the numerator is free from singularities in the lower half-plane, and if $L(t)$ satisfies the condition

$$\int_0^\infty |\,L(t)\,|\,dt < \infty, \tag{3.285}$$

it will be bounded in that half-plane. Thus $k(\omega)$ is of at most rational growth in the lower half-plane, in which it is free of singularities. If

$$k(\omega) = \int_{-\infty}^\infty e^{-i\omega t}\,dK_1(t), \tag{3.29}$$

then $K_1(t)$ is constant for negative arguments and is a possible $K(t)$.

3.3 Prediction and Filtering

Let us consider two cases in particular. To begin with, let

$$h(t) = \varphi(t+\alpha). \tag{3.30}$$

In this case,

$$\frac{1}{2\pi}\int_{-\infty}^\infty \frac{H(u)}{\overline{\Psi(u)}}e^{iut}\,du = \frac{1}{2\pi}\int_{-\infty}^\infty \frac{\Phi(u)}{\overline{\Psi(u)}}e^{iu\alpha}e^{iut}\,du$$
$$= \begin{cases}\psi(t+\alpha) & (t > -\alpha),\\ 0 & (t < -\alpha).\end{cases} \tag{3.305}$$

This will yield us

$$k(\omega) = \frac{\displaystyle\int_0^\infty \psi(t+\alpha)e^{-i\omega t}\,dt}{\Psi(\omega)}. \tag{3.31}$$

If $\Phi(\omega)$ is rational, this will be a rational expression. In fact, if

$$\Psi(\omega) = \sum_{m,n} A_{m,n}\frac{1}{(\omega-\omega_n)^m}, \tag{3.315}$$

we shall have

$$\psi(t) = \frac{1}{2\pi}\int_{-\infty}^\infty e^{i\omega t}\,d\omega \sum_{m,n} A_{m,n}\frac{1}{(\omega-\omega_n)^m}. \tag{3.32}$$

Now,

$$\frac{i^m}{(m-n)!}\int_0^\infty e^{-i\omega t}e^{i\omega_n t}t^{m-1}\,dt = \frac{1}{(\omega-\omega_n)^m} \quad (\mathrm{I}\{\omega_n\} > 0), \tag{3.325}$$

and hence

$$\frac{1}{2\pi}\int_{-\infty}^{\infty}\frac{e^{i\omega t}\,d\omega}{(\omega-\omega_n)^m} = \frac{i^m}{(m-1)!}t^{m-1}e^{i\omega_n t}\,(t>0). \tag{3.33}$$

Therefore

$$\psi(t) = \sum_{m,n} A_{m,n}\frac{i^m}{(m-1)!}t^{m-1}e^{i\omega_n t}, \tag{3.335}$$

and

$$\begin{aligned}\psi(t+\alpha) &= \sum_{m,n} A_{m,n}\frac{i^m}{(m-1)!}(t+\alpha)^{m-1}e^{i\omega_n(t+\alpha)} \\ &= \sum_{m,n} A_{m,n}\frac{e^{i\omega_n\alpha}i^m}{(m-1)!}\sum_{k=0}^{m-1}\frac{(m-1)!\alpha^k t^{m-1-k}}{k!(m-1-k)!}e^{i\omega_n t} \\ &= \sum_{m,n} A_{m,n}e^{i\omega_n\alpha}\sum_{k=0}^{m-1}\frac{(i\alpha)^k}{k!}\frac{i^{m-k}t^{m-1-k}}{(m-1-k)!}e^{i\omega_n t}\end{aligned} \tag{3.34}$$

As a consequence of this,

$$\int_0^\infty \psi(t+\alpha)e^{-i\omega t}\,dt = \sum_{m,n} A_{m,n}e^{i\omega_n\alpha}\sum_{k=0}^{m-1}\frac{(i\alpha)^k}{k!}\frac{1}{(\omega-\omega_n)^{m-k}}. \tag{3.345}$$

This gives us

$$k(\omega) = \frac{\displaystyle\sum_{m,n}\frac{1}{(\omega-\omega_n)^m}\sum_{\mu=m} A_{\mu,n}e^{i\omega_n\alpha}\frac{(i\alpha)^{\mu-m}}{(\mu-m)!}}{\displaystyle\sum_{m,n} A_{m,n}\frac{1}{(\omega-\omega_n)^m}}. \tag{3.35}$$

Again, let $H(\omega)$ be of the form $M(\omega)e^{i\alpha\omega}$, where $M(\omega)$ is rational and has no real zeros or singularities and has a denominator of higher degree than the numerator. Then we may write

$$\frac{M(\omega)}{\overline{\Psi(\omega)}} = \sum_{m,n}\frac{\alpha_{m,n}}{(\omega-\omega_n)^m} + \sum_{m,n}\frac{\beta_{m,n}}{(\omega-\omega_n')^m} \quad (\mathrm{I}\{\omega_n\} > 0, \mathrm{I}\{\omega_n'\} < 0); \tag{3.355}$$

then

$$\frac{1}{2\pi}\int_{-\infty}^{\infty}\frac{M(\omega)}{\overline{\Psi(\omega)}}e^{i\omega t}\,d\omega = \begin{cases}\displaystyle\sum_{m,n}\alpha_{m,n}\frac{i^m}{(m-1)!}t^{m-1}e^{i\omega_n t} & (t>0); \\ \displaystyle\sum_{m,n}\beta_{m,n}\frac{i^m}{(m-1)!}t^{m-1}e^{i\omega_n' t} & (t<0).\end{cases} \tag{3.36}$$

Let us call this function $w(t)$. Then, if α is positive,

$$\frac{1}{2\pi}\int_{-\infty}^{\infty}\frac{H(u)}{\overline{\Psi(u)}}e^{iut}\,du$$
$$=\sum_{m,n}\alpha_{m,n}\frac{i^{n}}{(m-1)!}(t+\alpha)^{m-1}e^{i\omega_n(t+\alpha)}\quad (t>0). \tag{3.365}$$

If, on the other hand, α is negative,

$$\frac{1}{2\pi}\int_{-\infty}^{\infty}\frac{H(u)}{\overline{\Psi(u)}}e^{iut}\,du$$
$$=\begin{cases}\displaystyle\sum_{m,n}\alpha_{m,n}\frac{i^{n}}{(m-1)!}(t+\alpha)^{m-1}e^{i\omega_n(t+\alpha)} & (t>-\alpha);\\ \displaystyle\sum_{m,n}\beta_{m,n}\frac{i^{m}}{(m-1)!}(t+\alpha)^{m-1}e^{i\omega_n'(t+\alpha)} & (0<t<-\alpha).\end{cases} \tag{3.37}$$

Thus the cases of lead and of lag must be treated differently. In the first case,

$$k(\omega)=\frac{\displaystyle\sum_{m,n}\frac{1}{(\omega-\omega_n)^m}\sum_{\mu=m+1}^{\infty}\alpha_{\mu,n}e^{i\omega_n\alpha}\frac{(i\alpha)^{\mu-m}}{(\mu-m)!}}{\Psi(\omega)}, \tag{3.375}$$

which is rational if $\Phi(\omega)$ is rational. In the second case, $k(\omega)$ will not be rational; and, if the function $k(\omega)$ is to be approximated as the voltage ratio of an electric circuit, a further investigation is needed.

3.4 The Error of Performance of a Filter; Long-lag Filters

In any case, the minimum of

$$-2\mathrm{R}\left\{\int_0^{\infty}h(\tau)\,d\overline{K(\tau)}\right\}+\int_0^{\infty}dK(\sigma)\int_0^{\infty}d\overline{K(\tau)}\varphi(\tau-\sigma) \tag{3.40}$$

will be given by

$$\frac{1}{2\pi}\int_{-\infty}^{\infty}\Phi(\omega)\left\{-\left|\frac{1}{2\pi\Psi(\omega)}\int_0^{\infty}e^{-i\omega t}\,dt\int_{-\infty}^{\infty}\frac{H(u)}{\overline{\Psi(u)}}e^{iut}\,du\right|^2\right\}d\omega$$
$$=\int_0^{\infty}\left\{-\left|\frac{1}{2\pi}\int_{-\infty}^{\infty}\frac{H(u)}{\overline{\Psi(u)}}e^{iut}\,du\right|^2\right\}dt. \tag{3.405}$$

If we put

$$\varphi_{ij}(\tau)=\frac{1}{2\pi}\int_{-\infty}^{\infty}\Phi_{ij}(\omega)e^{i\omega\tau}\,d\omega\quad (i,j=1,2), \tag{3.41}$$

the minimum of

$$\lim_{T\to\infty} \frac{1}{2T} \int_{-T}^{T} \left| f(t+\alpha) - \int_0^\infty [f(t-\tau) + g(t-\tau)]\, dK(\tau) \right|^2 dt \tag{3.415}$$

will be given by

$$= \frac{1}{2\pi} \int_0^\infty \left\{ \varphi_{11}(t) - \left| \frac{1}{2\pi} \int_{-\infty}^{\infty} \frac{H(u)}{\overline{\Psi(u)}} e^{iut}\, du \right|^2 \right\} dt. \tag{3.42}$$

If $g(t)$ vanishes, this will be

$$= \frac{1}{2\pi} \int_0^\alpha |\psi(t)|^2\, dt. \tag{3.425}$$

In general

$$\int_0^\infty \left| \sum_{m,n} A_{m,n} \frac{i^m}{(m-1)!} t^{m-1} e^{i\omega_n t} \right|^2 dt$$
$$= \sum_{m,n} \sum_{\mu,\nu} A_{m,n} A_{\mu,\nu} \frac{(-1)^\mu i^{(m+\mu-2)}}{(m-1)!(\mu-1)!} \left[\frac{1}{(\omega_n - \bar{\omega}_\nu)^{m+\mu-1}} \right], \tag{3.43}$$

and this may be used to determine the minimum of

$$\lim_{T\to\infty} \frac{1}{2T} \int_{-T}^{T} \left| f(t+\alpha) - \int_0^\infty [f(t-\tau) + g(t-\tau)]\, dK(\tau) \right|^2 dt \tag{3.435}$$

whenever $\alpha > 0$ and all the functions $\Phi_{ij}(\omega)$ are rational. If, on the other hand, $\alpha \to -\infty$, we shall have for the limit of the minimum

$$\lim_{\alpha\to-\infty} \frac{1}{2\pi} \int_{-\infty}^{\infty} \left\{ \Phi_{11}(\omega) - \Phi(\omega) \left| \frac{1}{2\pi\Psi(\omega)} \int_0^\infty e^{-i\omega t}\, dt \int_{-\infty}^{\infty} \frac{M(u) e^{iu(t+\alpha)}}{\overline{\Psi(u)}}\, du \right|^2 \right\} d\omega$$
$$= \frac{1}{2\pi} \int_{-\infty}^{\infty} \left\{ \Phi_{11}(\omega) - \frac{\Phi(\omega)\, |M(\omega)|^2}{|\Psi(\omega) \cdot \overline{\Psi(\omega)}|^2} \right\} d\omega$$
$$= \frac{1}{2\pi} \int_{-\infty}^{\infty} \left\{ \Phi_{11}(\omega) - \frac{|\Phi_{11}(\omega) + \Phi_{12}(\omega)|^2}{\Phi_{11}(\omega) + \Phi_{12}(\omega) + \overline{\Phi_{12}(\omega)} + \Phi_{22}(\omega)} \right\} d\omega$$
$$= \frac{1}{2\pi} \int_{-\infty}^{\infty} \frac{\Phi_{11}(\omega)\Phi_{22}(\omega) - |\Phi_{12}(\omega)|^2}{\Phi_{11}(\omega) + \Phi_{12}(\omega) + \overline{\Phi_{12}(\omega)} + \Phi_{22}(\omega)}\, d\omega. \tag{3.44}$$

In the particular case where $f(t)$ and $g(t)$ have zero cross-correlation under any finite lag, $\Phi_{12}(\omega)$ is identically zero, and the minimum value of (3.415) will tend to approach

$$\frac{1}{2\pi}\int_{-\infty}^{\infty}\frac{\Phi_{11}(\omega)\Phi_{22}(\omega)}{\Phi_{11}(\omega)+\Phi_{22}(\omega)}\,d\omega \tag{3.445}$$

as the lag $-\alpha$ becomes infinite. Whether the cross-correlation is zero or not, let it be noted that $f(t)$ and $g(t)$ occur symmetrically in these expressions.

We have now considered the performance of long-lag filters from the point of view of their "norm" or figure of merit. It is appropriate to consider the asymptotic form of their $k(\omega)$. This will be

$$k(\omega)=\frac{e^{i\omega\alpha}}{2\pi\Psi(\omega)}\int_{\alpha}^{\infty}e^{-i\omega t}\int_{-\infty}^{\infty}\frac{M(u)}{\Psi(u)}e^{iut}\,du; \tag{3.45}$$

and if we consider only the asymptotic value of $k(\omega)e^{-i\omega\alpha}$, this will be

$$\frac{M(\omega)}{\Phi(\omega)}=\frac{\Phi_{11}(\omega)+\Phi_{12}(\omega)}{\Phi(\omega)}. \tag{3.46}$$

3.5 Filters and Ergodic Theory

A fundamental theorem in the theory of the Brownian motion asserts that, if the responses of two linear resonators to Brownian inputs, whether the same, different, or partly the same and partly different, have a zero cross-correlation coefficient with respect to the parameter of distribution of the Brownian motion, then they are not merely linearly independent, but (as the parameters of the Brownian distributions vary) they have entirely independent distributions. Now the problem of optimum prediction is solved by reducing $f(t+\alpha)$ to a part linearly dependent on the past of $f(t)$ and a part uncorrelated with the past of $f(t)$. If the first part is

$$\int_0^{\infty}f(t-\tau)\,dK(\tau), \tag{3.50}$$

then the second will be

$$f(t+\alpha)-\int_0^{\infty}f(t-\tau)\,dK(\tau); \tag{3.505}$$

and by the ergodic theorem, the cross-correlation of the latter and $f(t+\alpha)$ with respect to the Brownian parameters of distribution will almost always be the average in time, or

$$\varphi(s+\alpha)-\int_0^{\infty}\varphi(s-\tau)\,dK(\tau), \tag{3.51}$$

which accordingly must vanish for positive s. This is the integral equation of prediction. The integral equation of filtering, or

$$\varphi_{11}(s+\alpha) - \int_0^\infty [\varphi_{11}(s-\tau) + \varphi_{22}(s-\tau)]\, dK(\tau) = 0 \quad (s > 0), \tag{3.52}$$

may be regarded similarly as the statement of the vanishing of a cross-correlation, and in the Brownian case will be found to assert that the error of performance of our optimum filter is wholly independent of known data and thus wholly unpredictable. Of course, its distribution will be known, but this will be a Gaussian distribution about zero determined exclusively by the functions $\varphi_{ij}(t)$ and will have nothing further to do with the function $f(t)$.

3.6 Computation of Specific Filter Characteristics

It may be of interest to work out several filter characteristics both for lead and for lag. We shall assume that $\varphi_{12}(t)$ is identically zero. As for $\varphi_{22}(t)$, we shall take a case which, although not formally contained in the theory we have given, constitutes a limiting case of it, and one of the greatest importance in practice. This is the case in which the noise input is due to a shot effect and has an equipartition of power in frequency. Theoretically, of course, this is not strictly realizable, as it would demand an infinite power; practically, as in the case of Planck's law in optics, it may hold within the limits of observation up to frequencies of a magnitude so great that they are no longer of interest for our particular problem. Thus we shall put $\Phi_{22}(\omega) = \epsilon^2$. As to $\Phi_{11}(\omega)$, it will depend on the particular problem considered. For an example, let us consider the case $\Phi_{11}(\omega) = 1/(1+\omega^2)$. Then

$$H(\omega) = \frac{e^{i\alpha\omega}}{1+\omega^2}; \tag{3.60}$$

$$\Phi(\omega) = \frac{1}{1+\omega^2} + \epsilon^2 = \frac{1+\epsilon^2+\epsilon^2\omega^2}{1+\omega^2}; \tag{3.601}$$

$$\Psi(\omega) = \frac{\sqrt{1+\epsilon^2} + \epsilon i\omega}{1+i\omega}; \tag{3.602}$$

$$\frac{H(\omega)}{\overline{\Psi(\omega)}} = \frac{e^{i\alpha\omega}}{(1+i\omega)(\sqrt{1+\epsilon^2} - \epsilon i\omega)}; \tag{3.603}$$

$$\frac{1}{2\pi}\int_{-\infty}^{\infty} \frac{H(u)}{\overline{\Psi(u)}} e^{iut}\, du = \frac{1}{2\pi}\int_{-\infty}^{\infty} \frac{e^{iu(t+\alpha)}}{(1+iu)(\sqrt{1+\epsilon^2} - \epsilon iu)}\, du$$

$$= \frac{1}{2\pi}\int_{-\infty}^{\infty} \frac{e^{iu(t+\alpha)}}{\epsilon+\sqrt{1+\epsilon^2}}\left(\frac{1}{1+iu} + \frac{\epsilon}{\sqrt{1+\epsilon^2}-\epsilon iu}\right) du$$

$$= \begin{cases} \dfrac{e^{-(t+\alpha)}}{(\epsilon + \sqrt{1+\epsilon^2})} & (t > -\alpha) \\ e\,\dfrac{\sqrt{1+\epsilon^2}(t+\alpha)}{\epsilon(\epsilon+\sqrt{1+\epsilon^2})} & (t < -\alpha). \end{cases} \tag{3.604}$$

If α is positive, this gives us

$$k(\omega) = \frac{1+i\omega}{\sqrt{1+\epsilon^2}+\epsilon i\omega}\int_0^{\infty} \frac{e^{-(t+\alpha)}e^{-i\omega t}}{\epsilon+\sqrt{1+\epsilon^2}}\,dt$$

$$= \frac{1+i\omega}{\sqrt{1+\epsilon^2}+\epsilon i\omega}\cdot\frac{e^{-\alpha}}{\epsilon+\sqrt{1+\epsilon^2}}\cdot\frac{1}{1+i\omega}$$

$$= \frac{e^{-\alpha}}{\epsilon+\sqrt{1+\epsilon^2}}\cdot\frac{1}{\sqrt{1+\epsilon^2}+\epsilon i\omega}\cdot \tag{3.605}$$

3.7 Lagging Filters

If α is negative, we must have recourse to approximations. For example, let us notice that

$$e^{i\alpha\omega} \cong \left(\frac{1+(\alpha i\omega/2\nu)}{1-(\alpha i\omega/2\nu)}\right)^{\nu} \tag{3.70}$$

Thus

$$\frac{H(\omega)}{\Psi(\omega)} \cong \left(\frac{1+(\alpha i\omega/2\nu)}{1-(\alpha i\omega/2\nu)}\right)^{\nu} \frac{1}{(1+i\omega)(\sqrt{1+\epsilon^2}-\epsilon i\omega)}$$

$$= \sum_1^N \frac{A_k}{[1-(\alpha i\omega/2\nu)]^k} + \frac{B}{1+i\omega} + \frac{C}{\sqrt{1+\epsilon^2}-\epsilon i\omega}, \tag{3.705}$$

where simultaneous equations for the A_k, B, and C may be obtained, for example, by substituting different values for ω in either the equation given or in its derivatives. Thus if ν is 1 and $\alpha \neq -2$, we have

$$(A_1 + B)\sqrt{1+\epsilon^2} + C = 1; \tag{3.71}$$

$$A_1(\sqrt{1+\epsilon^2}-\epsilon) + B\left(-\epsilon - \frac{\alpha}{2}\sqrt{1+\epsilon^2}\right) + C\left(1-\frac{\alpha}{2}\right) = \frac{\alpha}{2}; \tag{3.715}$$

$$-A_1\epsilon + \frac{B\alpha\epsilon}{2} - \frac{C}{2} = 0; \tag{3.72}$$

yielding

$$A_1 = \frac{\alpha^2/2}{\left(1+\frac{\alpha}{2}\right)\left(\frac{\alpha}{2}\sqrt{1+\epsilon^2}-\epsilon\right)}; \quad B = \frac{1-(\alpha/2)}{\left(1+\frac{\alpha}{2}\right)(\epsilon+\sqrt{1+\epsilon^2})}. \tag{3.73}$$

In general,

$$k(\omega) = \frac{1+i\omega}{\sqrt{1+\epsilon^2}+\epsilon i\omega}\left(\sum_1^{\nu}\frac{A_k}{(1-\alpha i\omega/2\nu)^k}+\frac{B}{1+i\omega}\right), \tag{3.74}$$

and in this special case

$$k(\omega) = \frac{\alpha^2/2}{\left(1+\frac{\alpha}{2}\right)\left(\frac{\alpha}{2}\sqrt{1+\epsilon^2}-\epsilon\right)}\cdot\frac{1+i\omega}{(\sqrt{1+\epsilon^2}+\epsilon i\omega)\left(1-\frac{\alpha i\omega}{2}\right)} + \frac{1-(\alpha/2)}{\left(1+\frac{\alpha}{2}\right)(\epsilon+\sqrt{1+\epsilon^2})(\sqrt{1+\epsilon^2}+\epsilon i\omega)}. \tag{3.75}$$

In this work, there is room for a considerable amount of freedom in selecting the rational operator which is taken to approximate $e^{i\alpha\omega}$. The problem of determining such an operator is the well-known one of designing pure delay networks and has been discussed at length in the *Bell System Technical Journal* and elsewhere. Given $\Phi(\omega)$, each such design leads to the determination of a rational filter characteristic with no singularities below the real axis. The realization of such real functions as the voltage-transfer-ratio characteristics of networks is also, at least in large measure, established engineering practice and may be found in such books as Guillemin's *Communication Networks.*

An alternative approach to filter design may be found in the use of Laguerre functions or similar combinations of algebraic and exponential functions to approximate to

$$\int_{-\infty}^{\infty}\frac{H(u)}{\Psi(u)}e^{iut}\,du \quad (0 \le t \le \infty). \tag{3.76}$$

Such a method will probably have no significant advantages over methods depending on an approximation to $e^{i\alpha\omega}$

The detailed design of a filter involves certain choices of constants which must be justified economically. In general, it does not pay to eliminate a small error from a quantity when there is a large irremovable error in it. The irremovable error in the mean square of the performance of a filter, as we have seen, is

$$\frac{1}{2\pi}\int_{-\infty}^{\infty}\frac{\Phi_{11}(\omega)\Phi_{22}(\omega)}{\Phi_{11}(\omega)+\Phi_{22}(\omega)}\,d\omega. \tag{3.77}$$

This minimum error is attainable only in a filter having an infinite time of delay. There is manifestly no use in increasing the delay time of a filter if the part of the error

$$\frac{1}{2\pi}\int_{0}^{\infty}\left\{\Phi_{11}(\omega)-\Phi(\omega)\left|\frac{1}{2\pi\Psi(\omega)}\int_{0}^{\infty}e^{i\omega t}\,dt\int_{-\infty}^{\infty}\frac{H(u)}{\overline{\Psi(\omega)}}e^{i\omega t}\,du\right|^{2}\right\}d\omega \tag{3.78}$$

due to the finiteness of this delay is already small compared with the intrinsic error.

3.8 The Determination of Lag and Number of Meshes in a Filter

If we thus fix on the reasonable delay for a filter, there is still another error dependent on the fact that the theoretically optimum design is not, in fact, realizable with a finite network of resistances, capacities, and inductances. This is the error implicit in our approximation to $e^{i\alpha u}$. Again, such an error may be estimated, and there is no point in reducing this part of the error to a level substantially below that of the error implicit in the delay already chosen. It is this last error which determines the number of meshes and parts appropriate to the complete filter network. Once this final error is decided upon, the rational voltage ratio characteristic of the network is determined, and its design is a matter of known technique.

The filter theory here presented, like our prediction theory, though it is adapted to the use of electrical circuits, may be carried out as a purely numerical computation, as may be indicated in certain studies of meteorological or geophysical time series, or may be applied to mechanical structures, as in the smoothing of the performance of hand-cranks and servo-mechanisms. In the latter cases, the large time scale of the process may put the use of purely passive electrical networks out of question, as really large inductances are not practical. On the other hand, such filters may be realized mechanically; or they may be realized electrically by the use of active networks, in which vacuum tubes or other amplifying devices appear, but in which inductances do not appear; or they may be realized by a combination of mechanical and electrical devices. In all cases, for a shot-effect noise, the information required for filter design consists in the spectrum of the message to be transmitted, the noise level, and the permissible delay.

3.9 Detecting Filters for High Noise Level

An especially interesting and important type of filter is that employed in the observation of very faint messages when the noise level is very high. Here certain simplifying assumptions may be made. We shall have

$$\Phi_{22}(\omega) = 1; \quad \Phi_{11}(\omega) = \epsilon F(\omega); \quad \Phi(\omega) = 1 + \epsilon F(\omega) \qquad (3.900)$$

if the noise is suitably normalized. To factor $\Phi(\omega)$, we introduce the function

$$G(\omega) = \frac{1}{2\pi}\int_0^\infty e^{-i\omega t}\,dt \int_{-\infty}^\infty F(u)e^{iut}\,du. \qquad (3.901)$$

Then approximately

$$\Phi(\omega) = |1 + \epsilon G(\omega)|^2; \qquad (3.9015)$$

and if we put

$$1 + \epsilon G(\omega) = \Psi(\omega), \qquad (3.902)$$

we shall have approximately

$$\frac{H(\omega)}{\overline{\Psi(\omega)}} = \frac{\epsilon F(\omega)e^{i\omega\alpha}}{1 + \epsilon\overline{G(\omega)}} = \epsilon F(\omega)e^{i\omega\alpha}[1 - \epsilon\overline{G(\omega)}]. \qquad (3.9025)$$

To a first degree of approximation,

$$k(\omega) = \frac{\epsilon}{2\pi}\int_0^\infty e^{-i\omega t}\,dt \int_{-\infty}^\infty F(u)e^{iu(t+\alpha)}\,du$$

$$= e^{i\omega\alpha}\frac{\epsilon}{2\pi}\int_\alpha^\infty e^{-i\omega t}\,dt \int_{-\infty}^\infty F(u)e^{iut}\,du. \qquad (3.903)$$

This means that, if we allow a long delay and neglect the effect of this delay, the frequency character of the desired filter will be $\epsilon F(\omega)$ itself.

The next degree of approximation is

$$k(\omega)e^{-i\omega\alpha} \cong \frac{\Phi_{11}(\omega)}{\Phi(\omega)} \cong \frac{\epsilon F(\omega)}{1 + \epsilon F(\omega)}. \qquad (3.9035)$$

It will be seen that this function is qualitatively very similar to $F(u)$, and that, except for a numerical factor, a filter with frequency characteristic $F(\omega)$ will not perform too badly for a considerable range of large noise intensities. Of course, the criterion which we have used for the performance of a filter is perhaps not so obviously natural at high noise levels as at low. Nevertheless, it is still not a valueless one.

3.91 Filters for Pulses

The problem of filtering at high noise levels arises chiefly in connection with the detection of extremely faint messages, at the threshold of detectability. A closely related problem is that of the proper pulse form for such messages. For many purposes, a message to be detected by such means must be accurately localizable in time and, for the avoidance of incoming and outgoing interference, must also be accurately localizable in frequency. Let the time form of such a pulse be $f(t)$, and its Fourier transform (the frequency form of the pulse) be $g(u)$. Here

$$g(u) = \frac{1}{\sqrt{2\pi}} \underset{A\to\infty}{\text{l.i.m.}} \int_{-A}^{A} f(t)e^{-iut}\,dt. \tag{3.910}$$

A reasonable measure of the time spread of such a pulse will be

$$\frac{\int_{-\infty}^{\infty} t^2 \,|\, f(t) \,|^2 \,dt}{\int_{-\infty}^{\infty} |\, f(t) \,|^2 \,dt}. \tag{3.9105}$$

Thus a reasonable measure of the combined time and frequency spread will be

$$\frac{\int_{-\infty}^{\infty} t^2 \,|\, f(t) \,|^2 \,dt}{\int_{-\infty}^{\infty} |\, f(t) \,|^2 \,dt} + \frac{\lambda^2 \int_{-\infty}^{\infty} u^2 \,|\, g(u) \,|^2 \,du}{\int_{-\infty}^{\infty} |\, \overline{g(u)} \,|^2 \,du}$$

$$= \frac{\int_{-\infty}^{\infty} \left[t^2 \,|\, f(t) \,|^2 + \lambda^2 \left| \frac{df(t)}{dt} \right|^2 \right] dt}{\int_{-\infty}^{\infty} |\, f(t) \,|^2 \,dt}. \tag{3.911}$$

Assuming the total energy of the pulse to be fixed, the problem minimizing the combined time and frequency spreads reduces to that of minimizing

$$\int_{-\infty}^{\infty} \left([t^2 f(t)]^2 + \lambda^2 \left[\frac{df(t)}{dt} \right]^2 \right) dt, \tag{3.9115}$$

while $\int_{-\infty}^{\infty} [f(t)]^2 \,dt$ is constant. If we now vary $f(t)$ by adding to it $\delta f(t)$, we see that we should have

$$0 = \int_{-\infty}^{\infty} t^2 f(t) \delta f(t) \,dt + \lambda^2 \int_{-\infty}^{\infty} \frac{df(t)}{dt} \frac{d}{dt} [\delta f(t)] \,dt$$

$$= \int_{-\infty}^{\infty} \left[t^2 f(t) - \lambda^2 \frac{d^2 f(t)}{dt^2} \right] \delta f(t) \,dt \tag{3.912}$$

whenever

$$\int_{-\infty}^{\infty} f(t)\delta f(t)\,dt = 0. \tag{3.9125}$$

This leads to the Euler-Lagrange equation,

$$t^2 f(t) - \lambda^2 \frac{d^2 f(t)}{dt^2} = \mu f(t), \tag{3.913}$$

and the L^2 solution of this, with smallest characteristic value, will be

$$f(t) = \exp\left(\frac{-t^2}{2\lambda}\right). \tag{3.9135}$$

This is accordingly the desired pulse shape, and the desired $k(\omega)e^{-i\omega\alpha}$ will be

$$F(\omega) = \text{const}\exp(-\lambda\omega^2). \tag{3.914}$$

For detection of faint pulses of this form, we accordingly wish a filter with characteristic approximating $e^{-\lambda\omega^2}$. Fortunately, the time characteristic of such a filter will be nearly

$$(\text{const})\exp\left[\frac{-(t-\tau)^2}{4\lambda}\right], \tag{3.9145}$$

and this may be approximated by taking

$$t^n \exp\left(-t\sqrt{\frac{n}{2\lambda}}\right) \cong (\text{const})\exp\left[-\frac{(t-\text{const})^2}{4\lambda}\right]. \tag{3.915}$$

On the frequency scale this yields us

$$(\text{const})\frac{1}{\left(\sqrt{\frac{n}{2\lambda}} + i\omega\right)^n}, \tag{3.9155}$$

and our problem is completely solved.

3.92 Filters Having Characteristics Linearly Dependent on Given Characteristics

Let us now explore the problem of the filter consisting in a linear combination of fixed operators with variable coefficients. Let us seek to minimize

$$M = \lim_{T\to\infty}\frac{1}{2T}\int_{-T}^{T}\left|f(t+\alpha) - \sum_1^n a_k \int_0^{\infty}[f(t-\tau)+g(t-\tau)]\,dK_k(\tau)\right|^2 dt$$

$$= \varphi_{11}(0) - 2\mathrm{R}\left\{\int_0^\infty [\varphi_{11}(\alpha+\tau) + \varphi_{12}(\alpha+\tau)] \sum_1^n \bar{a}_k\, d\overline{K_k(t)}\right\}$$

$$+ \int_0^\infty \sum_1^n a_k\, dK_k(\sigma) \int_0^\infty \sum_1^n \bar{a}_k\, d\overline{K_k(\tau)}\varphi(\tau-\sigma). \qquad (3.920)$$

Let us use the notation

$$\int_0^\infty e^{-i\omega t}\, dK_k(t) = k_k(\omega) \quad (k = 1, \cdots, n). \qquad (3.9205)$$

Then, using reductions now familiar to us, we may write

$$M = \frac{1}{2\pi}\int_{-\infty}^\infty \Phi_{11}(\omega)\, d\omega$$

$$- 2\mathrm{R}\left\{\sum_1^n \bar{a}_k \frac{1}{2\pi}\int_{-\infty}^\infty [\Phi_{11}(\omega) + \Phi_{12}(\omega)]e^{i\alpha\omega}\overline{k_k(\omega)}\, d\omega\right\}$$

$$+ \frac{1}{2\pi}\sum_1^n a_j \sum_1^n \bar{a}_k \int_{-\infty}^\infty \Phi(\omega)k_j(\omega)\overline{k_k(\omega)}\, d\omega. \qquad (3.921)$$

If our $k_k(\omega)$ are so normalized that

$$\frac{1}{2\pi}\int_{-\infty}^\infty \Phi(\omega)k_j(\omega)\overline{k_k(\omega)}\, d\omega = \begin{cases} 0 \text{ if } j \neq k; \\ 1 \text{ if } j = k; \end{cases} \qquad (3.9215)$$

this gives us

$$M = \frac{1}{2\pi}\int_{-\infty}^\infty \Phi_{11}(\omega)\, d\omega -$$

$$2\mathrm{R}\left\{\sum_1^n \bar{a}_k \frac{1}{2\pi}\int_{-\infty}^\infty [\Phi_{11}(\omega) + \Phi_{12}(\omega)]e^{i\alpha\omega}\overline{k_k(\omega)}\, d\omega\right\} + \sum_1^n |a_k|^2$$

$$= \frac{1}{2\pi}\int_{-\infty}^\infty \Phi_{11}(\omega)\, d\omega - \sum_1^n \left|\frac{1}{2\pi}\int_{-\infty}^\infty [\Phi_{11}(\omega) + \Phi_{12}(\omega)]e^{i\alpha\omega}\overline{k_k(\omega)}\, d\omega\right|^2$$

$$+ \sum_1^n \left|a_k - \frac{1}{2\pi}\int_{-\infty}^\infty [\Phi_{11}(\omega) + \Phi_{12}(\omega)]e^{i\alpha\omega}\overline{k_k(\omega)}\, d\omega\right|^2. \qquad (3.922)$$

This will be a minimum when and only when

$$a_k = \frac{1}{2\pi}\int_{-\infty}^\infty [\Phi_{11}(\omega) + \Phi_{12}(\omega)]e^{i\alpha\omega}\overline{k_k(\omega)}\, d\omega, \qquad (3.9225)$$

and the minimum will be

$$\frac{1}{2\pi}\int_{-\infty}^\infty \Phi_{11}(\omega)\, d\omega - \sum_1^n \left|\frac{1}{2\pi}\int_{-\infty}^\infty [\Phi_{11}(\omega) + \Phi_{12}(\omega)]e^{i\alpha\omega}\overline{k_k(\omega)}\, d\omega\right|^2. \qquad (3.923)$$

Under these conditions,

$$\sum_1^n a_k k_k(\omega) = \tag{3.9235}$$

$$\frac{1}{\Psi(\omega)}\left[\sum_1^n k_k(\omega)\Psi(\omega)\frac{1}{2\pi}\int_{-\infty}^{\infty}\frac{\Phi_{11}(\omega)+\Phi_{22}(\omega)}{\overline{\Psi(\omega)}}e^{i\alpha\omega}\overline{k_k(\omega)\Psi(\omega)}\,d\omega\right]$$

which is the formal development of $\dfrac{\Phi_{11}(\omega)+\Phi_{22}(\omega)}{\overline{\Psi(\omega)}}e^{i\alpha\omega}$ in the functions $k_k(\omega)\cdot\Psi(\omega)$, multiplied by $1/\Psi(\omega)$. If the set of functions $k_k(\omega)\cdot\Psi(\omega)$, all of which are free from singularities in the half-plane below the real axis, is complete in the set of such functions [and if it is not, it may be made so by adjoining other functions $k_k(\omega)\cdot\Psi(\omega)$, which may be taken so as to satisfy (3.9215)], we obtain as the formal limit for $\sum_1^n a_k k_k(\omega)$ the expression

$$\frac{1}{2\pi\Psi(\omega)}\int_0^{\infty}e^{-i\omega t}\,dt\int_{-\infty}^{\infty}\frac{\Phi_{11}(u)+\Phi_{12}(u)}{\overline{\Psi(u)}}e^{i\alpha u}\,du, \tag{3.924}$$

which we have already seen to solve the filter problem as stated, without any reference to a specific set of functions $k_k(\omega)$.

Let it be noted that we have here assumed the factorability of $\Phi(\omega)$ into $|\Psi(\omega)|^2$, where $\Psi(\omega)$ is free from singularities in the lower half-plane. If this is not the case, as we have seen,

$$\int_{-\infty}^{\infty}\frac{\log|\Phi(\omega)|}{1+\omega^2}\,d\omega = -\infty. \tag{3.9245}$$

Under these circumstances the set of functions $\sqrt{\Phi(\omega)}k(\omega)$, where $k(\omega)$ is free from singularities in the lower half-plane, is closed L^2. Otherwise there will exist a function $\Phi(\omega)l(\omega)$ of class L^2, not equivalent to zero, such that

$$\int_{-\infty}^{\infty}\Phi(\omega)k(\omega)l(\omega)\,d\omega = 0 \tag{3.925}$$

for every $k(\omega)$ for which

$$\int_{-\infty}^{\infty}\Phi(\omega)\,|k(\omega)|^2\,d\omega < \infty \tag{3.9255}$$

and which has no singularity in the lower half-plane. This will certainly be the case for every function $k(\omega)$ of class L^2 having no singularity in the lower half-plane. It will follow at once that

$$\Phi(\omega)\cdot l(\omega) \tag{3.926}$$

will be orthogonal to every function of class L^2 vanishing for negative arguments, and will itself vanish for positive arguments. Thus

$$\int_{-\infty}^{\infty} \frac{\log |\Phi(\omega)| + \log |l(\omega)|}{1+\omega^2} d\omega > -\infty. \tag{3.9265}$$

This would mean, by (3.9245), that

$$\int_{-\infty}^{\infty} \frac{\log |\sqrt{\Phi(\omega)}l(\omega)|}{1+\omega^2} d\omega = \infty, \tag{3.927}$$

which is manifestly false, as the function $\sqrt{\Phi(\omega)}l(\omega)$ belongs to L^2. Thus no such $l(\omega)$ exists, and the set of functions $\sqrt{\Phi(\omega)}k(\omega)$ is closed. In other words, we may choose the functions $\sqrt{\Phi(\omega)}k_n(\omega)$ in such a way that

$$\sqrt{\Phi(\omega)} \sum_1^{\infty} a_k k_k(\omega) = \frac{\Phi_{11}(\omega) + \Phi_{12}(\omega)}{\sqrt{\Phi(\omega)}} e^{i\alpha\omega}, \tag{3.9275}$$

and that formally

$$\sum_1^{\infty} a_k k_k(\omega) = \frac{\Phi_{11}(\omega) + \Phi_{12}(\omega)}{\Phi(\omega)} e^{i\alpha\omega}. \tag{3.928}$$

This means that in such a case the performance of a filter for a finite delay may be made to approximate as nearly as we wish the performance of a filter for an infinite delay. This is quite reasonable, since for such messages and noises the entire future of the message-plus-noise is determined by its past, and nothing new ever happens.

Let it be noted that the situation depends on the factorability or non-factorability of $\Phi(\omega)$, which involves both the message and the noise, and not on the factorability or non-factorability of a term containing the message alone. Even with a perfectly predictable message, the presence of an imperfectly predictable noise makes the filtering problem a significant one.

The problem which we have just solved is that of the design of a filter having a character which is the sum of fixed operators with adjustable coefficients. The functions $k_k(\omega)$ may be any functions obtained by normalizing in the proper sense the functions $(1 + ai\omega)^n$ $(n = 0, 1, 2, \cdots)$, for then they themselves [and *a fortiori* the functions $k_k(\omega)$] will be closed in the set of all functions of L^2 which are free from singularities in the half-plane below the real axis. Since we have an algorithm for obtaining the coefficients, the filter-design problem is solved. Such adjustable filters are of the greatest value in experimental installations, in which the adjustability may actually be realized by the turning of rheostats or potentiometers, or even in permanent installations in which the variety of work to be undertaken is very great. Of course, they will ordinarily be more complicated than the fixed-constant sets having the minimum number of elements for the same performance.

3.93 Computation of Filter: Résumé

Let us then sum up the mathematical stages in the design of a filter of fixed or variable characteristic. The first stage is the computation of the even function $\Phi_{11}(\omega)$. On the basis of this, with a proper choice of the scale constant P, the coefficients

$$\int_{-\infty}^{\infty} \Phi_{11}(\omega) e^{2ni \tan^{-1}\omega P} \frac{2P}{1+\omega^2P^2} d\omega \qquad (3.930)$$

are then computed. Then the Cesaro sum

$$\frac{1}{2\pi} \sum_{\nu=-N}^{N} \left(1 - \frac{|\nu|}{N}\right) e^{-2\nu i \tan^{-1}\omega P} \int_{-\infty}^{\infty} \Phi_{11}(u) e^{2\nu i \tan^{-1} uP} du \qquad (3.9305)$$

is computed as an approximate value for $\Phi_{11}(\omega)$. This is then written in the quotient form

$$\frac{(\text{polynomial in } \omega^2P^2)}{(1+\omega^2P^2)^{N-1}} \qquad (3.931)$$

and the numerator and denominator are factored into linear factors. In factoring the numerator, algebraic equations of high order may have to be solved for their complex roots, and the use of a device such as the Isograph* of the Bell Telephone Laboratories is indicated. Then, by selecting in both numerator and denominator only those roots with positive imaginary part, the function $\Psi(\omega)$ is determined. In the case of a lead filter, all is plain sailing from here on. In the case of a lag filter, through considerations such as those we have already indicated, the proper lag is determined, as well as the degree to which it is worth while to imitate $e^{i\alpha\omega}$ in determining the proper approximation to this lag. Then we call on the existing technique of delay-mechanism design to realize this approximation in terms of the simplest rational characteristic possible. When this approximation is known, we have already given the formulae which determine the final filter characteristic. Finally, whether for lead or for lag, we have to realize the (now determined) characteristic by a network, which is then subject to the many known tricks of network transformation.

* See Industrial Mathematics, by T. C. Fry, *Bell System Technical Journal*, Vol. 20, No. 3, July, 1941, p. 276. There are also well-known computational-algebraic methods of achieving the same result.

So much for the fixed filter. For the variable filter, we are faced instead with the choice of a suitable closed set of functions $1/(i\omega + a)^k$ and of their orthogonalization with respect to the function $\Phi(\omega)$. This alone determines the structure of the filter, while the determination of the numbers a_k in terms of α gives the setting of the apparatus for a desired lead or lag.

Much less important, though of real interest, is the problem of the numerical filter for statistical work, as contrasted with the filter as a physically active piece of engineering apparatus. In the case of continuous data, there is little new to say of this, except that the $k(\omega)$ already obtained must be translated into a $K(t)$ from which we may evaluate $\int_0^\infty f(t - \tau)\, dK(\tau)$. In that particular subcase of the discrete case in which f and g are independent, we follow the lines of the prediction theory of the previous chapter and define the function $\Phi(\omega)$ in terms of the auto-correlation coefficient

$$\varphi_{11\nu} = \lim_{N\to\infty} \frac{1}{2N+1} \sum_{\mu=-N}^{N} f_{\nu+\mu} f_{\mu}, \tag{3.9315}$$

where the f_ν constitute the time series with which we are working. We further put

$$\Phi_{11}(\omega) = \sum_{-\infty}^{\infty} \varphi_{11\nu} e^{-i\nu\omega}. \tag{3.932}$$

This will be a periodic function of period 2π. Similarly,

$$\varphi_{22\nu} = \lim_{N\to\infty} \frac{1}{2N+1} \sum_{\mu=-N}^{N} g_{\nu+\mu}\bar{g}_{\mu}; \quad \Phi_{22}(\omega) = \sum_{-\infty}^{\infty} \varphi_{22\nu} e^{-i\nu\omega}. \tag{3.9325}$$

As an approximation, we may use Cesaro methods as before and may put

$$\sum_{-N}^{N} \left(1 - \frac{|\nu|}{N}\right) (\varphi_{11\nu} + \varphi_{22\nu}) e^{-i\nu\omega} = \left| \sum_{0}^{N} \psi_\nu e^{i\nu\omega} \right|^2, \tag{3.933}$$

where the factoring is so carried out that

$$\Psi(\omega) = \sum_{0}^{N} \psi_\nu e^{i\nu\omega} \neq 0 \quad (\mathrm{I}\{\omega\} < 0). \tag{3.9335}$$

Here again the isograph may be used. We then put

$$k(\omega) = \frac{1}{2\pi\Psi(\omega)} \sum_{0}^{\infty} e^{-i\nu\omega} \int_{-\pi}^{\pi} \frac{\Phi_{11}(u) e^{i\alpha u}}{\overline{\Psi(u)}} e^{iu\nu}\, du \quad (\alpha \text{ an integer}). \tag{3.934}$$

If now we write

$$K_\nu = \frac{1}{2\pi} \int_{-\pi}^{\pi} e^{i\nu\omega} k(\omega)\, d\omega, \tag{3.935}$$

we shall minimize

$$\lim_{N\to\infty} \frac{1}{2N+1} \sum_{-N}^{N} \left| f_{\nu+\alpha} - \sum_{\mu=0}^{\infty} (f_{\nu-\mu} + g_{\nu-\mu}) K_\mu \right|^2. \tag{3.936}$$

The minimum of this expression will be

$$\frac{1}{2\pi} \int_{-\pi}^{\pi} \{\Phi_{11}(\omega) - [\Phi_{11}(\omega) + \Phi(\omega)] \mid k(\omega) \mid^2\}\, d\omega. \tag{3.937}$$

Control System Synthesis by Root Locus Method

WALTER R. EVANS

MUCH insight into the properties of a single-input single-output linear system can be obtained from a plot of the poles and zeros of its transfer function. For a single-loop system, the closed-loop transfer function is given by

$$G(s) = \frac{KN(s)}{D(s) + KN(s)}, \qquad (1)$$

where $KN(s)/D(s)$ is the open loop transfer function and K is the gain. The closed-loop zeros are the zeros of $N(s)$ and the poles are the roots of the equation

$$D(s) + KN(s) = 0 \qquad (2)$$

In the late 1940s it was still a tedious task to compute roots of polynomials. As an extension of a technique he had developed earlier in [2], and using homotopy ideas, Evans presented in this 1950 paper a clever method to find the locus of roots of (2) when the gain varies. He observed that the roots of (2) approach the roots of $D(s)$ and $N(s)$ as $K \to 0$ or ∞. For small gains the closed-loop poles are close to the open-loop poles, and for large values of the gain $\deg N$ of the closed-loop poles migrate to the zeros and the remaining $\deg \mathrm{D} - \deg N$ closed-loop poles go towards infinity in a symmetric star configuration. Evans developed simple methods to sketch the root loci based on initial directions, asymptotes, multiple roots, and behavior close to multiple roots.

The presentation of the method in the paper focuses on the graphical construction. The paper even describes how to make a simple "spirule" (Fig. 8 of the paper) to facilitate the construction. Many of the basic root loci rules as they are usually elaborately listed and emphasized in modern textbooks may be found in this paper, but they are introduced informally, often through an example. They include the start and end points of the loci, their asymptotes, portions on the real axis, and the presence of break-away points.

The paper demonstrates convincingly how root loci are used to establish important properties of closed-loop systems such as their response as determined by the dominant closed-loop poles. It does not stop at this, but also illustrates how root loci may be used to design corrective networks by anticipating and analyzing the effect of introducing additional open-loop poles and zeros. In Evans's own words, "[the] particular advantage [of the root locus method] is in retaining at all times a clear picture of the relationships between the overall network parameters and the parameters of individual circuit elements."

The root locus quickly became a favorite tool of control engineers, whose intuition was enriched by its clear visualization of feedback system properties. It has had a lasting impact on engineering education. Every present-day undergraduate control textbook includes a detailed discussion of the root locus method. The root locus is still a very effective way to quickly assess how the properties of a system change with the gain. Every control systems software package offers root loci as a standard tool. Today, we still plot root loci but the graphical construction is rarely used any more, because using efficient software packages one can easily compute the roots with much greater accuracy.

Root locus ideas contributed to the development of more advanced feedback system concepts, including multivariable root locus generalizations and high gain properties displayed by root locus asymptotes. The fundamental property that closed-loop poles tend to system zeros as the gain becomes very high is the essence of the modern concept of "zero dynamics," not only in linear, but also in a large class of nonlinear systems [1]. This property points to the obstacles to achieving high performance in systems with unstable zero dynamics (non-minimum phase systems). Feedback limitations were analyzed using the root locus asymptotes of optimal systems in [3]. A design which assigns the rates with which roots move along their loci was developed in [4]. These and similar recent results show that the impact of the root locus method was broader than just being a useful graphical tool.

References

[1] C.I. Byrnes and A. Isidori, "A frequency domain philosophy for for nonlinear systems," *Proc. 23th IEEE Conf. Decision and Control*, pp. 1569–1573, Las Vegas, NV, 1984.

[2] W.R. Evans, "Graphical analysis of control systems," *Trans. Amer. Inst. Electric. Engineers*, **67**:547–551, 1948.

[3] H. KWAKERNAAK AND R. SIVAN, "The maximally achievable accuracy of linear optimal regulators and linear optimal filters," *IEEE Trans. Aut. Contr.*, **17**:79–86, 1972.

[4] A. SABERI AND P. SANNUTI, "Time-scale structure assignment in linear multivariable systems using high-gain feedback," *Internat. J. Contr.*, **49**:2191–2213, 1989.

H.K.

Control System Synthesis by Root Locus Method

WALTER R. EVANS
MEMBER AIEE

Synopsis: The root locus method determines all of the roots of the differential equation of a control system by a graphical plot which readily permits synthesis for desired transient response or frequency response. The base points for this plot on the complex plane are the zeros and poles of the open loop transfer function, which are readily available. The locus of roots is a plot of the values of s which make this transfer function equal to -1 as loop gain is increased from zero to infinity. The plot can be established in approximate form by inspection and the significant parts of the locus calculated accurately and quickly by use of a simple device. For multiple loop systems, one solves the innermost loop first, which then permits the next loop to be solved by another root locus plot. The resultant plot gives a complete picture of the system, which is particularly valuable for unusual systems or those which have wide variations in parameters.

THE root locus method is the result of an effort to determine the roots of the differential equation of a control system by using the concepts now associated with frequency response methods.[1] The roots are desired, of course, because they describe the natural response of the system. The simplifying feature of the control system problem is that the open loop transfer function is known as a product of terms. Each term, such as $1/(1+Ts)$, can be easily treated in the same manner as an admittance such as $1/(R+jx)$. It is treated as a vector in the sense used by electrical engineers in solving a-c circuits. The phase shift and attenuation of a signal of the form ϵ^{st} being transmitted is represented by $1/(1+Ts)$ in which s in general is a complex number. The key idea in the root locus method is that the values of s which make transfer function around the loop equal to -1 are roots of the differential equation of the system.

The opening section in this paper, Background Theory, outlines the over-all pattern of analysis. The following section on Root Locus Plot points out the great usefulness of knowing factors of the open loop transfer function in finding the roots.

The graphical nature of the method requires that specific examples be used to demonstrate the method itself under the topics: Single Loop Example, Multiple Loop System, and Corrective Networks. The topic Correlation with Other Methods suggests methods by which experience in frequency methods can be extended to this method. The topic Other Applications includes the classic problem of solving an nth degree polynomial. Finally, the section on Graphical Calculations describes the key features of a plastic device called a "Spirule", which permits calculations to be made from direct measurement on the plot.

Background Theory

The over-all pattern of analysis can be outlined before explaining the technique of sketching a root locus plot. Thus consider the general single loop system shown in Figure 1.

Note that each transfer function is of the form $KG(s)$ in which K is a static gain constant and $G(s)$ is a function of the complex number. In general, $G(s)$ has both numerator and denominator known in factored form. The values of s which make the function zero or infinite can therefore be seen by inspection and are called zeros and poles respectively. The closed loop transfer function can be expressed directly from Figure 1 as given in equation 1

$$\frac{\theta_0}{\theta_i}(s)=\frac{K_\mu G_\mu(s)}{1+K_\mu G_\mu(s)K_\beta G_\beta(s)} \tag{1}$$

The problem of finding the roots of the differential equation here appears in the form of finding values of s which make the denominator zero. After these values are determined by the root locus method, the denominator can be expressed in factored form. The zeros of the function θ_0/θ_i can be seen from equation 1 to be the zeros of $G_\mu(s)$ and the poles of $G_\beta(s)$. The function can now be expressed as shown in equation 2

$$\frac{\theta_0}{\theta_i}(s)=K_c s^\gamma\frac{(1-s/q_1)(1-s/q_2)\ldots}{(1-s/r_1)(1-s/r_2)\ldots.} \tag{2}$$

The constant K_c and the exponent γ depend upon the specific system but for control systems γ is often zero and K_c is often 1.

The full power of the Laplace Transform[2] or an equivalent method now can be used. The transient response of the output for a unit step input, for example, is given by equation 3

$$\theta_0(t)=1-\sum_{i=1}^{i=n}A_i\epsilon^{r_it} \tag{3}$$

The amplitude A_i is given by equation 4

$$A_i=\left[\frac{\theta_0}{\theta_i}(s)(1-s/r_i)\right]_{s=r_i} \tag{4}$$

The closed loop frequency response, on the other hand, can be obtained by substituting $s=j\omega$ into equation 2. Fortunately, the calculation in finding A_i or $\theta_0/\theta_i(j\omega)$ involves the same problem of multiplying vectors that arises in making a root locus plot, and can be calculated quickly from the resultant root locus plot.

Paper 50-11, recommended by the AIEE Feedback-Control Systems Committee and approved by the AIEE Technical Program Committee for presentation at the AIEE Winter General Meeting, New York, N. Y., January 30–February 3, 1950. Manuscript submitted November 15, 1948; made available for printing November 22, 1949.

WALTER R. EVANS is with North American Aviation, Inc., Downey, Calif.

The author wishes to express his appreciation for the assistance given by his fellow workers, K. R. Jackson and R. M. Osborn, in the preparation of this paper. In particular, Mr. Osborn contributed the circuit analysis example.

Reprinted with permission from *Transactions of the American Institute of Electrical Engineers*, Walter R. Evans, "Control System Synthesis by Root Locus Method," Vol. 69, 1950, pp. 66–69.

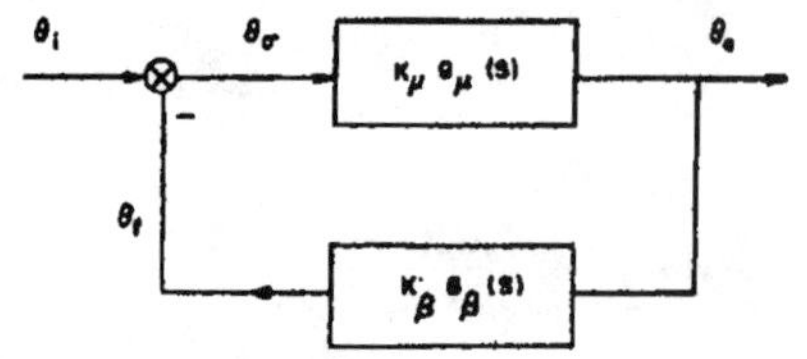

Figure 1 (left). General block diagram

Figure 3 (right). Single loop root locus

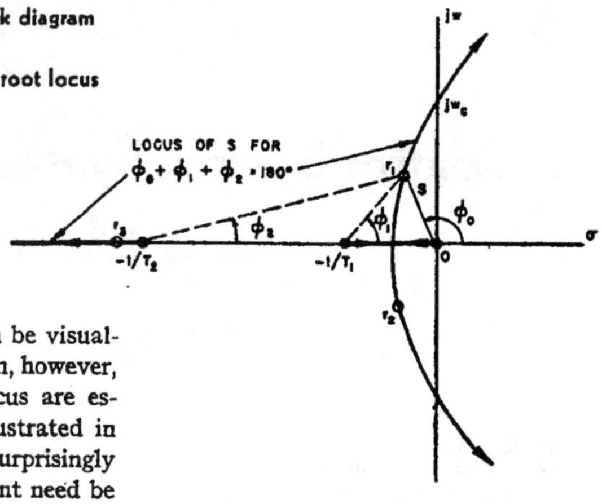

Root Locus Plot

The open loop transfer function is typically of the form given in equation 5.

$$K_\mu G_\mu(s)K_\beta G_\beta(s) = \frac{K(1+T_2s)[\sigma_3^2+\omega_3^2]}{s(1+T_1s)[(s+\sigma_3)^2+\omega_3^2]} \quad (5)$$

The parameters such as T_1 are constant for a given problem, whereas s assumes many values; therefore, it is convenient to convert equation 5 to the form of equation 6.

$$K_\mu G_\mu(s)K_\beta G_\beta(s) = \frac{K(1/T_2+s)T_2[\sigma_3^2+\omega_3^2]}{s(1/T_1+s)T_1[(s+\sigma_3+j\omega_3)(s+\sigma_3-j\omega_3)]} \quad (6)$$

The poles and zeros of the function are plotted and a general value of s is assumed as shown in Figure 2.

Note that poles are represented as dots, and zeros as crosses. All of the complex terms involved in equation 6 are represented by vectors with heads at the general point s and tails at the zeros or poles. The angle of each vector is measured with respect to a line parallel to the positive real axis. The magnitude of each vector is simply its length on the plot.

In seeking to find the values of s which make the open loop function equal to -1, the value -1 is considered as a vector whose angle is 180 degrees $\pm$ n 360 degrees, where n is an integer, and whose magnitude is unity. Then one can consider first the problem of finding the locus of values for which the angle condition alone is satisfied. In general, one pictures the exploratory s point at various positions on the plane, and imagines the lines from the poles and zeros to be constructed so that the angles in turn can be visualized. For any specific problem, however, many special parts of the locus are established by inspection as illustrated in examples in later sections. Surprisingly few trial positions of the s point need be assumed to permit the complete locus to be sketched.

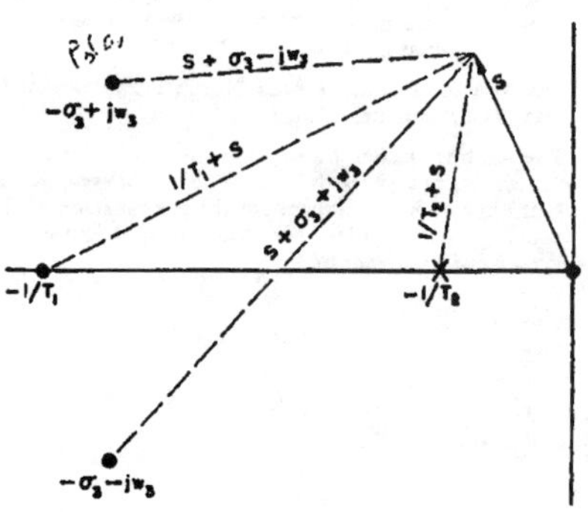

Figure 2. Root locus plot

After the locus has been determined, one considers the second condition for a root, that is, that the magnitude of $K_\mu G_\mu(s)K_\beta G_\beta(s)$ be unity. In general, one selects a particular value of s along the locus, estimates the lengths of the vectors, and calculates the static gain $K_\mu K_\beta = 1/G_\mu(s)G_\beta(s)$. After acquiring some experience, one usually can select the desired position of a dominant root to determine the allowable loop gain. The position of roots along other parts of the locus usually can be determined with less than two trials each.

An interesting fact to note from equation 6 is that for very low gain, the roots are very close to the poles in order that corresponding vectors be very small. For very high gain, the roots approach infinity or terminate on a zero.

Single Loop Example

Consider a single loop system such as shown in Figure 1 in which the transfer functions are given in equation 7.

$$K_\mu G_\mu(s) = \frac{K}{(1+T_1s)(1+T_2s)s}; K_\beta G_\beta(s) = 1 \quad (7)$$

The poles of the open loop function are at 0, $-1/T_1$ and $-1/T_2$ as represented by dots in Figure 3.

The locus along the real axis is determined by inspection because all of the angles are either 0 degrees or 180 degrees. An odd number of angles must therefore be 180 degrees as shown by the intervals between 0 and $-1/T_1$, and from $-1/T_2$ to $-\infty$. Along the $j\omega$ axis, ϕ_0 is 90 degrees so that ϕ_2 must be the complement of ϕ_1, as estimated at $s=j\omega_c$. For very large values of s, all angles are essentially equal so the locus for the complex roots finally approaches radial lines at ± 60 degrees.

The point where the locus breaks away from the real axis is found by considering a value of s just above the real axis. The decrease in ϕ_0 from 180 degrees can be made equal to the sum of ϕ_1 and ϕ_2 if the reciprocal of the length from the trial point to the origin is equal to the sum of the reciprocals of lengths from the trial point to $-1/T_1$ and $-1/T_2$. If a damping ratio of 0.5 for the complex roots is desired, the roots r_1 and r_2 are fixed by the intersection with the locus of radial lines at ± 60 degrees with respect to negative real axis.

In calculating K for $s=r_1$, it is convenient to consider a term $(1+T_1s)$ as a ratio of lengths from the pole $-1/T_1$ to the s point and from s to the origin respectively. After making gain $K=1/[G(s)]_{s=r_1}$ a good first trial for finding r_3 is to assume that it is near $-1/T_2$ and solve for $(1/T_2+s)$. After the roots are determined to the desired accuracy, the over-all transfer function can be expressed as given in equation 8.

$$\frac{\theta_0}{\theta_i} = \frac{1}{\left(1-\frac{s}{r_1}\right)\left(1-\frac{s}{r_2}\right)\left(1-\frac{s}{r_3}\right)} \quad (8)$$

The procedure in handling a multiple loop system now can be explained.

Multiple Loop System

Consider a multiple loop system in which the single loop system just solved is the forward path of another loop, as shown in Figure 4.

θ_o/θ_i is given in factored form by equation 8 so the roots of the inner loop now serve as base points for the new locus plot. For convenience, however, neglect the effect of the term $(1-s/r_3)$ so that the locus for the outer loop is shown in Figure 5.

The locus for the outer loop would be a circle about the $-1/T$ point as a center if the effect of θ_o/θ_i were completely neglected. Actually, the vectors from the points r_1 and r_2 introduce net angles so that the locus is modified as shown. The

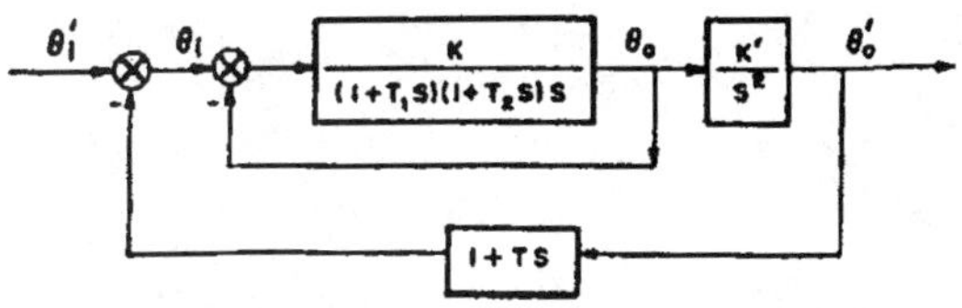

Figure 4 (above). Multiple loop block diagram

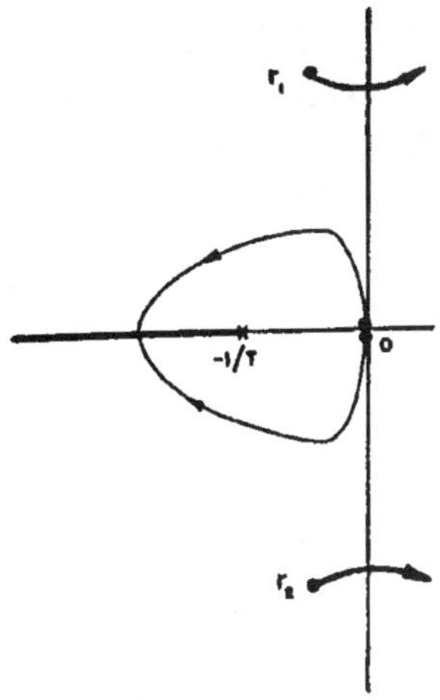

Figure 5 (right). Multiple loop root locus

angle at which the locus emerges from r_1 can be found by considering a value of s close to the point r_1, and solving for the angle of the vector $(s-r_1)$.

Assume that the static loop gain desired is higher than that allowed by the given system. The first modification suggested by the plot is to move the r_1 and r_2 points farther to the left by obtaining greater damping in the inner loop. If these points are moved far to the left, the loci from these points terminate in the negative real axis and the loci from the origin curve back and cross the $j\omega$ axis. Moving the $-1/T$ point closer to the origin would then be effective in permitting still higher loop gain. The next aspect of synthesis involves adding corrective networks.

Corrective Networks

Consider a somewhat unusual system which arises in instrument servos whose open loop transfer function is identified by the poles p_1 and p_2 in Figure 6(A). As loop gain is increased from zero, the roots which start from p_1 and p_2 move directly toward the unstable half plane. These roots could be made to move away from the $j\omega$ axis if 180 degrees phase shift were added. A simple network to add is three lag networks in series, each having a time constant T such that 60 degrees phase shift is introduced at p_1. The resultant locus plot is shown in Figure 6(B).

The gain now is limited only by the requirement that the new pair of roots do not cross the $j\omega$ axis. A value of gain is selected to obtain critical damping of these roots and the corresponding positions of all the roots are shown in Figures 6(A) and 6(B) as small circles.

Actually, greater damping could be achieved for roots which originate at p_1 and p_2 if a phase shifting bridge were used rather than the 3-lag networks. Its transfer function is $(3-Ts)/(1+Ts)$ and is of the "nonminimum phase" type of circuit.

Since these types of correction are somewhat unusual, it is perhaps well to point out that the analysis has been verified by actual test and application.

These examples serve to indicate the reasoning process in synthesizing a control system by root locus method. An engineer draws upon all of his experience, however, in seeking to improve a given system; therefore, it is well to indicate the correlation between this method and other methods.

Correlation with Other Methods

The valuable concepts of frequency response methods[1] are in a sense merely extended by the root locus system. Thus a transfer function with s having a complex value rather than just a pure imaginary value corresponds to a damped sinusoid being transmitted rather than an undamped one. The frequency and gain for which the Nyquist plot passes through the -1 point are exactly the same values for which the root locus crosses the $j\omega$ axis. Many other correlations appear in solving a single problem by both methods.

The results of root locus analysis can be easily converted to frequency response data. Thus one merely assumes values of s along the $j\omega$ axis, estimates the phase angles and vector lengths to the zeros and poles, and calculates the sum of the angles for total phase shift and the product of lengths for attenuation. The inverse problem of determining zeros and poles from experimental data is the more difficult one. Many techniques are already available, however, such as drawing asymptotes to the logarithmic attenuation curve. For unusual cases, particularly those in which resonant peaks are involved, the conformal mapping technique originated by Dr. Profos of Switzerland is recommended.[3]

The transient response is described by the poles of the transfer function. The inverse problem in this case is to locate the poles from an experimental transient response. One might use dead time, maximum build-up rate, overshoot, natural frequency of oscillation, and the damping rate as effective clues in solving this problem.

Other Applications

Many systems require a set of simultaneous equations to describe them and are said to be multicoupled. The corresponding block diagrams have several inputs to each loop so that the root locus method cannot be applied immediately. One should first lay out the diagram so that the main line of action of the signals forms the main loop with incidental coupling effects appearing as feedbacks and feed forwards. One then proceeds to isolate loops by replacing a signal which comes from within a loop by an equivalent signal at the output, replacing a signal entering a loop by an equivalent signal at the input. One can and should keep the physical picture of the equivalent system in mind as these manipulations are carried out.

The techniques of the root locus method can be used effectively in analyzing electric circuits. As a simple example, consider the lead-lag network of Figure 7(A).

It can be shown that the transfer function of this network is as given in equation 9

$$\frac{V_0}{V_i}=\frac{(1+R_1C_1s)(1+R_2C_2s)R_3}{(1+R_1C_1s)(1+R_2C_2s)R_3+R_1[1+(R_2+R_3)C_2s]} \tag{9}$$

The denominator can be factored algebraically by multiplying out and finding the zeros of the resulting quadratic. As an alternative, it will be noted that the zeros of the denominator must satisfy equation 10

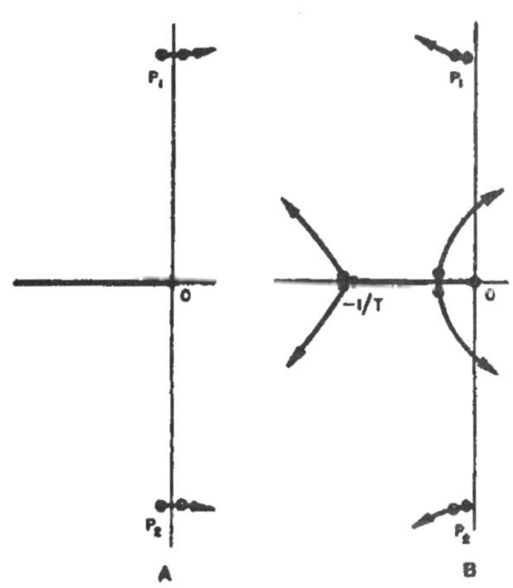

Figure 6. (A) Basic system. (B) Corrected system

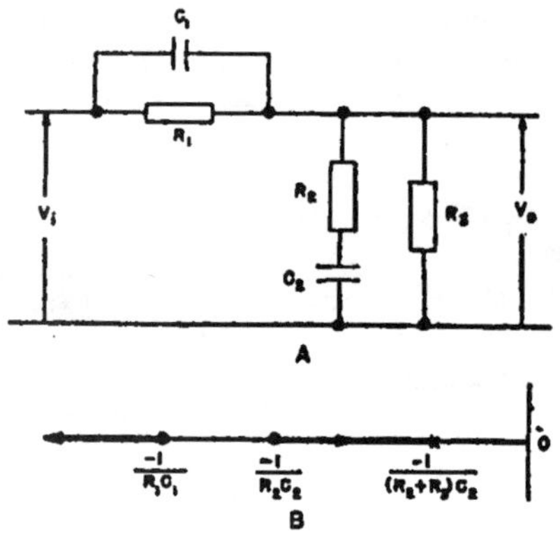

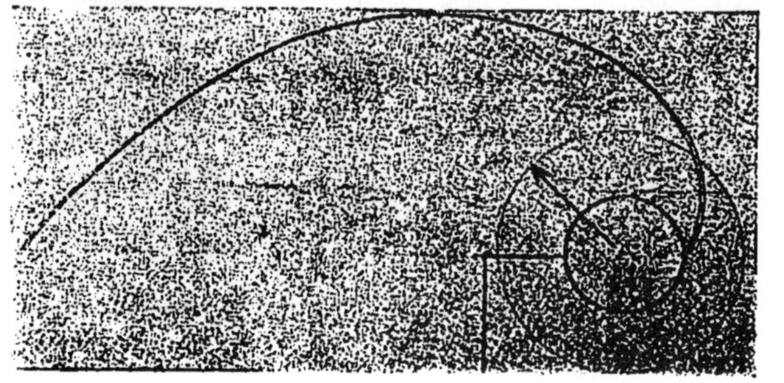

Figure 7 (left). (A) Circuit diagram. (B) Root locus

Figure 8 (right). Spirule

$$\frac{(1/R_1C_1+s)(1/R_2C_2+s)R_3(R_1C_1)(R_2C_2)}{[1/(R_2+R_3)C_2+s](R_2+R_3)C_2R_1} = -1 \quad (10)$$

The vectors in this expression are represented according to the root locus scheme in Figure 7(B). The two roots are thereby bounded as shown by the two dots and the cross. Their exact locations could be estimated or accurately determined by graphical methods.

The locus of roots now is simply intervals along the negative real axis between the open loop zeros and poles as shown in Figure 7 (B). The exact location of the roots along these intervals is determined in the usual way. Note that the constant in equation 10 is of the form $R'C_1$, in which R' is the effective value of R_2 and R_3 in parallel.

In more complicated networks, the advantages of the root locus concept over algebraic methods becomes greater; its particular advantage is in retaining at all times a clear picture of the relationships between the over-all network parameters and the parameters of individual circuit elements.

In the classical problem of finding roots, the differential equation is given in the form of a sum of terms of successively higher order. This can be converted to the form shown in equation 10

$$s^n+as^{n-1}+bs^{n-2}+\ldots+m = [(s+a)s+b]s+\ldots+m \quad (11)$$

This corresponds to a block diagram with another loop closed for each higher order term. Solve for the roots of the first loop which corresponds to the quantities in brackets above and proceed as before for the multiple loop system. If the roots close to the origin are of most interest, substitute $s=1/x$ first and solve for root values of x. Other combinations are, of course, possible because a single root locus basically determines the factors of the sum of two terms.

The root locus method is thus an analytical tool which can be applied to other problems than control system synthesis for which it was developed. But in attacking a new problem one would probably do well to try first to develop a method of analysis which is natural for that problem rather than seek to apply any existing methods.

Graphical Calculations

The root locus plot is first established in approximate form by inspection. Any significant point on the locus then can be checked by using the techniques indicated in this section. Note that only two calculations are involved, adding angles and multiplying lengths. Fortunately, all of these angles and lengths can be measured at the s point. Thus angles previously pictured at the zeros and poles also appear at the s point but between a horizontal line to the left and lines to the zeros and poles. A piece of transparent paper or plastic pivoted at the s point can be rotated successively through each of these angles to obtain their sum.

The reader can duplicate the "spirule" with two pieces of transparent paper, one for the disk and the other for the arm. Several procedures are possible, but the over-all purpose is to successively rotate the arm with respect to the disk through each of the angles of interest. Thus for adding phase angles, the disk is held fixed while the arm is rotated from a pole to the horizontal, whereas the two move together in getting aligned on the next pole. For multiplying lengths, the disk is held fixed while the arm is rotated from the position where a pole is on the straight line to the position where the pole is on the logarithmic curve. Rotations are made in the opposite directions for zeros than they are for poles.

Conclusions

The definite opinion of engineers using this method is that its prime advantage is the complete picture of a system which the root locus plot presents. Changing an open loop parameter merely shifts a point and modifies the locus. By means of the root locus method, all of the zeros and poles of the over-all function can be determined.

Any linear system is completely defined by this determination, and its response to any particular input function can be determined readily by standard mathematical or graphical methods.

References

1. PRINCIPLES OF SERVOMECHANISMS (book), G. S. Brown, D. P. Campbell. John Wiley and Sons, New York, N. Y., 1948.

2. TRANSIENTS IN LINEAR SYSTEMS (book), M. F. Gardner, J. L. Barnes. John Wiley and Sons, New York, N. Y., 1942.

3. GRAPHICAL ANALYSIS OF CONTROL SYSTEMS, W. R. Evans. *AIEE Transactions*, volume 67, 1948, pages 547–51.

No Discussion

The Structure of Dynamic Programming Processes

RICHARD BELLMAN

IF anything, dynamic programming, at its advent in the 1950s, had a greater impact on the engineering community than did the maximum principle (and a lesser impact on the mathematical community). While the maximum principle emerged from its womb in the calculus of variations as a fully formulated necessary condition of optimality for open-loop optimal control problems, dynamic programming emerged theoretically restricted, requiring strong smoothness assumptions not often satisfied. Nonetheless dynamic programming stimulated research enormously providing, uniquely, three things. Firstly, it gave the research community a powerful *conceptual framework* for formulating feedback control and decision problems, deterministic and stochastic; it remains today the only general methodology for handling optimal feedback control problems when stochastic disturbances are present [11]. Bellman's book [1] and the textbook [2] give some impression of the wide range of feedback problems addressed: multi-stage allocation and decision processes, inventory and stock control, production processes, control of queues, Markovian decision problems, combinatorial problems as well as the problems, more familiar to us, of controlling deterministic and stochastic systems described by difference or differential equations. Secondly, it provided a *constructive method* for solving these problems via the recurrence relation for the value function which, for deterministic, continuous-time problems, is the familiar Hamilton-Jacobi equation; complexity limits the use of this approach but it nonetheless provides a means for solving the linear quadratic and H_∞ optimal control problems where the value function can be finitely parameterized. The much used Viterbi algorithm for maximum likelihood decoding is an excellent example of dynamic programming. Finally, it provides, as does Hamilton-Jacobi theory (Caratheodory's method) in the deterministic case, *sufficient* conditions for optimality; it was used very early [9] for this purpose. Since dynamic programming, unlike the maximum principle, addressed feedback that is bread and butter to the control engineer, it was enthusiastically welcomed. The excitement that it generated was considerable; it provided the machinery to consider a whole host of problems previously considered inaccessible.

The conceptual tools provided by dynamic programming are well described in the review by Mitter [11] in the context of continuous-time stochastic systems. In discrete-time, the recurrence relation gives the value function at time k in terms of the value function at time $k+1$; if uncertainty is present, the value function at time k is simply the minimum (with repect to the control at time k) of the expected value of the sum of the path cost and the value function at time $k+1$, *conditional* on the data available. If the state is known, the modification to the deterministic recurrence relation is small. If the state is only partially observable, the problem is transformed into a problem with perfect information in which the hyper-state is the (unnormalized) conditional density of the original state. From these relations emerged solutions to the deterministic, stochastic, and partially observable feedback optimal control problems [5], [6], [8], [10], [13].

In the 1960s dynamic programming was widely judged to lack a rigorous foundation; it aimed to provide a general approach to the solution of optimal control problems yet, as originally formulated, depended on regularity properties of the value function which frequently it did not possess. The conceptual breakthroughs, required to elevate the status of dynamic programming from that of a useful heuristic tool to a cornerstone of modern optimal control, did not occur until the 1970s. They involved a complete re-appraisal of what was meant by a 'solution' to the Hamilton-Jacobi-Bellman (HJB) equation.

The first breakthrough was the introduction in [3] of the viscosity solution of the HJB equation; under certain hypotheses (including uniform continuity, but not differentiability, of the value function), the value function is the unique, uniformly continuous, generalized solution of the HJB equation (in the viscosity sense). This paper provides the first characterization of the value function as a unique solution to the HJB equation and gives continuous-time dynamic programming a rigourous foundation. The concepts have been very useful in stochastic control and have led to effective numerical techniques for solving HJB equations. They also provided a clean and more direct definition of the value of a differential game and existence proofs for solutions of the Hamilton-Jacobi-Isaacs equation [7] under very mild conditions.

However, in problems with terminal constraints, the cost functions are extended valued, discontinuous functions, and give rise to extended valued, discontinuous value functions. An important subsequent advance was to characterize discontinuous value functions as solutions to Hamilton-Jacobi-Bellman equations, appropriately defined. This was accomplished by refinement of the concept of viscosity solutions and by other methods based on techniques of non-smooth analysis. Overviews are provided in [4], [12].

Recent research in Caratheodory's method has been largely concerned with the inverse problem: the existence of verification functions that enable optimality of a candidate optimal control to be established. The main result is that, under a mild degeneracy hypothesis, there always exists a Lipschitz continuous verification function even when there are terminal constraints and when the value function (the obvious choice for a verification function) is not even continuous!

Dynamic programming, in addition to the powerful conceptual tools it has provided the control engineer, is now also, despite its slow start, an expanding and active area of theoretical research!

References

[1] R. Bellman, *Dynamic Programming*, Princeton University Press (Princeton), 1957.

[2] D.P. Bertsekas, *Dynamic Programming*, Prentice Hall (Englewood Cliffs, NJ), 1987.

[3] M.G. Crandall and P.L. Lions, "Viscosity solutions of Hamilton-Jacobi equations," *Trans. Amer. Math. Soc.*, **277**:1–42, 1983.

[4] W.H. Fleming and H.M. Soner, *Controlled Markov processes and viscosity solutions*, Springer-Verlag (New York), 1993.

[5] J.J. Florentin, Optimal control of continuus time, Markov, stochastic systems," *Journal of Electronics and Control*, **10**:473–488, 1961.

[6] J.J. Florentin, "Partial observability and optimal control," *Journal of Electronics and Control*, **13**:263–279, 1962.

[7] R. Isaacs, *Differential Games*, 2nd ed., Kruger (Huntington, N.Y.), 1975 (1st ed., Wiley, NY, 1965).

[8] P.D. Joseph and T.T. Tou, On linear control theory. *Transactions AIEE, Part III*, 80(18):193–196, 1961.

[9] R.E. Kalman, "Contributions to the theory of optimal control," *Boletin Sociedad Matematica Mexicana*, **5**:102–119, 1960.

[10] H.J. Kushner, "Optimal stochastic control," *IRE Trans. Automat. Contr.*, pp. 120–122, October 1962.

[11] S.K. Mitter, Filtering and stochastic control: A historical perspective. *IEEE Control Systems (Magazine)*, 16(3):67–76, 1996.

[12] R.B. Vinter, *Optimal Control*, Birkhäuser (Boston), 2000.

[13] W.M. Wonham, "Random differential equations in control theory," in *Probabilistic Methods in Applied Mathematics*, A. T. Bharucha-Reid, editor, **2**:131–212, Academic Press (New York), 1970.

D.Q.M.

CHAPTER III

The Structure of Dynamic Programming Processes

§ 1. Introduction

In this chapter we wish to examine and compare the essential features of the two processes we have considered in some detail in the first and second chapters. Disparate as these processes may seem at first glance, one being of deterministic type with a stochastic version and the other of a stochastic type with no deterministic version, we shall see that from an abstract point of view they are examples of the same general type of process. It is therefore no accident that they are governed by functional equations of a similar form.

After a discussion and analysis of these similarities, we shall consider the formulation of the more general decision processes and from these derive a number of functional equations possessing a common structure. We could, if we so desired, condense these into one all-embracing functional equation. However, since extreme generality is only gained at the expense of fine detail, it seems decidedly better, from both a conceptual and analytic point of view, to consider separately a number of important sub-categories of processes, each of which possesses certain distinctive mathematical and physical features.

We shall close the chapter with a further discussion of the concept of approximation in function space, which we have already encountered in the previous chapters, and a demonstration of its most important property, that of monotone convergence.

§ 2. Discussion of the two preceding processes

Let us begin by observing that the processes discussed in Chapters I and II have the following features in common:

a. In each case we have a physical system characterized at any stage by a small set of parameters, the *state variables*.
b. At each stage of either process we have a choice of a number of decisions.
c. The effect of a decision is a transformation of the state variables.
d. The past history of the system is of no importance in determining future actions.

Reprinted with permission from *Dynamic Programming* (Princeton, NJ: Princeton University Press, 1957), Richard Bellman, "The Structure of Dynamic Programming Processes (Chapter 3)" pp. 81–89.

e. The purpose of the process is to maximize some function of the state variables.

We have purposely left the description a little vague, since it is the spirit of the approach to these processes that is significant rather than the letter of some rigid formulation. It is extremely important to realize that one can neither axiomatize mathematical formulation nor legislate away ingenuity. In some problems, the state variables and the transformations are forced upon us; in others there is a choice in these matters and the analytic solution stands or falls upon this choice; in still others, the state variables and sometimes the transformations must be artificially constructed. Experience alone, combined with often laborious trial and error, will yield suitable formulations of involved processes.

Let us now identify the two processes discussed in the foregoing chapters with the description given above.

In the unbounded multi-stage allocation process, the state variables are x, the quantity of resources, and z the return obtained up to the current stage. The decision at any stage consists of an allocation of a quantity y to the first activity where $0 \leq y \leq x$. This decision has the effect of transforming x into $ay + b(x - y)$ and z into $z + g(y) + h(x - y)$. The purpose of the process is to maximize the final value of z.

In the stochastic gold-mining process, the state variables are x and y, the present levels of the two mines, and z the amount of gold mined to date. The decision at any stage consists of a choice of Anaconda or Bonanza. If Anaconda is chosen, (x, y) goes into $((1 - r_1)x, y)$ and z into $z + r_1 x$, and if Bonanza, (x, y) goes into $(x, (1 - r_2)y)$ and z into $z + r_2 y$. The purpose of the process is to maximize the expected value of z obtained before the machine is defunct.

In the finite versions of both processes, we have the additional parameter of time, manifesting itself in the form of the number of stages remaining in the process. It is, however, very useful to keep this state variable distinct from the others, since, as usual, time plays a unique role.

Let us now agree to the following terminology: A *policy* is any rule for making decisions which yields an allowable sequence of decisions; and an *optimal policy* is a policy which maximizes a preassigned function of the final state variables. A more precise definition of a policy is not as readily obtained as might be thought. Although not too difficult for deterministic processes, stochastic processes require more care. For any *particular* process, it is not difficult to render the concept exact. The key word is, of course, "allowable".

A convenient term for this preassigned function of the final state variables is *criterion function.* In many applications, the determination of a proper criterion function is a matter of some difficulty. From the analytic point of view, a solution may be quite easy to obtain for one criterion function, and quite difficult for a closely related one. It is well, consequently, to retain a certain degree of flexibility in the choice of such functions.

§ 3. The principle of optimality

In each process, the functional equation governing the process was obtained by an application of the following intuitive:

PRINCIPLE OF OPTIMALITY. *An optimal policy has the property that whatever the initial state and initial decision are, the remaining decisions must constitute an optimal policy with regard to the state resulting from the first decision.*

The mathematical transliteration of this simple principle will yield all the functional equations we shall encounter throughout the remainder of the book. A proof by contradiction is immediate.

§ 4. Mathematical formulation—I. A discrete deterministic process

Let us now consider a *deterministic* process, by which we mean that the outcome of a decision is uniquely determined by the decision, and assume that the state of the system, apart from the time dependence, is described at any stage by an M-dimensional vector $p = (p_1, p_2, \ldots, p_M)$, constrained to lie within some region D. Let $T = \{T_q\}$ where q runs over a set S which may be finite, enumerable, composed of continua, or a combination of sets of this type, be a set of transformations with the property that $p \varepsilon D$ implies that $T_q(p) \varepsilon D$ for all $q \varepsilon S$, which is to say that any transformation T_q carries D into itself.

The term "discrete" signifies here that we have a process consisting of a finite or denumerably infinite number of stages.

A policy, for the finite process which we shall consider first, consists of a selection of N transformations in order, $P = (T_1, T_2, \ldots, T_N)$,[1] yielding successively the sequence of states

$$
\begin{aligned}
&p_1 = T_1(p), \\
&p_2 = T_2(p_1), \\
&\quad\vdots \\
&p_N = T_N(p_{N-1}).
\end{aligned} \tag{1}
$$

These transformations are to be chosen to maximize a given function, R, of the final state p_N.

[1] where we write T_1 for T_{q_1}, T_2 for T_{q_2}, and so on.

There are a number of cases in which it is easy to see that a maximum will exist, in which case an optimal policy exists. The simplest is that where there are only a finite number of allowable choices for q at each stage. Perhaps next in order of simplicity is where we assume that D is a finite closed region, with $R(p)$ continuous in p for $p \varepsilon D$, $T_q(p)$ jointly continuous in p and q for all $p \varepsilon D$ and all q belonging to a finite closed region S.

These two cases cover the most important of the finite processes, while their limiting forms account for the unbounded processes.

Observe that the maximum value of $R(p_N)$, as determined by an optimal policy, will be a function only of the initial vector p and the number of stages N. Let us then define our basic auxiliary functions

$$f_N(p) = \operatorname*{Max}_P R(p_N) \tag{2}$$

= the N-stage return obtained starting from an initial state p and using an optimal policy.

This sequence is defined for $N = 1, 2, \ldots,$ and for $p \varepsilon D$.

Simple as this step is, it represents a fundamental principle in analysis, the *principle of continuity*. In order to solve our original problem involving one initial vector, p, and a multi-stage process of a definite number of stages, N, we consider the entire set of maximization problems arising from arbitrary values of p and from an arbitrary number of stages.

The original process has thus been imbedded within a family of similar processes. In place of attempting to determine the characteristics of an optimal policy for an isolated process, we shall attempt to deduce the common properties of the set of optimal policies possessed by the members of the family.

This procedure will enable us to resolve the original problem in a number of cases where direct methods fail.

To derive a recurrence relation connecting the members of the sequence $\{f_N(p)\}$, let us employ the principle of optimality stated above in 3. Assume that we choose some transformation T_q as a result of our first decision, obtaining in this way a new state vector, $T_q(p)$. The maximum "return"[2] from the following $(N-1)$ stages is, by definition, $f_{N-1}(T_q(p))$. It follows that if we wish to maximize the total N-stage return q must now be chosen so as to maximize this $N-1$ stage return. The result is the basic recurrence relation

$$f_N(p) = \operatorname*{Max}_{q \varepsilon S} f_{N-1}(T_q(p)), \tag{3}$$

for $N \geq 2$, with

[2] i.e. the value of the criterion function.

$$f_1(p) = \underset{q \epsilon S}{\text{Max}}\, R(T_q(p)) \tag{4}$$

Observe that $f_N(p)$ is unique, but that the q which maximizes is not necessarily so. Thus the maximum return is uniquely determined, but there may be many optimal policies which yield this return.

For the case of an unbounded process, the sequence $\{f_N(p)\}$ is replaced by a single function $f(p)$, the total return obtained using an optimal policy starting from state p, and the recurrence relation is replaced by the functional equation

$$f(p) = \underset{q}{\text{Max}}\, f(T_q(p)). \tag{5}$$

§ 5. Mathematical formulation—II. A discrete stochastic process

Let us once again consider a discrete process, but one in which the transformations which occur are *stochastic* rather than deterministic.

A decision now results in a distribution of transformations, rather than a single transformation. The initial vector p is transformed into a stochastic vector z with an associated distribution function $dG_q(p, z)$, dependent upon p and the choice q.

Two distinct types of processes arise, depending upon whether we assume that z is known after the decision has been made and before the next decision has to be made, or whether we assume that only the distribution function is known. We shall only consider processes of the first type in this volume, since processes of the second type require in general the concept of functions of functions, which is to say *functionals.*

It is clear, as we have stated several times before, that it is now on the whole meaningless to speak of maximizing *the* return. Rather we must agree to measure the value of a policy in terms of some average value of the function of the final state. Let us call this expected value the *return.*

Beginning with the case of a finite process, we define $f_N(p)$ as in (4.2). If z is a state resulting from any initial transformation T_q, the return from the last $N-1$ stages will be $f_{N-1}(z)$, upon the employment of an optimal policy. The expected return as a result of the initial choice of T_q is therefore

$$\int_{z \epsilon D} f_{N-1}(z)\, dG_q(p, z) \tag{1}$$

Consequently, the recurrence relation for the sequence $\{f_N(p)\}$ is

$$f_N(p) = \underset{q \epsilon S}{\text{Max}} \int_{z \epsilon D} f_{N-1}(z)\, dG_q(p, z),\ N \geq 2, \tag{2}$$

with

(3) $$f_1(p) = \operatorname*{Max}_{q \in S} \int_{z \in D} R(z)\, dG_q(p, z).$$

Considering the unbounded process, we obtain the functional relation

(4) $$f(p) = \operatorname*{Max}_{q \in S} \int_{z \in D} f(z)\, dG_q(p, z)$$

§ 6. Mathematical formulation—III. A continuous deterministic process

There are a number of interesting processes that require that decisions be made at each point of a continuum, such as a time interval. The simplest examples of processes of this character are furnished by the calculus of variations. As we shall see in Chapter IX below, this conception of the calculus of variations leads to a new view of various parts of this classical theory.

Let us define

(1) $f(p; T)$ = the return obtained over a time interval $[0, T]$ starting from the initial state p and employing an optimal policy.

Although we consider the process as one consisting of choices made at each point t on $[0, T]$, it is better to begin with the concept of choosing policies, which is to say functions, over intervals, and then pass to the limit as these intervals shrink to points. The analogue of (4.3) is

(2) $$f(p; S + T) = \operatorname*{Max}_{D[0, S]} f(p_D; T)$$

where the maximum is taken over all allowable decisions made over the interval $[0, S]$.

As soon as we consider infinite processes, occurring as the result of either unbounded sequences of operations, or because of choices made over continua, we are confronted with the difficulty of establishing the existence of an actual maximum rather than a supremum. In general, therefore, in the discussion of processes of continuous type, it is better to use initially the equation

(3) $$f(p; S + T) = \operatorname*{Sup}_{D} f(p_D; T),$$

which is usually easy to establish, and then show, under suitable assumptions that the maximum is actually attained.

As we shall see in Chapter IX, the limiting form of (2) as $S \to 0$ is a nonlinear partial differential equation. This is the important form for actual analytic utilization. For numerical purposes, S is kept non-zero but small.[3]

[3] We shall show, in Chapter IX, that it is possible to avoid many of the quite difficult rigorous details involved in this limiting procedure if we are interested only in the computational solution of variational processes.

§ 7. Continuous stochastic processes

An interesting and challenging question which awaits further exploration is the formulation and solution of general classes of continuous stochastic decision processes of both one-person and two-person variety. Although we shall discuss a particular process in Chapter VIII, we shall not discuss the general formulation of continuous stochastic decision processes here, since a rigorous treatment requires delicate and involved argumentation based upon sophisticated concepts.

§ 8. Generalizations

It will be apparent to the reader that the functional equations we have derived above for the case where the state variables and the decision variables were constrained to finite dimensional Euclidean spaces can be extended to cover the case where the state variables and decision variables are elements of more general mathematical spaces, such as Banach spaces.

Rather than present this extension abstractly we prefer to wait until a second volume where we will discuss examples of these more general processes. The theory of integral equations and variational problems involving functions of several variables, as well as more general stochastic processes, all afford examples of processes which escape the finite dimensional formulation to which we have restricted ourselves in this volume, and require for their formulation in the foregoing terms the theory of functionals and operations.

§ 9. Causality and optimality

Consider a multi-stage process involving no decisions, say one generated by the system of differential equations,

$$dx_i/dt = g_i(x_1, x_2, \ldots, x_N),\ x_i(0) = c_i;\ i = 1, 2, \ldots, N, \tag{1}$$

which may, more compactly, be written in vector form

$$dx/dt = g(x),\ x(0) = c. \tag{2}$$

The state of the system at time t, taking for granted existence and uniqueness of the solution, is a function only of c and t, thus we may write

$$x(t) = f(c, t). \tag{3}$$

The uniqueness of the solution leads to the functional equation

$$f(c, s + t) = f(f(c, s), t), \tag{4}$$

for $s, t \geq 0$, an analytical transliteration of the law of causality. This equation expresses the fundamental semi-group property of processes of this type.

Comparing (4) above with (6.2), we see that we may regard multi-stage decision processes as furnishing a natural extension of the theory of semi-groups. Any further discussion here along these lines would carry us beyond our self-imposed limits, and we shall consequently content ourselves with the above observation.

§ 10. Approximation in policy space

In solving a functional equation such as (4.4) or (5.3), we shall in Chapter IV make use of that general factotum of analysis, the method of successive approximations. The method very briefly, consists of choosing an initial function $f_0(p)$, and then determining a sequence of functions, $\{f_N(p)\}$, by means of the algorithm

$$(1) \qquad f_N(p) = \operatorname*{Max}_q f_{N-1}(T_q(p)),\ N = 1, 2, \ldots$$

as, for instance, in (4.4) We have already employed this method in dealing with the equations of Chapters I and II.

In many important cases, this method after a suitable preliminary preparation of the equation actually leads to a convergent sequence whose limit yields the solution of the functional equation.[4] We shall make extensive use of it in the following chapter.

In the theory of dynamic programming, however, we have an alternate method of approximation which is equally important in its own right, a method which we call "approximation in policy space".

Before discussing this method of approximation, let us observe that there is a natural duality existing in dynamic programming processes between the function $f(p)$ measuring the overall return, and the optimal policy (or policies) which yields this return. Each can be used to determine the other, with the additional feature that a knowledge of $f(p)$ yields *all* optimal policies, since it determines all maximizing indices q in an equation such as (4.4), while a knowledge of *any* particular optimal policy yields $f(p)$.

The maximizing index q can be considered to be a function of p. If the index is not unique, we have a multi-valued function. Whereas we call $f(p)$ an element in function space, let us call $q = q(p)$ an element of *policy space*. Both spaces are, of course, function spaces, but it is worth distinguishing between them, since their elements are quite different in meaning.

It follows now that we have two ways of making an initial approxima-

[4] It is interesting to observe that in many theories, as, for example, partial differential equations, the preliminary transformation of the equation is of such a nature that the principal difficulty of the existence proof resides in the demonstration that the limit function actually satisfies the *original* equation.

tion. We may approximate to $f(p)$, as we do ordinarily in the method of successive approximation, or we may, and this is a feature of the functional equations belonging to dynamic programming processes, approximate initially in policy space.[5]

Choosing an initial approximation $q_0 = q_0(p)$, we compute the return from this policy by means of the functional equation

$$f_0(p) = f_0(T_{q_0}(p)). \tag{2}$$

We have already given an example of this in § 11 of Chapter I.

There are now two ways we can proceed. Taking the function of q, $f_0(T_q(p))$, we can determine a function $q(p)$ which maximizes. Call this function $q_1(p)$. Using this new policy, we determine $f_1(p)$, the new return, by means of the functional equation

$$f_1(p) = f_1(T_{q_1}(p)). \tag{3}$$

This equation is solved iteratively, as in (11.3) and (11.4) of Chapter I.

Continuing in this way, we obtain two sequences $\{f_N(p)\}$ and $\{q_N(p)\}$.

In place of this procedure, we can define

$$f_1(p) = \operatorname*{Max}_q f_0(T_q(p)), \tag{4}$$

and then continue inductively, employing the usual method of successive approximations,

$$f_{N+1}(p) = \operatorname*{Max}_q f_N(T_q(p)). \tag{5}$$

It is immediate that $f_1 \geq f_0$ and thus that the sequence $\{f_N\}$ is monotone increasing. We shall discuss the convergence of this process in the next chapter.

The first procedure, although a more natural one, seems more difficult to treat rigorously and we shall not consider it here. In dealing with various types of continuous processes, such as those furnished by the calculus of variations, it would seem, however, that this technique is required for successive approximations. We shall discuss this topic again in Chapter IX.

[5] Actually this type of approximation is tacitly encountered in other branches of analysis as, for instance, in the theory of differential equations, where a differential equation is frequently replaced by a difference equation for approximation purposes. This replaces the space of general functions by the subspace of step-functions.

Optimal Regulation Processes

L. S. PONTRYAGIN

THE maximum principle that had such a dramatic effect on the development of the theory of control was introduced to the mathematical and engineering communities through this paper, and a series of other papers [3], [8], [2] and the book [15]. The paper selected for this volume was the first to appear (in 1961) in an English translation. The maximum principle, sometimes referred to as the Pontryagin maximum principle because of Pontryagin's role as leader of the research group at the Steklov Institute, rose to prominence largely through the book. The first proof of the maximum principle is attributed by some to Boltyanskii [2]. This paper by Pontryagin commences with these words: "In this paper will be found an account of results obtained by my students V. G. Boltyanskii, R. V. Gamkrelidze and myself."

Surprisingly, the initial impact on mathematicians of the new theory was small. The unenthusiastic reception at the 1958 Congress of Mathematicians to the announcement of the maximum principle by the Soviet group is described by Markus in [12]. Many believed that the new theory was, through its introduction of inequality constraints, a minor addition to the calculus of variations. Sussmann and Willems [17], on the other hand, emphasize the conceptual advance made with the discovery of the maximum principle; this advance was impeded in the calculus of variations literature where the differential equation for the 'dynamic system' has the very special form $\dot{x} = u$ which eliminates the adjoint variable in the Euler-Lagrange equation and obscures the fact that the Hamiltonian is maximized.

In marked contrast, the impact on the control engineering community was dramatic. The emergence, almost simultaneously, of the maximum principle, dynamic programming, and Kalman filtering created no less than a revolutionary change in the way control problems were formulated and solved. A graduate student at the time was acquainted with the classical frequency response approach to control, with selection of controller parameters to minimize a quadratic performance index subject, possibly, to a quadratic constraint on control energy [14] and with Wiener filtering theory (sometimes used, via reformulation, to solve linear, quadratic optimal control problems). Time-optimal control problems had already been solved by Bushaw [5] who obtained, *inter alia*, the switching curves for second-order oscillatory systems. But generally the field was static, constrained perhaps by inadequate tools, an illustrative example being the introduction of time-varying transforms to analyze time-varying linear systems. In this atmosphere the impact of the maximum principle, dynamic programming, and Kalman filtering enabling a powerful time-domain perspective of nonlinear and time-varying systems, was overwhelming. A whole set of new tools, and new problems, was suddenly available, injecting new life into graduate schools. Conferences and workshops multiplied to comprehend, use, and extend these new results. Every researcher who lived in this period was invigorated, and every research student had a stimulating and open field in which to work.

The maximum principle excited considerable attention. This, after all, was the aerospace era in which open-loop problems suddenly became meaningful; Goddard's problem of maximizing altitude given a fixed amount of fuel, for example, posed in 1919, was solved in 1951 by Tsien and Evans [18] using the calculus of variations. Like all revolutions, there were excesses. Linear, quadratic optimal control problems were solved in textbooks and papers via the maximum principle instead of using the sufficiency conditions provided by Hamilton-Jacobi (dynamic programming) theory. The importance of robustness, and other lessons from the past, were forgotten. But optimal control prospered. In aerospace, algorithms for determining optimal flight paths were developed, linear quadratic optimal control became a powerful design tool for a wide range of problems, and H_∞ optimal control was developed after a concern for robustness re-emerged. Model predictive control, a widely applied form of control in the process industries where constraints are significant, makes direct use of open-loop optimal control. Many books were written for the engineering community; that by Bryson and Ho [4] captured well the excitement and breadth of the new theories. The initial slow impact on the mathematical community disappeared with the entry into the field of Western mathematicians who wrote many influential papers and authoritative texts such as [1], [7], [10], [11], [20] that helped cement the field. The paper by Halkin [9] was particularly widely read. There was also a voluminous Russian literature the flavor of which is conveyed in the recent book by Milyutin and Osmolovskii [13].

The approach adopted by Pontryagin and his collaborators, and developed in the literature quoted above, may be roughly summarized as follows. Consider a fixed-time optimal control problem where the system is $\dot{x} = f(x, u), x := (y, z)$ with given initial state $x(0) = x_0 = (0, z_0)$. The control objective is to minimize $y(1)$ subject to $z(1) = 0$ and $u(t) \in \mathbb{U}$ for all $t \in [0, 1]$. It is the terminal equality constraint that makes determination of necessary conditions of optimality difficult. The set of reachable states at $t = 1$ is W. Suppose $(u^0(\cdot), x^0(\cdot))$ is an optimal control-state pair. The set of 'better' terminal states is $S = \{x \mid y < y^0(1), z = 0\}$. Optimality implies W and S are disjoint. A local approximation $\hat{W}$ to W at $x^0(1)$ is obtained using an approximation to the differential equation (that is, linear in x and nonlinear in u) and a set of variations of the optimal control; the set $\hat{W}$ is convex. It is then shown (and this is the hard part) that disjointness of W and S implies linear separation of $\hat{W}$ and S. The latter is easily shown to imply the maximum principle. This powerful, and intuitively appealing, approach is still being extended [16].

It is a sign of the continuing vitality of the field that, in addition, a very different approach has been recently developed. The starting point is that the maximum principle is easily derived for an optimal control with no terminal equality constraints. Clarke [6] uses an exact penalty function (which is non-smooth) to remove the troublesome constraints, formulates a perturbed problem (which is smooth), and obtains necessary conditions for this problem. Necessary conditions are then obtained for the original problem by passage to the limit. An exposition of the many ramifications of this approach appears in the forthcoming book [19].

With interest in optimal control, from both theoretical and applied researchers, as high as ever, the substantial impact that the introduction of the maximum principle has had and continues to have can be in no doubt.

References

[1] L.D. Berkovitz, *Optimal Control Theory*, Springer-Verlag (New York), 1974.

[2] V.G. Boltyanskii, "The maximum principle in the theory of optimal processes," *Dokl. Akad. Nauk SSSR*, **119**:1070–1073, 1958.

[3] V.G. Boltyanskii, R.V. Gramkrelidze and L.S. Pontryagin, "On the theory of optimal processes," *Dokl. Akad. Nauk SSSR*, **110**:7–10, 1956.

[4] A.E. Bryson and Y.-C. Ho, *Applied Optimal Control—Optimization, Estimation, and Control*, Hemisphere Publishing Corp. (Washington), 1969.

[5] D.W. Bushaw, "Optimal discontinuous forcing terms," in *Contributions to the theory of nonliner oscillations*, **4**:29–52. Princeton Univ. Press, 1958.

[6] F.H. Clarke, "The maximal principle under minimal hypotheses," *SIAM J. Contr. Optimiz.*, **14**:1078–1091, 1976.

[7] W.H. Fleming and R.W. Rishel, *Deterministic and Stochastic Optimal Control*, Springer-Verlag (New York, Heidelberg, Berlin), 1975.

[8] R.V. Gamkrelidze, "On the theory of optimal processes in linear systems," *Dokl. Akad. Nauk SSSR*, **116**:9–11, 1957.

[9] H. Halkin, "On the necessary condition for the optimal control of nonlinear systems," *J. Analyse Math.*, **12**:1–82, 1964.

[10] H. Hermes and J.P. LaSalle, *Functional Analysis and Time Optimal Control*, Academic Press (New York), 1969.

[11] E.B. Lee and L. Markus, *Foundations of Optimal Control Theory*, Wiley (New York), 1967.

[12] L. Markus, In *Differential Equations, Dynamical Systems, and Control Science: A Festschrift in honor of Lawrence Markus*, K. D. Elworthy, W. N. Everitt, and E. B. Lee, editors, Marcel Dekker, Inc. (New York), 1994.

[13] A.A. Milyutin and N.P. Osmolovskii, *Calculus of Variations and Optimal Control*, American Mathematical Society (Providence, RI), 1998. (Translations of mathematical monographs.)

[14] G. Newton, L.A. Gould and J.F. Kaiser, *Analytical Design of Linear Feedback Controls*, Wiley (New York), 1957.

[15] L.S. Pontryagin, V.G. Boltyanskii, R.V. Gamkrelidze, and E.F. Mishchenko, *The Mathematical Theory of Optimal Processes*, Interscience, Wiley (New York), 1962.

[16] H.J. Sussmann, "Geometry and optimal control," In *Mathematical Control Theory*, J. Baillieul and J.C. Willems, editors, pp. 140–194. Springer-Verlag (New York), 1999.

[17] H.J. Sussmann and J.C. Willems, "300 years of optimal control: from the brachystochrone to the maximum principle," *IEEE Control Systems (Magazine)*, **17**(3):32–44, 1997.

[18] H.S. Tsien and R.C. Evans, "Optimum thrust programming for a sounding rocket," *J. American Rocket Soc.*, **21**(5):99–107, 1951.

[19] R.B. Vinter, *Optimal Control*, Birkhaüser (Boston), 2000.

[20] J. Warga, *Optimal Control of Differential Equations and Functional Equations*, Academic Press (New York), 1972.

D.Q.M.

OPTIMAL REGULATION PROCESSES

L. Ś. PONTRYAGIN

In this paper will be found an account of results obtained by my students V. G. Boltyanskiĭ and R.V. Gamkrelidze and myself (see [1], [2], [3]).

§ 1. Statement of the problem. Let Ω be some topological space.* We shall say that a *control process* is given, if one has a system of ordinary differential equations

$$\frac{dx^i}{dt} = f^i(x^1, \ldots, x^n; u) = f^i(\mathrm{x}; u) \quad (i = 1, \ldots, n), \tag{1}$$

or, in vector form,

$$\frac{d\mathrm{x}}{dt} = \mathrm{f}(\mathrm{x}; u), \tag{2}$$

where $x^1, \cdots, x^n$ are real functions of the time t, $\mathrm{x} = (x^1, \cdots, x^n)$ is a vector of the n-dimensional vector space R, $u \in \Omega$, and

$$f^i(\mathrm{x}; u) \qquad (i = 1, \ldots, n)$$

are functions given and continuous for all values of the pairs $(\mathrm{x}, u) \in R \times \Omega$. We assume further that the partial derivatives

$$\frac{\partial f^i}{\partial x^j} \qquad (i, j = 1, \ldots, n)$$

are also defined and continuous in the entire space $R \times \Omega$.

In order to find a solution of equation (2), defined on the interval $t_0 \le t \le t_1$, it suffices to exhibit a *control function* $u(t)$ on the segment $t_0 \le t \le t_1$, and the initial value x_0 of the solution for $t = t_0$. In accordance with this we shall say that we are given a *control*

$$U = (u(t), t_0, t_1, \mathrm{x}_0) \tag{3}$$

of equation (2), if we are given a function $u(t)$, the segment of its definition $t_0 \le t \le t_1$, and the initial value x_0 of the solution $\mathrm{x}(t)$. In what follows we shall consider a piecewise-continuous control function $u(t)$, admitting discontinuities of the first order, and *continuous* solutions of equation (2). Here we shall suppose that the controls $u(t)$ are continuous in the initial point t_0 and semicontinuous from the left, i.e., the condition $u(t-0) = u(t)$, $t > t_0$ is satisfied. We shall say that the control (3) carries the point x_0 into the point x_1, if the corresponding solution $\mathrm{x}(t)$ of equation (2), satisfying the initial condition $\mathrm{x}(t_0) = \mathrm{x}_0$, satisfies as well the end condition: $\mathrm{x}(t_1) = \mathrm{x}_1$.

* In cases interesting in applications Ω is a closed region of a finite-dimensional linear space.

Reprinted with permission from *American Math Society Translations*, L. S. Pontryagin, "Optimal Regulations Processes," Series 2, Vol. 18, 1961, pp. 321–339.

Now suppose that $f^0(x^1, \cdots, x^n; u) = f^0(\mathbf{x}, u)$ is a function defined and continuous along with its partial derivatives $\frac{\partial f^0}{\partial x^j}$ $(j = 1, \ldots, n)$ on the whole space $R \times \Omega$. To each control (3) corresponds then the number

$$L(U) = \int_{t_0}^{t_1} f^0(\mathbf{x}(t), u(t))\, dt.$$

Thus, L is a functional of the control (3). The control U will be said to be *optimal*, if, for any control

$$U^* = (u^*(t), t_0^*, t_1^*, \mathbf{x}_0),$$

which carries the point $\mathbf{x}_0$ into the point $\mathbf{x}_1$, the inequality $L(U) \leq L(U^*)$ holds.

Remark 1. If (3) is an optimal control of the equation (2), $\mathbf{x}(t)$ the solution of equation (2) corresponding to it, and $t_2 < t_3$ are two points of the interval $t_0 \leq t \leq t_1$, then $U' = (u(t), t_2, t_3, \mathbf{x}(t_2))$ is also an optimal control.

Remark 2. If (3) is an optimal control of the equation (2), carrying the point $\mathbf{x}_0$ into the point $\mathbf{x}_1$, and τ is any number, then

$$U'' = (u(t - \tau);\ t_0 + \tau,\ t_1 + \tau,\ \mathbf{x}_0)$$

is also an optimal control carrying the point $\mathbf{x}_0$ into $\mathbf{x}_1$.

A particularly important case is that of a function $f^0(\mathbf{x}; u)$ which satisfies the equation

$$f^0(\mathbf{x}, u) \equiv 1. \qquad (4)$$

In this case we have: $L(U) = t_1 - t_0$, and the optimality of the control U means that *the time of transition from the position* $\mathbf{x}_0$ *to the position* $\mathbf{x}_1$ *is minimal.*

A case which is important in the applications is that in which Ω is a closed region of some r-dimensional Euclidean space E; then $u = (u^1, \cdots, u^r)$, and the one controlling parameter u goes over into a system of numerical parameters $u^1, \cdots, u^r$. In the case that Ω is an open set of the space E, the variational problem formulated here turns out to be a particular case of the problem of Lagrange ([4], p. 225) and the fundamental result presented below (the maximum principle) coincides with the known Weierstrass criterion. However it is important in the applications to consider the case in which the controlling parameters satisfy inequalities, including the possibility of equality, for example: $|u^i| \leq 1$ $(i = 1, \cdots, r)$. In this case the Weierstrass criterion obviously does not hold, amd the result presented below is new.

§2. Necessary conditions for optimality (maximum principle).

In order to formulate a necessary condition for optimality we introduce the vector

$$\tilde{\mathbf{x}} = (x^0, x^1, \cdots, x^n)$$

of $(n+1)$-dimensional Euclidean space S into the discussion, and we consider the control process

$$\frac{dx^i}{dt} = f^i(x^1, \ldots, x^n, u) = f^i(\mathbf{x}, u) = f^{i'}(\mathbf{x}, \tilde{u}) \quad (i = 0, 1, \ldots, n), \tag{5}$$

or, in vector form,

$$\frac{d\tilde{\mathbf{x}}}{dt} = \tilde{\mathbf{f}}(\tilde{\mathbf{x}}, u), \tag{6}$$

where $f^0(\mathbf{x},u)$ is the function which defines the functional L. In order, knowing the control (3) of equation (2), to obtain the control of equation (6), it suffices, beginning with the initial value

$$\mathbf{x}_0 = (x_0^1, \cdots, x_0^n),$$

to set down the initial value $\tilde{\mathbf{x}}_0$ of equation (6). We define the vector $\tilde{\mathbf{x}}_0$, writing

$$\tilde{\mathbf{x}}_0 = (0, x_0^1, \cdots, x_0^n).$$

In this way the control (3) of equation (2) uniquely defines a control of equation (6), and we will say for simplicity that (3) is a control of equation (6). If now the control (3) carries the initial value $\tilde{\mathbf{x}}_0$ into the terminal value

$$\tilde{\mathbf{x}}_1 = (x_1^0, x_1^1, \cdots, x_1^n),$$

then we have

$$L(U) = x_1^0,$$

and thus is determined a connection between equation (6) and the variational problem formulated above.

Along with the contravariant vector $\tilde{\mathbf{x}}$ of the space S we consider an auxiliary covariant vector

$$\tilde{\psi} = (\psi_0, \cdots, \psi_n)$$

of that space, and we set up the function

$$K(\tilde{\psi}, \tilde{\mathbf{x}}, u) = (\tilde{\psi}, \tilde{\mathbf{f}}(\tilde{\mathbf{x}}, u))$$

(the right side is the scalar product of the vectors $\tilde{\psi}$ and $\tilde{\mathbf{f}}$).

For fixed values of the quantitites $\tilde{\psi}$ and $\tilde{\mathbf{x}}$, the function K is a function of the parameter u; the upper bound of the values of this function will be denoted by $N(\tilde{\psi}, \tilde{\mathbf{x}})$. We set up, further, the Hamiltonian system of equations

$$\frac{dx^i}{dt} = \frac{\partial K}{\partial \psi_i} \quad (i = 0, \ldots, n); \tag{7}$$

$$\frac{d\psi_i}{dt} = -\frac{\partial K}{\partial x^i} \quad (i = 0, \ldots, n). \tag{8}$$

It is immediately evident that the system (7) coincides with (5), while the system (8) is:

$$\frac{d\psi_0}{dt} = 0, \quad \frac{d\psi_j}{dt} = -\sum_{i=0}^{n} \psi_i \frac{\partial f^i(\tilde{\mathbf{x}}, u)}{\partial x^j} \quad (j = 1, \ldots, n). \tag{9}$$

Theorem 1. *Suppose that* (3) *is an optimal control of equation* (2) *and* $\mathbf{x}(t)$ *is the solution of equation* (2) *corresponding to it. We complete the vector* $\mathbf{x}(t)$ *to a vector* $\tilde{\mathbf{x}}(t)$, *writing:*

$$x^0(t) = \int_{t_0}^{t} f^0(\tilde{x}(t), u(t))\, dt.$$

There exists then a non-zero continuous vector-function $\tilde{\psi}(t)$, *such that*

$$K(\tilde{\psi}(t_0), \tilde{x}(t_0), u(t_0)) = 0, \qquad \psi_0(t_0) \leqslant 0, \tag{10}$$

and the functions

$$\tilde{\psi}(t), \tilde{\mathbf{x}}(t), u(t)$$

constitute a solution of the Hamiltonian system (7), (8), *while*

$$K(\tilde{\psi}(t), \tilde{\mathbf{x}}(t), u(t)) = N(\tilde{\psi}(t), \tilde{\mathbf{x}}(t)); \tag{11}$$

furthermore it turns out that the function $K(\tilde{\psi}(t), \tilde{\mathbf{x}}(t), u(t))$ *is constant, so that*

$$K(\tilde{\psi}(t), \tilde{\mathbf{x}}(t), u(t)) \equiv 0. \tag{12}$$

In order to formulate the necessary condition in the case when one is dealing with the problem of minimum time, we set up the Hamiltonian function

$$H(\psi, \mathbf{x}, u) = (\psi, \mathbf{f}(\mathbf{x}, u)).$$

For fixed values of ψ and $\mathbf{x}$ the function $H(\psi, \mathbf{x}, u)$ is a function of the parameter u. The upper bound of the values of this function will be denoted by $M(\psi, \mathbf{x})$. We set up, further, the Hamiltonian system

$$\frac{dx^j}{dt} = \frac{\partial H}{\partial \psi_j} \qquad (j = 1, \ldots, n); \tag{13}$$

$$\frac{d\psi_j}{dt} = -\frac{\partial H}{\partial x^j} \qquad (j = 1, \ldots, n). \tag{14}$$

Obviously the system (13) coincides with the system (1), and the system (14) is:

$$\frac{d\psi_j}{dt} = -\sum_{k=1}^{n} \psi_k \frac{\partial f^k(x, u)}{\partial x^j} \qquad (j = 1, \ldots, n). \tag{15}$$

Theorem 2. *Suppose that* (3) *is a control of the equation* (2) *which is optimal for the functional* (4), *and that* $\mathbf{x}(t)$ *is the solution of equation* (2) *corresponding to this control. Then there exists a non-zero continuous vector function* $\psi(t) = (\psi_1(t), \ldots, \psi_n(t))$ *such that*

$$H(\psi(t_0), \mathbf{x}(t_0), u(t_0)) \geq 0,$$

and the functions

$$\psi(t), \mathbf{x}(t), u(t)$$

satisfy the Hamiltonian system of equations (13), (14), *while*

$$H(\psi(t), \mathbf{x}(t), u(t)) \doteq M(\psi(t), \mathbf{x}(t)). \tag{16}$$

It turns out, moreover, that the function $H(\psi(t), \mathbf{x}(t), u(t))$ *is constant, so that*

$$H(\psi(t), \mathbf{x}(t), u(t)) \geq 0. \tag{17}$$

Theorem 2 follows immediately from Theorem 1.

The substance of Theorems 1 and 2 is the equalities (11) and (16). Therefore theorem 2, which was first published as a hypothesis in the note [1], is called a *maximum principle*. In the same sense, it is natural to confer upon Theorem 1 the designation *of maximum principle.*

§3. Proof of the maximum principle (Theorems 1 and 2).

We shall prove Theorem 1. In the proof we shall use certain constructions of McShane [5]. Suppose that (3) is some control of equation (6) and that $\tilde{x}(t)$ is the solution of equation (6) corresponding to it. The system of equations in the variations for system (5) near the solution $\tilde{x}(t)$ is written, as is well known, in the form

$$\frac{dy^i}{dt} = \sum_{j=0}^{n} \frac{\partial f^i(\tilde{x}(t),\ u(t))}{\partial x^j} y^j \qquad (i = 0,\ 1,\ \ldots,\ n). \tag{18}$$

Writing the solution of the system (18) in vector form, we obtain the vector

$$\tilde{y}(t) = (y^0(t),\ \cdots,\ y^n(t)).$$

In what follows we shall consider only continuous solutions $\tilde{y}(t)$. The system of equations in the variations, as is well known, may be interpreted in the following way. Let $\tilde{y}_0$ be an arbitrary vector of the space S. We set down the initial condition*

$$\tilde{x}_0 + \varepsilon \tilde{y}_0 + \varepsilon o(\varepsilon)$$

for the solution of the equation (6). Then the same solution of equation (6) with this initial value may be written in the form

$$\tilde{x}(t) + \varepsilon \tilde{y}(t) + \varepsilon o(\varepsilon),$$

where $\tilde{y}(t)$ is a solution of the system (18), taken with the initial value $\tilde{y}_0$. We shall say that the solution $\tilde{y}(t)$ of system (18) is the *transport* of the vector $\tilde{y}_0$, given at the initial point $\tilde{x}_0$ of the trajectory $\tilde{x}(t)$, along the whole trajectory. In the same sense we may say that the solution $\tilde{y}(t)$ is the transport of the vector $\tilde{y}(\tau)$, given at the point $\tilde{x}(\tau)$ of the trajectory $\tilde{x}(t)$, along the whole trajectory.

Along with the contravariant vector $\tilde{y}(t)$ which is the solution of the system (18), we consider the covariant vector $\tilde{\psi}(t)$, which is the solution of the system (8). One verifies immediately that

$$\frac{d}{dt}(\tilde{\psi}(t),\ \tilde{y}(t)) = 0,$$

so that

$$(\tilde{\psi}(t),\ \tilde{y}(t)) = \text{const.} \tag{19}$$

If one interprets the covariant vector $\tilde{\psi}(t)$ as a plane passing through the point $\tilde{x}(t)$, then one may say that the plane $\tilde{\psi}(t)$ is the transport of the plane $\tilde{\psi}(\tau)$, given at the point $\tilde{x}(\tau)$ of the trajectory $\tilde{x}(t)$, along the whole trajectory.

A *variation* of the control (3) will be a control

$$U^* = U^*(\varepsilon,\ \alpha) = (u^*(t),\ t_0,\ t_1 + \alpha\varepsilon,\ \tilde{x}_0),$$

depending on the parameter ϵ and the real number α, defined for all sufficiently small positive values of the parameter ϵ and satisfying the following condition:

*In what follows the symbol $o(\epsilon)$ is used as a typical notation for quantities tending to zero along with ϵ.

The solution $\tilde{x}^*(t)$ of equation (6), corresponding to the control U^*, may be written, at the point $t = t_1 + \epsilon\alpha$, in the form:

$$\tilde{x}(t_1) + \varepsilon\tilde{\delta}(U^*) + \varepsilon o(\varepsilon),$$

where $\tilde{\delta}(U^*)$ does not depend on ϵ.

The family Δ of variations of one and the same control (3) will be said to be *admissible*, if along with each two variations $U_1^*(\epsilon, \alpha_1)$ and $U_2^*(\epsilon, \alpha_2)$ in it, there exists for any nonnegative γ_1, γ_2 a third variation $U^*(\epsilon, \gamma_1\alpha_1 + \gamma_2\alpha_2)$, satisfying the condition:

$$\tilde{\delta}(U^*) = \gamma_1\tilde{\delta}(U_1^*) + \gamma_2\tilde{\delta}(U_2^*). \tag{20}$$

We now construct a special variation

$$U^*(\varepsilon, \alpha) = V(\varepsilon, \alpha, \tau, \sigma, u^*),$$

depending on the point τ of the half-interval $t_0 < t \leq t_1$ (while in the case $\alpha < 0$ we have to take $\tau < t_1$), on the nonnegative number σ, and the point u^* of the space Ω. We define the variation $V(\epsilon, \alpha, \tau, \sigma, u^*)$ by giving the function $u^*(t)$ by the relations

$$u^*(t) = \begin{cases} u(t) & \text{for } t_0 \leqslant t \leqslant \tau - \varepsilon\sigma, \\ u^* & \text{for } \tau - \varepsilon\sigma < t \leqslant \tau, \\ u(t) & \text{for } \tau < t \leqslant t_1, \\ u(t_1) & \text{for } t_1 < t \leqslant t_1 + \varepsilon\alpha \ (\text{if } \alpha > 0). \end{cases} \tag{21}$$

It is easy to construct an admissible family Δ containing all the variations of type (21). This family will be taken as fundamental in the further constructions.

To each variation U^* of the admissible family Δ there corresponds a vector $\tilde{\delta}(U^*)$ issuing from the point $\tilde{x}_1$. The set of all these vectors fills out a convex cone Π with vertex at the point $\tilde{x}_1$ (see (20)). Let

$$\tilde{v} = (-1, 0, \cdots, 0)$$

be a vector issuing from the point $\tilde{x}_1$ and going in the direction of the negative x^0-axis in the space S. If the cone Π contains the end of the vector $\tilde{v}$ as an interior point, then the control U is not optimal. Suppose, indeed, that $U^* \in \Delta$ is that variation of the control U for which

$$\tilde{\delta}(U^*) = \tilde{v}.$$

We denote by $\tilde{x}_1^*$ the point into which the point moves under the control U^*. We obtain:

$$\tilde{x}_1^* = \tilde{x}_1 + \varepsilon\tilde{v} + \varepsilon o(\varepsilon).$$

Breaking this equation up into a scalar equation for the null coordinate and a vector equation for the remaining coordinates, we obtain:

$$L(U^*) = x_1^{0*} = x_1^0 - \varepsilon + \varepsilon o(\varepsilon) = L(U) - \varepsilon + \varepsilon o(\varepsilon),$$

$$x_1^* = x_1 + \varepsilon o(\varepsilon).$$

Thus, the functional goes down by a quantity of order ϵ, and the endpoint of the trajectory differs from that desired by the quantity $\varepsilon o(\varepsilon)$. By making this construction more exact, we are led to a variation $U^{\#}\varepsilon\Delta$, for which the endpoint $\tilde{x}_1^{\#}$ of the trajectory $\tilde{x}^{\#}$ satisfies the exact equality $\tilde{x}_1^{\#} = \tilde{x}_1 - s\tilde{v}$, and this contradicts the assumption that the control U was optimal.

So, assuming that the control U is optimal, we will suppose in what follows that the vector $\tilde{v}$ is not interior to the cone Π. Since the cone Π is convex, then there exists a supporting plane Γ, such that the whole cone lies in one (closed) halfspace defined by this space, and the vector $\tilde{v}$ in the other. Denoting by $\tilde{\psi}_1$ the covariant vector corresponding to the plane Γ, taken with the appropriate sign, we obtain:

$$(\tilde{\psi}_1, \tilde{\delta}(U^*)) \leqslant 0, \quad U^* \in \Delta, \tag{22}$$

$$(\tilde{\psi}_1, \tilde{v}) \geqslant 0. \tag{23}$$

From inequality (23) there follows at once the inequality

$$\psi_1^0 \leq 0. \tag{24}$$

We denote by $\tilde{\psi}(t)$ the covariant vector obtained by the transport of the vector $\tilde{\psi}_1$, given at the point $\tilde{x}_1$ along the whole trajectory $\tilde{x}(t)$. We shall show that the vector-function $\tilde{\psi}(t)$ is the one whose existence was asserted in Theorem 1.

Let $V(\varepsilon, 0, \tau, \sigma, u^*)$ be any special variation (see (21)) of the family Δ and $\tilde{x}^*(t)$ the solution of equation (6) corresponding to it. A simple calculation yields:

$$\tilde{x}^*(\tau) = \tilde{x}(\tau) + \varepsilon[\tilde{f}(\tilde{x}(\tau), u^*) - \tilde{f}(\tilde{x}(\tau), u(\tau))] + \varepsilon o(\varepsilon).$$

We denote by $\tilde{y}(t)$ the vector obtained from the vector

$$\tilde{y}(\tau) = \tilde{f}(\tilde{x}(\tau), u^*) - \tilde{f}(\tilde{x}(\tau), u(\tau)),$$

given at the point $\tilde{x}(\tau)$ by transport along the trajectory $\tilde{x}(t)$. Then we have:

$$\tilde{x}^*(t_1) = \tilde{x}_1 + \varepsilon\tilde{y}(t_1) + \varepsilon o(\varepsilon).$$

Since the vector $\tilde{y}(t_1)$ belongs to the cone Π, then from inequality (22) we obtain:

$$(\tilde{\psi}_1, \tilde{y}(t_1)) \leqslant 0.$$

Hence, from (19), we get:

$$(\tilde{\psi}(\tau), \tilde{f}(\tilde{x}(\tau), u^*) - \tilde{f}(\tilde{x}(\tau), u(\tau))) \leqslant 0.$$

Rewriting the last inequality in the notation of the function K, we obtain the inequality

$$K(\tilde{\psi}(\tau), \tilde{x}(\tau), u(\tau)) \geqslant K(\tilde{\psi}(\tau), \tilde{x}(\tau), u^*),$$

equivalent to the equality (11).

Now let $U^* = V(\varepsilon, \alpha, \tau, 0, u^*)$. The solution of equation (6) corresponding to this control U^*, will be denoted by $\tilde{x}^*(t)$. We have obviously:

$$\tilde{x}^*(t_1 + \alpha\varepsilon) = \tilde{x}_1 + \varepsilon\tilde{\delta}(U^*) + \varepsilon o(\varepsilon),$$

where

$$\tilde{\delta}(U^*) = \alpha\tilde{f}(\tilde{x}_1, u(t_1)).$$

Since the vector $\tilde{\delta}(U^*)$ belongs to the cone Π, then from inequality (22) we obtain:

$$\alpha(\tilde{\psi}_1, \tilde{f}(\tilde{x}_1, u(t_1))) \leqslant 0.$$

Taking account of the fact that α is an arbitrary real number, the last inequality is possible only under the condition

$$(\tilde{\psi}_1, \tilde{f}(\tilde{x}_1, u(t_1))) = 0,$$

i.e., only for

$$K(\tilde{\psi}(t_1), \tilde{x}(t_1), u(t_1)) = 0. \tag{25}$$

We shall prove finally that the function $K(t) = K(\tilde{\psi}(t), \tilde{x}(t), u(t))$ of the variable t is constant. Suppose that $t_0 \leq t_2 < t_3 \leq t_1$, while on the semi-interval $t_2 < t \leq t_3$ the function $u(t)$ is continuous. We shall show that the function $K(t)$ is constant on that semi-interval. Choose two arbitrary points τ_0 and τ_1 of the semi-interval $t_2 < t \leq t_3$. In view of (11), we have:

$$K(\tilde{\psi}(\tau_0), \tilde{x}(\tau_0), u(\tau_0)) - K(\tilde{\psi}(\tau_0), \tilde{x}(\tau_0), u(\tau_1)) \geqslant 0,$$

$$-K(\tilde{\psi}(\tau_1), \tilde{x}(\tau_1), u(\tau_1)) + K(\tilde{\psi}(\tau_1), \tilde{x}(\tau_1), u(\tau_0)) \leqslant 0.$$

Adding the difference $K(\tau_1) - K(\tau_0)$ to both sides of this inequality, we obtain the inequality

$$K(\tilde{\psi}(\tau_1), \tilde{x}(\tau_1), u(\tau_0)) - K(\tilde{\psi}(\tau_0), \tilde{x}(\tau_0), u(\tau_0)) \leqslant$$
$$\leqslant K(\tau_1) - K(\tau_0) \leqslant K(\tilde{\psi}(\tau_1), \tilde{x}(\tau_1), u(\tau_1)) - K(\tilde{\psi}(\tau_0), \tilde{x}(\tau_0), u(\tau_1)). \tag{26}$$

Further, since the function $K(\tilde{\psi}(t), \tilde{x}(t), u(\tau))$ of the variable t on the segment $t_2 < t \leq t_3$ is continuous and has a derivative, equal to zero in view of (7) and (8), then the outside terms of the inequality (26) disappear. Thus, $K(\tau_1) - K(\tau_0) = 0$, i.e., $K(t) = \text{const.}$ on the semi-interval $t_2 < t \leq t_3$.

Now suppose that τ_0 is a jump point for the function $u(t)$ and that $\tau_1 > \tau_0$ is a point close to τ_0. If $K(\tau_0) > K(\tau_1)$, then for sufficiently small $\tau_1 - \tau_0$ we have:

$$K(\tilde{\psi}(\tau_1), \tilde{x}(\tau_1), u(\tau_0)) > K(\tilde{\psi}(\tau_1), \tilde{x}(\tau_1), u(\tau_1)),$$

which contradicts equality (11). If now $K(\tau_0) < K(\tau_1)$, then for sufficiently small $\tau_1 - \tau_0$ we have:

$$K(\tilde{\psi}(\tau_0), \tilde{x}(\tau_0), u(\tau_1)) > K(\tilde{\psi}(\tau_0), \tilde{x}(\tau_0), u(\tau_0)),$$

which also contradicts equality (11). Thus, $K(\tau_0) = K(\tau_0 + 0)$.

It follows from what has been proved (see (25)) that equality (12) holds for the entire interval $t_0 \leq t \leq t_1$, which, in particular, proves the first of relations (10). The second of relations (10) follows from inequality (24) on taking account of the first equation of (9).

Thus, Theorem 1 is completely proved.

Remark to Theorem 1. Theorem 1 remains valid also in the case that the class of admissible control functions is taken to be the class of *measurable* bounded functions; here equality (11) for the optimal equation is satisfied almost everywhere.

§4. Optimality in the sense of fast motion of a linear control. As an important system in the applications, and an excellent illustration of the general results, one considers the example of a linear control system

$$\frac{dx^i}{dt} = \sum_{j=1}^{n} a^i_j x^j + \sum_{k=1}^{r} b^i_k u^k \qquad (i = 1, \ldots, n),$$

where $\mathbf{u} = (u^1, \cdots, u^r)$ is a point of a convex closed bounded polyhedron Ω lying in a linear space E with coordinates $u^1, \cdots, u^r$. In vector form this system may be written as follows:

$$\frac{d\mathbf{x}}{dt} = A\mathbf{x} + B\mathbf{u}, \tag{27}$$

where A is a linear operator in the space R of the variables $x^1, \cdots, x^n$ and B a linear operator from the space E into the space R. We shall consider here only the problem of minimizing the functional $\int_{t_0}^{t_1} dt$, i.e., the problem of the minimization of the time of passage.

In order to obtain certain results of a uniqueness character we shall impose on the control equation (27) the conditions following below, A) and B), whose roles will be clear in what follows:

A) Let $\mathbf{w}$ be some vector whose direction is that of one of the edges of the polyhedron Ω; then the vector $B\mathbf{w}$ does not lie in any proper subspace of the space R which is invariant under the operator A; thus, the vectors

$$B\mathbf{w},\ AB\mathbf{w},\ \cdots,\ A^{n-1}B\mathbf{w} \tag{28}$$

are linearly independent in the space R whenever $\mathbf{w}$ is a vector having as its direction one of the directions of the edges of the polyhedron Ω.

B) The origin of coordinates of the space E is an interior point of the polyhedron Ω.

The function $H(\psi, \mathbf{x}, \mathbf{u})$ in our case has the form

$$H = (\psi, A\mathbf{x}) + (\psi, B\mathbf{u}), \tag{29}$$

and the system (15) may be written in the form

$$\frac{d\psi_j}{dt} = -\sum_{i=1}^{n} a_j^i \psi_i \qquad (j = 1, \ldots, n),$$

or in vector form

$$\frac{d\psi}{dt} = -A^*\psi. \tag{30}$$

Obviously the function H, considered as a function of the variable $\mathbf{u} \in \Omega$, admits a maximum simultaneously with the function

$$(\psi, B\mathbf{u}).$$

Accordingly we shall denote the maximum of the function $(\psi, B\mathbf{u})$ by $P(\psi)$, the function being considered as a function of the variable $\mathbf{u} \in \Omega$. It follows from Theorem 2 therefore that if

$$U = (\mathbf{u}(t), t_0, t_1, \mathbf{x}_0)$$

is an optimal control of (27), then there exists a solution $\psi(t)$ of equation (30) such that

$$(\psi(t), B\mathbf{u}(t)) = P(\psi(t)). \tag{31}$$

Since equation (30) does not contain the unknown functions $\mathbf{x}(t)$ and $\mathbf{u}(t)$, then all the solutions of equation (30) may be found easily, and in the same way, under condition (31), one may easily find also all optimal controls $\mathbf{u}(t)$ of the equation (27). The question as to how uniquely condition (31) determines the control $\mathbf{u}(t)$ in terms of the function $\psi(t)$, is solved by the theorem following below.

Theorem 3. *If condition* A) *is satisfied, then for any given nontrivial solution* $\psi(t)$ *of equation* (30), *the relation* (31) *uniquely determines the control function* $\mathbf{u}(t)$: *here it results that the function* $\mathbf{u}(t)$ *is piecewise-continuous and its values can be only vertices of the polyhedron* Ω.

Proof. Since the function

$$(\psi(t), B\mathbf{u}), \tag{32}$$

considered as a function of the vector $\mathbf{u}$, is linear, then it is either constant, or it takes on its maximum on the boundary of the polygon Ω. The same consideration may be applied to each face of the polyhedron Ω. Thus, either the function (32) takes on its maximum only on one vertex of the polyhedron Ω, or else it takes it on on a whole face of the polyhedron Ω. We shall show that it follows from condition A) that the latter case is possible only for a finite number of values t. Suppose that the function (32) takes on its maximum (or is constant) on some face Γ of the polyhedron Ω. Let $\mathbf{w}$ be a vector whose direction is that of some edge of the face Γ. Since the function (32) is constant on the face Γ we have:

$$(\psi(t), B\mathbf{w}) = 0.$$

If now this relation were to hold for an infinite set of values of the variable t, then it would hold identically in t, and, differentiating this successively with respect to t, we would obtain

$$\left.\begin{aligned} (\psi(t), B\mathbf{w}) &= 0, \\ (A^*\psi(t), B\mathbf{w}) = (\psi(t), AB\mathbf{w}) &= 0, \\ (A^{*2}\psi(t), B\mathbf{w}) = (\psi(t), A^2B\mathbf{w}) &= 0, \\ \cdots\cdots\cdots\cdots & \\ (A^{*n-1}\psi(t), B\mathbf{w}) = (\psi(t), A^{n-1}B\mathbf{w}) &= 0, \end{aligned}\right\} \qquad (33)$$

and, since from condition A), the vectors (28) form a basis of the space R, then it would follow from relations (33) that $\psi(t) = 0$, which contradicts the supposition that the solution $\psi(t)$ was nontrivial.

§5. Theorem of uniqueness for linear controls. We solve equation (27), as an inhomogeneous equation, by the method of variation of constants. For this purpose we denote by

$$\phi_1(t), \cdots, \phi_n(t) \qquad (34)$$

a fundamental system of solutions of the homogeneous equation

$$\frac{d\mathbf{x}}{dt} = A\mathbf{x},$$

satisfying the initial condition $\phi_j^i(t_0) = \delta_j^i$, and by

$$\psi^1(t), \cdots, \psi^n(t)$$

a fundamental system of solutions of the homogeneous equation (30), satisfying the initial conditions $\psi_i^j(t_0) = \delta_i^j$. We shall seek the general solution of equation (27) in the form

$$\mathbf{x}(t) = \sum_{i=1}^{n} \varphi_i(t)\, c^i(t).$$

Substituting this solution in equation (27), we obtain:

$$\sum_{i=1}^{n} \varphi_i(t) \frac{dc^i(t)}{dt} = B\mathbf{u}(t).$$

Multiplying this last relation by the scalar ψ^j and taking account of the fact that $(\psi^j(t), \phi_i(t)) = \delta_i^j$, we obtain:

$$\frac{dc^i(t)}{dt} = (\psi^i(t), B\mathbf{u}(t)). \qquad (35)$$

Thus, the solution of equation (27) for an arbitrary control $U = (\mathbf{u}(t), t_0, t_1, \mathbf{x}_0)$ may be written in the form:

$$\mathbf{x}(t) = \sum_{i=1}^{n} \varphi_i(t) (x_0^i + \int_{t_0}^{t} (\psi^i(t), B\mathbf{u}(t))\, dt). \qquad (36)$$

Theorem 4. *Suppose that equation* (27) *satisfies condition* A), *and let*

$$U_1 = (\mathbf{u}_1(t), t_0, t_1, \mathbf{x}_0), \qquad U_2 = (\mathbf{u}_2(t), t_0, t_2, \mathbf{x}_0)$$

be two optimal controls for equation (27), *carrying the point* x_0 *into the same point* x_1. *Then these controls coincide:*

$$t_1 = t_2, \quad u_1(t) \equiv u_2(t).$$

Proof. Since both controls U_1 and U_2 are optimal, then $t_1 = t_2$; otherwise, if, for example, $t_1 < t_2$, then the control U_2 would not be optimal. We thus have the equality

$$x_1 = \sum_{i=1}^{n} \varphi_i(t_1)\left(x_0^i + \int_{t_0}^{t_1} (\psi^i(t), Bu_1(t))\, dt\right) = \sum_{i=1}^{n} \varphi_i(t_1)\left(x_0^i + \int_{t_0}^{t_1} (\psi^i(t), Bu_2(t))\, dt\right).$$

Since the vectors $\phi_1(t_1), \cdots, \phi_n(t_1)$ are linearly independent, then it follows from the last equation that

$$\int_{t_0}^{t_1} (\psi^i(t), Bu_1(t))\, dt = \int_{t_0}^{t_1} (\psi^i(t), Bu_2(t))\, dt \qquad (i = 1, \ldots, n). \tag{37}$$

From Theorem 3, it follows that there corresponds to the optimal control U_1 a vector function $\psi(t)$, which is a solution of equation (30). The initial value of this function for $t = t_0$ will be denoted by

$$\psi_0 = (\psi_{10}, \cdots, \psi_{n0});$$

then the solution $\psi(t)$ may be written in the form

$$\psi(t) = \sum_{i=1}^{n} \psi_{i0}\psi^i(t). \tag{38}$$

Multiplying relation (37) by ψ_{i0} and summing on i, we obtain:

$$\int_{t_0}^{t} (\psi(t), Bu_1(t))\, dt = \int_{t_0}^{t} (\psi(t), Bu_2(t))\, dt. \tag{39}$$

From Theorem 3 the function $u_1(t)$ satisfies the condition

$$(\psi(t), Bu_1(t)) = P(\psi(t))$$

and is determined uniquely by this condition. If now the function $u_2(t)$ did not coincide with the function $u_1(t)$, then the condition

$$(\psi(t), Bu_2(t)) \equiv P(\psi(t)),$$

would not be satisifed, and therefore the function $(\psi(t), Bu_2(t))$, not exceeding the function $(\psi(t), Bu_1(t))$ anywhere, will on some interval be less than it. Thus, if on the segment $t_0 \leq t \leq t_1$ the identity $u_1(t) \equiv u_2(t)$ did not hold, then equality (39) would be impossible.

Thus, Theorem 4 is proved.

We shall call the control

$$U = (u(t), t_0, t_1, x_0)$$

extremal, if it satisfies condition (31), where $\psi(t)$ is some nontrivial solution of equation (30).

In order to find all optimal controls carrying the point x_0 into the point x_1, one may find first all extremal controls carrying the point x_0 into the point x_1 and then chose from their number the unique one which carries out this passage in the shortest time. The question arises, as to whether there may be several extremal controls carrying the point x_0 into the point x_1. Generally speaking, there may be several of them. The theorem following below indicates an important case of unicity.

Theorem 5. *Suppose that equation* (27) *satisfies conditions* A) *and* B), *and let*

$$U_1 = (u_1(t),\ t_0,\ t_1,\ x_0),\qquad U_2 = (u_2(t),\ t_0,\ t_2,\ x_0)$$

be two extremal controls, carrying the point x_0 *into the origin of coordinates* $x_1 = 0$ *of the space R; then the controls* U_1 *and* U_2 *coincide:*

$$t_1 = t_2,\qquad u_1(t) \equiv u_2(t).$$

Proof. By hypothesis, we have the equality

$$\left.\begin{aligned}\sum_{i=1}^{n} \varphi_i(t_1)\Big(x_0^i + \int_{t_0}^{t_1} (\psi^i(t),\ Bu_1(t))\,dt\Big) = 0,\\ \sum_{i=1}^{n} \varphi_i(t_2)\Big(x_0^i + \int_{t_0}^{t_2} (\psi^i(t),\ Bu_2(t))\,dt\Big) = 0.\end{aligned}\right\} \tag{40}$$

Since the vectors (34) are linearly independent for any t, then from equation (40) it follows that

$$-x_0 = \int_{t_0}^{t_1} (\psi^i(t),\ Bu_1(t))\,dt = \int_{t_0}^{t_2} (\psi^i(t),\ Bu_2(t))\,dt. \tag{41}$$

We assume for definiteness that $t_1 > t_2$, and take $\psi(t)$ to be that solution of equation (30) for which the identity

$$(\psi(t),\ Bu_1(t)) \equiv P(\psi(t))$$

holds, defining the function $u_1(t)$. As in the proof of Theorem 4, the function $\psi(t)$ may be written in the form (38). Multiply relation (41) by ψ_{i0} and sum on i. We obtain:

$$\int_{t_0}^{t_1} (\psi(t),\ Bu_1(t))\,dt = \int_{t_0}^{t_2} (\psi(t),\ Bu_2(t))\,dt.$$

We now observe that it follows from condition B) that

$$P(\psi(t)) \geq 0. \tag{42}$$

Indeed, since zero is an interior point of the convex body Ω, then the function $(\psi(t),\ Bu)$, as a function of the variable u, is either identically zero or else may take on both negative and positive values.

In view of (42) we have the inequality

$$\int_{t_0}^{t_2} (\psi(t),\ B\mathbf{u}_1(t))\, dt \leqslant \int_{t_0}^{t_2} (\psi(t),\ B\mathbf{u}_2(t))\, dt.$$

Hence, just as in the proof of Theorem 4, we obtain:

$$\mathbf{u}_1(t) \equiv \mathbf{u}_2(t) \text{ for } t_0 \leq t \leq t_2.$$

Further, since the equation $P(\psi(t)) = 0$ holds only for isolated values t, then we must have $t_1 = t_2$.

Thus, Theorem 5 is proved.

§6. **Existence of optimal controls for linear systems.**

Theorem 6. *If there exists at least one control of the equation* (27), *carrying the point* $\mathbf{x}_0$ *into the point* $\mathbf{x}_1$, *then there exists an optimal control of the equation* (27), *carrying the point* $\mathbf{x}_0$ *into the point* $\mathbf{x}_1$.

Proof. The set of all controls of the form

$$U = (\mathbf{u}(t),\ 0,\ t,\ \mathbf{x}_0), \tag{43}$$

carrying the point $\mathbf{x}_0$ into the point $\mathbf{x}_1$, will be denoted by $\Delta_{\mathbf{x}_0, \mathbf{x}_1}$. To each control (43) there corresponds a time of passage t. We denote the lower bound of all such times for $U \in \Delta_{\mathbf{x}_0, \mathbf{x}_1}$ by t^*, and we shall prove that there exists a control $U^* = (\mathbf{u}^*(t),\ 0,\ t^*,\ \mathbf{x}_0)$, carrying the point $\mathbf{x}_0$ into the point $\mathbf{x}_1$.

We choose from the set $\Delta_{\mathbf{x}_0, \mathbf{x}_1}$ an infinite sequence of controls

$$U_k = (\mathbf{u}_k(t),\ 0,\ t_k,\ \mathbf{x}_0) \quad (k = 1, 2, \cdots),$$

for which the equality $\lim_{k\to\infty} t_k = t^*$ holds. Obviously, equality

$$\lim_{k\to\infty} \sum_{i=1}^{n} \varphi_i(t^*) \left(x_0^i + \int_0^{t^*} (\psi^i(t),\ B\mathbf{u}_k(t))\, dt \right) = \mathbf{x} \tag{44}$$

holds.

Consider the Hilbert space L_2 of all measurable functions with integrable squares, given on the interval $0 \leq t \leq t^*$. The control $\mathbf{u}_k(t)$ is a vector-function; the i-th coordinate of this function will be denoted by $u_k^i(t)$. The function $u_k^i(t)$, considered on the segment $0 \leq t \leq t^*$, lies in the space L_2. The set of all functions $u_k^i(t)$ $(k = 1, 2, \cdots)$, obviously lies in some sphere of the space L_2 and therefore we may select a weakly convergent subsequence from it. For simplicity we suppose that the original sequence

$$u_1^i(t),\ u_2^i(t),\ \cdots,\ u_k^i(t),\ \cdots \tag{45}$$

converges weakly to some function $u^i(t)$ $(i = 1, \cdots, r)$.

We shall show that the vector-function

$$\mathbf{u}^*(t) = (u^1(t),\ \cdots,\ u^r(t))$$

satisfies, for almost all values t, the condition

$$\mathbf{u}^*(t) \in \Omega.$$

Let

$$b(\mathbf{u}) = \sum_{i=1}^{r} b_i u^i = b$$

be the equation of a hyperplane carrying one of the $(r-1)$-dimensional faces of the polyhedron Ω, while the polyhedron Ω lies in the halfspace

$$b(\mathbf{u}) \leq b.$$

Let m be the set of all values t of the segment $[0, t^*]$, for which $b(\mathbf{u}^*(t)) > b$, and $v(t)$ the characteristic function of the set m. We then have

$$\lim_{k\to\infty} \int_0^{t^*} v(t)\,[b(\mathbf{u}^*(t)) - b(\mathbf{u}_k(t))]\,dt = 0$$

in view of the weak convergence of the sequence (45), and since

$$b(\mathbf{u}^*(t)) - b(\mathbf{u}_k(t)) > 0$$

on the set m, then the measure of m is zero.

Thus, changing the vector-function $\mathbf{u}^*(t)$ on a set of measure zero, we obtain a new function, which we shall again denote by $\mathbf{u}^*(t)$, which satisfies the condition $\mathbf{u}^*(t) \in \Omega$, $0 \leq t \leq t^*$.

It follows from relation (44) and the weak convergence of the sequence (45) that:

$$\sum_{i=1}^{n} \varphi_i(t^*) \left(x_0^i + \int_0^{t^*} (\psi^i(t),\ B\mathbf{u}^*(t))\,dt \right) = \mathbf{x}_1.$$

Thus, $U^* = (\mathbf{u}^*(t), 0, t^*, \mathbf{x}_0)$ is a measurable optimal control, carrying the point $\mathbf{x}_0$ into the point $\mathbf{x}_1$.

In view of the remark to Theorem 1, by changing the control $\mathbf{u}^*(t)$ on a set of measure zero, we may convert it into a control satisfying the maximum principle, i.e., in our case the condition

$$(\psi(t),\ B\mathbf{u}^*(t)) = P(\psi(t)).$$

It follows obviously from this condition that the function $\mathbf{u}^*(t)$ is piecewise continuous.

Thus, Theorem 6 is proved.

Theorem 7. *If equation* (27) *satisfies conditions* A) *and* B) *and the operator A is stable, i.e., if all of its eigenvalues have negative real parts, then to each point $\mathbf{x}_0 \in R$ there corresponds an optimal control carrying that point into the origin of coordinates $0 \in R$.*

Proof. We shall first of all prove that there exists a neighborhood V of the point 0 in R, each point $\mathbf{x}_0$ of which may be carried by some control into 0.

Choose in Ω a vector $\mathbf{v}$ such that the vector $-\mathbf{v}$ belongs to Ω and such that the vector

$$\mathbf{b} = B\mathbf{v}$$

does not belong to any proper subspace of the space R which is invariant under the operator A. From conditions A) and B) it follows that there is such a vector $\mathbf{v}$. For a sufficiently small positive ϵ the operators A and $e^{-\epsilon A}$ have coincident invariant subspaces, and therefore the vectors

$$e^{-\varepsilon A}\mathbf{b},\ e^{-2\varepsilon A}\mathbf{b},\ \ldots,\ e^{-n\varepsilon A}\mathbf{b}$$

are linearly independent.

Let $\chi(t)$ be any real function, defined on some segment $0 \leq t \leq t_1$ and not exceeding unity in absolute value; then

$$U = (\mathbf{v}\chi(t),\ 0,\ t_1,\ \mathbf{x}_0)$$

is a control of equation (27), and that control carries the point $\mathbf{x}_0$ into the point (see (36))

$$\mathbf{x}_1 = e^{t_1 A}\left(\mathbf{x}_0 + \int_0^{t_1} e^{-tA}\mathbf{b}\chi(t)\,dt\right). \tag{46}$$

Now we choose the function $\chi(t)$, depending on the parameters $\xi^1, \cdots, \xi^n$ in such a way that the point (46), which we denote by $\mathbf{x}_1(\mathbf{x}_0;\ \xi^1, \cdots, \xi^n)$, satisfies the following conditions:

$$\mathbf{x}_1(0;\ 0, \cdots, 0) = 0,$$

and the Jacobian

$$\left.\frac{\partial(x_1^1, \ldots, x_1^n)}{\partial(\xi^1, \ldots, \xi^n)}\right|_{\mathbf{x}_0=0,\ \xi^1=0,\ \ldots,\ \xi^n=0}$$

is nonzero. Constructing such a function $\chi(t)$, we show that the equation $\mathbf{x}_1(\mathbf{x}_0;\ \xi^1, \cdots, \xi^n) = 0$ may be solved for $\xi^1, \cdots, \xi^n$ for all values $\mathbf{x}_0$ lying in some neighborhood V of the origin 0.

First of all we define a function $\sigma(t, \tau, \xi)$ of the variable t, $0 \leq t \leq t_1$, where $0 < \tau < t_1$ and ξ is a parameter. The function $\sigma(t, \tau, \xi)$, as a function of the variable t, is equal to zero everywhere outside the interval $[\tau, \tau + \xi)$, and on that interval it is equal to sign ξ. Now put

$$\chi(t) = \sum_{k=1}^{n} \sigma(t,\ k\varepsilon,\ \xi^k).$$

A simple calculation shows that the point $\mathbf{x}_1(\mathbf{x}_0,\ \xi^1, \cdots, \xi^n)$, with this choice of the function $\chi(t)$, satisfies the stated conditions.

Now let $\mathbf{x}_0$ be any point of the space R. Suppose that it first moves under the control $\mathbf{u}(t) \equiv 0$. Since all eigenvalues of the operator A have negative real parts, then after the expiration of some time the point will come into the neighborhood, after which, as we have proved, it must be carried into the origin of coordinates.

Hence, from Theorem 6, it follows that there exists an *optimal* control, carrying the point $\mathbf{x}_0$ into the origin.

Thus Theorem 7 is proved.

§7. **Synthesis of a linear optimal control.** The problem of *synthesizing* an optimal control has a sense for any control system (1); however here we shall treat it only in the case of a linear control system (27), satisfying conditions A) and B), with a stable operator A. For such a system one has the theorems on existence and uniqueness (Theorems 7 and 5), thanks to which the problem of synthesis is in principle solved. The considerations presented here give a constructive method of solution of the problem. The practicability of this method in each concrete case requires, however, a series of constructions. The synthesis of an optimal control for the linear system (27) was carried out, in its entirety, by other methods up until now only for the case of one control parameter (i.e., for $r = 1$). Fel'dbaum [6], in the case that the operator A had real roots, and Bushaw [7], in the case that $n = 2$ and the eigenvalues of the operator A were complex.

We shall suppose that the equation (27) satisfies conditions A) and B) and has a stable operator A. Then for any point $\mathbf{x}_0 \in R$ there exists one (and only one) optimal control

$$U_{\mathbf{x}0} = (\mathbf{u}_{\mathbf{x}0}(t),\ t_0,\ t_1,\ \mathbf{x}_0), \tag{47}$$

which carries the point $\mathbf{x}_0$ into the origin of coordinates $0 \in R$. There is uniqueness, finally, up to a time translation (see remark 2 to the statement of the problem). The quantity $\mathbf{u}_{\mathbf{x}0}(t_0)$ depends, thus, only on the point $\mathbf{x}_0$, and not on the particular origin of the time reading t_0, and therefore one may put: $\mathbf{v}(\mathbf{x}_0) = \mathbf{u}_{\mathbf{x}0}(t_0)$. Let $\mathbf{x}(t)$ be a solution of equation (27), corresponding to the control (47); then $U_{\mathbf{x}(\tau)} = (\mathbf{u}_{\mathbf{x}0}(t),\ \tau,\ t_1,\ \mathbf{x}(\tau))$ (see remark 1 to the statement of the problem), and therefore $\mathbf{u}_{\mathbf{x}0}(\tau) = \mathbf{v}(\mathbf{x}(\tau))$. Thus,

$$\frac{d}{dt}\mathbf{x}(t) = A\mathbf{x}(t) + B\mathbf{v}(\mathbf{x}(t)),$$

and we see that the solution of the equation

$$\frac{d\mathbf{x}}{dt} = A\mathbf{x} + B\mathbf{v}(\mathbf{x}) \tag{48}$$

for arbitrary initial conditions $\mathbf{x}(t_0) = \mathbf{x}_0$ gives a law of optimal motion of the point $\mathbf{x}_0$ into the origin of coordinates. In this sense the function $\mathbf{v}(\mathbf{x})$ *synthesizes* an optimal solution, carrying any point $\mathbf{x}_0$ into the origin.

We now present a method of construction of the function $\mathbf{v}(\mathbf{x})$. Let $\psi(t)$ be that solution of equation (30) which under Theorem 2 corresponds to the control (47), so that

$$\frac{d\psi(t)}{dt} = -A^*\psi(t), \tag{49}$$

and the function $\mathbf{u}_{\mathbf{x}0}(t)$ is defined from the equation

$$(\psi(t), Bu_{x_0}(t)) = P(\psi(t)). \tag{50}$$

Suppose further that $x(t)$ is a solution of equation (27), satisfying the initial condition

$$x(t_0) = x_0 \tag{51}$$

and the end condition

$$x(t_1) = 0, \tag{52}$$

so that

$$\frac{dx(t)}{dt} = Ax(t) + Bu_{x_0}(t). \tag{53}$$

Then the function $v(x)$ satisfies the condition:

$$(\psi(t_0), Bv(x(t_0))) = P(\psi(t_0)). \tag{54}$$

From the existence and uniqueness theorems it follows that there exists one, and only one (up to a translation of time), pair of functions $u_{x_0}(t)$, $x(t)$, given on the segment $t_0 \leq t \leq t_1$ and satisfying conditions (49)–(53). In view of the possibility of translating the time, the numbers t_0 and t_1 are not determined uniquely by these conditions, but the number $t_1 - t_0$ is.

It is not perfectly clear how to find the functions $u_{x_0}(t)$ and $x(t)$, satisfying all the conditions (49)–(53), but it is easy to find all the functions $u_{x_0}(t)$, $x(t)$, satisfying only conditions (49), (50), (52), and (53). To do this we proceed as follows: in view of the possibility of an arbitrary translation of time, we fix the number t_1, putting $t_1 = 0$. Now let χ be any covariant vector, different from zero, and $\psi(t, \chi)$ the solution of equation (49), satisfying the initial condition:

$$\psi(0, \chi) = \chi$$

and defined for $t \leq 0$. Further, we define a function $u(t, \chi)$ from the condition

$$(\psi(t, \chi), Bu(t, \chi)) = P(\psi(t, \chi)), \quad t \leq 0,$$

and the function $x(t, \chi)$ from the equation

$$\frac{dx(t, \chi)}{dt} = Ax(t, \chi) + Bu(t, \chi).$$

From what has been said above, the function $v(x)$ is defined by the relation:

$$(\psi(t, \chi), Bv(x(t, \chi))) = P(\psi(t, \chi)). \tag{55}$$

It follows from the existence theorem (Theorem 7) that the point $x(t, \chi)$ sweeps out the whole space R, as t runs through negative values and the vector χ changes arbitrarily. Thus, relation (55) defines the value of the function $v(x)$ for any point x of the space R.

BIBLIOGRAPHY

[1] V. G. Boltyanskiĭ, R. V. Gamkrelidze and L. S. Pontryagin, *On the theory of optimal processes*, Dokl. Akad. Nauk SSSR 110 (1956), 7–10. (Russian)

[2] R. V. Gamkrelidze, *On the theory of optimal processes in linear systems*,

Dokl. Akad. Nauk SSSR 116 (1957), 9–11. (Russian)

[3] V. G. Boltyanskiĭ, *The maximum principle in the theory of optimal processes*, Dokl. Akad. Nauk SSSR 119 (1958), 1070–1073. (Russian)

[4] G. A. Bliss, *Lectures on the calculus of variations*, University of Chicago Press, Chicago, Ill., 1946. [The reference in the text is to the Russian edition, and probably corresponds to the first few pages of Chapter VII in the English edition.]

[5] E. J. McShane, *On multipliers for Lagrange problems*, Amer. J. Math. 61 (1939), 809–819.

[6] A. A. Fel'dbaum, *On the design of optimal systems by means of phase space*, Avtomat. i Telemeh. 16 (1955), 129–149. (Russian)

[7] D. W. Bushaw, *Experimental towing tank*, Stevens Institute of Technology, Report 469, Hoboken, N. J., 1953.

Translated by:
J. M. Danskin, Jr.

Contributions to the Theory of Optimal Control

R. E. KALMAN

THIS is one of the two ground-breaking papers by Kalman that appeared in 1960—with the other one (discussed next) being the filtering and prediction paper. This first paper, which deals with linear-quadratic feedback control, set the stage for what came to be known as LQR (Linear-Quadratic-Regulator) control, while the combination of the two papers formed the basis for LQG (Linear-Quadratic-Gaussian) control. Both LQR and LQG control had major influence on researchers, teachers, and practitioners of control in the decades that followed.

The idea of designing a feedback controller such that the integral of the square of tracking error is minimized was first proposed by Wiener [17] and Hall [8], and further developed in the influential book by Newton, Gould and Kaiser [12]. However, the problem formulation in this book remained unsatisfactory from a mathematical point of view, but, more importantly, the algorithms obtained allowed application only to rather low order systems and were thus of limited value. This is not surprising since it basically took until the H_2-interpretation in the 1980s of LQG control before a satisfactory formulation of least squares feedback control design was obtained. Kalman's formulation in terms of finding the least squares control that evolves from an arbitrary initial state is a precise formulation of the optimal least squares transient control problem.

The paper introduced the very important notion of *controllability*, as the possibility of transfering any initial state to zero by a suitable control action. It includes the necessary and sufficient condition for controllability in terms of the positive definiteness of the *Controllability Grammian*, and the fact that the linear time-invariant system with n states,

$$\frac{d}{dt}x = Fx + Gu$$

is controllable if and only if the matrix $[G, FG, \ldots, F^{n-1}G]$ has rank n. As is well known, this concept of controllability, its implications (e.g., in pole placement and stabilization), and generalizations to nonlinear or infinite-dimensional systems became one of the main *leitmotivs* in control research. Controllability is indeed one of the compelling notions that is truly endogenous to the field of control.

The paper also introduced the notion of *observability*, but as a mere "dual" of controllability. Contemporaneously Kalman provided an alternative, more satisfactory definition in [10], where observability is defined in a more intrinsic way in terms of the possibility of deducing the state trajectory from input/output measurements.

Kalman actually states (p. 102, fourth paragraph) that he views the introduction of the notions of controllability and observability, and their exploitation in the regulator problem, as the principal contribution of the present paper. Controllability and observability are shown in the paper to be of central importance in the analysis of the least squares control problem over an infinite horizon. They are also used to obtain the asymptotic properties of the Riccati differential equation [Equation (6.3) of the paper—henceforth referred to as RDE], and the stability properties of its limiting solution. This paper was in fact the first to introduce the RDE as an algorithm for computing the state feedback gain of the optimal controller for a general linear system with a quadratic performance criterion. RDE had emerged earlier in the study of the second variations in the calculus of variations, but its use in general linear systems, where the optimal trajectory needs to be generated by a control input, was new.

The analysis throughout the paper concentrates on time-varying systems, and uses the Hamilton-Jacobi theory to arrive at RDE and to deduce optimality of the LQ control gain. We now know, however, that an alternative way to prove optimality in least squares is by showing how RDE allows one to "complete the square" (see, e.g., [5], [18]).

Almost immediately after its appearance, the LQ-problem was included in influential textbooks [2], [5], [11], [1], and extended in a number of directions. For example, the case of indefinite cost is treated in [18], which requires a more delicate analysis, and later had many applications, in particular, in H_∞-control; and extensions to zero-sum and nonzero-sum differential games are discussed in [9] and [16], where again RDE-based feedback policies arise, albeit with more general structures [3]. Kalman's paper actually also motivated and led the way to a great deal of research on RDE, particularly on its algebraic version, and algorithms for solving it appeared very soon [13]. Most computer packages that aim at linear systems and control implement today a Riccati equation solver. A wealth of information and references on the work on the LQ problem can be found in [15] and [4].

Kalman's paper deals only with state feedback. The formulation of the output feedback version of the LQ problem requires either introducing, as in LQG, stochastic disturbances, or a formulation in terms of the H_2-norm. It is worth noting, however, that Kalman's paper contains (p. 104) an informal, unproven but quite precise statement of the separation theorem and the certainty equivalence principle, which are further discussed in this volume in the preamble to Feldbaum's papers.

The LQ problem and its sister, LQG, received enormous attention, and were sometimes presented as a panacea for solving linear control problems. Critics [14], [16] pointed to the excessive bandwidth and poor robustness properties of the LQG controller. In fact, Doyle's paper [6] about guaranteed margin of LQG-controllers had the abstract "*There are none*"—fortuitously making reading of the article superfluous. This lack of robustness led to the H_∞ control problem which was formulated by Zames [19] in part as a reaction to the LQG problems (see the comments in the preamble to [19], the last paper in this volume). However, it turns out that also in the H_∞-problem it was the Riccati equation that finally prevailed [7]! The introduction of the Riccati equation, along with the notions of controllability and observability, are among the many gems in the crown of original ideas in this seminal paper by Kalman.

References

[1] B.D.O. Anderson and J.B. Moore, *Linear Optimal Control*, Prentice Hall (Englewood Cliffs, NJ), 1971.

[2] M. Athans and P.L. Falb, *Optimal Control*, McGraw-Hill (New York), 1966.

[3] T. Başar and G.J. Olsder, *Dynamic Noncooperative Game Theory*, Classics in Applied Mathematics, SIAM (Philadelphia), 1999.

[4] S. Bittanti, A.J. Laub, and J.C. Willems, edts, *The Riccati Equation*, Springer Verlag (Berlin), 1991.

[5] R.W. Brockett, *Finite Dimensional Linear Systems*, Wiley (New York), 1970.

[6] J.C. Doyle, "Guaranteed margins for LQG regulators," *IEEE Trans. Automat. Contr.*, **AC-23**:756–757, 1978.

[7] J.C. Doyle, K. Glover, P.P. Khargonekar and B.A. Francis, "State-space solutions to standard H_2 and H_∞ control problems," *IEEE Trans. Automat. Contr.*, **AC-34**:831–847, 1989.

[8] A.C. Hall, *The Analysis and Synthesis of Linear Servomechanisms*, The Technology Press, M.I.T. (Cambridge, MA), 1943.

[9] Y.C. Ho, A.E. Bryson, Jr. and S. Baron, "Differential games and optimal pursuit-evasion strategies," *IEEE Trans. Automat. Contr.*, **AC-10**(4):385–389, 1965.

[10] R.E. Kalman, "On the general theory of control systems," in *Proc. First Internat. Congress Automat. Contr.*, pp. 481–491, Moscow, 1960.

[11] H. Kwakernaak and R. Sivan, *Linear Optimal Control Systems*, Wiley (New York), 1972.

[12] G.C. Newton, Jr., L.A. Gould and J.F. Kaiser, *Analytical Design of Linear Feedback Controls*, Wiley (New York), 1957.

[13] J.E. Potter, "Matrix quadratic solutions," *SIAM J. Appl. Math.*, **14**:496–501, 1964.

[14] H.H. Rosenbrock and P.D. McMorran, "Good, bad, or optimal?," *IEEE Trans. Automat. Contr.*, **AC-16**(6):552–554, 1971.

[15] *Special Issue on the Linear-Quadratic-Gaussian Estimation and Control Problem*, *IEEE Trans. Automat. Contr.*, **AC-16**(6), 1971.

[16] A.W. Starr and Y.C. Ho, "Nonzero-sum differential games," *J. Optimiz. Theory Appl.*, **3**(3):184–206, 1969.

[17] N. Wiener, *Extrapolation, Interpolation, and Smoothing of Stationary Time Series*, MIT Press (Cambridge, MA), 1949.

[18] J.C. Willems, "Least squares stationary optimal control and the algebraic Riccati equation," *IEEE Trans. Automat. Contr.*, **AC-16**:621–634, 1971.

[19] G. Zames, "Feedback and optimal sensitivity: Model reference transformations, multiplicative seminorms, and approximate inverses," *IEEE Trans. Automat. Contr.*, **AC-26**:301–320, 1981.

W.S.L. & J.C.W.

CONTRIBUTIONS TO THE THEORY OF OPTIMAL CONTROL

By R. E. Kalman

1. Introduction

The purpose of this paper is to give an account of recent research on a classical problem in the theory of control: the design of linear control systems so as to minimize the integral of a quadratic function evaluated along motions of the system. This problem dates back in its modern form to Wiener and Hall at about 1943 ([1], [2]). In spite of its relatively long history, the problem has never been formulated rigorously from a mathematical point of view. Even the most up-to-date expositions of the subject (see, e.g., [3]) are inaccessible to the mathematician due to the lack of precisely stated conditions and results.

The problem is quite broad, and there are many unsettled questions. This paper will be concerned with only the simplest case, the so-called *regulator problem*. For other aspects of the problem, the reader may consult [4]–[8] which, while devoted primarily to questions of theoretical engineering, contain precise mathematical results.

The conventional theory of the regulator problem is based largely on Fourier and Laplace transforms. By contrast, the approach of this paper is direct and uses the well-known theory of ordinary differential equations. We have also drawn on Lyapunov's theory of stability. While our earlier treatments ([5], [6], [8]) followed the point of view of dynamic programing, here we utilize classical tools of the calculus of variations, in particular the Hamilton-Jacobi equation.

The principal contribution of the paper lies in the introduction and exploitation of the concepts of *controllability* and *observability* ([8]), with the aid of which we give, for the first time, a complete theory of the regulator problem. In particular, we prove existence and stability theorems for the regulator problem and study in some detail stability properties of the matrix Riccati equation, which arises as a special case of the Hamilton-Jacobi equation.

A careful discussion of the conceptual aspects of the control problems has been included as an aid to persons not familiar with the field of control. Some mathematical arguments, in particular the review of the calculus of variations, are more leisurely than usual in order to render the paper reasonably self-contained.

2. Notation and terminology

We use standard vector-matrix notation, with the following conventions: small Greek letters are scalars; small Latin letters are vectors, capitals are matrices. The unit matrix is I. Exceptions: i, j, m, n, p are integers; l, E, L, V are scalars; ϕ, ψ, ξ are vectors. The inner product is $[x, y]$. The transpose of a matrix is denoted by the prime. The norm is $\| x \| = [x, x]^{\frac{1}{2}}$. The norm $\| A \|$ of a matrix A is sup $\| Ax \|$ over $\| x \| = 1$. Special conventions: $\| x \|_P^2 = [x, Px]$ where P is

Reprinted with permission from *Boletin de la Sociedad Matematica Mexicana*, R. E. Kalman, "Contributions to the Theory of Optimal Control," Vol. 5, 1960, pp. 102–119.

symmetric, nonnegative definite. If A, B are symmetric, $A > B[A \geq B]$ means $A - B$ is positive [nonnegative] definite. The letters t, σ, τ are arbitrary real numbers which always denote the *time*; we use $t_0 \leq t_1 \leq t_2$ to denote fixed, ordered values of t. All scalars, vectors, and matrices are real throughout. For a scalar function $L(u)$ of the vector u, L_u is the gradient vector and L_{uu} is the jacobian matrix.

We shall study the system represented by the equations

$$dx/dt = F(t)x + G(t)u(t) \tag{2.1}$$

$$y(t) = H(t)x(t) \tag{2.2}$$

where: u is an m-vector, x is an n-vector, y is a p-vector; $F(t)$, $G(t)$, as well as $H(t)$ are rectangular matrices continuous in t, either of which may be singular.

In view of the physical motivation of our problem, we adopt the following terminology: Equations (2.1–2) are the *plant* (or *model*); x is the *state* of; $u(t)$ is the *control function* or *input* to; and $y(t)$ is the *output* of the plant. The plant is *constant* if F, G, H are constants. If $u(t) = 0$ or $G(t) = 0$, the plant is *free*.

The behavior of the plant is described by the solution of the differential equation (2.1) which will exist for all t and be unique if, say, $u(t)$ is Lebesgue integrable. As is well known ([9]), the general solution has the form[1]

$$x(t) = \Phi(t, t_0)x(t_0) + \int_{t_0}^{t} \Phi(t, \tau)G(\tau)u(\tau)\, d\tau \tag{2.3}$$

where $\Phi(t, \tau)$, defined for all t, τ, is a fundamental matrix ([9]) of solutions of the free system (2.1), satisfying the additional requirement that

$$\Phi(t, t) = I \quad \text{for all } t \tag{2.4}$$

In view of (2.4), we call Φ the *transition matrix* of (2.1)—a terminology borrowed from the theory of Markov processes ([7], [8], [10]).

The solution (2.3) is conveniently regarded as the *motion* of the state of (2.1); this leads to the notation

$$x(t) = \phi_u(t; x, t_0) \tag{2.5}$$

Read: the motion of (2.1) starting at initial state x at time t_0 and observed at time t, and influenced by the *fixed* control function $u(t)$ defined in the interval $[t_0, t]$. Since (2.5) holds for all t, t_0, we have in particular the identity: $x = \phi_u(t; x, t)$ for all t, x, u. Free motions are denoted by ϕ_f. We observe also that (2.1) has an *equilibrium state* x^* at 0, in other words, a state for which $x^* = \phi_f(t; x^*, t_0)$ for all t, t_0.

3. Statement of problem

In the simplest applications, the object of a control system is the following: *Given any state x of the plant (2.1) at any time t_0, "generate" a control function*

[1] The function (2.3) will satisfy the differential equation (2.1) *almost everywhere*.

$u(t)$, defined for $t \geq t_0$ and depending on x, t_0, which causes x to be "transferred" to the equilibrium state 0. In other words, $u(t)$ is chosen so as to assure

$$\lim_{t\to\infty} \phi_u(t; x, t_0) = 0 \tag{3.1}$$

For technological reasons, the function $u(t)$ must be generated from actual measurements of the behavior of the plant. To describe how the control system is to be *physically realized*, one must therefore provide an algorithm for computing the number $u(t_1)$ from the knowledge of $y(t)$ for $t \leq t_1$. This is usually referred to as the *Feedback Principle.*

One may separate the problem of physical realization into two stages:

(A) Computation of the "best approximation" $\hat{x}(t_1)$ of the state $x(t_1)$ from knowledge of $y(t)$ for $t \leq t_1$.

(B) Computation of $u(t_1)$ given $\hat{x}(t_1)$.

In the engineering literature one often makes the simplifying assumption of treating the two problems separately, i.e., simply regarding $\hat{x}(t)$ as though it were $x(t)$. We are concerned here only with Problem (B) and therefore always assume that $x(t)$ is known exactly. Somewhat surprisingly, the theory of Problem (A), which includes as a special case Wiener's theory of the filtering and prediction of time series, turns out to be analogous to the theory of Problem (B) developed in this paper. This assertion follows from the *duality theorem* discovered by the author ([7], [8]); this theorem can be used to show also that the separation of Problems (A) and (B) is indeed legitimate.

Assuming $x(t)$ is known exactly and taking into account the Feedback Principle, the problem of generating $u(t)$ reduces to specifying the *control law*

$$u(t) = k(x(t), t) \tag{3.2}$$

From the definition of state it is clear that nothing would be gained by letting $u(t)$ depend also on values of the state prior to time t. To assure that (2.1) with (3.2) has a unique solution, it suffices to have $k \in C^1$. If (3.2) does not depend explicitly on t, we say the control law is *constant.*

To arrive at the control law "rationally," we now add the further desideratum that *the integral of a nonnegative function of the state along any motion ϕ_u should be minimized by the choice of $u(t)$*. Stating this requirement with some care, we shall see that it uniquely determines $u(t)$ and hence also the control law (3.2), and even implies (3.1). We call the resulting control system *optimal.*

Let us now state precisely the

(3.3) OPTIMAL REGULATOR PROBLEM. *Find a control law* (3.2) *for which*

$$V^0(x, t_0, t_1) = \inf_{u(t)} \left\{ \nu(\phi_u(t_1; x, t_0)) + \int_{t_0}^{t_1} L(\phi_u(t; x, t_0)\ u(t), t)\, dt \right\} \tag{3.4}$$

is attained for all x, t_0, and t_1, where ν, L are nonnegative scalar functions of class C^2 in all arguments.

The class of functions $u(t)$ which are admitted to competition in taking the infimum in (3.4) are to be of class D^0 (i.e., continuous except at isolated points at which $u(t)$ has finite left- and right-hand limits).

It is well known in the calculus of variations ([11], p. 196) that the condition $L_{uu} \geq 0$ or $L_{uu} \leq 0$ is necessary for the existence of even local extremals. To avoid complications due to the equality sign, we assume from the outset that

$$L_{uu}(x, u, t) > 0 \quad \text{for all } x, u, t \tag{3.5}$$

which is equivalent to assuming that L is strictly convex in u.

In engineering language, one calls t_1 the *terminal time* (it may be infinity!), the integral is the *performance index*, and (at least when it does not depend on u) L is the *error criterion* or more generally the *loss function*. The function ν is added for greater generality. We use the notation $V(x, t_0, t_1; u)$ for the value of the integral in (3.4) for some specified, fixed $u(t)$. The superscript o identifies "optimal".

4. Relations with the calculus of variations

In this section we transcribe some well-known results ([11], Ch. 12) of the local problem of the calculus of variations into a form best suited to our problem. Let us first solve (3.3) in a very special case.

(4.1) LEMMA (Carathéodory). *Let $k(x, t)$ be an m-vector function of class C^1 in x, t. Write $u^o = k(x, t)$. Assume $\nu = 0$ and that the function L in (3.4) satisfies the following conditions for all x and all $t_0 \leq t \leq t_1$:*

(a) $L(x, u^o, t) = 0$

(b) $L(x, u, t) > 0$ *for all* $u \neq u^o$

Then the optimal performance index V^o is identically zero for all x and is attained by using the optimal control law given by

$$u^o = k(x(t), t) \tag{4.2}$$

Proof. Let ϕ_{u^o} denote the motion of the plant under control law (4.2); similarly, let ϕ_{u^1} be the motion corresponding to some fixed control function $u^1(t)$. By (b) above and since the integrand of (3.4) is in class D^0 in t, it follows that $V(x, t_0, t_1; u^1) > 0$ unless $u^1(t) = u^o(t)$ at every continuity point of $u^1(t)$ in the interval (t_0, t_1). On the other hand, by (a) we see that $V^o(x, t_0, t_1)$ vanishes identically in x.[2] Q. E. D.

Anticipating the final result (4.14), let $V^o(x, t, t_1)$ be an arbitrary scalar function of class C^2 in x, t, (t_1 being a fixed number) and subject also to

$$V^o(x, t_1, t_1) = \nu(x) \tag{4.3}$$

Let us replace L by

[2] It is clear that two optimal controls $u^o(t)$ and $u^1(t)$ can differ only on a set of measure zero.

$$L^*(x, u, t) = L(x, u, t) + V_t^\circ(x, t, t_1) + [V_x^\circ(x, t, t_1), F(t)x + G(t)u]$$

The integral of the last two terms along any motion between the limits t_0, t_1 is

$$\nu(\phi_u(t_1 ; x, t_0)) - V^\circ(x, t_0, t_1) \tag{4.4}$$

Since the second term in (4.4) does not depend on $u(t)$, it follows that the two variational problems,

$$\inf_{u(t)} \int_{t_0}^{t_1} L^*(\phi_u(t; x, t_0), u(t), t)\, dt \tag{4.5}$$

and (3.3) are equivalent in that they have the same minimizing function $u(t)$ (if such exists).

Now we try to find functions $V^\circ(x, t, t_1)$ and $k(x, t, t_1)$ for which the hypotheses of the Lemma are satisfied when L is replaced by L^*.

In order that $L^*(x, u, t)$ have a minimum with respect to u at $u = u^\circ = k(x, t, t_1)$, it is necessary that all first partial derivatives of L^* with respect to u vanish at u°. This and the condition $L^*(x, u^\circ, t) = 0$ give

$$G'(t)V_x^\circ = -L_u(x, u^\circ, t) \tag{4.6}$$

$$-V_t^\circ = L(x, u^\circ, t) + [V_x^\circ, F(t)x + G(t)u^\circ] \tag{4.7}$$

These equations are called by Carathéodory the *fundamental equations* of the variational problem (3.3).

From assumption (3.5) it follows at once (by the strict convexity[3] of $L(x, u, t)$ in u) that (4.6) can be solved for u°; more precisely, there exists a function ψ of class C^1 such that

$$u^\circ = \psi(x, G'(t)V_x^\circ(x, t, t_1), t) = k(x, t, t_1) \tag{4.8}$$

which is the desired optimal control law.

To check condition (b) of the Lemma, we write, using (4.6–7),

$$\begin{aligned} L^*(x, u, t) &= L(x, u, t) - L(x, u^\circ, t) - [u - u^\circ, L_u(x, u^\circ, t)] \\ &= E(x, u, u^\circ, t) \end{aligned} \tag{4.9}$$

which is the well-known *Weierstrass E-function*. It is clear by inspection that E is the quadratic remainder in the Taylor series of L at $u = u^\circ$. Using the well-known estimate for the remainder, we have

$$E(x, u, u^\circ, t) = \| u - u^\circ \|^2_{L_{uu}(x, u+\theta(u^\circ-u), t)} \qquad (0 \le \theta \le 1) \tag{4.10}$$

which is nonnegative and in view of (3.5) vanishes if and only if $u = u^\circ$.

Hence if there is a function V° satisfying (4.3, 4.6–7) and in addition (3.5)

[3] A scalar function $\alpha(x)$ of a vector x is convex in x if and only if for all x_1, x_2 the function $\beta(\lambda) = \alpha(\lambda x_1 + (1 - \lambda)x_2)$ is convex in λ over the interval $0 \le \lambda \le 1$. This will be the case if and only if $d^2\beta/d\lambda^2 = [x_1 - x_2, \alpha_{xx}(x_1 - x_2)] \ge 0$. Hence $\alpha(x)$ is convex if and only if $\alpha_{xx} \ge 0$; similarly, $\alpha(x)$ is strictly convex if and only if $\alpha_{xx} > 0$.

holds, we can apply the Lemma and find that V^o is just the left-hand side of (3.4). The optimal control law is (4.8).

We now define the so-called *conjugate variable* ξ by

$$\xi = V_x^o \tag{4.11}$$

and write $\psi = \psi(x, G'(t)\xi, t)$. We define the *Hamiltonian* as

$$\mathcal{H}(x, \xi, t) = L(x, \psi, t) + [\xi, F(t)x + G(t)\psi] \tag{4.12}$$

It is easily shown that if L is of class C^2, so is also $\mathcal{H}$.

Now if $V^o(x, t, t_1)$ is any solution of class C^2 of (4.6–7), it follows by substitution that V^o is a solution of the Hamilton-Jacobi partial differential equation of the first order:

$$V_t^o + \mathcal{H}(x, V_x^o, t) = 0 \tag{4.13}$$

Conversely, let $V^o(x, t, t_1)$ (t_1 = parameter) be any solution of (4.13) of class C^2. Defining u^o by means of (4.8), it follows that V^o satisfies the fundamental equations (4.6–7).

In summary, we have

(4.14) THEOREM. *If there exists a solution $V^o(x, t, t_1)$ of class C^2 of the Hamilton-Jacobi equation (or, equivalently, of (4.6–7)) which satisfies $V^o(x, t_1, t_1) = \nu(x)$ and if (3.5) holds, then V^o is the optimal performance index for the regulator problem (3.3), and the corresponding optimal control law is given by (4.8).*

5. Controllability

The purpose of this section is to impose conditions on the plant (2.1) to assure that the problem posed by (3.3) is meaningful in the limit $t_1 = \infty$. Guided by physical intuition, we introduce the

(5.1) DEFINITION. A state x is said to be *controllable at time* t_0 if there exists a control function $u^1(t)$, depending on x and t_0 and defined over some finite closed interval $[t_0, t_1]$, such that $\phi_{u^1}(t_1; x, t_0) = 0$. If this is true for every state x, we say that the plant is *completely controllable at time* t_0; if this is true for every t_0, we say simply that the plant is *completely controllable.*

The following equivalent characterization of controllability is useful:

(5.2) PROPOSITION. *A plant is completely controllable at time* t (i) *if and* (ii) *only if the symmetric matrix*

$$W(t_0, t_1) = \int_{t_0}^{t_1} \Phi(t_0, t)G(t)G'(t)\Phi'(t_0, t)\, dt \tag{5.3}$$

is positive definite for some $t_1 > t_0$.

Proof. (i) Set

$$u^1(t) = -G'(t)\Phi'(t_0, t)W^{-1}(t_0, t_1)x \tag{5.4}$$

Substitution into (2.3) shows that $\phi_{u^1}(t_1; x, t_0) = 0$.

(ii) Suppose there exists some $x \neq 0$ such that $\| x \|^2_{W(t_0, t_1)} = 0$. Define

$$u^2(t) = -G'(t)\Phi'(t_0, t)x$$

which implies that

$$\| x \|^2_{W(t_0,t_1)} = \int_{t_0}^{t_1} \| u^2(t) \|^2 \, dt = 0$$

Since $u^2(t)$ is continuous in t, it is therefore identically zero in the interval $[t_0, t_1]$.

On the other hand, if the plant is completely controllable at t_0, there exists a control function $u^1(t)$ as required by (5.1) which satisfies the relation

$$x = -\int_{t_0}^{t_1} \Phi(t_0, t)G(t)u^1(t) \, dt$$

and therefore

$$\| x \|^2 = -\int_{t_0}^{t_1} [u^1(t), u^2(t)] \, dt = 0$$

contradicting the assumption that $x \neq 0$. Q. E. D.

(5.5) COROLLARY. *A constant plant is completely controllable* (i) *if and* (ii) *only if*

$$\text{rank } [G, FG, \cdots, F^{n-1}G] = n \tag{5.6}$$

(*where the square brackets denote a composite matrix of n rows and mn columns*) *in which case one may choose $t_1 - t_0 > 0$ as small as desired.*

Proof. Because of stationarity, controllability does not depend on t_0. Hence take $t_0 = 0$.

(i) By (5.2), it suffices to prove that $W(0, t_1)$ is positive definite no matter how small $t_1 > 0$. Let $g^1, \cdots, g^m$ be the columns of G. If $W(0, t_1)$ is semidefinite, then proceeding as in part (ii) of the proof of (5.2) we conclude that there is a vector $x \neq 0$ such that

$$[x, e^{Ft}g^i] = 0 \quad \text{for all} \quad 0 \leq t \leq t_1 \text{ and } i = 1, \cdots, m$$

Differentiating j times with respect to t, and then setting $t = 0$, we get

$$[x, F^j g^i] = 0 \quad \text{for all} \quad i = 1, \cdots, m \text{ and } j = 0, \cdots, n-1 \tag{5.7}$$

If (5.6) holds, this implies that x is orthogonal to a set of generators of E^n, contradicting the assumption that $x \neq 0$.

(ii) Assume the plant is completely controllable but (5.6) is false. Then there is a vector $x \neq 0$ which satisfies (5.7). By the Cayley-Hamilton theorem

$$[x, e^{Ft}g^i] = \left[x, \left(\sum_{j=0}^{\infty} (Ft)^j/j!\right) g^i\right] = \left[x, \left(\sum_{j=0}^{n-1} \alpha_j (Ft)^j\right) g^i\right], \qquad i = 1, \cdots, m$$

It follows that $\| x \|^2_{W(0,t_1)} = 0$ for all t_1, contradicting the assumption of complete controllability. Q. E. D.

Condition (5.6) has been used as a technical device in several recent papers in the theory of control ([12]–[14]), without reference to the "physical" interpretation (5.1).

(5.8) *Remark.* Let x be the state of the plant at t_0 and y the "desired" state at t_1. It follows easily by a slight extension of the preceding arguments that y is *reachable* from x (i.e., there exists a motion ϕ_{u^1} which meets x at t_0 and y at t_1) if and only if the equation

$$x - \Phi(t_0, t_1)y = W(t_0, t_1)v \tag{5.9}$$

has a solution, in which case

$$u^1(t) = -G'(t)\Phi'(t_0, t)v \tag{5.10}$$

is the appropriate control function.

Moreover, elementary methods of the calculus of variations show (see also [15]) that the minimum *control energy* required to achieve the transfer is

$$\mathcal{E}(x, t_0; y, t_1) = \int_{t_0}^{t_1} \| u^1(t) \|^2 \, dt = \| x - \Phi(t_0, t_1)y \|^2_{W^{-1}(t_0,t_1)} \tag{5.11}$$

Clearly, the required "energy" is zero if and only if the free motion going through x at t_0 intersects y at t_1.

Equation (5.9) may have a solution for some but not all x, y. Then W^{-1} does not exist and it is convenient to replace it with the *generalized inverse* $W\dagger$ in the sense of Penrose ([16], [17]). (See Appendix). With this convention, (5.11) *is the minimum energy required for transferring x as close to y as possible.*

If $W(t_0, t_1)$ is invertible, then (5.9) always has a solution; we see that *a plant is completely controllable at time t_0 if and only if starting from the origin at time t_0 any state x can be reached in a finite length of time by applying an appropriate control function $u(t)$.* In other words, there is a noteworthy "symmetry" between sending x to 0 and sending 0 to x.

(5.12) *Remark.* Using the generalized inverse, we may replace (5.10) by a control law defined in $[t_0, t_1]$:

$$u^1(t) = -G'(t)W\dagger(t, t_1)[x - \Phi(t, t_1)y]$$

Even if the plant is stationary, this control law is not. In fact, a stationary control law can be obtained in this case only by letting $u^1(t)$ be discontinuous ([8]).

The following definition is designed to single out a class of nonstationary plants which are in a sense "quasi-stationary". This will play an important role in the sequel.

(5.13) Definition. A plant is *uniformly completely controllable* if the following relations hold for all t:

(i) $0 < \alpha_0(\sigma)I \leq W(t, t + \sigma) \leq \alpha_1(\sigma)I$

(ii) $0 < \beta_0(\sigma)I \leq \Phi(t + \sigma, t)W(t, t + \sigma)\Phi'(t + \sigma, t) \leq \beta_1(\sigma)I$

where σ is a fixed constant. In other words, one can always transfer x to 0 and 0 to x in a finite length σ of time; moreover, such a transfer can never take place using an arbitrarily small amount (or requiring an arbitrarily large amount) of control energy.

Definition (5.13) has surprisingly far-flung consequences. We mention some of these; the proofs are elementary.

First of all, if (i–ii) hold, then, for all t,

$$\sqrt{\beta_0(\sigma)/\alpha_1(\sigma)} \leq \|\Phi(t + \sigma, t)\| \leq \sqrt{\beta_1(\sigma)/\alpha_0(\sigma)} \tag{5.14}$$

which is equivalent to

$$\sqrt{\alpha_0(\sigma)/\beta_1(\sigma)} \leq \|\Phi(t, t + \sigma)\| \leq \sqrt{\alpha_1(\sigma)/\beta_0(\sigma)} \tag{5.15}$$

(5.16) From formulas (5.3) and (5.14–15) we see that (i–ii) hold also for the constant $\sigma' = 2\sigma$; this implies further that (i–ii) hold for any $\sigma' \geq \sigma$.

(5.17) Using (5.16), we see that (i–ii) actually imply the following stronger bound on the transition matrix:

$$\|\Phi(t, \tau)\| \leq \alpha_3(|t - \tau|) \quad \text{for all } t, \tau \tag{5.18}$$

(5.19) It is now clear that if any two of the relations (5.18), (i), and (ii) hold, the remaining relation is also true.

The bound (5.18) obviously restricts the class of dynamical systems (2.1). Some such restriction appears to be an unavoidable consequence of any "reasonable" definition of uniform complete controllability. For instance, if only (i) holds, the following peculiar situation may arise. Consider the scalar system:

$$dx/dt = -tx + \sqrt{2(t - 1)e^{-t+1/2}}\,u(t)$$

(defined only for $t \geq 1$). We find easily that

$$\phi(t, \tau) = e^{(\tau^2 - t^2)/2}$$

which does not satisfy (5.18); furthermore,

$$w(t, t + \sigma) = e^{2(\sigma-1)t + (\sigma-1)^2} - e^{-2t+1}$$

and it is clear that w does not satisfy (i) unless $\sigma = 1$, while (ii) is never satisfied. In other words, to transfer x to 0 over an interval of time shorter than 1 may require an arbitrarily large amount of control energy, whereas doing the same job over an interval of time longer than 1 may require only a vanishingly small amount of energy as $t_0 \to \infty$. Transferring 0 to x will require more and more energy as $t_0 \to \infty$.

Finally, let us note a well-known and readily verifiable condition for (5.18) (easily proved using the Gronwall-Bellman lemma):

$$\int_{t_1}^{t_2} \|F(\tau)\| d\tau \leq \gamma(t_2 - t_1) \quad \text{for all } t_1, t_2 \tag{5.20}$$

We now seek to characterize a plant according to its "output" properties. This is most conveniently done as follows. Let $t^* = -t$ and $F^*(t^*) = F'(t)$, $G^*(t^*) = H'(t)$, and $H^*(t^*) = G'(t)$. Then

$$\begin{aligned} dx^*/dt^* &= F^*(t^*)x^* + G^*(t^*)u^*(t^*) \\ y^*(t^*) &= H^*(t^*)x^*(t^*) \end{aligned} \tag{5.21}$$

where x^*, u^*, y^* are n, p, and m vectors respectively, is the *dual plant* of (2.1–2). We shall not discuss the significance of this concept in detail (for which see [8]), except for pointing out that (i) the duality relations are reflexive if $t_0^{**} = t_0$; (ii) the transition matrix of (5.21) satisfies the relation

$$\Phi^*(t^*, \tau^*) = \Phi'(\tau, t) \qquad \text{for all } t, \tau \tag{5.22}$$

It is convenient to introduce the:

(5.23) Definition. A plant (2.1–2) is *uniformly completely observable* if its dual is uniformly completely controllable.

It follows easily from (5.23) that the explicit expression for W^* corresponding to (5.3) is

$$\begin{aligned} W^*(t^*, t_0^* + \sigma^*) &= W^*(t_0, t_0 - \sigma^*) \\ &= \int_{t_0-\sigma^*}^{t_0} \Phi'(t, t_0)H'(t)H(t)\Phi(t, t_0)\, dt \end{aligned} \tag{5.24}$$

Using W^* defined by (5.24), we can now state (5.23) explicitly. To avoid any possibility of confusion, the constants α, β, σ occurring in (i–ii) are to be replaced by α^*, β^*, σ^*.

6. Solution of the linear regulator problem

The point of view of the classical calculus of variations outlined in Section 4 is purely "local." At present, there are few global results and just about none in the theory of control. In the "local" (linear) case, however, the ideas of the preceding section lead to (what is hoped to become) a definitive theory of the regulator problem. This is the subject of the remainder of the paper.

To get the linear case of the regulator problem, it is not enough to have a linear model (2.1) for the plant but we need also the assumption:

(A_1) $\quad L(x, u, t) = \tfrac{1}{2}\{\| H(t)x \|^2_{Q(t)} + \| u \|^2_{R(t)}\}, \qquad \nu(x) = \tfrac{1}{2}\| x \|^2_A$

where A is symmetric, nonnegative definite while $Q(t)$, $R(t)$ are symmetric, positive definite and of class C^2 in t.

In view of (A_1), the Hamiltonian function (4.12) is

$$\mathcal{H}(x, \xi, t) = \tfrac{1}{2}\{\| H(t)x \|^2_{Q(t)} + 2[F(t)x, \xi] - \| G'(t)\xi \|^2_{R^{-1}(t)}\} \tag{6.1}$$

With this choice of $\mathcal{H}$, the function

$$V^\circ(x, t, t_1) = \tfrac{1}{2}\| x \|^2_{P(t, t_1)} \tag{6.2}$$

(t_1 = parameter) is a solution of the Hamilton-Jacobi equation (4.13) if and only if $P(t, t_1)$ is a solution of the following ordinary nonlinear differential equation of the Riccati type:

$$-\frac{dP}{dt} = F'(t)P + PF(t) - PG(t)R^{-1}(t)G'(t)P + H'(t)Q(t)H(t) \qquad (6.3)$$

It is clear that (6.2) determines P only up to a constant, skew-symmetric matrix (constancy follows from the fact that dP/dt is symmetric). Henceforth, to avoid trivia, we always assume that P is symmetric.

Given any symmetric, nonnegative definitive matrix A, (6.3) has a unique solution $\Pi(t; A, t_1)$ which takes on the value A at $t = t_1$. This solution is known to exist only in some neighborhood of t_1; without further analysis we cannot conclude existence for all t. (Because of the phenomenon of *finite escape time*, for which see [10], Example 3.)

Nonetheless, $\Pi(t; A, t_1)$ does exist for all $t \leq t_1$. We prove this indirectly as follows:

(6.4) EXISTENCE THEOREM. (i) *For all t_1 and all symmetric, nonnegative definite A,* (6.3) *has a unique solution $\Pi(t; A, t_1)$ defined for all $t \leq t_1$.* (ii) *Under Assumption* (A_1) *the optimal performance index for Problem* (3.3) *is given by*

$$V^{\circ}(x, t_0, t_1) = \| x \|^2_{\Pi(t_0; A, t_1)}$$

Moreover, the optimal performance index is attained if and only if the control law is given by

$$u^{\circ}(t) = R^{-1}(t)G'(t)\Pi(t; A, t_1)x(t) \qquad (6.5)$$

Proof. If we assume (i), then (ii) follows immediately from (4.14). Therefore, if $\Pi(t; A, t_1)$ exists, it must necessarily satisfy the relation

$$\| x \|^2_{\Pi(t_0; A, t_1)} \leq \int_{t_0}^{t_1} \| H(t)\Phi(t, t_0) x \|^2_{Q(t)} \, dt + \| \Phi(t_1, t_0)x \|^2_A$$
$$\leq \alpha(t_1, t_0) \| x \|^2$$

which follows by setting $u(t) \equiv 0$ in (3.4). Since $\alpha(t_1, t_0)$ is finite for all pairs t_1, t_0, it is clear that $\Pi(t; A, t_1)$ (if it exists) is contained in a compact region for all $t \in [t_0, t_1]$. Including this fact in the standard proof of the existence theorem for differential equations proves (i). Q. E. D.

In order to study the case $t_1 \to \infty$, we first define a particular solution of (6.3) which is of central significance for the ensuing development.

(6.6) PROPOSITION. *If the plant is completely controllable, then*

$$\lim_{t_1 \to \infty} \Pi(t; 0, t_1) = \bar{P}(t)$$

(i) *exists for all t and* (ii) *is a solution of* (6.3).

Proof. (i) Suppose the plant is completely controllable at $t = t_0$. Then for every x there exists a control function $u^1(t)$, given by (5.4), which transfers x to 0 at or before $t = t_2(x, t_0)$. We set $u^1(t) = 0$ for $t > t_2$. Then

$$\| x \|_{\Pi(t_0;0,t_1)} = V^o(x, t_0, t_1) \leq V(x, t_0, t_2; u^1) = V(x, t_0, \infty; u^1) \leq \alpha(t_0) \| x \|^2$$

which shows that $\| \Pi(t_0; 0, t_1) \|$ is bounded for all $t_1 \geq t_0$. On the other hand, (3.4) shows that $\| \Pi(t_0; 0, t_1) \|$ is nondecreasing as $t_1 \to \infty$. Hence the desired limit exists for arbitrary $t = t_0$. Q. E. D.

(ii) Using the continuity of solutions of (6.3) with respect to initial conditions, we have

$$\bar{P}(t) = \lim_{t_2 \to \infty} \Pi(t; 0, t_2) = \lim_{t_2 \to \infty} \Pi(t; \Pi(t_1; 0, t_2), t_1)$$

$$= \Pi(t; \lim_{t_2 \to \infty} \Pi(t_1; 0, t_2), t_1) = \Pi(t; \bar{P}(t_1), t_1)$$

which shows that $\bar{P}(t)$ is a solution of (6.3) which is defined for all t. Q. E. D.

(6.7) EXISTENCE THEOREM. *Assuming* (A_1), $\nu(x) = 0$, *and* $t_1 = \infty$, *the optimal performance index for Problem* (3.3) *is* $\| x \|^2_{\bar{P}(t)}$ *and the optimal control law is*

$$u^o(t) = R^{-1}(t)G'(t)\bar{P}(t)x(t) \tag{6.8}$$

Proof. Assume throughout the $\nu(x) = 0$. First we show: *If* $u(t)$ *is determined by the control law* (6.7), *the corresponding performance index is*

$$V(x, t_0, \infty; u^o) = \lim_{t_1 \to \infty} V(x, t_0, t_1; u^o) = \| x \|^2_{\bar{P}(t_0)}$$

We see from (6.4) and (6.6) that

$$V(x, t_0, t_1; u^o) = \| x \|^2_{\bar{P}(t_0)} - \| \phi_{u^o}(t_1; x, t_0) \|^2_{\bar{P}(t_1)} \leq \| x \|^2_{\bar{P}(t_0)}$$

On the other hand,

$$V(x, t_0, t_1; u^o) \geq V^o(x, t_0, t_1) = \| x \|^2_{\Pi(t_0;0,t_1)} \geq \| x \|^2_{\bar{P}(t_0)} - \epsilon$$

where $\epsilon \to 0$ as $t_1 \to \infty$, which proves (6.9). Hence

$$V^o(x, t_0, \infty) \leq V(x, t_0, \infty; u^o)$$

The inequality sign cannot arise. For if $V(x, t_0, \infty; u^o) - V^o(x, t_0, \infty) \geq \eta > 0$, there is some control function u^1 such that $V(x, t_0, \infty; u^o) - V(x, t_0, \infty, u^1) \geq \eta/2$. For t_1 sufficiently large, we then have

$$V(x, t_0, \infty; u^o) = V^o(x, t_0, t_1) + \eta/4 \geq V(x, t_0, t_1; u^1) + \eta/2$$

which is a contradiction and everything is proved. Q. E. D.

In the engineering literature it is often assumed (tacitly and incorrectly) that a system with optimal control law (6.8) is necessarily stable. We now give rigorous sufficient conditions insuring uniform asymptotic stability and point out in the process of proof some trivial but interesting parallels between the calculus of variations and the second method of Lyapunov.

The following definition is standard [10], [18]: The system (2.1) is *uniformly asymptotically stable* if (i) $\| \Phi(t, t_0) \| \leq \alpha$ and (ii) $\| \Phi(t, t_0) \| \to 0$ with $t \to \infty$ uniformly in t_0. It can be shown ([10], Theorem 3] that uniform asymptotic stability in the linear case is equivalent to *exponential asymptotic stability,* which is defined by the condition $(\alpha, \beta > 0)$

$$\| \Phi(t, t_0) \| \leq \alpha \exp[-\beta(t - t_0)] \quad \text{for all } t_0 \text{ and all } t \geq t_0 .$$

(6.10) STABILITY THEOREM. *Consider a plant with control law* (6.8) *which is uniformly completely controllable and uniformly completely observable. In addition to* (A_1), *assume also*

$$(\mathrm{A}_2) \qquad Q(t) \geq \alpha_4 I > 0, \qquad R(t) \geq \alpha_5 I > 0$$

$$(\mathrm{A}_3) \qquad Q(t) \leq \alpha_6 I, \qquad R(t) \leq \alpha_7 I$$

Then the controlled plant is uniformly asymptotically stable and $V^o(x, t, \infty)$ *is one of its Lyapunov functions.*

Proof. As is well known, it suffices to prove that (a) V^o is bounded from above and (b) below by increasing functions of $\| x \|$ independent of t, (c) the derivative $\dot{V}^o$ of V^o along optimal motions of the plant is negative definite [10, 18], and (d) $V^o \to \infty$ with $\| x \| \to \infty$.

(a) By uniform complete controllability, let $u^1(t)$ be the control function, depending on x, t_0 and defined in $[t_0, t_0 + \sigma]$ (σ = positive constant), which transfers x to 0 at or before $t = t_0 + \sigma$. In accordance with the remarks following (5.13), there is no loss of generality in taking the constants σ and σ^* (occurring in the definition of uniform complete controllability and uniform complete observability) to be the same. Having set $\sigma = \sigma^*$, we let $t_1 = t_0 + \sigma$. If $u^1(t)$ is defined explicitly by means of (5.4), then

$$\begin{aligned} \phi_{u^1}(t; x, t_0) &= \Phi(t, t_0)[I - W(t_0, t)W^{-1}(t_0, t_1)]x \\ &= \Phi(t, t_1)z(t) \end{aligned} \tag{6.11}$$

From the definition of W (see (5.3)), it follows easily[4] that the norm of the bracketed term above is less than or equal to 1. Using (5.18) then gives

$$\| z(t) \| \leq \alpha_8 \| x \|$$

where α_8 depends only on σ and it is therefore constant. By (6.11) and (A_3) we get

$$V^o(x, t_0, \infty) \leq \int_{t_0}^{t_1} \{\alpha_6 \| H(t)\Phi(t, t_1)z(t) \|^2 + \alpha_7 \| u^1(t) \|^2\} \, dt \tag{6.12}$$

[4] We need to show only that if $B > 0$ and $B \geq A \geq 0$, then $\| AB^{-1} \| \leq 1$. Now $\| AB^{-1} \|^2 = \lambda_{\max}(B^{-1}A^{-2}B^{-1}) = \lambda_{\max}(A^2B^{-2})$. By the well-known theorem about simultaneous diagonalization of a positive definite and a symmetric matrix, we have the representation $A^2 = T'\Lambda T$, $B^2 = T'T$, where T is nonsingular and Λ is diagonal. $B \geq A \geq 0$ implies $1 \geq \lambda(\Lambda) \geq 0$. Hence $\lambda_{\max}(A^2B^{-2}) = \lambda_{\max}(\Lambda) \leq 1$.

Making use of the foregoing and of the elementary inequality

$$\| Ax \|^2 \leq \| A \|^2 \cdot \| x \|^2 = \lambda_{\max}(A'A) \| x \|^2 \leq (\operatorname{tr} A'A) \| x \|^2$$

(valid for any matrix A and any vector x), (6.12) becomes

$$V^{o}(x, t_0, \infty) \leq \alpha_6 \alpha_8 (\operatorname{tr} W^*(t_1, t_0)) \| x \|^2 + \alpha_7 \| x \|^2_{W^{-1}(t_0, t_1)}$$

and by uniform complete controllability and observability we have finally

$$V^{o}(x, t_0, \infty) \leq [n \alpha_6 \alpha_8 \overset{*}{\alpha}_1(\sigma) + \alpha_7 \alpha_0(\sigma)] \| x \|^2 = \alpha_9 \| x \|^2.$$

(b) In view of (6.7), we can define

$$\inf_x V^{o}(x, t_0, \infty) = \alpha_{10}(t_0) \| x \|^2$$

We show that $\alpha_{10}(t_0) \geq \alpha_{11} > 0$. In fact, in the contrary case we can make ϵ, defined by

$$\| x \|^2 \epsilon(x, t) = \int_{t_0}^{\infty} \| u^{o}(t) \|^2 \, dt \leq \alpha_5^{-1} \int_{t_0}^{\infty} \| u^{o}(t) \|^2_{R(t)} \, dt \leq V^{o}(x, t_0, \infty)$$

as small as desired by suitable choice of x, t_0. We introduce the abbreviation

$$z(t) = \int_{t_0}^{t_1} \Phi(t_0, t) G(t) u^{o}(t) \, dt$$

and note that, by the Schwarz inequality,

$$\| z(t) \|^2 \leq \left(\int_{t_0}^{t_1} \| \Phi(t_0, t) G(t) \|^2 \, dt \right) \left(\int_{t_0}^{t_1} \| u^{o}(t) \|^2 \, dt \right)$$

and by uniform complete controllability,

$$\| z(t) \|^2 \leq n \alpha_1(\sigma) \epsilon(x, t_0) \| x \|^2.$$

Utilizing this estimate, we find with the aid of (A_2):

$$V^{o}(x, t_0, \infty) \geq \int_{t_0}^{t_1} \alpha_4 \| H(t) \Phi(t, t_0) [x + z(t)] \|^2 \, dt$$

$$\geq \int_{t_0}^{t_1} \alpha_4 \{ \| H(t) \Phi(t, t_0) x \|^2 - \| H(t) \Phi(t, t_0) z(t) \|^2 \} \, dt$$

$$\geq \alpha_4 \{ \| \Phi(t_1, t_0) x \|^2_{W^*(t_1, t_0)} - n \alpha_1(\sigma) \epsilon(x, t_0) [\operatorname{tr} W^*(t_1, t_0)] \| x \|^2 \}$$

By (5.18) and uniform complete observability, this reduces to

$$V^{o}(x, t_0, \infty) \geq \alpha_4 [\alpha_3^{-2}(\sigma) \overset{*}{\alpha}_0(\sigma) - n^2 \alpha_1(\sigma) \overset{*}{\alpha}_1(\sigma) \epsilon(x, t_0)] \| x \|^2$$

$$\geq [\alpha_{13} - \alpha_{14} \epsilon(x, t_0)] \| x \|^2$$

which contradicts the assumption that α_{12} (and hence ϵ) can be made arbitrarily small by suitable choice of x, t_0.

(c) Since G and H are allowed to be singular, we cannot prove of course that V is negative definite. However, inspection of the last step of (b) yields the further inequality,

$$V^o(\phi_{u^o}(t_1\,;x,t_0),t_1,\infty) - V^o(x,t_0,\infty) \leq -[\alpha_{13} - \alpha_{14}\,\epsilon\,(x,t_0)]\|\,x\,\|^2$$

and we have simultaneously also the further inequality

$$V^o(\phi_{u^o}(t_1\,;x,t_0),t_1,\infty) - V^o(x,t_0,\infty) \leq \alpha_5\,\epsilon\,(x,t_0)\|\,x\,\|^2$$

which follows immediately by (A_2) and the definition of V^o. Setting

$$\alpha_{15} = \alpha_5\alpha_{13}/(\alpha_5 + \alpha_{14}) > 0$$

we have finally that

$$V^o(\phi_{u^o}(t_0+\sigma;x,t_0),t_0+\sigma,\infty) - V^o(x,t_0,\infty) \leq -\alpha_{15}\,\|\,x\,\|^2$$

which shows that V^o is strictly decreasing along any interval of time of length σ_2, unless $x = 0$. Taking account of this fact, the proof of Lyapunov's theorem on uniform asymptotic stability ([10]) goes through as usual.

(d) This is trivial in view of $V^o \geq \alpha_{12}\,\|\,x\,\|^2$. Q. E. D.

It is of some interest to observe that if we have merely complete controllability, part (a) does not go through but we have nevertheless proved (nonuniform) asymptotic stability.

7. Stability of the Riccati equation

We now turn again to (6.3) and examine briefly its stability properties. Let $\delta P(t) = P(t) - \bar{P}(t)$ denote the deviation of a given motion $P(t)$ of (6.3) from $\bar{P}(t)$. Substituting into (6.3) shows that

$$d(\delta P)/dt = -\bar{F}'(t)\delta P - \delta P\bar{F}(t) - \delta PG(t)R^{-1}(t)G'(t)\delta P \tag{7.1}$$

where

$$\bar{F}(t) = F(t) - G(t)R^{-1}(t)G'(t)\bar{P}(t)$$

For simplicity, we temporarily drop the argument t in G, H, P, Q, R.

(7.2) STABILITY THEOREM. *Let*

$$\mathcal{V}(\delta P, t) = \tfrac{1}{2}\,\mathrm{tr}\,(\delta P\bar{P}^{-1})^2. \tag{7.3}$$

Then (i) *the derivative of* $\mathcal{V}$ *along motions of* (7.1) *is*

$$\begin{aligned}&\dot{\mathcal{V}}(\delta P, t)\\ &= tr\,\{(P^{\frac{1}{2}}GR^{-1}G'P^{\frac{1}{2}})\cdot(P^{\frac{1}{2}}\bar{P}^{-1}P^{\frac{1}{2}} - 1)^2 + (\bar{P}^{-\frac{1}{2}}H'QH\bar{P}^{-\frac{1}{2}})(\bar{P}^{-\frac{1}{2}}P\bar{P}^{-\frac{1}{2}} - 1)^2\}\end{aligned} \tag{7.4}$$

provided $P \geq 0$;

(ii) *If* $A \geq 0$, *then under the hypotheses of* (6.10), *all solutions* $\Pi(t; A, t_1)$ *of* (6.3) *are uniformly asymptotically stable relative to* $\bar{P}(t)$ *as* $t \to -\infty$, *and* $\mathcal{V}$ *is an appropriate Lyapunov function.*

Proof. (i) This is established by lengthy, elementary calculations. The square root of P exists by assumption and that of $\bar{P}$ by part (b) of (6.10).

(ii) Clearly, $\mathcal{V}$ vanishes if and only if $\delta P = 0$. We recall the fact that for any symmetric, $n \times n$ matrices A, B,

$$\lambda_{\min}(B)\lambda_i(A^2) \leq \lambda_i(ABA) \leq \lambda_{\max}(B)\lambda_i(A^2) \qquad (i = 1, \cdots, n) \quad (7.5)$$

where the λ_i are eigenvalues. This is a consequence of the Fischer-Courant variational description of eigenvalues (for which see [20], p. 115 Theorem 3 and p. 120, Exercise 9). Using (7.5) and the results of (6.10) it follows easily that

$$0 < \alpha \operatorname{tr} (\delta P)^2 \leq \mathcal{V}(\delta P, t) \leq \beta \operatorname{tr} (\delta P)^2, \qquad \delta P \neq 0$$

Moreover, (7.4) being the trace of a nonnegative definite matrix, $\dot{\mathcal{V}}$ is clearly nonnegative. By arguments analogous to part (c) of the proof of (6.10), it follows then also that $\mathcal{V}$ is uniformly decreasing along any motion of (6.3) as as $t \to -\infty$.

(7.6) Corollary. *The motion $\bar{P}(t)$ is unstable (as $t \to \infty$).*

Proof. Immediate consequence of part (ii) of the proof of (7.2).

(7.7) *Remark.* If the problem is stationary, i.e., F, G, H, Q, R are constants, $P(t, t_1) = P(t + h, t_1 + h)$ which shows that $dP(t, t_1)/dt_1 = -dP(t, t_1)/dt$. Hence in this case one can compute $P(t) =$ const. from (6.3) by replacing t by $-t$; for any initial $A \geq 0$, this computation is asymptotically stable in the large.

(7.8) *Remark.* Because of the Corollary, in the nonstationary case (at least one of F, G, H, Q, R not constant), one cannot compute $P(t)$ as $t \to \infty$ from the knowledge of $P(t_0)$.

8. General solution of the Riccati equation

Consider the canonic (Hamiltonian) differential equations associated with (6.1):

$$dx/dt = \mathcal{H}_\xi(x, \xi, t) = F(t)x - G(t)R^{-1}(t)G'(t)\xi \qquad (8.1)$$

$$d\xi/dt = -\mathcal{H}_x(x, \xi, t) = -H'(t)Q(t)H(t)x - F'(t)\xi \qquad (8.2)$$

Let $P(t)$ be a solution of (6.3), defined in some interval $U = (-\infty, t_2)$. In view of (4.11) and (6.2), we assume that the initial conditions of (8.1–2) at time t_1 are related by ($t_1 < t_2$!)

$$\xi(t_1) = V_x^o(x(t_1), t_1) = P(t_1)x(t_1)$$

Then the same relation will hold between solutions of (8.1–2) corresponding to these initial conditions, for all t that $P(t)$ exists:

$$\xi(t) = V_x^o(x(t), t) = P(t)x(t), \qquad t \in U \quad (8.3)$$

We can also verify (8.3) directly by substituting (6.3) into (8.1–2).

Now let $X(t)$, $\Xi(t)$ be a pair of matrix solutions of (8.1–2) satisfying the initial conditions $X(t_1) = I$, $\Xi(t_1) = P(t_1)$. By (8.3) we have obviously

$$\Xi(t) = P(t)X(t), \qquad t \in U \quad (8.4)$$

which shows that $X(t)$ is a solution of the matrix differential equation

$$dX/dt = [F(t) - G(t)R^{-1}(t)G'(t)P(t)]X, \qquad t \in U$$

Setting $\nu(x) = \| x \|^2_{P(t_1)}$ in (3.4), we see from (6.4–5) that $X(t)$ is the transition matrix $\Phi^o(t, t_1)$ of the optimally controlled plant corresponding to this choice of ν. Since $\Phi^o(t, t_1)$ exists for all $t \in U$, we have

$$P(t) = \Xi(t)\Phi^o(t_1, t), \qquad t \in U \quad (8.5)$$

To obtain an explicit expression for $P(t)$, let

$$\Theta(t, t_1) = \begin{pmatrix} \Theta_{11}(t, t_1) & \Theta_{12}(t, t_1) \\ \Theta_{21}(t, t_1) & \Theta_{22}(t, t_1) \end{pmatrix}$$

be the transition matrix of the system (8.1–2). We get the following formula valid for $t \in U$:

$$P(t) = [\Theta_{21}(t, t_1) + \Theta_{22}(t_1, t)P(t_1)][\Theta_{11}(t, t_1) + \Theta_{12}(t, t_1)P(t_1)]^{-1}$$

This procedure is very well known in the calculus of variations ([21; 11, Ch. 15]) and is being periodically rediscovered ([22], [23]).

RIAS, Baltimore, Maryland

Bibliography

[1] N. Wiener, The Extrapolation, Interpolation, and Smoothing of Stationary Time Series, Wiley, New York, N. Y., 1949.

[2] A. C. Hall, The Analysis and Synthesis of Linear Servomechanisms, The Technology Press, M. I. T., Cambridge, Mass., 1943.

[3] G. C. Newton, Jr., L. A. Gould, and J. F. Kaiser, Analytical Design of Linear Feedback Controls, Wiley, New York, N. Y., 1957.

[4] R. E. Kalman and J. E. Bertram, *General synthesis procedure for computer control of single and multi-loop linear systems*, Trans. Am. Inst. Elect. Engr., 77 II (1958) 602–609.

[5] R. E. Kalman and R. W. Koepcke, *Optimal synthesis of linear sampling control systems using generalized performance indexes*, Trans. Am. Soc. Mech. Engr., 80 (1958) 1820–1826.

[6] R. E. Kalman and R. W. Koepcke, *The role of digital computers in the dynamic optimization of chemical reactions*, Proc. Western Joint Comp. Conf., 1959, 107–116.

[7] R. E. Kalman, *A new approach to linear filtering and prediction problems*, J. Basic Engr., 82 D (1960) 35–45.

[8] R. E. Kalman, *On the general theory of control systems*, Proc. First Internat'l Congress on Automatic Control, Moscow, 1960.

[9] E. A. Coddington and N. Levinson, Theory of Ordinary Differential Equations, McGraw-Hill, New York, N. Y., 1955.

[10] R. E. Kalman and J. E. Bertram, *Control system analysis and design via the 'second method' of Lyapunov. I. Continuous-time systems*, J. Basic Engr., 82 D (1960).

[11] C. Carathéodory, Variationsrechnung und Partielle Differentialgleichungen Erster Ordnung, Teubner, Leipzig, 1935.

[12] L. S. PONTRYAGIN, *Optimal control processes* (in Russian), Uspekh. Mat. Nauk, 14 (1959) 3–20.
[13] N. N. KRASOVSKII, *On a problem of optimal control of nonlinear systems*, Prikl. Math. Mekh., 23 (1959) 209–229 (translation pp. 303–332).
[14] J. P. LASALLE, *Time optimal control systems*, Proc. Nat. Academy of Sci. USA, 45 (1959) 573–577.
[15] J. E. BERTRAM AND P. E. SARACHIK, *On optimal computer control*, Proc. First Internat'l Congress on Automatic Control, Moscow, 1960.
[16] R. PENROSE, *A generalized inverse for matrices*, Proc. Cambridge Phil. Soc., 51 (1955) 406–413.
[17] R. PENROSE, *On best approximate solutions of linear matrix equations*, Proc. Cambridge Phil. Soc., 52 (1956) 17–19.
[18] W. HAHN, Theorie und Anwendung der Direkten Methode von Lyapunov, Ergebnisse der Mathematik No. 22, Springer, Berlin, 1959.
[19] R. BELLMAN, Stability Theory of Differential Equations, McGraw-Hill, New York, N. Y., 1953.
[20] R. BELLMAN, Introduction to Matrix Analysis, McGraw-Hill, New York, N. Y., 1960.
[21] J. RADON, *Zum Problem von Lagrange*, Abh. Math. Sem. Univ. Hamburg, 6 (1928) 273–299.
[22] W. T. REID, *A matrix differential equation of the Riccati type*, Am. J. Math., 68 (1946) 237–246.
[23] J. J. LEVIN, *On the matrix Riccati equation*, Trans. Am. Math. Soc., 10 (1959) 519–524.

Appendix: The generalized inverse of a matrix

Following Penrose ([16]), the *generalized inverse* of an arbitrary square matrix A is a matrix $A\dagger$ satisfying the relations:

$$\text{(i) } AA\dagger A = A, \qquad \text{(ii) } A\dagger AA\dagger = A\dagger,$$

$$\text{(iii) } (A\dagger A)' = A\dagger A, \qquad \text{(iv) } (AA\dagger)' = AA\dagger$$

It can be shown that $A\dagger$ always exists and is uniquely determined by these relations. Examples: (1) If D is diagonal, then the elements of its generalized inverse are

$$\begin{aligned} d_{ii}^{\dagger} &= d_{ii}^{-1} \quad \text{if } d_{ii} \neq 0 \\ &= 0 \text{ otherwise} \end{aligned}$$

(2) If A is symmetric, there is an orthogonal transformation T such that $A = T'DT$. Then $A\dagger = T'D\dagger T$.

Consider now the linear equation $Ax = y$. Penrose proves ([17]) that the "best approximate solution" $x^o = A\dagger y$ of this equation has the properties:

(i) $\| Ax - y \| \geq \| Ax^o - y \|$ for all x

(ii) If $\| Ax - y \| = \| Ax^o - y \|$, then $\| x \| \geq \| x^o \|$

A New Approach to Linear Filtering and Prediction Problems

R. E. KALMAN

FILTERING is the process of estimating the current value of a discrete- or continuous-time stochastic signal, using the history and the current value of another (observed) stochastic process (so-called measurement process) which is correlated with it; *prediction* refers to a similar process where now the objective is to estimate the value of the stochastic process at a future point in time. In the 1950s, in the years that preceded the publication of this paper, the prevailing approach to filtering and prediction was the one provided by the Wiener-Kolmogorov theory, developed in the 1940s (see the preamble to chapter III of [13], included in this volume). The Wiener-Kolmogorov theory assumed that the signal and the measurement were jointly stationary stochastic processes, and it involved the solution of a Wiener-Hopf equation (an integral equation for the continuous-time problem, and an appropriately defined counterpart in the discrete time), which was not that easy to solve, even computationally. Hence, it was not a useful tool for engineers, to apply to practical problems. There was a need for a more practical theory, which would handle stationary as well as nonstationary (time-varying) signals, and allow for real-time recursive computation of the estimators.

This paper by Kalman was one of the first few papers to respond to this need, and could easily be considered as **the** paper that marked the beginning of a new era in filtering and prediction. Besides filtering and prediction, it also had a big impact on the **control** of dynamic systems under uncertainty and using imperfect state measurements. The paper was written during a period when the "state space approach" took over control theory, with its two distinctive features, namely its focus on the time domain, and the use of optimization as the principal tool for design. And, this movement is definitely reflected in the paper, both in the formulation of the problem and in the solution process. Motivated by the earlier work of Bode and Shannon [1], Kalman focused (in the discrete time) on signals generated by finite-dimensional linear dynamic systems excited by white noise, namely the discrete-time filtering and prediction problems with linear state equation [see Equation (16) in the paper]

$$x(t+1) = \Phi(t+1; t)x(t) + u(t),$$

and linear measurement equation [see Equation (17) of the paper]

$$y(t) = M(t)x(t),$$

where $u(t)$ is a white Gaussian noise, and $\Phi(\cdot, \cdot)$ and $M(\cdot)$ are known matrices, with the dimension (p) of y generally lower than that (n) of x. It is noteworthy that the formulation takes the nonstationary case head-on, but that the measurement equation does not include the white noise measurement error term that has become standard afterwards. In the discrete-time case it is, of course, not necessary to include this term (as any formulation with such a term can be transformed into the form above by introducing additional state variables), in contrast to the continuous-time case where it is essential.

In the paper, Kalman introduces orthogonal projection as a basic tool in linear optimal estimation—an idea that was little known in the systems and control community at the time. This leads to a recursive implementation of the optimal filter (least-squares estimator of the state) in the form of a linear dynamic system of the same dimension as x, driven by the measurement process, and an off-line recursive computation of the error covariance matrix, which determines the 'gain' in the filter equation. As noted by Sorenson in his 1970 historical treatise of least-squares estimation [9], the recursive filter of Kalman in this paper can also be "rightfully regarded as an efficient computational solution of the least-squares method" of Gauss [3].

The paper does not address the issue of stability of the optimal filter. Kalman points out, however, the duality of the filtering problem and the regulation problem, which in later papers (see, e.g., [5], [6]) he astutely exploits to establish his well-known stability and convergence results. There is also an interesting paragraph (Remark *j* on p. 23) about the continuous-time problem that apparently Kalman attempted to study as a limiting case of the discrete-time problem, which he later pursued further in [6]. He states that the continuous-time optimal filter has $n - p$ state variables rather than n. We now know that this is because of the missing white measurement noise that makes the continuous-time problem singular. The regular (nonsingular) continuous-time filtering problem was addressed by Kalman and Bucy a year later in [7], using a direct approach, which of course leads to the full (n) dimensional filter, with the 'gain' generated again off-line by solving an n-dimensional Riccati differential equation (or algebraic Riccati equation, in the stationary case). Another inspiring derivation of the same result was given later by Kailath using the innovations approach [4], which also found extensions to *nonlinear* filtering [2].

Recursive filtering results of the same type as Kalman's were obtained independently at about the same time by Swerling [12] in the United States and Stratonovic [11] in the USSR. Swerling's work, in fact, predates that of Kalman, but it is more application oriented, with the main motivation being the need to come up with a practical method for smoothing satellite data. What he came up with in this specific context, however, was essentially what is known today as the extended Kalman filter (as he did not start with a linear model, but considered a Taylor series expansion with respect to the estimate). In spite of these parallel works, this 1960 paper by Kalman and its 1961 continuous-time counterpart by Kalman and Bucy [7] are the two that attained instant fame, mainly because of the appealing state space modeling they have introduced for the stochastic signals (using difference or differential equations), and the computationally efficient schemes they have developed for optimal filters and predictors. They both are widely cited today for establishing the modern version of filtering, estimation, and prediction of stochastic systems. Their impact on the fields of linear systems theory, statistics, signal processing, identification, feedback control, and adaptive systems has been enormous, not only on the application side (it is noteworthy that many regard Kalman filtering an indispensible tool contributing to the success of the moon mission in 1970), but also in motivating new formulations of feasible filtering problems outside the linear domain. For example in nonlinear filtering, state space models involving nonlinear stochastic differential equations driven by Brownian motion have constituted a standard paradigm, with optimal filters sought as dynamic sytems driven by the innovations process [8]. Several other extensions and applications of Kalman filtering can be found in the 1985 IEEE Press volume [10] (of collection of papers) devoted to that topic.

References

[1] H.W. Bode and C.E. Shannon, "A simplified derivation of linear least-squares smoothing and prediction theory," *Proc. IRE*, **38**:417–425, 1950.

[2] P. Frost and T. Kailath, "An innovations approach to least-squares estimation. Part III: Nonlinear estimation in additive white noise," *IEEE Trans. Aut. Contr.*, **AC-16**(6):217–226, 1971.

[3] K.G. Gauss, *Theory of Motion of the Heavenly Bodies*, Dover (New York), 1963. (English translation of *Theoria Motus Corporum Coelestium*, 1809.)

[4] T. Kailath, "An innovations approach to least-squares estimation. Part I: Linear filtering in additive white noise," *IEEE Trans. Aut. Contr.*, **AC-13** (6):646–654, 1968.

[5] R.E. Kalman, "On the general theory of control systems," in *Proc. 1st IFAC Congress on Automatic Control*, Moscow, USSR, 1960.

[6] R.E. Kalman, "New methods of Wiener filtering theory," in *Proc. 1st Symp. Engineering Appl. Random Function Theory and Probability*, J.L. Bogdanoff and F. Kozin, edts., Wiley (New York), 1963, pp. 270–385.

[7] R.E. Kalman and R.S. Bucy, "New results in linear filtering and prediction theory," *Trans. ASME, J. Basic Engineering*, **83D**(1):95–108, March 1961.

[8] S.K. Mitter, "Filtering and stochastic control: a historical perspective," *IEEE Control Systems (Magazine)*, **16**(3):67–76, June 1996.

[9] H.W. Sorenson, "Least-squares estimation: from Gauss to Kalman," *IEEE Spectrum*, **7**:63–68, July 1970.

[10] H.W. Sorenson, ed., *Kalman Filtering: Theory and Application*, IEEE Press (New York), 1985.

[11] R.L. Stratonovich, "Application of the theory of Markoff processes in optimal signal detection," *Radio Eng. Electron. Phys.*, **1**:1–19, 1960.

[12] P. Swerling, "A proposed stagewise differential correction procedure for satellite tracking and prediction," *J. Astronaut. Sci.*, vol. 6, 1959. (First appeared as Rept. P-1292, RAND Corp., Santa Monica, CA, January 1958.)

[13] N. Wiener, *Extrapolation, Interpolation, and Smoothing of Stationary Time Series*, MIT Press (Cambridge, MA), 1949.

H.K. & T.B.

A New Approach to Linear Filtering and Prediction Problems[1]

R. E. KALMAN
Research Institute for Advanced Study,[2]
Baltimore, Md.

The classical filtering and prediction problem is re-examined using the Bode-Shannon representation of random processes and the "state-transition" method of analysis of dynamic systems. New results are:

(1) The formulation and methods of solution of the problem apply without modification to stationary and nonstationary statistics and to growing-memory and infinite-memory filters.

(2) A nonlinear difference (or differential) equation is derived for the covariance matrix of the optimal estimation error. From the solution of this equation the coefficients of the difference (or differential) equation of the optimal linear filter are obtained without further calculations.

(3) The filtering problem is shown to be the dual of the noise-free regulator problem.

The new method developed here is applied to two well-known problems, confirming and extending earlier results.

The discussion is largely self-contained and proceeds from first principles; basic concepts of the theory of random processes are reviewed in the Appendix.

Introduction

An important class of theoretical and practical problems in communication and control is of a statistical nature. Such problems are: (i) Prediction of random signals; (ii) separation of random signals from random noise; (iii) detection of signals of known form (pulses, sinusoids) in the presence of random noise.

In his pioneering work, Wiener [1][3] showed that problems (i) and (ii) lead to the so-called Wiener-Hopf integral equation; he also gave a method (spectral factorization) for the solution of this integral equation in the practically important special case of stationary statistics and rational spectra.

Many extensions and generalizations followed Wiener's basic work. Zadeh and Ragazzini solved the finite-memory case [2]. Concurrently and independently of Bode and Shannon [3], they also gave a simplified method [2] of solution. Booton discussed the nonstationary Wiener-Hopf equation [4]. These results are now in standard texts [5–6]. A somewhat different approach along these main lines has been given recently by Darlington [7]. For extensions to sampled signals, see, e.g., Franklin [8], Lees [9]. Another approach based on the eigenfunctions of the Wiener-Hopf equation (which applies also to nonstationary problems whereas the preceding methods in general don't), has been pioneered by Davis [10] and applied by many others, e.g., Shinbrot [11], Blum [12], Pugachev [13], Solodovnikov [14].

In all these works, the objective is to obtain the specification of a linear dynamic system (Wiener filter) which accomplishes the prediction, separation, or detection of a random signal.[4]

---

[1] This research was supported in part by the U. S. Air Force Office of Scientific Research under Contract AF 49 (638)-382.

[2] 7212 Bellona Ave.

[3] Numbers in brackets designate References at end of paper.

[4] Of course, in general these tasks may be done better by nonlinear filters. At present, however, little or nothing is known about how to obtain (both theoretically and practically) these nonlinear filters.

Contributed by the Instruments and Regulators Division and presented at the Instruments and Regulators Conference, March 29–April 2, 1959, of The American Society of Mechanical Engineers.

Note: Statements and opinions advanced in papers are to be understood as individual expressions of their authors and not those of the Society. Manuscript received at ASME Headquarters, February 24, 1959. Paper No. 59—IRD-11.

Present methods for solving the Wiener problem are subject to a number of limitations which seriously curtail their practical usefulness:

(1) The optimal filter is specified by its impulse response. It is not a simple task to synthesize the filter from such data.

(2) Numerical determination of the optimal impulse response is often quite involved and poorly suited to machine computation. The situation gets rapidly worse with increasing complexity of the problem.

(3) Important generalizations (e.g., growing-memory filters, nonstationary prediction) require new derivations, frequently of considerable difficulty to the nonspecialist.

(4) The mathematics of the derivations are not transparent. Fundamental assumptions and their consequences tend to be obscured.

This paper introduces a new look at this whole assemblage of problems, sidestepping the difficulties just mentioned. The following are the highlights of the paper:

(5) *Optimal Estimates and Orthogonal Projections.* The Wiener problem is approached from the point of view of conditional distributions and expectations. In this way, basic facts of the Wiener theory are quickly obtained; the scope of the results and the fundamental assumptions appear clearly. It is seen that all statistical calculations and results are based on first and second order averages; no other statistical data are needed. Thus difficulty (4) is eliminated. This method is well known in probability theory (see pp. 75–78 and 148–155 of Doob [15] and pp. 455–464 of Loève [16]) but has not yet been used extensively in engineering.

(6) *Models for Random Processes.* Following, in particular, Bode and Shannon [3], arbitrary random signals are represented (up to second order average statistical properties) as the output of a linear dynamic system excited by independent or uncorrelated random signals ("white noise"). This is a standard trick in the engineering applications of the Wiener theory [2–7]. The approach taken here differs from the conventional one only in the way in which linear dynamic systems are described. We shall emphasize the concepts of *state* and *state transition;* in other words, linear systems will be specified by systems of first-order difference (or differential) equations. This point of view is

Reprinted with permission from *Transactions of the ASME, Journal of Basic Engineering*, R. E. Kalman, "A New Approach to Linear Filtering and Prediction Problems," Vol. 82D, March 1960, pp. 35–45.

natural and also necessary in order to take advantage of the simplifications mentioned under (5).

(7) *Solution of the Wiener Problem.* With the state-transition method, a single derivation covers a large variety of problems: growing and infinite memory filters, stationary and nonstationary statistics, etc.; difficulty (3) disappears. Having guessed the "state" of the estimation (i.e., filtering or prediction) problem correctly, one is led to a nonlinear difference (or differential) equation for the covariance matrix of the optimal estimation error. This is vaguely analogous to the Wiener-Hopf equation. Solution of the equation for the covariance matrix starts at the time t_0 when the first observation is taken; at each later time t the solution of the equation represents the covariance of the optimal prediction error given observations in the interval (t_0, t). From the covariance matrix at time t we obtain at once, without further calculations, the coëfficients (in general, time-varying) characterizing the optimal linear filter.

(8) *The Dual Problem.* The new formulation of the Wiener problem brings it into contact with the growing new theory of control systems based on the "state" point of view [17–24]. It turns out, *surprisingly*, that the Wiener problem is the *dual* of the noise-free optimal regulator problem, which has been solved previously by the author, using the state-transition method to great advantage [18, 23, 24]. The mathematical background of the two problems is identical—this has been suspected all along, but until now the analogies have never been made explicit.

(9) *Applications.* The power of the new method is most apparent in theoretical investigations and in numerical answers to complex practical problems. In the latter case, it is best to resort to machine computation. Examples of this type will be discussed later. To provide some feel for applications, two standard examples from nonstationary prediction are included; in these cases the solution of the nonlinear difference equation mentioned under (7) above can be obtained even in closed form.

For easy reference, the main results are displayed in the form of theorems. Only Theorems 3 and 4 are original. The next section and the Appendix serve mainly to review well-known material in a form suitable for the present purposes.

Notation Conventions

Throughout the paper, we shall deal mainly with *discrete* (or *sampled*) dynamic systems; in other words, signals will be observed at equally spaced points in time (*sampling instants*). By suitable choice of the time scale, the constant intervals between successive sampling instants (*sampling periods*) may be chosen as unity. Thus variables referring to time, such as t, t_0, τ, T will always be integers. The restriction to discrete dynamic systems is not at all essential (at least from the engineering point of view); by using the discreteness, however, we can keep the mathematics rigorous and yet elementary. Vectors will be denoted by small bold-face letters: $\mathbf{a}, \mathbf{b}, \ldots, \mathbf{u}, \mathbf{x}, \mathbf{y}, \ldots$ A *vector* or more precisely an *n-vector* is a set of n numbers $x_1, \ldots x_n$; the x_i are the *co-ordinates* or *components* of the vector $\mathbf{x}$.

Matrices will be denoted by capital bold-face letters: $\mathbf{A}, \mathbf{B}, \mathbf{Q}, \mathbf{\Phi}, \mathbf{\Psi}, \ldots$; they are $m \times n$ arrays of elements $a_{ij}, b_{ij}, q_{ij}, \ldots$ The *transpose* (interchanging rows and columns) of a matrix will be denoted by the prime. In manipulating formulas, it will be convenient to regard a vector as a matrix with a single column.

Using the conventional definition of matrix multiplication, we write the *scalar product* of two n-vectors $\mathbf{x}$, $\mathbf{y}$ as

$$\mathbf{x}'\mathbf{y} = \sum_{i=1}^{n} x_i y_i = \mathbf{y}'\mathbf{x}$$

The scalar product is clearly a scalar, i.e., not a vector, quantity. Similarly, the *quadratic form* associated with the $n \times n$ matrix $\mathbf{Q}$ is

$$\mathbf{x}'\mathbf{Q}\mathbf{x} = \sum_{i,j=1}^{n} x_i q_{ij} x_j$$

We define the expression $\mathbf{x}\mathbf{y}'$ where $\mathbf{x}$ is an m-vector and $\mathbf{y}$ is an n-vector to be the $m \times n$ matrix with elements $x_i y_j$.

We write $E(\mathbf{x}) = E\mathbf{x}$ for the expected value of the random vector $\mathbf{x}$ (see Appendix). It is usually convenient to omit the brackets after E. This does not result in confusion in simple cases since constants and the operator E commute. Thus $E\mathbf{x}\mathbf{y}'$ = matrix with elements $E(x_i y_j)$; $E\mathbf{x}E\mathbf{y}'$ = matrix with elements $E(x_i)E(y_j)$.

For ease of reference, a list of the principal symbols used is given below.

Optimal Estimates

t	time in general; present time.
t_0	time at which observations start.
$x_1(t)$, $x_2(t)$	basic random variables.
$y(t)$	observed random variable.
$x_1^*(t_1\|t)$	optimal estimate of $x_1(t_1)$ given $y(t_0), \ldots, y(t)$.
L	loss function (nonrandom function of its argument).
ϵ	estimation error (random variable).

Orthogonal Projections

$\mathcal{Y}(t)$	linear manifold generated by the random variables $y(t_0), \ldots, y(t)$.
$\bar{x}(t_1\|t)$	orthogonal projection of $x(t_1)$ on $\mathcal{Y}(t)$.
$\tilde{x}(t_1\|t)$	component of $x(t_1)$ orthogonal to $\mathcal{Y}(t)$.

Models for Random Processes

$\mathbf{\Phi}(t+1; t)$	transition matrix
$\mathbf{Q}(t)$	covariance of random excitation

Solution of the Wiener Problem

$\mathbf{x}(t)$	basic random variable.
$\mathbf{y}(t)$	observed random variable.
$\mathcal{Y}(t)$	linear manifold generated by $\mathbf{y}(t_0), \ldots, \mathbf{y}(t)$.
$\mathcal{Z}(t)$	linear manifold generated by $\tilde{\mathbf{y}}(t\|t-1)$.
$\mathbf{x}^*(t_1\|t)$	optimal estimate of $\mathbf{x}(t_1)$ given $\mathcal{Y}(t)$.
$\tilde{\mathbf{x}}(t_1\|t)$	error in optimal estimate of $\mathbf{x}(t_1)$ given $\mathcal{Y}(t)$.

Optimal Estimates

To have a concrete description of the type of problems to be studied, consider the following situation. We are given signal $x_1(t)$ and noise $x_2(t)$. Only the sum $y(t) = x_1(t) + x_2(t)$ can be observed. Suppose we have observed and know exactly the values of $y(t_0), \ldots, y(t)$. What can we infer from this knowledge in regard to the (unobservable) value of the signal at $t = t_1$, where t_1 may be less than, equal to, or greater than t? If $t_1 < t$, this is a *data-smoothing* (*interpolation*) problem. If $t_1 = t$, this is called *filtering*. If $t_1 > t$, we have a *prediction* problem. Since our treatment will be general enough to include these and similar problems, we shall use hereafter the collective term *estimation*.

As was pointed out by Wiener [1], the natural setting of the estimation problem belongs to the realm of probability theory and statistics. Thus signal, noise, and their sum will be random variables, and consequently they may be regarded as random processes. From the probabilistic description of the random processes we can determine the probability with which a particular sample of the signal and noise will occur. For any given set of measured values $\eta(t_0), \ldots, \eta(t)$ of the random variable $y(t)$ one can then also determine, in principle, the probability of simultaneous occurrence of various values $\xi_1(t)$ of the random variable $x_1(t_1)$. This is the conditional probability distribution function

$$Pr[x_1(t_1) \leqq \xi_1 | y(t_0) = \eta(t_0), \ldots, y(t) = \eta(t)] = F(\xi_1) \quad (1)$$

Evidently, $F(\xi_1)$ represents all the information which the measurement of the random variables $y(t_0), \ldots, y(t)$ has conveyed about the random variable $x_1(t_1)$. Any statistical estimate of the random variable $x_1(t_1)$ will be some function of this distribution and therefore a (nonrandom) function of the random variables $y(t_0), \ldots, y(t)$. This statistical estimate is denoted by $X_1(t_1|t)$, or by just $X_1(t_1)$ or X_1 when the set of observed random variables or the time at which the estimate is required are clear from context.

Suppose now that X_1 is given as a fixed function of the random variables $y(t_0), \ldots, y(t)$. Then X_1 is itself a random variable and its actual value is known whenever the actual values of $y(t_0), \ldots, y(t)$ are known. In general, the actual value of $X_1(t_1)$ will be different from the (unknown) actual value of $x_1(t_1)$. To arrive at a rational way of determining X_1, it is natural to assign a *penalty* or *loss* for incorrect estimates. Clearly, the loss should be a (i) positive, (ii) nondecreasing function of the *estimation error* $\epsilon = x_1(t_1) - X_1(t_1)$. Thus we define a *loss function* by

$$L(0) = 0$$

$$L(\epsilon_2) \geqq L(\epsilon_1) \geqq 0 \quad \text{when} \quad \epsilon_2 \geqq \epsilon_1 \geqq 0 \quad (2)$$

$$L(\epsilon) = L(-\epsilon)$$

Some common examples of loss functions are: $L(\epsilon) = a\epsilon^2, a\epsilon^4, a|\epsilon|, a[1 - \exp(-\epsilon^2)]$, etc., where a is a positive constant.

One (but by no means the only) natural way of choosing the random variable X_1 is to require that this choice should minimize the average loss or risk

$$E\{L[x_1(t_1) - X_1(t_1)]\} = E[E\{L[x(t_1) - X_1(t_1)]|y(t_0), \ldots, y(t)\}] \quad (3)$$

Since the first expectation on the right-hand side of (3) does not depend on the choice of X_1 but only on $y(t_0), \ldots, y(t)$, it is clear that minimizing (3) is equivalent to minimizing

$$E\{L[x_1(t_1) - X_1(t_1)]|y(t_0), \ldots, y(t)\} \quad (4)$$

Under just slight additional assumptions, optimal estimates can be characterized in a simple way.

Theorem 1. *Assume that L is of type (2) and that the conditional distribution function $F(\xi)$ defined by (1) is:*

(A) symmetric about the mean $\bar{\xi}$:

$$F(\xi - \bar{\xi}) = 1 - F(\bar{\xi} - \xi)$$

(B) convex for $\xi \leqq \bar{\xi}$:

$$F(\lambda\xi_1 + (1 - \lambda)\xi_2) \leqq \lambda F(\xi_1) + (1 - \lambda)F(\xi_2)$$

for all $\xi_1, \xi_2 \leqq \bar{\xi}$ and $0 \leqq \lambda \leqq 1$

Then the random variable $x_1^*(t_1|t)$ which minimizes the average loss (3) is the conditional expectation

$$x_1^*(t_1|t) = E[x_1(t_1)|y(t_0), \ldots, y(t)] \quad (5)$$

Proof: As pointed out recently by Sherman [25], this theorem follows immediately from a well-known lemma in probability theory.

Corollary. *If the random processes $\{x_1(t)\}$, $\{x_2(t)\}$, and $\{y(t)\}$ are gaussian, Theorem 1 holds.*

Proof: By Theorem 5, (*A*) (see Appendix), conditional distributions on a gaussian random process are gaussian. Hence the requirements of Theorem 1 are always satisfied.

In the control systems literature, this theorem appears sometimes in a form which is more restrictive in one way and more general in another way:

Theorem 1-a. *If $L(\epsilon) = \epsilon^2$, then Theorem 1 is true without assumptions (A) and (B).*

Proof: Expand the conditional expectation (4):

$$E[x_1^2(t_1)|y(t_0), \ldots, y(t)] - 2X_1(t_1)E[x_1(t_1)|y(t_0), \ldots, y(t)] + X_1^2(t_1)$$

and differentiate with respect to $X_1(t_1)$. This is not a completely rigorous argument; for a simple rigorous proof see Doob [15], pp. 77–78.

Remarks: (*a*) As far as the author is aware, it is not known what is the most general class of random processes $\{x_1(t)\}$, $\{x_2(t)\}$ for which the conditional distribution function satisfies the requirements of Theorem 1.

(*b*) Aside from the note of Sherman, Theorem 1 apparently has never been stated explicitly in the control systems literature. In fact, one finds many statements to the effect that loss functions of the general type (2) cannot be conveniently handled mathematically.

(*c*) In the sequel, we shall be dealing mainly with vector-valued random variables. In that case, the estimation problem is stated as: Given a vector-valued random process $\{\mathbf{x}(t)\}$ and observed random variables $\mathbf{y}(t_0), \ldots, \mathbf{y}(t)$, where $\mathbf{y}(t) = \mathbf{M}\mathbf{x}(t)$ ($\mathbf{M}$ being a singular matrix; in other words, not all co-ordinates of $\mathbf{x}(t)$ can be observed), find an estimate $\mathbf{X}(t_1)$ which minimizes the expected loss $E[L(\|\mathbf{x}(t_1) - \mathbf{X}(t_1)\|)]$, $\| \; \|$ being the norm of a vector.

Theorem 1 remains true in the vector case also, provided we require that the conditional distribution function of the n co-ordinates of the vector $\mathbf{x}(t_1)$,

$$Pr[x_1(t_1) \leqq \xi_1, \ldots, x_n(t_1) \leqq \xi_n|\mathbf{y}(t_0), \ldots, \mathbf{y}(t)] = F(\xi_1, \ldots, \xi_n)$$

be symmetric with respect to the n variables $\xi_1 - \bar{\xi}_1, \ldots, \xi_n - \bar{\xi}_n$ and convex in the region where all of these variables are negative.

Orthogonal Projections

The explicit calculation of the optimal estimate as a function of the observed variables is, in general, impossible. There is an important exception: The processes $\{x_1(t)\}$, $\{x_2(t)\}$ are gaussian.

On the other hand, if we attempt to get an optimal estimate under the restriction $L(\epsilon) = \epsilon^2$ and the additional requirement that the estimate be a linear function of the observed random variables, we get an estimate which is identical with the optimal estimate in the gaussian case, without the assumption of linearity or quadratic loss function. This shows that results obtainable by linear estimation can be bettered by nonlinear estimation only when (i) the random processes are nongaussian and even then (in view of Theorem 5, (*C*)) only (ii) by considering at least third-order probability distribution functions.

In the special cases just mentioned, the explicit solution of the estimation problem is most easily understood with the help of a geometric picture. This is the subject of the present section.

Consider the (real-valued) random variables $y(t_0), \ldots, y(t)$. The set of all linear combinations of these random variables with real coefficients

$$\sum_{i=t_0}^{t} a_i y(i) \quad (6)$$

forms a *vector space* (*linear manifold*) which we denote by $\mathcal{Y}(t)$. We regard, abstractly, any expression of the form (6) as "point" or "vector" in $\mathcal{Y}(t)$; this use of the word "vector" should not be confused, of course, with "vector-valued" random variables, etc. Since we do not want to fix the value of t (i.e., the total number of possible observations), $\mathcal{Y}(t)$ should be regarded as a finite-dimensional subspace of the space of all possible observations.

Given any two vectors u, v in $\mathcal{Y}(t)$ (i.e., random variables expressible in the form (6)), we say that u and v are *orthogonal* if $Euv = 0$. Using the Schmidt orthogonalization procedure, as described for instance by Doob [15], p. 151, or by Loève [16], p. 459, it is easy to select an *orthonormal basis* in $\mathcal{Y}(t)$. By this is meant a set of vectors $e_{t_0}, \ldots, e_t$ in $\mathcal{Y}(t)$ such that any vector in $\mathcal{Y}(t)$ can be expressed as a unique linear combination of $e_{t_0}, \ldots, e_t$ and

$$\left.\begin{aligned} Ee_ie_j = \delta_{ij} &= 1 \quad \text{if} \quad i = j \\ &= 0 \quad \text{if} \quad i \neq j \end{aligned}\right\} (i, j = t_0, \ldots, t) \tag{7}$$

Thus any vector $\bar{x}$ in $\mathcal{Y}(t)$ is given by

$$\bar{x} = \sum_{i=t_0}^{t} a_i e_i$$

and so the coefficients a_i can be immediately determined with the aid of (7):

$$E\bar{x}e_j = E\left(\sum_{i=t_0}^{t} a_i e_i\right) e_j = \sum_{i=t_0}^{t} a_i E e_i e_j = \sum_{i=t_0}^{t} a_i \delta_{ij} = a_j \tag{8}$$

It follows further that any random variable x (not necessarily in $\mathcal{Y}(t)$) can be uniquely decomposed into two parts: a part $\bar{x}$ in $\mathcal{Y}(t)$ and a part $\tilde{x}$ orthogonal to $\mathcal{Y}(t)$ (i.e., orthogonal to every vector in $\mathcal{Y}(t)$). In fact, we can write

$$x = \bar{x} + \tilde{x} = \sum_{i=t_0}^{t} (Exe_i)e_i + \tilde{x} \tag{9}$$

Thus $\bar{x}$ is uniquely determined by equation (9) and is obviously a vector in $\mathcal{Y}(t)$. Therefore $\tilde{x}$ is also uniquely determined; it remains to check that it is orthogonal to $\mathcal{Y}(t)$:

$$E\tilde{x}e_i = E(x - \bar{x})e_i = Exe_i - E\bar{x}e_i$$

Now the co-ordinates of $\bar{x}$ with respect to the basis $e_{t_0}, \ldots, e_t$ are given either in the form $E\bar{x}e_i$ (as in (8)) or in the form Exe_i (as in (9)). Since the co-ordinates are unique, $Exe_i = E\bar{x}e_i$ ($i = t_0, \ldots, t$); hence $E\tilde{x}e_i = 0$ and $\tilde{x}$ is orthogonal to every base vector e_i and therefore to $\mathcal{Y}(t)$. We call $\bar{x}$ the *orthogonal projection* of x on $\mathcal{Y}(t)$.

There is another way in which the orthogonal projection can be characterized: $\bar{x}$ is that vector in $\mathcal{Y}(t)$ (i.e., that *linear* function of the random variables $y(t_0), \ldots, y(t)$) which minimizes the quadratic-loss function. In fact, if $\bar{w}$ is any other vector in $\mathcal{Y}(t)$, we have

$$E(x - \bar{w})^2 = E(\tilde{x} + \bar{x} - \bar{w})^2 = E[(x - \bar{x}) + (\bar{x} - \bar{w})]^2$$

Since $\tilde{x}$ is orthogonal to every vector in $\mathcal{Y}(t)$ and in particular to $\bar{x} - \bar{w}$ we have

$$E(x - \bar{w})^2 = E(x - \bar{x})^2 + E(\bar{x} - \bar{w})^2 \geqq E(x - \bar{x})^2 \tag{10}$$

This shows that, if $\bar{w}$ also minimizes the quadratic loss, we must have $E(\bar{x} - \bar{w})^2 = 0$ which means that the random variables $\bar{x}$ and $\bar{w}$ are equal (except possibly for a set of events whose probability is zero).

These results may be summarized as follows:

Theorem 2. *Let* $\{x(t)\}$, $\{y(t)\}$ *random processes with zero mean (i.e.,* $Ex(t) = Ey(t) = 0$ *for all* t*). We observe* $y(t_0), \ldots, y(t)$.

If either

(*A*) *the random processes* $\{x(t)\}$, $\{y(t)\}$ *are gaussian; or*

(*B*) *the optimal estimate is restricted to be a linear function of the observed random variables and* $L(\epsilon) = \epsilon^2$;

then

$$x^*(t_1|t) = \text{optimal estimate of } x(t_1) \text{ given } y(t_0), \ldots, y(t)$$
$$= \text{orthogonal projection } \bar{x}(t_1|t) \text{ of } x(t_1) \text{ on } \mathcal{Y}(t). \tag{11}$$

These results are well-known though not easily accessible in the control systems literature. See Doob [15], pp. 75–78, or Pugachev [26]. It is sometimes convenient to denote the orthogonal projection by

$$\bar{x}(t_1|t) \equiv x^*(t_1|t) = \hat{E}[x(t_1)|\mathcal{Y}(t)]$$

The notation $\hat{E}$ is motivated by part (*b*) of the theorem: If the stochastic processes in question are gaussian, then orthogonal projection is actually identical with conditional expectation.

Proof. (*A*) This is a direct consequence of the remarks in connection with (10).

(*B*) Since $x(t)$, $y(t)$ are random variables with zero mean, it is clear from formula (9) that the orthogonal part $\tilde{x}(t_1|t)$ of $x(t_1)$ with respect to the linear manifold $\mathcal{Y}(t)$ is also a random variable with zero mean. Orthogonal random variables with zero mean are uncorrelated; if they are also gaussian then (by Theorem 5 (*B*)) they are independent. Thus

$$\begin{aligned} 0 = E\tilde{x}(t_1|t) &= E[\tilde{x}(t_1|t)|y(t_0), \ldots, y(t)] \\ &= E[x(t_1) - \bar{x}(t_1|t)|y(t_0), \ldots, y(t)] \\ &= E[x(t_1)|y(t_0), \ldots, y(t)] - \bar{x}(t_1|t) = 0 \end{aligned}$$

Remarks. (*d*) A rigorous formulation of the contents of this section as $t \to \infty$ requires some elementary notions from the theory of Hilbert space. See Doob [15] and Loève [16].

(*e*) The physical interpretation of Theorem 2 is largely a matter of taste. If we are not worried about the assumption of gaussianness, part (*A*) shows that the orthogonal projection is the optimal estimate for all reasonable loss functions. If we do worry about gaussianness, even if we are resigned to consider only linear estimates, we know that orthogonal projections are *not* the optimal estimate for many reasonable loss functions. Since in practice it is difficult to ascertain to what degree of approximation a random process of physical origin is gaussian, it is hard to decide whether Theorem 2 has very broad or very limited significance.

(*f*) Theorem 2 is immediately generalized for the case of vector-valued random variables. In fact, we define the linear manifold $\mathcal{Y}(t)$ generated by $\mathbf{y}(t_0), \ldots, \mathbf{y}(t)$ to be the set of all linear combinations

$$\sum_{i=t_0}^{t} \sum_{j=1}^{m} a_{ij} y_j(i)$$

of all m co-ordinates of each of the random vectors $\mathbf{y}(t_0), \ldots, \mathbf{y}(t)$. The rest of the story proceeds as before.

(*g*) Theorem 2 states in effect that the optimal estimate under conditions (*A*) or (*B*) is a linear combination of all previous observations. In other words, the optimal estimate can be regarded as the output of a linear filter, with the input being the actually occurring values of the observable random variables; Theorem 2 gives a way of computing the impulse response of the optimal filter. As pointed out before, knowledge of this impulse response is not a complete solution of the problem; for this reason, no explicit formulas for the calculation of the impulse response will be given.

Models for Random Processes

In dealing with physical phenomena, it is not sufficient to give an empirical description but one must have also some idea of the underlying causes. Without being able to separate in some sense causes and effects, i.e., without the assumption of causality, one can hardly hope for useful results.

It is a fairly generally accepted fact that primary macroscopic sources of random phenomena are independent gaussian processes.[5] A well-known example is the noise voltage produced in a resistor due to thermal agitation. In most cases, *observed* random phenomena are not describable by independent random variables. The statistical dependence (correlation) between random signals observed at different times is usually explained by the presence of a dynamic system between the primary random source and the observer. *Thus a random function of time may be thought of as the output of a dynamic system excited by an independent gaussian random process.*

An important property of gaussian random signals is that they remain gaussian after passing through a linear system (Theorem 5(A)). Assuming independent gaussian primary random sources, if the observed random signal is also gaussian, we may assume that the dynamic system between the observer and the primary source is *linear*. This conclusion may be forced on us also because of lack of detailed knowledge of the statistical properties of the observed random signal: Given any random process with known first and second-order averages, we can find a gaussian random process with the same properties (Theorem 5(C)). Thus gaussian distributions and linear dynamics are natural, mutually plausible assumptions particularly when the statistical data are scant.

How is a dynamic system (linear or nonlinear) described? The fundamental concept is the notion of the *state*. By this is meant, intuitively, some quantitative information (a set of numbers, a function, etc.) which is the least amount of data one has to know about the past behavior of the system in order to predict its future behavior. The dynamics is then described in terms of *state transitions*, i.e., one must specify how one state is transformed into another as time passes.

A *linear* dynamic system may be described in general by the vector differential equation

$$\left.\begin{aligned} dx/dt &= \mathsf{F}(t)\mathbf{x} + \mathsf{D}(t)\mathbf{u}(t) \\ \text{and}\qquad \mathbf{y}(t) &= \mathsf{M}(t)\mathbf{x}(t) \end{aligned}\right\} \tag{12}$$

where $\mathbf{x}$ is an n-vector, the *state* of the system (the components x_i of $\mathbf{x}$ are called *state variables*); $\mathbf{u}(t)$ is an m-vector ($m \leq n$) representing the *inputs* to the system; $\mathsf{F}(t)$ and $\mathsf{D}(t)$ are $n \times n$, respectively, $n \times m$ matrices. If all coefficients of $\mathsf{F}(t)$, $\mathsf{D}(t)$, $\mathsf{M}(t)$ are constants, we say that the dynamic system (12) is *time-invariant* or *stationary*. Finally, $\mathbf{y}(t)$ is a p-vector denoting the outputs of the system; $\mathsf{M}(t)$ is an $n \times p$ matrix; $p \leq n$.

The physical interpretation of (12) has been discussed in detail elsewhere [18, 20, 23]. A look at the block diagram in Fig. 1 may be helpful. This is not an ordinary but a matrix block diagram (as revealed by the fat lines indicating signal flow). The integrator in Fig. 1 actually stands for n integrators such that the output of each is a state variable; $\mathsf{F}(t)$ indicates how the outputs of the integrators are fed back to the inputs of the integrators. Thus $f_{ij}(t)$ is the coefficient with which the output of the jth integrator is fed back to the input of the ith integrator. It is not hard to relate this formalism to more conventional methods of linear system analysis.

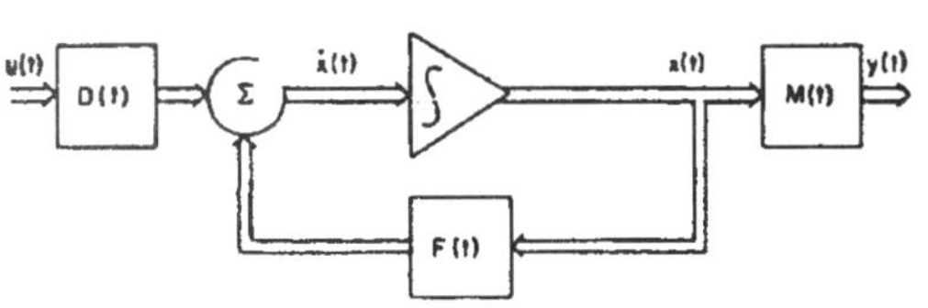

Fig. 1 Matrix block diagram of the general linear continuous-dynamic system

If we assume that the system (12) is stationary and that $\mathbf{u}(t)$ is constant during each sampling period, that is

$$\mathbf{u}(t + \tau) = \mathbf{u}(t); \quad 0 \leq \tau < 1, \quad t = 0, 1, \ldots \tag{13}$$

then (12) can be readily transformed into the more convenient discrete form

$$\mathbf{x}(t + 1) = \mathbf{\Phi}(1)\mathbf{x}(t) + \mathbf{\Delta}(1)\mathbf{u}(t); \quad t = 0, 1, \ldots$$

where [18, 20]

$$\mathbf{\Phi}(1) = \exp \mathsf{F} = \sum_{i=0}^{\infty} \mathsf{F}^i/i! \quad (\mathsf{F}^0 = \text{unit matrix})$$

and

$$\mathbf{\Delta}(1) = \left(\int_0^1 \exp \mathsf{F}\tau d\tau\right) \mathsf{D}$$

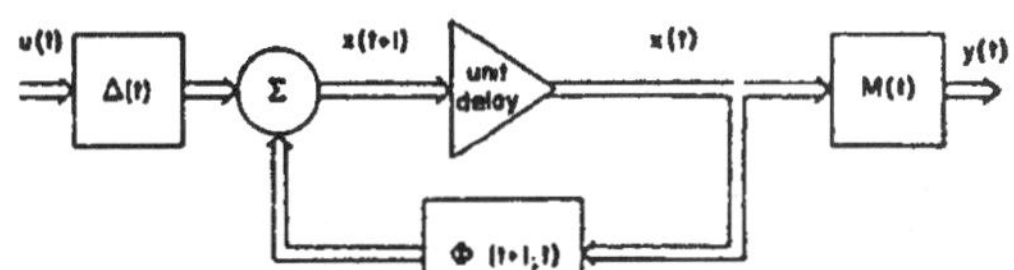

Fig. 2 Matrix block diagram of the general linear, discrete-dynamic system

See Fig. 2. One could also express $exp\ \mathsf{F}\tau$ in closed form using Laplace transform methods [18, 20, 22, 24]. If $\mathbf{u}(t)$ satisfies (13) but the system (12) is nonstationary, we can write analogously

$$\left.\begin{aligned} \mathbf{x}(t + 1) &= \mathbf{\Phi}(t + 1;\ t) + \mathbf{\Delta}(t)\mathbf{u}(t) \\ \mathbf{y}(t) &= \mathsf{M}(t)\mathbf{x}(t) \end{aligned}\right\} \quad t = 0, 1, \ldots \tag{14}$$

but of course now $\mathbf{\Phi}(t + 1;\ t)$, $\mathbf{\Delta}(t)$ cannot be expressed in general in closed form. Equations of type (14) are encountered frequently also in the study of complicated sampled-data systems [22]. See Fig. 2.

$\mathbf{\Phi}(t + 1;\ t)$ is the *transition matrix* of the system (12) or (14). The notation $\mathbf{\Phi}(t_2;\ t_1)$ (t_2, t_1 = integers) indicates transition from time t_1 to time t_2. Evidently $\mathbf{\Phi}(t;\ t) = \mathsf{I} =$ unit matrix. If the system (12) is stationary then $\mathbf{\Phi}(t + 1;\ t) = \mathbf{\Phi}(t + 1 - t) = \mathbf{\Phi}(1) =$ const. Note also the product rule: $\mathbf{\Phi}(t;\ s)\mathbf{\Phi}(s;\ r) = \mathbf{\Phi}(t;\ r)$ and the inverse rule $\mathbf{\Phi}^{-1}(t;\ s) = \mathbf{\Phi}(s;\ t)$, where t, s, r are integers. In a stationary system, $\mathbf{\Phi}(t;\ \tau) = \exp \mathsf{F}(t - \tau)$.

As a result of the preceding discussion, we shall represent random phenomena by the model

$$\mathbf{x}(t + 1) = \mathbf{\Phi}(t + 1;\ t)\mathbf{x}(t) + \mathbf{u}(t) \tag{15}$$

where $\{\mathbf{u}(t)\}$ is a vector-valued, independent, gaussian random process, with zero mean, which is completely described by (in view of Theorem 5 (C))

$$E\mathbf{u}(t) = 0 \text{ for all } t;$$

$$E\mathbf{u}(t)\mathbf{u}'(s) = 0 \quad \text{if} \quad t \neq s$$

$$E\mathbf{u}(t)\mathbf{u}'(t) = \mathsf{Q}(t).$$

Of course (Theorem 5 (A)), $\mathbf{x}(t)$ is then also a gaussian random process with zero mean, but it is no longer independent. In fact, if we consider (15) in the steady state (assuming it is a stable system), in other words, if we neglect the initial state $\mathbf{x}(t_0)$, then

---

[5] The probability distributions will be gaussian because macroscopic random effects may be thought of as the superposition of very many microscopic random effects; under very general conditions, such aggregate effects tend to be gaussian, regardless of the statistical properties of the microscopic effects. The assumption of independence in this context is motivated by the fact that microscopic phenomena tend to take place much more rapidly than macroscopic phenomena; thus primary random sources would appear to be independent on a macroscopic time scale.

$$\mathbf{x}(t) = \sum_{r=-\infty}^{t-1} \Phi(t;\ r+1)\mathbf{u}(r).$$

Therefore if $t \geqq s$ we have

$$E\mathbf{x}(t)\mathbf{x}'(s) = \sum_{r=-\infty}^{s-1} \Phi(t;\ r+1)Q(r)\Phi'(s;\ r+1).$$

Thus if we assume a linear dynamic model and know the statistical properties of the gaussian random excitation, it is easy to find the corresponding statistical properties of the gaussian random process $\{\mathbf{x}(t)\}$.

In real life, however, the situation is usually reversed. One is given the covariance matrix $E\mathbf{x}(t)\mathbf{x}'(s)$ (or rather, one attempts to estimate the matrix from limited statistical data) and the problem is to get (15) and the statistical properties of $\mathbf{u}(t)$. This is a subtle and presently largely unsolved problem in experimentation and data reduction. As in the vast majority of the engineering literature on the Wiener problem, we shall find it convenient to start with the model (15) and regard the problem of obtaining the model itself as a separate question. To be sure, the two problems *should* be optimized jointly if possible; the author is not aware, however, of any study of the *joint* optimization problem.

In summary, the following assumptions are made about random processes:

Physical random phenomena may be thought of as due to primary random sources exciting dynamic systems. The primary sources are assumed to be independent gaussian random processes with zero mean; the dynamic systems will be linear. The random processes are therefore described by models such as (15). The question of how the numbers specifying the model are obtained from experimental data will not be considered.

Solution of the Wiener Problem

Let us now define the principal problem of the paper.

Problem I. *Consider the dynamic model*

$$\mathbf{x}(t+1) = \Phi(t+1;\ t)\mathbf{x}(t) + \mathbf{u}(t) \tag{16}$$

$$\mathbf{y}(t) = \mathbf{M}(t)\mathbf{x}(t) \tag{17}$$

where $\mathbf{u}(t)$ is an independent gaussian random process of n-vectors with zero mean, $\mathbf{x}(t)$ is an n-vector, $\mathbf{y}(t)$ is a p-vector ($p \leqq n$), $\Phi(t+1;\ t)$, $\mathbf{M}(t)$ are $n \times n$, resp. $p \times n$, matrices whose elements are nonrandom functions of time.

Given the observed values of $\mathbf{y}(t_0), \ldots, \mathbf{y}(t)$ find an estimate $\mathbf{x}^*(t_1|t)$ of $\mathbf{x}(t_1)$ which minimizes the expected loss. (See Fig. 2, where $\Delta(t) = \mathbf{I}$.)

This problem includes as a special case the problems of filtering, prediction, and data smoothing mentioned earlier. It includes also the problem of reconstructing all the state variables of a linear dynamic system from noisy observations of some of the state variables ($p < n!$).

From Theorem 2-a we know that the solution of Problem I is simply the orthogonal projection of $\mathbf{x}(t_1)$ on the linear manifold $\mathcal{Y}(t)$ generated by the observed random variables. As remarked in the Introduction, this is to be accomplished by means of a linear (not necessarily stationary!) dynamic system of the general form (14). With this in mind, we proceed as follows.

Assume that $\mathbf{y}(t_0), \ldots, \mathbf{y}(t-1)$ have been measured, i.e., that $\mathcal{Y}(t-1)$ is known. Next, at time t, the random variable $\mathbf{y}(t)$ is measured. As before let $\tilde{\mathbf{y}}(t|t-1)$ be the component of $\mathbf{y}(t)$ orthogonal to $\mathcal{Y}(t-1)$. If $\tilde{\mathbf{y}}(t|t-1) \equiv 0$, which means that the values of all components of this random vector are zero for almost every possible event, then $\mathcal{Y}(t)$ is obviously the same as $\mathcal{Y}(t-1)$ and therefore the measurement of $\mathbf{y}(t)$ does not convey any additional information. This is not likely to happen in a physically meaningful situation. In any case, $\tilde{\mathbf{y}}(t|t-1)$ generates a linear manifold (possibly 0) which we denote by $\mathcal{Z}(t)$. By definition, $\mathcal{Y}(t-1)$ and $\mathcal{Z}(t)$ taken together are the same manifold as $\mathcal{Y}(t)$, and every vector in $\mathcal{Z}(t)$ is orthogonal to every vector in $\mathcal{Y}(t-1)$.

Assuming by induction that $\mathbf{x}^*(t_1-1|t-1)$ is known, we can write:

$$\begin{aligned}\mathbf{x}^*(t_1|t) &= \hat{E}[\mathbf{x}(t_1)|\mathcal{Y}(t)] = \hat{E}[\mathbf{x}(t_1)|\mathcal{Y}(t-1)] + \hat{E}[\mathbf{x}(t_1)|\mathcal{Z}(t)]\\ &= \Phi(t+1;\ t)\mathbf{x}^*(t_1-1|t-1) + \hat{E}[\mathbf{u}(t_1-1)|\mathcal{Y}(t-1)]\\ &\quad + \hat{E}[\mathbf{x}(t_1)|\mathcal{Z}(t)]\end{aligned} \tag{18}$$

where the last line is obtained using (16).

Let $t_1 = t + s$, where s is any integer. If $s \geqq 0$, then $\mathbf{u}(t_1-1)$ is independent of $\mathcal{Y}(t-1)$. This is because $\mathbf{u}(t_1-1) = \mathbf{u}(t+s-1)$ is then independent of $\mathbf{u}(t-2), \mathbf{u}(t-3), \ldots$ and therefore by (16–17), independent of $\mathbf{y}(t_0), \ldots, \mathbf{y}(t-1)$, hence independent of $\mathcal{Y}(t-1)$. Since, for all t, $\mathbf{u}(t_0)$ has zero mean by assumption, it follows that $\mathbf{u}(t_1-1)$ ($s \geqq 0$) is orthogonal to $\mathcal{Y}(t-1)$. Thus if $s \geqq 0$, the second term on the right-hand side of (18) vanishes; if $s < 0$, considerable complications result in evaluating this term. We shall consider only the case $t_1 \geqq t$. Furthermore, it will suffice to consider in detail only the case $t_1 = t + 1$ since the other cases can be easily reduced to this one.

The last term in (18) must be a linear operation on the random variable $\tilde{\mathbf{y}}(t|t-1)$:

$$\hat{E}[\mathbf{x}(t+1)|\mathcal{Z}(t)] = \Delta^*(t)\tilde{\mathbf{y}}(t|t-1) \tag{19}$$

where $\Delta^*(t)$ is an $n \times p$ matrix, and the star refers to "optimal filtering."

The component of $\mathbf{y}(t)$ lying in $\mathcal{Y}(t-1)$ is $\bar{\mathbf{y}}(t|t-1) = \mathbf{M}(t)\mathbf{x}^*(t|t-1)$. Hence

$$\tilde{\mathbf{y}}(t|t-1) = \mathbf{y}(t) - \bar{\mathbf{y}}(t|t-1) = \mathbf{y}(t) - \mathbf{M}(t)\mathbf{x}^*(t|t-1). \tag{20}$$

Combining (18–20) (see Fig. 3) we obtain

$$\mathbf{x}^*(t+1|t) = \Phi^*(t+1;\ t)\mathbf{x}^*(t|t-1) + \Delta^*(t)\mathbf{y}(t) \tag{21}$$

where

$$\Phi^*(t+1;\ t) = \Phi(t+1;\ t) - \Delta^*(t)\mathbf{M}(t) \tag{22}$$

Thus optimal estimation is performed by a linear dynamic system of the same form as (14). The state of the estimator is the previous estimate, the input is the last measured value of the observable random variable $\mathbf{y}(t)$, the transition matrix is given by (22). Notice that physical realization of the optimal filter requires only (i) the model of the random process (ii) the operator $\Delta^*(t)$.

The estimation error is also governed by a linear dynamic system. In fact,

$$\begin{aligned}\tilde{\mathbf{x}}(t+1|t) &= \mathbf{x}(t+1) - \mathbf{x}^*(t+1|t)\\ &= \Phi(t+1;\ t)\mathbf{x}(t) + \mathbf{u}(t) - \Phi^*(t+1;\ t)\mathbf{x}^*(t|t-1)\\ &\quad - \Delta^*(t)\mathbf{M}(t)\mathbf{x}(t)\end{aligned}$$

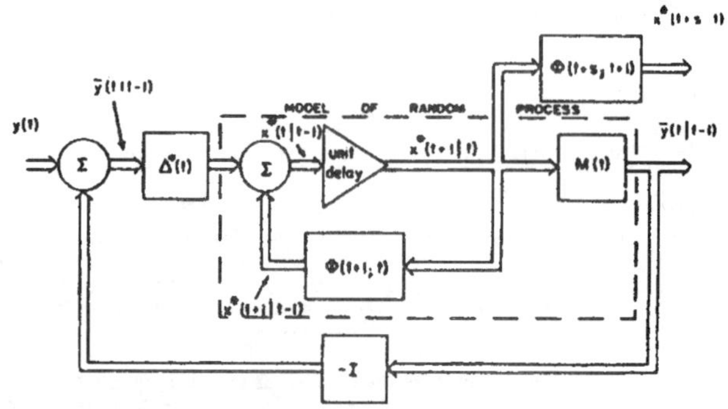

Fig. 3 Matrix block diagram of optimal filter

$$= \Phi^*(t+1;\ t)\tilde{x}(t|t-1) + u(t) \quad (23)$$

Thus Φ^* is also the transition matrix of the linear dynamic system governing the error.

From (23) we obtain at once a recursion relation for the covariance matrix $P^*(t)$ of the optimal error $\tilde{x}(t|t-1)$. Noting that $u(t)$ is independent of $x(t)$ and therefore of $\tilde{x}(t|t-1)$, we get

$$\begin{aligned} P^*(t+1) &= E\tilde{x}(t+1|t)\tilde{x}'(t+1|t) \\ &= \Phi^*(t+1;\ t)E\tilde{x}(t|t-1)\tilde{x}'(t|t-1)\Phi^{*\prime}(t+1;\ t) + Q(t) \\ &= \Phi^*(t+1;\ t)E\tilde{x}(t|t-1)\tilde{x}'(t|t-1)\Phi'(t+1;\ t) + Q(t) \\ &= \Phi^*(t+1;\ t)P^*(t)\Phi'(t+1;\ t) + Q(t) \end{aligned} \quad (24)$$

where $Q(t) = Eu(t)u'(t)$.

There remains the problem of obtaining an explicit formula for Δ^* (and thus also for Φ^*). Since,

$$\tilde{x}(t+1|\mathcal{Z}(t)) = x(t+1) - \hat{E}[x(t+1)|\mathcal{Z}(t)]$$

is orthogonal to $\tilde{y}(t|t-1)$, it follows by (19) that

$$\begin{aligned} 0 &= E[x(t+1) - \Delta^*(t)\tilde{y}(t|t-1)]\tilde{y}'(t|t-1) \\ &= Ex(t+1)\tilde{y}'(t|t-1) - \Delta^*(t)E\tilde{y}(t|t-1)\tilde{y}'(t|t-1). \end{aligned}$$

Noting that $\tilde{x}(t+1|t-1)$ is orthogonal to $\mathcal{Z}(t)$, the definition of $P(t)$ given earlier, and (17), it follows further

$$\begin{aligned} 0 &= E\tilde{x}(t+1|t-1)\tilde{y}'(t|t-1) - \Delta^*(t)M(t)P^*(t)M'(t) \\ &= E[\Phi(t+1;\ t)\tilde{x}(t|t-1) + u(t|t-1)]\tilde{x}'(t|t-1)M'(t) \\ &\qquad - \Delta^*(t)M(t)P^*(t)M'(t). \end{aligned}$$

Finally, since $u(t)$ is independent of $x(t)$,

$$0 = \Phi(t+1;\ t)P^*(t)M'(t) - \Delta^*(t)M(t)P^*(t)M'(t).$$

Now the matrix $M(t)P^*(t)M'(t)$ will be positive definite and hence invertible whenever $P^*(t)$ is positive definite, provided that none of the rows of $M(t)$ are linearly dependent at any time, in other words, that none of the observed scalar random variables $y_1(t)$, $\ldots$, $y_m(t)$ is a linear combination of the others. Under these circumstances we get finally:

$$\Delta^*(t) = \Phi(t+1;\ t)P^*(t)M'(t)[M(t)P^*(t)M'(t)]^{-1} \quad (25)$$

Since observations start at t_0, $\tilde{x}(t_0|t_0-1) = x(t_0)$; to begin the iterative evaluation of $P^*(t)$ by means of equation (24), we must obviously specify $P^*(t_0) = Ex(t_0)x'(t_0)$. Assuming this matrix is positive definite, equation (25) then yields $\Delta^*(t_0)$; equation (22) $\Phi^*(t_0+1;\ t_0)$, and equation (24) $P^*(t_0+1)$, completing the cycle. If now $Q(t)$ is positive definite, then all the $P^*(t)$ will be positive definite and the requirements in deriving (25) will be satisfied at each step.

Now we remove the restriction that $t_1 = t+1$. Since $u(t)$ is orthogonal to $\mathcal{Y}(t)$, we have

$$x^*(t+1|t) = \hat{E}[\Phi(t+1;\ t)x(t) + u(t)|\mathcal{Y}(t)] = \Phi(t+1;\ t)x^*(t|t)$$

Hence if $\Phi(t+1;\ t)$ has an inverse $\Phi(t;\ t+1)$ (which is always the case when Φ is the transition matrix of a dynamic system describable by a differential equation) we have

$$x^*(t|t) = \Phi(t;\ t+1)x^*(t+1|t)$$

If $t_1 \geqq t+1$, we first observe by repeated application of (16) that

$$\begin{aligned} x(t+s) &= \Phi(t+s;\ t+1)x(t+1) \\ &\qquad + \sum_{r1}^{s-1} \Phi(t+s;\ t+r)u(t+r) \quad (s \geqq 1) \end{aligned}$$

Since $u(t+s-1), \ldots, u(t+1)$ are all orthogonal to $\mathcal{Y}(t)$,

$$\begin{aligned} x^*(t+s|t) &= \hat{E}[x(t+s)|\mathcal{Y}(t)] \\ &= \hat{E}[\Phi(t+s;\ t+1)x(t+1)\,|\mathcal{Y}(t)] \\ &= \Phi(t+s;\ t+1)x^*(t+1|t) \quad (s \geqq 1) \end{aligned}$$

If $s < 0$, the results are similar, but $x^*(t-s|t)$ will have $(1-s)(n-p)$ co-ordinates.

The results of this section may be summarized as follows:

Theorem 3. *(Solution of the Wiener Problem)*

Consider Problem I. The optimal estimate $x^*(t+1|t)$ of $x(t+1)$ given $y(t_0), \ldots, y(t)$ is generated by the linear dynamic system

$$x^*(t+1|t) = \Phi^*(t+1;\ t)x^*(t|t-1) + \Delta^*(t)y(t) \quad (21)$$

The estimation error is given by

$$\tilde{x}(t+1|t) = \Phi^*(t+1;\ t)\tilde{x}(t|t-1) + u(t) \quad (23)$$

The covariance matrix of the estimation error is

$$\text{cov }\tilde{x}(t|t-1) = E\tilde{x}(t|t-1)\tilde{x}'(t|t-1) = P^*(t) \quad (26)$$

The expected quadratic loss is

$$\sum_{i=1}^{n} E\tilde{x}_i^2(t|t-1) = \text{trace } P^*(t) \quad (27)$$

The matrices $\Delta^*(t)$, $\Phi^*(t+1;\ t)$, $P^*(t)$ are generated by the recursion relations

$$\left.\begin{aligned} \Delta^*(t) &= \Phi(t+1;\ t)P^*(t)M'(t)[M(t)P^*(t)M'(t)]^{-1} \quad (28) \\ \Phi^*(t+1;\ t) &= \Phi(t+1;\ t) - \Delta^*(t)M(t) \quad (29) \\ P^*(t+1) &= \Phi^*(t+1;\ t)P^*(t)\Phi'(t+1;\ t) \\ &\qquad + Q(t) \quad (30) \end{aligned}\right\} t \geqq t_0$$

In order to carry out the iterations, one must specify the covariance $P^*(t_0)$ of $x(t_0)$ and the covariance $Q(t)$ of $u(t)$. Finally, for any $s \geqq 0$, if Φ is invertible

$$\begin{aligned} x^*(t+s|t) &= \Phi(t+s;\ t+1)x^*(t+1|t) \\ &= \Phi(t+s;\ t+1)\Phi^*(t+1;\ t)\Phi(t;\ t+s-1) \\ &\qquad \times x^*(t+s-1|t-1) \\ &\quad + \Phi(t+s;\ t+1)\Delta^*(t)y(t) \end{aligned} \quad (31)$$

so that the estimate $x^*(t+s|t)$ $(s \geqq 0)$ is also given by a linear dynamic system of the type (21).

Remarks. (*h*) Eliminating Δ^* and Φ^* from (28–30), a nonlinear difference equation is obtained for $P^*(t)$:

$$\begin{aligned} P^*(t+1) &= \Phi(t+1;\ t)\{P^*(t) - P^*(t)M'(t)[M(t)P^*(t)M'(t)]^{-1} \\ &\qquad \times P^*(t)M(t)\}\Phi'(t+1;\ t) + Q(t) \quad (t \geqq t_0) \end{aligned} \quad (32)$$

This equation is linear only if $M(t)$ is invertible but then the problem is trivial since all components of the random vector $x(t)$ are observable $P^*(t+1) = Q(t)$. Observe that equation (32) plays a role in the present theory analogous to that of the Wiener-Hopf equation in the conventional theory.

Once $P^*(t)$ has been computed via (32) starting at $t = t_0$, the explicit specification of the optimal linear filter is immediately available from formulas (29-30). Of course, the solution of Equation (32), or of its differential-equation equivalent, is a much simpler task than solution of the Wiener-Hopf equation.

(*i*) The results stated in Theorem 3 do not resolve completely Problem I. Little has been said, for instance, about the physical significance of the assumptions needed to obtain equation (25), the convergence and stability of the nonlinear difference equation (32), the stability of the optimal filter (21), etc. This can actually be done in a completely satisfactory way, but must be left to a future paper. In this connection, the principal guide and tool turns out to be the duality theorem mentioned briefly in the next section. See [29].

(*j*) By letting the sampling period (equal to one so far) approach zero, the method can be used to obtain the specification of a differential equation for the optimal filter. To do this, i.e., to pass from equation (14) to equation (12), requires computing the logarithm $\mathbf{F}^*$ of the matrix $\mathbf{\Phi}^*$. But this can be done only if $\mathbf{\Phi}^*$ is nonsingular—which is easily seen *not* to be the case. This is because it is sufficient for the optimal filter to have $n - p$ state variables, rather than n, as the formalism of equation (22) would seem to imply. By appropriate modifications, therefore, equation (22) can be reduced to an equivalent set of only $n - p$ equations whose transition matrix is nonsingular. Details of this type will be covered in later publications.

(*k*) The dynamic system (21) is, in general, nonstationary. This is due to two things: (1) The time dependence of $\mathbf{\Phi}(t+1; t)$ and $\mathbf{M}(t)$; (2) the fact that the estimation starts at $t = t_0$ and improves as more data are accumulated. If $\mathbf{\Phi}$, $\mathbf{M}$ are constants, it can be shown that (21) becomes a stationary dynamic system in the limit $t \to \infty$. This is the case treated by the classical Wiener theory.

(*l*) It is noteworthy that the derivations given are not affected by the nonstationarity of the model for $\mathbf{x}(t)$ or the finiteness of available data. In fact, as far as the author is aware, the only explicit recursion relations given before for the growing-memory filter are due to Blum [12]. However, his results are much more complicated than ours.

(*m*) By inspection of Fig. 3 we see that the optimal filter is a feedback system, and that the signal after the first summer is white noise since $\tilde{\mathbf{y}}(t|t-1)$ is obviously an orthogonal random process. This corresponds to some well-known results in Wiener filtering, see, e.g., Smith [28], Chapter 6, Fig. 6–4. However, this is apparently the first *rigorous* proof that every Wiener filter is realizable by means of a feedback system. Moreover, it will be shown in another paper that such a filter is always *stable*, under very mild assumptions on the model (16–17). See [29].

The Dual Problem

Let us now consider another problem which is conceptually very different from optimal estimation, namely, the noise-free regulator problem. In the simplest cases, this is:

Problem II. *Consider the dynamic system*

$$\mathbf{x}(t+1) = \hat{\mathbf{\Phi}}(t+1;\ t)\mathbf{x}(t) + \hat{\mathbf{M}}(t)\mathbf{u}(t) \qquad (33)$$

where $\mathbf{x}(t)$ *is an n-vector,* $\mathbf{u}(t)$ *is an m-vector* ($m \leq n$), $\hat{\mathbf{\Phi}}$, $\hat{\mathbf{M}}$ *are* $n \times n$ *resp.* $n \times m$ *matrices whose elements are nonrandom functions of time. Given any state* $\mathbf{x}(t)$ *at time t, we are to find a sequence* $\mathbf{u}(t), \ldots, \mathbf{u}(T)$ *of control vectors which minimizes the performance index*

$$V[\mathbf{x}(t)] = \sum_{\tau=t}^{T+1} \mathbf{x}'(\tau)\mathbf{Q}(\tau)\mathbf{x}(\tau)$$

where $\hat{\mathbf{Q}}(t)$ *is a positive definite matrix whose elements are nonrandom functions of time.* See Fig. 2, where $\mathbf{\Delta} = \hat{\mathbf{M}}$ and $\mathbf{M} = \mathbf{I}$.

Probabilistic considerations play no part in Problem II; it is implicitly assumed that every state variable can be measured exactly at each instant $t, t+1, \ldots, T$. It is customary to call $T \geq t$ the *terminal time* (it may be infinity).

The first general solution of the noise-free regulator problem is due to the author [18]. The main result is that the optimal control vectors $\mathbf{u}^*(t)$ are nonstationary linear functions of $\mathbf{x}(t)$. After a change in notation, the formulas of the Appendix, Reference [18] (see also Reference [23]) are as follows:

$$\mathbf{u}^*(t) = -\hat{\mathbf{\Delta}}^*(t)\mathbf{x}(t) \qquad (34)$$

Under optimal control as given by (34), the "closed-loop" equations for the system are (see Fig. 4)

$$\mathbf{x}(t+1) = \hat{\mathbf{\Phi}}^*(t+1;\ t)\mathbf{x}(t)$$

and the minimum performance index at time t is given by

$$V^*[\mathbf{x}(t)] = \mathbf{x}'(t)\mathbf{P}^*(t-1)\mathbf{x}(t)$$

The matrices $\hat{\mathbf{\Delta}}^*(t)$, $\hat{\mathbf{\Phi}}^*(t+1;\ t)$, $\hat{\mathbf{P}}^*(t)$ are determined by the recursion relations:

$$\left.\begin{aligned}
&\hat{\mathbf{\Delta}}^*(t) = [\hat{\mathbf{M}}'(t)\hat{\mathbf{P}}^*(t)\hat{\mathbf{M}}(t)]^{-1}\hat{\mathbf{M}}'(t)\hat{\mathbf{P}}^*(t)\hat{\mathbf{\Phi}}(t+1;\ t) && (35)\\
&\hat{\mathbf{\Phi}}^*(t+1;\ t) = \hat{\mathbf{\Phi}}(t+1;\ t) - \hat{\mathbf{M}}(t)\hat{\mathbf{\Delta}}^*(t) && (36)\\
&\hat{\mathbf{P}}^*(t-1) = \hat{\mathbf{\Phi}}'(t+1;t)\hat{\mathbf{P}}^*(t)\hat{\mathbf{\Phi}}^*(t+1;\ t) + \hat{\mathbf{Q}}(t) && (37)
\end{aligned}\right\} t \leq T$$

Initially we must set $\hat{\mathbf{P}}^*(T) = \hat{\mathbf{Q}}(T+1)$.

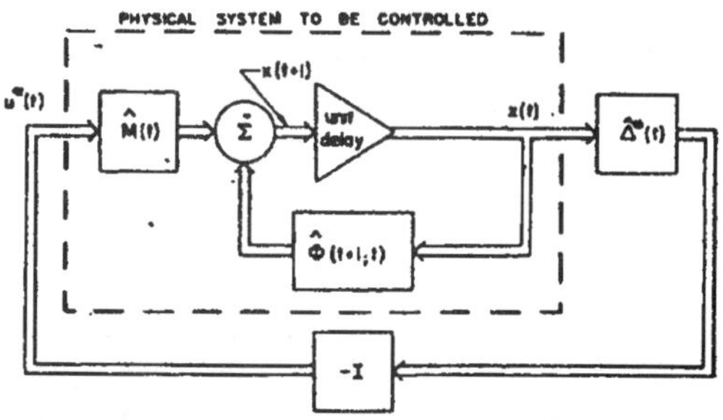

Fig. 4 Matrix block diagram of optimal controller

Comparing equations (35–37) with (28–30) and Fig. 3 with Fig. 4 we notice some interesting things which are expressed precisely by

Theorem 4. (*Duality Theorem*) *Problem I and Problem II are duals of each other in the following sense:*

Let $\tau \geq 0$. *Replace every matrix* $\mathbf{X}(t) = \mathbf{X}(t_0 + \tau)$ *in* (28–30) *by* $\hat{\mathbf{X}}'(t) = \hat{\mathbf{X}}'(T - \tau)$. *Then one has* (35–37). *Conversely, replace every matrix* $\hat{\mathbf{X}}(T - \tau)$ *in* (35–37) *by* $\mathbf{X}'(t_0 + \tau)$. *Then one has* (28–30).

Proof. Carry out the substitutions. For ease of reference, the dualities between the two problems are given in detail in Table 1.

Table 1

	Problem I	Problem II
1	$\mathbf{x}(t)$ (unobservable) state variables of random process.	$\mathbf{x}(t)$ (observable) state variables of plant to be regulated.
2	$\mathbf{y}(t)$ observed random variables.	$\mathbf{u}(t)$ control variables.
3	t_0 first observation.	T last control action.
4	$\mathbf{\Phi}(t_0 + \tau + 1;\ t_0 + \tau)$ transition matrix.	$\hat{\mathbf{\Phi}}(T - \tau + 1;\ T - \tau)$ transition matrix.
5	$\mathbf{P}^*(t_0 + \tau)$ covariance of optimized estimation error.	$\hat{\mathbf{P}}^*(T - \tau)$ matrix of quadratic form for performance index under optimal regulation.
6	$\mathbf{\Delta}^*(t_0 + \tau)$ weighting of observation for optimal estimation.	$\hat{\mathbf{\Delta}}^*(T - \tau)$ weighting of state for optimal control.
7	$\mathbf{\Phi}^*(t_0 + \tau + 1;\ t_0 + \tau)$ transition matrix for optimal estimation error.	$\hat{\mathbf{\Phi}}^*(T - \tau + 1;\ T - \tau)$ transition matrix under optimal regulation.
8	$\mathbf{M}(t_0 + \tau)$ effect of state on observation.	$\hat{\mathbf{M}}(T - \tau)$ effect of control vectors on state.
9	$\mathbf{Q}(t_0 + \tau)$ covariance of random excitation.	$\hat{\mathbf{Q}}(T - \tau)$ matrix of quadratic form defining error criterion.

Remarks. (*n*) The *mathematical* significance of the duality between Problem I and Problem II is that both problems reduce to the solution of the Wiener-Hopf-like equation (32).

(*o*) The *physical* significance of the duality is intriguing. Why are observations and control dual quantities?

Recent research [29] has shown that the essence of the Duality Theorem lies in the duality of constraints at the output (represented by the matrix $\hat{\mathbf{M}}(t)$ in Problem I) and constraints at the input (represented by the matrix $\hat{\mathbf{M}}(t)$ in Problem II).

(*p*) Applications of Wiener's methods to the solution of noise-free regulator problem have been known for a long time; see the recent textbook of Newton, Gould, and Kaiser [27]. However, the connections between the two problems, and in particular the duality, have apparently never been stated precisely before.

(*q*) The duality theorem offers a powerful tool for developing more deeply the theory (as opposed to the computation) of Wiener filters, as mentioned in Remark (*i*). This will be published elsewhere [29].

Applications

The power of the new approach to the Wiener problem, as expressed by Theorem 3, is most obvious when the data of the problem are given in numerical form. In that case, one simply performs the numerical computations required by (28–30). Results of such calculations, in some cases of practical engineering interest, will be published elsewhere.

When the answers are desired in closed analytic form, the iterations (28–30) may lead to very unwieldy expressions. In a few cases, Δ^* and Φ^* can be put into "closed form." Without discussing here how (if at all) such closed forms can be obtained, we now give two examples indicative of the type of results to be expected.

Example 1. Consider the problem mentioned under "Optimal Estimates." Let $x_1(t)$ be the signal and $x_2(t)$ the noise. We assume the model:

$$x_1(t+1) = \phi_{11}(t+1;\ t)x_1(t) + u_1(t)$$

$$x_2(t+1) = u_2(t)$$

$$y_1(t) = x_1(t) + x_2(t)$$

The specific data for which we desire a solution of the estimation problem are as follows:

1 $t_1 = t + 1;\ t_0 = 0$
2 $Ex_1^2(0) = 0$, i.e., $x_1(0) = 0$
3 $Eu_1^2(t) = a^2,\ Eu^2(t) = b^2,\ Eu_1(t)u_2(t) = 0$ (for all t)
4 $\phi_{11}(t+1;\ t) = \phi_{11} =$ const.

A simple calculation shows that the following matrices satisfy the difference equations (28–30), for all $t \geqq t_0$:

$$\Delta^*(t) = \begin{bmatrix} \phi_{11}C(t) \\ 0 \end{bmatrix}$$

$$\Phi^*(t+1;\ t) = \begin{bmatrix} \phi_{11}[1 - C(t)] & 0 \\ 0 & 0 \end{bmatrix}$$

$$\mathbf{P}^*(t+1) = \begin{bmatrix} a^2 + \phi_{11}^2 b^2 C(t) & 0 \\ 0 & b^2 \end{bmatrix}$$

where $$C(t+1) = 1 - \frac{b^2}{a^2 + b^2 + \phi_{11}^2 b^2 C(t)} \quad t \geqq 0 \qquad (38)$$

Since it was assumed that $x_1(0) = 0$, neither $x_1(1)$ nor $x_2(1)$ can be predicted from the measurement of $y_1(0)$. Hence the measurement at time $t = 0$ is useless, which shows that we should set $C(0) = 0$. This fact, with the iterations (38), completely determines the function $C(t)$. The nonlinear difference equation (38) plays the role of the Wiener-Hopf equation.

If $b^2/a^2 \ll 1$, then $C(t) \cong 1$ which is essentially pure prediction. If $b^2/a^2 \gg 1$, then $C(t) \cong 0$, and we depend mainly on $x_1^*(t|t-1)$ for the estimation of $x_1^*(t+1|t)$ and assign only very small weight to the measurement $y_1(t)$; this is what one would expect when the measured data are very noisy.

In any case, $x_2^*(t|t-1) = 0$ at all times; one cannot predict independent noise! This means that ψ^*_{12} can be set equal to zero. The optimal predictor is a first-order dynamic system. See Remark (*j*).

To find the stationary Wiener filter, let $t = \infty$ on both sides of (38), solve the resulting quadratic equation in $C(\infty)$, etc.

Example 2. A number of particles leave the origin at time $t_0 = 0$ with random velocities; after $t = 0$, each particle moves with a constant (unknown) velocity. Suppose that the position of one of these particles is measured, the data being contaminated by stationary, additive, correlated noise. What is the optimal estimate of the position and velocity of the particle at the time of the last measurement?

Let $x_1(t)$ be the position and $x_2(t)$ the velocity of the particle; $x_3(t)$ is the noise. The problem is then represented by the model,

$$x_1(t+1) = x_1(t) + x_2(t)$$

$$x_2(t+1) = x_2(t)$$

$$x_3(t+1) = \phi_{33}(t+1;\ t)x_3(t) + u_3(t)$$

$$y_1(t) = x_1(t) + x_3(t)$$

and the additional conditions

1 $t_1 = t;\ t_0 = 0$
2 $Ex_1^2(0) = Ex_2(0) = 0,\ Ex_2^2(0) = a^2 > 0;$
3 $Eu_3(t) = 0,\ Eu_3^2(t) = b^2.$
4 $\phi_{33}(t+1;\ t) = \phi_{33} =$ const.

According to Theorem 3, $\mathbf{x}^*(t|t)$ is calculated using the dynamic system (31).

First we solve the problem of predicting the position and velocity of the particle one step ahead. Simple considerations show that

$$\mathbf{P}^*(1) = \begin{bmatrix} a^2 & a^2 & 0 \\ a^2 & a^2 & 0 \\ 0 & 0 & b^2 \end{bmatrix} \quad \text{and} \quad \Delta^*(0) = \begin{bmatrix} 0 \\ 0 \\ 1 \end{bmatrix}$$

It is then easy to check by substitution into equations (28–30) that

$$\mathbf{P}^*(t) = \frac{b^2}{C_1(t-1)} \times \begin{bmatrix} t^2 & t & -\phi_{33}t(t-1) \\ t & 1 & -\phi_{33}(t-1) \\ -\phi_{33}t(t-1) & -\phi_{33}(t-1) & \phi_{33}^2(t-1)^2 + C_1(t-1) \end{bmatrix}$$

is the correct expression for the covariance matrix of the prediction error $\tilde{\mathbf{x}}(t|t-1)$ for all $t \geqq 1$, provided that we define

$$C_1(0) = b^2/a^2$$
$$C_1(t) = C_1(t-1) + [t - \phi_{33}(t-1)]^2,\ t \geqq 1$$

It is interesting to note that the results just obtained are valid also when ϕ_{33} depends on t. This is true also in Example 1. In conventional treatments of such problems there *seems* to be an essential difference between the cases of stationary and nonstationary noise. This misleading impression created by the conventional theory is due to the very special *methods* used in solving the Wiener-Hopf equation.

Introducing the abbreviation

$$C_2(0) = 0$$
$$C_2(t) = t - \phi_{33}(t-1),\ t \geqq 1$$

and observing that

$$\text{cov}\ \tilde{\mathbf{x}}(t+1|t) = \mathbf{P}^*(t+1)$$
$$= \Phi(t+1;\ t)[\text{cov}\ \tilde{\mathbf{x}}(t|t)]\Phi'(t+1;\ t) + \mathbf{Q}(t)$$

the matrices occurring in equation (31) and the covariance matrix of $\tilde{\mathbf{x}}(t|t)$ are found after simple calculations. We have, for all $t \geqq 0$,

$$\mathbf{\Phi}(t;\ t+1)\mathbf{\Delta}^*(t) = \frac{1}{C_1(t)}\begin{bmatrix} tC_2(t) \\ C_2(t) \\ C_1(t) - tC_2(t) \end{bmatrix}$$

$$\mathbf{\Phi}(t;\ t+1)\mathbf{\Phi}^*(t+1;\ t)\mathbf{\Phi}(t+1;\ t)$$

$$= \frac{1}{C_1(t)}\begin{bmatrix} C_1(t) - tC_2(t) & C_1(t) - tC_3(t) & -\phi_{33}tC_2(t) \\ -C_2(t) & C_1(t) - C_2(t) & -\phi_{33}C_2(t) \\ -C_1(t) + tC_2(t) & -C_1(t) + tC_2(t) & +\phi_{33}tC_2(t) \end{bmatrix}$$

and

$$\text{cov}\ \tilde{\mathbf{x}}(t|t) = E\tilde{\mathbf{x}}(t|t)\tilde{\mathbf{x}}'(t|t) = \frac{b^2}{C_1(t)}\begin{bmatrix} t^2 & t & -t^2 \\ t & 1 & -t \\ -t^2 & -t & t^2 \end{bmatrix}$$

To gain some insight into the behavior of this system, let us examine the limiting case $t \to \infty$ of a large number of observations. Then $C_1(t)$ obeys approximately the differential equation

$$dC_1(t)/dt \cong C_2^2(t) \qquad (t \gg 1)$$

from which we find

$$C_1(t) \cong (1 - \phi_{33})^2t^3/3 + \phi_{33}(1 - \phi_{33})t^2 + \phi_{33}^2t + b^2/a^2 \qquad (t \gg 1) \quad (39)$$

Using (39), we get further

$$\mathbf{\Phi}^{-1}\mathbf{\Phi}^*\mathbf{\Phi} \cong \begin{bmatrix} 1 & 1 & 0 \\ 0 & 1 & 0 \\ -1 & -1 & 0 \end{bmatrix} \text{ and } \mathbf{\Phi}^{-1}\mathbf{\Delta}^* \cong \begin{bmatrix} 0 \\ 0 \\ 1 \end{bmatrix} \qquad (t \gg 1)$$

Thus as the number of observations becomes large, we depend almost exclusively on $x_1^*(t|t)$ and $x_2^*(t|t)$ to estimate $x_1^*(t+1|t+1)$ and $x_3^*(t+1|t+1)$. Current observations are used almost exclusively to estimate the noise

$$x_3^*(t|t) \cong y_1(t) - x_1^*(t|t) \qquad (t \gg 1)$$

One would of course expect something like this since the problem is analogous to fitting a straight line to an increasing number of points.

As a second check on the reasonableness of the results given, observe that the case $t \gg 1$ is essentially the same as prediction based on continuous observations. Setting $\phi_{33} = 0$, we have

$$E\tilde{x}_1^2(t|t) \cong \frac{a^2b^2t^2}{b^2 + a^2t^3/3} \qquad (t \gg 1;\ \phi_{33} = 0)$$

which is identical with the result obtained by Shinbrot [11], Example 1, and Solodovnikov [14], Example 2, in their treatment of the Wiener problem in the finite-length, continuous-data case, using an approach entirely different from ours.

Conclusions

This paper formulates and solves the Wiener problem from the "state" point of view. On the one hand, this leads to a very general treatment including cases which cause difficulties when attacked by other methods. On the other hand, the Wiener problem is shown to be closely connected with other problems in the theory of control. Much remains to be done to exploit these connections.

References

1 N. Wiener, "The Extrapolation, Interpolation and Smoothing of Stationary Time Series," John Wiley & Sons, Inc., New York, N. Y., 1949.

2 L. A. Zadeh and J. R. Ragazzini, "An Extension of Wiener's Theory of Prediction," *Journal of Applied Physics*, vol. 21, 1950, pp. 645–655.

3 H. W. Bode and C. E. Shannon, "A Simplified Derivation of Linear Least-Squares Smoothing and Prediction Theory," *Proceedings IRE*, vol. 38, 1950, pp. 417–425.

4 R. C. Booton, "An Optimization Theory for Time-Varying Linear Systems With Nonstationary Statistical Inputs," *Proceedings IRE*, vol. 40, 1952, pp. 977–981.

5 J. H. Laning and R. H. Battin, "Random Processes in Automatic Control," McGraw-Hill Book Company, Inc., New York, N.Y., 1956.

6 W. B. Davenport, Jr., and W. L. Root, "An Introduction to the Theory of Random Signals and Noise," McGraw-Hill Book Company, Inc., New York, N. Y., 1958.

7 S. Darlington, "Linear Least-Squares Smoothing and Prediction, With Applications," *Bell System Tech. Journal*, vol. 37, 1958, pp. 1221–1294.

8 G. Franklin, "The Optimum Synthesis of Sampled-Data Systems," Doctoral dissertation, Dept. of Elect. Engr., Columbia University, 1955.

9 A. B. Lees, "Interpolation and Extrapolation of Sampled Data," *Trans. IRE Prof. Group on Information Theory*, IT-2, 1956, pp. 173–175.

10 R. C. Davis, "On the Theory of Prediction of Nonstationary Stochastic Processes," *Journal of Applied Physics*, vol. 23, 1952, pp. 1047–1053.

11 M. Shinbrot, "Optimization of Time-Varying Linear Systems With Nonstationary Inputs," Trans. ASME, vol. 80, 1958, pp. 457–462.

12 M. Blum, "Recursion Formulas for Growing Memory Digital Filters," *Trans. IRE Prof. Group on Information Theory*, IT-4, 1958, pp. 24–30.

13 V. S. Pugachev, "The Use of Canonical Expansions of Random Functions in Determining an Optimum Linear System," *Automatics and Remote Control* (USSR), vol. 17, 1956, pp. 489–499; translation pp. 545–556.

14 V. V. Solodovnikov and A. M. Batkov, "On the Theory of Self-Optimizing Systems (in German and Russian)," Proc. Heidelberg Conference on Automatic Control, 1956, pp. 308–323.

15 J. L. Doob, "Stochastic Processes," John Wiley & Sons, Inc., New York, N. Y., 1955.

16 M. Loève, "Probability Theory," Van Nostrand Company, Inc., New York, N. Y., 1955.

17 R. E. Bellman, I. Glicksberg, and O. A. Gross, "Some Aspects of the Mathematical Theory of Control Processes," RAND Report R-313, 1958, 244 pp.

18 R. E. Kalman and R. W. Koepcke, "Optimal Synthesis of Linear Sampling Control Systems Using Generalized Performance Indexes," Trans. ASME, vol. 80, 1958, pp. 1820–1826.

19 J. E. Bertram, "Effect of Quantization in Sampled-Feedback Systems," *Trans. AIEE*, vol. 77, II, 1958, pp. 177–182.

20 R. E. Kalman and J. E. Bertram, "General Synthesis Procedure for Computer Control of Single and Multi-Loop Linear Systems," *Trans. AIEE*, vol. 77, II, 1958, pp. 602–609.

21 C. W. Merriam, III, "A Class of Optimum Control Systems," *Journal of the Franklin Institute*, vol. 267, 1959, pp. 267–281.

22 R. E. Kalman and J. E. Bertram, "A Unified Approach to the Theory of Sampling Systems," *Journal of the Franklin Institute*, vol. 267, 1959, pp. 405–436.

23 R. E. Kalman and R. W. Koepcke, "The Role of Digital Computers in the Dynamic Optimization of Chemical Reactors," Proc. Western Joint Computer Conference, 1959, pp. 107–116.

24 R. E. Kalman, "Dynamic Optimization of Linear Control Systems, I. Theory," to appear.

25 S. Sherman, "Non-Mean-Square Error Criteria," *Trans. IRE Prof. Group on Information Theory*, IT-4, 1958, pp. 125–126.

26 V. S. Pugachev, "On a Possible General Solution of the Problem of Determining Optimum Dynamic Systems," *Automatics and Remote Control* (USSR), vol. 17, 1956, pp. 585–589.

27 G. C. Newton, Jr., L. A. Gould, and J. F. Kaiser, "Analytical Design of Linear Feedback Controls," John Wiley & Sons, Inc., New York, N. Y., 1957.

28 O. J. M. Smith, "Feedback Control Systems," McGraw-Hill Book Company, Inc., New York, N. Y., 1958.

29 R. E. Kalman, "On the General Theory of Control Systems," Proceedings First International Conference on Automatic Control, Moscow, USSR, 1960.

APPENDIX

RANDOM PROCESSES: BASIC CONCEPTS

For convenience of the reader, we review here some elementary definitions and facts about probability and random processes. Everything is presented with the utmost possible simplicity; for greater depth and breadth, consult Laning and Battin [5] or Doob [15].

A *random variable* is a function whose values depend on the outcome of a chance event. The *values* of a random variable may be any convenient mathematical entities; real or complex numbers, vectors, etc. For simplicity, we shall consider here only real-valued random variables, but this is no real restriction. Random variables will be denoted by $x, y, \ldots$ and their values by $\xi, \eta, \ldots$. Sums, products, and functions of random variables are also random variables.

A random variable x can be explicitly defined by stating the probability that x is less than or equal to some real constant ξ. This is expressed symbolically by writing

$$Pr(x \leq \xi) = F_x(\xi); \quad F_x(-\infty) = 0, F_x(+\infty) = 1$$

$F_x(\xi)$ is called the *probability distribution function* of the random variable x. When $F_x(\xi)$ is differentiable with respect to ξ, then $f_x(\xi) = dF_x(\xi)/d\xi$ is called the *probability density function* of x.

The *expected value* (*mathematical expectation, statistical average, ensemble average, mean*, etc., are commonly used synonyms) of any nonrandom function $g(x)$ of a random variable x is defined by

$$Eg(x) = E[g(x)] = \int_{-\infty}^{\infty} g(\xi)dF_x(\xi) = \int_{-\infty}^{\infty} g(\xi)f_x(\xi)d\xi \tag{40}$$

As indicated, it is often convenient to omit the brackets after the symbol E. A sequence of random variables (finite or infinite)

$$\{x(t)\} = \ldots, x(-1), x(0), x(1), \ldots \tag{41}$$

is called a *discrete* (or *discrete-parameter*) *random* (or *stochastic*) *process*. One particular set of observed values of the random process (41)

$$\ldots, \xi(-1), \xi(0), \xi(1), \ldots$$

is called a *realization* (or a *sample function*) of the process. Intuitively, a random process is simply a set of random variables which are indexed in such a way as to bring the notion of time into the picture.

A random process is *uncorrelated* if

$$Ex(t)x(s) = Ex(t)Ex(s) \qquad (t \neq s)$$

If, furthermore,

$$Ex(t)x(s) = 0 \qquad (t \neq s)$$

then the random process is *orthogonal*. Any uncorrelated random process can be changed into orthogonal random process by replacing $x(t)$ by $x'(t) = x(t) - Ex(t)$ since then

$$\begin{aligned} Ex'(t)x'(s) &= E[x(t) - Ex(t)]\cdot[x(s) - Ex(s)] \\ &= Ex(t)x(s) - Ex(t)Ex(s) = 0. \end{aligned}$$

It is useful to remember that, if a random process is orthogonal, then

$$E[x(t_1) + x(t_2) + \ldots]^2 = Ex^2(t_1) + Ex^2(t_2) + \ldots \quad (t_1 \neq t_2 \neq \ldots)$$

If $\mathbf{x}$ is a vector-valued random variable with components $x_1, \ldots, x_n$ (which are of course random variables), the matrix

$$\begin{aligned} [E(x_i - Ex_i)(x_j - Ex_j)] &= E(\mathbf{x} - E\mathbf{x})(\mathbf{x}' - E\mathbf{x}') \\ &= \operatorname{cov} \mathbf{x} \end{aligned} \tag{42}$$

is called the *covariance matrix* of x.

A random process may be specified explicitly by stating the probability of simultaneous occurrence of any finite number of events of the type

$x(t_1) \leq \xi_1, \ldots, x(t_n) \leq \xi_n$; $(t_1 \neq \ldots \neq t_n)$, i.e.,

$$Pr[(x(t_1) \leq \xi_1, \ldots, x(t_n) \leq \xi_n)] = F_{x(t_1),\ldots,x(t_n)}(\xi_1, \ldots, \xi_n) \tag{43}$$

where $F_{x(t_1),\ldots,x(t_n)}$ is called the *joint probability distribution function* of the random variables $x(t_1), \ldots, x(t_n)$. The *joint probability density function* is then

$$f_{x(t_1),\ldots,x(t_n)}(\xi_1, \ldots, \xi_n) = \partial^n F_{x(t_1),\ldots,x(t_n)}/\partial\xi_1, \ldots, \partial\xi_n$$

provided the required derivatives exist. The expected value $Eg[x(t_1), \ldots, x(t_n)]$ of any nonrandom function of n random variables is defined by an n-fold integral analogous to (40).

A random process is *independent* if for any finite $t_1 \neq \ldots \neq t_n$, (43) is equal to the product of the first-order distributions

$$Pr[x(t_1) \leq \xi_1] \ldots Pr[x(t_n) \leq \xi_n]$$

If a set of random variables is independent, then they are obviously also uncorrelated. The converse is not true in general. For a set of more than 2 random variables to be independent, it is not sufficient that any pair of random variables be independent.

Frequently it is of interest to consider the probability distribution of a random variable $x(t_{n+1})$ of a random process given the actual values $\xi(t_1), \ldots, \xi(t_n)$ with which the random variables $x(t_1), \ldots, x(t_n)$ have occurred. This is denoted by

$$\begin{aligned} Pr[x(t_{n+1}) &\leq \xi_{n+1} | x(t_1) = \xi_1, \ldots, x(t_n) = \xi_n] \\ &= \frac{\int_{-\infty}^{\xi_{n+1}} f_{x(t_1),\ldots,x(t_{n+1})}(\xi_1, \ldots, \xi_{n+1})d\xi_{n+1}}{f_{x(t_1),\ldots,x(t_n)}(\xi_1, \ldots, \xi_n)} \end{aligned} \tag{44}$$

which is called the *conditional probability distribution function* of $x(t_{n+1})$ given $x(t_1), \ldots, x(t_n)$. The *conditional expectation*

$$E\{g[x(t_{n+1})]|x(t_1), \ldots, x(t_n)\}$$

is defined analogously to (40). The conditional expectation is a random variable; it follows that

$$E[E\{g[x(t_{n+1})]|x(t_1), \ldots, x(t_n)\}] = E\{g[x(t_{n+1})]\}$$

In all cases of interest in this paper, integrals of the type (40) or (44) need never be evaluated explicitly; only the *concept* of the expected value is needed.

A random variable x is *gaussian* (or *normally distributed*) if

$$f_x(\xi) = \frac{1}{[2\pi E(x - Ex)^2]^{1/2}} \exp\left[-\frac{1}{2}\frac{(\xi - Ex)^2}{E(x - Ex)^2}\right]$$

which is the well-known bell-shaped curve. Similarly, a random vector $\mathbf{x}$ is *gaussian* if

$$f_x(\xi) = \frac{1}{(2\pi)^{n/2}(\det \mathbf{C})^{1/2}} \exp\left[-\frac{1}{2}(\xi - E\mathbf{x})'\mathbf{C}^{-1}(\xi - E\mathbf{x})\right]$$

where $\mathbf{C}^{-1}$ is the inverse of the covariance matrix (42) of $\mathbf{x}$. A *gaussian random process* is defined similarly.

The importance of gaussian random variables and processes is largely due to the following facts:

Theorem 5. *(A) Linear functions (and therefore conditional expectations) on a gaussian random process are gaussian random variables.*

(B) Orthogonal gaussian random variables are independent.

(C) Given any random process with means $Ex(t)$ and covariances $Ex(t)x(s)$, there exists a unique gaussian random process with the same means and covariances.

Dual Control Theory

A. A. FELDBAUM

In broad terms, a *stochastic optimum control problem* involves a stochastic dynamical system whose controlled input is to be selected based on *a priori* known statistical information on underlying random quantities (such as initial state, system noise, actuator noise, and parameter uncertainty) as well as on on-line noise-corrupted measurements of the system state and possibly of other random variables, so as to minimize the expected value of a performance index (equivalently, cost function, or loss function), with the planning horizon being finite or infinite. If the multiple inputs of the dynamic system do not all have access to the same measurement channel, then we are in the realm of *stochastic dynamic teams*, and if furthermore the stations (decision makers) controlling these different inputs have different performance indices to optimize, then what we have is a *stochastic dynamic game*.

In the 1950s, when the excitement in the development of modern control theory was high, stochastic optimum control theory was still in its infancy—this being the case in spite of the existence of powerful tools already in place in several neighboring fields and topics. Among these were Wiener's theory of filtering and estimation [18]; Wald's theory of sequential decision making [17]; Von Neumann's theory of games [16], also extended to stochastic frameworks [4]; Shannon's theory of information transmission [12]; and Bellman's dynamic programming [3]. Stochastic control problems carried many elements in common with these existing theories, but none of them alone would provide the necessary tools to address the underlying challenges. What was also needed was a precise, conceptually consistent, and mathematically satisfying formulation of control problems under statistical uncertainty, which would clearly be the first step toward obtaining a comprehensive theory of stochastic optimum control.

The four-part paper by Feldbaum (whose first two parts are included here) constitutes the first attempt to address that need, by introducing a conceptual framework within which a stochastic optimum control problem can be formulated, and some key features of an optimum solution can be identified. In this work, Feldbaum adopts a discrete-time framework (recognizing that a formulation in continuous time would require a more elaborate set of mathematical tools) and studies the case of finite horizon and stage-additive cost. Furthermore, he works with fairly simple models, where the system to be controlled is scalar and memoryless, which, however, does not lead to much loss of generality for the main message he wants to convey. Part IV of the paper [5] and Feldbaum's follow-up book [6] discuss extensions to multi-input, multi-output systems with memory.

The single most important result of the paper, and one that has had impact on research in stochastic and adaptive control, is the recognition that control generally has a *dual* role in stochastic optimum control problems. One of these is the "action" role, which is what one would have in deterministic optimum control problems, where the control tries to move the system state toward a desired target value while optimizing a given performance index. Because of the memoryless assumption, this task in fact reduces to that of independent minimization of the corresponding stagewise cost at each stage. The second role of control is the "investigative" one (the terminology used by Feldbaum), which is more commonly known in the current literature as *probing*. Recognizing that "higher quality" information acquired by the controller will generally lead to better performance, the controller will attempt to *shape* the signals carried to future stages in such a way that the information content of the measurement received by the control at future stages will be enhanced. This *probing* role of control is in general conflicting with its *action* role, and hence an optimal controller is one that achieves an "optimal trade-off" between these two conflicting objectives. In Feldbaum's work, and for the class of models considered, this trade-off is made explicit by showing that the total risk (expected value of a stage-additive loss function) can be written as the sum of two risks, *action risk* and *investigation* (*probing*) *risk*. Within the class of memoryless systems, Feldbaum shows that this decomposition is a genuine property of *closed-loop* (or *feedback*) systems, where the output is fed back into the system through a noisy channel, whereas in *open-loop* (or what is known today as *feedforward*) systems, the second component (investigation risk) is absent. In both cases, Feldbaum makes effective use of "dynamic programming" to obtain formulas for the optimum control laws, which he shows can be taken as *deterministic* maps, rather than stochastic rules which had been at the time a "rule of the land" in the neighboring fields of statistical decision theory [17] and game theory [4].

Stochastic control problems which do not exhibit a "dual" role for control are known today as *neutral* problems, where the conditional probability distribution of the state vector given past and present measurements, past control actions, and past control laws (or policies) is independent of the control laws [20]. The main implication of this property is that the "quality" of the information carried to future stages cannot be affected by the choice of the control policies in the past, thus allowing for a two-step derivation of the optimum controller: First determine the conditional probability distribution (*cpd*) of the state, and express the (action) risk function in terms of this quantity and the control (yet to be determined), and subsequently minimize the new risk function over all control laws as functions of the *cpd*, which provides *sufficient statistics* for the stochastic control problem. This property of neutral stochastic control problems is known in the literature as the *separation* of estimation and control (or loosely referred to as the *separation principle*), and is implicit in the work of Feldbaum, though shown only for a restricted class of problems.

A more comprehensive treatment of neutral stochastic control problems has later been provided by Striebel [14] in the discrete time, and Wonham [21] in the continuous time, among others. An important subclass of these problems is the linear-quadratic-Gaussian (LQG) problem where the system dynamics are linear in the state, control, and system noise, the measurement is linear in the state and measurement noise, the loss function is quadratic in the state and the control, and the system and measurement noises are Gaussian and independent, and so is the initial state. For this class, an even stronger version of the separation principle applies, where the minimization-relevant part of the risk as a function of the *cdp* depends only on the conditional mean, and hence the minimization problem faced by the controller is identical to the one where all random quantities are replaced by their mean values. This feature, where the control and estimation tasks are completely separated and therefore can be carried out independently, is known as *certainty equivalence*—a term first coined by Simon [13], when he encountered a similar feature in the solution of a simpler scalar model (with perfect state measurement) motivated by inventory control. Extension to higher dimensional models, but again under perfect state measurements, was carried out shortly after by Theil [15], but it took another four years before its counterpart was derived for the noisy state measurement case, by Joseph and Tou [9], and Gunckel and Franklin [7], with the most complete and rigorous derivation of the solution to the LQG problem in the early 60's provided by Shaw [11]. Of course, the availability of the Kalman filter [10] in the early 60's (see the preamble to [10] which is included in this volume) has facilitated the development of a complete solution here, since it essentially solved the estimation part of the design.

If a stochastic control problem is not neutral, then the *dual* role of control (as elucidated by Feldbaum) becomes dominant. Such problem formulations arise in various contexts, such as when control has limited or no memory, when system dynamics depend on some parameters which are correlated across stages (what is now more commonly known as *stochastic adaptive control*; see the preamble to [1] included in this volume), or stochastic teams where different control stations have access to different but correlated measurement channels. The well-known counterexample of Witsenhausen [19] provides perhaps the simplest model (two-stage, scalar, LQG, but memoryless controller) that depicts eloquently the conflicting role of the control at the first stage, between "action now" and "maximum information transmission" to the next stage for the benefit of the action then. Even though explicit solutions to some non-neutral stochastic control and team problems have been obtained in the past (see, for instance, [8], [2]), a general theory is still lacking, and the concept of *dual control* introduced by Feldbaum in these papers will undoubtedly continue to guide research efforts in stochastic control in the twenty-first century.

References

[1] K.J. Åström and B. Wittenmark, "On self-tuning regulators," *Automatica*, **9**:185–199, 1973.

[2] T. Başar, "Solutions to a class of nonstandard stochastic control problems with active learning," *IEEE Trans. Automat. Cont.*, **AC-33**(12):1122–1129, 1988.

[3] R. Bellman, *Dynamic Programming*, Princeton Univ. Press (Englewood Cliffs, NJ), 1957.

[4] D. Blackwell and M.A. Girshick, *Theory of Games and Statistical Decisions*, Wiley (New York), 1954.

[5] A.A. Feldbaum, "The theory of dual control, III & IV," *Automation and Remote Control*, **22**:1–12 & 109–121, 1961.

[6] A.A. Feldbaum, *Optimal Control Systems*, Academic Press (New York), 1965.

[7] T.F. Gunckel and G.F. Franklin, "A general solution for linear sampled data control," *Trans ASME, J. Basic Eng.*, **85-D**:197–201, 1963.

[8] Y.-C. Ho, "Team decision theory and information structures," *Proc. IEEE*, **68**:644–654, 1980.

[9] P.D. Joseph and J.T. Tou, "On linear control theory," *Trans AIEE*, Part III, **80**(18):193–196, 1961.

[10] R.E. Kalman, "A new approach to linear filtering and prediction problems," *Trans. ASME, J. Basic Eng.*, **82-D**(1):35–45, 1960.

[11] L. Shaw, "Optimum stochastic control," in *Disciplines and Techniques of Systems Control*, J. Peshcon, editor, chapter IV, Blaisdell, 1965, pp. 125–185.

[12] C.E. Shannon, "A mathematical theory of communication," *Bell System Techn. J.*, **27**(3), 1948.

[13] H.A. Simon, "Dynamic programming under uncertainty with a quadratic criterion function," *Econometrica*, **24**:74–81, 1956.

[14] C. Striebel, "Sufficient statistics in the optimum control of stochastic systems," *J. Math. Anal. Appl.*, **12**:576–592, 1965.

[15] H. Theil, "A note on certainty equivalence in dynamic planning," *Econometrica*, **25**:346–349, 1957.

[16] J. von Neumann and O. Morgenstern, *Theory of Games and Economic Behavior*, Princeton University Press (Princeton, NJ), 1947.

[17] A. Wald, *Statistical Decision Functions*, Wiley (New York), 1950.

[18] N. Wiener, *Extrapolation, Interpolation, and Smoothing of Stationary Time Series*, MIT Press (Cambridge, MA), 1949.

[19] H.S. Witsenhausen, "A counterexample in stochastic optimum control," *SIAM J. Contr.*, **6**(1):131–147, 1968.

[20] H.S. Witsenhausen, "Separation of estimation and control for discrete time systems," *Proceedings of the IEEE*, **59**(11):1557–1566, 1971.

[21] W.M. Wonham, "On the separation thorem of stochastic control," *SIAM J. Contr.*, **6**(2):312–326, 1968.

T.B.

DUAL-CONTROL THEORY. I

A. A. Fel'dbaum

Moscow

Translated from Avtomatika i Telemekhanika, Vol. 21, No. 9, pp. 1240-1249, September, 1960

Original article submitted March 23, 1960

Some fundamental problems in communication theory and control theory are compared. The problem of designing an optimum (in the statistical sense) closed-loop dual-control system, is formulated. Its solution, as well as examples and some generalizations, will be given in parts II, III, and IV.

Introduction

A general block diagram of signal transmission, as investigated in communication theory is shown in Fig. 1. The transmitted signal x^* proceeds from the transmitting device A to the communication channel H^*. The mixing of signal and interference (noise) h^* now takes place. The resultant signal y^* represents a mixture of the transmitted signal and the interference. The resultant signal proceeds to the input of receiver B. The optimum receiver problem consists in obtaining the signal $\underline{x}$ at its output, such that it is, in a specified sense, closest to the transmitted signal x^* or to some transformation of the signal x^*. The mathematical side of the problems related to such systems has been the subject of important investigations by A. N. Kolmogorov [1], N. Wiener [2], C. E. Shannon [3], and A. Wald [4]. This type of system was investigated in the works on communication theory by V. A. Kotel'nikov [5], D. Middleton, D. Van Meter [6], and others. The cited works differ in their various approaches to the problem, but are all basically concerned with the investigation of the scheme represented by the block diagram in Fig. 1. The results obtained in the above-cited works, and in particular the Kolmogorov-Wiener theory, have proved useful in formulating the statistical theory of automatic-control systems. This theory has been expounded in the books of V. S. Pugachev [7], J. H. Laning, Jr., and R. H. Battin [8], and others. The fullest consideration has been given to the theory of linear systems. If a system is linear, then whatever the closed-loop system, it is easy to obtain an open-loop system equivalent to it. That is why the automatic-control systems, which, as a rule, are closed-loop systems, enable one to use the scheme as in Fig. 1, provided that the system is linear. Some complications and difficulties arise when the interference does not appear at the input of the system, but at the input of the controlled object — the latter being inside the closed-loop network. This also creates difficulties which are not, however, of a fundamental nature. More serious difficulties arise due to the bounds to which the power of the system's signals are subjected. The problem becomes more involved when the controlled object is nonlinear or it is required that an optimum control system, which often proves to be nonlinear, is designed. It is not always possible to proceed from an open nonlinear system to an equivalent closed-loop one; furthermore, this is an extremely involved process. In such a case, the open-loop scheme depicted in Fig. 1 cannot be used in practice. A number of attempts have been made to reduce approximately nonlinear systems to equivalent linear ones (see, for example, the paper of I. E. Kazakov [9]). Such studies are of considerable practical value but do not, in their present state, provide a means of estimating how close the obtained approximation is to the true solution; neither do they enable one to synthesize the optimum system.

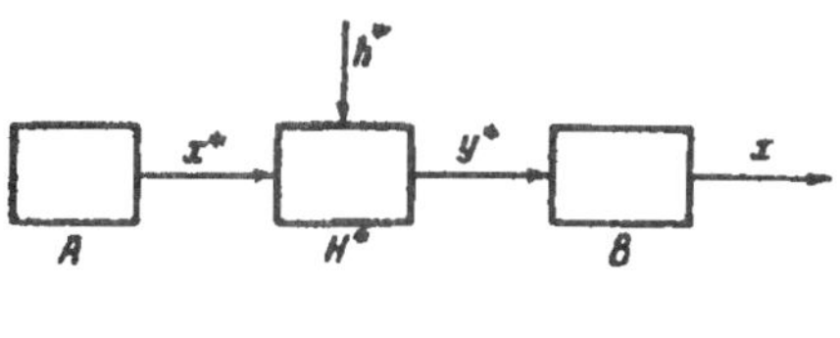

Fig. 1.

In order to be able to solve optimum problems of the control theory, a fundamentally different approach is required. Firstly, a different block diagram replacing that depicted in Fig. 1 is needed. Before selecting a common scheme to be used in automatic-control theory, it seems advisable to have a preliminary survey of certain basic concepts of the theory.

Figure 2, a shows the controlled object B with $\underline{x}$ as its output, $\underline{u}$ as the controller, and $\underline{z}$ as the disturbance (interference). When the system has several inputs and outputs, one can regard $\underline{x}$, $\underline{u}$, and $\underline{z}$ as vector quantities.

The output $\underline{x}$ depends on $\underline{u}$ and $\underline{z}$. This dependence can be described either by a linear or a nonlinear operator, and in a particular case of memoryless systems, only by a function. The interference $\underline{z}$ is generally a function

Reprinted with permission from *Automation and Remote Control*, A. A. Feldbaum, "Dual Control Theory, Parts I and II," Vol. 21, April 1961, pp. 874–880 and May 1961, pp. 1033–1039.

of time. Thus, since a change in the system's characteristics can be considered a particular result of interference (e.g., the parametric effect), then hereafter everything in the system's characteristics that changes with time will be attributed to interference. If, for example, x depends on u as in

$$x = [a_0 + f_0(t)]\, u^2 + [a_1 + f_1(t)]\, u + a_2 + f_2(t), \quad (1)$$

then the vector

$$z = \{z_0, z_1, z_2\},$$
$$z_0 = f_0(t), \; z_1 = f_1(t), \; z_2 = f_2(t) \quad (2)$$

gives the interference or disturbance, and the formula

$$x = [a_0 + z_0]\, u^2 + [a_1 + z_1]\, u + a_2 + z_2 \quad (3)$$

represents a particular operator.

If the object has memory and if, for example, its motion can be described by a differential equation of the nth order, then its state is considered as one of the characteristics of the object as described by the value of the vector x in the n-dimensional phase space.

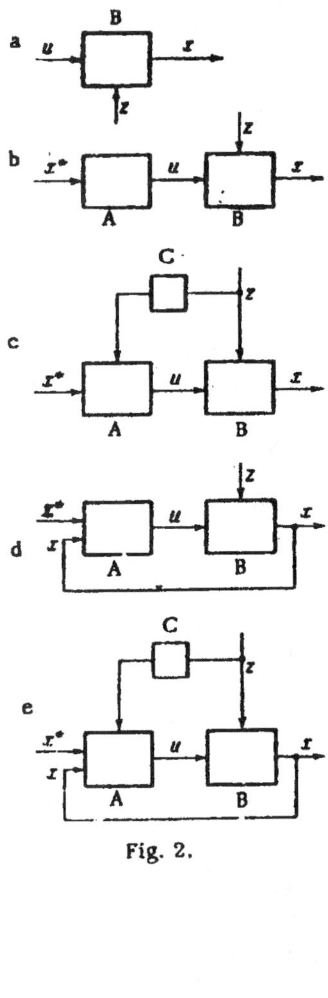

Fig. 2.

Complete information about a system thus consists of information about its operator, interference (noise), and the state of the system. The controlling device may be given and is considered as known. The open-loop systems are automatic-control systems of a simple type. A block diagram of an open-loop system is shown in Fig. 2, b and is of the same character as the one given in Fig. 1. The exciting quantity x* enters the input of the regulating member A, determining how the output quantity x should vary. The output u of the regulating device enters the input of the object B under control. The output x of the controlled object B does not proceed in this case to the regulating member A.

The required rule of change in x can only be implemented when full information about the controlled system is available, i.e., when its operator and state x are known, at least at the initial moment of time, as well as the interference z. The latter should be known a priori at all moments of time, including future moments. The required rule of change in x must be one of the admissible ones, such that it can be implemented for the given class of initial states of the system and for a class of controlling motions u staying within acceptable bounds.

The above conditions, and, in particular, the a priori full knowledge of the controlled system, cannot be satisfied in practice. This is why an accurate implementation of the required control rule cannot be obtained. Sometimes the interference z is not known a priori, but it is possible to measure it with a device C (see Fig. 2, c) and to introduce the outcome of the measurement into the controlling member A. One can then find in the latter the required controlling rule u. Such a scheme is also an open-loop one. But the scheme depicted in Fig. 2,c differs in some ways from the scheme with complete a priori information about the system, as now future magnitudes of the interference z remain unknown. Because of this, the exact implementation of the required rule of variation of the controlling quantity x is not always feasible.

When the state x of the system is not known, then, generally speaking, it is not possible to implement the required rule of the change in x. To be able to attain the required variation in x, or one near to it, a feedback network is needed to feed the output quantity x to the input of the controlling member A (see Fig. 2, d). Having compared x and x*, the controlling member generates the regulating action u, bringing x to its required value. The block diagram of Fig. 2, d is a closed-loop scheme, and is of the utmost importance in the automatic-control theory.

A closed-loop network offers far-reaching possibilities not available in an open-loop system. For example, it may be possible for a class of objects B of control to obtain a process x close to the one required even when the interference z remains unknown and incapable of measurement. Let, for example, the interference z, together with the controlling action u, be applied at the input of the controlled object B, the latter representing an inertia member. If the quantity $x^* = x$ is to be attained, the regulating member can be implemented in the form of an amplifier of high gain $k >> 1$, the difference $x^* - x$ being sent to its input.

It is not difficult to see that the requirement $x^* = x$ will be satisfied approximately, whatever the continuously varying interference z, provided k is sufficiently high and the bounds of variability of u are such that the interference z can be compensated by u.

This principle of neutralizing the interference can be generalized and applied to cases considerably more involved, combining the system's accuracy with its stability. A detailed analysis of the applicability of this principle was carried out by M. V. Meerov in his monograph [10].

When the interference z can be measured, it is possible to implement a combined system (see Fig. 2, e) of measurement of the state of the controlled system x as well as of its interference. Such systems are of considerable practical value. We shall not, however, concern ourselves with them but shall limit the study to the "pure" type of closed-loop systems, considering them as being of primary importance.

The input quantity x^* may be previously unknown; usually, neither do we have any prior knowledge of the interference z. Consequently, these processes become random, and, in a favorable case, the a priori information is limited to our knowledge of their statistical characteristics. Such processes may be regarded as belonging to a class of curves $x^*(\lambda)$ and $z(\mu)$ where λ and μ are parameter vectors $(\lambda_1, \ldots, \lambda_q)$ and $(\mu_1, \ldots, \mu_m)$, respectively, with their probability distributions either known or unknown.

In communication channels connecting the blocks of a system, the errors of measurement or noise can be regarded as subsidiary random processes as well, with either known or unknown characteristics. Thus, the analysis of a control system and the synthesis of the regulating member can be regarded as problems of a statistical nature. The problem should be solved for an over-all block diagram in which all the above features of an automatic-control system are reflected. Such a block diagram is depicted in Fig. 3; it is the subject of the present paper as well as that of further papers in this series.

The input quantity x^* proceeds to the input of the controlling member A through channel H^* where it becomes mixed with noise h^*. Thus, the quantity y^* entering the input of A is generally not equal to the actual value of the input quantity x^*. There also exists a class of systems with the external input x^* altogether absent. Generally speaking, however, it cannot be neglected. A similar mixing takes place of the state x of controlled object B and noise h in channel H; quantity y entering A will not, as a rule, be equal to x. The regulating action u proceeds next from A to controlled object B having previously passed through channel G where it was mixed with noise g. The quantity v proceeding to the controlled object is not, as a rule, equal to u.

Dual Control

One cannot neutralize, in a general case, the interference z by a regulation u if the interference z is not known. Its direct measurement is not, as a matter of fact, often possible. In such a case, an open-loop system is useless. But the closed system in Fig. 3 shows how z can be indirectly determined by measuring the input and the output, in and out of object B, by studying its characteristics. The input of controlling member A enters both the input v and the output x of the object or, in any case, the quantities u and y related to v and x. The examination of the quantities u and y provides information on the characteristics of object B. It should be understood that this information is never complete, as the noises g and h render an exact measurement of B's characteristics impossible; if the actual form of the object's operator is not known either, a full determination of its characteristics would not be possible even in the absence of noise, unless the determination time is infinitely great. The lack of complete information on the disturbance z can assume the form of an a posteriori probability distribution of its parameters. Although the latter does not provide precise values of the parameters, it is more accurate than an a priori distribution, as the former reflects the real character of the interference.

If the random process can be measured directly, one is able eventually to specify its statistical characteristics more accurately. The method which provides such improvement with the aid of dynamic programming was discussed in examples by R. Bellman and R. Kalaba [11, 12] and also by M. Freimer [13]. One is able to find the characteristics of the process x^* more accurately in the open part of the block diagram in Fig. 3, or in a similar scheme in Fig. 1.

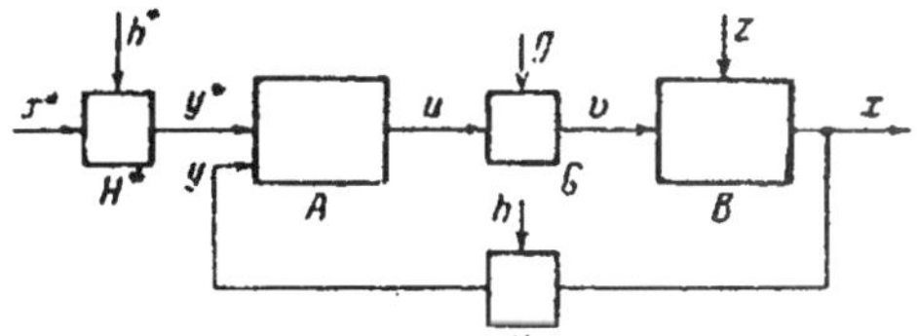

Fig. 3.

This formulation of the problem is characteristic for an open system. In a closed system its formulation becomes totally different. It is shown that some processes in the system of Fig. 3 may occur which have no counterpart in open-loop systems. Whereas open systems can only be studied by passive observations, the study may develop into an active one in closed systems. In order to improve the investigation one may vary the signals u (or v) which act on the controlled object B. The object is, as it were, "reconnoitered" by signals of an enquiring character whose purpose it is to promote a more rapid and more accurate study of the object's characteristics and of the methods of controlling it.

However, the controlling movements are necessary not only to study or to learn the characteristics of the object or the ways of controlling it, but also to implement the regulation, to direct the object to the required state. Thus, the controlling effects in the general block diagram in Fig. 3 must be twofold: they must, to a certain extent, be investigating as well as directing.

The control whose regulating effects are of this twofold character will in the sequel be called dual control; the papers in the present series will be devoted to the theory of dual control.

Dual control is particularly useful and even indispensible in cases where the operator and the interference z in the object B are complex, and the object is thus distinguished either by its complexity or by the variability of its characteristics. Some typical examples of systems with dual control are to be found in automatic search systems, in particular, in automatic optimization systems (see, for example, [14 and 15]). In these systems, the investigating or "trial" part can usually be separated easily from the controlling or "operating" part of the signal, either by the difference in their frequency ranges or because they interweave in time. Such a separation, however, need not always take place; an effect can be twofold in character by virtue of being partly diagnostic and partly regulating.

Thus, in dual-control systems, there is a conflict between the two sides of the controlling process, the investigational and the directional. An efficient control can only be effected by a well-timed action on the object. A delayed action weakens the control process. But the control can only be effective when the properties of the object are sufficiently well known; one needs, however, more time to become familiar with them. A too "hasty" controlling member will carry out the operational movement without making proper use of the results of trial investigations performed on the object. A too "cautious" system will bide its time unnecessarily long and process the received information without directing the object to its required state at the right time. In each case, the control process may not prove the best one and may not even prove to be up to the mark. Our problem is to find out, one way or another, which combination of these two sides of the regulation would prove to be most suitable. The operations must be so selected as to maximize a criterion of the control's quality.

As shown above, the incomplete information about the object will be expressed by the presence of the probability distributions of potentially possible characteristics of the object. The regulating member compares, as it were, the various hypotheses on the object, with probability of its occurrence being attached to each hypothesis. These probabilities vary with time. There may be a control method such that the most probable hypothesis will always be selected and, therefore, assuming that it is valid, the optimum control method will be attained. Such a control system is not generally optimum in the absolute meaning of the word as the complete information on the object has not been utilized. The probability distribution of the different hypotheses extracted from the experiments is distorted as the probability 1 was ascribed to one of them and the probability 0 to others. A better control method will be one whereby the probabilities of all the hypotheses would be taken into account.

The probability distribution of hypotheses will vary with time, the higher probabilities concentrating more and more in the region of those hypotheses which approach the true characteristics of the object. The pace of concentration and, therefore, the success of the subsequent regulating movements, depends on the character of the preceding regulating movements, on how well they have "sounded" the object. Thus, two factors should be taken into account by the controlling member which decides the specific amount of regulating movement at any given moment of time:

(a) The loss occurring in the value of the quality criterion due to the fact that the outcome of the operation at a given moment, and at subsequent moments of time, will cause a deviation of the object either from the required state or from the best attainable one. The average value of this loss shall be called the action risk.

(b) The loss occurring in the value of the quality criterion due to the fact that the magnitude of the controlling action has not proved the best to obtain information on the characteristics of the object; in view of this, the subsequent actions will not be the best posssible ones either. The average value of this loss shall be called the investigation risk.

It will be shown that for a certain class of systems, the total risk will be equal to the sum of the action and investigation risks.

All systems of automatic search (see [14]) are characterized by trial actions. Dual control, therefore, is applicable to all systems of automatic search and, in particular, to automatic optimization systems. It can also be applied to other types of closed-loop systems which do not belong to the automatic search class at all. To illustrate the difference between the two types of dual-control systems, a few examples will be given.

Figure 4,a shows a system which operates as follows: the main regulating member A implements the control of object B, either in an open- or in a closed-loop network (the closed one is indicated by a dashed line). The

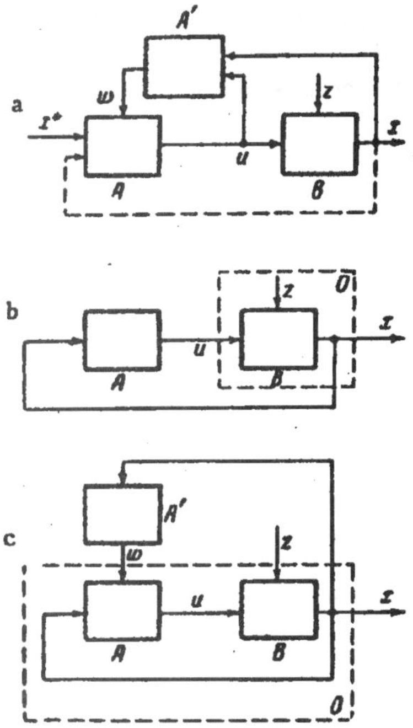

Fig. 4.

regulating movements u are of the investigational and action type simultaneously. The quantities u and x, from the input and the output of object B, respectively, enter an additional controlling member A'. The latter receives the characteristics of object B from the results of the investigations; subsequently, in accordance with an algorithm given in advance and fed into the device from outside, the parameters of controlling member A are so computed that its controlling action is optimal. Having the results of the computation, the additional regulating action w establishes the computed optimum parameters in the main controlling member A. This process may repeat itself periodically.

Such systems contain investigational movements; however, the automatic search is absent. The parameters of member A are established, not via automatic search, but from an algorithm given in advance, from a function of the determined characteristics of object B. There is no investigational component in the operation w. The channel w is not usually found in a closed network, as the change in the coefficients of A has no effect on the coefficients of B.

A block diagram of an automatic optimization system is presented in Fig. 4,b. Here the action u is dual in character, investigating object B as well as directing it to its optimum mode of action. The latter corresponds to an extremum of quantity Q dependent on x. The optimum mode is found by means of automatic search. The latter is conducted in such a way that the information received from the investigating action u and from the output x of the system is analyzed in controlling member A. This permits determination of the regulating part of the same action u on the same object B whose input and output were investigated, so that the same quantity x which was being investigated can be changed in the right direction.

The combination of the investigating and directing operations has, so far, not constituted the whole search, but only one of its distinct features. There is no search in the system depicted in Fig. 4,a; it takes place, however, in the system of automatic optimization shown in Fig. 4,b. But both systems are of the dual-control type.

In Fig. 4,b the controlled object is inside a dashed rectangle denoted by O. In this case, it does not differ from object B. The dual control, however, can also be applied to control the entire automatic system, considered as a complex object. For example, in Fig. 4,c the complex object O inside the dashed rectangle comprises the controlling member A and the object B of control. The auxiliary controlling member A' investigates the process x and, with the aid of the controlling process w, can vary the algorithm of regulation implemented by member A. The w processes are twofold in character. The investigation of changes in the algorithm of member A and their effect on the process x results in regulating processes w, bringing the algorithm of member A to such a form that the process x will either prove admissible, favorable, or optimum, depending for what purpose the system will be used. Here an automatic search takes place in the closed network of processes $w \to x \to u$.

Statement of the Synthesis Problem of an Optimum System of Dual Control

The problem of designing an optimum, in the specified meaning of the word, controlling member A, as shown in Fig. 3, is formulated below. It is advisable when formulating the problem to make use of certain concepts of the theory of games and of A. Wald's [4] theory of statistical decisions (see D. Blackwell and M. A. Girshick [16], and also Chow [17]). In solving the variational problem as stated later in the present series of papers, use is made of the concepts of R. Bellman's dynamic programming (see, for example, [18]). In the subsequent parts of this series, the mathematical exposition may appear somewhat cumbersome but is actually quite simple. The main contents of the papers deals with further development of the concepts of automatic control briefly described above.

Consider the scheme presented in Fig. 3. The following limitations of the statement of the problem are introduced.

1) A discrete-continuous system is investigated in which the time but not the level is quantized. All magnitudes occurring in the system are considered at discrete moments of time $t = 0, 1, 2, \ldots, n$ only. Any

magnitude at the sth moment of time will carry the index s. Thus, the considered quantities are x^*_s, x_s, y_s, v_s, g_s, etc.

Such limitation enables one to simplify the computation. Moreover, in many cases this actually occurs. The transition to the continuous time can in some cases be accomplished in an intuitive manner by making the time interval between the discrete values approach zero (see Part III). One meets with considerable difficulties in more fully examining the passage to the limit.

2) The time interval, or the number of cycles n within which the process is being investigated, is assumed to be a fixed constant. In certain cases no major difficulties arise when proceeding to the limit with $n \to \infty$. A wider generalization relating to a variable number n of cycles not known beforehand would be of interest, but will not be tackled in the present paper.

3) A Bayesian problem, in which a priori densities of random variables are given, is considered. Other formulations, for example, minimax, are also of considerable interest, but far more difficult to solve. This problem could also be formulated in relation to the concept of the so-called "inductive probability" (see, for example, the paper of L. S. Schwartz, B. Harris, and A. Hauptschein [19]).

We assume that h^*_s, h_s, g_s, are sequences of independent random variables with identical distribution densities $P(h^*_s)$, $P(h_s)$, $P(g_s)$. Further, let $z_s = z(s, \mu)$, and $x^*_s = x(s, \lambda)$, where μ and λ are random parameter vectors with coordinates μ_i and λ_i, respectively:

$$\mu = (\mu_1, \ldots, \mu_m), \qquad \lambda = (\lambda_1, \ldots, \lambda_q). \tag{4}$$

The a priori probability densities $P(\mu)$ and $P(\lambda)$ are given.

4) The object B is assumed to be memoryless; in other words, the values x_s of its output depend only on the values of the input quantities z_s and v_s at the same moment of time:

$$x_s = F_0(z_s, v_s). \tag{5}$$

The functions F_0 and z_s are assumed to be finite and single-valued, continuous and differentiable.

A generalization relating to objects with memory and with x_s depending on x_r, z_r, v_r $(r < s)$ will be given in Part IV. It should be pointed out that memoryless objects are of great practical value. Namely, if the input data (initial conditions or values of parameters) are given for a certain model, and one is able to carry out experiments using this model and also to register the results, then such an object becomes equivalent to a memoryless one.

5) A simple criterion W of quality is introduced.

Let the partial loss function corresponding to the sth time moment be of the form

$$W_s = W(s, x_s, x^*_s). \tag{6}$$

Moreover, let the total loss function W for the total time be equal to the sum of partial loss functions (such a criterion shall be called a simple one):

$$W = \sum_{s=0}^{s=n} W(s, x^*_s, x_s). \tag{7}$$

The smaller the mathematical expectation of W, the better is the system. It shall be called optimum when its average risk R (i.e., the mathematical expectation M of the quantity W) is minimal. The amount of risk is given by the formula

$$R = M\{W\} = \\ = M\left\{\sum_{s=0}^{s=n} W(s, x^*_s, x_s)\right\} = \sum_{s=0}^{s=n} M\{W_s\} = \sum_{s=0}^{s=n} R_s. \tag{8}$$

Each $R_s = M\{W_s\}$ will be called a partial risk due to the sth cycle.

There may be many types of simple criteria, for example,

$$W_s = \alpha(s)\,[x_s - x^*_s]^2 \tag{9}$$

Criteria of practical importance need not always be simple, and generalizations relating to other criteria would therefore be of interest.

The formulation of the optimum strategy problem in terms of risks is not the only one in existence. There exist a number of studies in which closed systems are investigated from the point of view of the information theory (see, for example, R. L. Dobrushin's paper [20]). As the primary aim of a control system does not lie in transmitting information but in designing required processes, the formulation of the problem in the language of statistical decisions fits in better with the intrinsic nature of the problem.

6) All the quantities occurring in the sth cycle will be regarded as scalar. Our object, therefore, has only a single input v and a single output x. The exposition becomes more involved with generalizations relating to objects with several inputs and outputs (see Part IV).

7) We assume that the manner by which the signal and the noise are combined in H^*, H, or G blocks is

known and invariable, and that the blocks are memoryless. Thus,

$$\begin{aligned} v &= v(u, g), \\ y^* &= y^*(h^*, x^*), \\ y &= y(h, x). \end{aligned} \tag{10}$$

Therefore, the conditional probabilities $P(y^*|x^*)$ and $P(y|x)$ and $P(v|u)$ make sense.

8) We assume that the controlling member A generally possesses a memory and that, moreover, for the sake of generality, the algorithm of its action is a random one, i.e., the part A exhibits random strategy.

We introduce the vectors $(0 \leq s \leq n)$:

$$\begin{aligned} \mathbf{u}_s &= (u_0, u_1, \ldots, u_s), \\ \mathbf{y}_s^* &= (y_0^*, y_1^*, \ldots, y_s^*), \\ \mathbf{y}_s &= (y_0, y_1, \ldots, y_s). \end{aligned} \tag{11}$$

The controlling member can now be characterized by the probability densities

$$P_s(u_s) = \Gamma_s(u_s, \mathbf{u}_{s-1}, \mathbf{y}_s^*, \mathbf{y}_{s-1}) \quad (0 \leqslant s \leqslant n). \tag{12}$$

The problem consists in finding a sequence of functions Γ_s such that the average risk R (see [8]) becomes minimal.

LITERATURE CITED

1. A. N. Kolmogorov "Interpolation and extrapolation of stationary random sequences," Izvest. AN SSSR Ser. Matem. 5, 1 (1941).
2. N. Wiener, Extrapolation, Interpolation, and Smoothing of Stationary Time Series (J. Wiley and Sons, New York, 1949).
3. C. E. Shannon, "A mathematical theory of communication," Bell System Techn. J. 27, 3 (1948).
4. A. Wald, Statistical Decision Functions (J. Wiley and Sons, New York; Chapman and Hall, London, 1950).
5. V. A. Kotel'nikov, Theory of Potential Noise Stability [in Russian] (Gosénergoizdat, 1956).
6. D. Van Meter and D. Middleton, Modern Statistical Approaches to Reception in Communication Theory. Trans. IRE, IT-4 (Sept., 1954).
7. V. S. Pugachev, Theory of Random Functions and Its Applications to Automatic Control [in Russian] (Gostekhizdat, 1957).
8. J. H. Laning, Jr. and R. H. Battin, Random Processes in Automatic Control (McGraw-Hill, New York, 1956).
9. I. E. Kazakov, "An approximate statistical analysis of accuracy of essentially nonlinear systems," Avtomat. i Telemekh. 17, 5 (1956).*
10. M. V. Meerov, The Synthesis of Networks of Automatic Control Systems of High Accuracy [in Russian] (Fizmatgiz, 1959).
11. R. Bellman and R. Kalaba, On Communication Processes Involving Learning and Random Duration. IRE National Convention Record, Part 4 (1959).
12. R. Bellman and R. Kalaba, On Adaptive Control Processes. IRE National Convention Record, Part 4 (1959).
13. M. Freimar, A Dynamic Programming Approach to Adaptive Control Processes. IRE National Convention Record, Part 4 (1959).
14. A. A. Fel'dbaum, Computers in Automatic Systems [in Russian] (Fizmatgiz, 1959).
15. A. A. Fel'dbaum, Problems of statistical theory of automatic optimization, Proc. of the 1st. International Congress of Automatic Control (IFAC) [in Russian] (Moscow, 1960).
16. D. Blackwell and M. A. Girshick, Theory of Games and Statistical Decisions [Russian translation] (IL, 1959).
17. C. K. Chow, An Optimum Character Recognition System Using Decision Functions. IRE Trans. EC-6, No. 4 (1957).
18. R. Bellman, "Dynamic programming and stochastic control processes," Information and Control 1, 3 (Sept., 1958).
19. L. S. Schwartz, B. Harris, and A. Hauptschein, Information Rate from the Viewpoint of Inductive Probability. IRE National Convention Record, Part 4 (1959).
20. R. L. Dobrushin, "Transmission of information in channels with feedback," Teor. Ver. i ee Prim. 3, 4 (1958).

* See English translation.

DUAL CONTROL THEORY. II

A. A. Fel'dbaum

Moscow

Translated from Avtomatika i Telemekhanika, Vol. 21, No. 11, pp. 1453-1464 , November, 1960

Original article submitted March 23, 1960

Basic formulas are derived and the optimum control algorithm is determined in the general case, first for an open-loop and then for a closed-loop nonlinear system of dual control [1].

Similarities and differences between the solutions for open and closed systems are indicated.

1. Derivation of the Risk Formula in Open-Loop Systems

We start first by deriving the formulas for an open system shown in Fig. 1. This is done to demonstrate certain methods which have led to the obtained results, also to compare certain characteristics of the open and closed systems. The open system is basically more simple than the closed one, and this makes the derivation of the formulas easier in the former case.

The problem is stated as follows [1]: all quantities are functions of discrete time at the time moments 0, 1, ..., s, ..., n, where n is fixed. The input is

$$x_s^* = x_s^*(s, \lambda), \tag{1}$$

where λ is the parameter vector

$$\lambda = (\lambda_1, \ldots, \lambda_q) \tag{2}$$

with an a priori probability density $P_0(\lambda) = P(\lambda)$. The input x^* becomes mixed with noise h^* in the channel or system H^*, at whose output y^* is obtained. Statistical properties of the noise, and the method of combining the signal and the noise in H^* are known. Consequently the conditional probability $P(y_s^*/x_s^*)$ is also known, being identical for all s, as the probability density $P(h_s^*)$ of the noise is assumed as not varying with s*.

The characteristic of the controlled object B is given by the formula

$$x_s = F_0(z_s, v_s), \tag{3}$$

where F_0 is a known function, and

$$z_s = z_s(s, \mu), \tag{4}$$

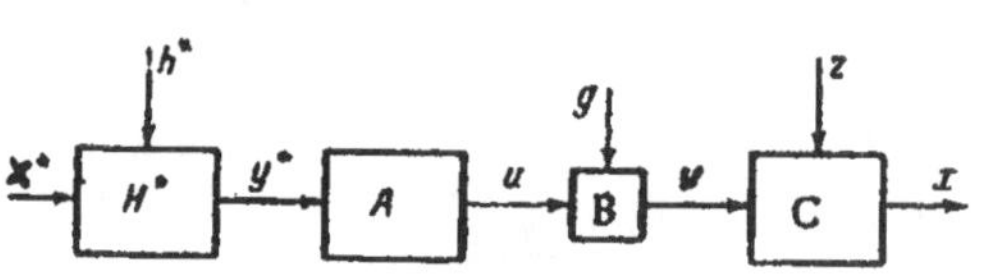

Fig. 1

in which μ is the parameter vector

$$\mu = (\mu_1, \mu_2, \ldots, \mu_m) \tag{5}$$

with given a priori probability density $P_0(\mu) = P(\mu)$. The quantity v_s is obtained at the output of the channel G, in which quantity u_s is mixed in a known manner with the noise g_s, whose probability density $P(g_s)$ does not vary with s. Thus, the conditional probability density $P(v_s | u_s)$ is also known. It is now required to find such a sequence of probability densities $\Gamma_s(u_s | \mathbf{y}_{s-1}^*)$, that the average risk, that is, the mathematical expectation of the quantity

$$W = \sum_{s=0}^{s=n} W_s(s, x_s^*, x_s), \tag{6}$$

will be minimum.

Therefore, the minimum of the quantity

$$R = M\{W\} = \sum_{s=0}^{s=n} M\{W_s\} = \sum_{s=0}^{s=n} R_s \tag{7}$$

is required. The function Γ_s is the required algorithm of the controlling member A.

Further notation is introduced. Let $P(x_s | u_s)$ be the conditional probability density of x_s with given u_s. This function can be computed from the formulas (3) and (4), when the probability density $P(\mu)$ is known. In addition, let $P(\mathbf{y}_{s-1}^* | \mathbf{x}_{s-1}^*)$ be the conditional probability density of the vector

$$\mathbf{y}_{s-1}^* = (y_0^*, y_1^*, \ldots, y_{s-1}^*) \tag{8}$$

when the vector

$$\mathbf{x}_{s-1}^* = (x_0^*, x_1^*, \ldots, x_{s-1}^*) \tag{9}$$

is known.

It follows from the properties of the channel H^* that

$$P(\mathbf{y}_{s-1}^* | \mathbf{x}_{s-1}^*) = \prod_{i=0}^{i=s-1} P(y_i^* | x_i^*). \tag{10}$$

---

* All external noises are regarded as independent.

The vector $\mathbf{x}^*_{s-1}$ depends on s and $\boldsymbol{\lambda}$, and therefore $P(\mathbf{y}^*_{s-1}|\mathbf{x}^*_{s-1})$ depends also on s and $\boldsymbol{\lambda}$.

We denote by $\Omega(x_s, v_s, u_s, \mathbf{y}^*_{s-1})$ the region of variation of the parameters x_s, v_s, u_s, $\mathbf{y}^*_{s-1}$. An infinitely small element of this region is denoted by

$$d\Omega(x_s, v_s, u_s, y^*_{s-1}) = dx_s\, dv_s\, du_s\, dy^*_0 \ldots dy^*_{s-1} \quad (11)$$

At first we shall write down the expression for the conditional partial risk r_s at the sth moment, understanding by the latter the magnitude of the risk R_s when the vector $\mathbf{x}^*_s$ is kept fixed or, what amounts to the same, when the vector $\boldsymbol{\lambda}$ remains fixed. Then

$$r_s = M\{W_s | \mathbf{x}^*_s\} =$$

$$= \int\limits_{\Omega(x_s, v_s, u_s, y^*_{s-1})} W_s(s, x^*_s, x_s) P(x_s | v_s) \times P(v_s | u_s) \Gamma_s(u_s | y^*_{s-1}) P(y^*_{s-1} | x^*_{s-1}) d\Omega(x_s, v_s, u_s, y^*_{s-1}). \quad (12)$$

Let $\Omega(\boldsymbol{\lambda})$ be the region of variation of the vector $\boldsymbol{\lambda}$ and $d\Omega(\boldsymbol{\lambda})$ an infinitely small element of it. The partial risk R_s is given by the formula

$$R_s = \int\limits_{\Omega(\lambda)} r_s P(\lambda) d\Omega(\lambda) =$$

$$= \int\limits_{\Omega(x_s, u_s, v_s, y^*_{s-1}, \lambda)} W_s(s, x^*_s, x_s) P(x_s | v_s) \times P(v_s | u_s) \Gamma_s(u_s | y^*_{s-1}) P(y_{s-1} | x^*_{s-1}) P(\lambda) d\Omega(x_s, u_s, v_s, y_{s-1}, \lambda) =$$

$$= \int\limits_{\Omega(x_s, u_s, v_s, y_{s-1})} P(x_s | v_s) P(v_s | u_s) \Gamma_s(u_s | y^*_{s-1}) \left\{ \int\limits_{\Omega(\lambda)} W_s[s, x^*_s(s, \lambda), x_s] \times P(y^*_{s-1} | x^*_{s-1}) P(\lambda) d\Omega(\lambda) \right\} d\Omega(x_s, u_s, v_s, y^*_{s-1}) \quad (13)$$

The expression in the braces represents an integral over the region $\Omega(\boldsymbol{\lambda})$, as W_s and $P(\mathbf{y}^*_{s-1}|\mathbf{x}^*_{s-1})$ generally depend on $\boldsymbol{\lambda}$. Having performed the integration, we obtain a function ρ_s in the braces depending on x_s, s, $\mathbf{y}^*_{s-1}$. The dependence on s is shown in the index s, and therefore one can write

$$\rho_s = \rho_s(x_s, y^*_{s-1}) = \int\limits_{\Omega(\lambda)} W_s[s, x^*_s(s, \lambda), x_s] \cdot P(y^*_{s-1} | x^*_{s-1}) P(\lambda) d\Omega(\lambda) \quad (14)$$

Then

$$R_s = \int\limits_{\Omega(x_s, v_s, u_s, y^*_{s-1})} P(x_s | v_s) P(v_s | u_s) \Gamma_s(u_s | y^*_{s-1}) \times \rho_s(x_s, y^*_{s-1}) d\Omega(x_s, v_s, u_s, y^*_{s-1}). \quad (15)$$

From (7) the total risk R is determined from the expression

$$R = \sum_{s=0}^{s=n} R_s = \sum_{s=0}^{s=n} \int\limits_{\Omega(x_s, v_s, u_s, y^*_{s-1})} P(x_s | v_s) P(v_s | u_s) \times \Gamma_s(u_s | y^*_{s-1}) \rho_s(x_s, y^*_{s-1}) d\Omega(x_s, v_s, u_s, y^*_{s-1}). \quad (16)$$

The functions Γ_s are to be selected in such a way that the value of R is a minimum.

2. Determination of Optimum Strategy for Open-Loop Systems

It can be seen from the formula (16) that the selection of functions Γ_s for fixed s only affects the component R_s corresponding to the sth time moment. In this way the total risk is identical with action risk and one is allowed to select Γ_s such that it minimizes a single R_s in (15). As the function Γ_s represents probability density, we have

$$\int\limits_{\Omega(u_s)} \Gamma_s(u_s) d\Omega(u_s) = 1. \quad (17)$$

The expression for R(s) is rewritten as follows:

$$R_s = \int\limits_{\Omega(u_s, y^*_{s-1})} \Gamma_s(u_s | y^*_{s-1}) \left\{ \int\limits_{\Omega(x_s, v_s)} P(x_s | v_s) P(v_s | u_s) \times \rho_s(x_s, y^*_{s-1}) d\Omega(x_s, v_s) \right\} d\Omega(u_s, y^*_{s-1}). \quad (18)$$

The integral in the braces represents a function of u_s and $\mathbf{y}^*_{s-1}$ which we shall denote by $\xi_s(u_s, \mathbf{y}^*_{s-1})$, that is,

$$\xi_s(u_s, y^*_{s-1}) = \int\limits_{\Omega(x_s, v_s)} P(x_s | v_s) P(v_s | u_s) \rho_s(x_s, y^*_{s-1}) d\Omega(x_s, v_s). \quad (19)$$

Then

$$R_s = \int\limits_{\Omega(y^*_{s-1})} I(y^*_{s-1}) d\Omega(y^*_{s-1}),$$

where

$$I(y^*_{s-1}) = \int\limits_{\Omega(u_s)} \Gamma_s(u_s | y^*_{s-1}) \xi_s(u_s, y^*_{s-1}) d\Omega(u_s) = (\xi_s)_{av} \int\limits_{\Omega(u_s)} \Gamma_s(u_s | y^*_{s-1}) d\Omega(u_s) = (\xi_s)_{av} \geqslant (\xi_s)_{min}. \quad (20)$$

One can take the average value $(\xi_s)_{av}$ outside the integration in accordance with the integral mean value theorem. The function Γ_s represents probability density and hence its integral over the region $\Omega(u_s)$ equals unity.

It follows from the expression (20) that a minimum value of $I(y^*_{s-1})$ is $(\xi_s)_{min}$. It is not difficult to show that I attains this value if the function Γ_s is chosen accordingly. In fact, let u^*_s be the value of u_s corresponding to the minimum (for instance, the least of all local minima) of the function $\xi_s(u_s)$ in the region $\Omega(u_s)$. We assume that such a value exists either within or on the boundary of the region.

Consider the function

$$\Gamma_s(u_s) = \delta(u_s - u^*_s), \tag{21}$$

where δ is the unit impulse function (otherwise Dirac's δ-function). Formula (21) provides the optimum algorithm of the member A, as substituting (21) into the left-hand side of (20) and using a well-known property of δ-function, we find

$$I = \int_{\Omega(u_s)} \Gamma_s(u_s | y^*_{s-1}) \xi_s(u_s, y^*_{s-1}) d\Omega(u_s) =$$

$$= \int_{\Omega(u_s)} \delta(u_s - u^*_s) \xi_s(u_s, y^*_{s-1}) d\Omega(u_s) =$$

$$= \xi_s(u^*_s) = \min_{u_s \in \Omega(u_s)} \xi_s(u_s) = (I)_{\min}. \tag{22}$$

The minimum values of I for each y^*_{s-1} provide also the minimum of R_s.

Thus, the optimum strategy, as seen from the formula (21), does not prove random but regular. The value of u_s should be taken as equal to a u^*_s. As seen from the formula (20), the value u^*_s which minimizes the function $\xi_s(u_s, y^*_{s-1})$ depends on y^*_{s-1}. In other words, as

$$\xi_s(u^*_s, y^*_{s-1}) = \min, \tag{23}$$

the quantity u^*_s is a function of y^*_{s-1}:

$$u^*_s = u^*_s(y^*_{s-1}). \tag{24}$$

This makes also the function Γ_s in the formula (21) depend on y^*_{s-1}:

$$\Gamma_s = \delta(u_s - u^*_s) = \delta[u_s - u^*_s(y^*_{s-1})]. \tag{25}$$

Thus, the optimum strategy proves regular; it is determined in the general case from all the preceding values of y^*_i ($i = 0, \ldots, s-1$). It should be shown that the regularity of the optimum decisions occurs in a wide group of problems in Wald's general theory of statistical decisions.

The integral $\xi_s(u_s, y^*_{s-1})$ can be evaluated with the aid of a computer CO (Fig. 2). The block diagram of the optimum controlling member can then be represented as shown in Fig. 2. The automatic optimizer AO selects a value u_s such that ξ_s becomes least. This value is now sent to the output of the block A. Were it possible to determine the value of u^*_s analytically as a function of y^*_{s-1}, the block A could be constructed differently. However, the formula for $\xi_s(u_s, y^*_{s-1})$ is as a rule most difficult to obtain explicitly and that is why the computer CO must perform automatically the integrations with respect to x_s and v_s, as shown in the formula (18). In this way the minimization of the integral ξ_s can be achieved directly.

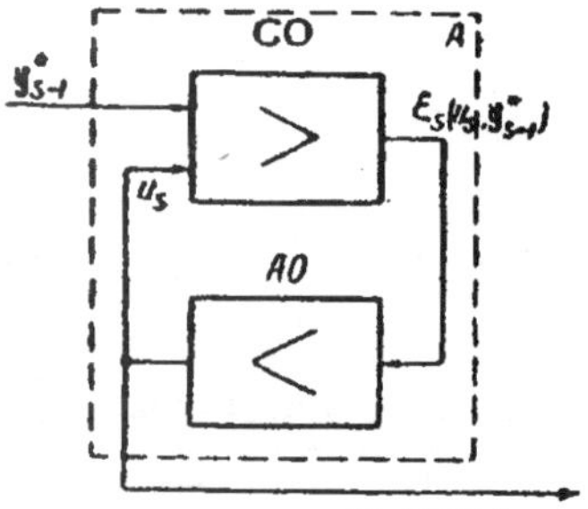

Fig. 2

3. Derivation of Risk Formula for Closed-Loop Systems

Consider now a block diagram of a closed system shown in Fig. 3. In order to simplify the exposition we neglect the interference h^* in the channel H^* (see, for example, Fig. 1 or Fig. 3 of [1]); x^* is thus assumed known. It was shown previously how the interference h^* should be taken into consideration, and should such a need arise, this can be done without any real difficulties. Our notation remains the same. In addition, there is also the feedback channel H with interference h_s which has a constant in time density distribution $P(h_s)$.

The way in which the signal and noise combine in the channel H is given and the conditional probability density $P(y_s | x_s)$ can therefore be found, and consequently the conditional probability density

$$P(\mathbf{y}_s | \mathbf{x}_s) = \prod_{i=0}^{i=s} P(y_s | x_s). \tag{26}$$

In a closed system it is necessary to take into account also the vectors

$$\begin{aligned} \mathbf{x}^*_s &= (x^*_0, x^*_1, \ldots, x^*_s), \\ \mathbf{y}_s &= (y_0, y_1, \ldots, y_s), \\ \mathbf{u}_s &= (u_0, u_1, \ldots, u_s), \\ \mathbf{v}_s &= (v_0, v_1, \ldots, v_s), \\ \mathbf{x}_s &= (x_0, x_1, \ldots, x_s). \end{aligned} \tag{27}$$

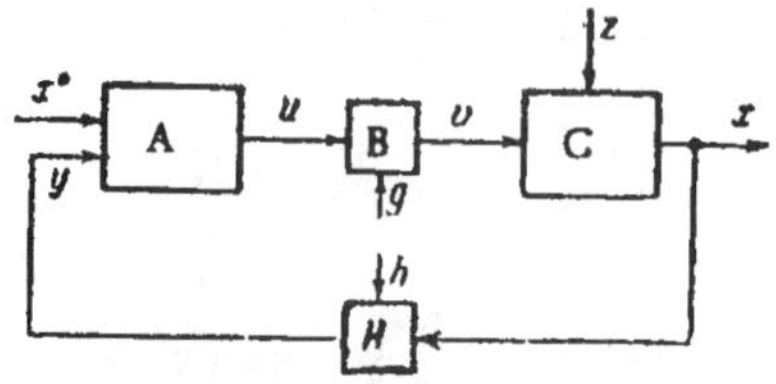

Fig. 3

The determination of the optimum controlling member A is reducible to the determination of its strategy, random in the general case, ensuring the minimizing of the risk R. The function $\Gamma_s(u_s \mid \mathbf{x}_s^*, \mathbf{y}_{s-1}, \mathbf{u}_{s-1})$, is assumed to represent random strategy, that is, the conditional probability density of u_s when the vectors $\mathbf{x}_s^*$, $\mathbf{y}_{s-1}$, and $\mathbf{u}_{s-1}$ remain fixed. Now Γ_s is a function of $\mathbf{x}_s^*$, $\mathbf{y}_{s-1}$, and $\mathbf{u}_s$.

First we shall write down the expression for the conditional partial risk r_s and we shall now understand by it the risk R_s when all the preceding motions of the system A are known, that is, when the vectors $\mathbf{x}_s^*$, $\mathbf{y}_{s-1}$, and $\mathbf{u}_{s-1}$ are fixed. Thus,

$$r_s = M\{W_s \mid \mathbf{x}_s^*, \mathbf{y}_{s-1}, \mathbf{u}_{s-1}\} =$$

$$= \int\limits_{\Omega(x_s, v_s, u_s)} W_s(s, x_s^*, x_s) P_s(x_s \mid v_s) \times$$

$$\times P(v_s \mid u_s) \Gamma_s(u_s \mid \mathbf{x}_s^*, \mathbf{y}_{s-1}, \mathbf{u}_{s-1}) \, d\Omega(x_s, v_s, u_s). \tag{28}$$

In this formula $P_s(x_s \mid v_s)$ is understood to represent the conditional probability density of the output $\underline{x}$ from the object B when its input $\underline{v}$ is known; as shown below, this probability need not be the same for different $\underline{s}$.

If one considers the values of r_s corresponding to different trials of the system, then generally speaking the vectors $\mathbf{u}_{s-1}$ and $\mathbf{y}_{s-1}$, not known beforehand, can for different trials assume different values. Therefore, in computing the average partial risk R_s it is necessary, when averaging r_s, to take into account the joint probability distributions $P(\mathbf{u}_{s-1}, \mathbf{y}_{s-1})$ of the vectors $\mathbf{u}_{s-1}$ and $\mathbf{y}_{s-1}$, which in the general case are not statistically independent of one another. Thus,

$$R_s = M\{r_s\} =$$

$$= \int\limits_{\Omega(x_s, v_s, u_s, y_{s-1})} W_s(s, x_s^*, x_s) P_s(x_s \mid v_s) P(v_s \mid u_s) \times$$

$$\times \Gamma_s(u_s \mid \mathbf{x}_{s-1}^*, \mathbf{u}_{s-1}, \mathbf{y}_{s-1}) P(\mathbf{u}_{s-1}, \mathbf{y}_{s-1})$$

$$d\Omega(x_s, v_s, \mathbf{u}_s, \mathbf{y}_{s-1}). \tag{29}$$

The values of the quantities u_i and y_i ($i = 0, \ldots, s-1$) are memorized in the controlling member A. Thus the input and the output of B are known — admittedly not exactly, but with some error due to the presence of $\underline{g}$ and $\underline{h}$, the input and the output interference respectively. This knowledge enables one to obtain a more accurate probability density of the parameter vector $\boldsymbol{\mu}$ on which in turn depends the interference z_s acting on the object B. The better $\boldsymbol{\mu}$ is known, the more fully determined is the method of regulating the object. One can derive a Bayesian formula in order to determine the a posteriori probability density of vector $\boldsymbol{\mu}$ in the $\underline{s}$th beat. Consider the probability density† $P(\boldsymbol{\mu}, \mathbf{u}_{s-1}, \mathbf{y}_{s-1})$ of a joint event consisting in a simultaneous occurrence of the vectors $\boldsymbol{\mu}$, $\mathbf{u}_{s-1}$ and $\mathbf{y}_{s-1}$. Let $P(\mathbf{u}_{s-1}, \mathbf{y}_{s-1} \mid \boldsymbol{\mu})$ be the conditional joint probability density of the vectors $\mathbf{u}_{s-1}$ and $\mathbf{y}_{s-1}$ when $\boldsymbol{\mu}$ is kept fixed and $P(\boldsymbol{\mu} \mid \mathbf{u}_{s-1}, \mathbf{y}_{s-1})$ is the conditional probability density of $\boldsymbol{\mu}$ when the vectors $\mathbf{u}_{s-1}$ and $\mathbf{y}_{s-1}$ are fixed. The latter is the required a posteriori probability density of the vector $\boldsymbol{\mu}$. Then, of course,

$$P(\boldsymbol{\mu}, \mathbf{u}_{s-1}, \mathbf{y}_{s-1}) = P(\mathbf{u}_{s-1}, \mathbf{y}_{s-1} \mid \boldsymbol{\mu}) P(\boldsymbol{\mu}) =$$

$$P(\boldsymbol{\mu} \mid \mathbf{u}_{s-1}, \mathbf{y}_{s-1}) P(\mathbf{u}_{s-1}, \mathbf{y}_{s-1}). \tag{30}$$

$P(\boldsymbol{\mu})$ denotes the a priori probability density of vector $\boldsymbol{\mu}$. Hence the formula is obtained giving the a posteriori probability density of vector $\boldsymbol{\mu}$:

$$P(\boldsymbol{\mu} \mid \mathbf{u}_{s-1}, \mathbf{y}_{s-1}) = \frac{P(\boldsymbol{\mu}) P(\mathbf{u}_{s-1}, \mathbf{y}_{s-1} \mid \boldsymbol{\mu})}{P(\mathbf{u}_{s-1}, \mathbf{y}_{s-1})} =$$

$$= \frac{P(\boldsymbol{\mu}) P(\mathbf{u}_{s-1}, \mathbf{y}_{s-1} \mid \boldsymbol{\mu})}{\int\limits_{\Omega(\boldsymbol{\mu})} P(\mathbf{u}_{s-1}, \mathbf{y}_{s-1} \mid \boldsymbol{\mu}) P(\boldsymbol{\mu}) \, d\Omega(\boldsymbol{\mu})}. \tag{31}$$

In the latter formula $\Omega(\boldsymbol{\mu})$ denotes the range of variability of vector $\boldsymbol{\mu}$ and $d\Omega(\boldsymbol{\mu})$ its infinitely small element.

In formula (31) a special role is played by the likelihood function $P(\mathbf{u}_{s-1}, \mathbf{y}_{s-1} \mid \boldsymbol{\mu})$. The probability densities of quantities are called likelihood functions when the vector $\boldsymbol{\mu}$, whose values are subject to hypothesis testing, is kept fixed. We obtain the likelihood function for the closed-loop system shown in Fig. 3. Obviously, we have

$$P(\mathbf{u}_{s-1}, \mathbf{y}_{s-1} \mid \boldsymbol{\mu}) = P(u_0, y_0 \mid \boldsymbol{\mu}) P(u_1, y_1 \mid \boldsymbol{\mu}, u_0, y_0) \times$$

$$\times P(u_2, y_2 \mid \boldsymbol{\mu}, \mathbf{u}_1, \mathbf{y}_1) \ldots P(u_{s-1}, y_{s-1} \mid \boldsymbol{\mu}, \mathbf{u}_{s-2}, \mathbf{y}_{s-2}). \tag{32}$$

A typical factor on the right-hand side is a probability density

$$P(u_i, y_i \mid \boldsymbol{\mu}, \mathbf{u}_{i-1}, \mathbf{y}_{i-1}) =$$

$$= P(y_i \mid \boldsymbol{\mu}, u_i, \mathbf{u}_{i-1}, \mathbf{y}_{i-1}) P(u_i \mid \boldsymbol{\mu}, \mathbf{u}_{i-1}, \mathbf{y}_{i-1}) =$$

$$= P(y_i \mid \boldsymbol{\mu}, i, u_i) P(u_i \mid \boldsymbol{\mu}, \mathbf{u}_{i-1}, \mathbf{y}_{i-1}) =$$

$$= P(y_i \mid \boldsymbol{\mu}, i, u_i) \Gamma_i(u_i \mid \mathbf{u}_{i-1}, \mathbf{y}_{i-1}). \tag{33}$$

Thus $P(y_i \mid \boldsymbol{\mu}, u_i, \mathbf{u}_{i-1}, \mathbf{y}_{i-1})$ depends neither on $\mathbf{u}_{i-1}$ nor on $\mathbf{y}_{i-1}$, but only on $\boldsymbol{\mu}$, on the beat index $\underline{i}$ and on the value of u_i. In addition, $P(u_i \mid \boldsymbol{\mu}, \mathbf{u}_{i-1}, \mathbf{y}_{i-1})$ does not depend on $\boldsymbol{\mu}$ but only on $\mathbf{u}_{i-1}$ and $\mathbf{y}_{i-1}$ and this is precisely the stochastic algorithm Γ_i of the controlling member A.

---

†In a number of subsequent formulas the dependence of the probability density on $\mathbf{x}_s^*$ is not stated explicitly but it is always implied.

By substituting the expression (33) into (32) we find that

$$P(\mathbf{u}_{s-1}, \mathbf{y}_{s-1} \mid \mu) = \left[\prod_{i=0}^{s-1} P(y_i \mid \mu, i, u_i)\right]$$

$$\left[\prod_{i=1}^{s-1} \Gamma_i(u_i \mid \mathbf{u}_{i-1}, \mathbf{y}_{i-1})\right] P_0(u_0). \tag{34}$$

Here $P_0(u_0)$ is the initial probability density of $\underline{u}$ which also should be given.

Now one is able to substitute the likelihood function obtained in (34) into the a posteriori probability density formula (31). We obtain

$$P_s(\mu) = P(\mu \mid \mathbf{u}_{s-1}, \mathbf{y}_{s-1}) =$$

$$P(\mu) \frac{\left[\prod_{i=0}^{s-1} P(y_i \mid \mu, i, u_i)\right] \prod_{i=0}^{s-1} \Gamma_i}{P(\mathbf{u}_{s-1}, \mathbf{y}_{s-1})}. \tag{35}$$

In the above formula the following notation has been introduced:

$$\Gamma_0 = P_0(u_0). \tag{36}$$

Knowing the a posteriori probability density $P_s(\mu)$ one is able to determine more precisely the expression of the probability density $P_s(x_s \mid v_s)$ which appears in the formula (29). From the formulas (3) and (4) we find that

$$x_s = F_0(z_s, v_s) = F(s, \mu, v_s). \tag{37}$$

The quantities $\underline{s}$ and v_s which appear in the function F are parameters. This formula enables one to find the conditional probability density $P_s(x_s \mid v_s)$ when $P_s(\mu)$ is known.

Since

$$\int_{(\Omega v_s)} P_s(x_s \mid v_s) P(v_s \mid u_s) d\Omega(v_s) = P_s(x_s \mid u_s) =$$

$$= \int_{\Omega(\mu)} P(x_s \mid \mu, u_s) P_s(\mu) d\Omega(\mu), \tag{38}$$

the expression (29) giving R_s can be re-written in the following way:

$$R_s = \int_{\Omega(x_s, \mathbf{u}_s, \mathbf{y}_{s-1})} W_s(s, x_s^{\bullet}, x_s)$$

$$\left[\int_{\Omega(v_s)} P_s(x_s \mid v_s) P(v_s \mid u_s) d\Omega(v_s)\right] \times$$

$$\times \Gamma_s(u_s \mid \mathbf{x}_{s-1}^{\bullet}, \mathbf{u}_{s-1}, \mathbf{y}_{s-1})$$

$$P(\mathbf{u}_{s-1}, \mathbf{y}_{s-1}) d\Omega(x_s, \mathbf{u}_s, \mathbf{y}_{s-1}) =$$

$$= \int_{\Omega(x_s, \mu, \mathbf{u}_s, \mathbf{y}_{s-1})} W_s(s, x_s^{\bullet}, x_s) P(x_s \mid \mu, u_s)$$

$$P_s(\mu) \Gamma_s(u_s \mid \mathbf{x}_{s-1}^{\bullet}, \mathbf{u}_{s-1}, \mathbf{y}_{s-1}) \times$$

$$\times P(\mathbf{u}_{s-1}, \mathbf{y}_{s-1}) d\Omega(x_s, \mu, \mathbf{u}_s, \mathbf{y}_{s-1}). \tag{39}$$

Substituting the formula (35) for $P_s(\mu)$ and cancelling $P(\mathbf{u}_{s-1}, \mathbf{y}_{s-1})$ we obtain

$$R_s = \int_{\Omega(x_s, \mu, \mathbf{u}_s, \mathbf{y}_{s-1})} W_s(s, x_s^{\bullet}, x_s) P(x_s \mid \mu, u_s) P(\mu) \times$$

$$\times \prod_{i=0}^{s-1} P(y_i \mid \mu, i, u_i) \prod_{i=0}^{s} \Gamma_i d\Omega(x_s, \mu, \mathbf{u}_s, \mathbf{y}_{s-1}). \tag{40}$$

The total risk is given by

$$R = \sum_{s=0}^{s=n} R_s = \sum_{s=0}^{s=n} \int_{\Omega(x_s, \mu, \mathbf{u}_s, \mathbf{y}_{s-1})}$$

$$W_s(s, x_s^{\bullet}, x_s) P(x_s \mid \mu, u_s) P(\mu) \times$$

$$\times \prod_{i=0}^{s-1} P(y_i \mid \mu, i, u_i) \prod_{i=0}^{s} \Gamma_i d\Omega(x_s, \mu, \mathbf{u}_s, \mathbf{y}_{s-1}). \tag{41}$$

Let us compare the expressions (16) and (41) of risks in open and closed systems. In the formula (16) only the $\underline{s}$th component R_s was influenced by the function Γ_s; the total risk associated with Γ_s was exclusively an action risk. In the formula (41), however, the function Γ_s influenced not only the term R_s (action risk) but also all the other components R_i where $i > s$. As regards Γ_s, the sum $\sum_{i=s+1}^{n} R_i$ represents the investigation risk. In fact, the selection of Γ_s influences the character of the process in the subsequent beats, as it causes either a better or a worse investigation of the characteristics of the object B; the latter amounts to the determination of the probability density $P_s(\mu)$. Thus the total risk regarded as a function of Γ_s $(0 \leq s \leq n)$ represents the sum of the action risk and the investigation risk. As regards Γ_n only a single action risk occurs.

When the object B has memory the selection of Γ_s influences not only R_s but also the terms R_i when $i > s$ even when the additional investigation of the object B has not taken place (see Paper IV in this series). Consequently, in this case the action risk is expressed by a more involved formula, and the interconnection between the action and the investigation risks proves also to be more involved.

4. Determination of Optimum Strategy in Closed Systems

It is required to select the functions Γ_i in the formula (41) such that the risk R be minimum. As Γ_i represents probability densities they must satisfy the condition (17).

To be able to select Γ_n we consider the last component R_n of R:

$$R_n = \int_{\Omega(x_n, \mu, \mathbf{u}_n, \mathbf{y}_{n-1})}$$

$$W_n(n, x_n^{\bullet}, x_n) P(x_n \mid \mu, u_n) P(\mu) \times$$

$$\times \prod_{i=0}^{n-1} P(y_i \mid \mu, i, u_i) \prod_{i=0}^{n-1} \Gamma_i \Gamma_n \, d\Omega(x_n, \mu, \mathbf{u}_n, \mathbf{y}_{n-1}). \tag{42}$$

We assume that $\Gamma_0, \ldots, \Gamma_{n-1}$ are given, and only Γ_n is to be selected such that R_n becomes minimum provided a constraint as in (17) is satisfied:

$$\int_{\Omega(u_n)} \Gamma_n(u_n, \mathbf{u}_{n-1}, \mathbf{y}_{n-1})\, d\Omega(u_n) = 1. \qquad (43)$$

In order to simplify the expression (42), put

$$\alpha_n = \alpha_n(\mathbf{u}_n, \mathbf{y}_{n-1}) =$$

$$= \int_{\Omega(x_n,\mu)} W_n(n, x_n, x_n^*)\, P(x_n \mid \mu, u_n)$$

$$P(\mu) \prod_{i=0}^{n-1} P(y_i \mid \mu, i, u_i)\, d\Omega(x_n, \mu). \qquad (44)$$

Further, put

$$\beta_k = \prod_{i=0}^{k} \Gamma_k; \qquad (45)$$

then

$$R_n = \int_{\Omega(u_n, \mathbf{u}_{n-1}, \mathbf{y}_{n-1})}$$

$$\alpha_n(u_n, \mathbf{u}_{n-1}, \mathbf{y}_{n-1})\, \beta_{n-1} \Gamma_n\, d\Omega(\mathbf{u}_n, \mathbf{y}_{n-1}) =$$

$$= \int_{\Omega(\mathbf{u}_{n-1}, \mathbf{y}_{n-1})} \beta_{n-1} \varkappa_n(\mathbf{u}_{n-1}, \mathbf{y}_{n-1})\, d\Omega(\mathbf{u}_{n-1}, \mathbf{y}_{n-1}), \qquad (46)$$

where

$$\varkappa_n(\mathbf{u}_{n-1}, \mathbf{y}_{n-1}) =$$

$$\int_{\Omega(u_n)} \alpha_n(u_n, \mathbf{u}_{n-1}, \mathbf{y}_{n-1})\, \Gamma_n(u_n, \mathbf{u}_{n-1}, \mathbf{y}_{n-1})\, d\Omega(u_n). \qquad (47)$$

If the Γ_n is selected such that for <u>any</u> set of vectors $\mathbf{u}_{n-1}$ and $\mathbf{y}_{n-1}$ the quantity κ_n remains least, then the integral (46) will also be minimum (bearing in mind that the quantity β_{n-1} is assumed known. By regarding the vectors $\mathbf{u}_{n-1}$ and $\mathbf{y}_{n-1}$ in the formulas (44) and (47) as parameters, one is able to find a value u_n^* such that the function α_n is minimum. We put

$$\gamma_n^* = \alpha_n(u_n^*, \mathbf{u}_{n-1}, \mathbf{y}_{n-1}) = \min_{u_n \in \Omega(u_n)} \alpha_n(u_n, \mathbf{u}_{n-1}, \mathbf{y}_{n-1}). \qquad (48)$$

The quantity u_n^*, of course, depends on $\mathbf{u}_{n-1}$ and $\mathbf{y}_{n-1}$:

$$u_n^* = u_n^*(\mathbf{u}_{n-1}, \mathbf{y}_{n-1}). \qquad (49)$$

Then the optimum function Γ_n satisfying the condition (43) is

$$\Gamma_n(u_n, \mathbf{u}_{n-1}, \mathbf{y}_{n-1}) = \delta(u_n - u_n^*), \qquad (50)$$

because in this case

$$\varkappa_n = \alpha_n(u_n^*, \mathbf{u}_{n-1}, \mathbf{y}_{n-1}) = (\alpha_n)_{\min} = (\varkappa_n)_{\min}. \qquad (51)$$

This follows from the considerations similar to those given above when the conditions (19), (20), and (21) were derived.

Now the optimum function Γ_{n-1} will be determined. Consider the sum of the last two terms R_{n-1} and R_n. First of all, we introduce the function

$$\alpha_k = \alpha_k(u_k, \mathbf{u}_{k-1}, \mathbf{y}_{k-1}) =$$

$$\int_{\Omega(x_k,\mu)} W_k(k, x_k^*, x_k)\, P(x_k \mid \mu, u_k) \times$$

$$\times P(\mu) \prod_{i=0}^{k-1} P(y_i \mid \mu, i, u_i)\, d\Omega(x_k, \mu) \qquad (0 \leqslant k \leqslant n). \qquad (52)$$

There the expression $\prod_{i=0}^{k-1} P(y_i \mid \mu, i, u_i)$ equals unity when $k = 0$.

Now the sum of the last two terms becomes

$$S_{n-1} = R_{n-1} + R_n =$$

$$= \int_{\Omega(\mathbf{u}_{n-1}, \mathbf{y}_{n-2})} \alpha_{n-1}\beta_{n-1}\, d\Omega(\mathbf{u}_{n-1}, \mathbf{y}_{n-2}) +$$

$$+ \int_{\Omega(\mathbf{u}_n, \mathbf{y}_{n-1})} \alpha_n\beta_n\, d\Omega(\mathbf{u}_n, \mathbf{y}_{n-1}). \qquad (53)$$

When $i \leq n-2$, the Γ_i are assumed to be known and Γ_n is always selected in accordance with the formula (50). Then

$$S_{n-1} = \int_{\Omega(\mathbf{u}_{n-2}, \mathbf{y}_{n-2})} \beta_{n-2} \Big\{ \int_{\Omega(u_{n-1})} \Gamma_{n-1}\alpha_{n-1}\, d\Omega(u_{n-1}) +$$

$$+ \int_{\Omega(u_{n-1}, y_{n-1})} \Gamma_{n-1}\alpha_n(u_n^*, \mathbf{u}_{n-1}, \mathbf{y}_{n-1})\, d\Omega(u_{n-1}, y_{n-1}) \Big\}. \qquad (54)$$

This quantity is least when the expression within the braces is least; we write the latter in full as

$$\varkappa_{n-1}(\mathbf{u}_{n-2}, \mathbf{y}_{n-2}) =$$

$$= \int_{\Omega(u_{n-1})} \Big\{ \Gamma_{n-1}\alpha_{n-1} + \int_{\Omega(y_{n-1})}$$

$$\Gamma_{n-1}\alpha_n(u_n^*, \mathbf{u}_{n-1}, \mathbf{y}_{n-1})\, d\Omega(y_{n-1}) \Big\}\, d\Omega(u_{n-1}) =$$

$$= \int_{\Omega(u_{n-1})} \Gamma_{n-1}(u_{n-1}, \mathbf{u}_{n-2}, \mathbf{y}_{n-2}) \times$$

$$\times \Big\{ \alpha_{n-1} + \int_{\Omega(y_{n-1})}$$

$$\alpha_n(u_n^*, \mathbf{u}_{n-1}, \mathbf{y}_{n-1})\, d\Omega(y_{n-1}) \Big\}\, d\Omega(u_{n-1}). \qquad (55)$$

Now consider the function

$$\gamma_{n-1} = \gamma_{n-1}(\mathbf{u}_{n-1}, \mathbf{y}_{n-2}) = \alpha_{n-1} +$$

$$+ \int_{\Omega(y_{n-1})} \alpha_n(u_n^*, \mathbf{u}_{n-1}, \mathbf{y}_{n-1})\, d\Omega(y_{n-1}). \qquad (56)$$

We find the value u_{n-1}^* which is the minimum value of this function. This value obviously depends on $\mathbf{u}_{n-2}$ and $\mathbf{y}_{n-2}$:

$$u_{n-1}^* = u_{n-1}^*(\mathbf{u}_{n-2}, \mathbf{y}_{n-2}). \qquad (57)$$

Then the most suitable function Γ_{n-1} such that it ensures the minimum of κ_{n-1} and satisfies the condition (17) will be

$$\Gamma_{n-1}(u_{n-1}, u_{n-2}, y_{n-2}) = \delta(u_{n-1} - u^*_{n-1}). \quad (58)$$

By applying analogous arguments one can determine the whole sequence of the functions Γ_i using the following procedure.

We introduce the function

$$\gamma_{n-k} = \alpha_{n-k} + \int\limits_{\Omega(y_{n-k})} \gamma_{n-k+1}(u^*_{n-k+1}, u_{n-k}, y_{n-k})\, d\Omega(y_{n-k}), \quad (59)$$

where $\gamma_n = \alpha_n$. Let the minimum of γ_{n-k} over u_{n-k} be denoted by γ^*_{n-k}:

$$\gamma^*_{n-k} = (\gamma_{n-k})_{u_{n-k}=u^*_{n-k}} =$$

$$= \min_{u_{n-k} \in \Omega(u_{n-k})} \gamma_{n-k}(u_{n-k}; u_{n-k-1}, y_{n-k-1});$$

$$u^*_{n-k} = u^*_{n-k}(u_{n-k-1}; y_{n-k-1}). \quad (60)$$

Then the optimum function Γ^*_{n-k} is given by the formula

$$\Gamma^*_{n-k} = \delta(u_{n-k} - u^*_{n-k}), \quad (61)$$

where δ is the unit impulse function. Then

$$(S_{n-k})_{\min} = \left(\sum_{i=0}^{i=k} R_{n-i}\right)_{\min} =$$

$$= \int\limits_{\Omega(u_{n-k-1}, y_{n-k-1})} \beta_{n-k-1}\gamma^*_{n-k}\, d\Omega(u_{n-k-1}, y_{n-k-1}). \quad (62)$$

In this way Γ_i are determined when $i = n, n-1, \ldots, 1$. As far as the determination of $\Gamma_0 = P_0(u_0)$ is concerned, this function may be given beforehand and then no selection is required. If the best Γ_0 can be chosen, the selection takes place in accordance with the procedure shown above. When $k = n-1$ we arrive at the formula

$$(S_1)_{\min} = \int\limits_{\Omega(u_0, y_0)} P_0(u_0)\gamma^*_1(u_0, y_0)\, d\Omega(u_0, y_0). \quad (63)$$

The total risk therefore when all Γ_i $(i = 1, \ldots, n)$ are optimal, is expressed by the formula

$$R = R_0 + (S_1)_{\min} =$$

$$= \int\limits_{\Omega(x_0, \mu, u_0)} W_0(0, x^*_0, x_0) P(x_0 \mid \mu, u_0) \times$$

$$\times P(\mu) P_0(u_0)\, d\Omega(x_0, \mu, u_0) +$$

$$+ \int\limits_{\Omega(u_0, y_0)} P_0(u_0)\gamma^*_1(u_0, y_0)\, d\Omega(u_0, y_0) =$$

$$= \int\limits_{\Omega(u_0)} P_0(u_0)\Big[\alpha_0(u_0) +$$

$$+ \int\limits_{\Omega(y_0)} \gamma^*_1(u_0, y_0)\, d\Omega(y_0)\Big] d\Omega(u_0), \quad (64)$$

where

$$x_0 = \alpha_0(u_0) =$$

$$= \int\limits_{\Omega(x_0, \mu)} W_0(0, x^*_0, x_0) P(x_0 \mid \mu, u_0) P(\mu)\, d\Omega(x_0, \mu). \quad (65)$$

We put

$$\gamma_0 = \gamma_0(u_0) = \alpha_0(u_0) + \int\limits_{\Omega(y_0)} \gamma^*_1(u_0, y_0)\, d\Omega(y_0). \quad (66)$$

It can be seen from the formula (64) that the quantity R becomes minimum when

$$\Gamma_0 = P_0(u_0) = \Gamma^*_0 = \delta(u_0 - u^*_0), \quad (67)$$

where the value u^*_0 makes the function $\gamma(u_0)$ minimum.

The value u^*_0 is dependent on the a priori probabilities, also generally on the given data, but not on observations which are not available at the initial moment of time.

Thus in the case of a closed system as well, the optimum strategy proves to be not random but a regular one. This conclusion is naturally valid within the bounds of the limitations accepted when formulating this theory.

The determination of sequence of functions γ_{n-k} may prove rather tedious in actual examples if electronic computers are not used. Even so, it has only been possible so far to solve only relatively simple problems in which the memory capacity necessary to memorize the functions is not too great.

The obtained system with the optimum control algorithm Γ^*_i is, generally speaking, a dual control system in which the u_i serve not only as directing but also as investigating operations. In the absence of investigation, the probability density $P_s(\mu)$ would remain equal to $P_0(\mu)$ and the investigation risk would be null.

It may happen that in a particular case of a dual control system the process of investigating the object proceeds in the same way, no matter what the values of the regulating operations within the admissible bandwidth; what is only required is that the regulating operations take place. Such a system (an example is given in the next paper) is called neutral. In a neutral system the investigation risk is independent of the quantity u_i, but $P_s(\mu)$ is not identical with $P_0(\mu)$ but varies with s.

LITERATURE CITED

1. A. A. Fel'dbaum, "Theory of dual control. I," Avtomat. i Telemekh. 21, No. 9 (1960). ‡

‡See English translation.

Absolute Stability of Nonlinear Systems of Automatic Control

V. M. POPOV

A CENTRAL specification imposed on a control mechanism is invariably its stability. Maxwell's paper "On Governors" [10], settled the problem of stability of systems described by third order linear differential equations in terms of the coefficients of the associated characteristic polynomial. Subsequent to this paper, the problem of stability of systems described by linear differential equations in terms of the coefficients of the characteristic polynomial was solved by Routh and Hurwitz. It is worth noting that the mathematical problem of determining when the roots of a polynomial lie in the left half of the complex plane had actually been effectively solved by Hermite [5] a decade before Maxwell posed the stability problem as a question in control.

In his thesis in 1895, Lyapunov put forward a method, now called 'the second method of Lyapunov', for determining the stability of an equilibrium of a system described by nonlinear differential equations. While Lyapunov was primarily interested in applications to mechanics, his method and stability concepts became very influential and widely applied in control. His ideas were especially popular in the Eastern European control literature, since it was customary in these countries to use differential equations as models for control mechanisms.

By contrast, in the Western literature, stability analysis in control had been centered on frequency-domain methods, propelled by the beauty, practicality, and generality (for linear systems) of the Nyquist criterion [11].

Each of these approaches has its advantages. Frequency-domain methods use very compact model specifications and are able to address robustness through gain and phase margins. However, the generalization to nonlinear systems of frequency-domain methods proved awkward. Differential equation models by and large proved a much better way of addressing nonlinearities.

Through the problem of determining conditions for the stability of systems with one nonlinear element, control witnessed a symbiosis between stability theory of Lyapunov methods and frequency-domain techniques. A key result that was the catalyst in this development was Popov's stability criterion, the subject of the paper that follows.

The history of the problem is as follows (see [2] for more details). In 1944, Lur'e and Postnikov [7] formulated the problem of *stability in the large* of a finite dimensional system with one nonlinear element $f : \mathbb{R} \to \mathbb{R}$ satisfying $\sigma f(\sigma) \geq 0$ for all $\sigma \in \mathbb{R}$. In subsequent publications, Lur'e obtained, using Lyapunov methods, a set of "resolving equations" that allowed to conclude stability. These ideas were further developed in [9], [16], [8], but the resulting conditions remained difficult to verify and to interpret.

This problem of stability of a system with one nonlinear element led also to the *Aizerman conjecture* [1]. This states that if there exist k_1, k_2 such that $k_1 < \frac{f(\sigma)}{\sigma} < k_2$ for all $\sigma \in \mathbb{R}$, and if the linear system obtained by replacing the nonlinear element by a linear one with gain k, is asymptotically stable for all $k \in [k_1, k_2]$, then the nonlinear system is also asymptotically stable in the large. Aizerman's conjecture is valid for second order systems, but was proven to be false for third and higher order systems [12] (see also [14], [15]).

By introducing the frequency response of a system, Popov obtained a sufficient condition for the stability of such systems. This criterion can be formulated as follows. Consider the feedback system described by (see Figure 3 in Popov's article):

$$\frac{d}{dt}x = Ax + bu; \quad y = c^T x; \quad u = -f(y)$$

with $A \in \mathbb{R}^{n\times n}$, $b, c \in \mathbb{R}^n$, and $f : \mathbb{R} \to \mathbb{R}$ a continuous nonlinearity with $f(0) = 0$. Denote by G the transfer function of the linear part, i.e., $G(s) := c^T(Is - A)^{-1}b$. Assume furthermore that A is a stability matrix [meaning that $t \in \mathbb{R} \mapsto \exp(At) \in \mathbb{R}^{n\times n}$ is bounded on $[0, \infty)$]. The matrix A need not have all its eigenvalues in the open left half of the complex plane. In fact, Popov assumes that A has all its eigenvalues in the open left half of the complex plane, except, possibly, one eigenvalue at the origin. The assumption that A need not be Hurwitz in fact complicates his analysis considerably.

Popov proves that the equilibrium trajectory $x = 0$ is globally asymptotically stable if the following conditions are satisfied:

1. $\sigma f(\sigma) > 0$ for all $0 \neq \sigma \in \mathbb{R}$,

2. there exists $q \geq 0$ such that $\text{Re}(1 + qj\omega)G(j\omega) \geq 0$ for all $\omega \in \mathbb{R}$.

This result is known as the *Popov criterion.* We have stated the criterion here for the case that the graph of f is contained in the first and third quadrant, but it is easily generalized to nonlinearities in an arbitrary conic sector. The above conditions on the frequency response can readily be verified using graphical methods.

There are two main techniques for proving this result. One is based on Lyapunov theory, and the other on studying an associated integral equation. The relevant Lyapunov equation consists of a quadratic plus an integral term,

$$V(x) = x^T P x + \int_0^{c^T x} \sigma f(\sigma)\, d\sigma,$$

with $P = P^T \in \mathbb{R}^{n \times n}$ a positive definite matrix that can be derived from what is now known as the *Positive-Real Lemma* [also known as the KYP (Kalman-Yakubovich-Popov) lemma]. As mentioned in the paper, Popov had developed the Lyapunov proof in earlier articles. Actually, the above type of Lyapunov function consisting of a quadratic form in the state plus an integral of the nonlinearity, was already used in [7]. The equivalence between Lur'e resolving equations, Popov's criterion, and the existence of a Lyapunov function of the quadratic form plus integral type was proven in [17] (see also [6]).

In the present paper, Popov presents a new proof that involves an integral equation, and in his analysis, he makes use of truncation operators. Truncation operators, and the associated extended $\mathfrak{L}_p$-spaces, are elegant and effective tools for stability analysis for feedback systems, used also for example by Zames [18] and Sandberg [13] (see also [15]).

The ideas that led to and emerged from the Popov criterion became of central importance in the field. It was a first instance of the use of 'multipliers' for feedback stability. It demonstrated the very fruitful interplay between time-domain and frequency-domain methods. It paved the way for important stability principles, such as the small gain and the passive operator theorem. The method of proof stimulated the very successful use of functional analysis ideas in stability analysis.

Popov's work immediately influenced many other leading researchers in the field; for example, Brockett [3], [4], Kalman [6], Yakubovich [17], and Zames [18]. It was one of key elements in the flowering of control theory in the sixties.

References

[1] M.A. AIZERMAN, "On the effect of nonlinear functions of several variables on the stability of automatic control systems," *Autom. i Telemekh.*, **8**:1, 1947.

[2] M.A. AIZERMAN AND F.R. GANTMACHER, *Absolute Stability of Regulator Systems*, Holden-Day, 1964 (translation of the 1963 Russian edition).

[3] R.W. BROCKETT AND J.L. WILLEMS, "Frequency domain stability criteria, Parts I and II," *IEEE Trans. Aut. Contr.*, **AC-10**:255–261 & 407–413, 1965.

[4] R.W. BROCKETT, "The status of stability theory for deterministic systems," *IEEE Trans. Aut. Contr.*, **AC-11**:596–606, 1966.

[5] C. HERMITE, "On the number of roots of an algebraic equation contained between two limits," extract of a letter from Mr. Ch Hermite to Mr. Borchardt of Berlin, *J. Reine Angew. Math.*, **52**:39–51, 1856. (Translation by P C Parks, *Int. J. Cont.*, **26**(2):183–196, 1977.)

[6] R.E. KALMAN, "Liapunov functions for the problem of Lur'e in automatic control," *Proc. Nat.Acad. Sci. U.S.*, **49**:2, 1963.

[7] A.I. LUR'E AND V.N. POSTNIKOV, "On the theory of stability of control systems," *Prikl. Mat. i Mekh.*, **8**:3, 1944.

[8] J.P. LASALLE AND S. LEFSCHETZ, *Stability by Liapunov's Direct Method with Applications*, Academic Press (New York), 1961.

[9] I.G. MALKIN, "On the theory of stability of control systems," *Prikl. Mat. i Mekh.*, **15**:1, 1951.

[10] J.C.MAXWELL, "On governors," *Proc. Royal Soc. London*, **16**:270–283, 1868.

[11] H. NYQUIST, "Regeneration theory," *Bell System Tech. J.*, **11**:126–147, 1932.

[12] V.A. PLISS, "On the Aizerman problem for a system of three differential equations," *Dokl. Acad. Nauk. SSSR*, **121**:3, 1958.

[13] I.W. SANDBERG, "A frequency-domain condition for the stability of feedback systems containing a single time-varying nonlinear element," *Bell System Tech. J.*, **43**:1601–1608, 1974.

[14] J.C. WILLEMS, "Perturbation theory for the analysis of instability in nonlinear feedback systems," *Proc. 4th Allerton Conf. on Circuit and System Theory*, pp. 836–848, 1966.

[15] J.C. WILLEMS, *The Analysis of Feedback Systems*, The MIT Press (Cambridge, MA), 1971.

[16] V.A. YAKUBOVICH, "On nonlinear differential equations for control systems with a single regulator," *Vest. LGU*, **2**:7, 1960.

[17] V.A. YAKUBOVICH, "The solution of certain matrix inequalities in automatic control theory," *Dokl. Akad. Nauk SSSR*, **143**(6):1304–1307, 1962. (English Translation: *Soviet Mathematics*, 1962, pp 620–623.)

[18] G. ZAMES, "On the input-output stability of time-varying nonlinear feedback systems. Part I: Conditions derived using concepts of loop gain, conicity, and positivity; Part II: Conditions involving circles in the frequency plane and sector nonlinearities," *IEEE Trans. Aut. Contr.*, **AC-11**:228–238 & 465–476, 1966.

J.C.W.

ABSOLUTE STABILITY OF NONLINEAR SYSTEMS OF AUTOMATIC CONTROL

V. M. Popov (Bucharest)

Translated from Avtomatika i Telemekhanika, Vol. 22, No. 8,
pp. 961-979, August, 1961
Original article submitted January 17, 1961

The problem of absolute stability of an "indirect control" system with a single nonlinearity is investigated by using a method which differs from the second method of Lyapunov. The main condition of the obtained criterion of absolute stability is expressed in terms of the transfer function of the system linear part. It is also shown that by forming the standard Lyapunov function -"a quadratic form plus the integral of the nonlinearity" it is not possible in the case considered here to obtain a wider stability domain than the one obtained from the presented criterion. Graphical criteria of absolute continuity are also given by means of the phase-amplitude characteristic or by what is known as "the modified phase-amplitude characteristic" of the system linear part.

In the present paper the absolute stability is investigated of nonlinear systems of "indirect control." The existing literature in this field (see [1] for example) deals exclusively with a direct application of Lyapunov's method. In this paper the solution is obtained by a different method, and this enables one to get new results.

It is assumed that the reader is not familiar with the author's previous publications. Therefore, not only the most general results are given here but also a very simple example shall be considered.

By using a new method the author has also investigated the absolute stability of other types of systems of differential equations (for example, of the system of "direct control"), as well as of other classes of nonlinear functions (for example, of functions whose graph is contained within a sector). In all these cases the absolute stability of the system with several nonlinearities is also studied (the case of systems with many controlling devices).

In his most recent papers now in press, the author has studied the stability in certain critical cases and also the stability of systems of differential equations with an "aftereffect".

1. Statement of the Problem

Systems of "indirect control" are considered which can be described by the following system of differential equations:

$$\frac{dx_l}{dt} = \sum_{k=1}^{n} a_{lk}x_k + b_l\varphi(\sigma) \qquad (l = 1, 2, \ldots, n), \tag{1.1}$$

$$\frac{d\xi}{dt} = \varphi(\sigma), \tag{1.2}$$

$$\sigma = \sum_{l=1}^{n} c_l x_l - \gamma\xi, \tag{1.3}$$

where a_{lk}, b_l, c_l, and γ are constant, and $\varphi(\sigma)$ is a function of class A*, that is, a continuous function satisfying the condition

$$\varphi(0) = 0 \tag{1.4}$$

and also the inequality

* It is assumed that the quantity σ_* introduced in [6] always vanishes. It should just be mentioned that if $\sigma_* \neq 0$, the trivial solution of the system (1.1)-(1.3) cannot be asymptotically stable.

Reprinted with permission from *Automation and Remote Control*, V. M. Popov, "Absolute Stability of Nonlinear Systems of Automatic Control," Vol. 22, February 1962, pp. 857–875.

$$\varphi(\sigma)\sigma > 0 \text{ for } \sigma \neq 0 \tag{1.5}$$

The system (1.1)-(1.3) admits the trivial solution

$$x_l = \xi = \sigma = 0, \tag{1.6}$$

whose stability is under investigation.

It is assumed that the trivial solution of the linear system with constant coefficients

$$\frac{dy_l}{dt} = \sum_{k=1}^{n} a_{lk} y_k \qquad (l = 1, 2, \ldots, n), \tag{1.7}$$

where a_{lk} are the same as in (1.1), is asymptotically stable, or (which is equivalent) that all real parts of the eigenvalues of the matrix (a_{lk}) are negative. The conditions are being sought which would be satisfied by the quantities a_{lk}, b_l, c_l, and γ in order that the trivial solution of the system (1.1)-(1.3) be asymptotically stable, whatever the function $\varphi(\sigma)$ of the class A (in other words, a condition of asymptotic absolute stability of the trivial solution). As we know to achieve this, it is necessary that*

$$\gamma > 0. \tag{1.8}$$

We shall therefore consider in the sequel the inequality (1.8) to be satisfied.

2. Introductory Definitions. Formulation of Criterion of Absolute Stability

Consider the functions $\psi_{lm}(t)$ $(l = 1, 2, \ldots, n;\ m = 1, 2, \ldots, n)$, defined when $t \geq 0$, and being the solution of the equations

$$\frac{d\psi_{lm}(t)}{dt} = \sum_{k=1}^{n} a_{lk} \psi_{km}(t) \qquad \begin{pmatrix} l = 1, 2, \ldots, n \\ m = 1, 2, \ldots, n \end{pmatrix} \tag{2.1}$$

together with the initial conditions

$$\psi_{lm}(0) = \delta_{lm} \quad (l = 1, 2, \ldots, n;\ m = 1, 2, \ldots, n), \tag{2.2}$$

where $\delta_{lm} = 0$ when $l \neq m$, and $\delta_{lm} = 1$ when $l = m$.

The functions $\psi_{lm}(t)$ form the fundamental system of solutions for the system (1.7).

Let $x_l(t)$, $\xi(t)$ be the solution of the equations (1.1)-(1.3)** which satisfies the initial conditions $x_l(0) = x_{l0}$, $\xi(0) = \xi_0$, and let $\varphi[\sigma(t)]$ be a function of t obtained by substituting the function $\sigma(t) = \sum c_l x_l(t) - \gamma\xi(t)$ in $\varphi(\sigma)$ [see (1.3)].

As a solution of the system (1.1) we obtain

$$x_l(t) = \sum_{m=1}^{n} \psi_{lm}(t) x_{m0} + \sum_{m=1}^{n} \int_0^t \psi_{lm}(t-\zeta) b_m \varphi(\sigma(\zeta))\, d\zeta. \tag{2.3}$$

It follows from (1.3) that

$$\sigma(t) = \sum_{m=1}^{n} \mu_m(t) x_{m0} - \int_0^t \nu(t-\zeta)\varphi(\sigma(\zeta))\, d\zeta - \gamma\xi(t), \tag{2.4}$$

*If $\gamma < 0$, the trivial solution is unstable when $\varphi(\sigma) = h\sigma$, $h>0$. If $\gamma = 0$, the condition $\xi(t) = 0$ is not satisfied for all solutions of the system.

**The existence of solutions is a consequence of the assumptions made in section 1. But their uniqueness is not assumed in the sequel. A solution can always be extended in view of the conditions for stability as formulated below.

where

$$\mu_m(t) = \sum_{l=1}^{n} c_l \psi_{lm}(t), \tag{2.5}$$

$$v(t) = -\sum_{l=1}^{n} \sum_{m=1}^{n} c_l \psi_{lm}(t)\, b_m. \tag{2.6}$$

In agreement with our assumptions the trivial solution of the system (1.7) is asymptotically stable, and therefore two positive constants K_0 and K_1 can be found such that [see (2:1) and (2.2)] for all $t \geq 0$ the inequality

$$|\psi_{lm}(t)| < K_1 e^{-K_0 t}, \qquad t > 0 \quad \begin{pmatrix} l = 1, 2, \ldots, n, \\ m = 1, 2, \ldots, n \end{pmatrix}. \tag{2.7}$$

takes place.

Therefore, for v (t) [see (2.6)] the inequality

$$|v(t)| < K_2 e^{-K_0 t} \quad (t > 0), \tag{2.8}$$

is valid where

$$K_2 = \sum_{l=1}^{n} |c_l| \sum_{l=1}^{n} |b_l| K_1.$$

We note that in view of the inequality (2.7), the Fourier transform of the function v (t) given by the formula

$$N(j\omega) = \int_0^\infty e^{-j\omega t} v(t)\, dt \qquad (j = \sqrt{-1}). \tag{2.9}$$

must exist.

We introduce the function

$$G(j\omega) = N(j\omega) + \frac{\gamma}{j\omega}, \tag{2.10}$$

which shall be called the transfer function of the system linear part (see Appendix 1).

Now the following theorem can be stated.

Theorem. If a nonnegative quantity q exists such that for all real w the inequality*

$$\mathrm{Re}\,(1 + j\omega q)\, G(j\omega) \geqslant 0, \tag{2.11}$$

takes place, then the trivial solution of the system (1.1)-(1.3) is asymptotically absolutely stable provided the assumptions made in Section 1 remain valid.

* ReX (or ImX) denotes the real (or, respectively, imaginary) part of a complex quantity X.

3. Proof of the Theorem in Section 2

Consider again the solution x_l (t), ξ (t), σ (t) of the system (1.1)-(1.3) as well as the function φ (σ (l)), corresponding to it. For any positive quantity T we define auxiliary functions

$$\varphi_T(t) = \begin{cases} \varphi(\sigma(t)) & \text{for} \quad 0 \leqslant t \leqslant T, \\ 0 & \text{for} \quad t > T, \end{cases} \tag{3.1}$$

$$\lambda_T(t) = -\int_0^t v(t-\zeta)\,\varphi_T(\zeta)\,d\zeta - q\int_0^t \frac{dv(t-\zeta)}{dt}\varphi_T(\zeta)d\zeta - q[v(0)+\gamma]\,\varphi_T(t), \tag{3.2}$$

where q is the quantity occurring in the theorem, and v (t) is given by the formula (2.6); it follows that [see (2.2)]

$$v(0) = -\sum_{l=1}^{n} c_l b_l. \tag{3.3}$$

It can easily be seen that the function λ_T (t) is bounded for all $0 \leq t \leq T$. When $t > T$, the inequality (see Appendix 1)

$$|\lambda_T(t)| < K_3 e^{-K_4 t} \quad (K_3 > 0, \ t > T), \tag{3.4}$$

takes place in which K_3 is independent of t. This guarantees the existence of the Fourier transform

$$L_T(j\omega) = \int_0^\infty e^{-j\omega t}\lambda_T(t)\,dt. \tag{3.5}$$

There exists also the transform

$$F_T(j\omega) = \int_0^\infty e^{-j\omega t}\varphi_T(t)\,dt. \tag{3.6}$$

In view of (2.6), (2.1),and (2.7) the Fourier transform of the function dv (t)/ dt exists and can be written as [se (2.9)]

$$\int_0^\infty e^{-j\omega t}\frac{dv(t)}{dt}dt = j\omega N(j\omega) - v(0), \tag{3.7}$$

Therefore, by taking the Fourier transform of (3.2) [see (3.5), (3.6),and (3.7)] we obtain

$$L_T(j\omega) = -[(1 + j\omega q)\,N(j\omega) + q\gamma]\,F_T(j\omega). \tag{3.8}$$

The following function of T is introduced:

$$\rho(T) = \int_0^\infty \lambda_T(t)\,\varphi_T(t)\,dt. \tag{3.9}$$

The Parseval formula can be applied in (3.9) because the function φ_T (t) is continuous for $0 \le t \le T$ and it vanishes for $t > T$, and the function λ_T (t) is also continuous for $0 \le t \le T$; and for $t > T$ the inequality (3.4) takes place. We now obtain

$$\rho(T) = \frac{1}{2\pi} \int_{-\infty}^{\infty} L_T(j\omega) \overline{F_T(j\omega)} \, d\omega, \tag{3.10}$$

where $\overline{F_T(j\omega)}$ is the conjugate complex of $F_T(j\omega)$.

By replacing $L_T(j\omega)$ by (3.8) in (3.10), and by taking into account that

$$F_T(j\omega) \overline{F_T(j\omega)} = |F_T(j\omega)|^2, \tag{3.11}$$

we obtain

$$\rho(T) = -\frac{1}{2\pi} \int_{-\infty}^{\infty} |F_T(j\omega)|^2 [(1 + j\omega q) N(j\omega) + q\gamma] \, d\omega. \tag{3.12}$$

By comparing their real parts we obtain

$$\rho(T) = -\frac{1}{2\pi} \int_{-\infty}^{\infty} |F_T(j\omega)|^2 \operatorname{Re} [(1 + j\omega q) N(j\omega) + q\gamma] \, d\omega. \tag{3.13}$$

Observing that [see (2.10) and (2.11)]

$$\operatorname{Re} [(1 + j\omega q) N(j\omega) + q\gamma] = \operatorname{Re} (1 + j\omega q)\left(N(j\omega) + \frac{\gamma}{j\omega}\right) = \operatorname{Re} (1 + j\omega q)\, g(j\omega) \geqslant 0, \tag{3.14}$$

gives the inequality

$$\rho(T) \leqq 0. \tag{3.15}$$

By substituting (3.1) into (3.9) we obtain

$$\rho(T) = \int_0^T \lambda_T(t) \varphi(\sigma(t)) \, dt \leqslant 0. \tag{3.16}$$

By substituting the expression for φ_T (t) into (3.2) and by applying the formula (2.4) as well as the formula

$$\frac{d\sigma(t)}{dt} = \sum_{m=1}^{n} \frac{d\mu_m(t)}{dt} x_{m0} - \int_0^t \frac{d\nu(t-\zeta)}{dt} \varphi(\sigma(\zeta)) \, d\zeta - [\nu(0) + \gamma] \, \varphi(\sigma(t)), \tag{3.17}$$

easily following from (2.4), we obtain within the interval $0 \le t \le T$

$$\lambda_T(t) = \sigma(t) + q \frac{d\sigma(t)}{dt} + \gamma \xi(t) - \sum_{m=1}^{n} \left(\mu_m(t) + q \frac{d\mu_m(t)}{dt}\right) x_{m0}. \tag{3.18}$$

By substituting this expression into (3.16) we obtain

$$\int_0^T \varphi(\sigma(t))\,\sigma(t)\,dt + q\int_0^T \varphi(\sigma(t))\frac{d\sigma(t)}{dt}\,dt + \gamma\int_0^T \varphi(\sigma(t))\,\xi(t)\,dt$$

$$-\sum_{m=1}^{n} x_{m0}\int_0^T \left(\mu_m(t) + q\frac{d\mu_m(t)}{dt}\right)\varphi(\sigma(t))\,dt \leqslant 0. \qquad (3.19)$$

Each term of the expression (3.19) shall be considered in turn. For the first term we have the inequality [see (1 and (1.5)]

$$-\int_0^T \varphi(\sigma(t))\,\sigma(t)\,dt < 0. \qquad (3.20)$$

The second term can be rewritten with the aid of the identity

$$\int_0^T \varphi(\sigma(t))\frac{d\sigma(t)}{dt}\,dt = F(\sigma(T)) - F(\sigma(0)), \qquad (3.21)$$

where

$$F(\sigma) = \int_0^\sigma \varphi(\sigma)\,d\sigma.$$

In the case when $\varphi(\sigma)$ belongs to the class A [see (1.4) and (1.5)], the condition

$$F(\sigma) \geq 0 \qquad (3.22)$$

is fulfilled, with the equality only taking place when $\sigma = 0$. As q is nonnegative, one has, of course, the inequality

$$-qF(\sigma(T)) \leq 0. \qquad (3.23)$$

The third term can be rewritten as follows [see (1.2)]:

$$\int_0^T \varphi(\sigma(t))\,\xi(t)\,dt = \int_0^T \frac{d\xi(t)}{dt}\,\xi(t)\,dt = \frac{1}{2}\xi^2(T) - \frac{1}{2}\xi_0^2. \qquad (3.24)$$

As $\gamma > 0$ [see (1.8)] we have

$$-\frac{1}{2}\gamma\xi^2(T) < 0. \qquad (3.25)$$

As far as the remaining terms are concerned we note [see Appendix 3] that there exists a positive quantity K_4 such that the inequality:

$$\left|\int_0^T \left(\mu_m(t) + q\frac{d\mu_m(t)}{dt}\right)\varphi(\sigma(t))\,dt\right| \leqslant \frac{1}{n}K_4 \sup_{0<\zeta<T} |\xi(\zeta)|, \quad (m = 1, 2, \ldots n). \qquad (3.26)$$

takes place.

Using the identities (3.21) and (3.24) as well as the inequalities (3.26), one is able to rewrite the inequality (3.19) as

$$\int_0^T \varphi(\sigma(t))\,\sigma(t)\,dt + qF(\sigma(T)) + \frac{1}{2}\gamma\xi^2(T)$$

$$-K_4 \max_{m=1,2,\dots,n}(|x_{m0}|)\sup_{0<\zeta<T}|\xi(\zeta)| \leqslant qF(\sigma(0)) + \frac{1}{2}\gamma\xi_0^2. \tag{3.27}$$

We shall show that the trivial solution of the system (1.1)-(1.3) is stable. To this end we combine the inequalities (3.27), (3.20), and (3.23)

$$\frac{1}{2}\gamma\xi^2(T) - K_4 \max_{m=1,2,\dots,n}(|x_{m0}|)\sup_{0<\zeta<T}|\xi(\zeta)| \leqslant qF(\sigma(0)) + \frac{1}{2}\gamma\xi_0^2. \tag{3.28}$$

Let t be an arbitrary positive quantity. The inequality (3.28) is valid for any T > 0. In particular, (3.25) remains valid for T (t) such that the inequality

$$0 \le T(t) \le t \tag{3.29}$$

takes place, and also the equality

$$|\xi(T(t))| = \sup_{0<\zeta<t}|\xi(\zeta)|. \tag{3.30}$$

Such a quantity T (t), of course, exists. It is easy to see that in view of (3.29) the inequalities

$$|\xi(T(t))| \leqslant \sup_{0<\zeta<T(t)}|\xi(\zeta)| \leqslant \sup_{0<\zeta<t}|\xi(\zeta)|. \tag{3.31}$$

take place.

By comparing (3.31) and (3.30) we obtain

$$|\xi(T(t))| = \sup_{0<\zeta<T(t)}|\xi(\zeta)|. \tag{3.32}$$

When considering (3.28) for T = T (t) and by using (3.32) we obtain

$$\frac{1}{2}\gamma\xi^2(T(t)) - K_4 \max_{m=1,2,\dots,n}(|x_{m0}|)|\xi(T(t))| - qF(\sigma(0)) - \frac{1}{2}\gamma\xi_0^2 \leqslant 0. \ (3.33) \tag{3.33}$$

The polynomial of the variable $|\xi(T(t))|$ in the lefthand side of the inequality (3.33) has real roots of opposite sign, when made equal to zero. The inequality (3.33) is equivalent to the inequality

$$|\xi(T(t))| \leqslant \frac{1}{\gamma}K_4 \max_{m=1,2,\dots,n}(|x_{m0}|)$$

$$+\frac{1}{\gamma}\sqrt{K_4^2\left(\max_{m=1,2,\dots,n}(|x_{m0}|)\right)^2 + 2\gamma\left(qF(\sigma(0)) + \frac{1}{2}\gamma\xi_0^2\right)}, \tag{3.34}$$

which is valid for all t > 0.

One can find two positive quantities K_5 and K_6 [see Appendix 4] such that for all t > 0 the inequality

$$|x_i(t)| \leqslant K_5 \max_{m=1,2,\dots,n}(|x_{m0}|) + K_6 \sup_{0<\zeta<t}|\xi(\zeta)|. \tag{3.35}$$

takes place.

By applying (3.30), (3.34), and (3.35) we finally obtain

$$|\xi(t)| < \frac{1}{\gamma} K_4 \max_{m=1,2,\ldots,n} (|x_{m0}|) + \frac{1}{\gamma} + \sqrt{K_4^2 (\max_{m=1,2,\ldots,n} (x_{m0}|))^2 + 2\gamma (qF(\sigma(0)) + \frac{1}{2}\gamma\xi_0^2)}, \tag{3.36}$$

$$|x_l(t)| < \left(K_5 + \frac{1}{\gamma} K_4 K_6\right) \max_{m=1,2,\ldots,n} (|x_{m0}|) + \frac{1}{\gamma} K_6 \sqrt{K_4^2 (\max_{m=1,2,\ldots,n} (|x_{m0}|))^2 + 2\gamma\left(qF(\sigma(0)) + \frac{1}{2}\gamma\xi_0^2\right)}. \tag{3.37}$$

Let ϵ be an arbitrary positive quantity. There exists than a $\delta > 0$ such that when $|x_{l0}| < \delta$ and $|\xi_0| < \delta$, the righthand sides of the inequalities (3.36) and (3.37) are less than ϵ;* and therefore the inequalities $|x_l(t)| < \epsilon$ and $|\xi(t)| < \epsilon$ take place. Consequently, the trivial solution of the system (1.1)-(1.3) is stable according to Lyapunov. Moreover, it follows from (3.36) and (3.37) that all the solutions of the system are bounded. This implies that all the solutions of the system (1.1)-(1.3) can be extended for all $t > 0$.

It shall be shown now that the trivial solution is also asymptotically stable for any $\varphi(\sigma)$ of the class A. We combine for this purpose the inequalities (3.27), (3.23), and (3.25):

$$\int_0^T \varphi(\sigma(t))\,\sigma(t)\,dt < K_4 \max_{m=1,2,\ldots,n} (|x_{m0}|) \sup_{0<\zeta<T} |\xi(\zeta)| + qF(\sigma(0)) + \frac{1}{2}\gamma\xi_0^2. \tag{3.38}$$

By making use of (3.36)

$$\int_0^T \varphi(\sigma(t))\,\sigma(t)\,dt < f_1(x_{m0}, \xi_0), \tag{3.39}$$

where $f_1(x_{m0}, \xi_0)$ is a specified function of the variables x_{m0} and ξ_0.

Taking into account the reaction [see (1.1), (1.3)]

$$\frac{d\sigma(t)}{dt} = \sum_{l=1}^n c_l \frac{dx_l}{dt} - \gamma\frac{d\xi}{dt} = \sum_{l=1}^n c_l\left(\sum_{k=1}^n a_{lk}x_k + b_l\varphi(\sigma)\right) - \gamma\varphi(\sigma), \tag{3.40}$$

[in the above formula x_k and ξ satisfy the inequalities (3.37) and (3.36), σ is determined by the equation (1.3) and $\varphi(\sigma)$ is a continuous function of σ], we obtain the inequality

$$\left|\frac{d\sigma(t)}{dt}\right| < f_2(x_{m0}, \xi_0), \tag{3.41}$$

where $f_2(x_{m0}, \xi_0)$ does not depend on t.

By using (3.39) and (3.41) we obtain [see Appendix 5]

$$\lim_{t\to\infty} \sigma(t) = 0 \tag{3.42}$$

* Here we make use of the fact that $F(\sigma(0))$ is a continuous function of $\sigma(0)$, vanishing when $\sigma(0) = 0$, and also of the equality $\sigma(0) = \sum_{l=1}^n c_l x_{l0} - \gamma\xi_0$ [see (1.3)].

and also

$$\lim_{t\to\infty} \varphi(\sigma(t)) = 0. \tag{3.43}$$

because $\varphi(\sigma)$ is a continuous function.

From (3.43) and from the equations (2.3) one obtains [see Appendix 6]

$$\lim_{t\to\infty} x_l(t) = 0. \tag{3.44}$$

The relation (1.3) implies [see (3.42) and '3.44)]

$$\lim_{t\to\infty} \xi(t) = \lim_{t\to\infty} \frac{1}{\gamma}\Big(\sum_{l=1}^{n} x_l(t) - \sigma t\Big) = 0. \tag{3.45}$$

The theorem has thus been proved.

4. Comparing the Results with the Results Obtainable with the Aid of Lyapunov Functions of Known Kind

It is of interest to compare the criteria (2.11) of absolute stability with the ones which can be obtained by constructing a Lyapunov function of the kind described as a "quadratic form plus the integral of the nonlinearity" [6-8]. We shall show that the latter are included in the criteria of the theorem in section 2, that is, if for the system under investigation there exists a Lyapunov function of the above kind, then a nonnegative quantity q also exists such that the inequality (2.11) takes place.

By differentiating the equation (1.3) we obtain a system equivalent to the (1.1)-(1.3):

$$\begin{aligned} \frac{dx_l}{dt} &= \sum_{k=1}^{n} a_{lk}x_k + b_l\varphi(\sigma) \qquad (l = 1, 2, \ldots, n), \\ \frac{d\sigma}{dt} &= \sum_{l=1}^{n} c_l\Big(\sum_{k=1}^{n} a_{lk}x_k + b_l\varphi(\sigma)\Big) - \gamma\varphi(\sigma). \end{aligned} \tag{4.1}$$

It is shown in Appendix 7 that the most general form of a negative definite Lyapunov function which is of the kind described as a "quadratic form plus the integral of the nonlinearity" and which can be formed for the system (4.1) takes the form

$$V = \sum_{l=1}^{n}\sum_{m=1}^{n} x_l r_{lm} x_m - \alpha\Big(\sigma - \sum_{l=1}^{n} c_l x_l\Big)^2 - 2\beta\int_0^{\sigma}\varphi(\sigma)\,d\sigma, \tag{4.2}$$

with the parameters r_{lm}, α, β satisfying the conditions:

$$r_{lm} = r_{ml},\ \alpha \geqslant 0,\ \beta \geqslant 0,\ \alpha + \beta > 0. \tag{4.3}$$

The derivative of this function which, in accordance with the system (1.4)*

$$\begin{aligned} \frac{1}{2}\frac{dV}{dt} = W(x_l, \sigma) &= \sum_{l=1}^{n}\sum_{m=1}^{n} x_l r_{lm}\Big(\sum_{k=1}^{n} a_{mk}x_k + b_m\varphi(\sigma)\Big) \\ &+ \alpha\gamma\varphi(\sigma)\Big(\sigma - \sum_{l=1}^{n} c_l x_l\Big) - \beta\varphi(\sigma)\Big(\sum_{l=1}^{n} c_l\Big(\sum_{k=1}^{n} a_{lk}x_k + b_l\,\varphi(\sigma)\Big) - \gamma\varphi(\sigma)\Big) \end{aligned} \tag{4.4}$$

must be positive definite for all functions $\varphi(\sigma)$ of the class A, and in particular when

* We note that $\frac{d}{dt}\big(\sigma - \sum_{l=1}^{n} c_l x_l\big) = -\gamma\varphi(\sigma)$.

$$\varphi(\sigma) = h\sigma, \; h>0. \tag{4.5}$$

By putting (4.5) in (4.4) we obtain a quadratic form of real variables x_l and σ which must be positive definite. This implies that for all complex values of x_l and σ not vanishing simultaneously the inequality

$$\begin{aligned}\mathrm{Re}\Big\{&\sum_{l=1}^{n}\sum_{m=1}^{n}\bar{x}_l r_{lm}\Big(\sum_{k=1}^{n}a_{mk}x_k + b_m h\sigma\Big) + \alpha\gamma h\bar{\sigma}\Big(\sigma - \sum_{l=1}^{n}c_l x_l\Big)\\ &-\beta h\bar{\sigma}\Big(\sum_{l=1}^{n}c_l\Big(\sum_{k=1}^{n}a_{lk}x_k + b_l h\sigma\Big) - \gamma h\sigma\Big)\Big\} > 0,\end{aligned} \tag{4.6}$$

is valid, the quantities $\bar{x}_l$ and $\bar{\sigma}$ being conjugate to x_l and σ. That this is so can be seen by putting in (4.6) $x_l = u_l + jv_l$, $\sigma = \mu + j\nu$, with u_l, v_l, μ, and ν real; the left-hand side of (4.6) assumes the form $\frac{1}{2}(W_0(u_l, \mu) + W_0(v_l, \nu))$, where W_0 is the quadratic form (4.4) with $\varphi(\sigma) = h\sigma$. Thus, if u_l, v_l, μ, and ν do not vanish simultaneousl we have the inequality (4.6).

The inequality (4.6) is satisfied in the particular case of

$$\begin{aligned}\sigma &= \frac{1}{h},\\ x &= M_l(j\omega) \qquad (l = 1, 2, \ldots, n),\end{aligned} \tag{4.7}$$

where $M_l(j\omega)$ satisfy the system of simultaneous equations

$$j\omega M_l(j\omega) = \sum_{k=1}^{n} a_{lk}M_k(j\omega) + b_l \qquad (l = 1, 2, \ldots, n). \tag{4.8}$$

The system (4.8) has a unique solution for any real ω as according to our assumptions (see section 1) the matrix (a_{lk}) has no purely imaginary eigenvalues.

By taking Fourier transforms of the system (2.1) including (2.2) we obtain

$$j\omega F\{\psi_{lm}(t)\} = \sum_{k=1}^{n} a_{lk}F\{\psi_{km}(t)\} + \delta_{lm} \qquad \begin{pmatrix} l = 1, 2, \ldots, n \\ m = 1, 2, \ldots, n\end{pmatrix}, \tag{4.9}$$

where

$$F\{\psi_{lm}(t)\} = \int_0^\infty e^{-j\omega t}\psi_{lm}(t)\,dt. \tag{4.10}$$

By comparing (4.9) and (4.8) we obtain

$$M_l(j\omega) = \sum_{m=1}^{n} F\{\psi_{lm}(t)\}\, b_m = \sum_{m=1}^{n}\int_0^\infty e^{-j\omega t}\psi_{lm}(t)\, b_m dt. \tag{4.11}$$

Substituting (4.7) into (4.6) and using the relations (4.8) we obtain

$$\begin{aligned}\mathrm{Re}\Big\{&\sum_{l=1}^{n}\sum_{m=1}^{n}\overline{M_l(j\omega)}\, r_{lm} j\omega M_m(j\omega) + \alpha\gamma\Big(\frac{1}{h} - \sum_{l=1}^{n}c_l M_l(j\omega)\Big)\\ &-\beta\Big(\sum_{l=1}^{n}c_l j\omega M_l(j\omega) - \gamma\Big)\Big\} > 0.\end{aligned} \tag{4.12}$$

But [see (4.11), (2.6), (2.9)]

$$\sum_{l=1}^{n} c_l M_l(j\omega) = \sum_{l=1}^{n}\sum_{m=1}^{n}\int_0^\infty e^{-j\omega t} c_l \Psi_{lm}(t)\, b_m dt = -\int_0^\infty e^{-j\omega t} v(t)\, dt = -N(j\omega). \tag{4.13}$$

In view of $r_{lm} = r_{ml}$, we have*

$$\mathrm{Re}\sum_{l=1}^{n}\sum_{m=1}^{n} \overline{M_l(j\omega)}\, r_{lm}\, j\omega M_m(j\omega) = \mathrm{Re}\left\{\frac{1}{2} j\omega \sum_{l=1}^{n}\sum_{m=1}^{n}\left(\overline{M_l(j\omega)}\, M_m(j\omega) + M_l(j\omega)\, \overline{M_m(j\omega)}\right) r_{lm}\right\} = 0. \tag{4.14}$$

The inequality (4.12) can therefore be rewritten as

$$\frac{\alpha\gamma}{h} + \mathrm{Re}\,(\alpha\gamma + j\omega\beta)\, N(j\omega) + \beta\gamma > 0 \tag{4.15}$$

or [see (2.10)]

$$\frac{\alpha\gamma}{h} + \mathrm{Re}\,(\alpha\gamma + j\omega\beta)\, G(j\omega) > 0. \tag{4.16}$$

The above consideration remain valid for any real ω. Thus, it has been shown that for a Lyapunov function of the considered kind to exist it is necessary that the inequality (4.16) be valid for all $h>0$ and for any ω.

If $\alpha \neq 0$, [see (4.3)] and thus if $\alpha\gamma > 0$ [see (1.8)], from (4.16) must follow the inequality

$$\mathrm{Re}\left(1 + j\omega\frac{\beta}{\alpha\gamma}\right) G(j\omega) \geqslant 0. \tag{4.17}$$

Indeed, should the inequality (4.17) be invalid for some $\omega = \omega_0$, then a positive quantity h exists such that the inequality (4.16) will not take place for $\omega = \omega_0$. The necessary condition (4.17) is identical with the inequality (2,11) if $q = \beta/\alpha\gamma > 0$.

If $\alpha = 0$, we obtain from (4.3) and (4.16) the inequalities

$$\beta > 0, \tag{4.18}$$

$$\mathrm{Re}\, j\omega G(j\omega) > 0. \tag{4.19}$$

We shall now consider certain properties of the function $N(j\omega)$ [see (2.9)]. The function $N(j\omega)$ is a continuous function of ω in view of the inequality (2.8). By the Riemann-Lebesque Lemma

$$\lim_{|\omega|\to\infty} N(j\omega) = 0. \tag{4.20}$$

Therefore, there exists a positive number P_1 such that for any ω

$$\mathrm{Re}N(j\omega) > -P_1. \tag{4.21}$$

By the Riemann-Lebesque Lemma we also obtain** [see (3.3)]

$$\lim_{|\omega|\to\infty} j\omega N(j\omega) = v(0) = -\sum_{l=1}^{n} c_l b_l. \tag{4.22}$$

---

*We note that $\mathrm{Im}\left(\overline{M_l(j\omega)}\, M_m(j\omega) + M_l(j\omega)\, \overline{M_m(j\omega)}\right) \equiv 0.$

**From the relations (2.6),(2.1) and (2.7) it is easy to find that the integral $\int_0^\infty \left|\frac{dv(t)}{dt}\right| dt$ must converge.

It follows from (4.21), (4.22) and (2.10) that

$$\operatorname{Re} G(j\omega) = \operatorname{Re} N(j\omega) > -P_1, \tag{4.23}$$

$$\lim_{|\omega|\to\infty} j\omega G(j\omega) = \lim_{|\omega|\to\infty} j\omega N(j\omega) + \gamma = -\sum_{i=1}^{n} c_i b_i + \gamma > 0. \tag{4.24}$$

The latter is a necessary condition and can be obtained from (4.6) and (4.18) by putting $x_l = 0$, $\sigma = 1/h$, $\alpha = 0$. It follows from (4.19), (4.24) and from the continuity of the function $G(j\omega)$ that a positive quantity P_2 can be found such that for any real ω the inequality

$$\operatorname{Re} j\omega G(j\omega) > P_2. \tag{4.25}$$

takes place.

By multiplying the inequality (4.25) by $2P_1/P_2$ and by adding the result to the inequality (4.23), we obtain

$$\operatorname{Re}\left(1 + 2\frac{P_1}{P_2} j\omega\right) G(j\omega) > -P_1 + 2P_1 > 0. \tag{4.26}$$

The latter implies that the inequality (2.11) is valid (as a strong inequality) when $q = 2P_1/P_2 > 0$.

Thus it has been proved that the inequality (2.11) suffices for the trivial solution of the system described in section 1 to be absolutely asymptotically stable and also to be a necessary condition for the existenec of a Lyapunov function of the considered type.

Remarks. (1) In order to construct a Lyapunov function of the "quadratic form" type (that is of the kind as in (4.2) with $\beta = 0$) it is necessary that the inequality (4.17) be valid with $\beta = 0$, that is that

$$\operatorname{Re} G(j\omega) \geq 0 \tag{4.27}$$

for any real ω.

The condition (4.27) is sufficient for the absolute asymptotic stability of the trivial solution of the system under investigation, as in this case the inequality (2.11) occurs for $q = 0$.

(2) Most of the Lyapunov functions so far constructed in [8] are of the kind as in (4.2), with $\alpha = 0$. That is why it is necessary that the inequality (4.19) be valid; the latter is also sufficient for the absolute stability of the system trivial solution.

The condition (4.19) can also be written for any real positive ω as

$$\operatorname{Im} G(j\omega) < 0. \tag{4.28}$$

This follows from the relation [see (2.9) and (2.10)]

$$\operatorname{Im} G(j\omega) = -\operatorname{Im} G(-j\omega) \tag{4.29}$$

and from the inequality [see (2.10), (3.7), and (1.8)].

$$[\operatorname{Re} j\omega G(j\omega)]_{\omega\to 0} = \gamma > 0. \tag{4.30}$$

(3) It would be interesting to find the general solution of the following inverse problem: If the condition (2.11) is satisfied is it always possible to construct a Lyapunov function of the kind as in (4.2) ? For some relatively simple cases the answer is in the affirmative.

5. Various Analytic and Graphic Forms of the (2.11) Criterion

The function $(1 + j\omega q)G(j\omega)$ can be written as

$$(1 + j\omega q) G(j\omega) = \frac{P(j\omega)}{Q(j\omega)}, \tag{5.1}$$

(see Appendix 1) with P (jω) and Q (jω) being polynomials of jω. The condition (2.11) can now be written as

$$\mathrm{Re} P(j\omega) Q(-j\omega) \geqslant 0 \tag{5.2}$$

or

$$R(x) \geqslant 0, \; x = \omega^2, \tag{5.3}$$

where R (x) is a polynomial of the variable x. The condition (2.11) is reduced, therefore, to a polynomial of x being nonnegative for x ≥ 0. The solution can be obtained by using standard algebraic methods.

An arbitrary nonnegative parameter q appears in the criterion, which can be selected in a suitable manner in every specific case.* The algebraic methods for obtaining the optimal values of q are quite straightforward.

The following graphical criteria of absolute asymptotic stability are of special interest in practical applications.

The locus of the points (u, v) in the plane of (u, v) such that

$$u(\omega) = \mathrm{Re}\, G(j\omega), \; v(\omega) = \omega \mathrm{Im}\, G(j\omega) \tag{5.4}$$

shall be called the "modified phase-amplitude characteristic" (M.P.A.C.). Directly from (2.11) we obtain the inquality

$$u(\omega) + q v(\omega) \geqslant 0, \; q \geqslant 0, \tag{5.5}$$

which means that the M.P.A.C. is in a half-plane.

A graphical criterion (Fig. 1). If there is a straight line situated either in the first and the third quadrants of the (u, v) plane or it is the ordinate axis**, and in addition it is such that the M.P.A.C. is "on the right" of this straight line, then the trivial solution of the investigated system is absolutely asymptotically stable. One may add that the M.P.A.C. is "on the right" of this straight line if any point of the M.P.A.C. is either on the straight line or it is in the half-plane bounded by this straight line and containing the point (+∞, 0).***

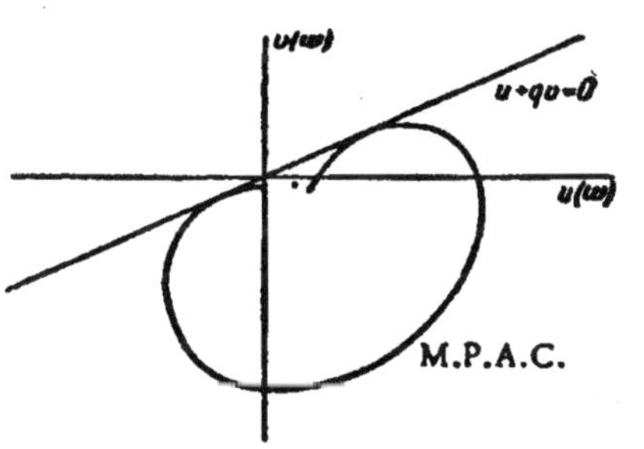

Fig. 1.

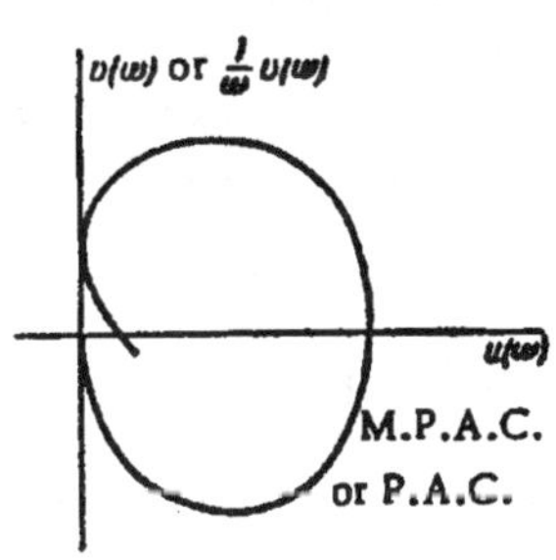

Fig. 2.

* We note that in the criteria (4.27) or (4.28) which described the results obtainable by the usual kind of Lyapunov function (see Remarks (1) and (2), there is no arbitrary parameter.

** Such a straight line obviously passes through the origin. Its equation is of the form u + qv = 0 with q ≥ 0.

*** Or, otherwise, that the inequality (5.5) takes place.

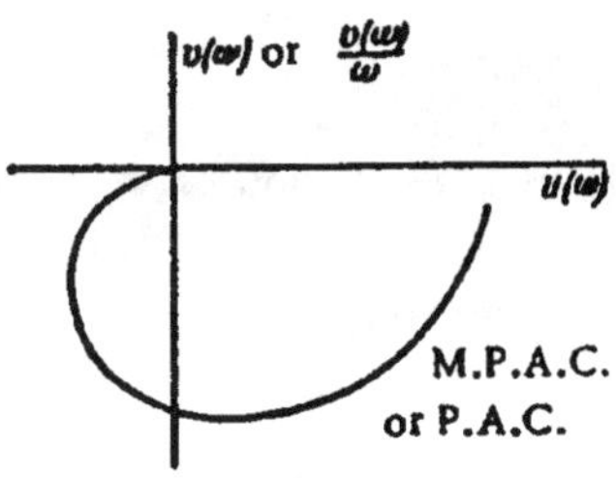

Fig. 3.

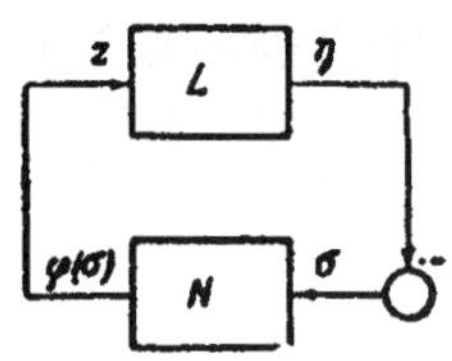

Fig. 4.

The ordinary phase-amplitude characteristic can also be made use of in order to obtain simplified graphical criteria of absolute stability.

The following graphical criterion is obtained from the condition (4.27).

Simplified graphical criterion No. 1. If all the points of the ordinary (or modified) phase-amplitude characteristic are situated "on the right" of the ordinate axis, the trival solution of the system under investigation is asymptotically absolutely stable. (Fig. 2)

It is necessary that this graphical criterion be satisfied in order that a Lyapunov function of the "quadratic form" kind may exist [see Remark (1)].

From the sufficient condition (4.28) of stability another simplified graphical criterion is obtained.

Simplified grapical criterion No. 2. If, when $\omega>0$, all the points of the ordinary (or modified) P.A.C. are situated in the third or the fourth quadrant or on the negative ordinate semi-axis, then the trivial solution of the system under investigation is asymptotically absolutely stable (Fig. 3).

It is necessary that this criterion be fulfilled in order that a Lyapunov function (5.3) with $\alpha = 0$ may exist [see Remark (2)].

We should like to mention that no simple method exists to express the general graphical criterion (Fig. 1) by means of the ordinary P.A.C. Knowing the ordinary P.A.C., one is able to obtain the modified P.A.C. by multiplying the ordinate of each point by the corresponding value of the variable.

6. Concluding Remarks

The majority of the arguments developed in the preceding sections can be applied with practically no alterations to more general cases mentioned in the introduction, and results of similar nature are obtained.

The fact that in the criteria only the transfer function of the system linear part appears, apart from the simple assumptions of section 1, seems to constitute the main characteristic of the achieved results. The latter need not be evaluated with the aid of the formula (2.9) but can be obtained by more direct methods which have been developed for linear systems of automatic control. The graphical criteria of absolute stability developed above are also applicable when nothing but its linearity and independence are known about the linear block of the system, its phase-amplitude characteristic being determined experimentally.

The author wishes to express his thanks to the collective of research-workers in the field of ordinary differential equations of the Mathematical Institute of the Academy of Sciences of the Romanian National Republic, in particular to Professor A. Khalan, for the interest they have shown and for their valuable remarks.

APPENDIX 1

The investigated system can always be represented in the form of a block diagram as in Fig. 4 where the linear block is denoted by L, the latter described by the system of equations

$$\frac{dx_l}{dt} = \sum_{k=1}^{n} a_{lk}x_k + b_l z \quad (l = 1, 2, \ldots, n),$$

$$\frac{d\xi}{dt} = z,$$

$$\eta = -\sum_{l=1}^{n} c_l x_l + \gamma\xi, \tag{I.1}$$

and N representing the nonlinear block. The input and the output quantities of the linear and nonlinear blocks are related by the equations

$$\eta = -\sigma; \quad z = \varphi(\sigma). \tag{I.2}$$

Let $z^0(t)$ be a known function such that the integral $\int_0^\infty |z^0(t)|$ exists, and let $L\{z^0(t)\}$ be its Laplace transform

$$L\{z^0(t)\} = \int_0^\infty e^{-st} z^0(t)\,dt, \quad \text{Re}\, s \geqslant 0. \tag{I.3}$$

Let further $x_l^0(t)$, $\xi^0(t)$, $\eta^0(t)$ be the solution of (I.1) when $z = z^0(t)$ and when the initial conditions are all nul. It follows from the assumptions in section 1 that $L\{x_l^0(t)\}$ exists at least for Re $s \geq 0$, and $L\{\xi^0(t)\}$ and $L\{\eta^0(t)\}$ exist for Re $s > 0$.

Taking Laplace transforms of (I.1) we obtain

$$sL\{x_l^0(t)\} = \sum_{k=1}^{n} a_{lk} L\{x_k^0(t)\} + b_l L\{z^0(t)\}, \quad \text{Re}\, s \geqslant 0 \quad (l = 1, 2, \ldots, n). \tag{I.4}$$

$$sL\{\xi^0(t)\} = L\{z^0(t)\}, \quad \text{Re}\, s > 0, \tag{I.5}$$

$$L\{\eta^0(t)\} = -\sum_{l=1}^{n} c_l L\{x_l^0(t)\} + \gamma L\{\xi^0(t)\}, \quad \text{Re}\, s > 0. \tag{I.6}$$

The system of simultaneous equations (I.4) has a unique solution when Re $s > 0$ (see the assumptions in section 1).

Now taking Laplace transforms of the system (2.1) with the initial conditions (2.2):

$$sL\{\psi_{lm}(t)\} = \sum_{k=1}^{n} a_{lk} L\{\psi_{km}(t)\} + \delta_{lm}, \quad \text{Re}\, s \geqslant 0 \quad \begin{pmatrix} l = 1, 2, \ldots, n \\ m = 1, 2, \ldots, n \end{pmatrix}. \tag{I.7}$$

By comparing (I.4) and (I.7) we obtain

$$L\{x_l^0(t)\} = \left(\sum_{m=1}^{n} L\{\psi_{lm}(t)\}\, b_m\right) L\{z^0(t)\}, \quad \text{Re}\, s \geqslant 0 \quad (l = 1, 2, \ldots, n). \tag{I.8}$$

Using (I.5), (I.6), and (I.80) gives

$$L\{\eta^0(t)\} = \tilde{G}(s)\, L\{z^0(t)\}, \quad \text{Re}\, s > 0, \tag{I.9}$$

where

$$\tilde{G}(s) = N(s) + \frac{\gamma}{s}, \quad \text{Re}\, s > 0. \tag{I.10}$$

$$N(s) = -\sum_{l=1}^{n} \sum_{m=1}^{n} c_l L\{\psi_{lm}(t)\}\, b_m, \quad \text{Re}\, s \geqslant 0. \tag{I.11}$$

The function $\tilde{G}(s)$ defined by the equations (I.10)-(I.11) is the transfer function of the system linear block. The function $\tilde{G}(s)$ is obtained in the form of a rational function of s. The above considerations are only valid when Re

$s > 0$; we shall say nevertheless that G (s) is a transfer function if it is a rational function defined in the whole s-pla and obtained as the analytic continuation of the $\tilde{G}$ (s) function.

It should be mentioned that the function L $\{\psi_{lm}$ (t)$\}$ exists when Re s = 0 and that when s = jω (ω real), then L$\{\psi_{lm}$ (t)$\}$ = F $\{\psi_{lm}$ (t)$\}$ [see (4.10)]. By comparing (I.11), with (2.6) and (2.9) we see that when s = jω the functic (I.11) is equal to the function of (2.9). Therefore, the function (2.10) is equal to the above defined function G (s) wi s = jω.

APPENDIX 2

By using (3.1) and (3.2) we obtain for t > T

$$\lambda_T(t) = -\int_0^T \left(\nu(t-\zeta) + q\frac{d\nu(t-\zeta)}{dt}\right)\varphi(\sigma(\zeta))\,d\zeta. \tag{II.1}$$

But [see (2.6), (2.1) and (2.7)]

$$\left|\nu(t) + q\frac{d\nu(t)}{dt}\right| < \sum_{l=1}^{n} |c_l|\left(1 + q\sum_{k=1}^{n}\sum_{m=1}^{n}|a_{km}|\right)\sum_{i=1}^{n}|b_i|K_1e^{-K_0t}. \tag{II.2}$$

Therefore,

$$|\lambda_T(t)| \leqslant \sup_{0<\zeta<T}|\varphi(\sigma(\zeta))|\int_0^T\left|\nu(t-\zeta) + q\frac{d\nu(t-\zeta)}{dt}\right|d\zeta$$

$$\leqslant \sup_{0<\zeta<T}|\varphi(\sigma(\zeta))|\sum_{l=1}^{n}|c_l|\left(1 + q\sum_{k=1}^{n}\sum_{m=1}^{n}|a_{km}|\right)\sum_{i=1}^{n}|b_i|K_1\int_0^T e^{-K_0(t-\zeta)}\,d\zeta. \tag{II.3}$$

This gives the inequality (3.4), where

$$K_3 = \sup_{0<\zeta<T}|\varphi(\sigma(\zeta))|\sum_{l=1}^{n}|c_l|\left(1 + q\sum_{k=1}^{n}\sum_{m=1}^{n}|a_{km}|\right)\sum_{i=1}^{n}|b_i|\frac{K_1}{K_0}(e^{K_0T} - 1). \tag{II.4}$$

For a specified solution of the system and for a given T, K_3 is a constant.

Therefore the Fourier transformation (3.5) exists.

APPENDIX 3

From the equation (1.2) we obtain

$$E = \int_0^T\left(\mu_m(t) + q\frac{d\mu_m(t)}{dt}\right)\varphi_T(\sigma(t))\,dt = \int_0^T\left(\mu_m(t) + q\frac{d\mu_m(t)}{dt}\right)\frac{d\xi(t)}{dt}\,dt. \tag{III.1}$$

The integration by parts gives

$$E = \left[\left(\mu_m(t) + q\frac{d\mu_m(t)}{dt}\right)\xi(t)\right]_0^T - \int_0^T\xi(t)\left(\frac{d\mu_m(t)}{dt} + q\frac{d^2\mu_m(t)}{dt^2}\right)dt, \tag{III.2}$$

and hence, in view of $|\xi(T)| \leqslant \sup_{0<\zeta<T}|\xi(\zeta)|$ и $|\xi(0)| \leqslant \sup_{0<\zeta<T}|\xi(\zeta)|$, it follows that

$$E| \leqslant \sup_{0<\zeta<T}|\xi(\zeta)|\left(2\sup_{0<t<T}\left|\mu_m(t) + q\frac{d\mu_m(t)}{dt}\right| + \int_0^T\left|\frac{d\mu_m(t)}{dt} + q\frac{d^2\mu_m(t)}{dt^2}\right|dt\right). \tag{III.3}$$

But [see (2.5), (2.1), and (2.7)]

$$\left|\mu_m(t) + q\frac{d\mu_m(t)}{dt}\right| < \sum_{l=1}^{n}|c_l|\left(1 + q\sum_{k=1}^{n}\sum_{m=1}^{n}|a_{km}|\right)K_1, \tag{III.4}$$

$$\left|\frac{d\mu_m(t)}{dt} + q\frac{d^2\mu_m(t)}{dt^2}\right| < \sum_{l=1}^{n}|c_l|\left(1 + q\sum_{k=1}^{n}\sum_{m=1}^{n}|a_{km}|\right)\sum_{k'=1}^{n}\sum_{m'=1}^{n}|a_{k'm'}|K_1e^{-K_0t}. \tag{III.5}$$

By substituting (III.4) and (III.5) into (III.3) and by using the inequality

$$\int_0^T e^{-K_0 t}\,dt < \int_0^\infty e^{-K_0 t}\,dt = \frac{1}{K_0}, \tag{III.6}$$

we obtain the inequality (3.26), where

$$\frac{1}{n}K_4 = \sum_{l=1}^{n} |c_l| \left(1 + q \sum_{k=1}^{n}\sum_{m=1}^{n} |a_{km}|\right)\left(2 + \frac{1}{K_0}\sum_{k'=1}^{n}\sum_{m'=1}^{n} |a_{k'm'}|\right)K_1. \tag{III.7}$$

APPENDIX 4

By replacing $\varphi(\sigma(\xi))$ in (2.3) by $d\xi(\zeta)/d\zeta$ [see (1.2)] and by integrating by parts we obtain

$$x_l(t) = \sum_{m=1}^{n} \psi_{lm}(t)\,x_{m0} + \sum_{m=1}^{n} b_m [\psi_{lm}(t-\zeta)\,\xi(\zeta)]_{\zeta=0}^{\zeta=t} - \sum_{m=1}^{n} b_m \int_0^t \xi(\zeta)\,\frac{d\psi_{lm}(t-\zeta)}{d\zeta}\,d\zeta. \tag{IV.1}$$

From (2.1) and (2.7) it follows that $|x_l(t)| < K_1 n \max(|x_{m0}|)$

$$+ \sup|\xi(\zeta)|\,K_1 \sum_{m=1}^{n} |b_m| \left(2 + \sum_{l=1}^{n}\sum_{k=1}^{n} |a_{lk}| \int_0^t e^{-K_0(t-\zeta)}\,d\zeta\right). \tag{IV.2}$$

As $\int_0^t e^{-K_0(t-\zeta)}\,d\zeta < \int_{-\infty}^t e^{-K_0(t-\zeta)}\,d\zeta = \frac{1}{K_0}$, we obtain the inequality (3.35) where

$$K_5 = nK_1,$$

$$K_6 = K_1 \sum_{m=1}^{n} |b_m| \left(2 + \frac{1}{K_0}\sum_{l=1}^{n}\sum_{k=1}^{n} |a_{lk}|\right)$$

APPENDIX 5

The result (3.42) could be obtained by making use of a lemma published by Barbalat [9]. Below we shall give a more direct proof.

We assume for this purpose that (3.42) is not true. Then a positive quantity δ and a sequence of quantities t_i must exist such that

$$|\sigma(t_i)| > \delta, \quad t_i > 0, \quad \lim_{i\to\infty} t_i = +\infty. \tag{V.1}$$

Let the solution in question [see (3.41), (3.36), (3.37), and (1.3)] be

$$\left|\frac{d\sigma(t)}{dt}\right| < M_1, \quad |\sigma(t)| < M_2. \tag{V.2}$$

The sequence t_k can be selected such that

$$t_k - t_{k-1} > \frac{\delta}{M_1}, \quad t_1 > \frac{\delta}{2M_1}. \tag{V.3}$$

When $|t - t_k| < \frac{\delta}{2M_1}$ we have

$$M_2 > |\sigma(t)| = \left|\sigma(t_k) + \int_{t_k}^{t} \frac{d\sigma(t)}{dt}\,dt\right| > |\sigma(t_k)| - M_1|t - t_k| > \frac{1}{2}\delta. \tag{V.4}$$

We also have [see (1.4) and (1.5)]

$$\int_0^T \varphi(\sigma(t))\,\sigma(t)\,dt \geqslant \sum_{k=1}^{N(T)} \int_{t_k - \frac{\delta}{2M_1}}^{t_k + \frac{\delta}{2M_1}} \varphi(\sigma(t))\,\sigma(t)\,dt, \tag{V.5}$$

where N (T) is an index such that $t_{N(T)} + \frac{\delta}{2M_1} \leqslant T$ and $t_{N(T)+1} + \frac{\delta}{2M_1} > T$.

Obviously

$$\lim_{T\to\infty} N(T) = \infty. \tag{V.6}$$

Let $\inf\limits_{\frac{\delta}{2}<|\sigma|<M_1} |\varphi(\sigma)| = m$. Then we obtain from (V.5) [see (1.5)]

$$\int_0^T \varphi(\sigma(t))\,\sigma(t)\,dt \geqslant N(T)\frac{\delta}{M_1}\cdot\frac{\delta}{2}m. \tag{V.7}$$

In view of (V.6)

$$\lim_{T\to\infty}\int_0^T \varphi(\sigma(t))\,\sigma(t)\,dt = \infty, \tag{V.8}$$

which contradicts (3.39), and so (3.42) is proved.

APPENDIX 6

In order to prove the relation (3.44) [see (2.3)] it is sufficient, in view of (2.7), to prove that

$$\overline{\lim_{t\to\infty}}\int_0^t \psi_{lm}(t-\zeta)\,\varphi(\sigma(\zeta))\,d\zeta = 0. \tag{VI.1}$$

follows from (3.44).

According to (2.7) we have

$$\left|\int_0^t \psi_{ml}(t-\zeta)\,\varphi(\sigma(\zeta))\,d\zeta\right| \leqslant K_1\int_0^t e^{-K_0(t-\zeta)}\,|\varphi(\sigma(\zeta))| \tag{VI.2}$$

By applying de l 'Hopital rule (see, for example, [10]) we obtain [see (3.43)]

$$\lim_{t\to\infty}\int_0^t e^{-K_0(t-\zeta)}\,|\varphi(\sigma(\zeta))|\,d\zeta = \lim_{t\to\infty}\frac{\int_0^t e^{K_0\zeta}\,|\varphi(\sigma(\zeta))|\,d\zeta}{e^{K_0\tau}} = \frac{1}{K_0}\lim_{t\to\infty}|\varphi(\sigma(t))| = 0, \tag{VI.3}$$

and hence follows (VI.1).

APPENDIX 7

It can easily be seen that the most general negative definite Lyapunov function of the type "a quadratic forr plus the integral of the nonlinearity" can be represented as

$$V = \sum_{l=1}^{n}\sum_{m=1}^{n} x_l r_{lm} x_m - \alpha\left(\sigma - \sum_{l=1}^{n} c_l x_l\right)^2 - 2\beta\int_0^\sigma \varphi(\sigma)\,d\sigma + 2\sigma\sum_{l=1}^{n} f_l x_l, \tag{VII.1}$$

with r_{lm}, α, β, f_l being constants. The derivative of the V function by virtue of (4.1) assumes the form

$$\frac{1}{2}\frac{dV}{dt} = W(x_l,\sigma) + \sigma\sum_{l=1}^{n} f_l\left(\sum_{k=1}^{n} a_{lk}x_k + b_l\varphi(\sigma)\right)$$
$$+\sum_{l=1}^{n} f_l x_l\left[\sum_{l=1}^{n} c_l\left(\sum_{k=1}^{n} a_{lk}x_k + b_l\varphi(\sigma)\right) - r\varphi(\sigma)\right], \tag{VII.2}$$

where $W(x_l,\sigma)$ is given by the expression (4.4).

We shall show that the equalities

$$f_l = 0 \quad (l = 1, 2, \ldots, n) \tag{VII.3}$$

are necessary. We note that the function $\frac{1}{2}\,dV/dt$ is positive when

$$x_l = x_l^0 = \varepsilon\int_0^\infty \psi_{lm}(t)\,dt \quad (l = 1, 2, \ldots, n) \tag{VII.4}$$

$$\varphi(\sigma) = \varepsilon^2\sigma, \tag{VII.5}$$

$$\sigma = 1, \tag{VII.6}$$

where m is one of the integers 1, 2, . . . , n; ψ_{lm} (t) the functions introduced in section 2, and ϵ an arbitrary (positive or negative) quantity. The function (VII.5) is of course of the class A. By integrating (2.1) we obtain, in view of $\psi_{lm}(t)$ [see also (2.2)],

$$-\psi_{lm}(0) = -\delta_{lm} = \sum_{k=1}^{n} a_{lk} \int_0^\infty \psi_{km}(t)\,dt. \tag{VII.7}$$

Consequently [see (VII.4)]

$$\sum_{k=1}^{n} a_{lk}x_k^0 = -\varepsilon\delta_{lm}. \tag{VII.8}$$

It is easily seen that for the values of (VII.4)-(VII.6) the function $\frac{1}{2}$ dV/dt can be written as

$$\frac{1}{2}\frac{dV}{dt} = \left(\sum_{l=1}^{n} f_l \sum_{k=1}^{n} a_{lk}x_k^0\right)\varepsilon + O(\varepsilon^2), \tag{VII.9}$$

in which $O(\epsilon^2)$ denotes terms with the property $|O(\epsilon^2)| < K\epsilon^2$, with K being a constant.

From (VII.9) using (VII.8) we obtain

$$\frac{1}{2}\frac{dV}{dt} = -f_m\varepsilon + O(\varepsilon^2). \tag{VII.10}$$

If $f_m \neq 0$, then for a sufficiently small ϵ the expression (VII.10) has the sign of $-f_m\epsilon$; the latter, however, is arbitrary because the sign of ϵ is also arbitrary. It is therefore necessary that $f_m = 0$ be true. As m was taken as arbitrary we see that the equalities (VII.3) must be true.

The function (4.2) is negative when $x_l = 0, \sigma = 1, \varphi(\sigma) = h\sigma, h > 0$. This gives the inequality

$$-\alpha - h\,\beta < 0 \text{ for any } h > 0, \tag{VII.11}$$

which proves finally the relations (4.3).

LITERATURE CITED

1. A. I. Lur'e and E. N. Rozenvasser, Methods of constructing Lyapunov functions of the nonlinear control systems - theory, Proc. of the First International Conference on Automatic Control, Acad. Sci. USSR Press (1961). Sci. Press (1961).
2. V. M. Popov, Criterii de stabilitate pentru sistemele neliniare de reglare automata pe utikizarea transformatei Laplace. Studii si Cercetarii de Energetica. Anul. IX. No. 4, 1959.
3. V. M. Popov, Criterii suficiente de stabilitate asimptotica in mare pentru sistemele neliniare cu mai multe organe oe executie. Studii si Cercetari de Energetica, Anul. IX, No. 4, 1959.
4. V. M. Popov, Nouveaux criteriums pe stabilite por les systemes automatiques non-lineaires. Revue d'Electrotechnique et d'Energetique, vol. V, No. 1, 1960.
5. V. M. Popov, Noi criterii grafice pentru stabilitatea starii stationare a sistemelor automati neliniare. Studii si Cerletari de Energetica. Anul. X, No. 3, 1960.
6. A. M. Letov, The stability of nonlinear systems under control, Gostekhizdat (1951).
7. A. I. Lur'e, Some nonlinear problems in the theory of automatic control, Gostekhizdat (1951).
8. V. A. Yakubovich, Nonlinear differential equations of automatic control systems with a single controlling device, Bull. Leningrad Univ., No. 7 (1960).
9. I. Barbalat, Systemes d'équations différentielles d'oscillations non linéaires. Revue de mathematiques pures et appliquées. Acad. R. P. R., vol. 4, No. 2, 1959.
10. Miron Nicolescu, Analiză Matematică, Vol. II, Bucaresti Editura Tehnică, 1958.

A Steepest-Ascent Method for Solving Optimum Programming Programs

A. E. BRYSON AND W. F. DENHAM

AEROSPACE problems influenced considerably the early development of optimal control and resulted in the formulation of trajectory optimization problems and methods for solving these problems well before the advent of dynamic programming and the maximum principle. As early as 1919 Goddard posed the problem of determining the thrust profile of a rocket to maximize altitude; an analytic solution for this problem, which requires a singular arc, was obtained in 1951 [8] and is described in [4], page 253. Hohmann [5] determined the optimal impulsive manoeuvre to transfer a space vehicle from one circular orbit to another in 1925. These early solutions were of necessity analytic; although the conceptual tools for numerical optimization were available, the means were not until the advent of the digital computer in the mid 1950s. The calculus of variations provided, at least implicitly, the concept of a function space gradient, and the method of steepest descent (or ascent) for optimization dates from Cauchy, but it could not be implemented on serious problems prior to the mid 1950s even though it had been well studied prior to this.

The stimulation provided by aerospace problems such as the determination of maximal range rocket trajectories, minimum-time trajectories for space flight from Earth to Mars and minimum-time trajectories for fighter aircraft to reach a high altitude was intense and resulted in a spate of papers in the late 1950s and early 1960s dealing with these problems. Among the first were the papers by Bryson and Ross [3] and Breakwell [1] that obtained necessary conditions of optimality for the maximal range rocket trajectory problem using the calculus of variations, and solved the resultant two point boundary value problem using a "shooting" algorithm, the initial value of the adjoint serving as the "decision" variable. Bryson and Ross used an IBM 650 computer made available by Garrett Birkhoff of the Mathematics Department at Harvard University and later applied the same method to the determination of optimal re-entry paths for an orbital glider. Lawden, Leitmann, Miele, and others also employed shooting algorithms to solve aerospace flight path problems in the late 1950s and early 1960s.

The extreme sensitivity of the shooting method motivated the development of gradient methods using the control function $u(\cdot)$ as the decision variable [6]. The problem studied in Kelley's paper and that of Bryson and Denham is the maximization of a terminal performance index $\phi(x(T))$ subject to constraints imposed by the differential equation

$$\dot{x} = f(x, u, t)$$

of the system, the initial condition $x(t_0) = x_0$, a terminal equality constraint $\psi(x(T)) = 0$, and a stopping condition $\Omega(x(T), T) = 0$ that determines the final time T. Kelley [6] replaces the terminal constraint by a penalty, making the problem unconstrained (apart from the differential equation); this simplifies the optimization procedure but causes numerical difficulties. Ignoring for simplicity the stopping condition (by regarding T as fixed), Bryson and Denham proceed as follows. Suppose that $u(\cdot)$ is the current control. The gradients $g_\phi(\cdot)$ and $g_\psi(\cdot)$ of the performance index and the equality constraint with respect to the control function $u(\cdot)$ are obtained by solving the adjoint differential equation

$$-\dot{\lambda} = f_x^T(x, u, t)\lambda$$

with terminal condition $\lambda(T) = \phi_x'(x(T))$ (to obtain $g_\phi(\cdot)$) and terminal condition $\lambda(T) = \psi_x'(x(T))$ (to obtain $g_\psi(\cdot)$). A control change $\delta u(\cdot)$ is selected that maximizes $\langle g_\phi(\cdot), \delta u(\cdot)\rangle_2$ subject to $\langle g_\phi(\cdot), \delta u(\cdot)\rangle_2 = a$ and $\|\delta u(\cdot)\|_2 \leq b$. Here a is the desired change in the value of the terminal constraint and b limits the step size. The successor control is $u'(\cdot) = u(\cdot) + \delta u(\cdot)$. The variables a and b are chosen manually to ensure convergence. Because both performance index and constraint change, the search direction is not steepest ascent; indeed the performance index may *decrease* in any given iteration.

What makes these early papers so exciting is the significant spacecraft and trajectory optimization problems that were solved. The first aircraft application of the gradient method appears in this paper by Bryson and Denham. One of the problems considered was the determination of the angle of attack profile that minimized time to achieve a specified altitude of 20 km and a specified speed of Mach 1 at level flight. The solution was surprising because it was considerably different from and better than the best previously achieved (by cut and try). The optimal path consisted of a rapid climb at a speed of just below Mach 1 to a height of 30,000 feet (to obtain an increase in potential energy) followed by a period in which height is *decreased* and speed increased (to gain kinetic energy). A final spurt to the specified height and speed concludes the manoeuvre. The optimal path was actually tested in January 1962 at the Patuxent River Naval Air Station; the time taken to reach the specified altitude was 338 seconds compared with 332 seconds predicted theoretically. Contrary to what might have been expected from such a beautiful demonstration of optimal control, the outcome was not immediately favorable; the two pilots who flew the optimal path were reprimanded and neither Harvard nor Raytheon received an increase in research funding!

However, the long-term effect on applications and research was significant. A wide range of spacecraft and aircraft application studies, well described by Bryson [2], were undertaken. But the work also stimulated theoretical research, particularly on the mathematical programming problems occurring in optimal control. The recent text by Polak [7] reveals the richness of the subject and the advances that have been made.

There are several features of the early research by Bryson and Denham and others that deserve comment. Firstly, the researchers used the calculus of variations rather than the maximum principle. This must have been in part owing to timing: the maximum principle was not readily available when the research began. There were other reasons: to this day there are only one or two algorithms that employ strong variations for the search "direction" and that generate accumulation points satisfying the maximum principle rather than weaker optimality conditions. Secondly, the researchers exercised considerable ingenuity in obtaining conditions of optimality for control problems involving state and state dependent control constraints, problems that are still being studied. Thirdly, the researchers developed novel algorithms to handle constraints, like that employed by Bryson and Denham. Nowadays the problem would probably use exact penalty functions that provide a single criterion in place of the two (performance index and constraint value) employed in this paper. Finally, it is clear that this early research inspired a long and fruitful activity, both theoretical and practical.

REFERENCES

[1] J.A. BREAKWELL, "The optimization of trajectories," *SIAM Journal*, **7**:215–247, 1959.

[2] A.E. BRYSON, "Optimal control—1950 to 1985," *IEEE Control Systems*, **16**(3):26–33, 1996.

[3] A.E. BRYSON AND S.E. ROSS, "Optimum rocket trajectories with aerodynamics drag," *Jet Propulsion*, 1958.

[4] A.E. BRYSON AND Y.-C. HO, *Applied Optimal Control—Optimization, Estimation, and Control*, Hemisphere Publishing Corp. (Washington), 1969.

[5] W. HOHMANN, *Die Erreichbarkeit der Himmelskörper*, Oldenbourg (Münich), 1925.

[6] H.J. KELLEY, "Gradient theory of optimal flight paths," *J. American Rocket Soc.*, **30**:948–954, 1960.

[7] E. POLAK, *Optimization: Algorithms and Consistent Approximations*, Springer Verlag (New York), 1997.

[8] H.S. TSIEN AND R.C. EVANS, "Optimum thrust programming for a sounding rocket," *J. American Rocket Soc.*, **21**(5):99–107, 1951.

D.Q.M.

A. E. BRYSON
Professor,
Division of Engineering
and Applied Physics,
Harvard University,
Cambridge, Mass.

W. F. DENHAM
Research Engineer,
Missiles & Space Division,
Raytheon Company,
Bedford, Mass.

A Steepest-Ascent Method for Solving Optimum Programming Problems

A systematic and rapid steepest-ascent numerical procedure is described for solving two-point boundary-value problems in the calculus of variations for systems governed by a set of nonlinear ordinary differential equations. Numerical examples are presented for minimum time-to-climb and maximum altitude paths for a supersonic interceptor and maximum-range paths for an orbital glider.

1 Summary

A systematic and rapid steepest-ascent numerical procedure is described for determining optimum programs for nonlinear systems with terminal constraints. The procedure uses the concept of local linearization around a nominal (non-optimum) path. The effect on the terminal conditions of a small change in the control variable program is determined by numerical integration of the adjoint differential equations for small perturbations about the nominal path. Having these adjoint (or influence) functions, it is then possible to determine the change in the control variable program that gives maximum increase in the pay-off function for a given mean-square perturbation of the control variable program while simultaneously changing the terminal quantities by desired amounts. By repeating this process in small steps, a control variable program that minimizes one quantity and yields specified values of other terminal quantities can be approached as closely as desired. Three numerical examples are presented: (*a*) The angle-of-attack program for a typical supersonic interceptor to climb to altitude in minimum time is determined with and without specified terminal velocity and heading. (*b*) The angle-of-attack program for the same interceptor to climb to maximum altitude is determined. (*c*) The angle-of-attack program is determined for a hypersonic orbital glider to obtain maximum surface range starting from satellite speed at 300,000 ft altitude.

2 Introduction

Optimum programming problems arise in connection with processes developing in time or space, in which one or more control variables must be programmed to achieve certain terminal conditions. The problem is to determine, out of all possible programs for the control variables, the one program that maximizes (or minimizes) one terminal quantity while simultaneously yielding specified values of certain other terminal quantities.

The calculus of variations is the classical tool for solving such problems. However, until quite recently, only rather simple problems had been solved with this tool owing to computational difficulties. Even with a high-speed digital computer these problems are quite difficult because, in the classical formulation, they are two-point boundary-value problems for a set of nonlinear ordinary differential equations. Numerical solution requires guessing the missing boundary conditions at the initial point, integrating the differential equations numerically to the terminal point, finding how badly the specified terminal boundary conditions are missed, and then attempting to improve the guess of the unspecified initial conditions. This process must be repeated over and over until all terminal conditions are satisfied. This process is not only tedious, expensive, and frustrating, it sometimes does not seem to work at all [1].[1] It is remarkably sensitive to small changes in the initial conditions; however, it can be made to work through great patience, good guessing, and second-order multiple interpolation [2, 3, 4].

Recently Kelley [5, 6] and the authors with several coworkers [7, 8] have revived a little-known procedure which offers a practical, straightforward method for finding numerical solutions to even the most complicated optimum programming problems. It is essentially a steepest-ascent method and it requires the use of a high-speed digital computer.

3 A Maximum Problem in Ordinary Calculus

In order to explain the steepest-ascent method it is helpful to consider its use in a simpler problem first; namely, the problem of finding the maximum of a nonlinear function of many variables subject to nonlinear constraints on these variables. This is a problem in the ordinary calculus. A quite general problem of this type can be stated as follows:

Determine α so as to maximize

$$\phi = \phi(\mathbf{x}), \tag{1}$$

subject to the constraints

$$\psi = \psi(\mathbf{x}) = 0, \tag{2}$$

$$f(\mathbf{x}, \alpha) + f_0 = 0 \tag{3}$$

where

$$\alpha = \begin{bmatrix} \alpha_1 \\ \cdot \\ \cdot \\ \cdot \\ \alpha_m \end{bmatrix}, \tag{4}$$

an $m \times 1$ matrix of *control variables*, which we are free to choose,

$$\mathbf{x} = \begin{bmatrix} x_1 \\ \cdot \\ \cdot \\ \cdot \\ x_n \end{bmatrix}, \tag{5}$$

an $n \times 1$ matrix of *state variables*, which result from the choice of α,

$$\psi = \begin{bmatrix} \psi_1 \\ \cdot \\ \cdot \\ \cdot \\ \psi_p \end{bmatrix}, \tag{6}$$

a $p \times 1$ matrix of *constraint functions*, each a known function of $\mathbf{x}$,

[1] Numbers in brackets designate References at end of paper.

A lecture presented at the Summer Conference of the Applied Mechanics Division, Chicago, Ill., June 14–16, 1961, of The American Society of Mechanical Engineers.

Discussion of this paper should be addressed to the Editorial Department, ASME, 345 East 47th Street, New York, N. Y., and will be accepted until July 20, 1962. Discussion received after the closing date will be returned. Manuscript received by ASME Applied Mechanics Division, September 6, 1961.

Reprinted with permission from *Transactions of the ASME, Journal of Applied Mechanics*, A. E. Bryson and W. F. Denham, "A Steepest-Ascent Method for Solving Optimum Programming Programs," April 1962, pp. 247–257.

ϕ is the *pay-off function*, a known function of $\mathbf{x}$, (7)

$$\mathbf{f} = \begin{bmatrix} f_1 \\ \cdot \\ \cdot \\ \cdot \\ f_n \end{bmatrix}, \text{ an } n \times 1 \text{ matrix of known functions of } \mathbf{x} \text{ and } \boldsymbol{\alpha}, \quad (8)$$

$$\mathbf{f}_0 = \begin{bmatrix} f_{10} \\ \cdot \\ \cdot \\ \cdot \\ f_{n0} \end{bmatrix}, \text{ an } n \times 1 \text{ matrix of constants} \quad (9)$$

4 Steepest-Ascent Method in Ordinary Calculus

The maximum problem stated in the preceding section can be solved systematically and rapidly on a high-speed digital computer using the steepest-ascent method. This method starts with a nominal control variable matrix $\boldsymbol{\alpha}^*$, and then improves this estimate by determining the direction of steepest ascent in the $\boldsymbol{\alpha}$ hyperspace; this determination is made by a linearization about the nominal point in the $\boldsymbol{\alpha}$ hyperspace. The method proceeds as follows:

(*a*) Guess some reasonable control variables $\boldsymbol{\alpha}^*$, and use them in equations (3) to calculate numerically the state variables $\mathbf{x}^*$ that correspond to this choice. In general, this nominal "point" will *not* satisfy the constraint conditions $\boldsymbol{\psi} = 0$, or yield the maximum value of ϕ.

(*b*) Consider small perturbations $d\boldsymbol{\alpha}$, about the nominal control variable point where

$$d\boldsymbol{\alpha} = \boldsymbol{\alpha} - \boldsymbol{\alpha}^* \quad (10)$$

These perturbations will cause perturbations $d\mathbf{x}$ in the state variables, where

$$d\mathbf{x} = \mathbf{x} - \mathbf{x}^* \quad (11)$$

Taking differentials of equations (3) we obtain to first order in the perturbations the linear set of equations for dx,

$$\mathbf{F}d\mathbf{x} + \mathbf{G}d\boldsymbol{\alpha} = 0 \quad (12)$$

where

$$\mathbf{F} = \begin{bmatrix} \left(\frac{\partial f_1}{\partial x_1}\right)^*, \ldots \left(\frac{\partial f_1}{\partial x_n}\right)^* \\ \cdot \quad \cdot \\ \cdot \quad \cdot \\ \left(\frac{\partial f_n}{\partial x_1}\right)^*, \ldots \left(\frac{\partial f_n}{\partial x_n}\right)^* \end{bmatrix} \quad (13)$$

$$\mathbf{G} = \begin{bmatrix} \left(\frac{\partial f_1}{\partial \alpha_1}\right)^*, \ldots \left(\frac{\partial f_1}{\partial \alpha_m}\right)^* \\ \cdot \quad \cdot \\ \cdot \quad \cdot \\ \left(\frac{\partial f_n}{\partial \alpha_1}\right)^*, \ldots \left(\frac{\partial f_n}{\partial \alpha_m}\right)^* \end{bmatrix} \quad (14)$$

and $(\ )^*$ indicates that the partial derivatives are evaluated at the nominal point. Using Lagrange multipliers (Appendix, section 1), we may write

$$d\phi = \boldsymbol{\lambda}_\phi'\mathbf{G}d\boldsymbol{\alpha} + \boldsymbol{\lambda}_\phi'd\mathbf{f}_0 \quad (15)$$

$$d\boldsymbol{\psi} = \boldsymbol{\lambda}_\psi'\mathbf{G}d\boldsymbol{\alpha} + \boldsymbol{\lambda}_\psi'd\mathbf{f}_0 \quad (16)$$

where the $\boldsymbol{\lambda}$ matrices are determined by the linear equations

$$\left(\frac{\partial \phi}{\partial \mathbf{x}}\right)^* + \boldsymbol{\lambda}_\phi'\mathbf{F} = 0, \quad (17)$$

$$\left(\frac{\partial \boldsymbol{\psi}}{\partial \mathbf{x}}\right)^* + \boldsymbol{\lambda}_\psi'\mathbf{F} = 0, \quad (18)$$

where

$$\frac{\partial \phi}{\partial \mathbf{x}} = \left[\frac{\partial \phi}{\partial x_1}, \ldots \frac{\partial \phi}{\partial x_n}\right], \quad \frac{\partial \boldsymbol{\psi}}{\partial \mathbf{x}} = \begin{bmatrix} \frac{\partial \psi_1}{\partial x_1}, \ldots \frac{\partial \psi_1}{\partial x_n} \\ \cdot \quad \cdot \\ \cdot \quad \cdot \\ \frac{\partial \psi_p}{\partial x_1}, \ldots \frac{\partial \psi_p}{\partial x_n} \end{bmatrix}, \quad (19)$$

and $(\ )'$ indicates the transpose of $(\ )$; i.e., rows and columns are interchanged.

Note that the $\boldsymbol{\lambda}$ are *influence numbers* since they tell how much ϕ or $\boldsymbol{\psi}$ is changed by small changes in the constraint levels $\mathbf{f}_0$.

For steepest ascent, we wish to find the $d\boldsymbol{\alpha}$ matrix that maximizes $d\phi$ in equation (15) for a given value of the positive definite quadratic form

$$(dP)^2 = d\boldsymbol{\alpha}'\mathbf{W}d\boldsymbol{\alpha}, \quad (20)$$

and given values of $d\boldsymbol{\psi}$ in equations (16). The values of $d\boldsymbol{\psi}$ are chosen to bring the nominal solution closer to the specified constraints, $\boldsymbol{\psi} = 0$. Choice of dP is made to insure that the perturbations $d\boldsymbol{\alpha}$ will be small enough for the linearization leading to equations (12) to be reasonable. $\mathbf{W}$ is an arbitrary non-negative definite $m \times m$ matrix of weighting numbers, essentially a metric in the $\boldsymbol{\alpha}$ hyperspace; it is at the disposal of the "optimizer" to improve convergence of the procedure.

The proper choice of $d\boldsymbol{\alpha}$ is derived in section 2 of the Appendix and the result is

$$d\boldsymbol{\alpha} = \pm\mathbf{W}^{-1}\mathbf{G}'(\boldsymbol{\lambda}_\phi - \boldsymbol{\lambda}_\psi\mathbf{I}_{\psi\psi}^{-1}\mathbf{I}_{\psi\phi})\left[\frac{(dP)^2 - d\boldsymbol{\beta}'\mathbf{I}_{\psi\psi}^{-1}d\boldsymbol{\beta}}{I_{\phi\phi} - \mathbf{I}_{\psi\phi}'\mathbf{I}_{\psi\psi}^{-1}\mathbf{I}_{\psi\phi}}\right]^{1/2} + \mathbf{W}^{-1}\mathbf{G}'\boldsymbol{\lambda}_\psi\mathbf{I}_{\psi\psi}^{-1}d\boldsymbol{\beta} \quad (21)$$

where

$$\begin{aligned} d\boldsymbol{\beta} &= d\boldsymbol{\psi} - \boldsymbol{\lambda}_\psi'd\mathbf{f}_0, \\ \mathbf{I}_{\psi\psi} &= \boldsymbol{\lambda}_\psi'\mathbf{G}\mathbf{W}^{-1}\mathbf{G}'\boldsymbol{\lambda}_\psi, \\ \mathbf{I}_{\psi\phi} &= \boldsymbol{\lambda}_\psi'\mathbf{G}\mathbf{W}^{-1}\mathbf{G}'\boldsymbol{\lambda}_\phi, \\ I_{\phi\phi} &= \boldsymbol{\lambda}_\phi'\mathbf{G}\mathbf{W}^{-1}\mathbf{G}'\boldsymbol{\lambda}_\phi, \end{aligned} \quad (22)$$

$(\ )^{-1}$ indicates inverse matrix, and the + sign is used if ϕ is to be increased, the − sign if ϕ is to be decreased. Note that the numerator under the square root in equation (21) can become negative if $d\boldsymbol{\beta}$ is chosen too large; thus there is a limit to the size of $d\boldsymbol{\beta}$ for a given dP. Since dP is chosen to insure valid linearization, the $d\boldsymbol{\beta}$ asked for must also be limited. The predicted change in ϕ for the change in control variables of equation (21) is

$$d\phi = \pm[((dP)^2 - d\boldsymbol{\beta}'\mathbf{I}_{\psi\psi}^{-1}d\boldsymbol{\beta})(I_{\phi\phi} - \mathbf{I}_{\psi\phi}'\mathbf{I}_{\psi\psi}^{-1}\mathbf{I}_{\psi\phi})]^{1/2} + \mathbf{I}_{\psi\phi}'\mathbf{I}_{\psi\psi}^{-1}d\boldsymbol{\beta} + \boldsymbol{\lambda}_\phi'd\mathbf{f}_0. \quad (23)$$

Notice if $d\boldsymbol{\psi} = 0$, $d\mathbf{f}_0 = 0$, equation (23) becomes

$$\frac{d\phi}{dP} = \pm(I_{\phi\phi} - \mathbf{I}_{\psi\phi}'\mathbf{I}_{\psi\psi}^{-1}\mathbf{I}_{\psi\phi})^{1/2} \quad (24)$$

which is the magnitude of the gradient in the $\boldsymbol{\alpha}$ hyperspace, since dP is the length of the step in the $\boldsymbol{\alpha}$ hyperspace. As the maximum is approached and the constraints are met ($d\boldsymbol{\psi} = 0$), this gradient must tend to zero, which results in

$$d\phi = \mathrm{I}_{\psi\phi}'\mathrm{I}_{\psi\psi}^{-1}d\psi + (\lambda_\phi' - \mathrm{I}_{\psi\phi}'\mathrm{I}_{\psi\psi}^{-1}\lambda_\psi')df_0 \tag{25}$$

This relation shows how much the maximum pay-off function changes for small changes in the constraint levels.

(c) A new control variable point is obtained as

$$\alpha_{NEW} = \alpha_{OLD} + d\alpha$$

where $d\alpha$ is obtained from equation (21). α_{NEW} is used in the original nonlinear equations (3) and the whole process is repeated several times until the $\psi = 0$ constraints are met and the gradient is nearly zero in equation (24). The maximum value of ϕ has then been obtained.

This process can be likened to climbing a mountain in a dense fog. We cannot see the top but we ought to be able to get there by always climbing in the direction of steepest ascent. If we do this in steps, climbing in one direction until we have traveled a certain horizontal distance, then reassessing the direction of steepest ascent, climbing in that direction, and so on, this is the exact analog of the procedure suggested here in a space of m-dimensions where ϕ is altitude and α_1, α_2 are co-ordinates in the horizontal plane, Fig. 1. There is, of course, a risk here in that we may climb a secondary peak and, in the fog, never become aware of our mistake.

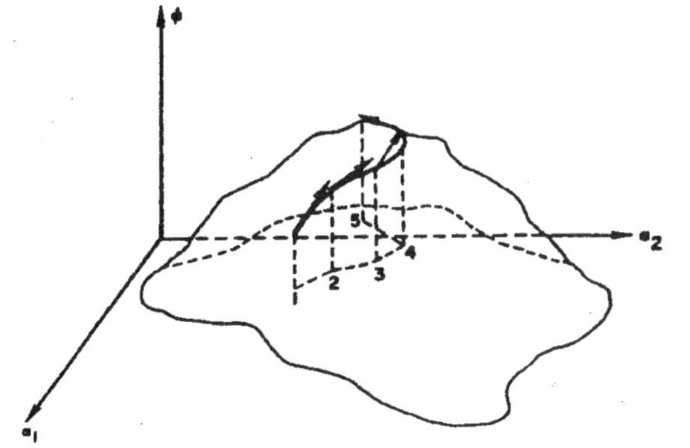

Fig. 1 Finding maximum of a function of two variables by steepest-ascent method

5 Optimum Programming, a Problem in the Calculus of Variations

An optimum programming problem of considerable generality can be stated as follows, Fig. 2:

Determine $\alpha(t)$ in the interval $t_0 \leq t \leq T$, so as to maximize

$$\phi = \phi(\mathbf{x}(T), T), \tag{26}$$ [3]

subject to the constraints,

$$\psi = \psi(\mathbf{x}(T), T) = 0, \tag{27}$$

$$\frac{d\mathbf{x}}{dt} = \mathbf{f}(\mathbf{x}(t), \alpha(t), t), \tag{28}$$

$$t_0 \text{ and } \mathbf{x}(t_0) \text{ given}, \tag{29}$$ [4]

$$T \text{ determined by } \Omega = \Omega(\mathbf{x}(T), T) = 0 \tag{30}$$

The nomenclature of the problem is as follows:

$$\alpha(t) = \begin{bmatrix} \alpha_1(t) \\ \cdot \\ \cdot \\ \cdot \\ \alpha_m(t) \end{bmatrix}, \text{ an } m \times 1 \text{ matrix of } \textit{control variable programs}, \text{ which we are free to choose,} \tag{31}$$

$$\mathbf{x}(t) = \begin{bmatrix} x_1(t) \\ \cdot \\ \cdot \\ \cdot \\ x_n(t) \end{bmatrix}, \text{ an } n \times 1 \text{ matrix of } \textit{state variable programs}, \text{ which result from a choice of } \alpha(t) \text{ and given values of } \mathbf{x}(t_0), \tag{32}$$

$$\psi = \begin{bmatrix} \psi_1 \\ \cdot \\ \cdot \\ \cdot \\ \psi_p \end{bmatrix}, \text{ a } p \times 1 \text{ matrix of } \textit{terminal constraint functions}, \text{ each of which is a known function of } \mathbf{x}(T) \text{ and } T, \tag{33}$$

[3] If an integral is to be maximized, simply introduce an additional state variable x_q and an additional differential equation $\dot{x}_q = q(\mathbf{x}, \alpha, t)$ where q is the integrand of the integral. $x_q(T)$ is then maximized with $x_q(t_0) = 0$.

[4] In some problems not all of the state variables are specified initially; in this case the unspecified state variables may be determined along with $\alpha(t)$ to maximize ϕ.

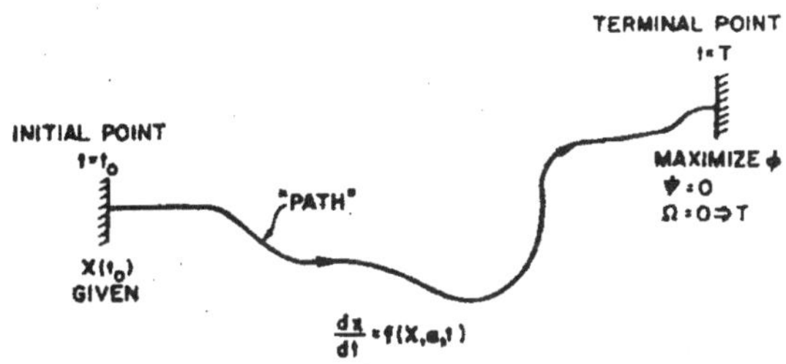

Fig. 2 Symbolic sketch of optimum programming problem

ϕ is the *pay-off function* and is known function of $\mathbf{x}(T)$ and T, (34)

$$\mathbf{f} = \begin{bmatrix} f_1 \\ \cdot \\ \cdot \\ \cdot \\ f_n \end{bmatrix}, \text{ an } n \times 1 \text{ matrix of known functions of } \mathbf{x}(t), \alpha(t), \text{ and } t, \tag{35}$$

$\Omega = 0$ is the *stopping condition that determines final time* T, and is a known function of $\mathbf{x}(T)$ and T (36)

The formulation of the necessary conditions for an extremal solution to this problem has been given by Breakwell [2] with the added complexity of inequality constraints on the control variables. The present paper is concerned with the efficient and rapid solution of such problems using a steepest-ascent procedure.

6 Steepest-Ascent Method in Calculus of Variations

The optimum programming problem stated in the preceding section can be solved systematically and rapidly on a high-speed digital computer using the steepest-ascent technique. This technique starts with a nominal control variable program $\alpha^*(t)$, and then improves this program in steps, using information obtained by a mathematical diagnosis of the program for the previous step. Conceptually it is a process of local linearization around the path of the previous step. The method proceeds as follows:

(a) Guess some reasonable control variable programs $\alpha^*(t)$, and use them with the initial conditions (29) and the differential equations (28) to calculate, numerically, the state variable programs $\mathbf{x}^*(t)$ until $\Omega = 0$. In general, this nominal "path" will *not* satisfy the terminal conditions $\psi = 0$, or yield the maximum possible value of ϕ.

(b) Consider small perturbations $\delta\alpha(t)$ about the nominal control variable programs, where

$$\delta \boldsymbol{\alpha} = \boldsymbol{\alpha}(t) - \boldsymbol{\alpha}^*(t) \tag{37}$$

These perturbations will cause perturbations in the state variable programs $\delta\mathbf{x}(t)$, where

$$\delta \mathbf{x} = \mathbf{x}(t) - \mathbf{x}^*(t) \tag{38}$$

Substituting these relations into the differential equations (28), we obtain, to first order in the perturbations, the linear differential equations for $\delta\mathbf{x}$,

$$\frac{d}{dt}(\delta\mathbf{x}) = \mathbf{F}(t)\delta\mathbf{x} + \mathbf{G}(t)\delta\boldsymbol{\alpha}, \tag{39}$$

where

$$\mathbf{F}(t) = \begin{bmatrix} \left(\frac{\partial f_1}{\partial x_1}\right)^*, \cdots\cdots & \left(\frac{\partial f_1}{\partial x_n}\right)^* \\ \cdot & \cdot \\ \cdot & \cdot \\ \cdot & \cdot \\ \left(\frac{\partial f_n}{\partial x_1}\right)^*, \cdots\cdots & \left(\frac{\partial f_n}{\partial x_n}\right)^* \end{bmatrix} \tag{40}$$

$$\mathbf{G}(t) = \begin{bmatrix} \left(\frac{\partial f_1}{\partial \alpha_1}\right)^*, \cdots\cdots & \left(\frac{\partial f_1}{\partial \alpha_m}\right)^* \\ \cdot & \cdot \\ \cdot & \cdot \\ \cdot & \cdot \\ \left(\frac{\partial f_n}{\partial \alpha_1}\right)^*, \cdots\cdots & \left(\frac{\partial f_n}{\partial \alpha_m}\right)^* \end{bmatrix} \tag{41}$$

and $(\ )^*$ indicates that the partial derivatives are evaluated along the nominal path. From the theory of adjoint equations, section 3 of the Appendix, we may write

$$d\phi = \int_{t_0}^{T} \boldsymbol{\lambda}_\phi'(t)\mathbf{G}(t)\delta\boldsymbol{\alpha}(t)dt + \boldsymbol{\lambda}_\phi'(t_0)\delta\mathbf{x}(t_0) + \dot{\phi}dT, \tag{42}$$

$$d\boldsymbol{\psi} = \int_{t_0}^{T} \boldsymbol{\lambda}_\psi'(t)\mathbf{G}(t)\delta\boldsymbol{\alpha}(t)dt + \boldsymbol{\lambda}_\psi'(t_0)\delta\mathbf{x}(t_0) + \dot{\boldsymbol{\psi}}dT, \tag{43}$$

$$d\Omega = \int_{t_0}^{T} \boldsymbol{\lambda}'_\Omega(t)\mathbf{G}(t)\delta\boldsymbol{\alpha}(t)dt + \boldsymbol{\lambda}_\Omega'(t_0)\delta\mathbf{x}(t_0) + \dot{\Omega}dT, \tag{44}$$

where the elements of the $\boldsymbol{\lambda}$ matrices are determined by numerical integration of the differential equations adjoint to equations (39); namely,

$$\frac{d\boldsymbol{\lambda}}{dt} = -\mathbf{F}'(t)\boldsymbol{\lambda} \tag{45}$$

with boundary conditions

$$\boldsymbol{\lambda}_\phi'(T) = \left(\frac{\partial \phi}{\partial \mathbf{x}}\right)^*_{t=T}, \quad \boldsymbol{\lambda}_\psi'(T) = \left(\frac{\partial \boldsymbol{\psi}}{\partial \mathbf{x}}\right)^*_{t=T},$$

$$\boldsymbol{\lambda}_\Omega'(T) = \left(\frac{\partial \Omega}{\partial \mathbf{x}}\right)^*_{t=T} \tag{46}$$

where

$$\frac{\partial \phi}{\partial \mathbf{x}} = \left[\frac{\partial \phi}{\partial x_1}, \cdots \frac{\partial \phi}{\partial x_n}\right], \quad \frac{\partial \boldsymbol{\psi}}{\partial \mathbf{x}} = \begin{bmatrix} \frac{\partial \psi_1}{\partial x_1} \cdots & \frac{\partial \psi_1}{\partial x_n} \\ \cdot & \cdot \\ \cdot & \cdot \\ \cdot & \cdot \\ \frac{\partial \psi_p}{\partial x_1} \cdot\cdot & \frac{\partial \psi_p}{\partial x_n} \end{bmatrix},$$

$$\frac{\partial \Omega}{\partial \mathbf{x}} = \left[\frac{\partial \Omega}{\partial x_1}, \cdots \frac{\partial \Omega}{\partial x_n}\right], \tag{47}$$

and

$$\dot{\phi} = \left(\frac{\partial \phi}{\partial t} + \frac{\partial \phi}{\partial \mathbf{x}}\mathbf{f}\right)^*_{t=T}, \quad \dot{\boldsymbol{\psi}} = \left(\frac{\partial \boldsymbol{\psi}}{\partial t} + \frac{\partial \boldsymbol{\psi}}{\partial \mathbf{x}}\mathbf{f}\right)^*_{t=T},$$

$$\dot{\Omega} = \left(\frac{\partial \Omega}{\partial t} + \frac{\partial \Omega}{\partial \mathbf{x}}\mathbf{f}\right)^*_{t=T}, \tag{48}$$

$(\ )'$ indicates the transpose of $(\ )$; i.e., rows and columns are interchanged.

Note that the $\boldsymbol{\lambda}$ are *influence functions* since they tell how much a certain terminal condition is changed by a small change in some initial state variable. Note also that the adjoint equations (47) must be integrated *backward* since the boundary conditions are given at the terminal point, $t = T$.

For steepest ascent we wish to find the $\delta\boldsymbol{\alpha}(t)$ programs that maximize $d\phi$ in equation (42) for a given value of the integral

$$(dP)^2 = \int_{t_0}^{T} \delta\boldsymbol{\alpha}'(t)\mathbf{W}(t)\delta\boldsymbol{\alpha}(t)dt, \tag{49}$$

given values of $d\boldsymbol{\psi}$ in equations (43) and $d\Omega = 0$ in equation (44). The values of $d\boldsymbol{\psi}$ are chosen to bring the nominal solution closer to the desired terminal constraints, $\boldsymbol{\psi} = 0$. Choice of dP is made to insure that the perturbations $\delta\boldsymbol{\alpha}(t)$ will be small enough for the linearization leading to equations (46) to be reasonable. $\mathbf{W}(t)$ is an arbitrary symmetric $m \times m$ matrix of weighting functions chosen to improve convergence of the steepest-ascent procedure; in some problems it is desirable to subdue $\delta\boldsymbol{\alpha}$ in certain highly sensitive regions in favor of larger $\delta\boldsymbol{\alpha}$ in the less sensitive regions.

The proper choice of $\delta\boldsymbol{\alpha}(t)$ is derived in section 4 of the Appendix and the result is

$$\delta\boldsymbol{\alpha}(t) = \pm\mathbf{W}^{-1}\mathbf{G}'(\boldsymbol{\lambda}_{\phi\Omega} - \boldsymbol{\lambda}_{\psi\Omega}\mathbf{I}_{\psi\psi}^{-1}\mathbf{I}_{\psi\phi})\left[\frac{(dP)^2 - d\boldsymbol{\beta}'\mathbf{I}_{\psi\psi}^{-1}d\boldsymbol{\beta}}{I_{\phi\phi} - \mathbf{I}_{\psi\phi}'\mathbf{I}_{\psi\psi}^{-1}\mathbf{I}_{\psi\phi}}\right]^{1/2}$$

$$+ \mathbf{W}^{-1}\mathbf{G}'\boldsymbol{\lambda}_{\psi\Omega}\mathbf{I}_{\psi\psi}^{-1}d\boldsymbol{\beta}, \tag{50}$$

where

$$d\boldsymbol{\beta} = d\boldsymbol{\psi} - \boldsymbol{\lambda}_{\psi\Omega}'(t_0)\delta\mathbf{x}(t_0),$$

$$\boldsymbol{\lambda}_{\phi\Omega} = \boldsymbol{\lambda}_\phi - \frac{\dot{\phi}}{\dot{\Omega}}\boldsymbol{\lambda}_\Omega,$$

$$\boldsymbol{\lambda}_{\psi\Omega} = \boldsymbol{\lambda}_\psi - \boldsymbol{\lambda}_\Omega\frac{\dot{\boldsymbol{\psi}}'}{\dot{\Omega}}$$

$$\mathbf{I}_{\psi\psi} = \int_{t_0}^{T} \boldsymbol{\lambda}_{\psi\Omega}'\mathbf{G}\mathbf{W}^{-1}\mathbf{G}'\boldsymbol{\lambda}_{\psi\Omega}dt, \tag{51}$$

$$\mathbf{I}_{\psi\phi} = \int_{t_0}^{T} \boldsymbol{\lambda}_{\psi\Omega}'\mathbf{G}\mathbf{W}^{-1}\mathbf{G}'\boldsymbol{\lambda}_{\phi\Omega}dt,$$

$$I_{\phi\phi} = \int_{t_0}^{T} \boldsymbol{\lambda}_{\phi\Omega}'\mathbf{G}\mathbf{W}^{-1}\mathbf{G}'\boldsymbol{\lambda}_{\phi\Omega}dt,$$

$(\ )^{-1}$ indicates inverse matrix, and the $+$ sign is used if ϕ is to be increased, the $-$ sign is used if ϕ is to be decreased. Note that the numerator under the square root in equation (50) can become negative if $d\boldsymbol{\beta}$ is chosen too large; thus there is a limit to the size of $d\boldsymbol{\beta}$ for a given dP. Since dP is chosen to insure valid linearization, the $d\boldsymbol{\beta}$ asked for must also be limited. The predicted change in ϕ for the change in control variable program (50) is

$$d\phi = \pm\{((dP)^2 - d\boldsymbol{\beta}'\mathbf{I}_{\psi\psi}^{-1}d\boldsymbol{\beta})(I_{\phi\phi} - \mathbf{I}_{\psi\phi}'\mathbf{I}_{\psi\psi}^{-1}\mathbf{I}_{\psi\phi})\}^{1/2}$$

$$+ \mathbf{I}_{\psi\phi}'\mathbf{I}_{\psi\psi}^{-1}d\boldsymbol{\beta} + \boldsymbol{\lambda}_{\phi\Omega}'(t_0)\delta\mathbf{x}(t_0) \tag{52}$$

Notice that if $d\boldsymbol{\psi} = 0$, $\delta\mathbf{x}(t_0) = 0$, equation (52) becomes

$$\frac{d\phi}{dP} = \pm(I_{\phi\phi} - \mathbf{I}_{\psi\phi}'\mathbf{I}_{\psi\psi}^{-1}\mathbf{I}_{\psi\phi})^{1/2} \tag{53}$$

which is a "gradient" in function space, since dP is the "length" of the step in the control variable programs. As the optimum program is approached and the terminal constraints are met ($d\psi = 0$), this gradient must tend to zero, which results in

$$d\phi = I_{\psi\phi}' I_{\psi\psi}^{-1} d\psi + [\lambda_{\phi\Omega}'(t_0) - I_{\psi\phi}' I_{\psi\psi}^{-1} \lambda_{\psi\Omega}'(t_0)]\delta \mathbf{x}(t_0) \quad (54)$$

This relation shows how much the maximum pay-off function changes for small changes in the terminal constraints and for small changes in the initial conditions.

(c) New control variable programs are obtained as

$$\alpha_{NEW}(t) = \alpha_{OLD}(t) + \delta\alpha(t) \quad (55)$$

where $\delta\alpha$ is given in equation (50). $\alpha_{NEW}(t)$ is used in the original nonlinear differential equations (28) and the whole process is repeated several times until the terminal constraints are met and the gradient is nearly zero in equation (53). The optimum program has then been obtained.

Note that *minimizing final time* T fits into this pattern conveniently. For this case we let $\phi = -t$ which implies $\dot{\phi} = -1$, $\lambda_\phi = 0$, and hence

$$\lambda_{\phi\Omega} = \frac{1}{\dot{\Omega}} \lambda_\Omega$$

$$d\phi = -dT = \int_{t_0}^{T} \frac{1}{\dot{\Omega}} \lambda_\Omega' \mathbf{G} \delta\alpha dt + \frac{1}{\dot{\Omega}} \lambda_\Omega'(t_0) \delta \mathbf{x}(t_0) \quad (56)$$

7 Computing Procedures

The computing procedures evolved over a period of time in solving many types of problems are summarized here. The computer used for all these problems was an IBM 704.

(a) Compute the nominal path by integrating the nonlinear physical differential equations with a nominal control variable program and appropriate (here assumed fixed) initial conditions and store the solution on tape.

(b) Compute the $\lambda_\phi, \lambda_{\psi_1}, \lambda_{\psi_2}, \ldots \lambda_{\psi_p}, \lambda_\Omega$ functions *all at the same time* by integrating the adjoint differential equations backward, evaluating the partial derivatives on the nominal path by reference to the tape in (a).

(c) Simultaneously with (b), calculate the quantities $\lambda_{\phi\Omega}, \lambda_{\psi_1\Omega}, \ldots \lambda_{\psi_p\Omega}$ and store $\lambda_{\phi\Omega}'\mathbf{G}$ and $\lambda_{\psi\Omega}'\mathbf{G}$ on tape.

(d) Also simultaneously with (b) and (c) perform the integrations (backward) leading to the numbers $I_{\phi\phi}, I_{\psi\phi}, I_{\psi\psi}$.

(e) Print out the values of $\phi, \psi_1, \psi_2, \ldots \psi_p$ achieved by the nominal path.

(f) Select desired terminal condition changes $d\psi_1, d\psi_2, \ldots d\psi_p$ to bring the next solution closer to the specified values $\psi = 0$ than were achieved by the nominal path.

(g) Select a reasonable value of $(dP)^2/(T - t_0)$, which is a mean-square deviation of the control variable programs from the nominal to the next step.

(h) Use the values of $d\psi$ and dP to calculate $(dP)^2 - d\psi' I_{\psi\psi}^{-1} d\psi$; if this quantity is negative, automatically scale down $d\psi$ to make this quantity vanish. If the quantity is positive, leave it as is.

(i) Using the values of dP and $d\psi$ (modified by (h) if necessary) calculate $\delta\alpha(t)$ from equation (50), ($\delta\mathbf{x}(t_0) = 0$).

(j) If final time, T, is not specified or being extremalized, compute the predicted change, dT, for the next step:

$$dT = -\frac{1}{\dot{\Omega}} \int_{t_0}^{T} \lambda_\Omega' \mathbf{G} \delta\alpha dt$$

If $|dT|$ is greater than a preselected maximum allowable value, scale down $\delta\alpha(t)$ to achieve this maximum value.

(k) Obtain a new nominal path by using $\alpha_{NEW} = \alpha_{OLD} + \delta\alpha$ and repeat processes (a) through (k) until the terminal constraints $\psi = 0$ are satisfied and the square of the gradient, $I_{\phi\phi} - I_{\psi\phi}' I_{\psi\psi}^{-1} I_{\psi\phi}$, tends to zero.

8 Example 1

Angle-of-Attack Program for a Supersonic Interceptor to Climb From Sea Level to a Given Altitude in Least Time. A typical supersonic interceptor is considered with lift, drag, and thrust characteristics as shown in Fig. 4. The vehicle is considered as a mass point (short period pitching motions are thus neglected) and the nomenclature used is shown in Fig. 3. The problem is to find the angle-of-attack program $\alpha(t)$, using maximum thrust, that takes the interceptor to a given final altitude in the least time, starting just after take-off at $M = 0.22$, $\gamma = 0$, at sea level ($h = 0$). First the problem is solved with no terminal constraints, then with $M = 0.9$ specified at the terminal point, then with $M = 0.9$ and $\gamma = 0$ at the terminal point. The differential equations for the interceptor path are

$$\dot{V} = \frac{F(h, M)}{m} \cos\alpha - \frac{D(h, M, \alpha)}{m} - g \sin\gamma \quad (57)$$

$$\dot{\gamma} = \frac{L(h, M, \alpha)}{mV} + \frac{F(h, M)}{mV} \sin\alpha - \frac{g\cos\gamma}{V} \quad (58)$$

$$\dot{h} = V \sin\gamma \quad (59)$$

$$\dot{x} = V \cos\gamma \quad (60)$$

$$\dot{m} = \dot{m}(h, M) \quad (61)$$

where $(\dot{\ })$ means $\frac{d}{dt}(\ )$, and

$F = F(h, M)$ is thrust, given as a tabular function, Fig. 4(a)

$D = C_D(\alpha, M) \frac{\rho V^2 S}{2}$ is drag

$L = C_L(\alpha, M) \frac{\rho V^2 S}{2}$ is lift

$C_D(\alpha, M)$, $C_L(\alpha, M)$ are given as tabular functions, Fig. 4(b)

$\rho = \rho(h)$ is air density, given as a tabular function

g = acceleration due to gravity (taken as constant here)

$M = \frac{V}{a}$ is Mach number

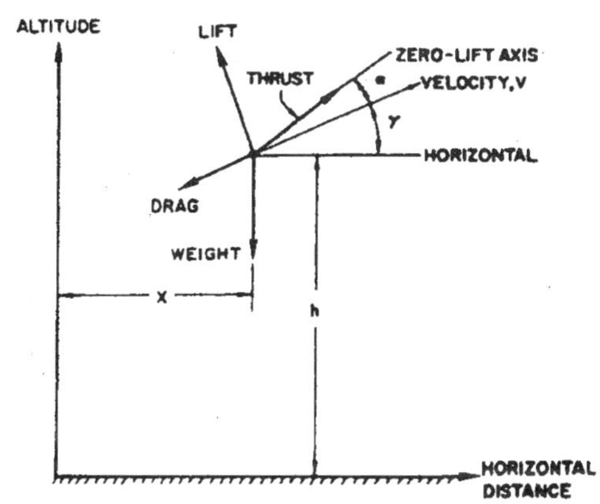

Fig. 3 Nomenclature used in analysis of supersonic interceptor

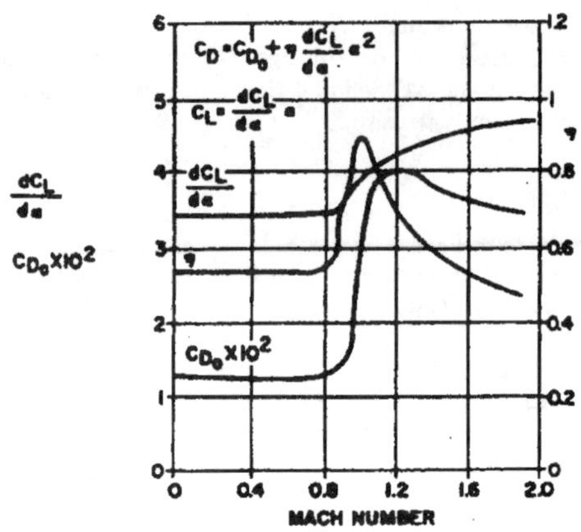

Fig. 4(a) Supersonic interceptor—lift and drag coefficient versus Mach number and angle-of-attack

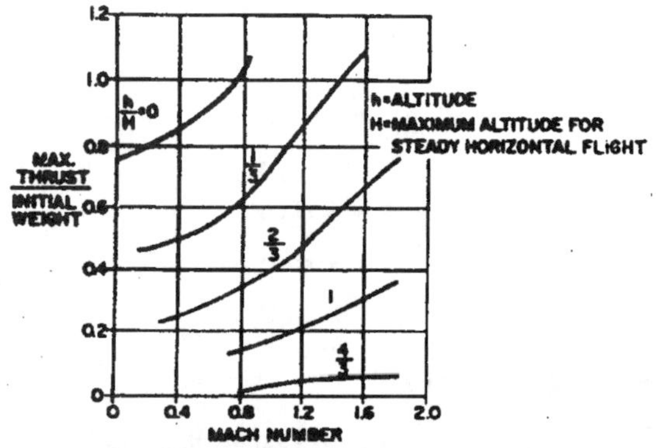

Fig. 4(b) Supersonic interceptor—thrust versus Mach number and altitude

$a = a(h)$ is speed of sound, given as a tabular function

$\dot{m}(h, M)$ is fuel consumption, given as a tabular function

m is mass of vehicle

The adjoint differential equations are:

$$\dot{\lambda}_V + \lambda_V \left(\frac{\cos\alpha}{ma}\frac{\partial F}{\partial M} - \frac{2D}{mV} - \frac{D}{ma}\frac{1}{C_D}\frac{\partial C_D}{\partial M} \right) + \lambda_\gamma \left(\frac{g\cos\gamma}{V^2} + \frac{L}{mV^2} + \frac{L}{mVa}\frac{1}{C_L}\frac{\partial C_L}{\partial M} - \frac{F\sin\alpha}{mV^2} + \frac{\sin\alpha}{mVa}\frac{\partial F}{\partial M} \right) + \lambda_h \sin\gamma + \lambda_x \cos\gamma + \lambda_m \frac{1}{a}\frac{\partial \dot{m}}{\partial M} = 0 \tag{62}$$

$$\dot{\lambda}_\gamma + \lambda_V(-g\cos\gamma) + \lambda_\gamma \left(\frac{g\sin\gamma}{V} \right) + \lambda_h V\cos\gamma - \lambda_x V\sin\gamma = 0 \tag{63}$$

$$\dot{\lambda}_h + \lambda_V \left(-\frac{D}{m}\frac{1}{\rho}\frac{d\rho}{dh} + \frac{D}{m}\frac{1}{C_D}\frac{\partial C_D}{\partial M}\frac{M}{a}\frac{da}{dh} + \frac{\cos\alpha}{m}\frac{\partial F}{\partial h} \right) + \lambda_\gamma \left(\frac{L}{mV}\frac{1}{\rho}\frac{d\rho}{dh} - \frac{L}{mV}\frac{1}{C_L}\frac{\partial C_L}{\partial M}\frac{M}{a}\frac{da}{dh} + \frac{\sin\alpha}{mV}\frac{\partial F}{\partial h} \right) + \lambda_m \frac{\partial \dot{m}}{\partial h} = 0 \tag{64}$$

$$\dot{\lambda}_x = 0 \tag{65}$$

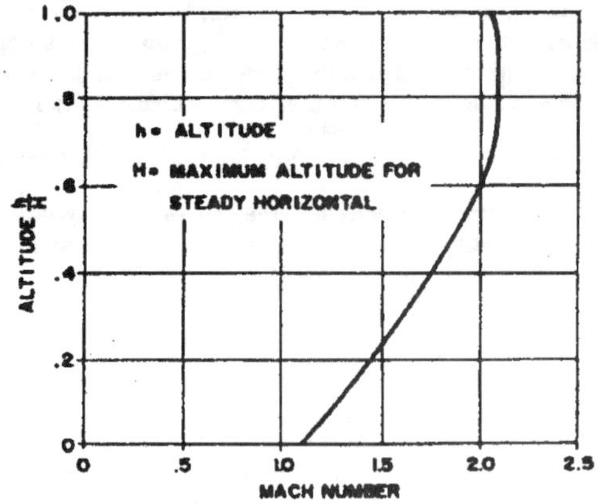

Fig. 4(c) Supersonic interceptor—altitude versus Mach number for steady level flight

$$\dot{\lambda}_m + \lambda_V \left(\frac{D}{m^2} - \frac{F\cos\alpha}{m^2} \right) + \lambda_\gamma \left(-\frac{L}{m^2V} - \frac{F\sin\alpha}{m^2V} \right) = 0 \tag{66}$$

In the nomenclature of Sections 5 and 6, three cases were calculated:

Case	ϕ	ψ_1	ψ_2	Ω
1	$-t,$	—,	—,	$h - H$
2	$-t,$	$M - 0.9,$	—,	$h - H$
3	$-t,$	$M - 0.9,$	$\gamma,$	$h - H$

(67)

where H is the final altitude desired, taken in this case as the maximum altitude for steady horizontal flight (the service ceiling). Initial conditions were

$$M = 0.22,\quad \gamma = 0,\quad h = 0,\quad x = 0,\quad m = \frac{W_0}{g} \tag{68}$$

where W_0 is the initial weight.

The results are shown in Fig. 5. The trajectories all show the "zoom" characteristic found by other investigators [9]; the airplane first climbs, then dives accelerating through sonic speed, then pulls up, trading kinetic for potential energy in the earth's gravitational field. This characteristic appears to be caused by the sharp transonic drag rise, the rapid thrust attenuation with altitude, and the thrust increase with Mach number. These surprising trajectories show the danger in using classical performance methods on high-performance airplanes. The quasi-steady analysis to determine local maximum rate-of-climb shows an almost constant slightly subsonic Mach number to be the best for climb. If the airplane is flown this way, it takes nearly *twice as long* to reach the altitude H as it does with the trajectories shown in Fig. 5. This quasi-steady flight path was used as the nominal path in the steepest-ascent method and the optimum path was reached in six or seven "steps."

Note that the addition of the $M = 0.9$ terminal constraint increased the minimum time to climb by 8 per cent. Further addition of the $\gamma = 0$ terminal constraint increased the minimum time to climb less than $^1/_2$ per cent. The angle-of-attack programs are not at all unusual and should not be difficult to approximate for actual flight situations.

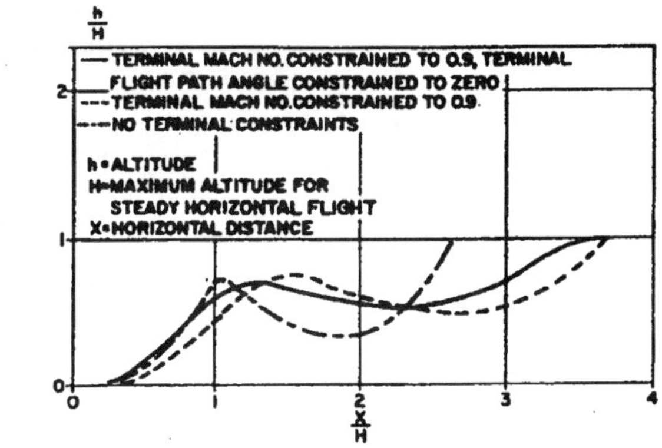

Fig. 5(a) Supersonic interceptor—flight path for minimum time to climb to service ceiling from sea level—altitude versus range

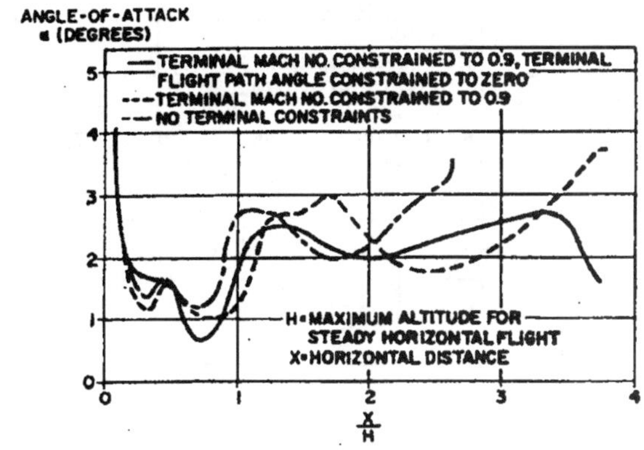

Fig. 5(c) Supersonic interceptor—flight path for minimum time to climb to service ceiling from sea level—angle-of-attack versus range

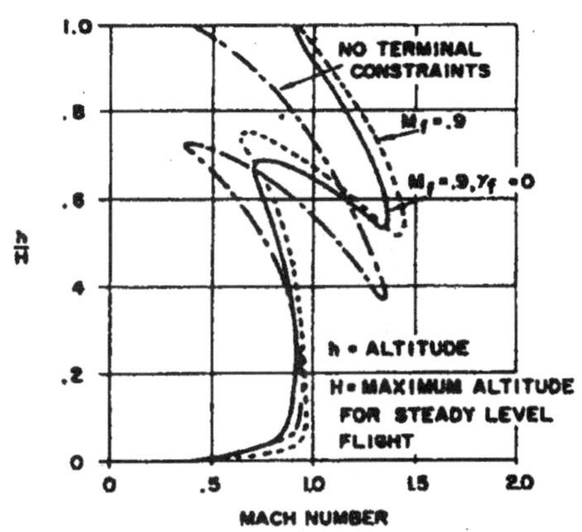

Fig. 5(b) Supersonic interceptor—flight path for minimum time to climb to service ceiling from sea level—altitude versus Mach number

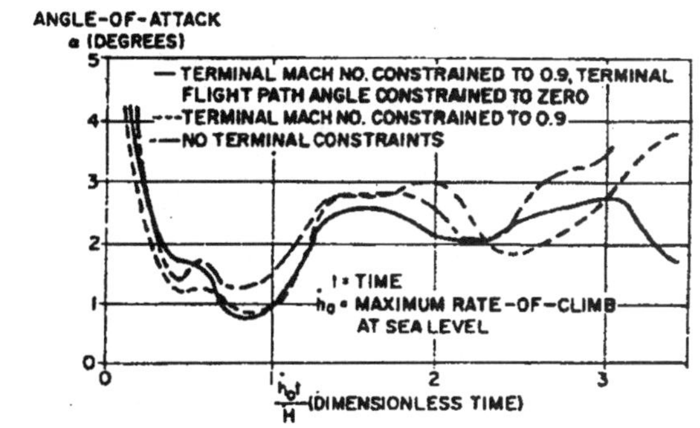

Fig. 5(d) Supersonic interceptor—flight path for minimum time to climb to service ceiling from sea level—angle-of-attack versus time

9 Example 2

Angle-of-Attack Program for a Supersonic Interceptor to Climb to Maximum Altitude. The same airplane considered in Section 8 is used here with the pay-off function, ϕ, being altitude, h. The stopping condition is $\Omega = \gamma = 0$; i.e., flight-path angle zero. The initial condition was taken as the maximum energy condition for steady horizontal flight, where energy per unit mass is $(V^2/2) + gh$; i.e., kinetic plus potential energy. This maximum energy condition occurred at

$$\frac{h}{H} = 0.81$$

$$\gamma = 0$$

$$M = 2.1$$

In this case no terminal constraints were used. The physical and adjoint equations are the same as those in Section 8.

The resulting flight path is shown in Fig. 6. It again contains a preliminary climb and dive, followed by a zoom to the maximum altitude which is 64 per cent higher than the service ceiling (maximum altitude for steady horizontal flight). If all the energy

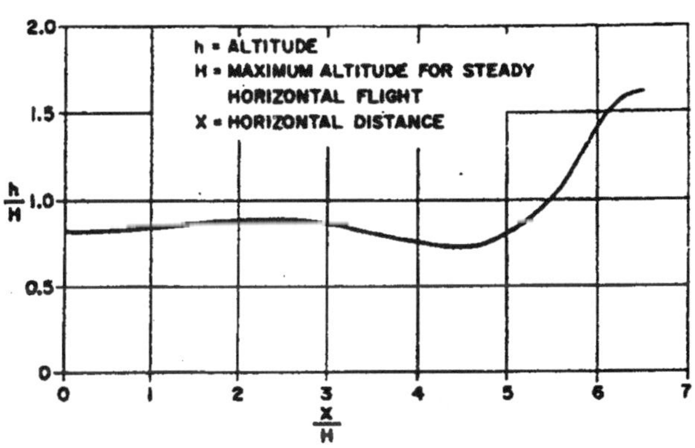

Fig. 6(a) Supersonic interceptor—flight path for maximum altitude—altitude versus range

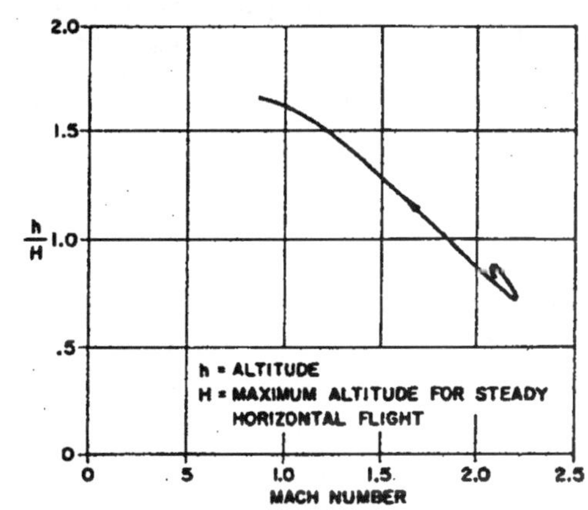

Fig. 6(b) Supersonic interceptor—flight path for maximum altitude—altitude versus Mach number

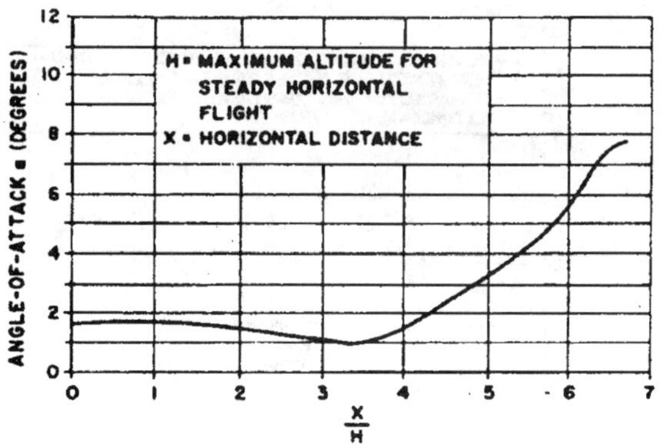

Fig. 6(c) Supersonic interceptor—flight path for maximum altitude—angle-of-attack versus range

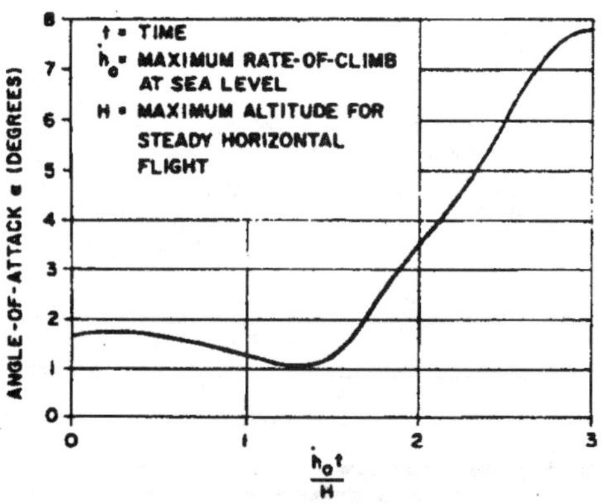

Fig. 6(d) Supersonic interceptor—flight path for maximum altitude—angle-of-attack versus time

could be converted into potential energy, the maximum altitude would be 97 per cent higher than the service ceiling. Note that the thrust essentially vanishes near $h/H = 4/3$; actually the turbojet engines will "blow out" somewhere near this latter altitude.

10 Example 3

Angle-of-Attack Program for a Hypersonic Orbital Glider to Achieve Maximum Range. The problem here is to find the angle-of-attack program $\alpha(t)$ for a hypersonic glider to achieve maximum range on the surface of the earth, starting from the point where it has been injected into a low-altitude satellite orbit by a rocket booster. The nomenclature for the analysis is shown in Fig. 7, and the lift-drag characteristics are shown in Fig. 8. The wing loading used was

$$\frac{mg_0}{S} = 27.3 \text{ lb ft}^{-2}$$

where

m = mass of vehicle
g_0 = acceleration of gravity at earth's surface
S = wing plan-form area

The initial conditions used were

$$\begin{aligned} V &= 25{,}920 \text{ ft sec}^{-1} \\ \gamma &= 0.18 \text{ deg} \\ h &= 300{,}000 \text{ ft} \end{aligned} \tag{69}$$

Owing to the wide range of velocities encountered in this problem (landing speeds were around 200 ft sec^{-1}) it was found

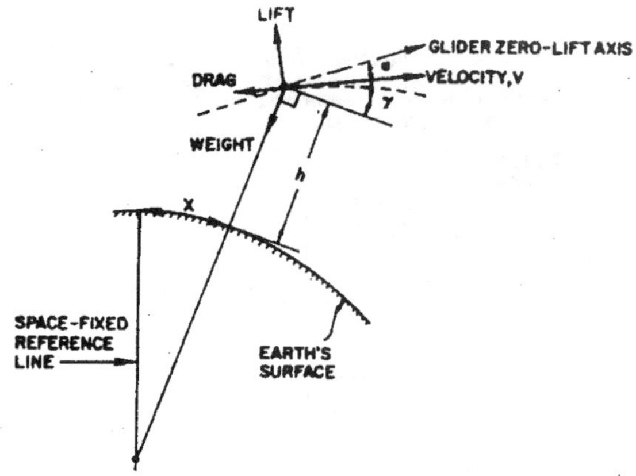

Fig. 7 Nomenclature used in analysis of hypersonic orbital glider

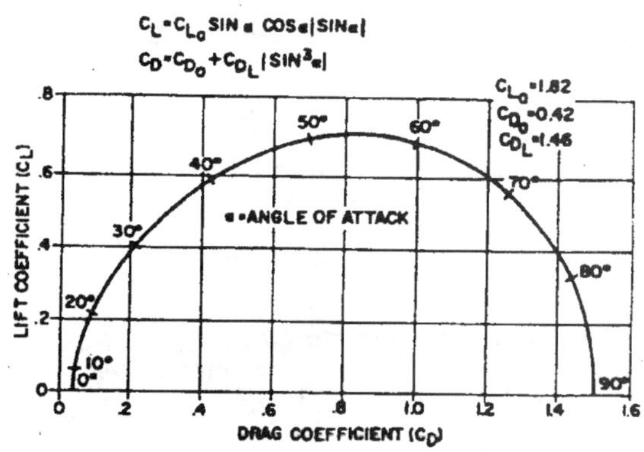

Fig. 8 Hypersonic orbital glider—lift and drag coefficients as functions of angle-of-attack

convenient to use distance along the flight path, s, as the independent variable in solving the differential equations. The physical differential equations used are:

$$\frac{dV}{ds} = -\frac{D(h, M, \alpha)}{mV} - \frac{g(h)\sin\gamma}{V} \tag{70}$$

$$\frac{d\gamma}{ds} = \frac{\cos\gamma}{R+h} + \frac{L(h, M, \alpha)}{mV^2} - \frac{g(h)\cos\gamma}{V^2} \tag{71}$$

$$\frac{dh}{ds} = \sin\gamma \tag{72}$$

$$\frac{dx}{ds} = \frac{\cos\gamma}{1 + \dfrac{h}{R}} \tag{73}$$

$$\frac{dt}{ds} = \frac{1}{V} \tag{74}$$

where

$g = g_0 \left(\dfrac{R}{R+h}\right)^2$, acceleration due to gravity
R = radius of earth (3440 nautical miles)
$D = C_D(\alpha)\,\dfrac{1}{2}\,\rho(h)V^2S$, drag force
$L = C_L(\alpha)\,\dfrac{1}{2}\,\rho(h)V^2S$, lift force

ρ = $\rho(h)$, density of atmosphere, a tabular function (ARDC standard atmosphere used)

C_D = $C_{D_0} + C_{D_L}|\sin^3 \alpha|$, $C_{D_0} = 0.042$, $C_{D_L} = 1.46$, Newtonian drag coefficient

C_L = $C_{L_0} \sin \alpha \, |\sin \alpha| \cos \alpha$, $C_{L_0} = 1.82$, Newtonian lift coefficient

The adjoint differential equations for the influence functions are:

$$\frac{d\lambda_w}{ds} + \lambda_w\left(\frac{2g \sin \gamma}{V^2}\right) + \lambda_\gamma\left(\frac{2g \cos \gamma}{V^2}\right) + \lambda_t\left(-\frac{1}{V}\right) = 0, \tag{75}$$

$$\frac{d\lambda_\gamma}{ds} + \lambda_w\left(-\frac{g \cos \gamma}{V^2}\right) + \lambda_\gamma\left(\frac{g \sin \gamma}{V^2} - \frac{\sin \gamma}{R + h}\right) - \lambda_x\left(\frac{\sin \gamma}{1 + \frac{h}{R}}\right) + \lambda_h \cos \gamma = 0, \tag{76}$$

$$\frac{d\lambda_h}{ds} + \lambda_w\left(\frac{2g \sin \gamma}{V^2(R + h)} - \frac{D}{mV^2}\frac{1}{\rho}\frac{d\rho}{dh}\right) + \lambda_\gamma\left(\frac{2g \cos \gamma}{V^2(R + h)} - \frac{\cos \gamma}{(R + h)^2} + \frac{L}{mV^2}\frac{1}{\rho}\frac{d\rho}{dh}\right) + \lambda_x\left(-\frac{\cos \gamma}{R\left(1 + \frac{h}{R}\right)^2}\right) = 0, \tag{77}$$

$$\frac{d\lambda_x}{ds} = 0, \tag{78}$$

$$\frac{d\lambda_t}{ds} = 0, \tag{79}$$

where

$$\lambda_w = V\lambda_V, \tag{80}$$

$w = \log \dfrac{V}{V_0}$ (V_0 some reference velocity)

The resulting flight path is shown in Fig. 9(a). For comparison the flight path for $\alpha = 20.5$ deg is as shown; this is the angle-of-attack for maximum lift-to-drag ratio, which is in this case $(L/D)_{\max} = 2.0$. It can be seen that the optimum $\alpha(t)$ program differs from the $\alpha = 20.5$ deg path most significantly in the first 10 min of the flight; this is truly the critical part of the flight. The optimum path achieves 15 per cent more range than the maximum L/D path. Again the optimum control variable program, Fig. 9(b) should not be difficult to approximate in practice.

References

1 C. R. Faulders, "Low Thrust Rocket Steering Program for Minimum Time Transfer Between Planetary Orbits," Society of Automotive Engineers, Paper 88A, October, 1958.

2 J. V. Breakwell, "The Optimization of Trajectories," *Journal of the Society of Industrial and Applied Mathematics*, vol. 7, 1959, pp. 215–247.

3 A. E. Bryson and S. E. Ross, "Optimum Trajectories With Aerodynamic Drag," *Jet Propulsion* (now *Journal of the American Rocket Society*), vol. 28, 1958, pp. 465–469.

4 L. J. Kulakowski and R. T. Stancil, "Rocket Boost Trajectories for Maximum Burnout Velocity," *Journal of the American Rocket Society*, vol. 30, 1960, pp. 612–619.

5 H. J. Kelley, "Gradient Theory of Optimal Flight Paths," *Journal of the American Rocket Society*, vol. 30, 1960, pp. 947–953.

6 H. J. Kelley, "Method of Gradients," chapter 6, "Optimization Techniques," edited by G. Leitmann, Academic Press (to be published).

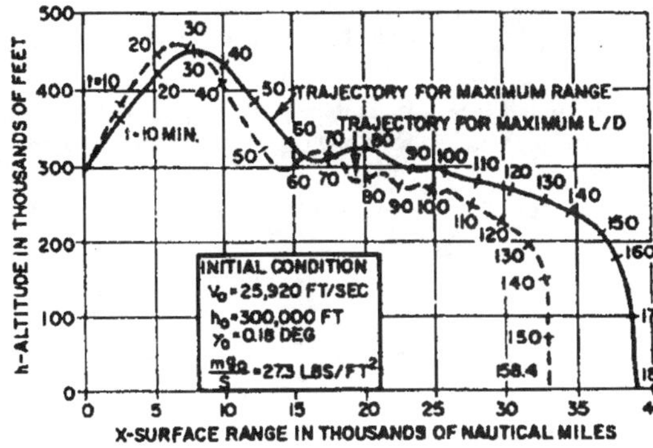

Fig. 9(a) Hypersonic orbital glider—altitude versus range

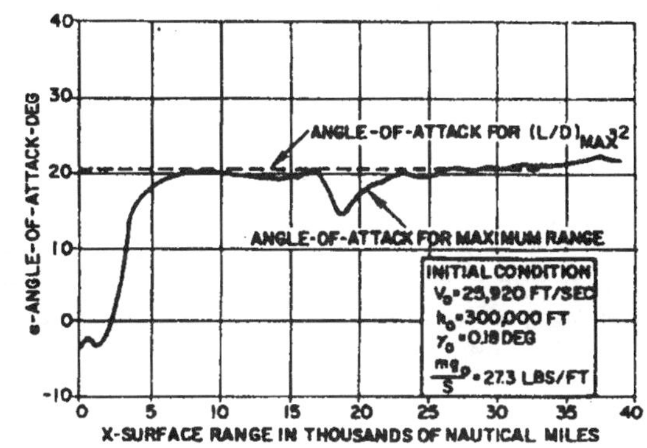

Fig. 9(b) Hypersonic orbital glider—angle-of-attack versus range

7 A. E. Bryson, W. F. Denham, F. J. Carroll, and K. Mikami, "Lift or Drag Programs That Minimize Re-entry Heating," *Journal of the Aerospace Sciences*, vol. 29, April, 1962, pp. 420–430.

8 A. E. Bryson, "A Gradient Method for Optimizing Multi-Stage Allocation Processes," Proceedings, Harvard University Symposium on Digital Computers and Their Applications, April, 1961 (to be published).

9 H. J. Kelley, "An Investigation of Optimal Zoom Climb Techniques," *Journal of Aerospace Sciences*, vol. 26, 1959, pp. 794–803.

10 G. A. Bliss, "Mathematics for Exterior Ballistics," John Wiley & Sons, Inc., New York, N. Y., 1944.

11 H. S. Tsien, "Engineering Cybernetics," chapter 13, "Control Design by Perturbation Theory," McGraw-Hill Book Company, Inc., New York, N. Y., 1954.

APPENDIX

1 Use of LaGrange Multipliers for Small Perturbations About a Given Control Variable Point.

For the maximum problem stated in Section 3 of the paper, we wish to determine the changes in ϕ and ψ for a small perturbation $d\boldsymbol{\alpha}$ in the control variables from a nominal point $\boldsymbol{\alpha}^*$. To do this consider first the quantity

$$\Phi = \phi(\mathbf{x}) + \lambda_\phi'(\mathbf{f}(\mathbf{x}, \boldsymbol{\alpha}) + \mathbf{f}_0) \tag{81}$$

where λ_ϕ' is a row matrix of Lagrange multipliers to be determined for our convenience. Note that the term multiplying λ_ϕ' is zero by equation (3) so that $\Phi = \phi$. Take the differential of this quantity:

$$d\Phi = \left(\frac{\partial \phi}{\partial \mathbf{x}} + \lambda_\phi' \frac{\partial \mathbf{f}}{\partial \mathbf{x}}\right) d\mathbf{x} + \lambda_\phi' \frac{\partial \mathbf{f}}{\partial \boldsymbol{\alpha}} d\boldsymbol{\alpha} + \lambda_\phi' \, d\mathbf{f}_0, \tag{82}$$

and evaluate the partial derivatives at the nominal point $\boldsymbol{\alpha}^* \rightarrow \mathbf{x}^*$:

$$d\Phi = d\phi = \left(\left(\frac{\partial\phi}{\partial x}\right)^* + \lambda_\phi'F\right)dx + \lambda_\phi'Gd\alpha + \lambda_\phi'df_0, \quad (83)$$

where the nomenclature is explained in Section 4 of the paper. Now let us *choose* λ_ϕ' so that the coefficient of dx vanishes; i.e.,

$$\left(\frac{\partial\phi}{\partial x}\right)^* + \lambda_\phi'F = 0, \quad (84)$$

or

$$\lambda_\phi' = -\left(\frac{\partial\phi}{\partial x}\right)^* F^{-1} \quad (85)$$

This reduces equation (83) to the expression

$$d\phi = \lambda_\phi'Gd\alpha + \lambda_\phi'\,df_0 \quad (86)$$

An exactly similar procedure yields

$$d\psi = \lambda_\psi'Gd\alpha + \lambda_\psi'df_0, \quad (87)$$

where

$$\left(\frac{\partial\psi}{\partial x}\right)^* + \lambda_\psi'F = 0 \quad (88)$$

2 Steepest-Ascent Method in Ordinary Calculus Using a Weighted Square Perturbation of the Control Variables to Determine Step Size.

The problem, as stated in Section 4 of the text, is to choose $d\alpha$ so as to maximize $d\phi$ for given values of $d\psi$, df_0 (usually zero), and dP, where

$$d\phi = \lambda_\phi'Gd\alpha + \lambda_\phi'df_0 \quad (89)$$

$$d\psi = \lambda_\psi'Gd\alpha + \lambda_\psi'df_0 \quad (90)$$

$$(dP)^2 = d\alpha'Wd\alpha \quad (91)$$

We use the process of Section 1 of the Appendix again; i.e., we consider a linear combination of the three foregoing equations:

$$d\phi = \lambda_\phi'Gd\alpha + \lambda_\phi'df_0 + v'(d\psi - \lambda_\psi'df_0 - \lambda_\psi'Gd\alpha) + \mu((dP)^2 - d\alpha'Wd\alpha) \quad (92)$$

where v' is a $1 \times p$ row matrix of constants, and μ is a constant, all to be determined for our convenience. Note that the quantities multiplying v' and μ are both zero, by equations (90) and (91). Take the second differential of this quantity:

$$d^2\phi = (\lambda_\phi'G - v'\lambda_\psi'G - 2\mu d\alpha'W)d^2\alpha \quad (93)$$

Thus the maximum of $d\phi$ occurs when the coefficient of $d^2\alpha$ vanishes in equation (93). This will be the case if

$$d\alpha = \frac{1}{2\mu}W^{-1}(G'\lambda_\phi - G'\lambda_\psi v) \quad (94)$$

Substituting (94) into (90) we have

$$d\beta = \frac{1}{2\mu}(I_{\psi\phi} - I_{\psi\psi}v) \quad (95)$$

where

$$\begin{aligned} d\beta &= d\psi - \lambda_\psi'df_0 \\ I_{\psi\phi} &= \lambda_\psi'GW^{-1}G'\lambda_\phi \\ I_{\psi\psi} &= \lambda_\psi'GW^{-1}G'\lambda_\psi \end{aligned} \quad (96)$$

Solving equation (95) for v we obtain

$$v = -2\mu I_{\psi\psi}^{-1}d\beta + I_{\psi\psi}^{-1}I_{\psi\phi} \quad (97)$$

Substituting (97) and (94) into (91), and solving for μ, we obtain

$$2\mu = \pm\left[\frac{I_{\phi\phi} - I_{\psi\phi}'I_{\psi\psi}^{-1}I_{\psi\phi}}{(dP)^2 - d\beta'I_{\psi\psi}^{-1}d\beta}\right]^{1/2}, \quad (98)$$

where

$$I_{\phi\phi} = \lambda_\phi'GW^{-1}G'\lambda_\phi \quad (99)$$

Substituting (97) and (98) into (94) and (89) we obtain the results given in Sections 4 of the text in equations (21) and (23), respectively.

These results have a simple geometric interpretation in the α-hyperspace. Equations (89) and (90) with $df_0 = 0$ may be written

$$d\phi = \nabla\phi d\alpha = (\nabla\phi W^{-1/2})W^{1/2}d\alpha \quad (1\ 0)$$

$$d\psi = \nabla\psi d\alpha = (\nabla\psi W^{-1/2})W^{1/2}d\alpha \quad (101)$$

i.e., $\lambda_\phi'G = \nabla\phi$ is the gradient of ϕ in the α-hyperspace, and $\lambda_\psi'G = \nabla\psi$ is the gradient of ψ. For the moment we will consider ψ to be a single scalar quantity rather than a column matrix. If W is the metric in the α-hyperspace then dP is the infinitesimal distance from the present nominal point to a neighboring point $\alpha + d\alpha$. We wish to choose $d\alpha$ to maximize $d\phi$, keeping $d\psi = 0$, for a given value of

$$dP = |W^{-1/2}d\alpha|.$$

Hence we must subtract the component of $\nabla\phi W^{-1/2}$ that is parallel to $\nabla\psi W^{-1/2}$ from $\nabla\phi W^{-1/2}$. Using the Pythagorean theorem we have simply then

$$\begin{aligned}\left(\frac{d\phi}{dP}\right)^2 &= (\nabla\phi W^{-1/2})(W^{-1/2}\nabla\phi') - \frac{[(\nabla\phi W^{-1/2})(W^{-1/2}\nabla\psi')]^2}{(\nabla\psi W^{-1/2})(W^{-1/2}\nabla\psi')} \\ &= \nabla\phi W^{-1}\nabla\phi - \frac{(\nabla\phi W^{-1}\nabla\psi')^2}{\nabla\psi W^{-1}\nabla\psi'} \\ &= I_{\phi\phi} - I_{\psi\phi}'I_{\psi\psi}^{-}I_{\psi\phi}\end{aligned} \quad (102)$$

which gives a geometric interpretation of equation (24). Clearly this quantity is positive unless $\nabla\phi$ is parallel to $\nabla\psi$, in which case it is zero.

3 Adjoint Differential Equations for Small Perturbations About Given Control Variable Programs.

For the optimum programming problem stated in Section 6 of the text, we wish to determine the changes in ϕ, ψ, and Ω for small perturbations, $\delta\alpha(t)$, in the control variable programs about given nominal control variable programs $\alpha^*(t)$, where $\Omega = 0$ is the stopping condition. To do this we consider the linear equations (39) describing small perturbations about the nominal path:

$$\frac{d}{dt}(\delta x) = F(t)\delta x + G(t)\delta\alpha \quad (103)$$

To determine the changes in ϕ, ψ, and Ω we introduce the linear differential equations adjoint to (103) defined as

$$\frac{d\lambda}{dt} = -F'(t)\lambda \quad (104)$$

where λ_ϕ is an $n \times 1$ matrix, λ_ψ is an $n \times p$ matrix, and λ_Ω is an $n \times 1$ matrix. If (103) is premultiplied by λ' and (104) is premultiplied by $(\delta x)'$ and the transpose of the second product is added to the first product, we have

$$\lambda'\frac{d(\delta x)}{dt} + \frac{d\lambda'}{dt}\delta x = \lambda'F\delta x - \lambda'F\delta x + \lambda'G\delta\alpha$$

or

$$\frac{d}{dt}(\lambda'\delta x) = \lambda'G\delta\alpha \quad (105)$$

If we integrate equations (105) from t_0 to T, the result is

$$(\lambda'\delta x)_{t=T} = \int_{t_0}^{T} \lambda' G \delta\alpha dt + (\lambda'\delta x)_{t=t_0} \tag{106}$$

If we let

$$\lambda_\phi'(T) = \left(\frac{\partial \phi}{\partial x}\right)^*_{t=T}; \quad \lambda_\psi'(T) = \left(\frac{\partial \psi}{\partial x}\right)^*_{t=T},$$

$$\lambda_\Omega'(T) = \left(\frac{\partial \Omega}{\partial x}\right)^* \tag{107}$$

where the nomenclature is given in equations (47), it is clear that

$$\delta\phi = (\lambda_\phi'\delta x)_{t=T}; \quad \delta\psi = (\lambda_\psi'\delta x)_{t=T}; \quad \delta\Omega = (\lambda_\Omega'\delta x)_{t=T} \tag{108}$$

For small perturbations the value of T will be changed only a small amount dT, so that

$$\begin{aligned} d\phi &= \delta\phi + \dot{\phi} dT \\ d\psi &= \delta\psi + \dot{\psi} dT \\ d\Omega &= \delta\Omega + \dot{\Omega} dT \end{aligned} \tag{109}$$

where the nomenclature is given in equations (48).

Substituting (107) and (109) into (106) we obtain equations (42), (43), and (44) of the text.

4 Steepest-Ascent Method in Calculus of Variations Using a Weighted Mean-Square Perturbation of Control Variables to Determine Step Size.

The problem as stated in Section 6 of the text is to choose $\delta\alpha(t)$ so as to maximize $d\phi$ for given values of $d\psi$, $\delta x(t_0)$ (usually zero), and dP, with $d\Omega = 0$ where

$$d\phi = \int_{t_0}^{T} \lambda_\phi' G \delta\alpha dt + \lambda_\phi'(t_0)\delta x(t_0) + \dot{\phi} dT \tag{110}$$

$$d\psi = \int_{t_0}^{T} \lambda_\psi' G \delta\alpha dt + \lambda_\psi'(t_0)\delta x(t_0) + \dot{\psi} dT \tag{111}$$

$$0 = d\Omega = \int_{t_0}^{T} \lambda_\Omega' G \delta\alpha dt + \lambda_\Omega'(t_0)\delta x(t_0) + \dot{\Omega} dT \tag{112}$$

$$(dP)^2 = \int_{t_0}^{T} \delta\alpha' W \delta\alpha dt \tag{113}$$

We use the analog of the process of Section 2 of the Appendix with only small differences arising from the additional term in dT in the foregoing equations. The first step is to eliminate dT from equation (112):

$$dT = -\frac{1}{\dot{\Omega}} \int_{t_0}^{T} \lambda_\Omega' G \delta\alpha dt - \frac{1}{\dot{\Omega}} \lambda_\Omega'(t_0)\delta x(t_0) \tag{114}$$

Substituting this into equations (110) and (111), we have the relations

$$d\phi = \int_{t_0}^{T} \lambda_{\phi\Omega}' G \delta\alpha dt + \lambda_{\phi\Omega}'(t_0)\delta x(t_0) \tag{115}$$

$$d\psi = \int_{t_0}^{T} \lambda_{\psi\Omega}' G \delta\alpha dt + \lambda_{\psi\Omega}'(t_0)\delta x(t_0) \tag{116}$$

where the nomenclature is given in equations (51). Next we consider a linear combination of (116) and (113) with (115); i.e.,

$$\begin{aligned} d\phi = &\int_{t_0}^{T} (\lambda_{\phi\Omega}' G - \nu' \lambda_{\psi\Omega}' G - \mu \delta\alpha' W) \delta\alpha dt \\ &+ (\lambda_{\phi\Omega}'(t_0) - \nu' \lambda_{\psi\Omega}'(t_0))\delta x(t_0) + \nu' d\psi + \mu (dP)^2 \end{aligned} \tag{117}$$

Now we take the variation of this relation (117):

$$\delta(d\phi) = \int_{t_0}^{T} (\lambda_{\phi\Omega}' G - \nu' \lambda_{\psi\Omega}' G - 2\mu \delta\alpha' W) \delta^2\alpha dt \tag{118}$$

where $\delta x(t_0)$, $d\psi$, and dP are considered constants. The maximum $d\phi$ will occur if the coefficient of $\delta^2\alpha$ in the integrand of (118) vanishes; i.e.,

$$\delta\alpha = \frac{1}{2\mu} W^{-1} G'(\lambda_{\phi\Omega} - \lambda_{\psi\Omega}\nu) \tag{119}$$

Substituting (119) back into (116) and then (113) we can solve for the constants ν and μ:

$$\nu = -2\mu I_{\psi\psi}^{-1} d\beta + I_{\psi\psi}^{-1} I_{\psi\phi} \tag{120}$$

$$2\mu = \pm \left[\frac{I_{\phi\phi} - I_{\psi\phi}' I_{\psi\psi}^{-1} I_{\psi\phi}}{(dP)^2 - d\beta' I_{\psi\psi}^{-1} d\beta}\right] \tag{121}$$

where

$$\begin{aligned} d\beta &= d\psi - \lambda_{\psi\Omega}'(t_0)\delta x(t_0) \\ I_{\psi\psi} &= \int_{t_0}^{T} \lambda_{\psi\Omega}' G W^{-1} G' \lambda_{\psi\Omega} dt \\ I_{\psi\phi} &= \int_{t_0}^{T} \lambda_{\psi\Omega}' G W^{-1} G' \lambda_{\phi\Omega} dt \\ I_{\phi\phi} &= \int_{t_0}^{T} \lambda_{\phi\Omega}' G W^{-1} G' \lambda_{\phi\Omega} dt \end{aligned} \tag{122}$$

Substituting (120) and (121) into (119) and (115) we obtain the results given in Section 6 of the text in equations (50) and (51), respectively.

The Solution of Certain Matrix Inequalities in Automatic Control Theory

V. A. YAKUBOVICH

IN 1961–62, Vladimir Yakubovich was well prepared to respond to the 1960–61 breakthrough of Vasile-Mihai Popov [14] with a breakthrough of his own. As a prominent young member of the Leningrad School, Yakubovich was investigating the solvability of Lurie's "resolving equations," that is, the problem of existence of Lyapunov functions consisting of a quadratic form plus the integral of a single sector nonlinearity. The nonlinearity was in feedback around the linear part of the closed-loop system. Motivated by actuator nonlinearities, this problem was formulated in 1944 by Lurie and Postnikov [11] and remained an open problem until it was solved by Popov in [14].

Popov's solution was a frequency domain condition that restricted the Nyquist plot of the linear part of the feedback system. Yakubovich interpreted Popov's condition using linear matrix inequalities-LMIs, further refined by Kalman [8]. Taken together, these results made a tremendous impact on the development of feedback control theory. Yakubovich's papers [16] set in motion a process that culminated in the computationally efficient LMI's method. Today's proponents of this method, Boyd et al. [2], credit Yakubovich as its father.

Yakubovich's four-page "doklad" (communication) is included in this collection because of its historical significance, although its laconic style is more challenging to the reader than the detailed presentation in Yakubovich's subsequent papers [16]. In the West, the most incisive reader of the papers by Popov and Yakubovich was Kalman, who employed his controllability-observability theory and the canonical structure theorem [8] to sharpen and clarify the LMI condition of Yakubovich. This led to a result which later became popular as the Kalman-Yakubovich-Popov (KYP) Lemma.

KYP Lemma [8]: Given a number $\gamma \geq 0$, two n-vectors b, c, and an $n \times n$ Hurwitz matrix A, if the pair (A, b) is completely controllable, then a vector q satisfying

$$A^T P + PA = -qq^T, \tag{1}$$

$$Pb - c = \sqrt{\gamma} q, \tag{2}$$

exists if and only if

$$\gamma + 2Re[c^T (j\omega I - A)^{-1} b] \geq 0 \tag{3}$$

for all real ω. Moreover, the set $\{x : x^T Px = 0\}$ is the unobservable subspace of the pair (A, c^T).

With a change of notation, and after some algebraic manipulations, the reader familiar with controllability-observability theory can deduce how this lemma evolved from Yakubovich's theorems in this paper. While (3) is a form of Popov's frequency domain condition, (1) and (2) represent the Lyapunov matrix inequality $A^T P + PA \leq 0$, restricted by a relationship connecting b, c and P. When $P > 0$ and $\gamma = 0$, this relationship is $Pb = c$, which implies $c^T b > 0$, that is, the transfer function $H(j\omega) = c^T (j\omega I - A)^{-1} b$ has relative degree one, and, by (3), its real part is nonnegative. In other words, $H(j\omega)$ belongs to the class of positive real functions extensively used in network theory for passive synthesis. This fact was recognized only a couple of years later independently by Popov [15] and Brockett [3]. With this positive real (passivity) property, the renamed *Positive-Real Lemma* soon led to the formulation of the circle criterion, as will be discussed in connection with the paper by Zames [17] included in this volume. Some of these developments are described in the 1966 survey by Brockett [4], while a comprehensive theory of multivariable positive real systems was presented by Anderson [1].

The impact of the PR lemma on model reference adaptive control was crucial as can be judged from the textbooks by Landau [10], Narendra and Annaswamy [13], and Ioannou and Sun [7]. It started with a proposal of Monopoli [12] (see also Fradkov [6]) that an unknown, minimum phase, relative degree one transfer function be made strictly positive real (SPR) by a feedback controller whose parameters are to be adaptively tuned to render the time derivative of a Lyapunov function nonpositive. The matrix P in the Lyapunov function was chosen to satisfy the SPR condition for a reference model. From adaptive control, the idea of achieving passivity by feedback spread to nonlinear control via Kokotović and Sussmann [9] and Byrnes et al. [5].

References

[1] B.D.O. ANDERSON, "A system theory criterion for positive real matrices," *SIAM J. Contr. Optimiz.*, **5**:171–182, 1967.

[2] S. Boyd, L. El Ghaoui, E. Feron, and V. Balakrishnan, *Linear Matrix Inequalities in System and Control Theory*, SIAM Studies in Applied Mathematics, Vol. 15, SIAM (Philadelphia, PA), 1994.

[3] R.W. Brockett, "On the stability of nonlinear feedback systems," *IEEE Trans. Applications and Industry*, **83**:443–448, 1964.

[4] R.W. Brockett, "The status of stability theory for deterministic systems," *IEEE Trans. Aut. Contr.*, **AC-11**:596–606, 1966.

[5] C.I. Byrnes, A. Isidori, and J.C. Willems, "Passivity, feedback equivalence, and global stabilization of minimum phase systems," *IEEE Trans. Aut. Contr.*, **AC-36**:1228–1240, 1991.

[6] A.L. Fradkov, "Quadratic Lyapunov functions in the adaptive stability problem of a linear dynamic target," *Siberian Math Journal*, pp. 341–348, 1976.

[7] P.A. Ioannou and J. Sun, *Robust Adaptive Control*, Prentice Hall (Englewood Cliffs, NJ), 1996.

[8] R.E. Kalman, "Canonical structure of linear dynamical systems," *Proc. National Academy of Sciences of the United States of America*, **48**:596–600, 1962.

[9] P.V. Kokotović and H.J. Sussmann, "A positive real condition for global stabilization of nonlinear systems," *Syst. Contr. Lett.*, **19**:177–185, 1989.

[10] I.D. Landau, *Adaptive Control: The Model Reference Approach*, Marcel Dekker, Inc. (New York), 1979.

[11] A.I. Lurie and V.N. Postnikov, "On the theory of stability of control systems," *Prikl. Mat. Meh.*, **8**:246–248, 1944.

[12] R.V. Monopoli, "The Kalman-Yakubovich lemma in adaptive control system design," *IEEE IEEE Trans. Aut. Contr.*, **AC-18**:527–529, 1973.

[13] K.S. Narendra and A.M. Annaswamy, *Stable Adaptive Systems*, Prentice Hall (Englewood Cliffs, NJ), 1989.

[14] V.M. Popov, "Absolute stability of nonlinear systems of automatic control," *Automation and Remote Control*, **22**:857–875, 1962. Translated from *Avtomatika i Telemekhanika*, **22**:961–979, 1961.

[15] V.M. Popov, "The solution of a new stability problem for controlled systems," *Automation and Remote Control*, **24**:1–23, 1963. Translated from *Avtomatika i Telemekhanika*, **24**:7–26, 1963.

[16] V.A. Yakubovich, "The matrix-inequality method in the theory of the stability of nonlinear control systems-Parts I–III," *Automation and Remote Control*, **25**:905–923, **26**:577–592, **26**:753–763, 1965. Translated from *Avtomatika i Telemekhanika*, **25**:1017–1029, 1964; **26**:577–590, **26**:753–763, 1965.

[17] G. Zames, "On the input-output stability of time-varying nonlinear feedback systems-Parts I and II," *IEEE Trans. Aut. Contr.*, **AC-11**:228–238 & 465–476, 1966.

P.V.K.

THE SOLUTION OF CERTAIN MATRIX INEQUALITIES IN AUTOMATIC CONTROL THEORY

V. A. JAKUBOVIČ

1. We shall denote square matrices by upper-case italic letters, column-vectors by lower-case italic letters, and numbers by Greek letters. An asterisk will denote the Hermitian conjugate, so that ab^* is a matrix and $b^*a = (a, b)$ is the scalar product. The notation $H > 0$ will signify that H is a Hermitian positive-definite matrix. I is the unit matrix.

Let us consider the following problems:

(I_ν). Given A, a, b; the eigenvalues of A lie in the left half-plane. For $H = H^*$ we define

$$G = -(A^*H + HA), \quad g = -(Ha + b). \tag{1}$$

We are required to find conditions whose satisfaction will indicate that the quadratic inequality

$$G - gg^* > 0 \tag{2}$$

with respect to the matrix $H = H^*$, has a solution.

(II_ν). Given B, $c \neq 0$, $d \neq 0$; the eigenvalues of B lie in the left half-plane. We are required to find conditions whose satisfaction will indicate that there exists $X = X^*$ satisfying the relations

$$-Y \equiv B^*X + XB < 0, \quad Xc + d = 0. \tag{3}$$

The index ν in both problems will denote the order of the corresponding vectors and matrices. Problem (I_ν) can be reduced to problem ($\mathrm{II}_{\nu+1}$) and vice versa by simple transformations (see below,

Reprinted with permission from *Soviet Mathematics (by American Math. Society)*, V. A. Yakubovich, "The Solution of Certain Inequalities in Automatic Control Theory," 1962, pp. 620–623.

§§7,8).

In applications the given problems (I_ν) and (II_ν) depend on parameters which must be chosen such that the corresponding problem has a solution. In connection with this, it is desirable to have rational solutions of problems (I_ν) and (II_ν), i.e., solutions which can be reduced to the verification of a finite number of conditions $\xi_\mu > 0$, $\eta_\mu = 0$, where ξ_μ and η_μ are polynomials associated with the given problems (i.e., associated with the real and imaginary parts of the elements of A, a, b, B, c, d).

For the sake of generality and ease of solution, we shall assume that the elements of the vectors and matrices are complex, although in practice they may be real.

2. To problems (I_ν) and (II_ν) can be reduced the problems of the determination of the global stability conditions, optimal in a defined sense, for nonlinear differential equations with one nonlinearity of class (A) [1–5]; problems with fixed nonlinearities of the type studied in [6]; and, some others. The method of Lur'e [1, 2] gives rational sufficient conditions for solving problem (I_ν). For $\nu \leq 2$ these conditions are also necessary [5], but it can be shown that for $\nu > 2$ these conditions do not coincide with the necessary ones.* Let us note the effective sufficient conditions of Lefschetz [7], which can be considered as sufficient conditions for the solvability of problem (I_ν). In [8] V. M. Popov deduced a special global stability condition for systems nonlinearities of class (A) encompassing all the conditions which can be obtained by means of Ljapunov functions of the type: "a quadratic form plus an integral of the nonlinearity." Theorem 1 (see below) with minor additions gives a new proof of Popov's condition, together with an answer to the converse problem posed in [8], p. 972. Namely, systems for which there exist Ljapunov functions of the indicated type will satisfy the special condition of Popov [8].

3. Let us introduce the following notations:

$$\begin{gathered} A_\omega = A - i\omega I, \quad a_\omega = A_\omega^{-1} a, \quad b_\omega = A_\omega^{*-1} b, \\ \varphi_{\mathrm{I}}(\omega) = 1 + 2\,\mathrm{Re}\,(a_\omega, b), \quad B_\omega = B - i\omega I, \quad c_\omega = B_\omega^{-1} c, \\ \varphi_{\mathrm{II}}(\omega) = \mathrm{Re}\,(c_\omega, d), \quad \varkappa = -(c, d), \quad \beta = \mathrm{Re}\,(Bc, d). \end{gathered} \tag{4}$$

The functions $\phi_{\mathrm{I}}(\omega)$, $\phi_{\mathrm{II}}(\omega)$ will be called the characteristics of problems (I_ν), (II_ν), respectively.

Theorem 1. *In order that inequality* (2) *have the solution* $H = H^*$ *it is necessary and sufficient that* $\phi_{\mathrm{I}}(\omega) > 0$ *for* $-\infty < \omega < +\infty$.

Theorem 2. *In order that there exist a matrix* $X = X^*$ *satisfying relations* (3) *it is necessary and sufficient that the following be satisfied:* 1) κ *is real;* 2) $\beta \neq 0$; 3) $\phi_{\mathrm{II}}(\omega) > 0$ *for* $-\infty < \omega < +\infty$.

Theorem 3. *In order that there exist a matrix* $X = X^*$ *satisfying relations* (3) *and the inequality* $Y > \epsilon I$ *for a given* $\epsilon > 0$, *it is necessary that the following be satisfied:* 1) κ *is real;* 2) $\epsilon_0 = 2\,\mathrm{Inf}\,\phi_{\mathrm{II}}(\omega)/|c_{\omega}^2| > 0$; 3) $\epsilon \leq \epsilon_0$; *and sufficient that the conditions* 1), 2), *and* 3') $\epsilon < \epsilon_0$, *be satisfied.*

Obviously Theorems 1, 2, and 3 give rational solutions.**

4. The necessity of the conditions of Theorems 1–3. We have $A_\omega^* H + HA_\omega = -G$, $(Ga_\omega, a_\omega) = -2\,\mathrm{Re}\,(Ha, a_\omega) = 2\,\mathrm{Re}\,(b, a_\omega) + 2\,\mathrm{Re}\,\zeta$, where $\zeta = (g, a_\omega)$. From (2) and the first of relations (1) we successively deduce $(Ga_\omega, a_\omega) > |\zeta|^2$, $\phi_{\mathrm{I}}(\omega) > |1 + \zeta^2|$.***

---

*Thus, the problem posed in [5], p. 129, has a negative solution.

**Other, less convenient, solutions of problems (I_ν) and (II_ν) were reported by the author in the spring of 1961 at the seminar of V. V. Nemyckiĭ at the Moscow State University.

***Let us note that the necessity of the condition $\phi_{\mathrm{I}}(\omega) > 0$ was essentially proved in [3, 5]. The necessary condition (3.2) of [5], for complex A, a, b in the notation here takes the form $\Gamma^2 = 1 + 2\,\mathrm{Re}\,(b, A^{-1}a) > 0$. As long as problem (I_ν) does not change when A_ω is substituted for A, we have from here that $\phi_{\mathrm{I}}(\omega) > 0$.

The necessity of the conditions of Theorems 2 and 3 follows from the relations $\kappa = (Xc, c)$, $\beta = (Yc, c)$, $2\phi_{II}(\omega) = (Yc_\omega, c_\omega) > \epsilon |c_\omega|^2$.

5. The sufficiency of the conditions of Theorem 3 follows from the sufficiency of Theorem 2 (for matrices of fixed order ν). Assuming $X = X_0 + \epsilon H_0$, where $B^* H_0 + H_0 B = -I$, we get that problem (II_ν) has the desired solution if there exists $X_0 = X_0^*$ satisfying the conditions $B^* X_0 + X_0 B \equiv -Y_0 < 0$, $X_0 c + d_0 = 0$, where $d_0 = \epsilon H_0 c + d$. The fulfillment of conditions 1), 2), 3) of Theorem 2 and, consequently, the existence of X_0, follow from conditions 1), 2), 3') of Theorem 3.

6. We shall write $\{A\} = \mu$ if the matrix A has μ eigenvalues in the left half-plane and does not have any purely imaginary or zero eigenvalues.

Lemma 1. *Let us consider the matrix of order* $\nu + 1$

$$\tilde{B} = \begin{bmatrix} A & p \\ q^* & \alpha \end{bmatrix}. \tag{5}$$

Let $\operatorname{Re}\alpha \neq 0$ *and let* e *be a vector with* $\nu + 1$ *components* $0, \cdots, 0, 1$. *Let us assume* $\chi_I(\omega) = ((A - i\omega)^{-1} p, q) - \alpha$, $\chi_{II}(\omega) = -((\tilde{B} - i\omega I)^{-1} e, e)$. *Then the conditions:* a_1) $\operatorname{Re}\chi_I(\omega) > 0$ *for* $-\infty < \omega < +\infty$, b_1) $\{A\} = \mu$, *are equivalent to the conditions:* a_2) $\operatorname{Re}\chi_{II}(\omega) > 0$ *for* $-\infty < \omega < +\infty$, b_2) $\{\tilde{B}\} = \mu + 1$.

Proof. Solving the equation $(\tilde{B} - i\omega I)\begin{bmatrix} x \\ \xi \end{bmatrix} = e$, we find $\xi = \chi_{II} = (\chi_I + i\omega)^{-1}$. Therefore a_2) follows from a_1). From a_1), b_1) and the relation $\chi_{II} = -\det(A - i\omega I)/\det(\tilde{B} - i\omega I)$ we conclude that $\tilde{B}$ has no purely imaginary eigenvalues. As $|\omega| \to \infty$ we have $\operatorname{Re}\chi_{II} = -\operatorname{Re}\alpha/\omega^2 - [1 + O(\omega^{-1})]$, i.e., $\operatorname{Re}\alpha < 0$ and $\Delta \operatorname{Arg}\chi_{II}(\omega)\,|_{-\infty}^{+\infty} = -\pi$. Therefore $\Delta \operatorname{Arg}\det(\tilde{B} - i\omega I)\,|_{-\infty}^{+\infty} = \pi[(\nu + 1) - (\mu + 1)]$, i.e., $\{\tilde{B}\} = \mu + 1$. The converse statement can be proved analogously.

7. The sufficiency of Theorem 1 follows from the sufficiency of Theorem 2. Let us consider problem (II_ν) for $B = \begin{bmatrix} A & a \\ b^* & -1/2 \end{bmatrix}$, $c = e$, $d = -e$. Applying Lemma 1 we find that the conditions of Theorem 2 are satisfied. Therefore, there exists $X = \begin{bmatrix} H & 0 \\ 0 & 1 \end{bmatrix}$ as the solution of problem ($II_{\nu+1}$). It is easy to verify that the matrix H is the solution of problem (I_ν).

8. Proof of sufficiency of Theorem 2. For $\nu = 1$ the assertion of Theorem 2 is easily verified. Let us assume that sufficiency has been proved for matrices of order ν. Let us consider problem ($II_{\nu+1}$) and assume that the conditions of Theorem 2 are satisfied. From the formula $\pi\kappa = \lim_{\Omega \to +\infty} \int_{-\Omega}^{\Omega} \phi_{II}(\omega)\, d\omega$ it follows that $\kappa > 0$. Therefore the matrix S whose last column equals c and whose first ν columns are some basis in a subspace orthonormal to d, is nonsingular. Multiplying the first relation in (3) on the left by S^* and on the right by S, and the second on the left by S^*, we pass to the equivalent problem

$$-\tilde{Y} \equiv \tilde{B}^* \tilde{X} + \tilde{X}\tilde{B} < 0, \quad \tilde{X} e = \kappa e, \tag{6}$$

where $\tilde{B} = S^{-1} BS$. Inasmuch as $\phi_{II}(\omega) = \frac{\beta}{\omega^2}[1 + O(\omega^{-1})]$ for $|\omega| \to \infty$, we have $\beta > 0$. Representing $\tilde{B}$ in the form (5), we find that $\operatorname{Re}\alpha = -\beta/\kappa < 0$, $\operatorname{Re}\chi_{II}(\omega) = \phi_{II}(\omega)/\kappa > 0$. By Lemma 1 we have $\operatorname{Re}\chi_I(\omega) > 0$, $\{A\} = \nu$. The second relation in (6) signifies that $\tilde{X}$ has the form $\tilde{X} = \begin{bmatrix} H & 0 \\ 0 & \kappa \end{bmatrix}$. From the first relation in (6) it follows that the desired matrix H is the solution of problem (I_ν), where $a = p/\sqrt{\beta}$, $b = \kappa q/\sqrt{\beta}$. The characteristic of this problem is $\phi_{\mathrm{I}}(\omega) = \frac{\kappa}{\beta} \operatorname{Re} \chi_1(\omega) > 0$.

Let us consider the first case, when $\operatorname{Re}(A_\omega^{-1} a, b) < 0$ for some ω. Then there exists a τ such that $\operatorname{Inf} \phi_{\mathrm{I}}(\omega) = \phi_{\mathrm{I}}(\tau)$. Let us consider the auxiliary problem

$$-Y_1 \equiv A^* X_1 + X_1 A < 0, \quad X_1 a_\tau + b_\tau - \delta H_0 a_\tau = 0, \tag{7}$$

where H_0 is the solution of the equation $A^* H_0 + H_0 A = -I$, $\delta > 0$, and a_τ, b_τ are determined from (4). According to the induction postulated in §4 we are justified in applying Theorem 3. The value of $\kappa_{\mathrm{ap}} = (A_\tau^{-1} a, b) + \delta(H_0 a_\tau, a_\tau)$ is real since $\phi_{\mathrm{I}}'(\tau) = 0$. The characteristic of the auxiliary problem can be rewritten in the form

$$\phi_{\mathrm{ap}} = \frac{[\phi_{\mathrm{I}}(\omega) - \phi_{\mathrm{I}}(\tau)]}{(\tau - \omega)^2} + \frac{\delta}{2} \mid A_\omega^{-1} a_\tau \mid^2.$$

Therefore $\epsilon_0 \geq \delta/2$ and by Theorem 3 there exists a matrix $X_1 = H$ such that $Y_1 > \frac{\delta}{3} I$. We will show that the matrix H is the solution of problem (I_ν) if the number $\delta > 0$ is chosen sufficiently small. After some computation we obtain, using the second relation in (7), $(G^{-1} g, g) = -2 \operatorname{Re}(a_\tau, b) + \delta^2 (G^{-1} h, h) < 1 - \phi_{\mathrm{I}}(\tau) + 3\delta \mid h \mid^2$, where $h = A_\tau^* H_0 a_\tau$. For $0 < \delta < \phi_{\mathrm{I}}(\tau)/3 \mid h \mid^2$ we get $(G^{-1} g, g) < 1$, which for the condition $G = Y_1 > 0$ is equivalent to (2). The proof is complete for the first case.

In the second case, when $\operatorname{Re}(a_\omega, b) > 0$, the auxiliary problem has the form $-Y_1 \equiv A^* X_1 + X_1 A < 0$, $X_1 a + b - \delta H_0 a = 0$, where $0 < \delta \mid H_0 a \mid^2 /3$. All the above arguments are simplified. The number $\kappa_{\mathrm{ap}} = -(a, b) - \delta(H_0 a, a)$ is real since as $\mid \omega \mid \to \infty$ from the condition $\operatorname{Re}(a_\omega, b) = -\operatorname{Im}(a, b)/\omega + O(\omega^{-2}) > 0$ it follows that $\operatorname{Im}(a, b) = 0$. The characteristic of the auxiliary problem has the form $\phi_{\mathrm{ap}} = \operatorname{Re}(a_\omega, b) + \frac{\delta}{2} \mid a_\omega \mid^2$, i.e., $\epsilon_0 \geq \delta/2$. By Theorem 3 there exists $X_1 = X_1^*$ such that $Y_1 > \frac{\delta}{3} I$. For $H = X_1$ we have $G = Y_1 > 0$, $-g = \delta H_0 a$, $(G^{-1} g, g) < 1$. Theorem 2 is proved.

Received 11/DEC/61

BIBLIOGRAPHY

[1] A. I. Lur'e, *Some nonlinear problems of the theory of automatic regulation*, GITTL, Moscow, 1951. (Russian)

[2] A. M. Letov, *Stability in nonlinear control systems*, GITTL, Moscow, 1955. (Russian)

[3] V. A. Jakubovič, Dokl. Akad. Nauk SSSR 117 (1957), 44.

[4] ———, ibid. 121 (1958), 984.

[5] ———, Vestnik Leningrad. Univ. (15) (1960), no. 7, 120.

[6] ———, Dokl. Akad. Nauk SSSR 135 (1960), 26 = Soviet Math. Dokl. 1 (1960), 1238.

[7] S. Lefschetz, RIAS Tech. Rep. 60–9, 1960.

[8] V. M. Popov, Avtomat. i Telemeh. 22 (1961), 961 = Automat. Remote Control 22 (1962), 857.

Translated by:
N. H. Choksy

Mathematical Description of Linear Dynamical Systems

R. E. KALMAN

THE contrast and relationship between models that explain phenomena by either (i) displaying the (internal) wiring diagram/event sequence in a system, or (ii) classifying the (external) input/output behavior, are central themes in many areas of science. Undoubtedly a good example is the modern computer, where "friendliness" is identified with the need to understand only the external behavior, but where design and efficiency require mastery of the internal structure. In the field of control, the input/output point of view dominated the development in the first half of the twentieth century, particularly in the western world. This is evident from the works of Nyquist and Bode (see their papers elsewhere in this volume) and that of other reserchers of the period. It is fair to say, in fact, that until the 1960s, systems and control was actually identified with input/output thinking; even nowadays researchers in other fields (such as biology and psychology) view the stimulus/response model as the essence of systems theory—a point of view strongly advocated, for example, by Wiener in his influential book on cybernetics [16].

The limitations of the input/output point of view became strikingly evident in the earlier work of Kalman ([8], [9]—both included in this volume), and others. By using state space descriptions for systems, very effective new algorithms were put forward for filtering and prediction on the one hand, and feedback control design on the other. The filtering algorithms turned out to be dramatic improvements over those obtained earlier by Wiener and Kolmogorov (see the preambles to [15] and [8], elsewhere in this volume), and filtering and optimal feedback control became practical as a result of these developments.

Soon after the appearance of these algorithms, the question was raised as to what the relationship was between state space and input/output models. In the paper that follows, Kalman examines this relationship, and especially the question of irreducibility, and the relevance of controllability and observability to the state space representation question. The results, while more general, appear to have been inspired by the work of Gilbert in a paper [5] that precedes Kalman's paper in the same issue of *SIAM Journal on Control*.

Kalman's paper discusses the relationship between the finite-dimensional linear state space system

$$\frac{d}{dt}x = F(t)x + G(t)u; \quad y = H(t)x \tag{1}$$

and the convolution integral

$$y(t) = \int_{t_0}^{t} S(t, \tau)u(\tau)\, d\tau \tag{2}$$

Special attention is given to the situation in which the matrices $F(t)$, $G(t)$, and $H(t)$ are constant (i.e., independent of t), and the input/output relationship (2) is shift-invariant, that is, $S(t, \tau) = S(t - \tau, 0)$. (In mathematics, the term *convolution* is usually reserved for the shift-invariant case.)

The paper contains a series of important results, among which are the following:

(i) The system (2) can be represented by a state space system (1) (assuming $x(t_0) = 0$) if and only if $S(t, \tau)$ is factorizable as $S(t, \tau) = P(t)Q(\tau)$, with $P(t)$ and $Q(t)$ finite matrices.

(ii) The system (1) is *irreducible* (meaning that among all systems (1) that represent, in the above sense, the same system (2), it has minimal state space dimension) if, and only if, it is completely controllable and completely observable.

(iii) Any two irreducible realizations are equivalent (in the sense that they differ only in the choice of the basis in the state space).

(iv) The system (1) can be decomposed into a controllable/observable subsystem, a controllable/unobservable subsystem, an uncontrollable/observable subsystem, and an uncontrollable/unobservable subsystem (see Figure 4 of the paper).

Important developments that were inspired by these results are generalizations to nonlinear systems (see, e.g., [14]) and to stochastic systems (see, e.g., [2]), and the associated algebraic theory using system descriptions in terms of matrices of polynomial or rational functions [3], [18]. Further developments were in automata theory [10] and coding theory [11].

The issues raised in this paper and the results obtained form the basis of what has later been called *realization theory*, one of the central themes in system theory. The one important element that was added later is the Hankel matrix [7], [13], which turned out to be the most effective algorithmic tool for the computation of models from input/output data. This is already evident in the algorithms in [13], but was further accentuated by the work on model reduction, in which the paper [6] stands out. Model reduction has become a very effective tool for reducing the complexity (in the sense of reducing the number of state variables) of a linear system. Two of the most widely used methods for model reduction are (i) balancing (which may be traced back at least to [12]) and (ii) finding a Hankel matrix of lower rank that optimally approximates a given Hankel matrix (often called AAK-model reduction [1]). These methods are based on analyzing the singular values of the Hankel matrix, by exploiting effectively the fact that the basis of the state space may be chosen freely, and by considering the controllability and observability gramians of the system to be reduced. A germ of these ideas is of course present in the above mentioned results in Kalman's paper.

The view that an external description of a system takes the form of an input/output map or a transfer function, and that an internal description leads to a state space model has recently been challenged by what is called the *behavioral approach* [17]. In this framework, a system is not viewed as an input/output object; instead all external ("manifest") variables are treated on an equal footing. Here irreducibility is equivalent to mere observability, since external behaviors can be uncontrollable (contrary to the system (2), where controllability is basically built in).

References

[1] V.M. Adamjan, D.Z. Arov, and M.G. Krein, "Analytic properties of Schmidt pairs for a Hankel operator and the generalized Schur-Takagi problem," *Math. USSR Sbornik*, **15**:31–73, 1971.

[2] H. Akaike, "Stochastic theory of minimal realization," *IEEE Trans. Autom. Contr.*, **AC-19(6)**:667–672, 1974.

[3] P.A. Fuhrmann, *Linear Systems and Operators in Hilbert Space*, McGraw-Hill (New York), 1981.

[4] F.R. Gantmacher, *The Theory of Matrices*, Chelsea, 1959.

[5] E.G. Gilbert, "Controllability and observability in multivariable control systems," *SIAM J. Control*, **1**:128–151, 1963.

[6] K. Glover, "All optimal Hankel-norm approximations of linear multivariable systems and their L^∞ bounds," *Int. J. Contr.*, **39**:1115–1193, 1984.

[7] B.L. Ho and R.E. Kalman, "Effective construction of linear state-variable models from input/output functions," in *Proc. Third Allerton Conf.*, pp. 449–459, 1965.

[8] R.E. Kalman, "A new approach to linear filtering and prediction problems," *Trans. ASME, J. Basic Engineering*, **82D**(1):35–45, 1960.

[9] R.E. Kalman, "Contributions to the theory of optimal control," *Bol. Soc. Mat. Mexicana*, **5**:102–119, 1960.

[10] R.E. Kalman, P.L. Falb, and M.A. Arbib, *Topics in Mathematical System Theory*, McGraw-Hill (New York), 1969.

[11] J.L. Massey and M.K. Sain, "Codes, automata, and continuous systems: Explicit interconnections," *IEEE Trans. Autom. Contr.*, **AC-12(6)**:644–650, 1967.

[12] B.C. Moore, "Principal component analysis in linear systems: controllability, observability, and model reduction," *IEEE Trans. Autom. Contr.*, **AC-26**:17–32, 1981.

[13] L.M. Silverman, "Realization of linear dynamical systems, *IEEE Trans. Autom. Contr.*, **AC-26**:554–567, 1971.

[14] H. Sussmann,"Existence and uniqueness of minimal realizations of nonlinear systems," *Math. Syst. Theory*, **10**:263–284, 1977.

[15] N. Wiener, *Extrapolation, Interpolation, and Smoothing of Stationary Time Series*, MIT Press (Cambridge, MA), 1949.

[16] N. Wiener, *Cybernetics*, MIT Press (Cambridge, MA), 1961.

[17] J.C. Willems, "Paradigms and puzzles in the theory of dynamical systems," *IEEE Trans. Automat. Contr.*, **AC-36**:259–294, 1991.

[18] W.A. Wolovich, *Linear Multivariable Systems*, Springer-Verlag (New York), 1974.

J.C.W.

J.S.I.A.M. Control
Ser. A, Vol. 1, No. 2
Printed in U.S.A., 1963

MATHEMATICAL DESCRIPTION OF LINEAR DYNAMICAL SYSTEMS*

R. E. KALMAN†

Abstract. There are two different ways of describing dynamical systems: (i) by means of state variables and (ii) by input/output relations. The first method may be regarded as an axiomatization of Newton's laws of mechanics and is taken to be the basic definition of a system.

It is then shown (in the linear case) that the input/output relations determine only one part of a system, that which is completely observable and completely controllable. Using the theory of controllability and observability, methods are given for calculating irreducible realizations of a given impulse-response matrix. In particular, an explicit procedure is given to determine the minimal number of state variables necessary to realize a given transfer-function matrix. Difficulties arising from the use of reducible realizations are discussed briefly.

1. Introduction and summary. Recent developments in optimal control system theory are based on vector differential equations as models of physical systems. In the older literature on control theory, however, the same systems are modeled by transfer functions (i.e., by the Laplace transforms of the differential equations relating the inputs to the outputs). Two different languages have arisen, both of which purport to talk about the same problem. In the new approach, we talk about state variables, transition equations, etc., and make constant use of abstract linear algebra. In the old approach, the key words are frequency response, pole-zero patterns, etc., and the main mathematical tool is complex function theory.

Is there really a difference between the new and the old? Precisely what are the relations between (linear) vector differential equations and transfer-functions? In the literature, this question is surrounded by confusion [1]. This is bad. Communication between research workers and engineers is impeded. Important results of the "old theory" are not yet fully integrated into the new theory.

In the writer's view—which will be argued at length in this paper—the difficulty is due to insufficient appreciation of the concept of a *dynamical system*. Control theory is supposed to deal with physical systems, and not merely with mathematical objects such as a differential equation or a transfer function. We must therefore pay careful attention to the relationship between physical systems and their representation via differential equations, transfer functions, etc.

* Received by the editors July 7, 1962 and in revised form December 9, 1962.

Presented at the Symposium on Multivariable System Theory, SIAM, November 1, 1962 at Cambridge, Massachusetts.

This research was supported in part under U. S. Air Force Contracts AF 49(638)-382 and AF 33(616)-6952 as well as NASA Contract NASr-103.

† Research Institute for Advanced Studies (RIAS), Baltimore 12, Maryland.

Reprinted with permission from *SIAM Journal on Control*, R. E. Kalman, "Mathematical Description of Linear Dynamical Systems" Vol. 1, 1963, pp. 152–192.

To clear up these issues, we need first of all a precise, abstract definition of a (physical) dynamical system. (See sections 2–3.) The axioms which provide this definition are generalizations of the Newtonian world-view of causality. They have been used for many years in the mathematical literature of dynamical systems. Just as Newtonian mechanics evolved from differential equations, these axioms seek to abstract those properties of differential equations which agree with the "facts" of classical physics. It is hardly surprising that under special assumptions (finite-dimensional state space, continuous time) the axioms turn out to be equivalent to a system of ordinary differential equations. To avoid mathematical difficulties, we shall restrict our attention to linear differential equations.

In section 4 we formulate the central problem of the paper:

Given an (experimentally observed) impulse response matrix, how can we identify the linear dynamical system which generated it?

We propose to call any such system a *realization* of the given impulse response. It is an *irreducible realization* if the dimension of its state space is minimal.

Section 5 is a discussion of the "canonical structure theorem" [2, 14] which describes abstractly the coupling between the external variables (input and output) and the internal variables (state) of any linear dynamical system. As an immediate consequence of this theorem, we find that *a linear dynamical system is an irreducible realization of an impulse-response matrix if and only if the system is completely controllable and completely observable.* This important result provides a link between the present paper and earlier investigations in the theory of controllability and observability [3–5].

Explicit criteria for complete controllability and complete observability are reviewed in a convenient form in section 6.

Section 7 provides a constructive computational technique for determining the canonical structure of a constant linear dynamical system.

In section 8 we present, probably for the first time, a complete and rigorous theory of how to define the state variables of a multi-input/multi-output constant linear dynamical system described by its transfer-function matrix. Since we are interested only in irreducible realizations, there is a certain unique, well-defined number n of state variables which must be used. We give a simple proof of a recent theorem of Gilbert [5] concerning the value of n. We give canonical forms for irreducible realizations in simple cases. We give a constructive procedure (with examples) for finding an irreducible realization in the general case.

Many errors have been committed in the literature of system theory by carelessly regarding transfer functions and systems as equivalent concepts. A list of these has been collected in section 9.

The field of research outlined in this paper is still wide open, except perhaps in the case of constant linear systems. Very little is known about irreducible realizations of nonconstant linear systems. It is not clear what additional properties—besides complete controllability and complete observability—are required to identify the stability type of a system from its impulse response. Nothing is known about nonlinear problems in this context.

Finally, the writer would like to acknowledge his indebtedness to Professor E. G. Gilbert, University of Michigan, whose work [5] predates this and whose results were instrumental in establishing the canonical structure theorem.

2. Axiomatic definition of a dynamical system. Macroscopic physical phenomena are commonly described in terms of cause-and-effect relationships. This is the "Principle of Causality". The idea involved here is at least as old as Newtonian mechanics. According to the latter, the motion of a system of particles is fully determined for all future time by the present positions and momenta of the particles and by the present and future forces acting on the system. How the particles actually attained their present positions and momenta is immaterial. Future forces can have no effect on what happens at present.

In modern terminology, we say that the numbers which specify the instantaneous position and momentum of each particle represent the *state* of the system. The state is to be regarded always as an abstract quantity. Intuitively speaking, the state is the minimal amount of information about the past history of the system which suffices to predict the effect of the past upon the future. Further, we say that the forces acting on the particles are the *inputs* of the system. Any variable in the system which can be directly observed is an *output*.

The preceding notions can be used to give a precise mathematical definition of a dynamical system [6]. For the present purposes it will be convenient to state this definition in somewhat more general fashion [14].

DEFINITION 1. A dynamical system is a mathematical structure defined by the following axioms:

(D_1) There is given a *state space* Σ and a set of values of *time* Θ at which the behavior of the system is defined; Σ is a topological space and Θ is an ordered topological space which is a subset of the real numbers.

(D_2) There is given a topological space Ω of functions of time defined on Θ, which are the admissible *inputs* to the system.

(D_3) For any initial time t_0 in Θ, any initial state x_0 in Σ, and any input u in Ω defined for $t \geqq t_0$, the future states of the system are determined by the transition function $\varphi: \Omega \times \Theta \times \Theta \times \Sigma \to \Sigma$, which is written as $\varphi_u(t; t_0, x_0) = x_t$. This function is defined only for $t \geqq t_0$. Moreover, any $t_0 \leqq t_1 \leqq t_2$ in Θ, any x_0 in Σ, and any fixed u in Ω defined over $[t_0, t_1] \cap \Theta$, the following relations hold:

(D₃-i) $$\varphi_u(t_0 ; t_0, x_0) = x_0,$$

(D₃-ii) $$\varphi_u(t_2 ; t_0, x_0) = \varphi_u(t_2 ; t_1, \varphi_u(t_1 ; t_0, x_0)).$$

In addition, the system must be *nonanticipatory*, i.e., if u, $v \in \Omega$ and $u \equiv v$ on $[t_0, t_1] \cap \Theta$ we have

(D₃-iii) $$\varphi_u(t; t_0, x_0) \equiv \varphi_v(t; t_0, x_0).$$

(D₄) Every *output* of the system is a function $\psi : \Theta \times \Sigma \to$ reals.

(D₅) The functions φ and ψ are continuous, with respect to the topologies defined for Σ, Θ, and Ω and the induced product topologies.

In this paper we will study only a very special subclass of dynamical systems: those which are *real, finite-dimensional, continuous-time,* and *linear*.

"Real, finite-dimensional" means that $\Sigma = R^n = n$-dimensional real linear space. "Continuous-time" means that $\Theta = R^1$ = set of real numbers. "Linear" means that φ is linear on $\Omega \times \Sigma$ and ψ is linear on Σ.

By requiring φ and ψ to be sufficiently "smooth" functions, we can deduce from the axioms a set of equations which characterize every real, finite-dimensional, continuous-time, and linear dynamical system. The proof of this fact is outside the scope of the present paper [14]. Here we shall simply assume that every such system is governed by the equations

$$\frac{dx}{dt} = F(t)x + G(t)u(t), \tag{2.1}$$

$$y(t) = H(t)x(t), \tag{2.2}$$

defined on the whole real line $-\infty < t < \infty$, where x, u, and y are n, m, and p-vectors* respectively, and the matrices $F(t)$, $G(t)$, and $H(t)$ are continuous functions of the time t.

We call (2.1–2) the *dynamical equations* of the system.

It is instructive to check whether the axioms are satisfied. (D_1) is obviously true; we have $\Sigma = R^n$, $\Theta = R^1$. The *state* of the system is the vector x. To satisfy (D_2), we must specify the class of all inputs, that is, a subclass of all vector functions $u(t) = (u_1(t), \cdots, u_m(t))$. To define Ω, we shall assume that these functions are piecewise continuous; this is sufficiently

* Vectors will be denoted by small Roman letters, matrices by Roman capitals. The components of a vector x are x_i, components of a matrix A are a_{ij}. On the other hand, $x^1, x^2, \cdots$, are vectors, and F^{AA}, F^{AB} are matrices. A' is the transpose of A.

general for most applications. We have exactly p observations on the system (the components of the vector y) and by (2.2) they are functions of t, x. Hence (D_4) is satisfied. To check (D_3), we recall that the general solution of (2.1) is given by

$$\varphi_u(t; t_0, x_0) \equiv x_t = \Phi(t, t_0)x_0 + \int_{t_0}^{t} \Phi(t, \tau)G(\tau)u(\tau)\, d\tau, \tag{2.3}$$

where $\Phi(t, \tau)$ is the transition matrix of the free differential equation defined by $F(t)$ [4, 7]†. Since (2.3) is valid for any $t \geqq t_0$ (in fact, also for $t < t_0$), φ is well defined. Property (D_3-i) is obvious. (D_3-ii) follows from the composition property [4, 7] of the transition matrix:

$$\Phi(t, \sigma) = \Phi(t, \tau)\Phi(\tau, \sigma), \tag{2.4}$$

which holds for every set of real numbers t, τ, σ. Indeed, (2.4) is simply the linear version of (D_3-ii). (D_3-iii) is obvious from formula (2.3). The continuity axiom (D_5) is satisfied by hypothesis.

Evidently φ given by (2.3) is linear on the cartesian product of Σ with the linear space of vector-valued piecewise continuous functions.

We call a linear dynamical system (2.1–2) *constant*, *periodic*, or *analytic* whenever F, G, and H are constant, periodic, or analytic in t.

It is often convenient to have a special name for the couple $(t, x) \mid \in \Theta \times \Sigma$. Giving a fixed value of (t, x) is equivalent to specifying at some time (t) the state (x) of the system. We shall call (t, x) a *phase* and $\Theta \times \Sigma$ the *phase space*. (Recall the popular phrase: "phases" of the Moon.)

To justify our claim—implicit in the above discussion—that equations (2.1–2) are a good model of physical reality, we wish to point out that these equations can be concretely simulated by a simple physical system: a general-purpose analog computer. Indeed, the numbers (or functions) constituting F, G, and H may be regarded as specifying the "wiring diagram" of the analog computer which simulates the system (2.1–2) (see, for instance, [8]).

3. Equivalent dynamical systems. The state vector x must always be regarded as an abstract quantity. By definition, it cannot be directly measured. On the other hand, the inputs and outputs of the system (2.1–2) have concrete physical meaning. Bearing this in mind, equations (2.1–2) admit two interpretations:

(a) They express relations involving the abstract linear transformations $F(t)$, $G(t)$, and $H(t)$.

(b) At any fixed time, we take an arbitrary but fixed coordinate system

† I.e., Φ is a solution of $d\Phi/dt = F(t)\Phi$, subject to the initial condition $\Phi(\tau, \tau) = I =$ unit matrix for all τ.

in the (abstract) vector space Σ. Then the symbol $x \equiv (x_1, \cdots, x_n)$ is interpreted as the numerical n-tuple consisting of the coördinates of the abstract state vector which is also denoted by x. F, G, and H are interpreted as the matrix representations of the abstract linear transformations denoted by the same letters under (a)..

To describe the behavior of a dynamical system in concrete terms, the second point of view must be used. Then we must also ask ourselves the question: To what extent does the description of a dynamical system depend on the arbitrary choice of the coordinate system in the state space? (No such arbitrariness occurs in the definition of the numerical vectors u, y since the input and output variables u_i and y_j are concrete physical quantities.) This question gives rise to the next definition.

Definition 2. Two linear dynamical systems (2.1–2), with state vectors x, $\bar{x}$, are *algebraically equivalent* whenever their numerical phase vectors are related for all t as

$$(t, \bar{x}) = (t, T(t)x), \tag{3.1}$$

where $T(t)$ is a $n \times n$ matrix, nonsingular for all t and continuously differentiable in t. In other words, there is a 1-1 differentiable correspondence between the phase spaces $\Theta \times \Sigma$ and $\Theta \times \bar{\Sigma}$.

Remark: We could generalize this definition of equivalence to $(\bar{t}, \bar{x}) = (\tau(t), T(t)x)$ where τ is an increasing function of t. But this involves distortion of the time scale which is not permitted in Newtonian physics.

Algebraic equivalence implies the following relations between the defining matrices of the two systems:

$$\begin{aligned} \bar{\Phi}(t, \tau) &= T(t)\Phi(t, \tau)T^{-1}(\tau), \\ \bar{F}(t) &= \dot{T}(t)T^{-1}(t) + T(t)F(t)T^{-1}(t), \\ \bar{G}(t) &= T(t)G(t), \\ \bar{H}(t) &= H(t)T^{-1}(t). \end{aligned} \tag{3.2}$$

In general, algebraic equivalence does not preserve the stability properties of a dynamical system [7, 9, 10]. For this it is necessary and sufficient to have *topological equivalence*: algebraic equivalence plus the condition

$$\| T(t) \| \leqq c_1 \quad \text{and} \quad \| T^{-1}(t) \| \leqq c_2, \tag{3.3}$$

where c_1 and c_2 are fixed constants, and $\| \quad \|$ is the euclidean norm*.

A nonconstant system may be algebraically and even topologically equivalent to a constant system. The latter case is called by Markus [11]

* Let Θ, Σ, and $\bar{\Sigma}$ have the usual topologies induced by the euclidean norm. Then the product topologies induced on $\Theta \times \Sigma$ and $\Theta \times \bar{\Sigma}$ are equivalent if and only if (3.3) holds.

"kinematic similarity". Moreover, two constant systems may be algebraically and topologically equivalent without $T(t)$ being a constant. To bypass these complications, we propose

DEFINITION 3. Two constant linear dynamical systems are *strictly equivalent* whenever their numerical phase vectors are related for all t as $(t, \bar{x}) = (t, Tx)$, where T is a nonsingular constant matrix.

Evidently strict equivalence implies topological equivalence.

4. The impulse-response matrix and its realization by a linear dynamical system.

Sections 2–3 were concerned with mathematics, that is, abstract matters. If we now take the point of view of physics, then a dynamical system must be "defined" in terms of quantities which can be directly observed. For linear dynamical systems, this is usually done in the following way.

We consider a system which is at rest at time t_0; i.e., one whose input and outputs have been identically zero for all $t \leqq t_0$. We apply at each input in turn a very sharp and narrow pulse. Ideally, we would take $u_i^{(j)}(t) = \delta_{ij}\delta(t - t_0)$, where δ is the Dirac delta function, δ_{ij} is the Kronecker symbol, and $1 \leqq i, j \leqq m$. We then observe the effect of each vector input $u^{(j)}(t)$ on the outputs, which are denoted by $u(t; j)$. The matrix $S(t, t_0) = [s_{ij}(t, t_0)] = [y_i(t; j)]$ so obtained is called the *impulse-response matrix* of the system. Since the system was at rest prior to $t = t_0$, we must define $S(t, t_0) \equiv 0$ for $t < t_0$. We also assume, of course, that S is continuous in t and t_0 for $t > t_0$.

With these conventions, the output of a linear system originally at rest is related to its input by the well-known convolution integral:

$$y(t) = \int_{t_0}^{t} S(t, \tau)u(\tau)\, d\tau. \tag{4.1}$$

In much of the literature of system theory [12] (and also at times in physics) formula (4.1) is the basic definition of a system. The Fourier transform of S is often called "the system function" [13, p. 92].

Unfortunately, this definition does not explain how to treat systems which are not "initially at rest". Hence we may ask, "To what extent, if any, are we justified in equating the physical definition (4.1) of a system with the mathematical one provided by (2.1–2)?"

Suppose that the system in question is actually (2.1–2). Then (2.3) shows that

$$\begin{aligned} S(t, \tau) &= H(t)\Phi(t, \tau)G(\tau), \qquad && t \geqq \tau, \\ &= 0, && t < \tau.\dagger \end{aligned} \tag{4.2}$$

† The right-hand side of the first equation (4.2) is defined also for $t < \tau$; then the left-hand side may be regarded as the "backward impulse response", whose physical interpretation is left to the reader.

Thus it is trivial to calculate the impulse-response matrix of a given linear dynamical system. The converse question, however, is non trivial and interesting. *When and how does the impulse-response matrix determine the dynamical equations of the system?*

This problem is commonly called the *identification* of the system from its impulse-response matrix.

Having been given an impulse-response matrix, suppose that we succeed in finding matrices F, G, and H such that (4.2) holds. We have then identified a physical system that may have been the one which actually generated the observed impulse-response matrix. We shall therefore call (2.12) a *realization* of $S(t, \tau)$. This terminology is justified because the axioms given in section 2 are patterned after highly successful models of classical macroscopic physics; in fact, the system defined by (2.1–2) can be concretely realized, actually built, using standard analog-computer techniques in existence today. In short, proceeding from the impulse-response matrix to the dynamical equations we get closer to "physical reality". But we are also left with a problem: Which one of the (possibly very many) realizations of $S(t, \tau)$ is the actual system that we are dealing with?

It is conceivable that certain aspects of a dynamical system cannot ever be identified from knowledge of its impulse response, as our knowledge of the physical world gained from experimental observation must always be regarded as incomplete. Still, it seems sensible to ask how much of the physical world can be determined from a given amount of experimental data.

The first clear problem statement in this complex of ideas and the first results appear to be due to the writer [2, 14].

First of all we note

THEOREM 1. *An impulse-response matrix $S(t, \tau)$ is realizable by a finite-dimensional dynamical system* (2.1–2) *if and only if there exist continuous matrices $P(t)$ and $Q(t)$ such that*

$$S(t, \tau) = P(t)Q(\tau) \quad \text{for all} \quad t, \tau. \tag{4.3}$$

Proof. Necessity follows by writing the right-hand side of (4.2) as $H(t)\Phi(t, 0)\Phi(0, \tau)G(\tau)$, with the aid of (2.4). Sufficiency is equally obvious. We set $F(t) = 0$, $G(t) = Q(t)$, and $H(t) = P(t)$. Then $\Phi(t, \tau) \equiv I$ and the desired result follows by (4.2).

A realization (2.1–2) of $S(t, \tau)$ is *reducible* if over some interval of time there is a proper (i.e., lower-dimensional) subsystem of (2.1–2) which also realizes $S(t, \tau)$. As will be seen later, a realization of S (particularly the one given in the previous paragraph) is often reducible.

An impulse-response matrix S is *stationary* whenever $S(t, \tau) = S(t + \sigma, \tau + \sigma)$ for all real numbers t, τ, and σ. S is *periodic* whenever the preceding relation holds for all t, τ, and some σ. An impulse-response matrix is *analytic* whenever S is analytic in t and τ; if (4.3) holds, then P and Q must be analytic in t.

The main result, whose proof will be discussed later, is the following [14]:

THEOREM 2. *Hypothesis: The impulse-response matrix S satisfies* (4.3) *and is either periodic (and continuous) or analytic.*

Conclusions: (i) *There exist irreducible realizations of S, all of which have the same constant dimension n and are algebraically equivalent.* (ii) *If S is periodic [analytic] so are its irreducible realizations.*

Topological equivalence cannot be claimed in general. It may happen that S has one realization which is asymptotically stable and another which is asymptotically unstable [15]. Hence it may be impossible to identify the stability of a dynamical system from its impulse response! This surprising conclusion raises many interesting problems which are as yet unexplored [15]. If S is not periodic or analytic, it may happen that the dimension $n(t)$ of an irreducible realization is constant only over finite time intervals.

In the stationary case, Theorem 2 can be improved [14].

THEOREM 3. *Every stationary impulse-response matrix $S(t, \tau) = W(t - \tau)$ satisfying* (4.3) *has constant irreducible realizations. All such realizations are strictly equivalent.*

In view of this theorem, we may talk indifferently about a stationary impulse-response matrix or the dynamical system which generates it—as has long been the practice in system theory on intuitive grounds. But note that we must require the realization to be irreducible. For nonconstant systems, such a conclusion is at present not justified. The requirement of irreducibility in Theorem 3 is essential; disregarding it can lead—and has led—to serious errors in modeling dynamical systems. (See section 9.)

In many practical cases, it is not the *weighting-function matrix* $W(t - \tau)$ (see Theorem 3) which is given, but its Laplace transform, the *transfer-function matrix* $Z(s) = \mathcal{L}[W(t)]$. Then condition (4.3) has an interesting equivalent form, which is often used as a "working hypothesis" in engineering texts:

THEOREM 4. *A weighting-function matrix $W(t - \tau)$ satisfies* (4.3) *if and only if its elements are linear combinations of terms of the type $t^i e^{s_j t}$* ($i = 0, 1, \cdots, n - 1, j = 1, \cdots, n$). *Hence every element of the transfer-function matrix is a ratio of polynomials in s such that the degree of the denominator polynomial always exceeds the degree of the numerator polynomial.*

This result is provedd in [14]. It implies that the realization of an impulse-response matrix is equivalent to expressing the elements of F, G, and H as functions of the coefficients of the numerator and denominator polynomials of elements of $Z(s)$. (See section 8.)

In the remainder of the paper, we wish to investigate two main problems arising in the theory sketched above:

(i) Explicit criteria for reducibility.

(ii) Construction of irreducible realizations.

Remark. Elementary expositions of system theory often contain the statement that the operator d/dt ($\equiv s$) is a "system." Is a it system in the same sense as that word is used here? The answer is no. To define such a system rigorously in accordance with the axioms introduced in section 2, one must proceed as follows. The output of the system, which by definition is the derivative of the input, is given by

$$y(t) = \frac{du(t)}{dt} = \psi(t, x(t)), \tag{3.4}$$

so that at any fixed t, $u(t)$ must be a *point* function of $(t, x(t))$. Therefore the state space Σ must include the space Ω of functions on which the operator d/dt is defined. It is simplest to let $\Sigma = \Omega$. Then Σ is usually infinite dimensional because Ω is. Thus we define the state $x \equiv x(t)$ as the function $u(\tau)$, defined for all $\tau \leqq t$. The mapping $\varphi_u(t; t_0, x_{t_0})$ assigns to the function x_0 defined for $\tau \leqq t_0$ the function x_t, which is equal to x_{t_0} on $\tau \leqq t_0$ and equal to u on $t_0 < \tau \leqq t$.

In this paper, the finite dimensionality of Σ is used in an essential way, which rules out consideration of the "system" d/dt in all but trivial cases.

5. Canonical structure of linear dynamical systems. The concept of irreducibility can be understood most readily with the help of the writer's "canonical structure theorem" for linear dynamical systems [2, 14].

Before presenting and illustrating this central result, it is necessary to recall some definitions and facts concerning the *controllability* and *observability* of linear dynamical systems.

DEFINITION 4. A linear dynamical system (2.1-2) is *completely controllable* at time t_0 if it is not algebraically equivalent, for all $t \geqq t_0$, to a system of the type

$$\begin{aligned} &\text{(a)} \quad dx^1/dt = F^{11}(t)x^1 + F^{12}(t)x^2 + G^1(t)u(t) \\ \text{(5.1)}\ &\text{(b)} \quad dx^2/dt = F^{22}(t)x^2 \\ &\text{(c)} \quad y(t) = H^1(t)x^1(t) + H^2(t)x^2(t). \end{aligned}$$

(In (5.1), x^1 and x^2 are vectors of n_1 and $n_2 = n - n_1$ components respectively.)

In other words, it is *not* possible to find a coördinate system in which the state variables x_i are separated into two groups, $x^1 = (x_1, \cdots, x_{n_1})$ and $x^2 = (x_{n_1+1}, \cdots, x_n)$, such that the second group is not affected either by the first group or by the inputs to the system. If one could find such a

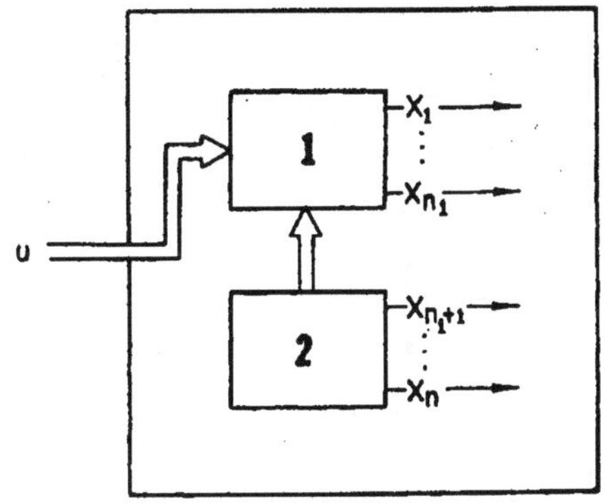

FIGURE 1.

coördinate system, we would have the state of affairs depicted schematically in Fig. 1.

Clearly, controllability is a system property which is completely independent of the way in which the outputs of the system are formed. It is a property of the couple $\{F(t), G(t)\}$.

The "dual" of controllability is observability, which depends only on the outputs but not on the inputs.

DEFINITION 5. A linear dynamical system (2.1–2) is *completely observable* at time t_0 if it is not algebraically equivalent, for all $t \leqq t_0$, to any system of the type

$$(5.2)\quad \begin{aligned} &(a) && dx^1/dt = F^{11}(t)x^1(t) + G^1(t)u(t) \\ &(b) && dx^2/dt = F^{21}(t)x^1(t) + F^{22}(t)x^2 + G^2(t)u(t) \\ &(c) && y(t) = H^1(t)x^1(t). \end{aligned}$$

(Again, x^1 is an n_1-vector and x^2 is an $(n - n_1)$-vector.)

In other words, it is not possible to find a coördinate system in which the state variables x_i are separated into two groups, such that the second group does not affect either the first group or the outputs of the system. If such a coördinate system could be found, we would have the state of affairs depicted in Fig. 2.

The above definitions show that controllability and observability are preserved under algebraic equivalence. These properties are coördinate-free, i.e., independent of the particular choice of basis in the state space.

The equivalence of the present definitions with other more abstract

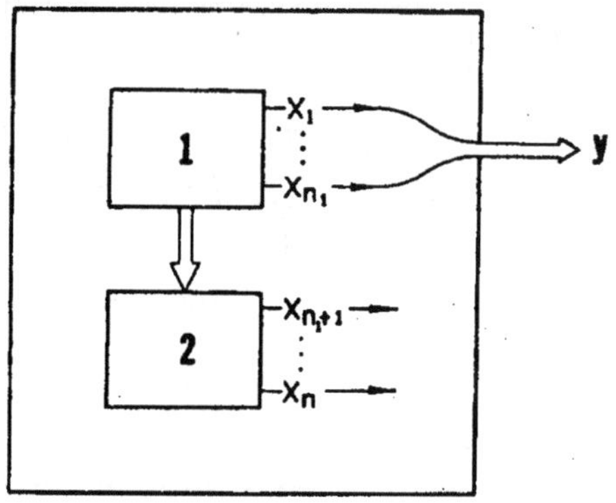

FIGURE 2.

definitions of controllability may be found in [4]. As to observability, we note that the *duality relations*

$$
(5.3)\quad
\begin{aligned}
&\text{(a)} && t - t_0 = t_0 - t',\\
&\text{(b)} && F(t - t_0) \Leftrightarrow F'(t_0 - t'),\\
&\text{(c)} && G(t - t_0) \Leftrightarrow H'(t_0 - t'),\\
&\text{(d)} && H(t - t_0) \Leftrightarrow G'(t_0 - t'),
\end{aligned}
$$

transform the system (5.2) into (5.1). Hence all theorems on controllability can be "dualized" to yield analogous results on observability.

It can be shown that in applying definitions 4–5 to constant systems it is immaterial whether we require algebraic or strict equivalence [14]. Hence—as one would of course expect—for constant systems the notions of complete controllability and complete observability do not depend on the choice of t_0.

EXAMPLE 1. A simple, well-known, and interesting case of a physical system which is neither completely controllable nor completely observable is the so-called constant-resistance network shown in Fig. 3.

Let x_1 be the magnetic flux in the inductor and x_2 the electric charge on the capacitor in Fig. 3, while $u_1(t)$ is a voltage source (zero short-circuit resistance) and $y_1(t)$ is the current into the network. The inductor and capacitor in the network may be time-varying, but we assume—this is the constant-resistance condition—that $L(t)$ and $C(t)$ are related by:

$$L(t)/C(t) = R^2 = 1 \qquad (L(t), C(t) > 0).$$

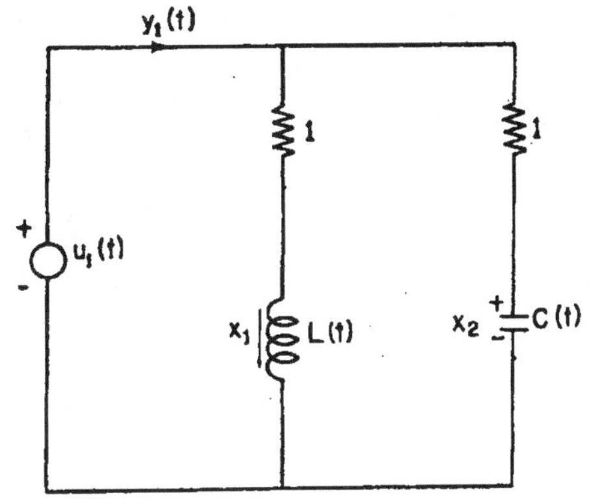

FIGURE 3.

The differential equations of the network are

$$dx_1/dt = -[1/L(t)]x_1 + u_1(t),$$
$$dx_2/dt = -[1/C(t)]x_2 + u_1(t),$$
$$y_1(t) = [1/L(t)]x_1 - [1/C(t)]x_2 + u_1(t).$$

If we let

$$\bar{x}_1 = (x_1 + x_2)/2,$$
$$\bar{x}_2 = (x_1 - x_2)/2,$$

the dynamical equations become

$$(5.4) \qquad \begin{aligned} d\bar{x}_1/dt &= -[1/L(t)]\bar{x}_1 + u_1(t), \\ d\bar{x}_2/dt &= -[1/L(t)]\bar{x}_2, \\ y_1(t) &= 2[1/L(t)]\bar{x}_2 + u_1(t).^* \end{aligned}$$

Here the state variable $\bar{x}_1$ is controllable but not observable, while $\bar{x}_2$ is observable but not controllable.

For obvious reasons, the subsystem (b) of (5.1) may be regarded as (completely) uncontrollable, while subsystem (b) of (5.2) is (completely) unobservable. In view of linearity, it is intuitively clear that it must be possible to arrange the components of the state vector—referred to a

* Note that this equation does not correspond to (2.2) but to $y(t) = H(t)x(t) + J(t)u(t)$. This is a minor point. In fact, Axiom (D_4) may be generalized to: "(D_4)' Every output is a function of t, $x(t)$, and $u(t)$." This entails only minor modifications as far as the results and arguments of the present paper are concerned.

suitable (possibly time-varying) coördinate system—into four mutally exclusive parts, as follows:

Part (A): Completely controllable but unobservable.

Part (B): Completely controllable and completely observable.

Part (C): Uncontrollable and unobservable.

Part (D): Uncontrollable but completely observable.

The precise statement of this idea is [2, 14]:

Theorem 5 (*Canonical Structure Theorem*). *Consider a fixed linear dynamical system* (2.1–2).

(i) *At every fixed instant t of time, there is a coördinate system in the state space relative to which the components of the state vector can be decomposed into four mutually exlusive parts*

$$x = (x^A, x^B, x^C, x^D),$$

which correspond to the scheme outlined above.

(ii) *This decomposition can be achieved in many ways, but the number of state variables $n_A(t), \cdots, n_D(t)$ in each part is the same for any such decomposition.*

(iii) *Relative to such a choice of coördinates, the system matrices have the canonical form*

$$F(t) = \begin{bmatrix} F^{AA}(t) & F^{AB}(t) & F^{AC}(t) & F^{AD}(t) \\ 0 & F^{BB}(t) & 0 & F^{BD}(t) \\ 0 & 0 & F^{CC}(t) & F^{CD}(t) \\ 0 & 0 & 0 & F^{DD}(t) \end{bmatrix},$$

$$G(t) = \begin{bmatrix} G^A(t) \\ G^B(t) \\ 0 \\ 0 \end{bmatrix},$$

and

$$H(t) = [0 \quad H^B(t) \quad 0 \quad H^D(t)].$$

In view of this theorem, we shall talk, somewhat loosely, about "Parts $(A), \cdots, (D)$ of the system." Thus the system (5.4) consists of Parts (A) and (D).

The canonical form of F, G, and H can be easily remembered by reference to the causal diagram shown on Fig. 4.

It is intuitively clear (and can be easily proved) that algebraically equivalent systems have the same canonical structure.

Unfortunately, the coördinate system necessary to display the canonical form of F, G, and H will not be continuous in time unless $n_A(t), \cdots, n_D(t)$ are constants. If these dimension numbers vary, we cannot call the various

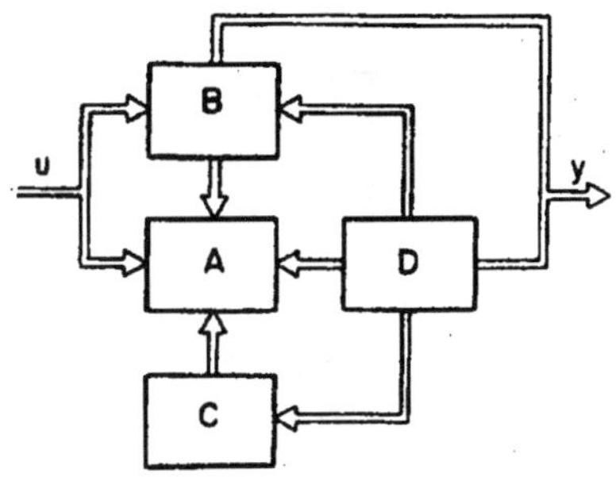

FIGURE 4.

parts of the canonical structure "subsystems." For constant systems this difficulty does not arise. More generally, we have:

THEOREM 6. *For a periodic or analytic linear dynamical system* (2.1–2) *the dimension numbers* $n_A, \cdots, n_D$ *are constants, and the canonical decomposition is continuous with respect to* t.

An illustration of the canonical structure theorem is provided by

EXAMPLE 2. Consider the constant system defined by

$$F = \begin{bmatrix} -3 & -3 & 0 & 1 \\ 26 & 36 & -3 & -25 \\ 30 & 39 & -2 & -27 \\ 30 & 43 & -3 & -32 \end{bmatrix},$$

$$G = \begin{bmatrix} 3 & 3 \\ -2 & -1 \\ 0 & 0 \\ 0 & 1 \end{bmatrix},$$

and

$$H = [-5 \quad -8 \quad 1 \quad 5].$$

We introduce new coördinates by letting $\bar{x} = Tx$, where

$$T = \begin{bmatrix} 2 & 3 & 0 & -2 \\ 1 & 1 & 0 & -1 \\ -2 & -3 & 0 & 3 \\ -6 & -9 & 1 & 6 \end{bmatrix},$$

and

$$T^{-1} = \begin{bmatrix} 0 & 3 & 1 & 0 \\ 1 & -2 & 0 & 0 \\ 3 & 0 & 0 & 1 \\ 1 & 0 & 1 & 0 \end{bmatrix}.$$

With respect to these new coordinates the system matrices assume the canonical form:

$$\bar{F} = TFT^{-1} = \begin{bmatrix} 2 & 4 & 1 & -1 \\ 0 & -1 & 0 & 1 \\ 0 & 0 & -3 & -2 \\ 0 & 0 & 0 & 1 \end{bmatrix},$$

$$\bar{G} = TG = \begin{bmatrix} 0 & 1 \\ 1 & 1 \\ 0 & 0 \\ 0 & 0 \end{bmatrix},$$

and

$$\bar{H} = HT^{-1} = [0 \quad 1 \quad 0 \quad 1].$$

On the other hand, if we define the new coördinates by

$$T = \begin{bmatrix} 3 & 4 & 0 & -3 \\ 1 & 1 & 0 & -1 \\ -5 & -7.5 & 0.5 & 6 \\ -6 & 9 & 1 & 6 \end{bmatrix},$$

$$T^{-1} = \begin{bmatrix} 0 & 3 & 1 & -0.5 \\ 1 & -3 & 0 & 0 \\ 3 & -3 & 0 & 1 \\ 1 & -1 & 1 & -0.5 \end{bmatrix},$$

then the system matrices become

$$\bar{F} = \begin{bmatrix} 2 & 1 & 1 & 0 \\ 0 & -1 & 0 & 1 \\ 0 & 0 & -3 & 0 \\ 0 & 0 & 0 & 1 \end{bmatrix},$$

$$\bar{G} = \begin{bmatrix} 1 & 2 \\ 1 & 1 \\ 0 & 0 \\ 0 & 0 \end{bmatrix},$$

and

$$\bar{H} = [0 \quad 1 \quad 0 \quad 1].$$

The numerical values of these two canonical forms are different, yet Theorem 5 is verified in both cases. In the second case the connections from Part (D) to Parts (A) and (C) are missing. This is not a contradiction since Theorem 5 does not require that all the indicated casual connections in Fig. 4 be actually present.

The transfer-function matrix of the system is easily found from the canonical representation. The coördinate transformations affect only the internal (state) variables, but not the external (input and output) variables; consequently the impulse response matrix is invariant under such transformations. We get by inspection:

$$Z(s) = \begin{bmatrix} \frac{1}{s+1} & \frac{1}{s+1} \end{bmatrix}.$$

It would be rather laborious to determine these transfer functions directly from the signal-flow graph [16] corresponding to F, G, and H.

EXAMPLE 3. A far less trivial illustration of the canonical decomposition theorem is provided by the following dynamical system, which occurs in the solution of a problem in the theory of statistical filtering [17]. Let A be an arbitrary positive function of t and define

$$F = \begin{bmatrix} -t^4/4A & 1 & 0 \\ -t^3/2A & 0 & 1 \\ -t^2/2A & 0 & 0 \end{bmatrix},$$

$$G = \begin{bmatrix} t^4/4A \\ t^3/2A \\ t^2/2A \end{bmatrix},$$

and

$$H = [0 \quad 1 \quad 0].$$

We introduce new state variables

$$\bar{x}(t) = T(t)x(t),$$

where

$$T(t) = \begin{bmatrix} 0 & 0 & 1 \\ 2 & -t & 0 \\ 0 & 1 & -t \end{bmatrix},$$

$$T^{-1}(t) = \begin{bmatrix} t^2/2 & 1/2 & t/2 \\ t & 0 & 1 \\ 1 & 0 & 0 \end{bmatrix}.$$

Then

$$\bar{F}(t) = T(t)F(t)T^{-1}(t) - \dot{T}(t)T^{-1}(t) = \left[\begin{array}{c|c|c} -t^4/4A & -t^3/4A & -t^2/4A \\ \hline 0 & 0 & 1 \\ \hline 0 & 0 & 0 \end{array}\right],$$

$$\bar{G}(t) = T(t)G(t) = \left[\begin{array}{c} t^2/2A \\ \hline 0 \\ \hline 0 \end{array}\right],$$

and

$$H(t) = H(t)T^{-1}(t) = [t \,|\, 0 \,|\, 1].$$

Hence the system consists of Parts $(B - D)$, with $n_B = n_C = n_D = 1$. It is interesting that the canonical decomposition is of constant dimension, even though the system may be neither periodic nor analytic.

The preceding examples illustrate special cases of a noteworthy general relationship which exists between the canonical structure of a dynamical system and irreducible realizations of an impulse-response matrix. The main facts here are the following:

THEOREM 7. (i) *The impulse-response matrix of a linear dynamical system* (2.1–2) *depends solely on Part* (*B*) *of the system and is given explicitly by*

$$S(t, \tau) = H^B(t)\Phi^{BB}(t, \tau)G^B(\tau), \tag{5.5}$$

where Φ^{BB} *is the transition matrix corresponding to* F^{BB}.

(ii) *Any two completely controllable and completely observable realizations of S are algebraically equivalent.*

(iii) *A realization of S is irreducible if and only if at all times it consists of Part* (*B*) *alone; thus every irreducible realization of S is completely controllable and completely observable.*

Proof. The first statement can be read off by inspection from Fig. 4. The second statement is proved in [14]. The necessity of the third statement follows from Theorem 5, while the sufficiency is implied by (ii).

It is clear that Theorem 2 is a consequence of Theorems 5–7.

We can now answer the question posed in section 4 in a definite way:

THEOREM 8 (*Main Result*). *Knowledge of the impulse-response matrix* $S(t, \tau)$ *identifies the completely controllable and completely observable part, and this part alone, of the dynamical system which generated it. This part ("B"* in Theorem 5) *is itself a dynamical system and has the smallest dimension among all realizations of S. Moreover, this part is identified by S uniquely up to algebraic equivalence.*

Using different words, we may say that an impulse-response matrix is a *faithful representation* of a dynamical system (2.1–2) if and only if the latter is completely controllable and completely observable.

Remark. It is very interesting to compare this result with Theorem 4 of E. F. Moore, in one of the early papers on finite automata [26]:

"*The class of all machines which are indistinguishable from a given strongly connected machine S by any single experiment has a unique (up to isomorphism) member with a minimal number of states. This unique machine, called the reduced form of S, is strongly connected and has the property that any two of its states are distinguishable.*"

"Indistinguishable machines" in Moore's terminology correspond in ours to alternate realizations of the same input/output relation. "Strongly connected" in his terminology means completely controllable in ours. "Indistinguishable states" in our terminology corresponds to states whose difference, not zero, is an unobservable state in the sense of [3].

Evidently the two theorems are concerned with the same abstract facts, each being stated in a different mathematical framework.

6. Explicit criteria for complete controllability and observability. The canonical structure theorem is so far merely an abstract result, since we have not yet given a constructive procedure for obtaining the coördinate transformation which exhibits the system matrices in canonical form. We shall do this in section 7. The method rests on the possibility of finding explicit criteria for complete controllability and complete observability. The following lemmas, proved in [4], play a central role:

LEMMA 1. $n_A(t_0) + n_B(t_0) = \text{rank } W(t_0, t_1)$ *for* $t_1 > t_0$ *sufficiently large, where*

$$W(t_0, t_1) = \int_{t_0}^{t_1} \Phi(t_0, \tau)G(\tau)G'(\tau)\Phi'(t_0, \tau)\, d\tau \tag{6.1}$$

or

$$dW/dt_0 = F(t_0)W + WF'(t_0) - G(t_0)G'(t_0),\ W(t_1) = 0. \tag{6.2}$$

LEMMA 2. $n_C(t_0) + n_D(t_0) = \text{rank } M(t_0, t_{-1})$ *for* $t_{-1} < t_0$ *sufficiently small, where*

$$M(t_0, t_{-1}) = \int_{t_{-1}}^{t_0} \Phi'(\tau, t_0)H'(\tau)H(\tau)\Phi(\tau, t_0)\, d\tau \tag{6.3}$$

or

$$-dM/dt_0 = F'(t_0)M + MF(t_0) - H'(t_0)H(t_0),\ M(t_{-1}) = 0. \tag{6.4}$$

For constant systems, the preceding lemmas can be considerably improved [4]:

LEMMA 3. *For a constant system,*

$$n_A + n_B = \text{rank } [G, FG, \cdots, F^{n-1}G]. \tag{6.5}$$

LEMMA 4. *For a constant system,*

$$n_C + n_D = \text{rank } [H', F'H', \cdots, (F')^{n-1}H']. \tag{6.6}$$

EXAMPLE 4. For F and G defined in Example 2, the matrix (6.5) is

$$\begin{bmatrix} 3 & 3 & -3 & -3 & 3 & 3 & -3 & -3 \\ -2 & -1 & 6 & 8 & 2 & 6 & 14 & 22 \\ 0 & 3 & 12 & 18 & 12 & 24 & 36 & 60 \\ 0 & 1 & 4 & 6 & 4 & 8 & 12 & 20 \end{bmatrix} \tag{6.7}$$

The rank of this matrix is 2, which checks with the fact that $n_A = 1$ and $n_B = 1$ in Example 2.

The determination of the rank of (6.7), while elementary, is laborious. For practical purposes it might be better to compute W; for instance, by solving the differential equation (6.2).

In the constant case, there is another criterion of complete controllability which is particularly useful in theoretical investigations. The most general form of this theorem (which may be found in [14]) is complicated; we state here a simplified version which is adequate for the present purposes:

LEMMA 5. *Hypothesis: The matrix F is similar to a diagonal matrix. In other words, there is a nonsingular coördinate transformation $\bar{x} = Tx$ with the property that in the new coördinate system F has the form*

$$\bar{F} = TFT^{-1} = \begin{bmatrix} \lambda_1 I_{q_1} & & 0 \\ & \ddots & \\ 0 & & \lambda_r I_{q_r} \end{bmatrix},$$

where I_{q_i} is a $q_i \times q_i$ unit matrix,

$$\sum_{i=1}^{r} q_i = n,$$

and the matrix G has the form

$$\bar{G} = TG = \begin{bmatrix} \bar{G}^{(1)} \\ \hline \vdots \\ \hline \bar{G}^{(r)} \end{bmatrix} \begin{matrix} \} \; q_1 \text{ rows} \\ \vdots \\ \} \; q_r \text{ rows.} \end{matrix}$$

Conclusion: The system is completely controllable if and only if

$$\text{rank } \bar{G}^{(1)} = q_1, \cdots, \text{rank } \bar{G}^{(r)} = q_r. \tag{6.8}$$

We leave it to the reader to dualize this result to complete observability.

EXAMPLE 5. Consider the special case $q_1 = \cdots = q_r = 1$ of Lemma 5. The eigenvalues of F are then distinct. If condition (6.8) is satisfied, every element of the one-column matrix $\bar{G}$ is nonzero; by a trivial transformation, all of these elements can be made equal to 1, without affecting $\bar{F}$. Thus we can choose a coordinate system in which F, G have the representation:

$$\bar{F} = \begin{bmatrix} \lambda_1 & & 0 \\ & \ddots & \\ 0 & & \lambda_n \end{bmatrix} (\lambda_i = \lambda_j \Rightarrow i = j), \; \bar{G} = \begin{bmatrix} 1 \\ \vdots \\ 1 \end{bmatrix}. \tag{6.9}$$

This is the canonical form of Lur'e [18]. It is closely related to the partial-fraction expansion of transfer functions. To illustrate this, consider the 1×1 transfer-function matrix:

$$z_{11}(s) = \frac{s+2}{(s+1)(s+3)(s+4)} = \frac{1/6}{s+1} + \frac{1/2}{s+3} - \frac{2/3}{s+4}.$$

This transfer function is realized by the system:

$$F = \begin{bmatrix} -1 & 0 & 0 \\ 0 & -3 & 0 \\ 0 & 0 & -4 \end{bmatrix}, \tag{6.10}$$

$$G = \begin{bmatrix} 1 \\ 1 \\ 1 \end{bmatrix}, \tag{6.11}$$

and

$$H = [\tfrac{1}{6} \quad \tfrac{1}{2} \quad -\tfrac{2}{3}] \tag{6.12}$$

which is in the canonical form of Lur'e.

By Lemma 5, (6.10–11) is completely controllable; by the dual of Lemma 5, (6.10–12) is completely observable.

We can double-check these facts by means of Lemmas 3–4. For (6.9) the matrix (6.5) is

$$\begin{bmatrix} 1 & \lambda_1 & \lambda_1^2 & \cdots & \lambda_1^{n-1} \\ 1 & \lambda_2 & \lambda_2^2 & \cdots & \lambda_2^{n-1} \\ \vdots & \vdots & \vdots & & \vdots \\ 1 & \lambda_n & \lambda_n^2 & \cdots & \lambda_n^{n-1} \end{bmatrix}, \tag{6.13}$$

where the λ_i are the diagonal elements (= eigenvalues) of F in (6.9). But the determinant of (6.13) is the well-known Vandermonde determinant. The latter is nonzero if and only if all the λ_i are distinct, which is what we have assumed.

7. Computation of the canonical structure. We show now how to determine explicitly the change of coordinates which reduces F, G, H to the canonical form. We consider only the constant case of (2.1–2). The computations are elementary; it is not necessary to diagonalize the matrix F or even to determine its eigenvalues.

The procedure is as follows:

(a) We compute the controllability matrix $W = W(0, 1)$* given by

* It can be shown [4, Theorem 10] that in the constant case one may choose any $t_1 > t_0$ in Lemma 1.

(6.1); for instance, by solving the differential equation (6.2). Then we find a nonsingular matrix T such that

$$T'WT = E = \begin{bmatrix} I_{n_1} & 0 \\ 0 & 0 \end{bmatrix} \tag{7.1}$$

where I_{n_1} is the $n_1 \times n_1$, $0 \leqq n_1 \leqq n$, unit matrix and the 0's are zero matrices of appropriate size. Clearly $n_1 = n_A + n_B$ is the number of controllable state variables.

The matrix T defines the change of coördinates

$$x = T\bar{x}; \tag{7.2}$$

in terms of the new coordinates, the system matrices are

$$\bar{F} = T^{-1}FT, \qquad \bar{G} = T^{-1}G, \qquad \bar{H} = HT, \qquad \bar{W} = E. \tag{7.3}$$

$$\bar{x} = \begin{bmatrix} \bar{x}^1 \\ \bar{x}^2 \end{bmatrix}, F = \begin{bmatrix} \bar{F}^{11} & \bar{F}^{12} \\ 0 & \bar{F}^{22} \end{bmatrix}, \bar{G} = \begin{bmatrix} \bar{G}^1 \\ 0 \end{bmatrix}, \text{ and } \; H = [\bar{H}^1 \;\; \bar{H}^2]. \tag{7.4}$$

This decomposition is trivial (and therefore omitted) if $n_1 = n$, i.e., when the system is completely controllable.

(b) Next we consider the two subsystems defined by

$$\begin{aligned} &\bar{F}^{11}, \bar{G}^1, \text{ and } \bar{H}^1; \\ &\bar{F}^{22}, 0, \text{ and } \bar{H}^2. \end{aligned} \tag{7.5}$$

We compute the observability matrices $\bar{M}^1 = \bar{M}^1(0, 1)$ and $\bar{M}^2 = \bar{M}^2(0, 1)$ given by (6.3) for both of these subsystems. Then we determine two nonsingular matrices $\bar{U}^1$, $\bar{U}^2$ such that

$$(\bar{U}^1)'\bar{M}^1\bar{U}^1 = \bar{E}^1 = \begin{bmatrix} 0 & 0 \\ 0 & I_{n_B} \end{bmatrix}, \tag{7.6}$$

and

$$(\bar{U}^2)'\bar{M}^2\bar{U}^2 = \bar{E}^2 = \begin{bmatrix} 0 & 0 \\ 0 & I_{n_d} \end{bmatrix}. \tag{7.7}$$

These results define another change of coördinates

$$\bar{x} = \begin{bmatrix} \bar{x}^1 \\ \bar{x}^2 \end{bmatrix} = \bar{U}\hat{x} = \begin{bmatrix} \bar{U}^1 & 0 \\ 0 & \bar{U}^2 \end{bmatrix} \cdot \begin{bmatrix} \hat{x}^1 \\ \hat{x}^2 \end{bmatrix}.$$

One or the other of these transformations is superfluous if $n_B = n_1$ or $n_d = n - n_1$.

After the coördinate changes (7.2) and (7.8), we obtain the following matrices

$$
(7.9) \quad \tilde{x} = \begin{bmatrix} x^A \\ x^B \\ -- \\ x^c \\ x^d \end{bmatrix}, \qquad \tilde{F} = \tilde{U}^{-1}\tilde{F}\tilde{U} = \left[\begin{array}{cc|cc} F^{AA} & F^{AB} & F^{Ac} & F^{Ad} \\ 0 & F^{BB} & \tilde{F}^{Bc} & F^{Bd} \\ \hline 0 & 0 & \tilde{F}^{Bc} & F^{cd} \\ 0 & 0 & 0 & F^{dd} \end{array}\right],
$$

$$
\tilde{U}^{-1}G = \tilde{G} = \begin{bmatrix} G^A \\ G^B \\ -- \\ 0 \\ 0 \end{bmatrix}, \qquad \tilde{H} = \tilde{H}\tilde{U} = [0 \quad H^B \quad 0 \quad H^d],
$$

$$
\tilde{M}^1 = \tilde{E}^1, \quad \tilde{M}^2 = \tilde{E}^2.
$$

Clearly, n_B is the number of state variables which are both controllable and observable. But, in general, $n_d < n_D$ and $n_c > n_C$.

(c) It remains to transform the element $\tilde{F}^{Bc}$ into 0, if this is not already the case. (If $\tilde{F}^{Bc} = 0$, then $n_c = n_C$, $n_d = n_D$ and (7.9) has the desired canonical structure.)

We consider the subsystem

$$
(7.10) \quad \tilde{F}^* = \left[\begin{array}{c|c} F^{BB} & \tilde{F}^{Bc} \\ \hline 0 & \tilde{F}^{cc} \end{array}\right], \quad \tilde{G}^* = \begin{bmatrix} G^B \\ --- \\ 0 \end{bmatrix}, \quad \text{and} \quad \tilde{H}^* = [H^B \mid 0].
$$

The corresponding observability matrix given by (6.3) is

$$
\tilde{M}^*(0,1) = \tilde{M}^* = \left[\begin{array}{c|c} I_{n_B} & A \\ \hline A' & Q \end{array}\right], \quad (Q = Q' \text{ nonnegative definite.})
$$

(The upper left element of $\tilde{M}^*$ is I_{n_B} in view of (7.9); all we know about the other elements is their symmetry properties.) Letting

$$
\tilde{V}^* = \left[\begin{array}{c|c} I_{n_B} & -A \\ \hline 0 & I_{n_c} \end{array}\right],
$$

we find that

$$
(\tilde{V}^*)'\tilde{M}^*\tilde{V}^* = \tilde{M}^* = \left[\begin{array}{c|c} I_{n_B} & 0 \\ \hline 0 & R \end{array}\right],
$$

where $R = Q - A'A$ is a symmetric, nonnegative-definite matrix.

Now let $\tilde{V}^{**}$ be a nonsingular matrix such that

$$(\tilde{V}^{**})'\tilde{M}^{**}\tilde{V}^{**} = \left[\begin{array}{c:cc} I_{n_B} & 0 & 0 \\ \hdashline 0 & 0 & 0 \\ 0 & 0 & I_{n_e} \end{array}\right],$$

where $n_e = \text{rank } R$. Let $\tilde{V} = \tilde{V}^* \tilde{V}^{**}$. Since $\tilde{V}^*$ and $\tilde{V}^{**}$ are upper triangular relative to the partitioning in (7.10), so is $\tilde{V}$, which will take $\tilde{F}^*$ into the upper triangular form

$$\tilde{V}^{-1}\tilde{F}^*\tilde{V} = \left[\begin{array}{c|cc} F^{BB} & F^{BC} & F^{Be} \\ \hline 0 & F^{CC} & F^{Ce} \\ 0 & F^{eC} & F^{ee} \end{array}\right].$$

where $n_C = n_c - n_e$. But these transformations decompose $\tilde{F}^*$ into a completely observable and an unobservable part. Hence $F^{BC} = F^{eC} = 0$. Moreover,

$$\tilde{H}^*\tilde{V} = [H^B \mid 0]\tilde{V} = [H^B \mid 0 \quad H^e]$$

THEOREM 9. *The explicit transformation which takes the constant matrices F, G, and H into the canonical form required by Theorem (5-iii) is given by* $x \to \tilde{V}^{-1}\tilde{U}^{-1}\tilde{T}^{-1}x$. *We partition*

$$F^{Ac} = [F^{AC} \quad F^{Ae}],$$

and partition

$$F^{cd} = \begin{bmatrix} F^{Cd} \\ F^{ed} \end{bmatrix}$$

Then we define $n_D = n_d + n_e$ *and find*

$$F^{AD} = [F^{Ae} \quad F^{Ad}],$$

$$F^{BD} = [F^{Be} \quad F^{Bd}],$$

$$F^{CD} = [F^{Ce} \quad F^{Cd}],$$

$$F^{DD} = \begin{bmatrix} F^{ee} & F^{ed} \\ 0 & F^{dd} \end{bmatrix},$$

$$H^D = [H^e \quad H^d].$$

8. Construction of irreducible realizations.

Now we give an explicit procedure for the construction of an irreducible realization of a weighting-function matrix $W(t - \tau)$. In view of Theorem 7, part (iii), we can do this in two stages:

(I) We construct a realization of W, then

(II-A) we prove, using Lemmas 1–5, that the resultant system is completely controllable and completely observable, hence irreducible; or

(II-B) we carry out explicitly the canonical decomposition and remove all parts other than (B).

Instead of the weighting-function matrix W, it is usually more convenient to deal with its Laplace transform Z.

Let us consider the problem with Method A in order of increasing difficulty.

Case 1. $m = p = 1$. This is equivalent to the problem of simulating a single transfer function on an analog computer. There are several well-known solutions. They may be found in textbooks on classical servomechanism theory or analog computation.

Without loss of generality (see Theorem 4) we may consider transfer functions of the form

$$z_{11}(s) = \frac{a_n s^{n-1} + \cdots + a_1}{s^n + b_n s^{n-1} + \cdots + b_1} = \frac{N(s)}{D(s)} \tag{8.1}$$

where the $a_n, \cdots, a_1; b_n, \cdots, b_1$ are real numbers. Of course, at least one of the a_i must be different from zero. We assume also that the numerator $N(s)$ and denominator $D(s)$ of $z_{11}(s)$ have no common roots.

There are two basic realizations of (8.1). See Figs. 5–6, where the standard signal-flow-graph notation [16] is used. In either case, one verifies almost by inspection that the transfer functions relating y_1 to u_1 are indeed given by z_{11}.

In Fig. 5, the system matrices are

$$F = \begin{bmatrix} 0 & 1 & 0 & \cdots & 0 & 0 \\ 0 & 0 & 1 & \cdots & 0 & 0 \\ \vdots & \vdots & \vdots & & \vdots & \vdots \\ 0 & 0 & 0 & \cdots & 0 & 1 \\ -b_1 & -b_2 & -b_3 & \cdots & -b_{n-1} & -b_n \end{bmatrix}, \tag{8.2}$$

$$G = \begin{bmatrix} 0 \\ 0 \\ \vdots \\ 0 \\ 1 \end{bmatrix}, \tag{8.3}$$

and

$$H = [a_1 \quad a_2 \quad \cdots \quad a_{n-1} \quad a_n]. \tag{8.4}$$

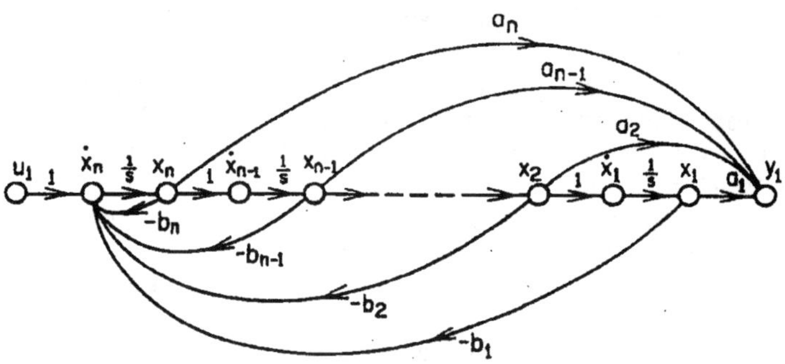

FIGURE 5.

In Fig. 6, the system matrices are.

$$(8.5)\qquad F = \begin{bmatrix} 0 & 0 & 0 & \cdots & 0 & -b_1 \\ 1 & 0 & 0 & \cdots & 0 & -b_2 \\ 0 & 1 & 0 & \cdots & 0 & -b_3 \\ \vdots & \vdots & \vdots & & \vdots & \vdots \\ 0 & 0 & 0 & \cdots & 1 & -b_n \end{bmatrix},$$

$$(8.6)\qquad G = \begin{bmatrix} a_1 \\ a_2 \\ a_3 \\ \vdots \\ a_n \end{bmatrix},$$

and

$$(8.7)\qquad H = [0 \quad 0 \quad 0 \quad \cdots \quad 1].$$

It is very easy to check by means of (6.5) and (6.6) that the system (8.2, 3) is completely controllable and (8.5, 7) is completely observable.

However, if we attempt to check the controllability of (8.5, 6) by means of (6.5) we get a matrix whose elements are complicated products of the coefficients of $N(s)$ and $D(s)$. To prove that the determinant of this matrix does not vanish, we have only one fact at our disposal: the assumption that $N(s)$ and $D(s)$ have no common roots. Guided by this observation, we find that the following is true:

LEMMA 7. *Suppose F has the form* (8.5) *and G has the form* (8.6). *Then* (i) *we have the relation*

$$(8.8)\qquad K(F, G) = [G \quad FG \quad \cdots — \quad F^{n-1}G] = N(F),$$

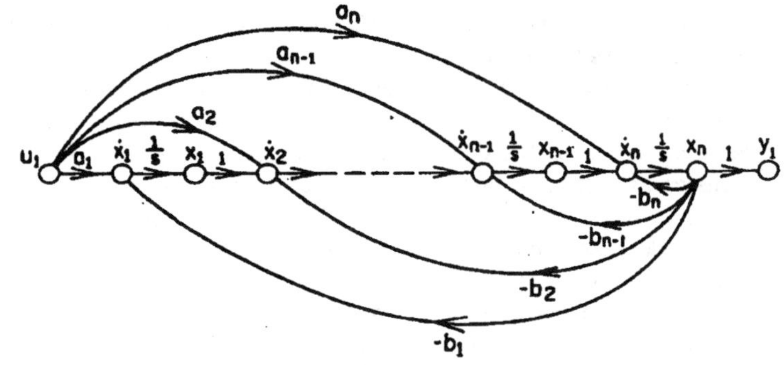

FIGURE 6.

and (ii) the polynomials $N(s)$ and $D(s)$ have no root in common if and only if $\det K(F, G) \neq 0$.

The main fact to be proved is (ii), for then the complete controllability of (8.5, 6) follows by Lemma 3. A straightforward way of establishing (ii) is to transform the standard Euler-Sylvester determinantal criterion [19, p. 84] for the nonexistence of common roots of $N(s)$ and $D(s)$ (the so-called *resolvent* of $N(s)$ and $D(s)$) into the form (8.8). This can be easily done, but the details are not very transparent. Therefore we prefer to give another

Proof. Let e_i, $i = 1, \cdots, n$, be the set of n-vectors in which the j-th component of e_i is δ_{ij}. Since F is given by (8.5), we see that $e_{i+1} = Fe_i$, $1 \leqq n - 1$, and $K(F, e_1) = [e_1, e_2, \cdots, e_n] = I$. Hence $K(F, e_i) = K(F, F^{i-1}e_1) = F^{i-1}K(F, e_1) = F^{i-1}$, when $1 \leqq i \leqq n$. Then (8.8) follows by linearity.

Let $\lambda_i[A]$, $i = 1, \cdots, n$ denote the eigenvalues (not necessarily distinct) of a square matrix A. Then

$$\det K(F, G) = \prod_{i=1}^{n} \lambda_i[N(F)] = \prod_{i=1} N(\lambda_i[F]),$$

where the second equality follows from (8.8) by a well-known identity in matrix theory. Thus $\det K(F, G) = 0$ if and only if $N(\lambda_i[F]) = 0$ for some i; that is, when an eigenvalue of F is a root of $N(\lambda)$. Since the eigenvalues of F are roots of $D(\lambda)$, this proves (ii).*

It is interesting that (8.8) provides a new representation for the resolvent, which is preferable in some respects to the Euler-Sylvester determinant. The latter is a $2n \times 2n$ determinant, whereas $\det K(F, G)$ is $n \times n$.

The complete observability of (8.2, 4) is proved similarly.

The systems given by (8.2–4) and (8.5–7) are duals of one another in

* The present proof of Lemma 6 was suggested by Drs. John C. Stuelpnagel and W. M. Wonham of RIAS.

the sense defined by (5.3). Fig. 6 is a reflection of Fig. 5 about the vertical axis, with all arrows reversed.

A third type of realization in common use is obtained from the partial-fraction expansion of $z_{11}(s)$ (see Example 5). Note, however, that this requires factorization of the denominator of $z_{11}(s)$, whereas the preceding realizations can be written down by inspection, using only the coefficients of $z_{11}(s)$.

These considerations may be summarized as the following result, which is a highly useful fact in control theory:

THEOREM 10. *Consider a linear constant dynamical system with $m = p = 1$, which is completely controllable and completely observable. Then one may always choose a basis in the state space so that F, G, H have the form* (8.2–4) *or (with respect to a different basis)* (8.5–7).

Proof. Let (8.1) be the transfer-function matrix of the given dynamical system. By Theorem 8, the given system is an irreducible realization of (8.1). So are the systems specified by (8.2–4) and (8.5–7). By Theorem (7-ii), all three systems are algebraically equivalent and by constancy (Theorem 3) they are even strictly equivalent.

Extensions of this theorem may be found in [14]. For an interesting application to the construction of Lyapunov functions, see [25].

The procedure described here may be generalized to the non-constant case. Assuming the factorization (4.3) of $S(t, \tau)$ is known (with $m = p = 1$), Batkov [20] shows how to determine the coefficients of the differential equation

$$
\begin{aligned}
(8.9)\qquad d^n y_1/dt^n + b_n(t)^{n-1}y_1/dt^{n-1} + \cdots + b_1(t)y_1& \\
= a_n(t)d^{n-1}u_1/dt^{n-1} + \cdots + a_1(t)u_1 .&
\end{aligned}
$$

Laning and Battin [21, p. 191–2] show how one converts (8.9) into a system of first-order differential equations (2.1) with variable coefficients. We shall leave to the reader the proof of the irreducibility of the realization so obtained.

Case 2-a. $m = 1$, $p > 1$. We have a single-input/multi-output system. We can realize $Z(s)$, without factoring the denominators of its transfer functions, by the following generalization of the procedure given by Fig. 5 and (8.2–4).

First, we find the smallest common denominator of the elements of $Z(s)$. (This can be done, of course, without factorization.) $Z(s)$ assumes the form

$$
Z(s) = \frac{1}{s^n + b_n s^{n-1} + \cdots + b_1}\begin{bmatrix} a_{1n} s^{n-1} + \cdots + a_{11} \\ \cdots \\ a_{pn} s^{n-1} + \cdots + a_{p1}\end{bmatrix}.
$$

Then the following dynamical system provides an irreducible realization of $Z(s)$: F and G are as in (8.2–3), while H given by (8.4) is generalized to

$$H = \begin{bmatrix} a_{11} & \cdots & a_{1n} \\ \vdots & & \vdots \\ a_{p1} & \cdots & a_{pn} \end{bmatrix}.$$

Complete controllability is trivial; complete observability is established by a straightforward generalization of Lemma 6.

In this case we form p linear functions of the state, rather than merely one as in Fig. 5.

Case 2-b. $m > 1$, $p = 1$. We can realize this multi-input/single-output system analogously to Case 2-a by generalizing the procedure given by Fig. 6 and (8.5–8.7). Let us write the elements of $Z(s)$ in terms of their smallest common denominator:

$$Z(s) = \left[\frac{a_{n1}s^{n-1} + \cdots a_{11}}{s^n + b_n s^{n-1} + \cdots + b_1} \cdots \frac{a_{nm}s^{n-1} + \cdots + a_{1m}}{s^n + b_n s^{n-1} + \cdots + b_1}\right].$$

Then the desired irreducible realization consists of F and H as defined by (8.5–6), while

$$G = \begin{bmatrix} a_{11} & \cdots & a_{1m} \\ \vdots & & \vdots \\ a_{n1} & \cdots & a_{nm} \end{bmatrix}.$$

This case is the dual of Case 2-a.

Even in Case 2, it is impractical to give a general formula which expresses the coefficients of F, G, and H in terms of the coefficients of the transfer functions in $Z(s)$ if the denominators are not all the same. When we pass to the general case, determination of F, G, and H often requires extensive numerical computation.

Case 3. m, p arbitrary. Here Method (A) is very complicated if any transfer function in $Z(s)$ has multiple poles [14]. In most practical applications, however, such complications are of no interest. Ruling them out, E. G. Gilbert gave an elegant and relatively simple solution [5].

Let $s_1, \cdots, s_q$ be distinct complex numbers corresponding to the poles of all the elements of $Z(s)$. Assume that all poles are simple. Then

$$R(k) = \lim_{s \to s_k} (s - s_k)Z(s), \; k = 1, \cdots, q$$

is the k-th residue matrix of $Z(s)$. If $s_l = \bar{s}_k$, then $R(s_l) = \bar{R}(s_k)$, where the bar denotes the complex conjugate. In terms of the residue matrices, the weighting-function matrix $W(t)$ corresponding to $Z(s)$ has the explicit form

$$W(t) = \mathcal{L}^{-1}[Z(s)] = \sum_{k=1}^{q} R(k)e^{s_k t}.$$

We have then:

THEOREM 11. (*Gilbert*). *Hypotheses: No element of the transfer-function matrix* $Z(s)$ *has multiple poles.* $Z(s)$ *has a total of* q *distinct poles* $s_1, \cdots, s_q$, *with corresponding residue matrices* $R(1), \cdots, R(q)$.

Conclusions: (i) *The dimension of irreducible realizations of* $Z(s)$ *is*

$$n = \sum_{k=1}^{q} r_k, \text{ where } r_k = \operatorname{rank} R(k). \tag{8.11}$$

(ii) *Write*

$$R(k) = H(k)G(k), \qquad k = 1, \cdots, q, \tag{8.12}$$

where $H(k)$ *is a* $p \times r_k$ *matrix and* $G(k)$ *is an* $r_k \times m$ *matrix, both of rank* r_k. *Then* $Z(s)$ *has the irreducible realization*

$$F = \begin{bmatrix} s_1 I_{r_k} & & 0 \\ & \ddots & \\ 0 & & s_q I_{r_q} \end{bmatrix}, \qquad (I_r = r \times r \text{ unit matrix}), \tag{8.13}$$

$$G = \begin{bmatrix} G(1) \\ \vdots \\ G(q) \end{bmatrix}, \tag{8.14}$$

and

$$H = [H(1) \quad \cdots \quad H(q)]. \tag{8.15}$$

Proof. This is one of the main results in [5]. With the aid of machinery developed here, we can give a shorter (though more abstract) demonstration. The factorization (8.12) is well known in linear algebra. We give in the Appendix various explicit formulae (which are easily machine-computable) for $G(k)$ and $H(k)$. Applying Lemma 5 shows that the dynamical system defined by (8.13–15) is completely controllable and completely observable. Hence it is irreducible, which implies formula (8.11). By elementary changes of variables, (8.13–15) can be transformed into matrices which have only real elements.

A serious disadvantage of Method (A), as expressed by Theorem 11, is that the denominators of the transfer functions in $Z(s)$ must be factored in order to determine the poles. This is not easily done numerically. Moreover, the residue matrices $R(k)$ corresponding to complex poles are complex, which makes the factorization (8.11) more complicated (see Appendix).

Now we turn to Method (B). This method does not require computation of eigenvalues, and it is not bothered by multiple poles. This is a decided advantage in numerical calculations. On the other hand, the method is not convenient for simple illustrative examples. Nor is it possible to display the elements of F, G, and H as simple functions of the coefficients in $z_{ij}(s)$.

An easy way of realizing $Z(s)$ (without guaranteeing irreducibility) is the following. Let α_i be the number of distinct poles (counting each pole with its maximum multiplicity) in the i-th row of $Z(s)$, and let β_i be the number of poles in the i-th column. Then the maximum number n_0 of state variables required to realize $Z(s)$ by repeatedly using the scheme given under Case 2-a or 2-b is

$$n_0 = \min\left\{ \sum_{i=1}^{p} \alpha_i , \quad \sum_{j=1}^{m} \beta_j \right\}.$$

As before, we can determine the α_i and β_j without factoring the transfer functions of $Z(s)$. There is in general no simple way in this method to determine the dimension $n \leqq n_0$ of irreducible realizations without performing the computations outlined in Section 7.

The two methods are best compared via an example. This example must be of fairly high order, since we wish to provide accurate numerical checks.

EXAMPLE 6. Consider the transfer-function matrix

$$Z(s) = \begin{bmatrix} \dfrac{3(s+3)(s+5)}{(s+1)(s+2)(s+4)} & \dfrac{6(s+1)}{(s+2)(s+4)} & \dfrac{2s+7}{(s+3)(s+4)} & \dfrac{2s+5}{(s+2)(s+3)} \\[2ex] \dfrac{2}{(s+3)(s+5)} & \dfrac{1}{(s+3)} & \dfrac{2(s-5)}{(s+1)(s+2)(s+3)} & \dfrac{8(s+2)}{(s+1)(s+3)(s+5)} \\[2ex] \dfrac{2(s^2+7s+18)}{(s+1)(s+3)(s+5)} & -\dfrac{2s}{(s+1)(s+3)} & \dfrac{1}{(s+3)} & \dfrac{2(5s^2+27s+34)}{(s+1)(s+3)(s+5)} \end{bmatrix}$$

Applying Method (A) first, we find that the residue matrices are:

$$R(1) = \begin{bmatrix} 8 & 0 & 0 & 0 \\ 0 & 0 & 4 & 1 \\ 3 & 1 & 0 & 3 \end{bmatrix}; \qquad r_1 = 3.$$

$$R(2) = \begin{bmatrix} -4.5 & -3 & 0 & 1 \\ 0 & 0 & -6 & 0 \\ 0 & 0 & 0 & 0 \end{bmatrix}; \qquad r_2 = 2.$$

$$R(3) = \begin{bmatrix} 0 & 0 & 1 & 1 \\ 1 & 1 & 2 & 2 \\ -3 & -3 & 1 & 1 \end{bmatrix}; \qquad r_3 = 2.$$

$$R(4) = \begin{bmatrix} -0.5 & 9 & 1 & 0 \\ 0 & 0 & 0 & 0 \\ 0 & 0 & 0 & 0 \end{bmatrix}; \qquad r_4 = 1.$$

$$R(5) = \begin{bmatrix} 0 & 0 & 0 & 0 \\ -1 & 0 & 0 & -3 \\ 2 & 0 & 0 & 6 \end{bmatrix}; \qquad r_5 = 1.$$

Thus $n = 9$.

Employing the procedure given in the Appendix, we find the following factors for matrices $R(k)$ (the products are accurate up to four places beyond the decimal point):

$$H(1) = \begin{bmatrix} 8.0000 & 0.0000 & 0.0000 \\ 0.0000 & 4.1231 & 0.0000 \\ 3.0000 & 0.7276 & 3.0774 \end{bmatrix},$$

$$G(1) = \begin{bmatrix} 1.0000 & 0.0000 & 0.0000 & 0.0000 \\ 0.0000 & 0.0000 & 0.9701 & 0.2425 \\ 0.0000 & 0.3249 & -0.2294 & 0.9175 \end{bmatrix};$$

$$H(2) = \begin{bmatrix} 5.5000 & 0.0000 \\ 0.0000 & 6.0000 \\ 0.0000 & 0.0000 \end{bmatrix},$$

$$G(2) = \begin{bmatrix} -0.8182 & -0.5455 & 0.0000 & 0.1818 \\ 0.0000 & 0.0000 & -1.0000 & 0.0000 \end{bmatrix};$$

$$H(3) = \begin{bmatrix} 1.3416 & 0.4472 \\ 3.1305 & -0.4472 \\ 0.0000 & 4.4721 \end{bmatrix},$$

$$G(3) = \begin{bmatrix} 0.2236 & 0.2236 & 0.6708 & 0.6708 \\ -0.6708 & -0.6708 & 0.2236 & 0.2236 \end{bmatrix};$$

$$H(4) = \begin{bmatrix} 9.0692 \\ 0.0000 \\ 0.0000 \end{bmatrix}, \qquad G(4) = [-0.0551 \quad 0.9924 \quad 0.1103 \quad 0.0000];$$

$$H(5) = \begin{bmatrix} 0.0000 \\ -3.1623 \\ 6.3246 \end{bmatrix}, \qquad G(5) = [0.3162 \quad 0.0000 \quad 0.0000 \quad 0.9487].$$

Using these numerical results, we find that the dynamical equations of the irreducible realization are given by

$$F = \begin{bmatrix} 1 & 0 & 0 & 0 & 0 & 0 & 0 & 0 & 0 \\ 0 & 1 & 0 & 0 & 0 & 0 & 0 & 0 & 0 \\ 0 & 0 & 1 & 0 & 0 & 0 & 0 & 0 & 0 \\ 0 & 0 & 0 & 2 & 0 & 0 & 0 & 0 & 0 \\ 0 & 0 & 0 & 0 & 2 & 0 & 0 & 0 & 0 \\ 0 & 0 & 0 & 0 & 0 & 3 & 0 & 0 & 0 \\ 0 & 0 & 0 & 0 & 0 & 0 & 3 & 0 & 0 \\ 0 & 0 & 0 & 0 & 0 & 0 & 0 & 4 & 0 \\ 0 & 0 & 0 & 0 & 0 & 0 & 0 & 0 & 5 \end{bmatrix},$$

$$G = \begin{bmatrix} 1.0000 & 0.0000 & 0.0000 & 0.0000 \\ 0.0000 & 0.0000 & 0.9701 & 0.2425 \\ 0.0000 & 0.3249 & -0.2294 & 0.9175 \\ -0.8182 & -0.5455 & 0.0000 & 0.1818 \\ 0.0000 & 0.0000 & -1.0000 & 0.0000 \\ 0.2236 & 0.2236 & 0.6708 & 0.6708 \\ -0.6708 & 0.6708 & 0.2236 & 0.2236 \\ -0.0551 & 0.9924 & 0.1103 & 0.0000 \\ 0.3162 & 0.0000 & 0.0000 & 0.9487 \end{bmatrix},$$

$$H = \begin{bmatrix} 8.0000 & 0.0000 & 0.0000 & 5.5000 & 0.0000 & 1.3416 & 0.4472 & 0.0692 & 0.0000 \\ 0.0000 & 4.1231 & 0.0000 & 0.0000 & 6.0000 & 3.1305 & -0.4472 & 0.0000 & -3.1623 \\ 3.0000 & 0.7276 & 3.0774 & 0.0000 & 0.0000 & 0.0000 & 4.4721 & 0.0000 & 6.3246 \end{bmatrix}.$$

Now we apply Method (B). First of all we note that $\alpha_1 = \alpha_2 = 4$, $\alpha_3 = 3$, while $\beta_1 = 5$, $\beta_2 = \beta_3 = 4$ (see p. 181). Hence it is best to choose for the preliminary realization three structures of the type discussed under Case 2-b. This will require $n_0 = p(\alpha_1 + \alpha_2 + \alpha_3) = 11$ dimensions.

Next, we find the least common denominator of the rows of $Z(s)$. See Fig. 7.

$$Z(s) = \begin{bmatrix} \dfrac{3(s^3 + 11s^2 + 39s + 45)}{s^4 + 10s^3 + 35s^2 + 50s + 24} & \dfrac{6(s^3 + 5s^2 + 7s + 3)}{\cdots} & \dfrac{2s^3 + 13s^2 + 25s + 14}{\cdots} & \dfrac{2s^3 + 15s^2 + 33s + 20}{\cdots} \\ \dfrac{2(s^2 + 3s + 2)}{s^4 + 11s^3 + 41s^2 + 61s + 30} & \dfrac{s^3 + 8s^2 + 17s + 10}{\cdots} & \dfrac{2(s^2 + 10s + 25)}{\cdots} & \dfrac{8(s^2 + 4s + 4)}{\cdots} \\ \dfrac{2(s^2 + 7s + 18)}{s^3 + 9s^2 + 23s + 15} & \dfrac{-2(s^2 + 5s)}{\cdots} & \dfrac{s^2 + 6s + 5}{\cdots} & \dfrac{2(5s^2 + 27s + 34)}{\cdots} \end{bmatrix}$$

FIGURE 7.

The desired realization of $Z(s)$ can be read off by inspection from Fig. 7, using (8.5) and (8.6):

$$F = \left[\begin{array}{cccc|cccc|ccc} 0 & 0 & 0 & -24 & & & & & & & \\ 1 & 0 & 0 & -50 & & 0 & & & & 0 & \\ 0 & 1 & 0 & -35 & & & & & & & \\ 0 & 0 & 1 & -10 & & & & & & & \\ \hline & & & & 0 & 0 & 0 & -30 & & & \\ & 0 & & & 1 & 0 & 0 & -61 & & 0 & \\ & & & & 0 & 1 & 0 & -41 & & & \\ & & & & 0 & 0 & 1 & -11 & & & \\ \hline & & & & & & & & 0 & 0 & -15 \\ & 0 & & & & 0 & & & 1 & 0 & -23 \\ & & & & & & & & 0 & 1 & -9 \end{array}\right],$$

$$
G = \begin{bmatrix}
135 & 18 & 14 & 20 \\
117 & 42 & 25 & 33 \\
33 & 30 & 13 & 15 \\
3 & 6 & 2 & 2 \\
\hdashline
4 & 10 & 50 & 32 \\
6 & 17 & 20 & 32 \\
2 & 8 & 2 & 8 \\
0 & 1 & 0 & 0 \\
\hdashline
36 & 0 & 5 & 68 \\
14 & -10 & 6 & 54 \\
2 & -2 & 1 & 10
\end{bmatrix},
$$

and

$$
H = \begin{bmatrix}
0 & 0 & 0 & 1 & 0 & 0 & 0 & 0 & 0 & 0 & 0 \\
\hdashline
0 & 0 & 0 & 0 & 0 & 0 & 0 & 1 & 0 & 0 & 0 \\
\hdashline
0 & 0 & 0 & 0 & 0 & 0 & 0 & 0 & 0 & 0 & 1
\end{bmatrix}.
$$

By virtue of its construction, this system is completely observable but we cannot tell by inspection whether or not it is completely controllable. (From the results obtained above with Method (A), we know that the system is not completely controllable since $11 = n_0 > n = 9$.) Therefore the canonical decomposition may contain Parts (B) and (D).

To see what the dimensions of these parts are, we compute numerically the decomposition of the system into completely controllable and uncontrollable parts according to the method described in Section 8. These calculations involve only the matrices F and G, but the resulting transformations must be applied also to the matrix H.

$$
\hat{F} = \left[\begin{array}{ccccccccc:cc}
-0.3346 & 0.1182 & 0.0139 & -0.0299 & 0.0097 & -0.0001 & -0.0663 & -0.0113 & 0.0000 & 0.0000 & 0.8334 \\
0.2455 & -0.2025 & -0.0115 & 0.0268 & -0.0101 & 0.0000 & 0.0734 & 0.0120 & 0.0001 & 0.0000 & 0.7189 \\
-0.8333 & -0.3850 & -0.1023 & 1.0535 & -0.2230 & -0.0005 & -0.9998 & 0.0237 & -0.0120 & 21.5463 & -9.9652 \\
0.2943 & 0.3032 & 0.0361 & 0.0022 & -0.0610 & -0.0044 & 0.1773 & -0.0569 & 0.0154 & -1.6475 & 0.8726 \\
-0.8896 & 0.8321 & 0.1838 & 0.4999 & -0.4287 & -0.0275 & 1.1199 & 0.0024 & 0.0089 & -1.1882 & 2.5562 \\
0.2477 & 1.3097 & 2.0439 & -0.3777 & -0.9685 & -0.4965 & -0.4046 & 0.3657 & 0.0199 & -462.2221 & -73.0068 \\
-0.1358 & 0.0429 & 0.0009 & -0.0321 & 0.0325 & 0.0016 & -0.3252 & -0.0152 & -0.0010 & 0.0000 & 0.4698 \\
1.1634 & -0.0290 & 0.1114 & 1.1645 & -0.4233 & -0.0196 & 0.4057 & -0.0480 & -0.0040 & -25.8717 & -2.1911 \\
-0.2787 & -0.6854 & 2.4604 & 2.4604 & -1.5802 & -0.2801 & 0.8992 & 0.8863 & -0.2649 & -710.5771 & -52.0706 \\
\hdashline
0.0000 & 0.0000 & 0.0000 & 0.0000 & 0.0000 & 0.0000 & 0.0000 & 0.0000 & 0.0000 & -0.5\text{[illegible]}2 & -0.0003 \\
0.0000 & 0.0000 & 0.0000 & 0.0000 & 0.0000 & 0.0000 & 0.0000 & 0.0000 & 0.0000 & -0.0401 & -0.2993
\end{array}\right] \times 10
$$

FIGURE 8.

$$\hat{G} = \left[\begin{array}{cccc} 1.2012 & -0.7886 & -0.3459 & -0.5320 \\ 1.6822 & 0.6039 & 0.3594 & 0.4745 \\ -1.7962 & 0.8640 & 2.5387 & -1.7927 \\ -0.1683 & -0.2420 & 2.2981 & 0.4039 \\ -0.0663 & 0.9596 & 1.4137 & 2.7344 \\ 1.0412 & 1.2734 & 1.5800 & -2.9382 \\ -0.1117 & -0.3946 & -0.0977 & 1.5554 \\ 1.3130 & -2.8589 & 1.9252 & -0.4989 \\ -2.8756 & -1.7637 & 0.9281 & -2.5137 \\ \hdashline 0.0000 & 0.0000 & 0.0000 & 0.0000 \\ 0.0000 & 0.0000 & 0.0000 & 0.0000 \end{array}\right] \times 10$$

$$\hat{H} = \left[\begin{array}{ccccccccc:cc} -0.2928 & 0.4076 & 0.0160 & -0.0373 & 0.0141 & 0.0000 & -0.0981 & -0.0167 & -0.0001 & 0.0000 & 1.0000 \\ -0.0286 & 0.0023 & -0.0116 & -0.0169 & 0.0373 & 0.0022 & -0.0830 & -0.0033 & -0.0007 & 1.0000 & 0.0000 \\ -0.0599 & 0.2173 & 0.0161 & 0.0354 & -0.0603 & -0.0039 & 0.6736 & 0.0178 & 0.0024 & 0.0000 & 0.0000 \end{array}\right]$$

FIGURE 9.

The final results may be seen in Figs. 8–9, which give the matrices $\hat{F}$, $\hat{G}$, and $\hat{H}$. Elements in the lower left-hand corner of $\hat{F}$ should be exactly zero. In fact, they are zero to at least the number of digits indicated in Fig. 8.

To check the accuracy of these two irreducible realizations of the transfer function matrix on p. 181, we have computed the corresponding weighting-function matrices $W^{(1)}(t)$ and $W^{(2)}(t)$. The equality $W^{(1)}(t) = W^{(2)}(t)$ was found to be correct to at least four significant digits.

9. Other applications to system theory. The literature of system theory contains many instances of errors, incomplete or misleading solutions of problems, etc., which can be traced to a lack of understanding of the issues discussed in this paper. This section presents some cases of this known to the writer; other examples may be found in the paper of Gilbert [5].

Analog computers. According to Theorem 8, a linear dynamical system (2.1–2) is a "faithful" realization of an impulse-response matrix if and only if it is irreducible. Suppose the dynamical equations (2.1–2) are programmed on an analog computer. (See [8].) Then it is clear from Theorem 8 that *the computer will simulate the impulse-response matrix correctly if and only if a minimal number of integrators are used.* Otherwise the system programmed on the analog computer will have, besides Part (B), at least one of the Parts (A), (C), or (D). Since the impulse-response matrix determines Part (B), and that alone, the nature of the redundant parts will depend not on the impulse-response matrix but on the particular method used to obtain the dynamical equations. It should be borne in mind that the canonical decomposition is an abstract thing; usually it is not possible to identify the redundant integrators without a change of variables.

The writer is not aware of any book or paper on analog computation where this is explicitly pointed out. But the facts of life seem to be well known (intuitively) to practitioners of the analog art.

That redundancy in the number of integrators used *can* cause positive harm is quite clear from the canonical structure theorem.

EXAMPLE 7. Let the simulated system consist of Parts (A) and (B) and suppose that Part (A) is unstable. Because of noise in the computer, Part (A) will be subject to perturbations; they will be magnified more and more, because of the instability. As long as assumptions of linearity hold exactly, the unstable (A) component of the state vector will not be noticed, but soon the computer will cease to function because its linear range will be exceeded.

Lur'e canonical form. In his book on the Lur'e problem, Letov implies [18; equation (2.4) and (2.23)] that every vector system

$$dx/dt = Fx + g\cdot\sigma \qquad (\sigma = \text{scalar}) \tag{9.1}$$

can be reduced to the canonical form

$$dx_i/dt = \lambda_i x_i + \sigma, \qquad i = 1, \cdots, n \tag{9.2}$$

whenever the eigenvalues λ_i of F are distinct. Since (9.2) is completely controllable, this assertion, if true, would imply that (9.1) is also completely controllable, which is false. In fact, the system defined by

$$F = \begin{bmatrix} \lambda & 0 \\ 0 & \mu \end{bmatrix}, \qquad g = \begin{bmatrix} 1 \\ 1 \end{bmatrix} \tag{9.3}$$

is obviously not equivalent to

$$F = \begin{bmatrix} \lambda & 0 \\ 0 & \mu \end{bmatrix}, \qquad g = \begin{bmatrix} 1 \\ 0 \end{bmatrix} \tag{9.4}$$

whenever $\lambda \neq \mu$.

In examining the derivation originally given by Lur'e for his canonical form [27; Chapter 1, §2–3], it is clear that the last step before equation (3.5) is valid if and only if det $[H_k(\lambda_p)] \neq 0$ (in the notation of Lur'e [27].) It is easy to show that this condition is equivalent to complete controllability, whenever the eigenvalues of F are distinct. Unfortunately, the condition det $[H_k(\lambda_p)] \neq 0$ was not emphasized explicitly by Lur'e [28] in the original publications.

We may thus conclude that *when F has distinct eigenvalues and there is a* *single control variable, the Lur'e-Letov canonical form exists if and only if the pair $\{F, g\}$ is completely controllable.*

It is interesting to note that (9.3) can be transformed into (9.4) when $\lambda = \mu$; in other words, when the eigenvalues are not distinct the Lur'e canonical form may exist even if the system is not completely controllable.

Cancellations in the transfer-function. When a mathematical model is derived from physical principles, the equations of the system are in or near the form (2.1–2). Regrettably, it has become widespread practice in system engineering to dispense with differential equations and to replace them by transfer functions $Z(s)$. Later, $Z(s)$ must be converted back into the form (2.1–2) for purposes of analog computation. In the process of algebraic manipulations, some transfer functions may have (exactly or very nearly) common factors in the numerator and denominator, which are then *canceled.* This is an indication that a part of the dynamics of the system is not represented by the transfer function.

Such cancellations are the basic idea of some elementary design methods in control theory. These methods do not bring the system under better control but merely "decouple" some of the undesirable dynamics. But then the closed-loop transfer function is no longer a faithful representation of the (closed-loop) dynamics. Stability difficulties may arise. Similar criticisms may be leveled against the large, but superficial, literature on "noninteracting" control system design.

EXAMPLE 8. Consider the system defined by the matrices

$$F = \begin{bmatrix} 0 & 1 & 0 \\ 5 & 0 & 2 \\ -2 & 0 & -2 \end{bmatrix}, \quad G = \begin{bmatrix} 0 \\ 0 \\ 0.5 \end{bmatrix}, \quad H = [-2 \quad 1 \quad 0]. \tag{9.5}$$

The transfer function relating y_1 to u_1 is the sum of two terms:

$$\begin{aligned} \frac{y_1(s)}{u_1(s)} &= -2\frac{x_1(s)}{u_1(s)} + \frac{x_2(s)}{u_1(s)} \\ &= -\frac{2}{s^3 + 2s^2 - 5s - 6} + \frac{s}{s^3 + 2s^2 - 5s - 6} \\ &= \frac{(s-2)}{(s+1)(s-2)(s+3)} = \frac{1}{(s+1)(s+3)}. \end{aligned} \tag{9.6}$$

Thus, by cancellation, the transfer function is reduced from the third to the second order. The system has an unstable "natural mode" (corresponding to $s_3 = 2$) about which the transfer functions gives no information.

Using (6.5) we see that the system (9.5) is completely controllable. By Theorem 5, the system cannot be completely observable: $n_B = 2$ from (9.6) and Case 1, section 8. The canonical structure consists of Parts (A) and (B). In canonical coördinates the system matrices can be taken as

$$\bar{F} = \begin{bmatrix} -1 & 0 & 0 \\ 0 & -3 & 0 \\ 0 & 0 & 2 \end{bmatrix}, \qquad \bar{G} = \begin{bmatrix} 1 \\ 1 \\ 1 \end{bmatrix}, \qquad \bar{H} = [0.5 \quad -0.5 \quad 0].$$

We can easily calculate the change of coördinates

$$\bar{x} = Tx$$

by the method of partial fractions discussed in [8]. First we find T^{-1}, then T. The results are

$$T^{-1} = \frac{1}{30}\begin{bmatrix} -5 & 3 & 2 \\ 5 & -9 & 4 \\ 10 & 6 & -1 \end{bmatrix}, \qquad T = \begin{bmatrix} -1 & 1 & 2 \\ 3 & -1 & 2 \\ 8 & 4 & 2 \end{bmatrix}.$$

Loss of controllability and observability due to sampling. Consider a single-input/single-output constant linear system. Suppose the output is observed only at the instants $t = kT$ (k = integer, $T > 0$), and that the input is constant over the intervals $kT \leqq t < (k + 1)T$. This situation is commonly called "sampling"; it arises when a digital computer is used in control or data processing. T is the *sampling period.* We can regard such a setup as a discrete-time dynamical system. We define here Θ (Axiom (D_1)) as the set of integers and replace (2.1) by a difference equation. All theorems carry over to this situation with small modifications.

The analysis of discrete-time systems by conventional techniques requires the computation of the so-called z-transform of $Z(s)$ [22]. The analysis using z-transforms then proceeds in close analogy with analysis based on Laplace transforms.

A constant linear system which is completely controllable and completely observable will retain these properties even after the introduction of sampling if and only if [4]

$$\text{(9.7)} \qquad \text{Re } s_i = \text{Re } s_j \quad \text{implies} \quad \text{Im } (s_i - s_j) \neq q\pi/T$$

where $i, j = 1, \cdots, n$ and q = positive integer.

If this condition is violated (the sampling process "resonates" with the system dynamics) then cancellations will take place in the z-transform. The z-transform will then no longer afford a faithful representation of the system, so that if (9.7) *is violated, results based on formal manipulation of z-transforms may be invalid.*

This point is not at all clear in the literature. True, Barker [23] has drawn attention to a related phenomenon and called it "hidden oscillation." The textbooks, however, dismiss the problem without providing real insight [22, §5-3; 24, §2.13].

A practical difficulty arises from the fact that near the "resonance" point given by (9.7) it is hard to identify the dynamical equations accurately from the z-transform. Small numerical errors in the computation of the z-transform may have a large effect on the parameters of the dynamical equations.

REFERENCES

[1] R. E. Kalman, *Discussion of paper by I. Flügge-Lotz*, Proc. 1st International Conference on Automatic Control, Moscow, 1960; Butterworths, London, 1961, Vol. 1, pp. 396–7.

[2] R. E. Kalman, *Canonical structure of linear dynamical systems*, Proc. Nat. Acad. Sci. USA, 48 (1962), pp. 596–600.

[3] R. E. Kalman, *On the general theory of control systems*, Proc. 1st International Congress on Automatic Control, Moscow, 1960; Butterworths, London, 1961, Vol. 1, pp. 481–492.

[4] R. E. Kalman, Y. C. Ho, and K. S. Narendra, *Controllability of linear dynamical systems*, (to appear in Contributions to Differential Equations, Vol. 1, John Wiley, New York.)

[5] E. G. Gilbert, *Controllability and observability in multivariable control systems*, J. Soc. Indust. Appl. Math. Ser. A: On Control, Vol. 1, No. 2 (1963), pp. 128–151.

[6] V. V. Nemitskii and V. V. Stepanov, *Qualitative Theory Of Differential Equations*, Princeton Univ. Press, Princeton, 1960.

[7] R. E. Kalman and J. E. Bertram, *Control system analysis and design via the 'second method' of Lyapunov*, J. Basic Engr. (Trans. A.S.M.E.), 82 D (1960), pp. 371–393.

[8] R. E. Kalman, *Analysis and design principles of second and higher-order saturating servomechanisms*, Trans. Amer. Inst. Elect. Engrs., 74, II (1955), pp. 294–310.

[9] W. Hahn, *Theorie und Anwendung der direkten Methode von Ljapunov*, Springer, Berlin, 1959.

[10] J. P. LaSalle and S. Lefschetz, *Stability By Lyapunov's Direct Method*, Academic Press, New York, 1961.

[11] L. Markus, *Continuous matrices and the stability of differential systems*, Math. Z., 62 (1955), pp. 310–319.

[12] L. A. Zadeh, *A general theory of linear signal transmission systems*, J. Franklin Inst., 253 (1952), pp. 293–312.

[13] D. Middleton, *An Introduction To Statistical Communication Theory*, McGraw-Hill, New York, 1960.

[14] R. E. Kalman, *On controllability, observability, and identifiability of linear dynamical systems*, (to appear).

[15] R. E. Kalman, *On the stability of time-varying linear systems*, Trans. I.R.E. Prof. Gr. Circuit Theory, (CT-9 (1962), pp. 420–422.).

[16] S. J. Mason, *Feedback theory: some properties of signal flow graphs*, Proc. I.R.E., 41 (1953), pp. 1144–56; *Further properties of signal flow graphs*, ibid., 44 (1956), pp. 920–926.

[17] R. E. Kalman, *New results in filtering and prediction theory*, RIAS Report 61-1, Research Institute for Advanced Studies (RIAS), Baltimore, 1961.

[18] A. M. Letov, *Stability In Nonlinear Control Systems*, Princeton Univ. Press, Princeton, 1961.

[19] B. L. VAN DER WAERDEN, *Modern Algebra*, Vol. 1, 2nd Ed., Ungar, New York, 1949.
[20] A. M. BATKOV, *On the problem of synthesis of linear dynamic systems with two parameters*, Avtomat. i Telemeh., 19 (1958), pp. 49–54.
[21] J. H. LANING, JR. AND R. H. BATTIN, *Random Processes In Automatic Control*, McGraw-Hill, New York, 1956.
[22] J. R. RAGAZZINI AND G. F. FRANKLIN, *Sampled-Data Control Systems*, McGraw-Hill, New York, 1958.
[23] R. H. BARKER, *The pulse transfer function and its application to sampling servo systems*, Proc. Inst. Elec. Engrs. 99 IV (1952), pp. 302–317.
[24] E. I. JURY, *Sampled-Data Control Systems*, John Wiley, New York, 1957.
[25] R. E. KALMAN, *Lyapunov functions for the problem of Lur'e in automatic control*, Proc. Nat. Acad. Sci. USA, 49, (1963), pp. 201–205.
[26] E. F. MOORE, *Gedanken-experiments on sequential machines*, *Automata Studies*, Princeton Univ. Press, Princeton, 1956.
[27] A. I. LUR'E, *Certain Nonlinear Problems in the Theory of Automatic Control*. (in Russian), Gostekhizdat, Moscow, 1951; German translation Akademie-Verlág, Berlin, 1957.
[28] Private communication, Academician A. I. Lur'e.

APPENDIX

Factorization of rectangular matrices. Given an arbitrary, real, $p \times m$ matrix R of rank $q \leqq \min(m, p)$. We wish to find a $p \times q$ matrix H and a $q \times m$ matrix G, both of rank q, such that $R = HG$. The existence of H and G follows almost immediately from the definition of rank. We describe below a constructive procedure for determining H and G numerically from numerical values of R.

Let $p \leqq m$. Form the $p \times p$ matrix $S = RR'$.

As is well known, there exists a nonsingular matrix T such that

$$TRR'T' = TST' = E, \tag{A-1}$$

where precisely q diagonal elements of E are 1, all other elements are 0. T can be calculated by steps similar to the gaussian elimination procedure.

Compute the generalized inverse $R^{\#}$ (in the sense of Penrose [4]) of R. $R^{\#}$ is an $m \times p$ matrix.

Using the properties of $R^{\#}$ ([4]) we obtain

$$R = RR^{\#}R = RR'R^{\#\prime} = SR^{\#\prime} = T^{-1}ET^{-1\prime}R^{\#\prime} = (T^{-1}E)(T^{-1}E)'R^{\#\prime}. \tag{A-2}$$

Now $T^{-1}E$ is a matrix which contains precisely $p - q$ zero columns. Deleting these columns, we obtain a $p \times q$ matrix $(T^{-1}E)^0 = H$. Similarly, deleting $p - q$ zero rows from $(T^{-1}E)'R^{\#\prime} = (R^{\#}T^{-1}E)'$ we obtain a $m \times q$ matrix $G' = (R^{\#}T^{-1}E)^0$. Evidently $R = HG$. Since the ranks of H and G are obviously less than or equal to q, both ranks must be exactly q for otherwise *rank* $R < q$, contrary to hypothesis.

Alternately, let T, U be nonsingular matrices such that

$$TRU = E;$$

then

$$R = (T^{-1}E)^0(EU^{-1})^0 \tag{A-3}$$

is the desired decomposition. However, the computation of (A-3) may require more steps than that of (A-2).

Suppose now that R is complex. Then $S = R\bar{R}' = RR^* = A + iB$ is complex hermitian; it corresponds to the $2n \times 2n$ nonnegative matrix

$$\Sigma = \begin{bmatrix} A & B \\ -B & A \end{bmatrix} \tag{A-4}$$

where $A = A'$ and $B = -B'$. In fact, if $z = x + iy$, the hermitian form z^*RR^*z (which is real-valued) is equal to the quadratic form

$$[x \quad y] \begin{bmatrix} A & B \\ -B & A \end{bmatrix} \begin{bmatrix} x \\ y \end{bmatrix}$$

As is well known, there exists a nonsingular *complex* matrix T such that $TST^* = E$. If $T = U + iV$, it follows further that

$$\begin{bmatrix} U & V \\ -V & U \end{bmatrix} \begin{bmatrix} A & B \\ -B & A \end{bmatrix} \begin{bmatrix} U' & -V' \\ V' & U' \end{bmatrix} = \begin{bmatrix} E & 0 \\ 0 & E \end{bmatrix}.$$

Hence the determination of the complex $n \times n$ matrix has been reduced to the determination of a real $2p \times 2p$ matrix. Similar remarks apply to the calculation of $R^{\#}$. Thus the problem of factoring complex $p \times m$ matrices can be embedded in the problem of factoring real $2p \times 2m$ matrices.

On the Input-Output Stability of Time-Varying Nonlinear Feedback Systems

G. ZAMES

IN the early sixties, a number of specific results regarding stability of feedback systems emerged. Especially important instances of such results are the Popov criterion (discussed elsewhere in this volume), and the circle criterion. The circle criterion has its roots in the work of Bongiorno [1] and Tsypkin [10], and was developed almost simultaneously by a number of authors. Popov's stability criterion and the circle criterion seem to be rather specific results with special purpose proofs, but they led researchers to discover interesting general underlying principles, and to develop a mathematical framework that allowed them to analyze stability of feedback systems.

Three such general and intuitively very pleasing stability principles are formulated in the first part of the article by Zames: the small loop gain theorem, the positive operator theorem, and the conic sector theorem. Intuitively, the small loop theorem states that if the open loop of a feedback system attenuates signals, then the feedback system is stable. Intuitively, the positive operator theorem states that if a feedback system can be viewed as the interconnection of two passive systems, then the feedback system dissipates, and will hence be stable. The conic operator theorem provides a way of unifying and generalizing both the small gain and the positive operator theorem. These principles are stated in the summary of Part 1, and the paper itself contains a formalization and very clear proof of these results.

These results were cast in terms of the properties of input/output relations on *extended spaces*. The concept of an extended space provided a rich and useful alternative to the Lyapunov methods which at the time were a more commonly used approach to stability. Extended spaces were introduced and formalized in these articles by Zames, with roots going back to his earlier work [14], [15]. They also occur in the work of Sandberg [8], [9]. Extended spaces allow one to analyze situations in which signals are not a priori assumed to belong to a normed space, but are a posteriori proven to belong to this normed space. The fact that all signals in a feedback loop belong to a normed space is what is called input/output stability. Often, the normed space may be $L_2[0, \infty]$, and the associated extended space $L_{2e}[0, \infty]$ is the space of signals whose time-truncations are square integrable. This technique of approaching stability is not unlike what is called Picard iteration in the theory of differential equations.

In the second part of the paper, these general stability principles are used to obtain proofs of the circle criterion and the Popov criterion.

The powerful general feedback stability principles discussed in Part I were further developed in a number of monographs [12], [7] and textbooks [5], [11] and led to a wide variety of stability criteria, often based on the use of multipliers, an instance of which is already used in Part II of the paper. Multipliers that are more flexible than the one used in the Popov criterion were used before in [2], and became a very effective technique for obtaining stability criteria. Many examples of such criteria are collected in the reprint volume [6]. This volume also contains an extensive bibliography related to the small gain theorem, the circle criterion, etc.

The circle criterion as stated in Part II of the paper is a beautiful generalization of the Nyquist criterion (as a stability criterion) to systems with one time-varying nonlinear element. It is worth pointing out, however, that the circle criterion for time-varying systems can sometimes give quite conservative results. For example (see [3]), it can be shown that the system $\frac{d^2}{dt^2}y + 2\frac{d}{dt}y + k(t)y = 0$ is asymptotically stable whenever k is in the sector $0 < k(t) < 11.6$, whereas the circle criterion only predicts asymptotic stability for $0 < k(t) < 4$. The circle criterion has an instability converse, proven by Brockett and Lee [4] in a Lyapunov setting. The combination of the stability and the instability parts of the circle criterion makes for a full generalization of the Nyquist criterion. The question of how to generalize the instability part of the circle criterion in an input/output stability setting proved to be non-trivial, and was carried out in [13].

These input/output stability principles had a deep impact on the development of control theory in the decades that followed. This influence was perhaps most striking in the work on robust control. Indeed, the small loop gain theorem led to the insight that in order to obtain an adequate robustness margin, the controller should be designed so as to moderate the $\mathcal{L}_2$-induced gain (the H_∞-norm) of the linear time invariant part of the system.

The small loop gain theorem was therefore instrumental for the development of H_∞ optimal control. The original paper [16] where the H_∞-control problem was first formulated will be discussed elsewhere in this volume.

References

[1] J. J. Bongiorno, Jr., "An extension of the Nyquist-Barkhausen stability criterion to linear lumped-parameter systems with time-varying elements," *IEEE Trans. Aut. Contr.*, **AC-8**:167–170, 1963.

[2] R. W. Brockett and J. L. Willems, "Frequency domain stability criteria, Parts I and II," *IEEE Trans. Aut. Contr.*, **AC-10**:255–261 & 407–413, 1965.

[3] R. W. Brockett, "The status of stability theory for deterministic systems," *IEEE Trans. Aut. Contr.*, **AC-11**:596–606, 1966.

[4] R. W. Brockett and H. B. Lee, "Instability criteria for time-varying and nonlinear systems," *IEEE Proceedings*, **55**:604–619, 1967.

[5] C. A. Desoer and M. Vidyasagar, *Feedback Systems: Input-Output Properties*, Academic Press (New York), 1975.

[6] A. G. J. MacFarlane, *Frequency-Response Methods in Control Systems*, IEEE Press (Piscataway, NJ), 1979.

[7] M. G. Safonov, *Stability and Robustness of Multivariable Feedback Systems*, The MIT press (Cambridge, MA), 1980.

[8] I. W. Sandberg, "On the properties of some systems that distort signals," *Bell Sys. Tech. J.*, **42**:2033, 1963, and **43**:91–112, 1964.

[9] I. W. Sandberg, "On the L_2-boundedness of solutions of nonlinear functional equations," *Bell Sys. Tech. J.*, **43**:1581–1599, 1964.

[10] Y. Z. Tsypkin, "On the stability in the large of nonlinear sampled-data systems," *Dokl. Akad. Nauk.*, **145**:52–55, 1962.

[11] M. Vidyasagar, *Nonlinear Systems Analysis*, Prentice Hall (Englewood Cliffs, NJ), 1978.

[12] J. C. Willems, *The Analysis of Feedback Systems*, The MIT Press (Cambridge, MA), 1971.

[13] J. C. Willems, "Stability, instability, invertibility and causality," *SIAM J. Contr.*, **7**:645–671, 1969.

[14] G. Zames, *Conservation of Bandwidth in Nonlinear Operations*, Report MIT-RLE No. 55, 1959.

[15] G. Zames, *Nonlinear Operators for System Analysis*, Report MIT-RLE No. 370, 1960.

[16] G. Zames, "Feedback and optimal sensitivity: Model reference transformations, multiplicative semi-norms and approximate inverses," *IEEE Trans. Aut. Contr.*, **AC-26**:301–320, 1981.

J.C.W

On the Input-Output Stability of Time-Varying Nonlinear Feedback Systems Part I: Conditions Derived Using Concepts of Loop Gain, Conicity, and Positivity

G. ZAMES, MEMBER, IEEE

Abstract—The object of this paper is to outline a stability theory for input-output problems using functional methods. More particularly, the aim is to derive open loop conditions for the boundedness and continuity of feedback systems, without, at the beginning, placing restrictions on linearity or time invariance.

It will be recalled that, in the special case of a linear time invariant feedback system, stability can be assessed using Nyquist's criterion; roughly speaking, stability depends on the amounts by which signals are amplified and delayed in flowing around the loop. An attempt is made here to show that similar considerations govern the behavior of feedback systems in general—that stability of nonlinear time-varying feedback systems can often be assessed from certain gross features of input-output behavior, which are related to amplification and delay.

This paper is divided into two parts: Part I contains general theorems, free of restrictions on linearity or time invariance; Part II, which will appear in a later issue, contains applications to a loop with one nonlinear element. There are three main results in Part I, which follow the introduction of concepts of *gain, conicity, positivity,* and *strong positivity:*

THEOREM 1: If the open loop gain is less than one, then the closed loop is bounded.

THEOREM 2: If the open loop can be factored into two, suitably proportioned, conic relations, then the closed loop is bounded.

THEOREM 3: If the open loop can be factored into two positive relations, one of which is strongly positive and has finite gain, then the closed loop is bounded.

Results analogous to Theorems 1–3, but with boundedness replaced by continuity, are also obtained.

I. Introduction

FEEDBACK, broadly speaking, affects a system in one of two opposing ways: depending on circumstances it is either degenerative or regenerative—either stabilizing or destabilizing. In trying to gain some perspective on the qualitative behavior of feedback systems we might ask: What are the kinds of feedback that are stabilizing? What kinds lead to a stable system? Can some of the effects of feedback on stability be described without assuming a very specific system representation?

Part I of this paper is devoted to the system of Fig. 1, which consists of two elements in a feedback loop.[1] This simple configuration is a model for many controllers, amplifiers, and modulators; its range of application will be extended to include multi-element and distributed systems, by allowing the system variables to be multi-dimensional or infinite-dimensional.

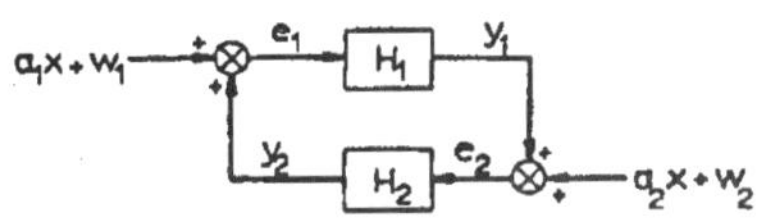

Fig. 1. A feedback loop with two elements.

The traditional approach to stability involves Lyapunov's method; here it is proposed to take a different course, and to stress the relation between input-output behavior and stability. An *input-output* system is one in which a function of time, called the output, is required to track another function of time, called the input; more generally the output might be required to track some function of the input. In order to behave properly an input-output system must usually have two properties:

1) Bounded inputs must produce bounded outputs—i.e., the system must be nonexplosive.

2) Outputs must not be critically sensitive to small changes in inputs—changes such as those caused by noise.

Manuscript received December 29, 1964; revised October 1, 1965; February 2, 1966. This work was carried out at the M.I.T. Electronic Systems Laboratory in part under support extended by NASA under Contract NsG-496 with the Center for Space Research. Parts of this paper were presented at the 1964 National Electronics Conference, Chicago, Ill., and at the 1964 International Conference on Microwaves, Circuit Theory, and Information Theory, Tokyo, Japan.

The author is with the Department of Electrical Engineering, Massachusetts Institute of Technology, Cambridge, Mass.

[1] The system of Fig. 1 has a *single* input x, multiplied by constants a_1 and a_2, and added in at *two* points. This arrangement has been chosen because it is symmetrical and thus convenient for analysis; it also remains invariant under some of the transformations that will be needed. Of course, a single input loop can be obtained by setting a_1 or a_2 to zero. The terms w_1 and w_2 are fixed bias functions, which will be used to account for the effects of initial conditions. The variables e_1, e_2, y_1, and y_2 are outputs.

Reprinted from *IEEE Transactions on Automatic Control*, Vol. AC-11, 1966, pp. 228–238, 465–476.

These two properties will form the basis of the definition of stability presented in this paper. It is desired to find conditions on the elements H_1 and H_2 (in Fig. 1) which will ensure that the overall loop will remain stable after H_1 and H_2 are interconnected. It is customary to refer to H_1 and H_2 prior to interconnection as the "open-loop" elements, and to the interconnected structure as the "closed loop." The problem to be considered here can therefore be described as seeking *open-loop conditions for closed-loop stability.*

Although the problem at hand is posed as a feedback problem, it can equally well be interpreted as a problem in networks; it will be found, for example, that the equations of the system of Fig. 1 have the same form as those of the circuit of Fig. 2, which consists of two elements in series with a voltage source, and in parallel with a current source.[2]

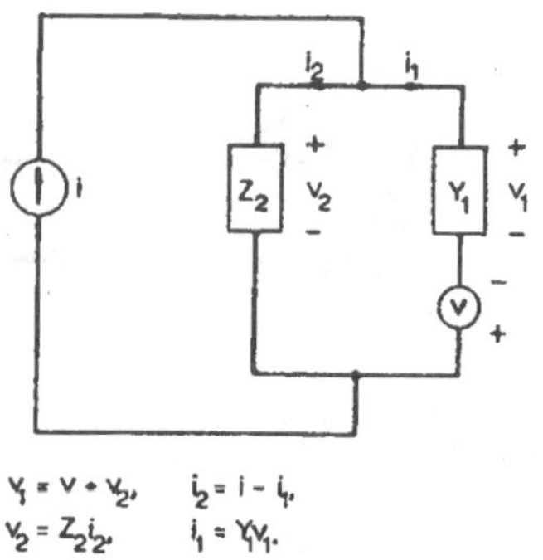

Fig. 2. A circuit equivalent to the loop of Fig. 1.

1.1 Historical Note

The problem of Lyapunov stability has a substantial history with which the names of Lur'e, Malkin, Yakubowitch, Kalman, and many others, are associated. On the other hand, functional methods for stability received less attention until relatively recently, although some versions of the well-known Popov [1] theorem might be considered as fitting into this category.

The present paper has its origin in studies [2a, b] of nonlinear distortion in bandlimited feedback loops, in which contraction methods were used to prove the existence and stability of an inversion scheme. The author's application of contraction methods to more general stability problems was inspired in part by conversations with Narendra during 1962–1963; using Lyapunov's method, Narendra and Goldwyn [3] later obtained a result similar to the circle condition of Part II of this paper.

The key results of this paper, in a somewhat different formulation, were first presented in 1964 [2d, e]. Many of these results are paralleled in the work of Brockett and Willems [4], who use Lyapunov based methods. Several others have obtained similar or related results by functional methods: Sandberg [5a] extended the nonlinear distortion theory mentioned above; later [5b] he obtained a stability theorem similar to Theorem 1 of this paper. Kudrewicz [6] has obtained circle conditions by fixed point methods. Contraction methods for incrementally positive operators have been developed by Zarantonello [7], Kolodner [8], Minty [9], and Browder [10]. A stability condition for linear time-varying systems has been described by Bongiorno [11].

2. Formulation of the Problem

There are several preliminaries to settle, namely, to specify a system model, to define stability, and to write feedback equations. What is a suitable mathematical model of a feedback element? A "black box" point of view towards defining a model will be taken. That is to say, only input-output behavior, which is a purely external property, will be considered; details of internal structure which underlie this behavior will be omitted. Accordingly, throughout Part I, a feedback element will be represented by an abstract relation, which can be interpreted as a mapping from a space of input functions into a space of output functions. More concrete representations, involving convolution integrals, characteristic graphs, etc., will be considered in Part II.

Some of the elementary notions of functional analysis will be used, though limitations of space prevent an introduction to this subject.[3] Among the concepts which will be needed and used freely are those of an abstract relation, a normed linear space, an inner product space, and the L_p spaces.

The practice of omitting the quantifier "for all" shall be utilized. For example, the statement

$$\text{“}-x^2 \leq x^2 (x \in X)\text{”}$$

is to be read:

$$\text{“for all } x \in X,\ -x^2 \leq x^2.\text{”}$$

CONVENTION: *Any expression containing a condition of the type "$x \in X$," free of quantifiers, holds* for all $x \in X$.

2.1 The Extended Normed Linear Space X_e

In order to specify what is meant by a system, a suitable space of input and output functions will first be defined.[4] Since unstable systems will be involved, this space must contain functions which "explode," i.e., which grow without bound as time increases [for example, the exponential exp (t)]. Such functions are not contained in the spaces commonly used in analysis, for example, in the L_p spaces. Therefore it is necessary to

[2] It is assumed that the source voltage v and the source current i are inputs, with $v = a_1 x + w_1$ and $i = a_2 x + w_2$; the currents and voltages in the two elements are outputs.

[3] A good reference is Kolmogorov and Fomin [12].

[4] The space of input functions will equal the space of output functions.

construct a special space, which will be called X_e. X_e, will contain both "well-behaved" and "exploding" functions, which will be distinguished from each other by assigning finite norms to the former and infinite norms to the latter. X_e will be an extension, or enlargement, of an associated normed linear space X in the following sense. Each finite-time truncation of every function in X_e will lie in X; in other words, the restriction of $x \in X_e$ to a finite time interval, say to $[0, t]$, will have a finite norm—but this norm may grow without limit as $t \to \infty$.

First a time interval T and a range of input or output values V will be fixed.

DEFINITION: *T is a given subinterval of the reals, of the type $[t_0, \infty)$ or $(-\infty, \infty)$. V is a given linear space.*

[For example, in the analysis of multielement (or distributed) networks, V is multidimensional (or infinite-dimensional).] Second, the notion of a truncated function is introduced.

NOTATION: *Let x be any function mapping T into V, that is, $x: T \to V$; let t be any point in T; then the symbol x_t denotes the* truncated function, *$x_t: T \to V$, which assumes the values $x_t(\tau) = x(\tau)$ for $\tau < t$ and $x_t(\tau) = 0$ elsewhere.*

(A truncated function is shown in Fig. 3.) Next, the space X is defined.

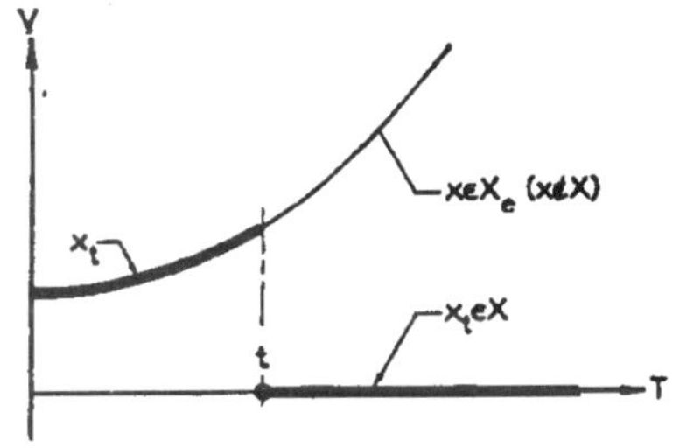

Fig. 3. A truncated function.

DEFINITION: *X is a space consisting of functions of the type $x: T \to V$; the following assumptions are made concerning X:*

(1) *X is a normed linear space; the norm of $x \in X$ is denoted by $\|x\|$.*
(2) *If $x \in X$ then $x_t \in X$ for all $t \in T$.*
(3) *If $x: T \to V$, and if $x_t \in X$ for all $t \in T$, then:*
   (a) *$\|x_t\|$ is a nondecreasing function of $t \in T$.*
   (b) *If $\lim_{t\to\infty} \|x_t\|$ exists, then $x \in X$ and the limit equals $\|x\|$.*

(For example, it can be verified that assumptions (1)–(3) are satisfied by the L_p spaces.) Next, X_e is defined.

DEFINITION: *The* extension *of X, denoted by X_e, is the space consisting of those functions whose truncations lie in X, that is, $X_e = \{x \mid x: T \to V$, and $x_t \in X$ for all $t \in T\}$.* (NOTE: X_e is a linear space.) *An* extended norm, *denoted $\|x\|_e$, is assigned to each $x \in X_e$ as follows: $\|x\|_e = \|x\|$ if $x \in X$, and $\|x\|_e = \infty$ if $x \notin X$.*

The point of assumptions (2)–(3) on X can now be appreciated; these assumptions make it possible to determine whether or not an element $x \in X_e$ has a finite norm, by observing whether or not $\lim_{t\to\infty} \|x_t\|$ exists. For example:

EXAMPLE 1: Let $L_2[0, \infty)$ be the normed linear space consisting of those real-valued functions x on $[0, \infty)$ for which the integral $\int_0^\infty x^2(t)dt$ exists, and let this integral equal $\|x\|^2$. Let $X = L_2[0, \infty)$, and let $L_{2e} = X_e$; that is, L_{2e} is the extension of $L_2[0, \infty)$. Let x be the function on $[0, \infty)$ given by $x(t) = \exp(t)$. Is $\|x\|_e$ finite, that is, is x in X? No, because $\|x_t\|$ grows without limit as $t \to \infty$, or in other words, $\|x\|_e = \infty$.

2.2 Input-Output Relations

The mathematical model of an input-output system will be a relation on X_e:

DEFINITION: *A* relation *H on X_e is any subset of the product space $X_e \times X_e$. If (x, y) is any pair belonging to H then y will be said to be H-*related *to x; y will also be said to be an* image *of x under H.*[5]

In other words, a relation is a set of pairs of functions in X_e. It will be convenient to refer to the first element in any pair as an input, and to the second element as an output, even though the reverse interpretation is sometimes more appropriate. A relation can also be thought of as a mapping, which maps some (not necessarily all) inputs into outputs. In general, a relation is multivalued; i.e., a single input can be mapped into many outputs.

The concept of state, which is essential to Lyapunov's method, will not be used here. This does not mean that initial conditions cannot be considered. One way of accounting for various initial conditions is to represent a system by a multi-valued relation, in which each input is paired with many outputs, one output per initial condition. Another possibility is to introduce a separate relation for each initial condition.

Note that the restrictions placed on X_e tend to limit, a priori, the class of allowable systems. In particular, the requirement that truncated outputs have finite norms means, roughly speaking, that only systems having infinite "escape times," i.e., systems which do not blow up in finite time, shall be considered.

Some additional nomenclature follows:

DEFINITION: *If H is a relation on X_e, then the* domain *of H denoted $Do(H)$, and the* range *of H denoted $Ra(H)$, are the sets,*

$$Do(H) = \{x \mid x \in X_e, \text{ and there exists } y \in X_e \text{ such that } (x, y) \in H\}$$

$$Ra(H) = \{y \mid y \in X_e, \text{ and there exists } x \in X_e \text{ such that } (x, y) \in H\}$$

[5] In general x can have *many* images.

NOTATION: *If H is a relation on X_e, and if x is a given element of X_e, then the symbol Hx denotes an* image *of x under H.*[5]

The idea here is to use a special symbol for an element instead of indicating that the element belongs to a certain set. For example, the statement, "there exists Hx having property P" is shorthand for "there exists $y \in \text{Ra}(H)$, such that y is an image of x, and y has property P."[6]

Observing that Hx is, according to the definitions used here, a function on T, the following symbol for the value of Hx at time t is adopted:

NOTATION: *The symbol $Hx(t)$ denotes the value assumed by the function Hx at time $t \in T$.*

Occasionally a special type of relation, called an operator, will be used:

DEFINITION: *An* operator *H is a relation on X_e which satisfies two conditions: 1) $Do(H) = X_e$. 2) H is single-valued; that is, if x, y, and z are elements of X_e, and if y and z are images of x under H, then $y = z$.*

2.3 The Class $\mathcal{R}$

DEFINITION: *$\mathcal{R}$ is the class of those relations H on X_e having the property that the zero element, denoted o, lies in $Do(H)$, and $Ho = o$.*

The assumption that H maps zero into zero simplifies many derivations; if this condition is not met at the outset, it can be obtained by adding a compensating bias to the feedback equations.

If H and K are relations in $\mathcal{R}$, and c is a real constant, then the *sum* $(H+K)$, the *product* cH, and the *composition product* KH of K *following* H, are defined in the usual way,[7] and are relations in $\mathcal{R}$. The *inverse of H* in $\mathcal{R}$, denoted by H^{-1}, *always exists*. The *identity operator* on X_e is denoted by I.

2.4 Input-Output Stability

The term "stable" has been used in a variety of ways, to indicate that a system is somehow well behaved. A system shall be called stable if it is well behaved in two respects: (1) It is bounded, i.e., not explosive. (2) it is continuous, i.e., not critically sensitive to noise.

DEFINITION: *A* subset *Y of X_e is* bounded *if there exists $A > 0$ such that, for all $y \in Y$, $\|y\|_e < A$. A* relation *H on X_e is* bounded[8] *if the image under H of every bounded subset of X_e is a bounded subset of X_e.*

DEFINITION: *A relation H on X_e is* continuous *if H has the following property: Given any $x \in X$ (that is, $\|x\|_e < \infty$), and any $\Delta > 0$, there exists $\delta > 0$ such that, for all $y \in X$, if $\|x - y\| < \delta$ then $\|Hx - Hy\| < \Delta$.*

DEFINITION: *A relation H on X_e is* input-output stable *if H is bounded and continuous.*

2.5 Feedback Equations

Although negative feedback loops will be of interest, the positive feedback configuration of Fig. 1 has been chosen because it is symmetrical.[1] The equations describing this system, to be known as the FEEDBACK EQUATIONS, are:

$$e_1 = w_1 + a_1 x + y_2 \tag{1a}$$

$$e_2 = w_2 + a_2 x + y_1 \tag{1b}$$

$$y_2 = H_2 e_2 \tag{2a}$$

$$y_1 = H_1 e_1 \tag{2b}$$

in which it is assumed that:

H_1 and H_2 are relations in $\mathcal{R}$
a_1 and a_2 are real constants
w_1 and w_2 are fixed biases in X
x in X_e is an input
e_1 and e_2 in X_e are (error) outputs
y_1 and y_2 in X_e are outputs.

(The biases are used to compensate for nonzero zero-input responses and, in particular, for the effects of initial conditions.) The *closed-loop relations* E_1, E_2, F_1, and F_2, are now defined as follows.

DEFINITION: *E_1 is the relation that relates e_1 to x or, more precisely, $E_1 = \{(x, e_1) \mid (x, e_1) \in X_e \times X_e$, and there exist e_2, y_1, y_2, $H_1 e_1$, and $H_2 e_2$, all in X_e, such that* (1) *and* (2) *are satisfied.$\}$ Similarly E_2 relates e_2 to x; F_1 relates y_1 to x; F_2 relates y_2 to x.*

All the prerequisites are now assembled for defining the problem of interest which is: *Find conditions on H_1 and H_2 which ensure that E_1, E_2, F_1, and F_2 are bounded or stable.* In general it will be enough to be concerned with E_1 and E_2 only, and to neglect F_1 and F_2, since every $F_2 x$ is related to some $E_1 x$ by the equation $F_2 x = E_1 x - a_1 x - w_1$, so that F_2 is bounded (or stable) whenever E_1 is, and similarly for F_1 vs. E_2.

It should be noted that by posing the feedback problem in terms of relations (rather than in terms of operators) all questions of existence and uniqueness of solutions are avoided. For the results to be practically significant, it must usually be known from some other source[9] that solutions exist and are unique (and have infinite "escape times").

[6] In keeping with the usual convention used here, any statement containing Hx free of quantifiers holds *for all x* in $\text{Ra}(H)$. For example, "$Hx > 1$ $(x \in X_e)$" means that "for all x in X_e, and for all Hx in $\text{Ra}(H)$, $Hx > 1$."

[7] In particular, $\text{Do}(H+K) = \text{Do}(H) \cap \text{Do}(K)$. Note that $\mathcal{R}$ is not a linear space; for example, if $\text{Do}(H) \neq \text{Do}(K)$ then $\text{Do}[(H+K) - K] \neq \text{Do}(H)$.

[8] This definition implies that inputs of finite norm produce outputs of finite norm. More than that, it implies that the sort of situation is avoided in which a bounded sequence of inputs, say $\|x_n\| < 1$ where $n = 1, 2, \cdots$, produces a sequence of outputs having norms that are finite but increasing without limit, say $\|Hx_n\| = n$.

[9] Existence and stability can frequently be deduced from entirely separate assumptions. For example, existence can often be deduced, by iteration methods, solely from the fact that (loosely speaking) the open loop delays signals; stability can not. (The connection between existence and generalized delay is discussed in G. Zames, "Realizability conditions for nonlinear feedback systems," *IEEE Trans. on Circuit Theory*, vol. CF-11, pp. 186–194, June 1964.)

3. Small Loop Gain Conditions

To secure a foothold on this problem a simple situation is sought in which it seems likely, on intuitive grounds, that the feedback system will be stable. Such a situation occurs when the open loop attenuates all signals. This intuitive idea will be formalized in Theorem 1; in later sections, a more comprehensive theory will be derived from Theorem 1.

To express this idea, a measure of attenuation, i.e., a notion of gain, is needed.

3.1 Gains

Gain will be measured in terms of the ratio of the norm of a truncated output to the norm of the related, truncated input.

DEFINITION: *The* gain *of a relation H in $\mathcal{R}$, denoted by $g(H)$, is*

$$g(H) = \sup \frac{\|(Hx)_t\|}{\|x_t\|} \tag{3}$$

where the supremum is taken over all x in $Do(H)$, all Hx in $Ra(H)$, and all t in T for which $x_t \neq 0$.

In other words, the supremum is taken over all possible input-output pairs, and over *all possible truncations*. The reason for using truncated (rather than whole) functions is that the norms of truncated functions are known to be finite *a priori*.

It can be verified that gains have all the properties of norms. In addition, if H and K belong to $\mathcal{R}$ then $g(KH) \leq g(K)g(H)$. Gains also satisfy the following inequalities:

$$\|(Hx)_t\| \leq g(H)\cdot\|x_t\| \qquad [x \in Do(H);\, t \in T] \tag{4}$$

$$\|Hx\|_e \leq g(H)\cdot\|x\|_e \qquad [x \in \mathrm{Do}(H)] \tag{5}$$

where (4) is implied by (3), and (5) is derived from (4) by taking the limit as $t \to \infty$.

If $g(H) < \infty$ then (5) implies that H is bounded. In fact, conditions for boundedness will be derived using the notion of gain and inequalities such as (5). In a similar way, conditions for continuity will be derived using the notion of incremental gain, which is defined as follows:

DEFINITION: *The* incremental gain *of any H in $\mathcal{R}$, denoted by $\tilde{g}(H)$, is*

$$\tilde{g}(H) = \sup \frac{\|(Hx)_t - (Hy)_t\|}{\|x_t - y_t\|} \tag{6}$$

where the supremum is taken over all x and y in $Do(H)$, all Hx and Hy in $Ra(H)$, and all t in T for which $x_t \neq y_t$.

Incremental gains have all the properties of norms, and satisfy the inequalities

$$\tilde{g}(KH) \leq \tilde{g}(K)\cdot\tilde{g}(H) \tag{7}$$

$$\|(Hx)_t - (Hy)_t\| \leq \tilde{g}(H)\cdot\|x_t - y_t\| \qquad [x, y \in \mathrm{Do}(H);\, t \in T] \tag{8}$$

$$\|Hx - Hy\|_e \leq \tilde{g}(H)\cdot\|x - y\|_e \qquad [x, y \in \mathrm{Do}(H)]. \tag{9}$$

In the Feedback Equations (1)–(2), the product $g(H_1)\cdot g(H_2)$ will be called the *open-loop gain-product*, and similarly, $\tilde{g}(H_1)\cdot\tilde{g}(H_2)$ will be called the *incremental open-loop gain-product*.

3.2 A Stability Theorem

Consider the Feedback Equations (1)–(2).

THEOREM 1:[10] *a) If $g(H_1)\cdot g(H_2) < 1$, then the closed loop relations E_1 and E_2 are bounded. b) If $\tilde{g}(H_1)\cdot\tilde{g}(H_2) < 1$, then E_1 and E_2 are input-output stable.*

Theorem 1 is inspired by the well known Contraction Principle.[11]

PROOF OF THEOREM 1: (a) Since eqs. (1)–(2) are symmetrical in the subscripts $_1$ and $_2$, it is enough to consider E_1. This proof will consist of showing that there are positive constants a, b, and c, with the property that any pair (x, e_1) belonging to E_1 [and so being a solution of eqs. (1)–(2)], satisfies the inequality

$$\|e_1\| \leq a\|w_1\| + b\|w_2\| + c\|x\|. \tag{10}$$

It will follow that if x is confined to a bounded region, say $\|x\| \leq A$, then e_1 will also be confined to a bounded region, in this case $\|e_1\| \leq a\|w_1\| + b\|w_2\| + cA$. Thus E_1 will be bounded.

PROOF OF INEQUALITY (10): If (x, e_1) belongs to E_1 then, after truncating eqs. (1a) and (1b), and using the triangle inequality to bound their norms, the following inequalities are obtained:

$$\|e_{1t}\| \leq \|w_{1t}\| + |a_1|\cdot\|x_t\| + \|y_{2t}\| \qquad (t \in T) \tag{10a}$$

$$\|e_{2t}\| \leq \|w_{2t}\| + |a_2|\cdot\|x_t\| + \|y_{1t}\| \qquad (t \in T) \tag{10b}$$

Furthermore, applying Inequality (4) to eqs. (2), the following is obtained, for each t in T:

$$\|y_{2t}\| \leq g(H_2)\cdot\|e_{2t}\| \tag{11a}$$

$$\|y_{1t}\| \leq g(H_1)\cdot\|e_{1t}\|. \tag{11b}$$

Letting $g(H_1) \triangleq \alpha$ and $g(H_2) \triangleq \beta$, and applying (11a) to (10a) and (11b) to (10b), the following inequalities are obtained:

$$\|e_{1t}\| \leq \|w_{1t}\| + |a_1|\cdot\|x_t\| + \beta\|e_{2t}\| \qquad (t \in T) \tag{12a}$$

$$\|e_{2t}\| \leq \|w_{2t}\| + |a_2|\cdot\|x_t\| + \alpha\|e_{1t}\| \qquad (t \in T). \tag{12b}$$

Applying (12b) to $\|e_{2t}\|$ in (12a), and rearranging,

$$(1 - \alpha\beta)\|e_{1t}\| \leq \|w_{1t}\| + \beta\|w_{2t}\| + (|a_1| + \beta|a_2|)\|x_t\| \qquad (t \in T). \tag{13}$$

[10] A variation of Theorem 1 was originally presented in [2d]. A related continuity theorem was used in [2c]. An independent, related result is Sandberg's [5b].

[11] If X is a complete space, if all relations are in fact operators, and if the hypothesis of Theorem 1b holds, then the Contraction Principle implies existence and uniqueness of solutions—a matter that has been disregarded here.

Since $(1-\alpha\beta)>0$ (as $\alpha\beta<1$, by hypothesis), Inequality (13) can be divided by $(1-\alpha\beta)$; after dividing and taking the limit of both sides as $t\to\infty$, the Inequality (10) remains. Q.E.D.

(b) Let (x', e_1') and (x'', e_1'') be any two pairs belonging to $\mathbf{E}_1$. Proceeding as in Part (a) an inequality of the form $\|e_1''-e_1'\|\leq c\|x''-x'\|$ is obtained, which implies that $\mathbf{E}_1$ is continuous. Moreover, since the hypothesis of Part (b) implies the hypothesis of Part (a), $\mathbf{E}_1$ is bounded too. Therefore $\mathbf{E}_1$ is input-output stable.

EXAMPLE 2: In eqs. (1)–(2) (and in Fig. 1) let one of the two relations, say $\mathbf{H}_1$, be the identity on L_{2e}. (L_{2e} is defined in Example 1.) Let the other relation, $\mathbf{H}_2$ on L_{2e}, be given by the equation $\mathbf{H}_2x(t)=kN[x(t)]$, where $k>0$ is a constant, and N is a function whose graph is shown in Fig. 4. *For what values of k are the closed loop relations (a) bounded? (b) stable?*

(a) First the gain is calculated.

$$g(\mathbf{H}_2) = \sup\left\{\int_0^\infty N^2[x(t)]\,dt \Big/ \int_0^\infty x^2(t)\,dt\right\}^{1/2}$$

$$= k \sup_{x \text{ real}} \left|\frac{N(x)}{x}\right| = k$$

where the first sup is over $[x\in\mathrm{Do}(\mathbf{H});\ \mathbf{H}x\in\mathrm{Ra}(\mathbf{H});\ t\in T,\ x_t\neq 0]$. That is, $g(\mathbf{H})$ is k times the supremum of the absolute slopes of lines drawn from the origin to points on the graph of N. Here $g(\mathbf{H})=k$, so Theorem 1 implies *boundedness for $k<1$*. This example is trivial in at least one respect, namely, in that H has no memory; examples with memory will be given in Part II.

(*b*) $\tilde{g}(\mathbf{H})$ can be worked out to be k times the supremum of the absolute *Lipschitzian* slopes of N, that is, $\tilde{g}(\mathbf{H})=k\ \sup_{x,y\ \mathrm{real}}|N(x)-N(y)/x-y|=2k$. *The closed loop is therefore stable for $k<1/2$.*

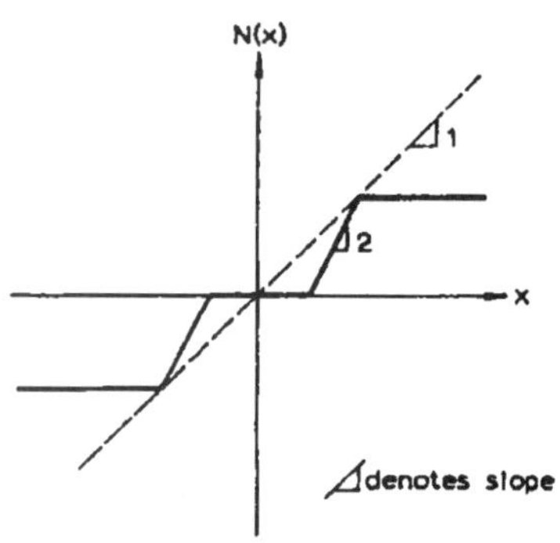

Fig. 4. Graph of the relation in Example 2.

4. Conditions Involving Conic Relations

The usefulness of Theorem 1 is limited by the condition that the open-loop gain-product be less than one—a condition seldom met in practice. However, a reduced gain product can often be obtained by transforming the feedback equations. For example, if $c\mathbf{I}$ is added to and subtracted from $\mathbf{H}_2$, as shown in Fig. 5, then e_2 remains unaffected; however, $\mathbf{H}_1$ is changed into a new relation $\mathbf{H}_1'$, as in effect $-c\mathbf{I}$ appears in feedback around $\mathbf{H}_1$. Under what conditions does this transformation give a gain product less than one? It will appear that a sufficient condition is that the input-output relations of the open loop elements be confined to certain "conic" regions in the product space $X_e\times X_e$.

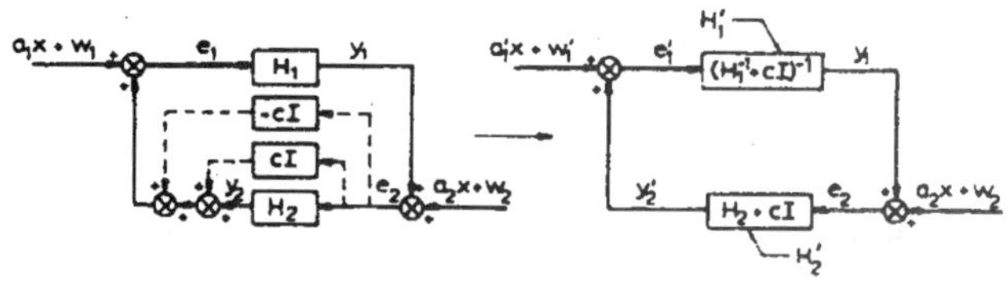

Fig. 5. A transformation.

RESTRICTION: *In the remainder of this paper, assume that X is an inner-product space, that $\langle x, y\rangle$ denotes the inner product on X, and that $\langle x, x\rangle=\|x\|^2$.*

This restriction is made with the intention of working mainly in the extended $L_2[0, \infty)$ norm,[12] with $\langle x, y\rangle = \int_0^\infty x(t)y(t)dt$.

4.1 Definitions of Conic and Positive Relations

DEFINITION: *A relation $\mathbf{H}$ in $\mathcal{R}$ is* interior conic *if there are real constants $r\geq 0$ and c for which the inequality*

$$\|(\mathbf{H}x)_t - cx_t\| \leq r\|x_t\| \quad [x\in\mathrm{Do}(\mathbf{H});\ t\in T] \quad (14)$$

is satisfied. $\mathbf{H}$ is exterior conic *if the inequality sign in* (14) *is reversed. $\mathbf{H}$ is* conic *if it is exterior conic or interior conic. The constant c will be called the* center parameter *of $\mathbf{H}$, and r will be* called the radius parameter.

The truncated output $(\mathbf{H}x)_t$ of a conic relation lies either inside or outside a sphere in X, with center proportional to the truncated input x_t and radius proportional to $\|x_t\|$. The region thus determined in $X_e\times X_e$ will be called a "cone," a term suggested by the following special case:

EXAMPLE 3: Let $\mathbf{H}$ be a relation on L_{2e} (see Example 1); let $\mathbf{H}x(t)$ be a function of $x(t)$, say $\mathbf{H}x(t)=N[x(t)]$, where N has a graph in the plane; then, as shown in Fig. 6, the graph lies inside or outside a conic sector of the plane, with a center line of slope c and boundaries of slopes $c-r$ and $c+r$. More generally, for $\mathbf{H}$ to be conic [without $\mathbf{H}x(t)$ necessarily being a function of $x(t)$, that is, if $\mathbf{H}$ has memory], it is enough for the point $[x(t), \mathbf{H}x(t)]$ to be confined to a sector of the plane. In this case, it will be said that $\mathbf{H}$ is *instantaneously confined to a sector of the plane.*

Inequality (14) can be expressed in the form $\|(\mathbf{H}x)_t-cx_t\|^2-r\|x_t\|^2\leq 0$. If norms are expressed in

[12] However, in engineering applications it is often more interesting to prove stability in the L_∞ norm. The present theory has been extended in that direction in the author's [2f]. The idea is [2f] is to transform L_2 functions into L_∞ functions by means of exponential weighting factors.

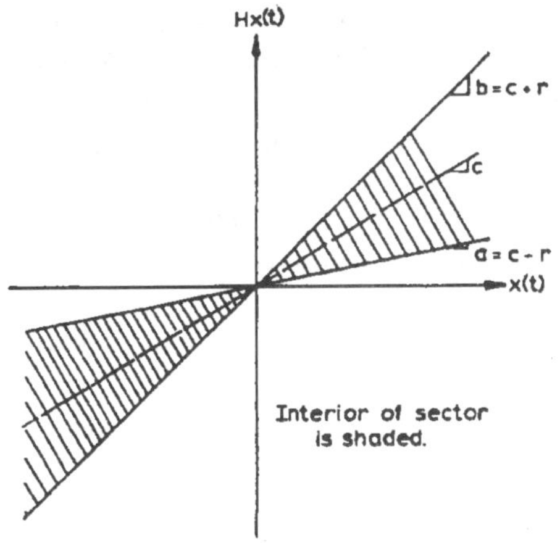

Fig. 6. A conic sector in the plane.

terms of inner products then, after factoring, there is obtained the equivalent inequality

$$\langle (Hx)_t - ax_t, (Hx)_t - bx_t \rangle \leq 0 \quad [x \in \text{Do}(H); t \in T] \qquad (15)$$

where $a=c-r$ and $b=c+r$. It will often be desirable to manipulate inequalities such as (15), and a notation inspired by Fig. 6 is introduced:

NOTATION: *A conic relation H is said to be* inside the sector $\{a, b\}$, *if $a \leq b$ and if Inequality (15) holds. H is* outside the sector $\{a, b\}$ *if $a \leq b$ and if (15) holds with the inequality sign reversed.*

The following relationship will frequently be used: If H is interior (exterior) conic with center c and radius r then H is inside (outside) the sector $\{c-r, c+r\}$. Conversely, if H is inside (outside) the sector $\{a, b\}$, then H is interior (exterior) conic, with center $(b+a)/2$ and radius $(b-a)/2$.

DEFINITION: *A relation H in $\mathcal{R}$ is* positive[11] *if*

$$\langle x_t, (Hx)_t \rangle \geq 0 \qquad [x \in \text{Do}(H); t \in T]. \qquad (16)$$

A positive relation can be regarded as degenerately conic, with a sector from 0 to ∞. [Compare (15) and (16).] For example, the relation H on L_{2e} is positive if it is instantaneously confined (see Example 3) to the first and third quadrants of the plane.

4.2 Some Properties of Conic Relations

Some simple properties will be listed. It will be assumed, in these properties, that H and H_1 are conic relations; that H is inside the sector $\{a, b\}$, with $b>0$; that H_1 is inside $\{a_1, b_1\}$ with $b_1>0$; and that $k \geq 0$ is a constant.

(i) *I is inside $\{1, 1\}$.*
(ii) *kH is inside $\{ka, kb\}$; $-H$ is inside $\{-b, -a\}$.*
(iii) SUM RULE: *$(H+H_1)$ is inside $\{a+a_1, b+b_1\}$.*
(iv) INVERSE RULE

CASE 1a: *If $a>0$ then H^{-1} is inside $\{1/b, 1/a\}$.*
CASE 1b: *If $a<0$ then H^{-1} is outside $\{1/a, 1/b\}$.*
CASE 2: *If $a=0$ then $(H^{-1}-(1/b)I)$ is positive.*

(v) *Properties (ii), (iii), and (iv) remain valid with the terms "inside $\{\ \}$" and "outside $\{\ \}$" interchanged throughout.*
(vi) *$g(H) \leq \max(|a|, |b|)$. Hence if H is in $\{-r, r\}$ then $g(H) \leq r$.*

The proofs are in Appendix A. One consequence of these properties is that it is relatively easy to estimate conic bounds for simple interconnections, where it might be more difficult, say, to find Lyapunov functions.

4.3 A Theorem on Boundedness

Consider the feedback system of Fig. 1, and suppose that H_2 is confined to a sector $\{a, b\}$. It is desirable to find a condition on H_1 which will ensure the boundedness of the closed loop. A condition will be found, which places H_1 inside or outside a sector depending on a and b, and which requires either H_1 or H_2 to be bounded away from the edge of its sector by an arbitrarily small amount, Δ or δ.

THEOREM 2a: *[In eqs. (1)–(2)] Let H_1 and H_2 be conic relations. Let Δ and δ be constants, of which one is strictly positive and one is zero. Suppose that*

(I) *$-H_2$ is inside the sector $\{a+\Delta, b-\Delta\}$ where $b>0$, and,*
(II) *H_1 satisfies one of the following conditions.*

CASE 1a: *If $a>0$ then H_1 is outside*

$$\left\{-\frac{1}{a} - \delta, -\frac{1}{b} + \delta\right\}.$$

CASE 1b: *If $a<0$ then H_1 is inside*

$$\left\{-\frac{1}{b} + \delta, -\frac{1}{a} - \delta\right\}.$$

CASE 2: *If $a=0$ then*

$$H_1 + \left(\frac{1}{b} - \delta\right)I$$

is positive; in addition, if $\Delta=0$ then $g(H_1) < \infty$.

Then E_1 and E_2 are bounded.

The proof of Theorem 2a is in Appendix B. Note that the minus sign in front of H_2 reflects an interest in *negative* feedback.

EXAMPLE 4: If H_1 and H_2 are relations on L_{2e} instantaneously confined to sectors of the plane (as in Example 3), then the closed loop will be bounded if the sectors are related as in Fig. 7. (More realistic examples will be discussed in Part II.)

[11] Short for "positive semidefinite." The terms "passive" and "nondissipative" have also been used.

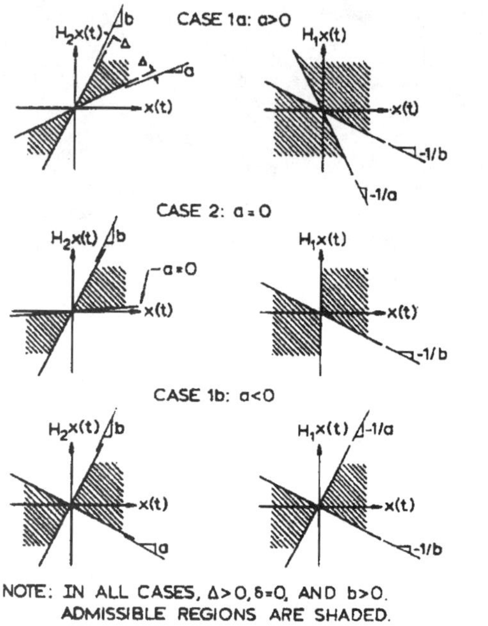

Fig. 7. Mutually admissible sectors for H_2 and H_1

4.4 *Incrementally Conic and Positive Relations*

Next, it is desired to find a stability result similar to the preceding theorem on boundedness. To this end the recent steps are repeated with all definitions replaced by their "incremental" counterparts.

DEFINITION: *A relation H in $\mathcal{R}$ is* incrementally interior (exterior) *conic if there are real constants $r>0$ and c for which the inequality*

$$\|(Hx - Hy)_t - c(x - y)_t\| \leq r\|(x - y)_t\| \quad [x, y \in \mathrm{Do}(H); t \in T] \quad (17)$$

is satisfied (with inequality sign reversed). An incrementally conic relation H is incrementally inside (outside) the sector $\{a, b\}$, *if $a \leq b$ and if the inequality*

$$\langle (Hx - Hy)_t - a(x - y)_t, (Hx - Hy)_t - b(x - y)_t \rangle \leq 0 \quad [x, y \in \mathrm{Do}(H); t \in T] \quad (18)$$

is satisfied (with inequality sign reversed). A relation H in $\mathcal{R}$ is incrementally positive[14] if

$$\langle (x - y)_t, (Hx - Hy)_t \rangle \geq 0 \quad [x, y \in \mathrm{Do}(H); t \in T]. \quad (19)$$

EXAMPLE 5: Consider the relation H on L_{2e}, with $Hx(t) = N[x(t)]$, where N is a function having a graph in the plane. If N is incrementally inside $\{a, b\}$ then N satisfies the Lipschitz conditions, $a(x-y) \leq N(x) - N(y) \leq b(x-y)$. Thus N lies in a sector of the plane, as in the nonincremental case (see Fig. 6), and in addition has upper and lower bounds to its slope.

Incrementally conic relations have properties similar to those of conic relations (see Section 4.2).

[14] The terms "monotone" and "incrementally passive" have also been used.

THEOREM 2b: *Let H_1 and H_2 be incrementally conic relations. Let Δ and δ be constants, of which one is strictly positive and one is zero. Suppose that,*

(I) *$-H_2$ is incrementally inside the sector $\{a+\Delta, b-\Delta\}$, where $b>0$, and,*

(II) *H_1 satisfies one of the following conditions:*

CASE 1a: *If $a>0$ then H_1 is incrementally outside*

$$\left\{-\frac{1}{a} - \delta, \; -\frac{1}{b} + \delta\right\}.$$

CASE 1b: *If $a<0$ then H_1 is incrementally inside*

$$\left\{-\frac{1}{b} + \delta, \; -\frac{1}{a} - \delta\right\}.$$

CASE 2: *If $a=0$ then*

$$H_1 + \left\{\frac{1}{b} - \delta\right\} I$$

is incrementally positive; in addition, if $\Delta = 0$ then $\tilde{g}(H_1) < \infty$.

Then E_1 and E_2 are input-output stable.

The proof is similar to that of Theorem 1a, and is omitted.

5. Conditions Involving Positive Relations

A special case of Theorem 2, of interest in the theory of passive networks, is obtained by, in effect, letting $a=0$ and $b \to \infty$. Both relations then become positive; also, one of the two relations becomes strongly positive, i.e.:

DEFINITION: *A relation H in $\mathcal{R}$ is* strongly (incrementally) positive *if, for some $\sigma > 0$, the relation $(H - \sigma I)$ is (incrementally) positive.*

The theorem, whose proof is in Appendix C, is:

THEOREM 3:[15] (a) [In eqs. (1)–(2)] If H_1 and $-H_2$ are positive, and if $-H_2$ is strongly positive and has finite gain, then E_1 and E_2 are bounded. (b) If H_1 and $-H_2$ are incrementally positive, and if $-H_2$ is strongly incrementally positive and has finite incremental gain, then E_1 and E_2 are input-output stable.

For example, if H_2 on L_{2e} is instantaneously confined to a sector of the plane, then, under the provisions of Theorem 3, the sector of H_2 lies in the first and third quadrants, and is bounded away from both axes.

5.1 *Positivity and Passivity in Networks*

A passive element is one that always absorbs energy. *Is a network consisting of passive elements necessarily stable?* An attempt will be made to answer this question for the special case of the circuit of Fig. 2.

First, an elaboration is given on what is meant by a

[15] A variation of this result was originally presented in [2d]. Kolodner [8] has obtained related results, with a restriction of linearity on one of the elements.

passive element. Consider an element having a current i and a voltage v; the absorbed energy is the integral $\int_0^\infty i(t)v(t)dt$, and the condition for passivity is that this integral be non-negative. Now, let $\boldsymbol{Z}$ be a relation mapping i into v; by analogy with the linear theory, it is natural to think of $\boldsymbol{Z}$ as an *impedance relation;* suppose $\boldsymbol{Z}$ is defined on L_{2e}, where the energy integral equals the inner product $\langle i, v\rangle$; then passivity of the element is equivalent to positivity of $\boldsymbol{Z}$. Similarly, if $\boldsymbol{Y}$ on L_{2e} is an admittance relation, which maps v into i, then passivity is equivalent to positivity of $\boldsymbol{Y}$.

Now consider the circuit of Fig. 2. This circuit consists of an element characterized by an impedance relation $\boldsymbol{Z}_2$, an element characterized by an admittance relation $\boldsymbol{Y}_1$, a voltage source v, and a current source i. The equations of this circuit are,

$$v_1 = v + v_2 \tag{20a}$$

$$i_2 = i - i_1 \tag{20b}$$

$$v_2 = \boldsymbol{Z}_2 i_2 \tag{21a}$$

$$i_1 = \boldsymbol{Y}_1 v_1. \tag{21b}$$

It is observed that these equations have the same form as the Feedback Equations, provided that the sources i and v are constrained by the equations $v=a_1x+w_1$, and $i=a_2x+w_2$. (By letting $a_1=0$ the familiar "parallel circuit" is obtained. Similarly, by letting $a_2=0$ the "series circuit" is obtained.) Thus there is a correspondence between the feedback system and the network considered here. Corresponding to the closed loop relation $\boldsymbol{E}_1$ there is a voltage transfer relation mapping v into v_1. Similarly, corresponding to $\boldsymbol{E}_2$ there is a current transfer relation mapping i into i_2. If Theorem 3 is now applied to eqs. (20)–(21) it may be concluded that: *If both elements are passive, and if, in addition, the relation of one of the elements is strongly positive and has finite gain, then the network transfer relations are bounded.*

6. Conclusions

The main result here is Theorem 2. This theorem provides sufficient conditions for continuity and boundedness of the closed loop, without restricting the open loop to be linear or time invariant. Theorem 2 includes Theorems 1 and 3 as special cases. However, all three theorems are equivalent, in the sense that each can be derived from any of the others by a suitable transformation.

There are resemblances between Theorem 2 and Nyquist's Criterion. For example, consider the following, easily derived, limiting form of Theorem 2: "If $\boldsymbol{H}_2=k\boldsymbol{I}$ then a sufficient condition for boundedness of the closed loop is that $\boldsymbol{H}_1$ be bounded away from the critical value $-(1/k)\boldsymbol{I}$, in the sense that

$$\left\|\left(\boldsymbol{H}_1 x - \frac{1}{k}x\right)_t\right\| \geq \delta\|x_t\|$$

for all x in X_e and t in T, where δ is an arbitrarily small positive constant." In fact, the conic sectors defined here resemble the disk-shaped regions on a Nyquist chart. However, Theorem 2 differs from Nyquist's Criterion in two important respects: (1) Unlike Nyquist's Criterion, Theorem 2 is not necessary, which is hardly surprising, since bounds on $\boldsymbol{H}_1$ and $\boldsymbol{H}_2$ are assumed in place of a more detailed characterization. (2) Nyquist's criterion assesses stability from observation of only the eigenfunctions exp (jwt), where Theorem 2 involves *all* inputs in X_e.

There is also a resemblance between the use of the notions of gain and inner product as discussed here, and the use of attenuation and phaseshift in the linear theory. A further discussion of this topic is postponed to Part II, where linear systems will be examined in some detail.

One of the broader implications of the theory developed here concerns the use of functional analysis for the study of poorly defined systems. It seems possible, from only coarse information about a system, and perhaps even without knowing details of internal structure, to make useful assessments of qualitative behavior.

Appendix

A. *Proofs of Properties (i–vi)*

Properties (i, ii). These two properties are immediately implied by the inequalities

$$\langle(\boldsymbol{I}x)_t - 1\cdot x_t, (\boldsymbol{I}x)_t - 1\cdot x_t\rangle = 0$$

$$\langle(c\boldsymbol{H}x)_t - cax_t, (c\boldsymbol{H}x)_t - cbx_t\rangle$$
$$= c^2\langle(\boldsymbol{H}x)_t - ax_t, (\boldsymbol{H}x)_t - bx_t\rangle \leq 0$$

in which c is a (positive or negative) real constant.

Property (iii). It is enough to show that $(\boldsymbol{H}+\boldsymbol{H}_1)$ has center $\frac{1}{2}(b+b_1+a+a_1)$ and radius $\frac{1}{2}(b+b_1-a-a_1)$; the following inequalities establish this:

$$\|[(\boldsymbol{H}+\boldsymbol{H}_1)x]_t - \tfrac{1}{2}(b+b_1+a+a_1)x_t\|$$
$$\leq \|(\boldsymbol{H}x)_t - \tfrac{1}{2}(b+a)x_t\|$$
$$+ \|(\boldsymbol{H}_1x)_t - \tfrac{1}{2}(b_1+a_1)\| \quad \text{(Triangle Ineq.)} \tag{A1a}$$
$$\leq \tfrac{1}{2}(b-a)\|x_t\| + \tfrac{1}{2}(b_1-a_1)\|x_t\|$$
$$= \tfrac{1}{2}(b+b_1-a-a_1)\|x_t\| \tag{A1b}$$

where eq. (A1b) follows from eq. (A1a) since $\boldsymbol{H}$ has center $\frac{1}{2}(b+a)$ and radius $\frac{1}{2}(b-a)$, and since $\boldsymbol{H}_1$ has center $\frac{1}{2}(b_1+a_1)$ and radius $\frac{1}{2}(b_1-a_1)$.

Property (iv).

Cases 1a and 1b: Here $a\neq 0$ and $b>0$, and

$$\left\langle(\boldsymbol{H}^{-1}x)_t - \frac{1}{b}x_t, (\boldsymbol{H}^{-1}x)_t - \frac{1}{a}x_t\right\rangle$$
$$= \left\langle y_t - \frac{1}{b}(\boldsymbol{H}y)_t, y_t - \frac{1}{a}(\boldsymbol{H}y)_t\right\rangle$$
$$= \frac{1}{ab}\langle(\boldsymbol{H}y)_t - ay_t, (\boldsymbol{H}y)_t - by_t\rangle$$

where $H^{-1}x=y$ and $x=Hy$. Since, by hypothesis, H is inside $\{a, b\}$ and $b>0$, the sign of the last expression is opposite to that of a. Thus the Inverse Rule is obtained.

CASE 2: Here $a=0$. The property is implied by the inequality,

$$\left\langle x_t, (H^{-1}x)_t - \frac{1}{b}x_t \right\rangle = \frac{1}{b}\langle (Hy)_t, by_t - (Hy)_t\rangle \geq 0.$$

Property (v). Simply reverse all the inequality signs.

Property (vi).

$$\begin{aligned} \|(Hx)_t\| &\leq \|(Hx)_t - \tfrac{1}{2}(b+a)x_t\| \\ &\quad + \|\tfrac{1}{2}(b+a)x_t\| \text{ (Triangle Ineq.)} \quad \text{(A2a)} \\ &\leq \tfrac{1}{2}(b-a)\|x_t\| + \tfrac{1}{2}|b+a|\cdot\|x_t\| \quad \text{(A2b)} \\ &= \max(|a|, |b|)\cdot\|x_t\| \end{aligned}$$

where eq (A2b) follows from eq (A2a) since, from the hypothesis, H has center $\frac{1}{2}(b+a)$ and radius $\frac{1}{2}(b-a)$. It follows that $g(H) \leq \max(|a|, |b|)$. Q.E.D.

B. Proof of Theorem 2a

The proof is divided into three parts: (1) The transformation of Fig. 5 is carried out, giving a new relation E_2'; E_2' is shown to contain E_2. (2) The new gain product is shown to be less than one. (3) E_2' is shown to be bounded, by Theorem 1; the boundedness of E_2 and E_1 follows.

Let $c=\frac{1}{2}(b+a)$ and $r=\frac{1}{2}(b-a)$.

B.1 Transformation of Eqs. (1)-(2)

The proof will be worked backwards from the end; the equations of the transformed system of Fig. 5 are,

$$e_1' = w_1' + a_1'x + y_2' \quad \text{(A3a)}$$
$$e_2 = w_2 + a_2x + y_1 \quad \text{(A3b)}$$
$$y_2' = H_2'e_2 \quad \text{(A4a)}$$
$$y_1 = H_1'e_1' \quad \text{(A4b)}$$

where

$$H_2' = (H_2 + cI) \quad \text{(A5a)}$$
$$H_1' = (H_1^{-1} + cI)^{-1}. \quad \text{(A5b)}$$

(It may be observed that these equations have the same form as eqs. (1)-(2), but H_1 is replaced by H_1' and H_2 is replaced by H_2'.) Let E_2' be the closed-loop relation that consists of all pairs (x, e_2) satisfying eqs. (A3)-(A4). It shall now be shown that E_2' contains E_2, that is, that any solution of eqs. (1)-(2) is also a solution of eqs. (A3)-(A4); thus boundedness of E_2' will imply boundedness of E_2.

In greater detail

(I) let (x, e_2) be any given element of E_2.

(II) Let e_1, y_1, y_2, H_1e_1, and H_2e_2 be fixed elements of X_e that satisfy eqs. (1)-(2) simultaneously with x and e_2.

(III) (Using Fig. 5 as a guide,) define two new elements of X_e,

$$y_2' = y_2 + ce_2 \quad \text{(A6a)}$$
$$e_1' = e_1 + cy_1. \quad \text{(A6b)}$$

It shall now be shown that there are elements $H_1'e_1'$ and $H_2'e_2'$ in X_e that satisfy eqs. (A3)-(A4) simultaneously with the elements defined in (I)-(III). Taking eqs. (A3)-(A4) one at a time:

Equation (A3a). Substituting eq. (1a) for e_1 in eq. (A6b), and eq. (1b) for y_1,

$$e_1' = (w_1 - cw_2) + (a_1 - ca_2)x + (y_2 + ce_2). \quad \text{(A7)}$$

If $w_1'=w_1-cw_2$ and $a_1'=a_1-ca_2$, then, with the aid of eq. (A6a), eq. (A7) reduces to eq. (A3a).

Equation (A3b): This is merely eq. (1b), repeated.

Equation (A4a): Recalling that $H_2'=H_2+cI$, it follows, from eqs. (A6a) and (2a), that there is an $H_2'e_2$ in X_e for which eq. (A4a) holds.

Equation (A4b): If eq. (A6b) is substituted for e_1 in eq. (2b), it is found that there exists $H_1(e_1'-cy_1)$ in X_e such that $y_1=H_1(e_1'-cy_1)$. Therefore,

(after inversion) $\quad H_1^{-1}y_1 = e_1' - cy_1$

(after rearrangement) $\quad (H_1^{-1} + cI)y_1 = e_1'$

(after inversion) $\quad y_1 = (H_1^{-1} + cI)^{-1}e_1'$.

That is, there exists $H_1'e_1'$ in X_e for which eq. (A4b) holds. Since eqs. (A3)-(A4) are all satisfied, (x, e_2) is in E_2'. Since (x, e_2) is an arbitrary element of E_2, E_2' contains E_2.

B.2 Boundedness of E_2'

It will be shown that $g(H_1')\cdot g(H_2') < 1$.

The Case $\Delta>0$, $\delta=0$: $g(H_2')$ will be bounded first. Since H_2 is in $\{-b+\Delta, -a-\Delta\}$ by hypothesis, (H_2+cI) is in $\{-b+\Delta+c, -a-\Delta+c\}$ by the Sum Rule of Section 4.2. Observing that $(H_2+cI)=H_2'$, that $(-b+c)=-r$, and that $(-a+c)=r$, it is concluded that H_2' is in $\{-r+\Delta, r-\Delta\}$. Therefore $g(H_2') \leq r-\Delta$.

Next, $g(H_1')$ will be bounded. In Case 1a, where $a>0$ and H_1 is outside

$$\left\{-\frac{1}{a}, -\frac{1}{b}\right\},$$

the Inverse Rule of Section 4.2 implies that H_1^{-1} is outside $\{-b, -a\}$; the same result is obtained in Cases 1b and 2. In all cases, therefore, the Sum Rule implies that $(H_1^{-1}+cI)$ is outside $\{-r, r\}$. By the Inverse Rule again, $(H_1^{-1}+cI)^{-1}$ is in

$$\left\{-\frac{1}{r}, \frac{1}{r}\right\}.$$

Therefore $g(H_1') \leq 1/r$.

Finally,

$$g(H_1')\cdot g(H_2') \leq \frac{r-\Delta}{r} < 1.$$

The Case $\Delta=0$, $\delta>0$: It shall be shown that this is a special case of the case $\Delta>0$, $\delta=0$. In other words, it will be shown that there are real constants a^*, b^*, and Δ^* for which the conditions of the case $\Delta>0$, $\delta=0$ are fulfilled, but with a replaced by a^*, b by b^*, and Δ by Δ^*.

Consider Case 1a, in which $a>0$. (Cases 1b and 2 have similar proofs, which will be omitted.) It must be shown that: (1) $-\boldsymbol{H}_2$ is in $\{a^*+\Delta,\ b^*-\Delta\}$. (2) $\boldsymbol{H}_1$ is outside

$$\left\{-\frac{1}{a^*},\ \frac{1}{b^*}\right\}.$$

Without loss of generality it can be assumed that δ is smaller than either $1/a$ or $1/b$. Choose a^* and b^* in the ranges

$$\frac{a}{1+a\delta} < a^* < a \quad\text{and}\quad b < b^* < \frac{b}{1-b\delta}\,.$$

Since $-\boldsymbol{H}_2$ is in $\{a,\ b\}$ by hypothesis, and since $a^*<a$ and $b^*>b$ by construction, there must be a $\Delta^*>0$ such that $\boldsymbol{H}_2$ satisfies condition (1). Since $\boldsymbol{H}_1$ is outside

$$\left\{-\frac{1}{a}-\delta,\ -\frac{1}{b}+\delta\right\}$$

by hypothesis, and since by construction

$$-\frac{1}{a^*} > -\frac{1}{a}-\delta \quad\text{and}\quad -\frac{1}{b^*} < -\frac{1}{b}+\delta,$$

condition (2) is satisfied. Hence this is, indeed, a special case of the previous one.

B.3 Conclusion of the Proof

Since $g(\boldsymbol{H}_1')\cdot g(\boldsymbol{H}_2')<1$, $\boldsymbol{E}_2'$ is bounded by Theorem 1, and so is $\boldsymbol{E}_2$, which is contained in $\boldsymbol{E}_2'$. Moreover, the boundedness of $\boldsymbol{E}_2$ implies the boundedness of $\boldsymbol{E}_1$; for, if $(x,\ e_1)$ is in $\boldsymbol{E}_1$ and $(x,\ e_2)$ is in $\boldsymbol{E}_2$, then

$$\|e_1\| \le \|w_1\| + |a_1|\cdot\|x\| + g(\boldsymbol{H}_2)\|e_2\|. \tag{A8}$$

Thus, if $\|x\|\le$const. and $\|e_2\|\le$const., then $\|e_1\|\le$const. (Inequality (A8) was obtained by applying the Triangle Inequality and Inequality (4) to eq. (1a), and taking the limit as $t\rightarrow\infty$. It may be noted that $g(\boldsymbol{H}_2)<\infty$, since $-\boldsymbol{H}_2$ is in $\{a,\ b\}$ by hypothesis.) Q.E.D.

C. Proof of Theorem 2

This shall be reduced to a special case of Theorem 2 [Case 2 with $\delta=0$]. In particular, it shall be shown that there are constants $b>0$ and $\Delta>0$ for which (I) $-\boldsymbol{H}_2$ is inside $\{\Delta,\ b-\Delta\}$, and, (II) the relation $[\boldsymbol{H}_1+(1/b)\boldsymbol{I}]$ is positive.

$[\boldsymbol{H}_1+(1/b)\boldsymbol{I}]$ is clearly positive for any $b>0$, since by hypothesis $\boldsymbol{H}_2$ is positive; the second condition is therefore satisfied. To prove the first condition it is enough to show that $\boldsymbol{H}_2$ is conic with center $-r$ and radius $r-\Delta$, where $r=b/2$. This is shown as follows: The hypothesis implies that, for some constant $\sigma>0$ and for any constant $\lambda>g(\boldsymbol{H}_2)$, the following inequalities are true

$$-\langle x_t,\ (\boldsymbol{H}_2x)_t\rangle \ge \sigma\|x_t\|^2 \tag{A9}$$

$$\|(\boldsymbol{H}_2x)_t\|^2 \le \lambda^2\|x_t\|^2 \tag{A10}$$

for any x in X_e and for any t in T. Hence, for any $r>0$,

$$\|(\boldsymbol{H}_2x)_t + rx_t\|^2 \le (\lambda^2 - 2r\sigma + r^2)\|x_t\|^2. \tag{A11}$$

Equation (A11) was obtained by expanding the square on its l.h.s., and applying eqs. (A9) and (A10). Constants λ, r, and Δ, are selected so that $\lambda>\sigma$, $r=\lambda^2/\sigma$, and $\Delta=r[1-\sqrt{1-(\sigma/\lambda)^2}]$. Now it can be verified that, for this choice of constants, the term $(\lambda^2-2r\sigma+r^2)$ in eq. (A11) equals $(r-\Delta)^2$; also, $0<\Delta<r$ since $(\sigma/\lambda)<1$); therefore eq. (A11) implies that $\boldsymbol{H}_2$ is conic with center $-r$ and radius $r-\Delta$. Q.E.D.

Acknowledgment

The author thanks Dr. P. Falb for carefully reading the draft of this paper, and for making a number of valuable suggestions concerning its arrangement and concerning the mathematical formulation.

References

[1] V. M. Popov, "Absolute stability of nonlinear systems of automatic control," *Automation and Remote Control*, pp. 857–875, March 1962. (Russian original in August 1961.)

[2] (a) G. Zames, "Conservation of bandwidth in nonlinear operations," M.I.T. Res. Lab. of Electronics, Cambridge, Mass., Quarterly Progress Rept. 55, pp. 98–109, October 15, 1959.
(b) ——, "Nonlinear operators for system analysis," M.I.T. Res. Lab. of Electronics, Tech. Rept. 370, September 1960.
(c) ——, "Functional analysis applied to nonlinear feedback systems," *IEEE Trans. on Circuit Theory*, vol. CT-10, pp. 392–404, September 1963.
(d) ——, "On the stability of nonlinear, time-varying feedback systems," *Proc. NEC*, vol. 20, pp. 725–730, October 1964.
(e) ——, "Contracting transformations—A theory of stability and iteration for nonlinear, time-varying systems," *Summaries, 1964 Internat'l Conf. on Microwaves, Circuit Theory, and Information Theory*, pp. 121–122.
(f) ——, "Nonlinear time varying feedback systems—Conditions for L∞-boundedness derived using conic operators on exponentially weighted spaces," *Proc. 1965 Allerton Conf.*, pp. 460–471.

[3] K. S. Narendra and R. M. Goldwyn, "A geometrical criterion for the stability of certain nonlinear nonautonomous systems," *IEEE Trans. on Circuit Theory (Correspondence)*, vol. CT-11, pp. 406–408, September 1964.

[4] R. W. Brockett and J. W. Willems, "Frequency domain stability criteria," pts. I and II, *1965 Proc. Joint Automatic Control Conf.*, pp. 735–747.

[5] (a) I. W. Sandberg, "On the properties of some systems that distort signals," *Bell Sys. Tech. J.*, vol. 42, p. 2033, September 1963, and vol. 43, pp. 91–112, January 1964.
(b) ——, "On the L_2-boundedness of solutions of nonlinear functional equations," *Bell. Sys. Tech. J.*, vol. 43, pt. II, pp. 1581–1599, July 1964.

[6] J. Kudrewicz, "Stability of nonlinear feedback systems," *Automatika i Telemechanika*, vol. 25, no. 8, 1964 (and other papers).

[7] E. H. Zarantonello, "Solving functional equations by contractive averaging," U. S. Army Math. Res. Ctr., Madison, Wis. Tech. Summary Rept. 160, 1960.

[8] I. I. Kolodner, "Contractive methods for the Hammerstein equation in Hilbert spaces," University of New Mexico, Albuquerque, Tech. Rept. 35, July 1963.

[9] G. J. Minty, "On nonlinear integral equations of the Hammerstein type," survey appearing in *Nonlinear Integral Equations*, P. M. Anselone, Ed. Madison, Wis.: University Press, 1964, pp. 99–154.

[10] F. E. Browder, "The solvability of nonlinear functional equations," *Duke Math. J.*, vol. 30, pp. 557–566, 1963.

[11] J. J. Bongiorno, Jr., "An extension of the Nyquist-Barkhausen stability criterion to linear lumped-parameter systems with time-varying elements," *IEEE Trans. on Automatic Control (Correspondence)*, vol. AC-8, pp. 166–170, April 1963.

[12] A. N. Kolmogorov and S. V. Fomin, *Functional Analysis*, vols. I and II. New York: Graylock Press, 1957.

On the Input-Output Stability of Time-Varying Nonlinear Feedback Systems–Part II: Conditions Involving Circles in the Frequency Plane and Sector Nonlinearities

G. ZAMES, MEMBER, IEEE

Abstract—The object of this paper is to outline a stability theory based on functional methods. Part I of the paper was devoted to a general feedback configuration. Part II is devoted to a feedback system consisting of two elements, one of which is linear time-invariant, and the other nonlinear.

An attempt is made to unify several stability conditions, including Popov's condition, into a single principle. This principle is based on the concepts of conicity and positivity, and provides a link with the notions of gain and phase shift of the linear theory.

Part II draws on the (generalized) notion of a "sector nonlinearity." A nonlinearity N is said to be INSIDE THE SECTOR $\{\alpha, \beta\}$ if it satisfies an inequality of the type $\langle(Nx-\alpha x)_t, (Nx-\beta x)_t\rangle \leq 0$. If N is memoryless and is characterized by a graph in the plane, then this simply means that the graph lies inside a sector of the plane. However, the preceding definition extends the concept to include nonlinearities with memory.

There are two main results. The first result, the CIRCLE THEOREM, asserts in part that: *If the nonlinearity is inside a sector $\{\alpha, \beta\}$, and if the frequency response of the linear element avoids a "critical region" in the complex plane, then the closed loop is bounded; if $\alpha>0$ then the critical region is a disk whose center is halfway between the points $-1/\alpha$ and $-1/\beta$, and whose diameter is greater than the distance between these points.*

The second result is a method for taking into account the detailed properties of the nonlinearity to get improved stability conditions. This method involves the removal of a "multiplier" from the linear element. The frequency response of the linear element is modified by the removal, and, in effect, the size of the critical region is reduced. Several conditions, including Popov's condition, are derived by this method, under various restrictions on the nonlinearity N; the following cases are treated:

(i) N is instantaneously inside a sector $\{\alpha, \beta\}$.
(ii) N satisfies (i) and is memoryless and time-invariant.
(iii) N satisfies (ii) and has a restricted slope.

1. Introduction

THE feedback system of Fig. 1 consists[1] of a linear time-invariant element H and a (not necessarily linear or time-invariant) element N. It will be supposed, for the moment, that N has no memory. These assumptions are among, the simplest which ensure that the system is both

(i) general enough to have many applications
(ii) complicated enough to exhibit such characteristic nonlinear phenomena as jump resonances, subharmonics, etc.

The object here is to find stability conditions for the closed-loop system. For practical reasons, it is desirable to express these conditions in terms of quantities that can be measured experimentally, such as frequency responses, transfer characteristics, etc. In particular, the following question is of interest: Imagine that the graph of N lies inside a sector of the plane, as shown in Fig. 2(a), and that the frequency response of H is plotted in the complex plane; can the complex plane be divided

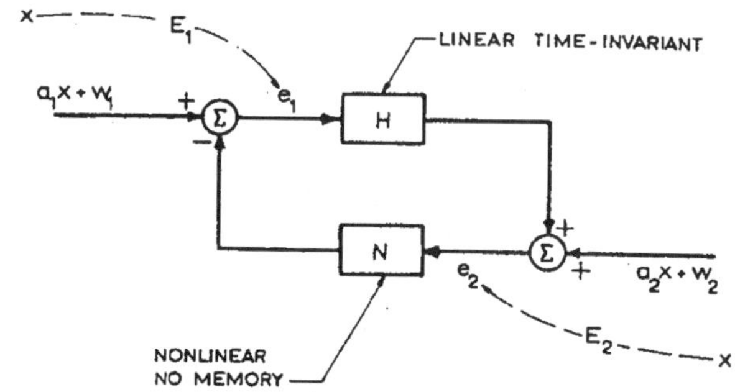

Fig. 1. A feedback system.

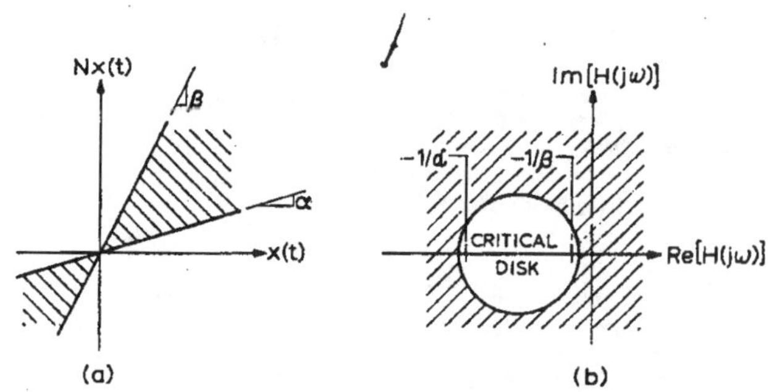

Fig. 2. If $Nx(t)$ vs. $x(t)$ and $H(j\omega)$ lie in the shaded regions, and if the Nyquist diagram of $H(j\omega)$ does not encircle the critical disk, then the closed loop is bounded.

Manuscript received December 29, 1964; revised October 1, 1965, and May 10, 1966. Parts of this paper were presented at the 1964 National Electronics Conference [1a], Chicago, Ill. This work was carried out in part under support extended by the National Aeronautics and Space Administration under Contract NsG-496 with the M.I.T. Center for Space Research.

The author is with the Department of Electrical Engineering, Massachusetts Institute of Technology, Cambridge, Mass.

[1] A single input x, multiplied by real constants a_1 and a_2, is added in at *two* points. By setting a_1 or a_2 to zero, it is possible to obtain a single-input system, in which the element closest to the input is either the linear element or the nonlinearity. The terms w_1 and w_2 are fixed bias functions, which will be used to account for the effects of initial conditions. The variables e_1 and e_2 are outputs.

into regions that are "safe" or "unsafe" as far as stability is concerned?

It will be shown that, with certain qualifications, such a division is possible. In fact it has already been shown in Part I that such regions, called "conic sectors," exist in a quite general sense. Here these general results will be applied to some concrete situations, involving frequency responses, etc. (Fig. 2, which illustrates the simplest of the results to be obtained here, gives some idea of what is being sought.)

2. Statement of the Problem

The purpose of this section is to define H and N, and to write feedback equations. H and N will be represented by input-output relations or by operators, in keeping with the theory outlined in Part I.

DEFINITION: *$R[0, \infty)$ is the space of real-valued functions on the interval $[0, \infty)$.*

L_p, where $p=1, 2, \cdots$, is the space consisting of those x in $R[0, \infty)$ for which the integral $\int_0^\infty |x(t)|^p dt$ is finite. In addition, for the case $p=2$, it is assumed that L_2 is an inner-product space, with inner-product

$$\langle x, y\rangle = \int_0^\infty x(t)y(t)dt$$

and norm $\|x\|_2$. The symbol $\|x\|$, without subscript, will often be used instead of $\|x\|_2$.

L_∞ is the space consisting of those functions x in $R[0, \infty)$ that are measurable and essentially bounded. L_∞ is assumed to be a normed linear space, with norm

$$\|x\|_\infty = \operatorname*{ess\,sup}_{t\geq 0} |x(t)|.$$

No distinction will be made between functions differing over sets of zero measure.

Those definitions which were introduced in Part I will only be summarized here. Following the convention of Part I, the subscripted symbol x_t denotes a function in $R[0, \infty)$ *truncated* after $[0, t)$. The space L_{pe}, where $p=1, 2, \cdots, \infty$, is the *extension* of L_p, i.e.,

$$L_{pe} = \{x \mid x \in R[0, \infty) \text{ and } x_t \in L_p \text{ for all } t \geq 0\}.$$

An *extended norm* $\|x\|_{pe}$ is defined on L_{pe}, where $\|x\|_{pe} = \|x\|_p$ if $x \in L_p$ and $\|x\|_{pe} = \infty$ if $x \notin L_p$. The symbol $\|x\|_{2e}$ will usually be abbreviated to $\|x\|_e$.

The concept of a *relation H on L_{pe}*, with *domain $Do(H)$* and *range $Ra(H)$* was introduced in Part I. A relation H on L_{pe} is *L_p-bounded* if H maps bounded subsets of L_{pe} into bounded subsets of L_{pe}. H is *L_p-continuous* if, given any x in $Do(H)$ and any $\Delta>0$, there is a $\delta>0$ such that, for any y in $Do(H)$, if $\|x-y\|_{pe}<\delta$ then $\|Hx-Hy\|_{pe}<\Delta$.

Part II will be devoted entirely to finding L_2 conditions (for boundedness and continuity), since these are easier to derive than the other L_p conditions. However, most of the results of this paper have been extended to the L_∞ norm, in [1b]. It has been found that, in most cases, the L_2 conditions imply L_∞-boundedness or continuity. For physical applications the most appropriate definitions of boundedness and continuity are, of course, obtained in the L_∞ norm.

DEFINITION: *Let $\mathcal{R}_0$ be the class of relations on L_{2e} having the property that the zero element, denoted o, is in $Do(H)$, and $Ho=o$.* An operator H on L_{2e} is any function of the type $H: L_{2e}\rightarrow L_{2e}$.

DEFINITION: *An operator H on L_{2e} is* TIME-INVARIANT *if it commutes with all delays. That is, for $t\geq 0$ let T_τ be the operator on L_{2e} given by: $T_\tau x(t) = x(t-\tau)$ for $t\geq\tau$, and $T_\tau x(t) = 0$ for $t<\tau$. Then $HT_\tau = T_\tau H$ for all $\tau\geq 0$.*

H is MEMORYLESS *if $Hx(t)$ is a function of $x(t)$ (i.e., only of $x(t)$) for all x in L_{2e} and for all $t\geq 0$.*

2.1. *The Operator Classes $\mathfrak{N}$ and $\mathfrak{L}$*

DEFINITION: *$\mathfrak{N}$ is the class of operators on L_{2e} having the following property: If N is in $\mathfrak{N}$ then there s a function, N: Reals→Reals, satisfying*[1]

$$Nx(t) = N(x(t)) \qquad (x \in L_{2e};\, t \geq 0) \tag{1}$$

and having the following properties: (i) $N(0)=0$, (ii) $|N(x)| \leq$ const. $|x|$, and (iii) for any real x, $\int_0^x N(x')dx'$ is finite.

An operator in $\mathfrak{N}$ is memoryless, time-invariant, not necessarily linear, and can be characterized by a graph in the plane. The letter N will indicate the graph of N.

DEFINITION. *$\mathfrak{L}$ is the class of those operators H on L_{2e} satisfying an equation of the type*[2]

$$Hx(t) = h_\infty x(t) + \int_0^t h(t-\tau)x(\tau)d\tau \quad (x \in L_{2e};\, t \geq 0) \tag{2}$$

in which h_∞ is a real constant, and the impulse response h is a function in L_1 with the property that, for some $\sigma_0<0$, $h(t)\exp(-\sigma_0 t)$ is also in L_1.

Operators in $\mathfrak{L}$ are linear and time-invariant.

2.2. *Feedback Equations*

Consider the feedback system of Fig. 1, but with two modifications: (i) N is not necessarily memoryless; (ii) a_1 and a_2 are operators on L_{2e}, multiplying x. (This amount of generality will be needed for some of the intermediate results; ultimately, the interesting case is that in which N has *no* memory, and a_1 and a_2 are *real* constants.) The equations of this system are

$$e_1 = a_1 x + w_1 - Ne_2 \tag{3a}$$

$$e_2 = a_2 x + w_2 + He_1 \tag{3b}$$

in which it is assumed that:

H is an operator in $\mathfrak{L}$
N is a relation in $\mathcal{R}_0$

[1] It can be verified that every mapping of the type $N: L_{2e}\rightarrow R[0, \infty)$ satisfying (1) is in fact an operator on L_{2e}. Similarly, every mapping $H: L_{2e}\rightarrow R[0, \infty)$ satisfying (2) is an operator on L_{2e} [see (B1) of Appendix B].

x in L_{2e} is an input
e_1 and e_2 in L_{2e} are outputs
w_1 and w_2 in L_2 are fixed biases
either a_1 and a_2 are real constants, or, more generally,[3] *a_1 is a relation on L_{2e} having the property that $\|a_1x\|_e \leq$const. $\|x\|_e$, and similarly for a_2.*

REMARK: If, to begin with, the linear element satisfies a state equation, then He_1 is set equal to the "zero-initial-condition response" of the state equation, and w_2 is set equal to the "zero-input response."

The closed-loop relations which map x into e_1 and e_2 will be denoted E_1 and E_2. The objective here is: *Find conditions on N and H which ensure that E_1 and E_2 are L_2-bounded and L_2-continuous.*

3. Conditions for Conicity and Positivity

This section has some preliminary results, which will be needed later in the analysis of stability. The following questions are fundamental to this analysis. Under what conditions is an operator conic or positive? Under what conditions is the composition product of two operators conic or positive?

The definitions of conicity and positivity were introduced in Part I. They are repeated here, for the special case of relations on L_{2e}.

DEFINITION: *Let H be a relation in $\mathcal{R}_0$. H is* INTERIOR CONIC *if c and $r \geq 0$ are real constants and the inequality*

$$\|(Hx)_t - cx_t\| \leq r\|x_t\| \qquad (x \in Do(H);\, t \geq 0) \qquad (4)$$

holds. H is exterior conic if (4) holds with the inequality sign reversed. c will be called the center parameter of H, and r will be called the radius parameter of H.

H is INSIDE (OUTSIDE) THE SECTOR $\{\alpha, \beta\}$ *if $\alpha \leq \beta$ and if the inequality*

$$\langle (Hx - \alpha x)_t, (Hx - \beta x)_t \rangle \leq 0 \quad (x \in Do(H);\, t \geq 0) \quad (5)$$

holds (with the inequality sign reversed).

H is POSITIVE *if it satisfies the inequality $\langle x_t, (Hx)_t \rangle \geq 0$ for all x in $Do(H)$ and all $t \geq 0$.*

In Part I, the concepts of conicity and positivity were subdivided into categories such as "instantaneous" conicity, "incremental" conicity, etc. The definitions of these terms are listed in Appendix A.

REMARK: The following conditions are equivalent: (i) H is interior conic with parameters c and r. (ii) H is inside $\{c-r, c+r\}$.

3.1. Memoryless, Time-Invariant Nonlinearities

Consider the operator class $\mathfrak{N}$; the conditions for N in $\mathfrak{N}$ to be conic, positive, etc., are simply the "instantaneous" conditions of Appendix A. Some of these conditions are illustrated in Fig. 3. In particular, N is inside the sector $\{\alpha, \beta\}$ if its graph lies in a sector of the plane bounded by lines of slopes α and β; N is incrementally inside $\{\alpha, \beta\}$ if, *in addition*, N satisfies the slope restrictions $\alpha \leq [N(x) - N(y)]/(x-y) \leq \beta$. N is positive if its graph lies in fhe first and third quadrants; N is incrementally positive if, *in addition*, N is nondecreasing.

[3] The more general assumption will be needed in Section 5 only.

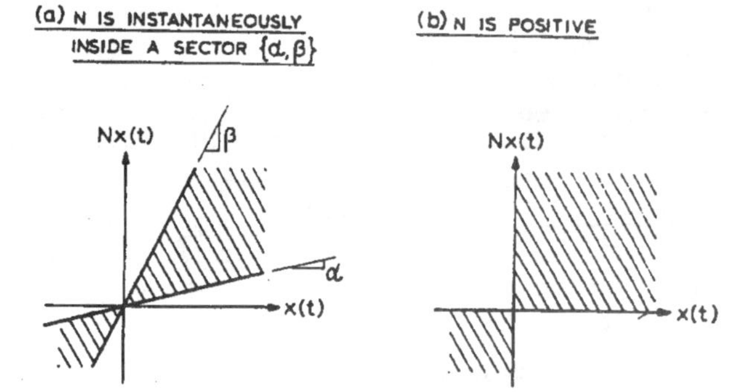

Fig. 3. Permissible regions (shaded) for instantaneously confined nonlinearities.

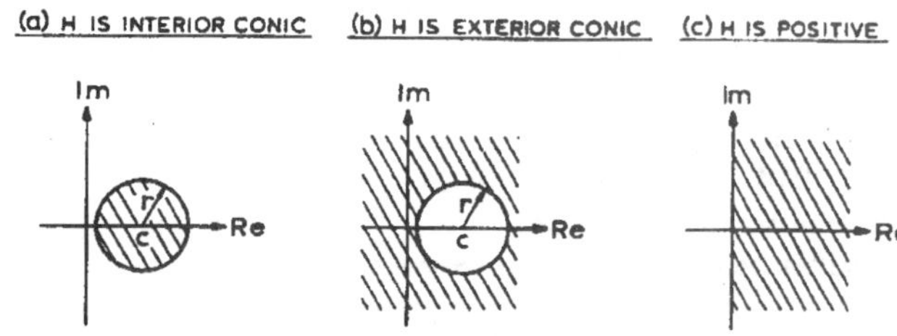

Fig. 4. Permissible regions (shaded) for the frequency response $H(j\omega)$.

3.2. Linear Time-Invariant Operators

Consider the operator class $\mathfrak{L}$; it will be shown, roughly speaking, that a conic sector has a counterpart in the frequency plane, in the form of a circular disk (see Fig. 4). This disk degenerates into a half-plane in the case of a positive operator.

DEFINITION: *Let $s = \sigma + j\omega$ denote a point in the complex plane. The* LAPLACE TRANSFORM *$H(s)$ of H in $\mathfrak{L}$ is*

$$H(s) = h_\infty + \int_0^\infty h(t) \exp(-st)dt \qquad (\sigma \geq 0) \qquad (6)$$

(The integral on the right-hand side of (6) exists and is analytic for $\sigma \geq 0$ [See (B1) of Appendix B].)

DEFINITION: *The* NYQUIST DIAGRAM *of $H(s)$ is a curve in the complex plane consisting of*: (i) *the image of the $j\omega$-axis under the mapping $H(s)$, and* (ii) *the point h_∞.*

LEMMA 1. *Let H be an operator in $\mathfrak{L}$, and let c and $r \geq 0$ be real constants.*

(a) *If $H(s)$ satisfies the inequality*

$$|H(j\omega) - c| \leq r \qquad (\omega \in (-\infty, \infty)) \qquad (7)$$

then H is incrementally interior conic with center parameter c and radius parameter r.

(b) *If $H(s)$ satisfies the inequality*

$$|H(j\omega) - c| \geq r \qquad (\omega \in (-\infty, \infty)) \tag{8}$$

and if the Nyquist diagram of $H(s)$ does not encircle the point $(c, 0)$, then ***H*** *is incrementally exterior conic with center parameter c and radius parameter r.*

(c) *If* $\operatorname{Re}\{H(j\omega)\} \geq 0$ *for* $\omega \in (-\infty, \infty)$ *then* ***H*** *is incrementally positive.*

The Proof of Lemma 1 is in Appendix B.

REMARK. The gains $g(\boldsymbol{H})$ and $\tilde{g}(\boldsymbol{H})$ were defined in Part I. It follows from Lemma 1(a) that if $|H(j\omega)| \leq c$, then $\tilde{g}(\boldsymbol{H}) = g(\boldsymbol{H}) \leq c$.

3.3. *Composition Products and Sector Products*

The composition product of two positive operators need not be positive. Those special cases in which the product is positive are of interest because they give a tighter bound on the composite behavior than would be obtained in general. (They form the basis of the factorization method of Section 5.2.)

Similarly, those special cases in which the product of two sector operators lies inside the "product sector" are of interest.

DEFINITION: *The* PRODUCT SECTOR $\{\alpha_1, \beta_1\} \times \{\alpha_2, \beta_2\}$ *is the sector* $\{\alpha, \beta\}$, *where* $[\alpha, \beta]$ *is the interval of the reals defined by* $[\alpha, \beta] = \{xy \mid x \in [\alpha_1, \beta_1] \text{ and } y \in [\alpha_2, \beta_2]\}$.

In other words, product sectors behave like pointwise products of the corresponding real intervals (see Fig. 5).

It is easy to show that if both operators are memoryless, say if both operators are in $\mathfrak{N}$, then their product has the above mentioned properties. [This can be shown by expressing the ratio $N_1(N_2(x))/x$ as the product $(N_1(y)/y) \times (N_2(x)/x)$, where $y \triangleq N_2(x)$.] More difficult are cases in which one operator is in $\mathfrak{N}$ and the other is in $\mathfrak{L}$, as in Lemmas 2 and 3.

3.4. *A Memoryless Nonlinearity and a First-Order Multiplier*

The following lemma is the basis for Popov's condition (Section 5.1).

LEMMA 2. *Let* ***N*** *be an operator in* $\mathfrak{N}$, ***K*** *be an operator in* $\mathfrak{L}$, *and let the Laplace transform of* ***K*** *be* $K(s) = k\lambda/s + \lambda$ *where* $k > 0$ *and* $\lambda > 0$.

(a) *If* ***N*** *is positive*[4] *then* ***NK*** *is positive.*

(b) *If* ***N*** *is inside*[5] *a sector* $\{\alpha, \beta\}$ *then* ***NK*** *is inside the product sector* $\{\alpha, \beta\} \times \{0, k\}$.

The proof of Lemma 2 is in Appendix C. (Note that ***K*** itself is positive and inside $\{0, k\}$, since $K(j\omega)$ lies entirely in the right-half plane and since $|K(j\omega) - \frac{1}{2}k| = \frac{1}{2}k$.)

[4] i.e., $xN(x) \geq 0$.
[5] i.e., $\alpha \leq N(x)/x \leq \beta$.

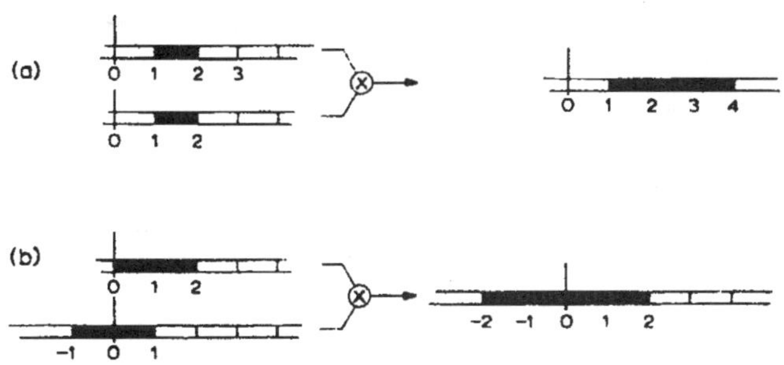

Fig. 5. Products of intervals.

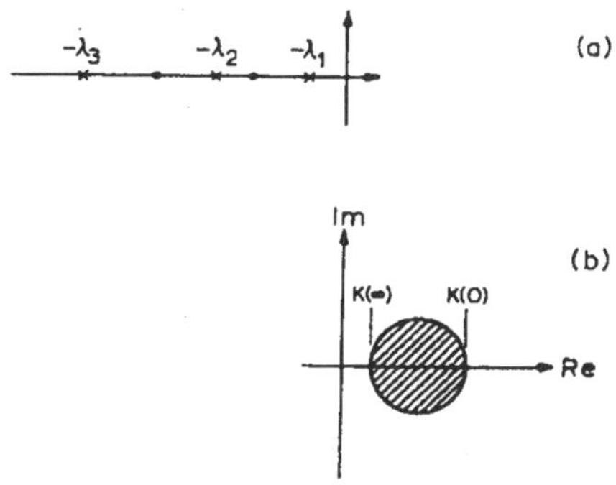

Fig. 6. $\mathfrak{RC}$ operators. (a) Typical pole-zero pattern. (b) Circle of confinement of $K(j\omega)$.

3.5. *A Memoryless Nonlinearity and an $\mathfrak{RC}$ Multiplier*

A situation resembling Lemma 2, but with ***N*** more restricted and ***K*** more general, is considered next. ***K*** is taken to be a sum of first-order terms, of a type that can be realized as the driving-point impedance of an *RC* network (see Guillemin [2], ch. 4).

DEFINITION: *Let* $\mathfrak{RC}$ *be the class of those operators* ***K*** *in* $\mathfrak{L}$ *having Laplace transforms of the form*

$$K(s) = \sum_{i=1}^{n} \frac{k_i \lambda_i}{s + \lambda_i} + K(\infty) \tag{9}$$

where $k_i \geq 0$, $\lambda_i > 0$, *and* $K(\infty) \geq 0$ *are real constants.*

An operator ***K*** in $\mathfrak{RC}$ has poles and zeros alternating on the negative-real axis, with a pole nearest but not at the origin. The frequency response of ***K*** lies inside a circle in the right-half plane, located as shown in Fig. 6(b); it follows that ***K*** is positive and inside the sector $\{K(\infty), K(0)\}$. (Observe that

$$K(0) = K(\infty) + \sum_{i=1}^{n} k_i \geq K(\infty).)$$

LEMMA 3. *Let* ***N*** *be an operator in* $\mathfrak{N}$, *and* ***K*** *be an operator in* $\mathfrak{RC}$.

(a) *If* ***N*** *is incrementally positive then* ***NK*** *is positive.*

(b) *If* ***N*** *is incrementally inside the sector* $\{\alpha, \beta\}$ *then* ***NK*** *is inside the product sector* $\{\alpha, \beta\} \times \{K(\infty), K(0)\}$.

In other words, multiplication by ***K*** affects the composite sector as if ***K*** had no memory. The proof of Lemma 3 is in Appendix D.

4. Circle Conditions for Stability

Consider now the main problem of this paper, namely, the problem of stability for the loop of Fig. 1. Suppose that N is a relation (which may or may not be memoryless) inside a sector $\{\alpha, \beta\}$. What conditions on the frequency response $H(j\omega)$ are sufficient to ensure boundedness of the closed loop? It will appear that the following "circle conditions" are sufficient:

Definition: *$H(j\omega)$ will be said to* satisfy the circle conditions for the sector $\{\alpha, \beta\}$, with offset δ, *where $\alpha \leq \beta$, $\beta > 0$, and $\delta \geq 0$ are real constants, if the following conditions hold:*

Case 1A. *If $\alpha > 0$, then*

$$\left| H(j\omega) + \frac{1}{2}\left(\frac{1}{\alpha} + \frac{1}{\beta}\right) \right| \geq \frac{1}{2}\left(\frac{1}{\alpha} - \frac{1}{\beta}\right) + \delta \qquad (\omega \in (-\infty, \infty)) \tag{10}$$

and the Nyquist diagram of $H(j\omega)$ does not encircle the point $-\frac{1}{2}(1/\alpha + 1/\beta)$.

Case 1B. *If $\alpha < 0$, then*

$$\left| H(j\omega) + \frac{1}{2}\left(\frac{1}{\alpha} + \frac{1}{\beta}\right) \right| \leq \frac{1}{2}\left(\frac{1}{\beta} - \frac{1}{\alpha}\right) - \delta \qquad (\omega \in (-\infty, \infty)). \tag{11}$$

Case 2. *If $\alpha = 0$, then* $\operatorname{Re}\{H(j\omega)\} \geq -(1/\beta) + \delta$ *for* $\omega \in (-\infty, \infty)$.

In other words, the complex plane is divided into two regions, shaped either like a circular disk and its complement, or like two half-planes. (The case $\alpha > 0$ is illustrated in Fig. 2.) One of the regions will be called "permissible" and the other will be called "critical." If $H(j\omega)$ does not enter or encircle the critical region, then the closed loop is bounded. If, in addition, N is *incrementally* inside $\{\alpha, \beta\}$, then the closed loop is continuous. These results are formalized in the following theorem:

A Circle Theorem. *Suppose that*

(I) *N is a relation in $\mathcal{R}_0$, (incrementally) inside the sector $\{\alpha + \Delta, \beta - \Delta\}$, where $\beta > 0$.*

(II) *H is an operator in $\mathcal{L}$, which satisfies the circle conditions for the sector $\{\alpha, \beta\}$ with offset δ.*

(III) *δ and Δ are non-negative constants, at least one of which is greater than zero.*

Then the closed-loop operators E_1 and E_2 are L_2-bounded (L_2-continuous).

The Circle Theorem is based on Theorem 2 of Part I. It was assumed in Theorem 2 that a_1 and a_2 were real constants. However, with only minor changes in the proof, it can be shown that Theorem 2 holds more generally if a_1 and a_2 are relations on L_{2e}, provided a_1 and a_2 satisfy inequalities of the type $\|a_1 x\|_t \leq \text{const.}\ \|x\|_t$. The Circle Theorem then follows immediately with the aid of Lemma 1 of Part II.

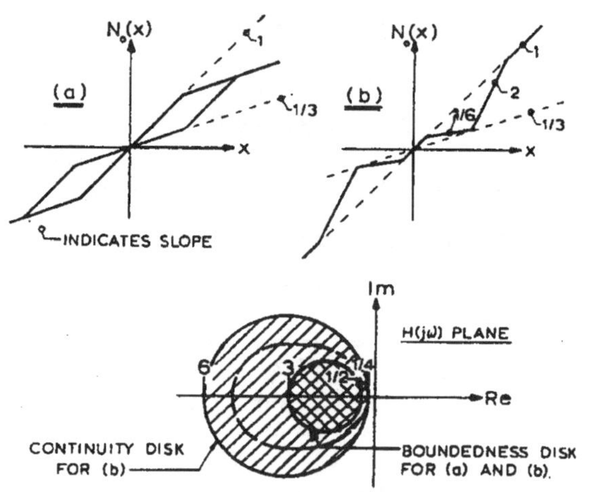

Fig. 7. Critical disks for Example 1. (Broken curve indicates edges of jump region in $H(j\omega)$ plane.)

The Circle Theorem can be viewed as a generalization of Nyquist's criterion,[6] in which a critical region replaces the critical point. For a given N there are two critical regions, one for boundedness and one for continuity. It can be shown that the continuity region always contains the boundedness region (see Example 1 and Fig. 7).

The Circle Theorem will serve as the generating theorem for the rest of this paper; i.e., the remaining results will be obtained as corollaries to the Circle Theorem by variously constraining the form of N. In particular, the following corollary is obvious.

4.1. A Circle Condition for Instantaneous Nonlinearities[7]

Corollary 1. *If (I) N in $\mathcal{R}_0$ is instantaneously (incrementally) inside the sector $\{\alpha + \Delta, \beta - \Delta\}$ where $\beta > 0$, and if conditions (II) and (III) of the Circle Theorem hold, then E_1 and E_2 are L_2-bounded (L_2-continuous).*

Example 1. (a) Let N_0 be the relation shown in Fig. 7(a), and N be the relation in $\mathcal{R}_0$ defined by the equation $Nx(t) = [1 + \sin^2(t)] \cdot N_0(x(t))$. Find the critical regions for boundedness and continuity. (b) Repeat for the function shown in Fig. 7(b).

(a) Observe that N_0 is inside the sector $\{1/3, 1\}$ and that the time-varying gain $[1 + \sin^2(t)]$ is inside the sector $\{1, 2\}$. It follows that N is inside the product sector, $\{1/3, 1\} \times \{1, 2\} = \{1/3, 2\}$. Corollary 1 therefore implies that the critical region for boundedness is a disk, as shown in Fig. 7. However, since N is multivalued, and therefore not incrementally in any sector, Corollary 1 provides no information about continuity.

(b) The same results as in (a) are obtained for boundedness. In addition, N is incrementally inside $\{1/6, 4\}$, and a continuity disk is obtained, as shown in Fig. 7.

[6] More accurately, of the sufficient part of Nyquist's criterion.

[7] Similar or closely related circle conditions were found independently by the author [1a], Sandberg [5], Narendra and Goldwyn [6], and Kudrewicz [7].

Observe that the nonlinearity N in Corollary 1 *can* be time-varying and *can* have memory. In fact, very little has been assumed about the detailed character of N. The price paid for this is that Corollary 1 is often conservative, i.e., the critical region is too large. This is especially true of the boundedness condition (see Example 2). The continuity condition probably gives a quite fair estimate of what to expect. In fact, an approximate analysis, based on the harmonic balance method (cf. Hatanaka [3]), suggests that continuity breaks down in the following way: There is a zone, inside the critical continuity disk, in which jump-resonance phenomena occur. The zone is not much smaller than the continuity disk. Furthermore, the magnitudes of jump resonances depend on the Nyquist diagram behavior inside the continuity disk.

5. Conditions with Transferred Multipliers

The next two corollaries can be viewed as attempts to reduce the size of the critical region, at the cost of added restrictions on N. In certain cases, it will be possible to remove a "multiplier" K from the linear element, *before* applying the Circle Theorem. The removal of K will shift the frequency response of the remainder, $H_1(j\omega)$, away from the critical region. Thus the effective size of the critical region will be reduced.

5.1. Popov's Condition

Consider the feedback system of Fig. 1, under the same conditions as in Corollary 1, but with the added constraint that N is a memoryless, time-invariant operator. The following condition for boundedness (*not* continuity) involves the removal of a first-order multiplier from H.

Corollary 2. *If*

(I) *N is an operator in $\mathfrak{N}$, inside the sector $\{\alpha, \beta\}$ where $\beta > 0$.*

(II) *H is an operator in $\mathfrak{L}$ that can be factored into a product $H = KH_1$, where H_1 and K are in $\mathfrak{L}$, and $K(s) = \lambda/(s+\lambda)$ where $\lambda > 0$.*

(III) *H_1 satisfies the circle conditions for the product sector $\{\alpha, \beta\} \times \{0, 1\}$ with offset δ, where $\delta > 0$.*

(IV) *$a_2 = 0$ and $\dot{w}_2$ is in L_2, where $\dot{w}_2$ denotes the derivative on $[0, \infty)$.*

Then E_1 and E_2 are L_2-bounded.

Remarks: (i) For $\alpha > 0$, Condition III simply means that $\mathrm{Re}\{(j\omega + \lambda)H(j\omega)\} \geq -\lambda/\beta + \delta$. (ii) Condition IV limits the result to that configuration in which the directions of flow is from the input to H to N.

Proof of Corollary 2. The feedback equations will be transformed, as illustrated in Fig. 8; i.e., H will be split into a product, $H = KH_1$, and the multiplier K will be transferred into a composition with N. It will then be shown, in Step 1, that the transformed equations are bounded, and, in Step 2, that they are equivalent to the original equations as far as stability is concerned.

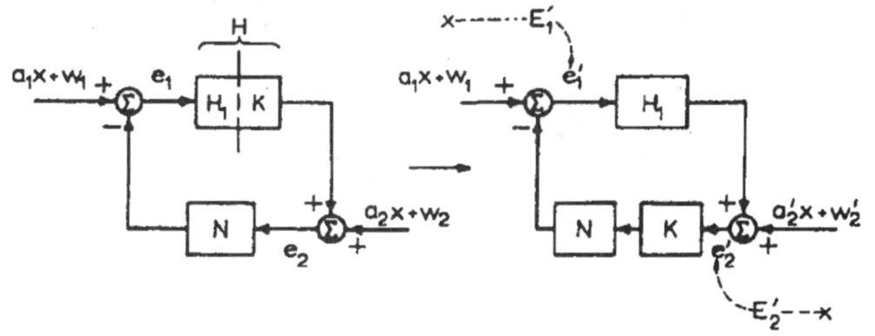

Fig. 8. A transformation.

Letting $w_2' = w_2 + \lambda^{-1}\dot{w}_2$, and recalling that $a_2 = 0$, consider the equations of the transformed system of Fig. 8,

$$e_1' = w_1 + a_1x - NKe_2' \tag{12a}$$

$$e_2' = w_2' + H_1e_1' \tag{12b}$$

Let E_1' and E_2' be the closed-loop relations for (12a)–(12b).

Step 1. E_1' and E_2' are L_2-bounded. This follows from the Circle Theorem whose hypotheses are satisfied because: w_2' is in L_2 by assumption IV; NK is in the product sector $\{\alpha, \beta\} \times \{0, 1\}$ by Lemma 2; and NK satisfies the appropriate circle conditions.

Step 2. It will be shown (below) that

$$E_1 = E_1' \tag{13a}$$

$$E_2 = KE_2' \tag{13b}$$

Since E_1' and E_2' have been proved L_2-bounded, and since K is certainly L_2-bounded, it follows that E_1 and E_2 are L_2-bounded.

To prove (13b), recall that E_2 and E_2' are subsets (of a product space), so that it is enough to establish that each contains the other. Suppose that (x, e_2) is an element of E_2; by definition of E_2, there is an e_1 in L_{2e} satisfying (3a)–(3b); let $e_1' = e_1$ and $e_2' = w_2' + H_1e_1$. Direct substitution shows that (x, e_1', e_2') satisfies (12a)–(12b), so that (x, e_2') belongs to E_2'. Substitution also shows that $e_2 = Ke_2'$, so that (x, e_2) is in KE_2'. Since (x, e_2) is an arbitrary element of E_2, it follows that KE_2' contains E_2. It can similarly be shown that E_2 contains KE_2', so that (13b) holds. The proof of (13a) is similar. Q.E.D.

Example 2. Let N be the operator in $\mathfrak{N}$ whose graph is shown in Fig. 9(a), and let $H(s) = k/(s+\lambda)(s+\mu)$. For what values of k is the closed-loop L_2-bounded? Compare Corollaries 1 and 2.

Here N is inside $\{0, 1\}$, so the critical region is a half-plane, $\mathrm{Re}\{\cdot\} \leq -1 + \delta$, in both corollaries. In Corollary 2 let $K(s) = \lambda/(s+\lambda)$ and $H_1(s) = k/\lambda(s+\mu)$; the following estimates are obtained:

Corollary 1: $-\lambda\mu < k < (\lambda+\mu)(\lambda+\mu+2\sqrt{\lambda\mu})$
Corollary 2: $-\lambda\mu < k$

Corollary 2 is less conservative than Corollary 1, as it shows the closed loop to be bounded for *all* positive k.

Figure 10 also illustrates the following point: For $a \leq 0$, both corollaries predict the same critical region; however, in many cases of interest, $H_1(j\omega)$ is further from the critical region than $H(j\omega)$.

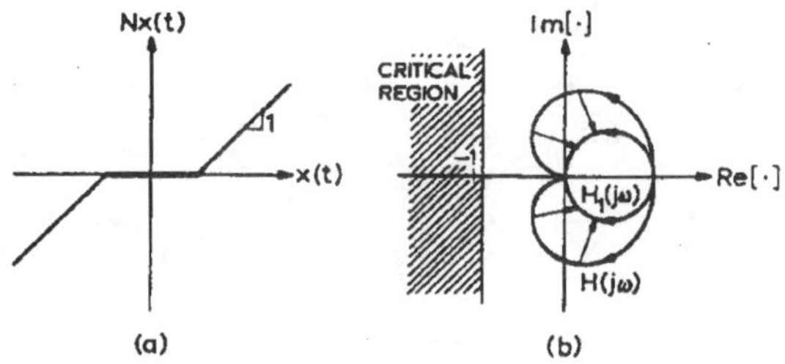

Fig. 9. An example of Popov's Condition. Arrows indicate shift away from critical region.

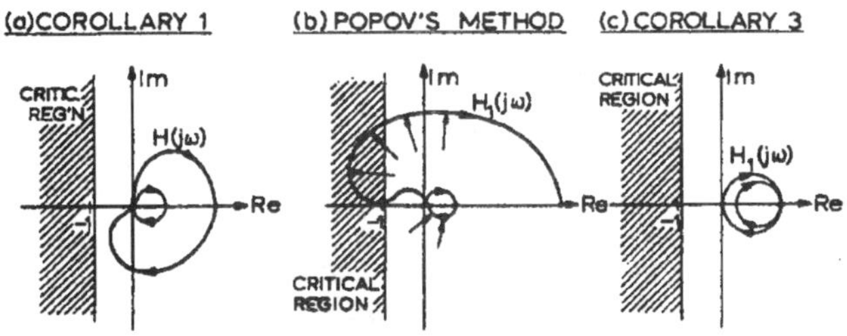

Fig. 10. Nyquist diagrams for Corollaries 1–3.

5.2. The Factorization Method

The proof of Corollary 2 suggests a method for generating a class of Popov-like conditions. The method consists of a factorization of H into $H=KH_1$, followed by the transformation of (3a)–(3b) into (12a)–(12b) followed by an application of the Circle Theorem. Various stability conditions are produced by variously choosing the multiplier K.

The method has two preconditions:

(Ia) Either K^{-1} exists or $a_2=0$.

(Ib) There is a w_2' in L_2 such that $K(w_2')=w_2$.

These preconditions ensure that (3) are transformable into (12). Note that if K^{-1} exists, then a_2 need not be zero; however, in that case, (12b) must be modified by the addition of a term $a_2'x$ where $a_2'x=K^{-1}a_2x$ (that is, a_2' is a *relation* on L_{2e}).

The method is worthwhile only if it gives a smaller effective critical region than Corollary 1. This happens if:

(IIa) NK lies in a sector not greater than the product of the sectors of N and K.

(IIb) KH_1 lies in a sector greater than the product of the sectors of K and H_1.

If Requirements (IIa)–(IIb) are satisfied, then it is advantageous to transfer K from a composition with H_1 into a composition with N. Requirement (IIa) usually means that the multiplier K has a very special form, and the difficulty in finding suitable multipliers is the main problem in applying this method. Once K is fixed, Requirement (IIb) defines a (limited) class of operators H for which this method is useful.

As an illustration of this method, a condition resembling Popov's is derived next.

5.3. A Slope-Restricted Nonlinearity and an $\mathcal{RC}$ Multiplier

Consider the feedback system of Fig. 1, under the conditions of Corollary 2, but with an added slope restriction on N, and a more general type of multiplier K.

COROLLARY 3.[8] *If:*

(I) *N is an operator in $\mathfrak{N}$, incrementally inside*[9] $\{\alpha, \beta\}$ *where* $\beta>0$.

(II) *H is an operator in $\mathcal{L}$, which can be factored into a product $H=KH_1$, where K is in $\mathcal{RC}$ and H_1 is in $\mathcal{L}$.*

(III) *H_1 satisfies the circle conditions for the product sector* $\{\alpha, \beta\}\times\{K(\infty), K(0)\}$ *with offset* δ, *where* $\delta>0$.

(IV) *There is*[10] *a* w_2' *such that* $K(w_2')=w_2$.

(V) *Either* $K(\infty)>0$ *or* $a_2=0$.

Then E_1 and E_2 are L_2-bounded.

Corollary 3 is obtained immediately by the factorization method, with the help of Lemma 3.

REMARK: For suitably restricted N, Corollary 3 has several advantages over Popov's method:

(i) The shift away from the critical region, which depends on $K(j\omega)$, can be controlled more flexibly as a function of ω. This is likely to be useful where a negative magnitude slope $d/d\omega|H(j\omega)|$ is followed by a positive slope at a larger ω.

(ii) a_2 need not be zero if $K(\infty)>0$.

(iii) If $\alpha>0$, the critical region predicted by Corollary 3 (a disk) is sometimes smaller than by Popov's method (which always gives a half-plane for $\alpha>0$).

EXAMPLE 3. Let N be the operator in $\mathfrak{N}$ whose graph is shown in Fig. 9(a) (the same as in Example 2), and let H have the Laplace transform

$$H(s) = kr^{-1}\left\{\frac{s+1}{s+r^{-1}}\right\} \cdot \left\{\frac{r^{-1}s}{(s+r^{-2})(s+r^{-1})} + \frac{r^2s}{(s+r)(s+r^2)}\right\}$$

where $r\gg1$. For what values of k is the closed loop bounded? Compare Corollaries 1, 2, and 3.

Figure 10(a) illustrates the significant features of the Nyquist diagram of $H(j\omega)$ (not drawn to scale). Observe that $H(j\omega)$ has two "pass bands," one for $r^{-2}<\omega<r^{-1}$ and the other for $r<\omega<r^2$; these "pass bands" produce the two loops in Fig. 10(a). Note that the critical region is the same half-plane in all corollaries, namely, $\mathrm{Re}\{\cdot\}\geq-1+\delta$.

[8] Corollary 3 and the factorization method, in a functional setting, were introduced by the author in Reference [1a]. A related method in a Liapunov setting has been exploited by Brockett, Willems, and Forys [4a]–[4b].

[9] i.e., $\alpha\leq\dfrac{N(x)-N(y)}{x-y}\leq\beta$.

[10] This condition is satisfied automatically if $K(\infty)>0$. If $K(\infty)=0$, then it is satisfied if $\dot{w}_2$ is in L_2.

Corollary 1 predicts boundedness for $-1<k<8$ approximately.

Popov's method is useless here. A comparison of Figs. 10(a) and 10(b) shows the effect of removing the multiplier $\lambda/s+\lambda$: $H_1(j\omega)$ is moved away from the critical region in the lower left half-plane (in the decaying edge of the lower pass band, $r^{-1}<\omega<1$); however, this improvement is more than offset by the bulge introduced in the upper left half-plane (in the rising edge of the upper pass band, $1<\omega<r$).

What is obviously needed here is a multiplier that acts like Popov's in the lower pass band, but has no effect in the upper pass band. The $\mathcal{RC}$ multiplier $K(s)=r^{-1}(s+1)(s+r^{-1})$ accomplishes just this. Its removal shifts $H_1(j\omega)$ entirely into the right half-plane [Fig. 10(c)]. Corollary 3 therefore implies that the closed loop is bounded for all positive k, in fact, for $k>-1$.

6. Comments and Conclusions

6.1. *Circle Conditions*

The main result here is the Circle Theorem. The Circle Theorem is a sufficient condition for closed-loop stability, which requires the nonlinearity N to lie inside a sector, but which leaves N free otherwise. The other conditions are all corollaries of the Circle Theorem.

Corollary 1 is probably the most useful result, since it roughs out the region of stability, with a minimum of restrictions on N. However, it is often conservative.

Corollaries 2 and 3 provide a tradeoff between limitations on N and limitations on $H(j\omega)$. Probably more significant than the actual conditions is the fact that there is a method of generating them, namely, the factorization method.

The results derived in Part II hold for nonzero initial conditions in the linear element, provided the "zero-input response" w_2 satisfies the indicated restrictions.

6.2. *Extensions of the Theory*

The theory has been extended in several directions (see [1b]), notably,

1) to L_∞,
2) to systems with a limited rate of time variation.

The extension to L_∞ involves the use of exponential weighting factors, which transform L_∞ functions into L_2 functions. The extension to time-varying systems involves the use of a shifted Nyquist diagram, $H(\sigma+j\omega)$, in which σ depends on the rate of time variation.

6.3. *Gain and Phase Shift in Relation to Nonlinear Time-Varying Systems*

The stability of a linear time-invariant feedback system depends on the amounts of gain and phase shift introduced by the open loop. Are similar considerations involved in nonlinear, time-varying problems? Of course the classical definitions of gain and phase shift, in terms of frequency response, have no strict meaning in nonlinear or time-varying systems. However, stability does seem to depend on certain measures of signal amplification and signal shift. Thus the norm ratio $\|Hx\|/\|x\|$ plays a role similar to the role of gain. Furthermore, the inner product $\langle x, Hx\rangle$, a measure of input-output cross-correlation, is closely related to the notion of phase shift. For example, for linear time-invariant operators in $\mathcal{L}$ the condition of positivity, $\langle x, Hx\rangle\geq 0$, is equivalent (by Lemma 1) to the phase condition,

$$|\operatorname{Arg}\{H(j\omega)\}| \leq 90^\circ.$$

It may be worthwhile to see what the theorems of Part I mean in terms of gain and phase shift. This can be done with the help of Lemma 1. Theorem 1 of Part I can be viewed as a generalization to nonlinear time-varying systems of the rule that, "if the open-loop gain is less than one, then the closed loop is stable." Theorem 3 can be viewed as the generalization of, "if the open-loop absolute phase shift is less than 180° then[11] the closed loop is stable." Theorem 2 places gain and phase shift in competition, permitting large gains at small phase shifts, etc.

6.4. *Conclusions*

Some of the salient features of the functional theory are:

(i) It provides an alternative to the method of Liapunov, an alternative resembling the classical Nyquist-Bode theory.

(ii) It is well suited to input-output problems.

(iii) It is free of state-space restrictions, and is therefore useful for distributed systems, hysteritic systems, etc. It also lends itself well to multivariable systems.

(iv) It unifies several results in stability theory. In particular, it is noteworthy that Popov's condition, the slope-restricted-N result, etc., can all be derived from the Circle Theorem.

(v) It has led to some new results, notably Corollary 3 and [1b].

The theory outlined here is probably still far from its definitive form. Nevertheless, it provides enough insight to make possible a reasonably systematic design of stabilizers.

Appendix A

Definitions of Conicity and Positivity

It will be assumed that H is a relation in $\mathcal{R}_0$ and that c, $r\geq 0$, and $\alpha\leq\beta$ are real constants.

Group I. *"Incremental" Conditions*

H is incrementally interior conic if

$$\|(Hx - Hy)_t - c(x-y)_t\| \leq r\|(x-y)_t\|;$$

[11] There are two positive elements in the open loop; each contributes an absolute phase shift of less than 90°; the open-loop absolute phase shift is therefore less than 180°.

H is incrementally inside the sector $\{\alpha, \beta\}$ if

$$\langle (Hx - Hy)_t - \alpha(x - y)_t, (Hx - Hy)_t - \beta(x - y)_t \rangle \leq 0;$$

H is incrementally positive if

$$\langle (x - y)_t, (Hx - Hy)_t \rangle \geq 0;$$

(where the inequalities of Group I hold for all x and y in $Do(H)$ and all $t \geq 0$).

The definitions of an operator that is "exterior" conic or "outside" a sector, are identical to the preceding ones except for a reversal of the inequality sign, and will therefore be omitted.

REMARK: If H is incrementally inside $\{\alpha, \beta\}$ then H is inside $\{\alpha, \beta\}$. Similarly, each inequality in Group I implies a corresponding inequality in Section 3.

GROUP II. *"Instantaneous" Conditions*

H is instantaneously inside the sector $\{\alpha, \beta\}$ if

$$\alpha \leq Hx(t)/x(t) \leq \beta \qquad (x \in L_{2e}; t \geq 0; x(t) \neq 0);$$

H is instantaneously positive if

$$x(t) \cdot Hx(t) \geq 0 \qquad (x \in L_{2e}; t \geq 0).$$

GROUP III. *"Instantaneous Incremental" Conditions*

H is instantaneously incrementally inside the sector $\{\alpha, \beta\}$ if

$$\alpha \leq \frac{Hx(t) - Hy(t)}{x(t) - y(t)} \leq \beta$$
$$(x \in L_{2e}; t \geq 0; x(t) - y(t) \neq 0).$$

H is instantaneously incrementally positive if

$$[Hx(t) - Hy(t)] \cdot [x(t) - y(t)] \geq 0 \qquad (x \in L_{2e}; t \geq 0).$$

REMARK: If H is instantaneously inside $\{\alpha, \beta\}$, then H is inside $\{\alpha, \beta\}$. Similarly, each inequality in Group II implies a corresponding inequality in Section 3. Also each inequality in Group III implies a corresponding inequality in Group II.

APPENDIX B

LEMMA 1

The proof of Lemma 1(b) will be based on the Principle of the Argument, a theorem of Paley and Wiener, and Parseval's theorem. The proofs of Lemmas 1(a) and 1(c), being straightforward applications of Parseval's theorem, will be omitted.

Some preliminary lemmas and properties will now be introduced.

DEFINITION: *If x is a function in L_2 then its* L.I.M. TRANSFORM *is*

$$X(s) = \underset{T\to\infty}{\text{l.i.m.}} \int_0^T x(t) \exp(-st)dt. \qquad (\sigma \geq 0) \qquad \text{(B1)}$$

The l.i.m. transform of $X(s)$ is

$$x_1(\tau) = \frac{1}{2\pi} \underset{W\to\infty}{\text{l.i.m.}} \int_{-W}^{W} X(j\omega) \exp(j\omega\tau)d\omega. \quad (\tau \text{ real}) \quad \text{(B2)}$$

The limits in the mean in (B1)–(B2) exist, and $x_1(\tau) = x(\tau)$.[12]

B.1. *Properties of Transforms*

In the following properties, H is an operator in $\mathcal{L}$ having a Laplace transform $H(s)$.

(A) The integral defining $H(s)$ [see (6)] converges and is bounded for $\sigma > \sigma_0$, since

$$\int_0^\infty h(t) \exp(-st)dt$$
$$\leq \int_0^\infty |h(t)| \exp(-\sigma_0 t)dt = \text{const.}$$

(B) $H(s)$ is analytic for $\sigma > \sigma_0$.[13]

(C) For $\sigma \geq 0$, $\lim_{|s|\to\infty} H(s) = h_\infty$.[14]

(D) If x is in L_2 then Hx is in L_2.[15]

(E) If the l.i.m. transforms of $x \in L_2$ and $Hx \in L_2$ are $X(s)$ and $Y(s)$ respectively, then[15]

$$Y(s) = H(s)X(s) \qquad \text{(B3)}$$

One consequence of Property D is that every mapping of the type defined by the right-hand side of (2) is an operator[16] on L_{2e}, and belongs to $\mathcal{R}_0$.

B.2. *Some Consequences of Nonencirclement*

A contour in the complex plane will be said to have Property N if it does not pass through or encircle the origin.

LEMMA 4. If the Nyquist diagram of $H(s)$ has Property N then: (a) $1/H(s)$ is analytic for $\sigma \geq 0$. (b) If the inequality $1/|H(\sigma + j\omega)| \leq M$ holds for $\sigma = 0$ then it holds for all $\sigma \geq 0$.

PROOF: (a) Since $H(s)$ is analytic for $\sigma \geq 0$, it is enough to show that $H(s) \neq 0$ for $\sigma \geq 0$ to prove (a). For this purpose several contours are defined: Let Γ_ω denote the $j\omega$-axis (as shown in Fig. 11); for $R \geq 0$ a constant, let Γ_R denote the clockwise contour bounding the semicircular region $|\sigma + j\omega| \leq R$ where $\sigma \geq 0$; let $\Gamma_\omega - \Gamma_R$ denote the difference contour; and let $H(\Gamma_\omega)$ and $H(\Gamma_\omega - \Gamma_R)$ denote images of the respective contours, each augmented with the point h_∞.

It will be shown that $H(\Gamma_R)$ has Property N for $R \geq R_0$; since by hypothesis $H(s)$ is analytic for $\sigma \geq 0$, and has no zeros on the $j\omega$-axis, Lemma 4(a) follows by the Principle of the Argument.

[12] See Widder [8], ch. II, Theorem 10.

[13] *Ibid.*, Theorem 5a.

[14] The special case $s = j\omega$ is implied by the Rieman-Lebesgue theorem (Titchmarsh [9a], Theorem 1). The general case follows from the special case and from Properties A and B by a theorem of Phragmén-Lindelöf (Titchmarsh [9b], sec. 5.64).

[15] This follows from Theorem 65 of Titchmarsh [9a], which implies that the convolution of an L_2 function with an L_1 function is in L_2, and has a transform of the type (B3).

[16] Suppose that x is in L_2; x_t is certainly in L_2, and $H(x_t)$ is in L_2 by Property D; since $[H(x_t)]_t = [Hx]_t$, it follows that $[Hx]_t$ is in L_2; i.e., Hx is in L_{2e}. Thus H maps L_{2e} into L_{2e}; since H also maps o into o, it follows that H is an operator on L_{2e} and in $\mathcal{R}_0$.

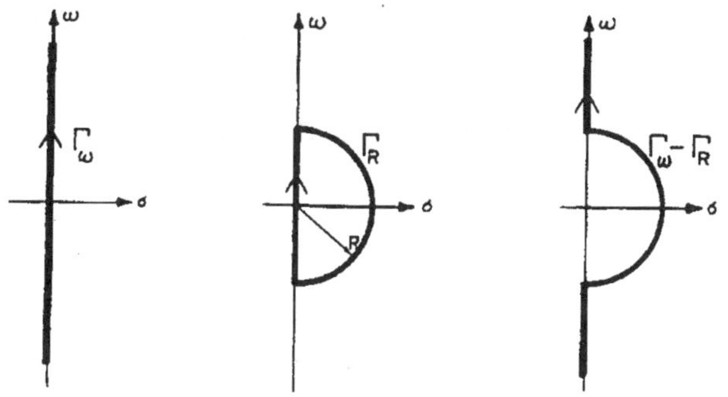

Fig. 11. Contours.

To prove that $H(\Gamma_R)$ has Property N observe that $H(\Gamma_R)=H(\Gamma_\omega)+H(\Gamma_R-\Gamma_\omega)$; since $H(\Gamma_\omega)$ has Property N by hypothesis, it is enough now to show that $H(\Gamma_R-\Gamma_\omega)$ has Property N. This can be accomplished by showing that $H(\Gamma_R-\Gamma_\omega)$ lies in a circle centered at h_∞ and not including the origin. The last assertion is a consequence of two facts.

(i) There is an $R_0>0$ for which, for $R\geq R_0$ and s in $(\Gamma_R-\Gamma_\omega)$, $|H(s)-h_\infty|\leq\frac{1}{2}|h_\infty|$.

(ii) $h_\infty\neq 0$.

(i) is obtained from Property C of Appendix B.1 for $|s|\geq R_0$, and therefore certainly holds for s in $(\Gamma_R-\Gamma_\omega)$. (ii) holds since

$$h_\infty = \lim_{\omega\to\infty} H(j\omega),$$

and since $H(j\omega)\neq 0$ by Property N.

(b) This is a special case of the Maximum Modulus Theorem of Phragmén-Lindelof.[17] The theorem implies that a function analytic in a half-plane, and bounded on the boundary, is bounded throughout the half-plane.

B.3. A Paley-Wiener Lemma

A complex-valued function $W(s)$ will be said to satisfy the Paley-Wiener conditions if

(PW1) $W(s)$ is analytic for $\sigma > 0$, and

(PW2) $\displaystyle\int_{-\infty}^{\infty} |W(\sigma + j\omega)|^2 d\omega \leq \text{const.} \qquad (\sigma > 0).$

The following lemma is a modification of Theorem 5 of Paley-Wiener [10], and is stated without proof.

Lemma 5. (a) If w is in L_2, $w_t=0$, and $W(s)$ is the l.i.m. transform of $w(\tau)$, then $W(s)$ exp (st) satisfies the Paley-Wiener conditions. (b) Conversely, if $t\geq 0$ and $W(s)$ exp (st) satisfies the Paley-Wiener conditions, then there is a function w in L_2 having the properties that $w_t=0$ and that $W(s)$ is the l.i.m. transform of w.

B.4. Proof of Lemma 1(b)

Let x in L_{2e} and $t\geq 0$ be given. Since H is linear, it is enough to show that H is conic with parameters c and r; that is,

$$\|(Hx - cx)_t\| \geq r\|x_t\|. \tag{B4}$$

For this purpose, let $y\triangleq(Hx-cx)$ and $\delta\triangleq H(x_t) - [Hx]_t$. Hence

$$y_t = (Hx - cx)_t = H(x_t) - cx_t - \delta. \tag{B5}$$

Now x_t is in L_2. Hence $H(x_t)$ is in L_2 by Property D of Appendix B.1. Since $[H(x_t)]_t=(Hx)_t$, it follows that $(Hx)_t$ is in L_2. Thus δ is in L_2, all terms in (B5) are in L_2, and by Property E,

$$Y_t(s) = H(s)X_t(s) - cX_t(s) - \Delta(s) \tag{B6}$$

where $Y_t(s)$, $X_t(s)$, and $\Delta(s)$ are the l.i.m. transforms of y_t, x_t, and δ. Hence

$$\{X_t(s)\} - \left\{\frac{\Delta(s)}{H(s) - c}\right\} = \left\{\frac{Y_t(s)}{H(s) - c}\right\}. \tag{B7}$$

Now the braced terms in (B7) are l.i.m. transforms of functions in L_2; for $X_t(s)$ this is true by definition; for the remaining terms, this can be proved by the reasoning given below in Assertion 1. Suppose that $\Delta(s)/H(s)-c$ is the l.i.m. transform of a function $q(\tau)$; it follows that

$$\|x_t - q\|^2 = \frac{1}{2\pi}\int_{-\infty}^{\infty}\left|x_t(j\omega) - \frac{\Delta(j\omega)}{H(j\omega) - c}\right|^2 d\omega$$

(Parseval's Theorem)

$$= \frac{1}{2\pi}\int_{-\infty}^{\infty}\left|\frac{Y_t(j\omega)}{H(j\omega) - c}\right|^2 d\omega \quad [\text{by (B7)}]$$

$$\leq \frac{1}{2\pi r^2}\int_{-\infty}^{\infty} |Y_t(j\omega)|^2 d\omega$$

[by inequality (8) of Section 3.2]

$$= \frac{1}{r}\|y_t\|^2 \quad \text{(Parseval's Theorem)} \tag{B8}$$

It will be shown, in Assertion 1, that $q(\tau)=0$ for almost all $\tau<t$. Therefore

$$\|x_t - q\|^2 = \|x_t\|^2 + \|q\|^2 \geq \|x_t\|^2. \tag{B9}$$

(B8) and (B9) imply (B4).

Assertion 1. The expression $\Delta(s)/[H(s)-c]$ is the l.i.m. transform of a function q in L_2; furthermore $q(\tau)=0$ for almost all $\tau<t$.

Proof of Assertion 1. By Lemma 5b, it is enough to show that $[\Delta(s)$ exp $(st)]/[H(s)-c]$ satisfies the Paley-Wiener conditions.

To prove (PW1), observe that the following three terms are analytic for $\sigma>0$: $\Delta(s)$, because it is the l.i.m. transform of an L_2 function; exp (st), because it is analytic throughout the plane; $[H(s)-c]^{-1}$, by Lemma 4 and the hypothesis on the Nyquist diagram. The product of these terms must therefore also be analytic for $\sigma>0$.

To prove (PW2), observe that

[17] Titchmarsh [9b], sec. 5.61.

$$\int_{-\infty}^{\infty}\left|\frac{\Delta(s)\exp(st)}{H(s)-c}\right|^2 d\omega \le \frac{1}{r^2}\int_{-\infty}^{\infty}|\Delta(s)\exp(st)|\,d\omega$$

$$\le \text{const.} \quad (\sigma > 0) \text{ [by inequality (8)]}$$

(Since $\delta \in L_2$ and $\delta_t = 0$ by construction, the last inequality is implied by Lemma 5(a).) Q.E.D.

Appendix C

Lemma 2

A preliminary assertion will be proved first.

Assertion 2. If N in $\mathfrak{N}$ is positive, K is in $\mathfrak{L}$, and $y = Kx$, then

$$\int_0^t \frac{dy}{d\tau}\cdot N(y(\tau))d\tau \ge 0. \qquad \text{(C1)}$$

Proof of Assertion 2.

$$\int_0^t \frac{dy}{d\tau}\cdot N(y(\tau))d\tau = \int_{y(0)}^{y(t)} N(y')dy'. \qquad \text{(C2)}$$

Since K is in $\mathfrak{L}$, $y(t)$ is given by a convolution integral, whose kernel is fixed for a fixed x, and whose limits of integration are 0 and t; therefore $y(0) = 0$. Furthermore, since N is positive, its graph lies in the first and third quadrants. It follows that the right-hand side of (C2) is non-negative. Q.E.D.

C.1. *Proof of Lemma 2*

Part a) It is required to show that, for any given x in L_{2e} and any given $t \ge 0$, the inequality

$$\int_0^t x(\tau)\cdot[NKx(\tau)]d\tau \ge 0 \qquad \text{(C3)}$$

holds. For this purpose, make the following substitutions: Write $y \triangleq Kx$, and observe that, since $K(s) = k\lambda/(s+\lambda)$, y is differentiable and

$$x(\tau) = \frac{1}{k\lambda}\frac{dy}{d\tau} + \frac{1}{k}y(\tau). \qquad \text{(C4)}$$

(C3) is therefore equivalent to

$$\int_0^t \left\{\frac{1}{k\lambda}\frac{dy}{d\tau} + \frac{1}{k}y(\tau)\right\}\cdot N(y(\tau))d\tau \ge 0 \text{ or}$$

$$\frac{1}{k\lambda}\int_0^t \frac{dy}{d\tau}\cdot N(y(\tau))d\tau + \frac{1}{k}\int_0^t y(\tau)\cdot N(y(\tau))d\tau \ge 0. \qquad \text{(C5)}$$

Now k and λ are positive by hypothesis; the first integral in (C5) is non-negative by Assertion 2; the second integral is non-negative, since N is a positive operator; therefore (C5) is true. Q.E.D.

Part b) It will be assumed, for simplicity, that $\beta > 0$.

Case A. Suppose $\alpha \ge 0$. It must be shown that NK is inside $\{\alpha, \beta\} \times \{0, k\}$. This is equivalent to saying that NK is inside $\{0, k\beta\}$, or that

$$\langle (NKx)_t), (NKx - k\beta x)_t\rangle \le 0. \qquad \text{(C6)}$$

(C6) is equivalent to the inequality

$$k\beta\langle x_t, (Ny)_t\rangle \ge \|(Ny)_t\|^2 \qquad \text{(C7)}$$

where $y = Kx$. Now recalling that $\langle x_t, (Ny)_t\rangle$ equals the left-hand side of (C5), we get

$$k\beta\langle x_t, (Ny)_t\rangle \ge \beta\int_0^t y(\tau)\cdot N(y(\tau))d\tau.$$

Observing that, for $\alpha \ge 0$, N satisfies the inequality $\beta y(\tau)\cdot N(y(\tau)) \ge [N(y(\tau))]^2$, we get

$$k\beta\langle x_t, (Ny)_t\rangle \ge \int_0^t [N(y(\tau))]^2 d\tau,$$

which implies (C7). Q.E.D.

Case B. Suppose $\alpha < 0$. Decompose $N(x)$ into two parts, $N(x) = N_+(x) + N_-(x)$; let $N_+(x) = N(x)$ for $N(x) \ge 0$ and $N_+(x) = 0$ elsewhere, and let $N_-(x)$ be similarly defined.

Since N_+ is clearly inside $\{0, \beta\}$, Case A implies that N_+K is inside $\{0, k\beta\}$. Similarly N_-K is inside $\{k\alpha, 0\}$. On summing the sectors of N_+K and N_-K (by the Sum Rule of Part I) it is found that NK is inside $\{k\alpha, k\beta\}$; that is, inside $\{\alpha, \beta\} \times \{0, k\}$. Q.E.D.

Appendix D

Lemma 3

Before proving Lemma 3, a few related assertions will be introduced.

Assertion 3. Let K be an operator in $\mathfrak{RC}$, x a fixed element of L_{2e}, and $y \triangleq Kx$. Then x has a "Foster expansion" in y;[18] that is, x can be expressed as a finite sum,

$$x = \sum_{i=0}^{m} F_i y,$$

in which F_i are operators mapping the image under K of L_{2e} into $R[0, \infty)$, as follows:

Case 1. $F_0 y = K^{-1}(0)\cdot y$.

Case 2. If $i = 1, 2, \cdots, (m-1)$, then F_i is in $\mathfrak{L}$ and has a Laplace Transform,

$$F_i(s) = h_i s/(s + \theta_i), \quad (h_i > 0, \quad \theta_i > 0).$$

Case 3. $F_m y = h_m \dot{y}$ if $K(\infty) = 0$ and $F_m y = 0$ otherwise, where $h_m \ge 0$.

Assertion 4. If N is incrementally positive, and $\langle x_t, [Ny]_t\rangle \ge 0$ then $\langle x_t, [N(x+y)]_t\rangle \ge 0$.

Proof of Assertion 4. It is enough to show that

$$\langle x_t, [N(x + y)_t]\rangle - \langle x_t, [Ny]_t\rangle \ge 0 \qquad \text{(D1)}$$

But the left-hand side of (D1) can be expressed as

$$\langle (x + y)_t - y_t, [N(x + y)]_t - (Ny)_t\rangle$$

which has the form $\langle x_{1t} - x_{2t}, Nx_{1t} - Nx_{2t}\rangle$, and is non-negative, since N is an incrementally positive operator. Therefore (D1) holds. Q.E.D.

[18] See Guillemin [2], p. 115.

ASSERTION 5. If N is an operator in $\mathfrak{N}$, incrementally inside a sector $\{\alpha, \beta\}$ where $\alpha<0$ and $\beta>0$, then N can be decomposed into $N=N_+ + N_-$, where N_- is inside $\{\alpha, 0\}$, and N_+ is inside $\{0, \beta\}$.

PROOF OF ASSERTION 5. Since N is incrementally inside a sector, its graph N is continuous and has bounded variation on every finite interval. Consequently N can be expressed as an integral, $N(x)=\int_0^x n(x')dx'$. Let $n_+(x)=n(x)$ if $n(x)\geq 0$, and $n_+(x)=0$ if $n(x)<0$; let $N_+(x)=\int_0^x n_+(x')dx'$. Clearly N_+ has the desired property. N_- is constructed similarly. Q.E.D.

D.1. Proof of Lemma 3

Part a) Let x be any given element of L_{2e}, and t any given point in $[0, \infty)$; it is required to show that

$$\langle x_t, (NKx)_t\rangle \geq 0. \tag{D2}$$

Letting $y \triangleq Kx$, and recalling that x can be expressed by the Foster expansion

$$x = \sum_{i=0}^{m} F_i y$$

(see Assertion 3), (D2) is equivalent to

$$\sum_{i=0}^{m} \langle (F_i y)_t, (Ny)_t\rangle \geq 0. \tag{D3}$$

It will be shown that each component on the left-hand side of (D3) is non-negative.

CASE 1. Here $F_0 y = K^{-1}(0)\cdot y$. Hence $\langle (F_0 y)_t, (Ny)_t\rangle = [K^{-1}(0)]\cdot\langle y_t, (Ny)_t\rangle$; this is non-negative since N is a positive operator, and since $K(0)$ is necessarily positive.

CASE 2. Here $F_i(s) = h_i s/(s+\theta_i)$. Let

$$z(t) = h_i^{-1}\int_0^t F_i y(\tau)d\tau.$$

It follows that $y = \dot{z} + \theta_i z$ almost everywhere, and that $F_i y = h_i \dot{z}$ almost everywhere. Hence

$$\langle (F_i y)_t, (Ny)_t\rangle = h_i\langle \dot{z}_t, [N(\dot{z} + \theta_i z)]_t\rangle. \tag{D4}$$

Now, observing that $\theta_i>0$, Assertion 2 implies that $\langle \dot{z}, [N(\theta_i z)]_t\rangle$ is non-negative. Observing that h_i is positive, the right-hand side of (D4) is non-negative by Assertion 4. Thus Case 2 is proved.

CASE 3. Here $F_m y = h_m \dot{y}$ if $K(\infty)=0$. Hence $\langle (F_m y)_t, (Ny)_t\rangle = h_m\langle \dot{y}_t, (Ny)_t\rangle$. Case 3 follows by Assertion 2.

Since the inner product is non-negative in all three cases, (D3) holds. Q.E.D.

Part b) Assume, for simplicity, that $\beta>0$.

CASE A. If $\alpha=0$ then, by reasoning similar to that used in Lemma 2(b), (C5)–(C7), the following inequality is obtained:

$$\langle x_t, (NKx)_t\rangle \geq [\beta K(0)]^{-1}\|(Ny)_t\|^2.$$

Hence NK is inside $\{0, \beta K(0)\}$, which equals $\{0, \beta\}\times\{0, K(0)\}$.

CASE B. If $\alpha>0$ then NK is decomposed into three parts

$$NK = \{[N - \alpha I]K\} + \{\alpha[K - K(\infty)\cdot I]\} + \{\alpha K(\infty)\cdot I\}. \tag{D5}$$

Now the three parts lie in the sectors $\{0, (\beta-\alpha)K(0)\}$, $\{0, \alpha[K(0)-K(\infty)]\}$, and $\{\alpha K(\infty), \alpha K(\infty)\}$, respectively. (The first two of these sectors are determined by the rule formed in Case A, after observing that $[N-\alpha I]$ is inside $\{0, \beta-\alpha\}$, and that $[K-K(\infty)\cdot I]$ is inside $\{0, [K(0)-K(\infty)]\}$; the third sector is simply the sector of a constant times the identity.) On summing the three sectors (by the Sum Rule of Part I), it is found that NK is inside $\{\alpha K(\infty), \beta K(0)\}$; that is, inside $\{\alpha, \beta\}\times\{K(\infty), K(0)\}$.

CASE C. If $\alpha<0$, N is decomposed into N_+ and N_-, as in Assertion 5. This case then follows by the reasoning used in Case B of Lemma 2. Q.E.D.

ACKNOWLEDGMENT

The author thanks Dr. P. Falb for correcting the manuscript, and for offering many valuable suggestions. He also thanks Dr. G. Kovatch and NASA's Electronic Research Center, Cambridge, Mass., for supporting the completion of the paper, and Mrs. Iris McDonald for typing it.

REFERENCES

[1a] G. Zames, "On the stability of nonlinear, time-varying feedback systems," *Proc. 1964 NEC*, vol. 20, pp. 725–730.

[1b] ——, "Nonlinear, time-varying feedback systems—Conditions for L_∞-boundedness derived using conic operators on exponentially weighted spaces," *Proc. 1965 Allerton Conf.*, pp. 460–471.

[1c] ——, "On the input-output stability of time-varying nonlinear feedback systems—Part I. Conditions derived using concepts of loop gain, conicity, and positivity," *IEEE Trans. on Automatic Control*, vol. AC-11, pp. 228–239, April 1966.

[2] E. A. Guillemin, *Synthesis of Passive Networks*. New York: Wiley, 1957.

[3] H. Hatanaka, "The frequency responses and jump-resonance phenomena of nonlinear feedback control systems," *Trans. ASME*, pp. 236–242, June 1963.

[4a] R. W. Brockett and J. W. Willems, "Frequency domain stability criteria, Parts I and II," *Proc. JACC*, pp. 735–747, 1965.

[4b] R. W. Brockett and L. J. Forys, "On the stability of systems containing a time-varying gain," *Proc. 1964 Allerton Conf.*, pp. 413–430.

[5] I. W. Sandberg, "A frequency domain condition for the stability of systems containing a single time-varying nonlinear element," *Bell Sys. Tech. J.*, vol. 43, p. 1601, 1964.

[6] K. S. Narendra, and R. M. Goldwyn, "A geometrical criterion for the stability of certain nonlinear, nonautonomous systems," *IEEE Trans. on Circuit Theory*, vol. CT-11, pp. 406–408, September 1964.

[7] J. Kudrewicz, "Stability of nonlinear feedback systems," *Automatika i Telemechanika*, vol. 25, no. 8, 1964.

[8] D. V. Widder, *The Laplace Transform*. Princeton, N. J.: Princeton University Press, 1946.

[9a] E. C. Titchmarsh, *Introduction to the Theory of Fourier Integrals*, 2nd ed. Oxford, England: University Press, 1962.

[9b] ——, *The Theory of Functions*, 2nd ed. Oxford, England: University Press, 1964.

[10] R. Paley and N. Wiener, *Fourier Transforms in the Complex Domain*. New York: Am. Math. Soc. Colloquium Publications, 1934.

An Invariance Principle in the Theory of Stability

JOSEPH P. LASALLE

THE stability theorems of Lyapunov have been among the oldest and strongest pillars of control theory. The centenary of Lyapunov's celebrated 1892 memoir was recently marked with its English translation [17], while the 1907 French translation was reprinted in 1949 by Princeton University Press [16].

Applications of Lyapunov stability concepts to control problems began to appear in Russia in the 1930s–1940s. Significant theoretical results from the 1950s were summarized in the books by Chetayev [3], Lurie [15], Malkin [18], Letov [14], Zubov [22], and Krasovskii [7]. In the post-1957 Sputnik era, English translations of these books, and of other Russian works, further increased the interest already stimulated by "Contributions to the Theory of Nonlinear Oscillations," a five volume series (1950, 1952, 1956, 1958, and 1960) edited by Lefschetz, and published by Princeton University Press. The fourth volume contained a detailed and rigorous survey of Lyapunov stability theory by Antosiewicz [1]. The 1960 survey by Kalman and Bertram [6] was more accessible and had a stronger impact on engineering audiences, as did the books by Lefschetz [12], Lefschetz and LaSalle [13], and a paper by LaSalle [10]. This activity continued in the early 1960s when several volumes of "Contributions to Differential Equations" were published, including the important results by Yoshizawa [20], whose 1966 book [21] presented a collection of advanced stability results. The status of stability theory in the pre-1965 period is summarized in the scholarly work by Hahn [4], which is the most comprehensive source covering that period.

Among the innovations from that period are the results which settle the issue of existence of Lyapunov functions. Particularly important among the results of this type are the "converse theorems" of Massera [19], Krasovskii [14] and Kurzweil [8], which found applications in diverse areas of control theory. Radially unbounded Lyapunov functions were introduced in 1952 by Barbashin and Krasovskii [2] to guarantee *stability in the large*, that is, global stability. The same Barbashin-Krasovskii paper and the book by Krasovskii [14] initiated another line of research which culminated in this paper by LaSalle, and its extended 1968 version [11].

This line of research was aimed at extracting from Lyapunov's stability theorem more information about the asymptotic behavior of the solutions $x(t) \in \mathbb{R}^n$ of

$$\dot{x} = f(x, t), \quad f(0, t) = 0 \quad \text{for all } t \tag{1}$$

With a positive definite function $V(x, t)$, a theorem of Lyapunov establishes asymptotic stability of the equilibrium $x = 0$, if $\dot{V}(x, t)$ is negative definite along the solutions $x(t)$ of (1). If $\dot{V}$ is only nonpositive, the same theorem guarantees stability, but does not reveal whether $x(t)$ converges to zero or to a set in $\mathbb{R}^n$. For autonomous systems, $\dot{x} = f(x)$, a theorem of Barbashin and Krasovskii [2] (see also Theorem 14.1 on p. 67 of [14]) examines the set E in which $\dot{V} = 0$. If this set does not contain any positive semi-trajectory (solution $x(t)$ for all $t \geq 0$) other than $x(t) \equiv 0$, then $x(t) \to 0$ as $t \to \infty$. This result gave a practical test for asymptotic stability when $\dot{V}$ is only nonpositive, instead of negative definite. In his 1960 paper [10], LaSalle extracted further information on the asymptotic behavior of $x(t)$. Using the limit sets and the largest invariant set M of $\dot{x} = f(x)$ contained in the set E where $\dot{V}$ is zero, he showed that $x(t)$ converges to the set M. This set need not be an equilibrium, but can be a compact limit set like a limit cycle.

This result of LaSalle, which he later termed *Invariance Principle*, had a significant connection with the then new concept of *observability*. For the system $\dot{x} = Ax$, a positive definite Lyapunov function $V = x'Px$ has the derivative $\dot{V} = A'P + PA = -x'Qx$. Suppose that Q is positive semi-definite, so that it can be expressed as $Q = C'C$ for some matrix C. It then follows that $\dot{V} = -y'y$, where $y = Cx$ can be treated as an output of $\dot{x} = Ax$. Clearly, the set E where $\dot{V} = 0$ is $y(t) \equiv 0$. If the pair (A, C) is completely observable, then $y(t) \equiv 0$ implies $x(t) = 0$. Hence, no nonzero positive semi-trajectory $x(t)$ is contained in E, which proves, via Barbashin-Krasovskii [2], that $\dot{x} = Ax$ is asymptotically stable. However, if the system is only stable with a pair of purely imaginary eigenvalues unobservable from $y = Cx$, then LaSalle's Principle shows that $x(t)$ converges to the periodic solution in the unobservable subspace.

The Invariance Principle was subsequently extended to periodic and almost periodic systems, but it does not hold for more general nonautonomous systems (1). To obtain similar information on the asymptotic behavior of $x(t)$, Yoshizawa [20] derived a set of conditions under which $\dot{V}(x, t) \leq W(x)$, where $W(x)$ is "positive definite with respect to a set M," implies that $x(t)$ converges to M. The main theorem in this paper by LaSalle modifies and improves this result of Yoshizawa, and provides the strongest convergence result for nonautonomous systems. To go beyond this result, further restrictions on $f(x, t)$ are needed. One of them is the so-called "persistency of excitation" condition in adaptive identification and control. Typically, an adaptive algorithm guarantees that $\dot{V} \leq -e^2$, where e is the scalar tracking error. The Yoshizawa-LaSalle theorem provides the conditions under which $e(t)$ converges to zero. It has thus become an indispensable tool in adaptive control design. It has also become instrumental in deducing stability from passivity properties as in feedback passivation and backstepping designs of nonlinear systems.

Other settings where LaSalle's *Invariance Principle* has been studied are infinite dimensional systems, as in the work of Hale [5], and stochastic systems governed by continuous-time Markov processes, as discussed by Kushner in [9].

REFERENCES

[1] H. ANTOSIEWICZ, "A survey of Lyapunov's second method," in *Contr. to Nonlinear Oscillations*, S. Lefschetz, Ed., **4**:141–166 (Princeton Univ. Press, Princeton NJ), 1958.

[2] E.A. BARBASHIN AND N.N. KRASOVSKII, "On the stability of motion in the large," *Dokl. Akad. Nauk USSR*, **86**:453–456, 1952.

[3] N.G. CHETAYEV, *The Stability of Motion*, Pergamon Press (Oxford), 1961. (Russian original, 1946.)

[4] W. HAHN, *Stability of Motion* (Springer-Verlag, New York), 1967.

[5] J.K. HALE, "Dynamical systems and stability," *J. Math. Anal. Appl.*, **26**:39–59, 1969.

[6] R.E. KALMAN AND J.E. BERTRAM, "Control system analysis and design via the 'second method' of Lyapunov, I Continuous-time systems," *J. Basic Engineering (Trans. ASME)*, **82D**:371–393, 1960.

[7] N.N. KRASOVSKII, *Stability of Motion*, Stanford Univ. Press (Stanford, CA), 1963. (Russian original, 1959.)

[8] J. KURZWEIL, "The converse second Liapunov's theorem concerning the stability of motion," *Czechoslovak Math. J.*, **6**(81):217–259 & 455–473, 1956.

[9] H.J. KUSHNER, "Stochastic stability," in *Stability of Stochastic Dynamical Systems*, R. Curtain, Ed., Lect. Notes in Math., Springer-Verlag (New York), **294**:97–124, 1972.

[10] J.P. LASALLE, "Some extensions of Liapunov's second method," *IRE Trans. Circuit Theory*, **CT-7**:520–527, 1960.

[11] J.P. LASALLE, "Stability theory for ordinary differential equations," *J. Differential Equations*, **4**:57–65, 1968.

[12] S. LEFSCHETZ, *Differential Equations: Geometric Theory*, Interscience, Wiley (New York), 1957.

[13] S. LEFSCHETZ AND J.P. LASALLE, *Stability by Liapunov's Direct Method, with Applications*, Academic Press (New York), 1961.

[14] A.M. LETOV, *Stability in Nonlinear Control Systems* (English translation), Princeton Univ. Press (Princeton, NJ), 1961. (Russian original, 1955.)

[15] A.E. LURIE, *Some Non-linear Problems in the Theory of Automatic Control* (English translation), H.M.S.O., London, 1957. (Russian original, 1951.)

[16] A.M. LYAPUNOV, "Probléme général de la stabilité du mouvement" (in French), *Ann. Fac. Sci. Toulouse*, **9**:203–474, 1907. Reprinted in *Ann. Math. Study*, No. 17, 1949, Princeton Univ. Press (Princeton, NJ).

[17] A.M. LYAPUNOV, "The general problem of the stability of motion" (translated into English by A.T. Fuller), *Int. J. Control*, **55**:531–773, 1992.

[18] I.G. MALKIN, *Theory of Stability of Motion*, AEC (Atomic Energy Commission) Translation 3352, Dept. of Commerce, United States, 1958. (Russian original, 1952.)

[19] J.L. MASSERA, "Contributions to stability theory," *Ann. Math.*, **64**:182–206, 1956.

[20] T. YOSHIZAWA, "Asymptotic behavior of solutions of a system of differential equations," in *Contributions to Differential Equations*, **1**:371–387, 1963.

[21] T. YOSHIZAWA, *Stability Theory by Liapunov's Second Method*, The Mathematical Society of Japan, Publication No. 9 (Tokyo), 1966.

[22] V.I. ZUBOV, *Mathematical Methods for the Study of Automatic Control Systems*, Pergamon Press (Oxford), 1962. (Russian original, 1957.)

P.V.K. & J.B.

An Invariance Principle in the Theory of Stability

JOSEPH P. LASALLE[1]
Center for Dynamical Systems
Brown University, Providence, Rhode Island

1. Introduction

The purpose of this paper is to give a unified presentation of Liapunov's theory of stability that includes the classical Liapunov theorems on stability and instability as well as their more recent extensions. The idea being exploited here had its beginnings some time ago. It was, however, the use made of this idea by Yoshizawa in [*7*] in his study of nonautonomous differential equations and by Hale in [*1*] in his study of autonomous functional-differential equations that caused the author to return to this subject and to adopt the general approach and point of view of this paper. This produces some new results for dynamical systems defined by ordinary differential equations which demonstrate the essential nature of a Liapunov function and which may be useful in applications. Of greater importance, however, is the possibility, as already indicated by Hale's results for functional-differential equations, that these ideas can be extended to more general classes of dynamical systems. It is hoped, for instance, that it may be possible to do this for some special types of dynamical systems defined by partial differential equations.

In Section 2 we present some basic results for ordinary differential equations. Theorem 1 is a fundamental stability theorem for nonautonomous systems and is a modified version of Yoshizawa's Theorem 6 in [*7*]. A simple example shows that the conclusion of this theorem is the best possible. However, whenever the limit sets of solutions are known to have an invariance property, then sharper results can be obtained. This "invariance principle" explains the title of this paper. It had its origin for autonomous and periodic systems in [*2*] and [*4*], although we present here improved versions of those results. Miller in [*5*] has established an invariance property

[1] This research was supported in part by the National Aeronautics and Space Administration under Grant No. NGR-40-002-015 and under Contract No. NAS8-11264, in part by the United States Air Force through the Air Force Office of Scientific Research under Grant No. AF-AFOSR-693-65, and in part by the United States Army Research Office, Durham, under Contract No. DA-31-124-ARO-D-270.

Reprinted with permission from *Differential Equations and Dynamical Systems* (New York: Academic Press, 1967), J. Hale and J. P. LaSalle, eds., Joseph P. LaSalle, "An Invariance Principle in the Theory of Stability," pp. 277–286.

for almost periodic systems and obtains thereby a similar stability theorem for almost periodic systems. Since little attention has been paid to theorems which make possible estimates of regions of attraction (regions of asymptotic stability) for nonautonomous systems results of this type are included. Section 3 is devoted to a brief discussion of some of Hale's recent results [1] for autonomous functional-differential equations.

2. Ordinary Differential Equations

Consider the system

$$\dot{x} = f(t, x) \tag{1}$$

where x is an n-vector, f is a continuous function on R^{n+1} to R^n and satisfies any one of the conditions guaranteeing uniqueness of solutions. For each x in R^n, we define $|x| = (x_1^2 + \cdots + x_n^2)^{1/2}$, and for E a closed set in R^n we define $d(x, E) = \text{Min}\,\{|x - y|;\ y \text{ in } E\}$. Since we do not wish to confine ourselves to bounded solutions, we introduce the point at ∞ and define $d(x, \infty) = |x|^{-1}$. Thus, when we write $E^* = E \cup \{\infty\}$, we shall mean $d(x, E^*) = \text{Min}\,\{d(x, E), d(x, \infty)\}$. If $x(t)$ is a solution of (1), we say that $x(t)$ approaches E as $t \to \infty$, if $d(x(t), E) \to 0$ as $t \to \infty$. If we can find such a set E, we have obtained information about the asymptotic behavior of $x(t)$ as $t \to \infty$. The best that we could hope to do is to find the smallest closed set Ω that $x(t)$ approaches as $t \to \infty$. This set Ω is called the *positive limit set* of $x(t)$ and the points p in Ω are called the *positive limit points* of $x(t)$. In exactly the same way, one defines $x(t) \to E$ as $t \to -\infty$, negative limit sets, and negative limit points. This is exactly Birkhoff's concept of limit sets. A point p is a positive limit point of $x(t)$, if and only if there is a sequence of times t_n approaching ∞ as $n \to \infty$ and such that $x(t_n) \to p$ as $n \to \infty$. In the above, it may be that the maximal interval of definition of $x(t)$ is $[0, \tau)$. This causes no difficulty since, in the results to be presented here, we need only, with respect to time t, replace ∞ by τ. We usually ignore this possibility and speak as though our solutions are defined on $[0, \infty)$ or $(-\infty, \infty)$.

Let $V(t, x)$ be a C^1 function on $[0, \infty) \times R^n$ to R, and let G be any set in R^n. We shall say that V is a *Liapunov function* on G for Eq. (1), if $V(t, x) \geqq 0$ and $\dot{V}(t, x) \leqq -W(x) \leqq 0$ for all $t > 0$ and all x in G, where W is continuous on R^n to R, and

$$\dot{V} = \frac{\partial V}{\partial t} + \sum_{i=1}^{n} \frac{\partial V}{\partial x_i} f_i. \tag{2}$$

We define ($\bar{G}$ is the closure of G)

$$E = \{x;\ W(x) = 0,\ x \text{ in } \bar{G}\}.$$

The following result is then a modified but closely related version of Yoshizawa's Theorem 6 in [7].

Theorem 1. If V is a Liapunov function on G for Eq. (1), then each solution $x(t)$ of (1) that remains in G for all $t > t_0 \geqq 0$ approaches $E^* = E \cup \{\infty\}$ as $t \to \infty$, provided one of the following conditions is satisfied:

(i) For each p in $\bar{G}$ there is a neighborhood N of p such that $|f(t, x)|$ is bounded for all $t > 0$ and all x in N.

(ii) W is C^1 and $\dot{W}$ is bounded from above or below along each solution which remains in G for all $t > t_0 \geqq 0$.

If E is bounded, then each solution of (1) that remains in G for $t > t_0 \geqq 0$ either approaches E or ∞ as $t \to \infty$.

Thus this theorem explains precisely the nature of the information given by a Liapunov function. A Liapunov function relative to a set G defines a set E, which under the conditions of the theorem contains (locates) all the positive limit sets of solutions which for positive time remain in G. The problem in applying the result is to find "good" Liapunov functions. For instance, the zero function $V = 0$ is a Liapunov function for the whole space R^n and condition (ii) is satisfied but gives no information since $E = R^n$. It is trivial but useful for applications to note that if V_1 and V_2 are Liapunov functions on G, then $V = V_1 + V_2$ is also a Liapunov function and $E = E_1 \cap E_2$. If E is smaller than either E_1 or E_2, then V is a "better" Liapunov function than either E_1 or E_2 and is always at least as "good" as either of the two.

Condition (i) of Theorem 1 is essentially the one used by Yoshizawa. We now look at a simple example, where condition (ii) is satisfied and condition (i) is not. The example also shows that the conclusion of the theorem is the best possible. Consider $\ddot{x} + p(t)\dot{x} + x = 0$, where $p(t) \geqq \delta > 0$. Define $2V = x^2 + y^2$, where $y = \dot{x}$. Then $\dot{V} = -p(t)y^2 \leqq -\delta y^2$ and V is a Liapunov function on R^2. Now $W = \delta y^2$ and $\dot{W} = 2\delta y\dot{y} = -2\delta(xy + p(t)y^2) \leq -2\delta xy$. Since all solutions are evidently bounded for all $t > 0$, condition (ii) is satisfied. Here E is the x-axis ($y = 0$) and for each solution $x(t)$, $y(t) = \dot{x}(t) \to 0$ as $t \to \infty$. Noting that the equation $\ddot{x} + (2 + \exp[t])\dot{x} + x = 0$ has a solution $x(t) = 1 + \exp(-t)$, we see that this is the best possible result without further restrictions on p.

In order to use Theorem 1, there must be some means of determining which solutions remain in G. The following corollary, which is an obvious consequence of Theorem 1, gives one way of doing this and also provides, for nonautonomous systems, a method for estimating regions of attraction.

Corollary 1. Assume that there exist continuous functions $u(x)$ and $v(x)$ on R^n to R such that $u(x) \leqq V(t, x) \leqq v(x)$ for all $t \geqq 0$. Define $Q_\eta^+ = \{x; u(x) < \eta\}$ and let G^+ be a component of Q_η^+. Let G denote the component of $Q_\eta = \{x; v(x) < \eta\}$ containing G^+.

If V is a Liapunov function on G for (1) and the conditions of Theorem 1 are satisfied, then each solution of (1) starting in G^+ at any time $t_0 \geqq 0$ remains in G for all $t > t_0$ and approaches E^* as $t \to \infty$. If G is bounded and $E^0 = E \cap G \subset G^+$, then E^0 is an attractor and G^+ is in its region of attraction.

In general we know that if $x(t)$ is a solution of (1)—in fact, if $x(t)$ is any continuous function on R to R^n—then its positive limit set is closed and connected. If $x(t)$ is bounded, then its positive limit set is compact. There are, however, special classes of differential equations where the limit sets of solutions have an additional invariance property which makes possible a refinement of Theorem 1. The first of these are the autonomous systems

$$\dot{x} = f(x). \tag{3}$$

The limit sets of solutions of (3) are invariant sets. If $x(t)$ is defined on $[0, \infty)$ and if p is a positive limit point of $x(t)$, then points on the solution through p on its maximal interval of definition are positive limit points of $x(t)$. If $x(t)$ is bounded for $t > 0$, then it is defined on $[0, \infty)$, its positive limit set Ω is compact, nonempty and solutions through points p of Ω are defined on $(-\infty, \infty)$ (i.e., Ω is invariant). If the maximal domain of definition of $x(t)$ for $t > 0$ is finite, then $x(t)$ has no finite positive limit points: That is, if the maximal interval of definition of $x(t)$ for $t > 0$ is $[0, \beta)$, then $x(t) \to \infty$ as $t \to \beta$. As we have said before, we will always speak as though our solutions are defined on $(-\infty, \infty)$ and it should be remembered that finite escape time is always a possibility unless there is, as for example in Corollary 2 below, some condition that rules it out. In Corollary 3 below, the solutions might well go to infinity in finite time.

The invariance property of the limit sets of solutions of autonomous systems (3) now enables us to refine Theorem 1. Let V be a C^1 function on R^n to R. If G is any arbitrary set in R^n, we say that V is a *Liapunov function on* G for Eq. (3) if $\dot{V} = (\text{grad } V) \cdot f$ does not change sign on G. Define $E = \{x; \dot{V}(x) = 0, x \text{ in } \bar{G}\}$, where $\bar{G}$ is the closure of G. Let M be the largest invariant set in E. M will be a closed set. The fundamental stability theorem for autonomous systems is then the following:

Theorem 2. If V is a Liapunov function on G for (3), then each solution $x(t)$ of (3) that remains in G for all $t > 0$ $(t < 0)$ approaches $M^* = M \cup \{\infty\}$ as $t \to \infty$ $(t \to -\infty)$. If M is bounded, then either $x(t) \to M$ or $x(t) \to \infty$ as $t \to \infty$ $(t \to -\infty)$.

This one theorem contains all of the usual Liapunov like theorems on stability and instability of autonomous systems. Here, however, there are no conditions of definiteness for V or $\dot{V}$, and it is often possible to obtain stability information about a system with these more general types of Liapunov functions. The first corollary below is a stability result which for applications has been quite useful, and the second illustrates how one obtains information on instability. Četaev's instability theorem is similarly an immediate consequence of Theorem 2 (see Section 3).

Corollary 2. Let G be a component of $Q_\eta = \{x; V(x) < \eta\}$. Assume that G is bounded, $\dot{V} \leqq 0$ on G, and $M^0 = \overline{M \cap G} \subset G$. Then M^0 is an attractor and G is in its region of attraction. If, in addition, V is constant on the boundary of M^0, then M^0 is a stable attractor.

Note that if M^0 consists of a single point p, then p is asymptotically stable and G provides an estimate of its region of asymptotic stability.

Corollary 3. Assume that relative to (3) that $V\dot{V} > 0$ on G and on the boundary of G that $V = 0$. Then each solution of (3) starting in G approaches ∞ as $t \to \infty$ (or possibly in finite time).

There are also some special classes of nonautonomous systems where the limit sets of solutions have an invariance property. The simplest of these are periodic systems (see [2])

$$\dot{x} = f(t, x), \qquad f(t + T, x) = f(t) \quad \text{for all } t \text{ and } x. \tag{4}$$

Here, in order to avoid introducing the concept of a periodic approach of a solution of (4) to a set and the concept of a periodic limit point, let us confine ourselves to solutions $x(t)$ of (4) which are bounded for $t > 0$. Let Ω be the positive limit set of such a solution $x(t)$, and let p be a point in Ω. Then there is a solution of (4) starting at p which remains in Ω for all t in $(-\infty, \infty)$; that is, if one starts at p at the proper time, the solution remains in Ω for all time. This is the sense now in which Ω is an invariant set. Let $V(t, x)$ be C^1 on $R \times R^n$ and periodic in t of period T. For an arbitrary set G of R^n we say that V is a *Liapunov function on* G for the periodic system (4) if $\dot{V}$ does not change sign for all t and all x in G. Define $E = \{(t, x); \dot{V}(t, x) = 0, x \text{ in } \bar{G}\}$ and let M be the union of all solutions $x(t)$ of (4) with the property that $(t, x(t))$ is in E for all t. M could be called "the largest invariant set relative to E." One then obtain the following version of Theorem 2 for periodic systems:

Theorem 3. If V is a Liapunov function on G for the periodic system (4), then each solution of (4) that is bounded and remains in G for all $t > 0$ $(t < 0)$ approaches M as $t \to \infty$ $(t \to -\infty)$.

In [5] Miller showed that the limit sets of solutions of almost periodic systems have a similar invariance property and from this he obtains a result quite like Theorem 3 for almost periodic systems. This then yields, for periodic and almost periodic systems, a whole chain of theorems on stability and instability quite similar to that for autonomous systems. For example, one has

Corollary 4. Let $Q_\eta^+ = \{x; V(t, x) < \eta, \text{ all } t \text{ in } [0, T]\}$, and let G^+ be a component of Q_η^+. Let G be the component of $Q_\eta = \{x; V(t, x) < \eta$ for some r in $[0, T]\}$ containing G^+. If G is bounded, $\dot{V} \leqq 0$ for all t and all x in G, and if $M^0 = \overline{M \cap G} \subset G^+$, then M^0 is an attractor and G^+ is an its region of attraction. If $V(t, x) = \varphi(t)$ for all t and all x on the boundary of M^0, then M^0 is a stable attractor.

Our last example of an invariance principle for ordinary differential equations is that due to Yoshizawa in [7] for "asymptotically autonomous" systems. It is a consequence of Theorem 1 and results by Markus and Opial (see [7] for references) on the limit sets of such systems. A system of the form

$$\dot{x} = F(x) + g(t, x) + h(t, x) \tag{5}$$

is said to be *asymptotically autonomous* if (i) $g(t, x) \to 0$ as $t \to \infty$ uniformly for x in an arbitrary compact set of R^n, (ii) $\int_0^\infty |h(t, \varphi(t))| \, dt < \infty$ for all φ bounded and continuous on $[0, \infty)$ to R^n. The combined results of Markus and Opial then state that the positive limit sets of solutions of (5) are invariant sets of $\dot{x} = F(x)$. Using this, Yoshizawa then improved Theorem 1 for asymptotically autonomous systems.

It turns out to be useful, as we shall illustrate in a moment on the simplest possible example, in studying systems (1) which are not necessarily asymptotically autonomous to state the theorem in the following manner:

Theorem 4. If, in addition to the conditions of Theorem 1, it is known that a solution $x(t)$ of (1) remains in G for $t > 0$ and is also a solution of an asymptotically autonomous system (5), then $x(t)$ approaches $M^* = M \cup \{\infty\}$ as $t \to \infty$, where M is the largest invariant set of $\dot{x} = F(x)$ in E.

It can happen that the system (1) is itself asymptotically autonomous, in which case the above theorem can be applied. However, as the following example illustrates, the original system may not itself be asymptotically autonomous, but it still may be possible to construct for each solution of (1) an asymptotically autonomous system (5) which it also satisfies.

Consider again the example

$$\begin{aligned} \dot{x} &= y \\ \dot{y} &= -x - p(t)y, \qquad 0 < \delta \leqq p(t) \leqq m \\ &\qquad\qquad \text{for all } t > 0. \end{aligned} \tag{6}$$

Now we have the additional assumption that $p(t)$ is bounded from above. Let $(\bar{x}(t), \bar{y}(t))$ be any solution of (6). As was argued previously below Theorem 1, all solutions are bounded and $\bar{y}(t) \to 0$ as $t \to \infty$. Now $(\bar{x}(t), \bar{y}(t))$ satisfies $\dot{x} = y$, $\dot{y} = -x - p(t)\,\bar{y}(t)$, and this system is asymptotically autonomous to (*) $\dot{x} = y$, $\dot{y} = -x$. With the same Liapunov function as before, E is the x-axis and the largest invariant set of (*) in E is the origin. Thus for (6) the origin is asymptotically stable in the large.

3. Autonomous Functional-Differential Equations

In this section we adopt completely the notations and assumptions introduced by Hale in his paper in these proceedings and present a few of the stability results that he has obtained for autonomous differential equations

$$\dot{x} = f(x_t). \tag{7}$$

A more complete account with numerous examples is given in [*1*]. For the extension to periodic and almost periodic functional-differential equations by Miller see [*6*].

We continue where Hale left off in Section 2 of his paper, except that we shall assume that the open set Ω is the whole state space C of continuous functions. We also confine ourselves to solutions x of (7) that are bounded and hence defined on $[-r, \infty)$. Except that we are in the state space C, the definition of the positive limit set of a trajectory x_t of (7) is essentially the same as for ordinary differential equations, and the notion of an invariant set is modified to take into account the fact that there is no longer uniqueness to the left. A set $M \subset C$ is invariant in the sense that if $\varphi \in M$, then $x(\varphi)$ is defined on $[-r, \infty)$, there is an extension on $(-\infty, -r]$, and $x_t(\varphi)$ remains in M for all t in $(-\infty, \infty)$. With these extensions of these geometric notions to the state space C, Hale then showed that the positive limit set of a trajectory of (7) bounded in the future is a nonempty, compact, connected, and invariant set in C. He was then able to obtain a theory of stability quite similar to that for autonomous ordinary differential equations.

Let V be a continuous function on C to R and define relative to (7)

$$\dot{V}(\varphi) = \overline{\lim_{\tau \to 0+}} \frac{1}{\tau} [V(x_\tau(\varphi)) - V(\varphi)]. \tag{8}$$

With G an arbitrary set in C, we say that V is a *Liapunov function* on G for (7) if $\dot{V}(\varphi) \leqq 0$ for all φ in G. Define $E = \{\varphi; \dot{V}(\varphi) = 0, \varphi \in \bar{G}\}$ and let M denote the largest invariant set of (7) in E.

Theorem 5. If V is a Liapunov function on G for (7) and x_t is a trajectory of (7) which remains in G and is bounded for $t > 0$, then $x_t \to M$ as $t \to \infty$.

Hale has also given the following more useful version of this result.

Corollary 5. Define $Q_\eta = \{\varphi; V(\varphi) < \eta\}$ and let G be Q_η or a component of Q_η. Assume that V is a Liapunov function on G for (7) and that either: (i) G is bounded or (ii) $|\varphi(0)|$ is bounded for φ in G. Then each trajectory starting in G approaches M as $t \to \infty$.

The following is an extension of Četaev's instability theorem. This is a somewhat simplified version of Hale's Theorem 4 in [*1*], which should have stated "$V(\varphi) > 0$ on U when $\varphi \neq 0$ and $V(0) = 0$" and at the end "... intersect the boundary of C_γ ...". This is clear from his proof and is necessary, since he wanted to generalize the usual statement of Četaev's theorem to include the possibility that the equilibrium point be inside U as well as on its boundary.

Corollary 6. Let $p \in C$ be an equilibrium point of (7) contained in the closure of an open set U and let N be a neighborhood of p. Assume that: (i) V is a Liapunov function on $G = U \cap N$, (ii) $M \cap G$ is either the empty set or p, (iii) $V(\varphi) < \eta$ on G when $\varphi \neq p$, and (iv) $V(p) = \eta$ and $V(\varphi) = \eta$ on that part of the boundary of G inside N. Then p is unstable. In fact, if N_0 is a bounded neighborhood of p properly contained in N, then each trajectory starting at a point of $G_0 = G \cap N_0$ other than p leaves N_0 in finite time.

Proof. By the conditions of the corollary and Theorem 6 each trajectory starting inside G_0 at a point other than p_0 must either leave G_0, approach its boundary or approach p. Conditions (i) and (iv) imply that it cannot reach or approach that part of the boundary of G_0 inside N_0 nor can it approach p as $t \to \infty$. Now (iii) states that there are no points of M on that part of the boundary of N_0 inside G. Hence each such trajectory must leave N_0 in finite time. Since p is either in the interior or on the boundary of G, each neighborhood of p contains such trajectories, and p is therefore unstable.

In [*1*] it was shown that the equilibrium point $\varphi = 0$ of

$$\dot{x}(t) = ax^3(t) + bx^3(t - r)$$

was unstable if $a > 0$ and $|b| < |a|$. Using the same Liapunov function and Theorem 6 we can show a bit more. With

$$V(\varphi) = -\frac{\varphi^4(0)}{4a} + \frac{1}{2}\int_{-r}^{0} \varphi^6(\theta)\, d\theta,$$

$$V(x_t) = -\frac{x^4(t)}{4a} + \frac{1}{2}\int_{t-r}^{t} x^6(\theta)\, d\theta,$$

and

$$\dot{V}(\varphi) = -\frac{1}{2}\left(\varphi^6(0) + 2\frac{b}{a}\varphi^3(0)\,\varphi^3(-r) + \varphi^6(-r)\right),$$

which is nonpositive when $|b| < |a|$ (negative definite with respect to $\varphi(0)$ and $\varphi(-r)$); that is, V is a Liapunov function on C and $E = \{\varphi; \varphi(0) = \varphi(-r) = 0\}$. Therefore M is simply the null function $\varphi = 0$. If $a > 0$, the region $G = \{\varphi; V(\varphi) < 0\}$ is nonempty, and no trajectory starting in G can have $\varphi = 0$ as a positive limit point nor can it leave G. Hence by Theorem 5, each trajectory starting in G must be unbounded. Since $\varphi = 0$ is a boundary point of G, it is unstable. It is also easily seen [*1*] that if $a < 0$ and $|b| < |a|$, then $\varphi = 0$ is asymptotically stable in the large.

In [*1*] Hale has also extended this theory for systems with infinite lag ($r = \infty$), and in that same paper gives a number of significant examples of the application of this theory.

REFERENCES

[*1*] Hale, J., Sufficient conditions for stability and instability of autonomous functional differential eauations, *J. Diff. Eqs.* **1**, 452–482 (1965).

[*2*] LaSalle, J., Some extensions of Liapunov's second method, *IRE Trans. Circuit Theory* **CT-7**, 520–527 (1960).

[3] LaSalle, J., Asymptotic stability criteria, *Proc. Symposia Applied Mathematics*, vol. 13, "Hydrodynamic Instability." 299–307. Amer. Math. Soc., Providence, Rhode Island, 1962.

[4] LaSalle, J., and Lefschetz, S., "Stability by Liapunov's Direct Method with Applications." Academic Press, New York, 1961.

[5] Miller, R., On almost periodic differential equations. *Bull. Amer. Math. Soc.* **70**, 792–795 (1964).

[6] Miller, R., Asymptotic behavior of nonlinear delay-differential equations, *J. Diff. Eqs.* **3**, 293–305 (1965).

[7] Yoshizawa, T., Asymptotic behavior of solutions of a system of differential equations. *Contrib. Diff. Eq.* **1**, 371–387 (1963).

Decoupling and Pole Assignment in Linear Multivariable Systems: A Geometric Approach

W. M. WONHAM AND A. S. MORSE

IN the theory of isolated dynamical systems (flows on manifolds, etc.) the notion of an invariant set serves a crucial role for classifying the nature of a system. A subset of the state space is said to be invariant if starting in it, the trajectory must remain in it for all time. For control systems, this property is not particularly useful. For example, if a system is controllable, in the sense that any initial state can be driven to any terminal state, then there are obviously no non-trivial invariant sets in this sense. However, if one modifies "must remain" into "can remain" in this definition, then a very useful notion is obtained. This is the key observation that underlies the notion of invariance introduced in the paper by Wonham and Morse (see also [1]).

Wonham and Morse consider finite dimensional linear systems. The linear subspace $\mathcal{V}$ of the state space $\mathcal{X}$ of $\frac{d}{dt}x = Ax + Bu$ is said to be controlled invariant [or $(A, B)-$ invariant, as they call it] if for any $x_0 \in \mathcal{X}$ there exists an input $u(\cdot)$ such that the resulting solution $x(\cdot)$ with $x(0) = x_0$ satisfies $x(t) \in \mathcal{V}$ for all $t \in \mathbb{R}$. It turns out that this condition is equivalent to $(A, B)-$invariance ($A\mathcal{V} \subseteq \mathcal{V} + \mathrm{im}(B)$) and to the possibility of making the subspace invariant by state feedback ($\exists F$ such that $(A + BF)\mathcal{V} \subseteq \mathcal{V}$).

This concept, together with the related notion of controllability subspace and the dual notions of conditional invariance, opened the road to the elegant "geometric" theory of linear systems and a wide range of applications, especially to problems of disturbance decoupling, non-interacting control, and regulation. Indeed, the decoupling problem was the main motivation for the development of the geometric approach. The early contributions to the study of this problem are reviewed in Wonham's seminal book [20, Sec. 9.10]. The problem of disturbance decoupling with output feedback was completely solved in the linear case in [16] and, incorporating stability, in [18].

Soon after the appearance of the notion of controlled invariance for finite dimensional linear systems, it was generalized in many directions, notably to infinite dimensional systems [4], to "almost" versions of these notions [19] (that made it possible to treat approximate decoupling problems and high gain feedback), and to nonlinear systems. It also sparked a totally new approach to discrete event systems; see, e.g., [15].

In the late 70s and in the early 80s the notion of controlled invariance was extended to nonlinear systems, and this extension opened the way to the systematic development of methods for the design of nonlinear feedback laws. In the context of nonlinear systems, the notion of invariant subspace generalizes in two distinct ways: invariant submanifolds and invariant distributions. Given a vector field f, an f-invariant submanifold is a "surface" with the property that any integral curve of f which intersects this surface is entirely contained in it, while an f-invariant distribution is, essentially, a partition of the state space into a "family of surfaces" with the property that the flow of f carries surfaces into surfaces (if, in particular, one of the surfaces of this family contains an equilibrium of f, then this surface is an invariant submanifold, otherwise it is not). In light of this, the extension of the concept of controlled invariance to nonlinear systems leads to the notions of controlled invariant submanifolds/distributions as objects that can be rendered invariant by means of suitable feedback laws. Thus, for instance, in a control system

$$\frac{d}{dt}x = f(x) + g(x)u$$

with state $x \in \mathbb{R}^n$ and $u \in \mathbb{R}$, a submanifold/distribution is controlled invariant if there exists a feedback law $u = \alpha(x)$ which renders it invariant for the "closed-loop" system

$$\frac{d}{dt}x = f(x) + g(x)\alpha(x)$$

In the nonlinear case it is the notion of controlled invariant distribution (in contrast to invariant linear manifold, around which the linear theory is centered) that plays a fundamental role in the analysis and solution of the problems of disturbance decoupling and noninteracting control. Decoupling a fixed output from a fixed disturbance input is achieved by imposing, by feedback, the existence of an invariant distribution that renders the influence of the disturbance unobservable by that particular output. This fruitful domain of research was opened by the works [9] and [6], and then continued by various authors (see, e.g., [13], [10], [5]). The problem of noninteracting control with stability required some extra effort in view of the existence of certain obstructions that do not have a counterpart in the case of linear systems, but eventually was solved in [8], [17], and [2].

The notion of invariant manifold led to the introduction of the concept of *zero dynamics*, the nonlinear analogue of the dynamics associated with the numerator of a transfer function. The zero dynamics of a nonlinear system is a dynamical system describing all (forced) state trajectories that are consistent with the constraint that the output is identically zero. As in the case of linear systems, it can be shown (under appropriate "regularity" hypotheses) that those trajectories are actually free trajectories of a feedback-modified system, whose state-space is a submanifold (a controlled invariant submanifold) of the original state space. If all such trajectories converge as time tends to ∞ to an equilibrium, the system has properties analogous to those of a linear *minimum-phase system*. This notion, developed in a few preliminary conference papers whose results are summarized in [3], had a significant impact in the analysis of the problem of output tracking and, above all, in the systematic design of adaptive feedback laws for systems affected by parameter uncertainties, as shown in the monographs [12], [11].

REFERENCES

[1] G. BASILE AND G. MARRO, "Controlled and conditioned invariant subspaces in linear system theory," *J. Optimization Th. & Appl.* **3**:306–315, 1969.

[2] S. BATTILOTTI, "A sufficient condition for noninteracting control with stability via dynamic state feedback," *IEEE Trans. Aut. Contr.*, **AC-36**:1033–1045, 1991.

[3] C.I. BYRNES AND A. ISIDORI, "Asymptotic stabilization of minimum-phase nonlinear systems," *IEEE Trans. Aut. Contr.*, **AC-36**:1122–1137, 1991.

[4] R.F. CURTAIN AND H. J. ZWART, *An Introduction to Infinite-Dimensional Linear Systems Theory*, Springer-Verlag (Berlin), 1995.

[5] W.P. DAYAWANSA, D. CHENG, T. J. TARN, AND W. M. BOOTHBY, "Global (f, g)-invariance of nonlinear systems," *SIAM J. Contr. Optimiz.*, **26**:1119–1132, 1988.

[6] R.M. HIRSCHORN, "(A, B)-invariant distributions and disturbance decoupling of nonlinear systems," *SIAM J. Contr. Optimiz.*, **19**:1–19. 1981.

[7] A. ISIDORI, *Nonlinear Control Systems: An Introduction*, 3rd ed., Springer-Verlag (Berlin), 1995.

[8] A. ISIDORI AND J. W. GRIZZLE, "Fixed modes and nonlinear noninteracting control with stability," *IEEE Trans. Aut. Contr.*, **AC-33**:907–914, 1988.

[9] A. ISIDORI, A. J. KRENER, C. GORI GIORGI, AND S. MONACO, "Nonlinear decoupling via feedback: A differential geometric approach," *IEEE Trans. Aut. Contr.*, **AC-26**:331–345, 1981.

[10] A.J. KRENER, "($\mathrm{Ad}_{f,g}$), ($\mathrm{ad}_{f,g}$) and locally ($\mathrm{ad}_{f,g}$)-invariant and controllability distributions," *SIAM J. Contr. Optimiz.*, **23**:523–549, 1985.

[11] M. KRSTIC, I. KANELLAKOPOULOS AND P. KOKOTOVIC, *Nonlinear Adaptive Control Design*, Wiley (New York), 1995.

[12] R. MARINO AND P. TOMEI, *Nonlinear Control Design: Geometric, Adaptive, Robust*, Prentice Hall (New York), 1995.

[13] H. NIJMEIJER AND A.J. VAN DER SCHAFT, "Controlled invariance for nonlinear systems," *IEEE Trans Aut. Contr.*, **AC-27**:904–914, 1982.

[14] H. NIJMEIJER AND A.J. VAN DER SCHAFT, *Nonlinear Dynamical Control Systems*, Springer-Verlag (New York), 1990.

[15] P.J. RAMADGE AND W.M. WONHAM, "On the supremal controllable sublanguage of a given language," *SIAM J. Contr. Optimiz.*, **25**:637–659, 1987.

[16] J.M. SCHUMACHER, "Compensator synthesis using (C, A, B)−pairs," *IEEE Trans. Aut. Contr.*, **AC-25**:1133–1138, 1980.

[17] K.G. WAGNER, "Nonlinear noninteraction with stability by dynamic state feedback," *SIAM J. Contr. Optimiz.*, **29**:609–622, 1991.

[18] J.C. WILLEMS AND C. COMMAULT, "Disturbance decoupling by measurement feedback with stability or pole placement," *SIAM J. Contr. Optimiz.*, **19**:490–504, 1981.

[19] J.C. WILLEMS, "Almost invariant subspaces: An approach to high gain feedback design—Part I: Almost controlled invariant subspaces, Part II: Almost conditionally invariant subspaces," *IEEE Trans. Aut. Contr.*, **AC-26**:235–252, 1981, and **27**:1071–1085, 1982.

[20] W.M. WONHAM, *Linear Multivariable Control: A Geometric Approach*, 1st edition, Lecture Notes in Economics and Mathematical Systems, Vol. 101, Springer-Verlag (New York), 1974; 2nd edition, *Applications in Mathematics series*, Volume 10, Springer-Verlag (New York), 1979.

A.I. & J.C.W

DECOUPLING AND POLE ASSIGNMENT IN LINEAR MULTIVARIABLE SYSTEMS: A GEOMETRIC APPROACH*

W. M. WONHAM† AND A. S. MORSE‡

1. Introduction. The current interest in linear multivariable control has led to several algebraic results with important applications to system synthesis. In particular, the problem of decoupling of individual system outputs by means of state variable feedback was studied by Rekasius [1], Falb and Wolovich [2] and Gilbert [3]; the problem of realizing arbitrary pole locations in the closed loop system transfer matrix was investigated by Wonham [4] and Heymann [5]. In the present article, new results are obtained along these lines. In § 3, the problem of neutralizing the effect of disturbances with respect to a specified group of output variables is solved. In § 4, the concept of a controllability subspace is introduced and its relation to pole assignability is investigated. This material is preliminary to the formulation of a general problem of output decoupling in § 5. In § 6 and § 7, necessary and sufficient conditions for decoupling are obtained in two special cases; the results of § 7 complement and extend those obtained previously in [1], [2] and [3]. In each case, the problem of pole assignment is solved completely.

Our viewpoint is that such problems are usefully treated in a geometric framework in which both definitions and results become intuitively transparent. In this way, entanglement at the outset in a thicket of algebraic calculations is avoided. Of course, for applications, it is necessary to translate the geometric criteria into matrix operations suitable for computation. This matter will be considered in a future article.

2. Notation. The control system of interest is specified by the differential equation

$$\dot{x}(t) = Ax(t) + Bu(t) \tag{2.1}$$

with x an n-vector, u an m-vector and A, B constant matrices of dimension, respectively, $n \times n$ and $n \times m$. Here and below, all vectors and matrices have real-valued elements. Script letters denote linear subspaces; $\mathscr{E}^n$ is real n-space; $\mathscr{V}^{\perp}$ is the orthogonal complement of the subspace $\mathscr{V}$; 0 denotes both the vector zero and the zero subspace.

If K is a matrix, $\{K\}$ or $\mathscr{K}$ is the range of K, and $\mathscr{N}(K)$ is the null space of K. If K is of dimension $\mu \times \nu$ and $\mathscr{V} \subset \mathscr{E}^{\mu}$, we write $K^{-1}\mathscr{V}$ for the subspace $\{z : z \in \mathscr{E}^{\nu}, Kz \in \mathscr{V}\} \subset \mathscr{E}^{\nu}$.

The *controllable subspace* of the pair (A, B), written $\{A|\mathscr{B}\}$, is defined as

$$\{A|\mathscr{B}\} = \mathscr{B} + A\mathscr{B} + \cdots + A^{n-1}\mathscr{B}.$$

* Received by the editors February 3, 1969, and in revised form June 4, 1969.

† Office of Control Theory and Application, NASA Electronics Research Center, Cambridge, Massachusetts 02139. The work of this author was supported by the National Aeronautics and Space Administration while he held an NRC postdoctoral resident research associateship.

‡ Office of Control Theory and Application, NASA Electronics Research Center, Cambridge, Massachusetts 02139.

Reprinted with permission from *SIAM Journal on Control*, W. M. Wonham and A. S. Morse, "Decoupling and Pole Assignment in Linear Multivariable Systems: A Geometric Approach" Vol. 8, 1970, pp. 1–18.

Thus, $\{A|\mathscr{B}\}$ is the largest subspace of $\mathscr{E}^n$ which the control $u(\cdot)$ in (2.1) can influence. Observe that $\{A|\mathscr{B}\}$ is an A-invariant subspace of $\mathscr{E}^n$.

With (2.1), we consider the auxiliary equation

$$y(t) = Hx(t), \tag{2.2}$$

where H is a constant $q \times n$ matrix. The vector y is the *output*.

Equations (2.1) and (2.2) play no essential role but serve to guide the investigation.

3. Localization of disturbances. In place of (2.1), consider the perturbed system

$$\dot{x}(t) = Ax(t) + Bu(t) + D\xi(t), \tag{3.1}$$

where D is a constant $n \times d$ matrix and $\xi(\cdot)$ is a disturbance input. If $u(t) = Cx(t) + v(t)$ (where $v(\cdot)$ is an external control input), then the output $y(\cdot)$ will be unaffected by all possible $\xi(\cdot)$ if and only if $\{A + BC|\mathscr{D}\} \subset \mathscr{N}(H)$. This suggests the problem: given A, B, $\mathscr{D} \subset \mathscr{E}^n$, $\mathscr{N} \subset \mathscr{E}^n$, under what conditions does there exist an $m \times n$ matrix C such that $\{A + BC|\mathscr{D}\} \subset \mathscr{N}$? If C exists, the effect of disturbances is, in an algebraic sense, localized to $\mathscr{N}$.

THEOREM 3.1. *There exists C such that $\{A + BC|\mathscr{D}\} \subset \mathscr{N}$ if and only if $\mathscr{D} \subset \mathscr{V}$, where $\mathscr{V}$ is the maximal subspace such that*

$$\mathscr{V} \subset \mathscr{N} \cap A^{-1}(\mathscr{B} + \mathscr{V}). \tag{3.2}$$

Furthermore $\mathscr{V}$ is given by $\mathscr{V} = \mathscr{V}^{(\nu)}$, where

$$\mathscr{V}^{(0)} = \mathscr{N}; \qquad \mathscr{V}^{(i)} = \mathscr{V}^{(i-1)} \cap A^{-1}(\mathscr{B} + \mathscr{V}^{(i-1)}), \qquad i = 1, 2, \cdots, \nu, \tag{3.3}$$

and $\nu = \dim \mathscr{N}$.

Here and below, "maximal" ("minimal") mean l.u.b. (g.l.b.) with respect to the usual partial ordering of subspaces by inclusion.

To prove the theorem we need two auxiliary facts.

LEMMA 3.1. *Let $x_i \in \mathscr{E}^n$, $u_i \in \mathscr{E}^m$, $i = 1, \cdots, N$, and write $X = (x_1, \cdots, x_N)$, $U = (u_1, \cdots, u_N)$. There exists an $m \times n$ matrix C such that $Cx_i = u_i, i = 1, \cdots, N$, if and only if $\mathscr{N}(X) \subset \mathscr{N}(U)$. C always exists if the x_i are linearly independent.*

The simple proof is omitted.

LEMMA 3.2. *Let $\mathscr{V} \subset \mathscr{E}^n$. There exists an $m \times n$ matrix C such that $(A + BC)\mathscr{V} \subset \mathscr{V}$ if and only if $A\mathscr{V} \subset \mathscr{B} + \mathscr{V}$.*

Proof. Necessity is clear. For sufficiency, let $v_1, \cdots, v_\mu$ be a basis of $\mathscr{V}$. Then $Av_i = Bu_i + w_i$ for some $u_i \in \mathscr{E}^m$ and $w_i \in \mathscr{V}$. Choose C, by Lemma 3.1, such that $Cv_i = -u_i$, $i = 1, \cdots, \mu$; then $(A + BC)v_i = w_i$.

Proof of Theorem 3.1. For sufficiency, (3.2) implies $\mathscr{V} \subset \mathscr{N}$ and $A\mathscr{V} \subset \mathscr{B} + \mathscr{V}$. By Lemma 3.2, there exists C such that $(A + BC)\mathscr{V} \subset \mathscr{V}$. Then

$$\{A + BC|\mathscr{D}\} \subset \{A + BC|\mathscr{V}\} = \mathscr{V} \subset \mathscr{N}.$$

The maximal property of $\mathscr{V}$ was not required.

For necessity write $\{A + BC|\mathscr{D}\} = \mathscr{W}$. Then

$$\mathscr{W} \subset \mathscr{N}, \qquad A\mathscr{W} \subset \mathscr{B} + \mathscr{W}. \tag{3.4}$$

If $\mathscr{W}$ is the class of all $\mathscr{W}' \subset \mathscr{E}^n$ which satisfy (3.4), then clearly $0 \in \mathscr{W}$ and $\mathscr{W}$ is closed under addition. Hence, $\mathscr{W}$ contains a (unique) maximal member $\hat{\mathscr{V}}$. Then $\mathscr{D} \subset \mathscr{W}' \subset \hat{\mathscr{V}}$ and $\hat{\mathscr{V}}$ satisfies (3.2).

To prove the second statement of the theorem, observe that $\mathscr{V}^{(0)} \supset \hat{\mathscr{V}}$, and if $\mathscr{V}^{(i-1)} \supset \hat{\mathscr{V}}$, then $\mathscr{V}^{(i)} \supset \mathscr{V} \cap A^{-1}(\mathscr{B} + \hat{\mathscr{V}}) = \hat{\mathscr{V}}$. Thus, $\mathscr{V}^{(i)} \supset \hat{\mathscr{V}}$ for all i; and since $\mathscr{V}^{(i)} \subset \mathscr{V}^{(i-1)}$, there is a least integer j such that $\mathscr{V}^{(i)} = \mathscr{V}^{(j)}$ if $i \geqq j$. Since $\mathscr{V}^{(j)} \supset \hat{\mathscr{V}}$ and $\mathscr{V}^{(j)}$ satisfies (3.4), $\mathscr{V}^{(j)} = \hat{\mathscr{V}}$. Clearly, $0 \leqq j \leqq \nu$; and if $\mathscr{D} \subset \hat{\mathscr{V}}$ we even have $0 \leqq j \leqq \nu - \dim \mathscr{D}$.

Remark 1. Theorem 3.1 depends essentially on the fact that the class $\mathscr{W}$ determined by (3.4), or equivalently

$$\mathscr{W} = \{\mathscr{W}' : \mathscr{W}' \subset \mathscr{V} \cap A^{-1}(\mathscr{B} + \mathscr{W}')\},$$

has a maximal element $\hat{\mathscr{V}}$. Furthermore, $\hat{\mathscr{V}}$ is defined constructively by means of (3.3). This fact will be used without special comment in the following sections.

4. Controllability subspaces. In regard to the system (2.1), suppose that a subspace $\mathscr{R} \subset \mathscr{E}^n$ is selected and that it is desired to modify the system in such a way that $\mathscr{R}$, but no larger subspace, is completely controllable. This aim is to be realized by feedback of state variables and by forming suitable linear combinations of control variables: that is, by setting $u = Cx + Kv$, where K is an $m \times m'$ matrix for some $m' \leqq m$. Then (2.1) becomes

$$\dot{x} = (A + BC)x + BKv$$

and we require

$$\{A + BC|\{BK\}\} = \mathscr{R}. \tag{4.1}$$

Condition (4.1) can be expressed more neatly by noting that $\{BK\} \subset \mathscr{B}$ and the following.

LEMMA 4.1. *If $\hat{\mathscr{B}} \subset \mathscr{B}$ and $\{A|\hat{\mathscr{B}}\} = \mathscr{R}$, then $\{A|\mathscr{B} \cap \mathscr{R}\} = \mathscr{R}$. Conversely, if $\{A|\mathscr{B} \cap \mathscr{R}\} = \mathscr{R}$, there exists a matrix K such that $\{A|\{BK\}\} = \mathscr{R}$.*

Proof. $\{A|\hat{\mathscr{B}}\} = \mathscr{R}$ implies $\hat{\mathscr{B}} \subset \mathscr{R}$, so $\hat{\mathscr{B}} \subset \mathscr{B} \cap \mathscr{R}$, and thus $\mathscr{R} = \{A|\hat{\mathscr{B}}\} \subset \{A|\mathscr{B} \cap \mathscr{R}\}$. Also, $A\mathscr{R} \subset \mathscr{R}$ implies $A(\mathscr{B} \cap \mathscr{R}) \subset \mathscr{R}$; by induction $A^j(\mathscr{B} \cap \mathscr{R}) \subset \mathscr{R}$, $j = 1, 2, \cdots$, and so $\{A|\mathscr{B} \cap \mathscr{R}\} \subset \mathscr{R}$.

For the converse, let b_i, $i = 1, \cdots, m$, be the ith column of B and let $\{r_j, j = 1, \cdots, m'\}$ be a basis of $\mathscr{B} \cap \mathscr{R}$. Then

$$r_j = \sum_{i=1}^{m} k_{ij} b_i, \qquad j = 1, \cdots, m',$$

for suitable k_{ij}, and we set $K = [k_{ij}]$. This completes the proof of the lemma.

By Lemma 4.1, we can pose the synthesis problem as follows:

Given A, B and $\mathscr{R}$, find conditions for the existence of C such that

$$\{A + BC|\mathscr{B} \cap \mathscr{R}\} = \mathscr{R}. \tag{4.2}$$

If such a C exists, we call $\mathscr{R}$ a *controllability subspace* of the pair (A, B). Observe that $\mathscr{R} = 0$ and $\mathscr{R} = \{A|\mathscr{B}\}$ are controllability subspaces.

Controllability subspaces can be characterized as follows.

THEOREM 4.1. *Let A, B, $\mathscr{R} \subset \mathscr{E}^n$ be fixed. $\mathscr{R}$ is a controllability subspace of (A, B) if and only if*

$$A\mathscr{R} \subset \mathscr{B} + \mathscr{R} \tag{4.3}$$

and

$$\mathscr{R} = \hat{\mathscr{R}}, \tag{4.4}$$

where $\hat{\mathscr{R}}$ is the minimal subspace such that

$$\hat{\mathscr{R}} = \mathscr{R} \cap (A\hat{\mathscr{R}} + \mathscr{B}). \tag{4.5}$$

Furthermore, $\hat{\mathscr{R}} = \mathscr{R}^{(\rho)}$, where $\rho = \dim \mathscr{R}$ and

$$\begin{aligned} &\mathscr{R}^{(0)} = 0, \\ &\mathscr{R}^{(i)} = \mathscr{R} \cap (A\mathscr{R}^{(i-1)} + \mathscr{B}), \qquad i = 1, 2, \cdots, n. \end{aligned} \tag{4.6}$$

Write $\mathbf{C}$ for the class of matrices C such that $(A + BC)\mathscr{R} \subset \mathscr{R}$. To prove the theorem we need two preliminary results.

LEMMA 4.2. *Let $\tilde{\mathscr{R}} \subset \mathscr{R}$. For all $C \in \mathbf{C}$,*

$$\mathscr{R} \cap \mathscr{B} + (A + BC)\tilde{\mathscr{R}} = \mathscr{R} \cap (A\tilde{\mathscr{R}} + \mathscr{B}).$$

Proof. Let $C \in \mathbf{C}$. Then $(A + BC)\tilde{\mathscr{R}} \subset \mathscr{R}$ and $A\tilde{\mathscr{R}} + \mathscr{B} = (A + BC)\tilde{\mathscr{R}} + \mathscr{B}$. By the modular distributive rule for subspaces,

$$\begin{aligned} \mathscr{R} \cap (A\tilde{\mathscr{R}} + \mathscr{B}) &= \mathscr{R} \cap [(A + BC)\tilde{\mathscr{R}} + \mathscr{B}] \\ &= (A + BC)\tilde{\mathscr{R}} + \mathscr{R} \cap \mathscr{B}. \end{aligned}$$

LEMMA 4.3. *If $C \in \mathbf{C}$ then*

$$\sum_{j=1}^{i} (A + BC)^{j-1}(\mathscr{B} \cap \mathscr{R}) = \mathscr{R}^{(i)}, \qquad i = 1, \cdots, n, \tag{4.7}$$

where the sequence $\mathscr{R}^{(i)}$ is defined by (4.6).

Proof. Equation (4.7) is true for $i = 1$. If it is true for $i = k - 1$, then by Lemma 4.2,

$$\begin{aligned} \sum_{j=1}^{k} (A + BC)^{j-1}(\mathscr{B} \cap \mathscr{R}) &= \mathscr{B} \cap \mathscr{R} + (A + BC)\mathscr{R}^{(k-1)} \\ &= \mathscr{R} \cap (A\mathscr{R}^{(k-1)} + \mathscr{B}) \\ &= \mathscr{R}^{(k)}. \end{aligned}$$

Proof of Theorem 4.1. By Lemma 3.2, $\mathbf{C}$ is nonempty if and only if (4.3) is true. Let

$$\mathscr{R} = \{A + BC | \mathscr{B} \cap \mathscr{R}\}. \tag{4.8}$$

Then $C \in \mathbf{C}$. By Lemma 4.3,

$$\mathscr{R} = \sum_{j=1}^{n} (A + BC)^{j-1}(\mathscr{B} \cap \mathscr{R}) = \mathscr{R}^{(n)} = \mathscr{R}^{(\rho)}.$$

Conversely, if $\mathscr{R} = \mathscr{R}^{(n)}$, then (4.8) is true for every $C \in \mathbf{C}$. It remains to show that (4.5) has the minimal solution $\mathscr{R}^{(\rho)}$. By induction on i in (4.6), it is seen that $\mathscr{R}^{(i)} \subset \hat{\mathscr{R}}$, $i = 1, 2, \cdots$, for every solution $\hat{\mathscr{R}}$ of (4.5), and that the sequence $\mathscr{R}^{(i)}$ is monotone nondecreasing. Hence, there is $\mu \leqq \rho$ such that $\mathscr{R}^{(i)} = \mathscr{R}^{(\mu)}$ for $i \geqq \mu$; in particular, $\mathscr{R}^{(\rho)} \subset \hat{\mathscr{R}}$ and $\mathscr{R}^{(\rho)}$ satisfies (4.5).

Remark 2. If $\mathscr{R}$ is a controllability subspace, then it was proved incidentally that

$$\mathscr{R} = \{A + BC | \mathscr{B} \cap \mathscr{R}\}$$

for every C such that $(A + BC)\mathscr{R} \subset \mathscr{R}$. This fact will be used later without special mention.

Consider now the problem of assigning the eigenvalues of the restriction of $A + BC$ to $\mathscr{R}$. It will be shown that there is complete freedom of assignment and that simultaneously the control v introduced earlier can be made a scalar; i.e., in (4.1) K can be made an m-vector ($m' = 1$). For this, recall [4] that a subspace $\mathscr{X}$ is *A-cyclic* if there exists $x \in \mathscr{X}$ such that $\{A | \{x\}\} = \mathscr{X}$; that is, if $\mathscr{X}$ contains a *generator* x. Thus we can take $m' = 1$ if and only if $\mathscr{R}$ can be made $(A + BC)$-cyclic and $\mathscr{R} \cap \mathscr{B}$ contains a generator.

THEOREM 4.2. *Let* (4.3) *and* (4.4) *hold, and let* $\alpha_1, \cdots, \alpha_\rho$ *be arbitrary real numbers* ($\rho = \dim \mathscr{R}$). *Then* C *can be chosen such that* (4.2) *is true and* $\mathscr{R}$ *is* $(A + BC)$-*cyclic with characteristic polynomial*

$$\lambda^\rho - \sum_{i=1}^{\rho} \alpha_i \lambda^{i-1}. \tag{4.9}$$

If $0 \neq b \in \mathscr{B} \cap \mathscr{R}$ *is arbitrary,* C *can be chosen so that, in addition,* b *generates* $\mathscr{R}$.

Proof. By Lemma 4.3 and Theorem 4.1, $\mathbf{C}$ is nonempty and

$$\{A + BC | \mathscr{B} \cap \mathscr{R}\} = \mathscr{R} \tag{4.10}$$

for every $C \in \mathbf{C}$. Choose $C_1 \in \mathbf{C}$ arbitrarily and write $A + BC_1 = A_1$. Let $b_1 = b \in \mathscr{R} \cap \mathscr{B}$ and let ρ_1 be the largest integer such that the vectors

$$b_1, A_1 b_1, \cdots, A_1^{\rho_1 - 1} b_1$$

are independent. Put $r_1 = b_1$ and $r_j = A_1 r_{j-1} + b_1$, $j = 2, \cdots, \rho_1$. Then $r_i \in \mathscr{R}$ and the r_i are independent. If $\rho_1 < \rho$, choose $b_2 \in \mathscr{R} \cap \mathscr{B}$ such that $r_1, \cdots, r_{\rho_1}, b_2$ are independent; such a b_2 exists by (4.7). Let ρ_2 be the greatest integer such that

$$b_1, \cdots, A_1^{\rho_1 - 1} b_1, b_2, \cdots, A_1^{\rho_2 - 1} b_2$$

are independent, and define

$$r_{\rho_1 + i} = A_1 r_{\rho_1 + i - 1} + b_2, \qquad i = 1, \cdots, \rho_2.$$

Then $r_1, \cdots, r_{\rho_2}$ are independent and in $\mathscr{R}$. Continuing thus, we obtain eventually $r_1, \cdots, r_\rho$ independent and in $\mathscr{R}$, with the property

$$r_{i+1} = A_1 r_i + \tilde{b}_i, \qquad i = 1, \cdots, \rho - 1,$$

where $\tilde{b}_i \in \mathscr{R} \cap \mathscr{B}$. Now let C_2 be chosen such that

$$BC_2 r_i = \tilde{b}_i, \qquad i = 1, \cdots, \rho,$$

where $\tilde{b}_\rho \in \mathscr{R} \cap \mathscr{B}$ is arbitrary. Since $\tilde{b}_i = Bu_i$ for suitable u_i, and the r_i are independent, Lemma 3.1 guarantees that C_2 exists. The situation now is that

$$r_{i+1} = (A_1 + BC_2)r_i, \qquad i = 1, \cdots, \rho - 1,$$

and

$$(A_1 + BC_2)r_\rho \in \mathscr{R}.$$

By independence of the r_i,

$$\{A_1 + BC_2|\{r_1\}\} = \mathscr{R};$$

that is, $\mathscr{R}$ is cyclic relative to $A + B(C_1 + C_2)$ with generator $r_1 = b_1 \in \mathscr{R} \cap \mathscr{B}$. It is well known [4] that now an n-vector c can be found such that $A + B(C_1 + C_2) + b_1c'$ (restricted to $\mathscr{R}$) has the characteristic polynomial (4.9). Setting $b_1 = Bg$ for suitable $g \in \mathscr{E}^m$, it follows that the matrix

$$C = C_1 + C_2 + gc'$$

has all the required properties.

Remark 3. The result that any nonzero vector in $\mathscr{B} \cap \mathscr{R}$ can serve as generator is an extension of the useful lemma in [5].

Remark 4. If $\mathscr{R} = \mathscr{E}^n$, (4.3) holds automatically and (4.4) amounts to $\{A|\mathscr{B}\} = \mathscr{E}^n$, i.e., complete controllability of (A, B). Then Theorem 4.2 yields the known result [4] that controllability implies pole assignability. The construction just used furnishes a simpler proof of this fact than that in [4].

It will be necessary later to compute the maximal controllability subspace contained in a given subspace $\mathscr{S}$. For this, let $\mathscr{V}$ be the maximal subspace of $\mathscr{S}$ which is $(A + BC)$-invariant for some C (recall Remark 1 following Theorem 3.1): and let $\mathbf{C}(\mathscr{V})$ be the class of C for which $(A + BC)\mathscr{V} \subset \mathscr{V}$.

Theorem 4.3. *If* $C \in \mathbf{C}(\mathscr{V})$, *the subspace*

$$\bar{\mathscr{R}} = \{A + BC|\mathscr{B} \cap \mathscr{V}\} \tag{4.11}$$

is the maximal controllability subspace in $\mathscr{S}$.

Proof. By (4.2) and Lemma 4.1, $\bar{\mathscr{R}}$ is a controllability subspace. Furthermore, by Lemma 4.3 with $\mathbf{C}(\mathscr{V})$ in place of $\mathbf{C}$, $\bar{\mathscr{R}}$ is independent of $C \in \mathbf{C}(\mathscr{V})$ and so is uniquely defined. Now suppose

$$\hat{\mathscr{R}} = \{A + B\hat{C}|\mathscr{B} \cap \hat{\mathscr{R}}\}, \qquad \hat{\mathscr{R}} \subset \mathscr{S}.$$

Since $\hat{\mathscr{R}}$ is $(A + B\hat{C})$-invariant and $\mathscr{V}$ is maximal, there follows $\hat{\mathscr{R}} \subset \mathscr{V}$. Let $\mathscr{V} = \hat{\mathscr{R}} \oplus \hat{\mathscr{V}}$. By the construction used in proving Lemma 3.2, a matrix C exists such that

$$Cx = \hat{C}x, \quad x \in \hat{\mathscr{R}}; \qquad (A + BC)\hat{\mathscr{V}} \subset \mathscr{V}.$$

Then $C \in \mathbf{C}(\mathscr{V})$, and

$$\begin{aligned} \hat{\mathscr{R}} &= \{A + BC|\mathscr{B} \cap \hat{\mathscr{R}}\} \\ &\subset \{A + BC|\mathscr{B} \cap \mathscr{V}\} \\ &= \bar{\mathscr{R}}; \end{aligned}$$

that is, $\bar{\mathscr{R}}$ is maximal.

5. Decoupling of output variables: Problem statement. Consider the output equation (2.2), with

$$H = \begin{bmatrix} H_1 \\ \vdots \\ H_k \end{bmatrix}, \tag{5.1}$$

where H_i is of dimension $q_i \times n$, $i = 1, \cdots, k$, $k \geqq 2$, $q_1 + \cdots + q_k = q$. Then (2.2) can be written

$$y_i = H_i x, \qquad i = 1, \cdots, k, \tag{5.2}$$

where y_i is a q_i-vector. The vectors y_i may be regarded as physically significant groups of scalar output variables. It may therefore be desirable to control completely each of the output vectors y_i individually, without affecting the behavior of the remaining $y_j, j \neq i$. This end is to be achieved by linear state-variable feedback together with the assignment of a suitable group of control inputs to each y_i. That is, in (2.1) we set

$$u = Cx + \sum_{i=1}^{k} K_i v_i. \tag{5.3}$$

For v_i to control y_i completely, we must have

$$H_i\{A + BC|\{BK_i\}\} = \mathscr{Y}_i, \tag{5.4}$$

where $\mathscr{Y}_i$ is the range of H_i. Since the ith control v_i is to leave the outputs y_j, $j \neq i$, unaffected, we require also

$$H_j\{A + BC|\{BK_i\}\} = 0, \qquad j \neq i. \tag{5.5}$$

Recalling the equivalence of (4.1) and (4.2), we can express conditions (5.4) and (5.5) more neatly as follows. Write $\mathscr{E}^n = \mathscr{E}$ and

$$\mathscr{N}(H_i) = \mathscr{N}_i, \qquad i = 1, \cdots, k. \tag{5.6}$$

Then our problem is: *Given A, B and $\mathscr{N}_1, \cdots, \mathscr{N}_k$, find a matrix C and controllability subspaces $\mathscr{R}_1, \cdots, \mathscr{R}_k$, with the properties*:

$$\mathscr{R}_i = \{A + BC|\mathscr{B} \cap \mathscr{R}_i\}, \qquad i = 1, \cdots, k, \tag{5.7}$$

$$\mathscr{R}_i + \mathscr{N}_i = \mathscr{E}, \qquad i = 1, \cdots, k, \tag{5.8}$$

$$\mathscr{R}_i \subset \bigcap_{j \neq i} \mathscr{N}_j, \qquad i = 1, \cdots, k. \tag{5.9}$$

Here (5.8) and (5.9) are equivalent, respectively, to (5.4) and (5.5).

The relations (5.7)–(5.9) provide a geometric formulation of the problem of simultaneous decoupling and complete control of the output vectors $y_i, \cdots, y_k$. Thus stated, the problem definition is both natural and intuitively transparent.

We observe that the output matrices H_i play no role beyond specification of the subspaces $\mathscr{N}_i$. Since the H_i need have no special structure, the $\mathscr{N}_i$ are similarly unrestricted. Nevertheless, we shall rule out trivialities by tacitly assuming:

(i) $\mathcal{N}_i \neq \mathcal{E}, i = 1, \cdots, k$.

(ii) The subspaces $\mathcal{N}_i^\perp$ are mutually independent.[1] In particular, the $\mathcal{N}_i$ are distinct and

$$\mathcal{N}_i \neq 0, \qquad i = 1, \cdots, k. \tag{5.10}$$

(iii) The pair (A, B) is completely controllable, i.e., $\{A|\mathcal{B}\} = \mathcal{E}$.

For if (i) fails, then for some i, $\mathcal{N}_i = \mathcal{E}$; that is, $H_i = 0$ and $y_i \equiv 0$. If (ii) fails, then for some i,

$$\mathcal{N}_i^\perp \cap \sum_{j \neq i} \mathcal{N}_j^\perp \neq 0$$

or, by taking orthogonal complements,

$$\mathcal{N}_i + \bigcap_{j \neq i} \mathcal{N}_j \neq \mathcal{E}$$

and (5.8) must fail. For (iii), if $\{A|\mathcal{B}\} = \mathcal{E}_1 \neq \mathcal{E}$ we can write $\mathcal{E} = \mathcal{E}_1 \oplus \mathcal{E}_2$ and (2.1) as

$$\dot{x}_1 = A_1 x_1 + A_3 x_2 + B_1 u,$$

$$\dot{x}_2 = A_2 x_2,$$

where $x_i \in \mathcal{E}_i$, $i = 1, 2$, and $\{A_1|\mathcal{B}_1\} = \mathcal{E}_1$. The problem is unrealistic unless A_2 is stable (i.e., the pair (A, B) is stabilizable [4]). Hence, we may assume $x_2(t) \equiv 0$ and take as starting point

$$\dot{x}_1 = A_1 x_1 + B_1 u.$$

The problem can then be reformulated with $\mathcal{E}_1$ in place of $\mathcal{E}$.

We turn now to the determination of necessary and sufficient conditions for the existence of a solution to (5.7)–(5.9) in two special, but interesting, cases.

In the following sections, $\bar{\mathcal{R}}_i$ denotes the maximal controllability subspace such that

$$\bar{\mathcal{R}}_i \subset \bigcap_{j \neq i} \mathcal{N}_j, \qquad i = 1, \cdots, k. \tag{5.11}$$

The $\bar{\mathcal{R}}_i$ are constructed according to Theorem 4.3.

6. Decoupling when rank $(H) = n$. Our assumption is equivalent to

$$\bigcap_{i=1}^{k} \mathcal{N}_i = 0. \tag{6.1}$$

That is, there is a one-to-one mapping of state variables into output variables.

THEOREM 6.1. *If* (6.1) *holds, then the problem* (5.7)–(5.9) *has a solution if and only if*

$$\bar{\mathcal{R}}_i + \mathcal{N}_i = \mathcal{E}, \qquad i = 1, \cdots, k. \tag{6.2}$$

[1] Equivalently, the row spaces of the H_i are mutually independent.

Proof. If the problem has a solution $\mathscr{R}_i$, $i = 1, \cdots, k$, then by maximality of the $\bar{\mathscr{R}}_i$, $i = 1, \cdots, k$, there follows $\mathscr{R}_i \subset \bar{\mathscr{R}}_i$, and (6.2) follows from (5.8).

Conversely, suppose (6.2) holds. The $\bar{\mathscr{R}}_i$ are mutually independent; for, by (5.11) and (6.1),

$$\bar{\mathscr{R}}_i \cap \sum_{\mu \neq i} \bar{\mathscr{R}}_\mu \subset \left[\bigcap_{j \neq i} \mathscr{N}_j\right] \cap \left[\sum_{\mu \neq i} \bigcap_{\nu \neq \mu} \mathscr{N}_\nu\right] \subset \bigcap_{j \neq i} \mathscr{N}_j \cap \mathscr{N}_i = 0.$$

Let C_i be chosen such that

$$\bar{\mathscr{R}}_i = \{A + BC_i | \mathscr{B} \cap \bar{\mathscr{R}}_i\}, \qquad i = 1, \cdots, k.$$

Since the $\bar{\mathscr{R}}_i$ are independent there exists, by Lemma 3.1, a matrix C such that $Cr = C_i r$ $(r \in \bar{\mathscr{R}}_i,\ i = 1, \cdots, k)$, i.e.,

$$(A + BC)r = (A + BC_i)r, \qquad r \in \bar{\mathscr{R}}_i, \quad i = 1, \cdots, k.$$

Then

$$\bar{\mathscr{R}}_i = \{A + BC | \mathscr{B} \cap \bar{\mathscr{R}}_i\}, \qquad i = 1, \cdots, k;$$

and C, together with the $\bar{\mathscr{R}}_i$, satisfy (5.7)–(5.9).

Remark 5. By Theorem 4.2, the C_i can be chosen so that $A + BC_i$, restricted to $\bar{\mathscr{R}}_i$, has any desired spectrum. Hence, the same is true for $A + BC$. Furthermore, there exists $b_i \in \mathscr{B} \cap \bar{\mathscr{R}}_i$ such that

$$\bar{\mathscr{R}}_i = \{A + BC | \{b_i\}\}, \qquad i = 1, \cdots, k.$$

7. Decoupling when rank $(B) = k$. Our assumption is equivalent to

$$\dim \mathscr{B} = k. \tag{7.1}$$

Here the situation has been simplified by narrowing the choice of generating subspaces $\mathscr{B} \cap \mathscr{R}_i$. The same assumption was made in [1], [2] and [3], with the additional restriction that the outputs y_i be scalars.

THEOREM 7.1. *If* (7.1) *holds, then the problem* (5.7)–(5.9) *has a solution if and only if*

$$\bar{\mathscr{R}}_i + \mathscr{N}_i = \mathscr{E}, \qquad i = 1, \cdots, k, \tag{7.2}$$

and

$$\mathscr{B} = \sum_{i=1}^{k} \mathscr{B} \cap \bar{\mathscr{R}}_i. \tag{7.3}$$

Furthermore, if $C, \mathscr{R}_1, \cdots, \mathscr{R}_k$ *is any solution, then*

$$\mathscr{R}_i = \bar{\mathscr{R}}_i, \qquad i = 1, \cdots, k. \tag{7.4}$$

Proof. Part 1. Suppose $C, \mathscr{R}_1, \cdots, \mathscr{R}_k$ is a solution. The necessity of (7.2) follows, as in the proof of Theorem 6.1. To verify (7.3), write

$$\mathscr{B} \cap \mathscr{R}_i = \mathscr{B}_i \oplus \left[\mathscr{B} \cap \mathscr{R}_i \cap \sum_{j \neq i} \mathscr{R}_j\right], \qquad i = 1, \cdots, k.$$

The $\mathscr{B}_i$ are mutually independent; in fact,

$$\mathscr{B}_i \cap \sum_{j \neq i} \mathscr{B}_j \subset \mathscr{B}_i \cap \sum_{j \neq i} \mathscr{B} \cap \mathscr{R}_j \subset \mathscr{B}_i \cap (\mathscr{B} \cap \mathscr{R}_i) \cap \sum_{j \neq i} \mathscr{R}_j = 0.$$

Recall that the $\mathscr{R}_i$ are $(A + BC)$-invariant. Then

$$\mathscr{R}_i = \{A + BC | \mathscr{B}_i\} + \bar{\mathscr{R}}_i,$$

where

$$\bar{\mathscr{R}}_i \subset \left\{ A + BC \middle| \sum_{j \neq i} \mathscr{R}_j \right\} \subset \sum_{j \neq i} \mathscr{R}_j \subset \sum_{j \neq i} \bigcap_{\mu \neq j} \mathscr{V}_\mu \subset \mathscr{V}_i.$$

Therefore, by (5.8),

$$\{A + BC | \mathscr{B}_i\} + \mathscr{V}_i = \mathscr{X},$$

and since $\mathscr{V}_i \neq \mathscr{X}$ there follows $\mathscr{B}_i \neq 0$, $i = 1, \cdots, k$. Therefore

$$\dim \sum_{i=1}^{k} \mathscr{B}_i = \sum_{i=1}^{k} \dim \mathscr{B}_i = k;$$

so

$$\mathscr{B} = \mathscr{B}_1 \oplus \cdots \oplus \mathscr{B}_k \tag{7.5}$$

and

$$\dim \mathscr{B}_i = 1, \qquad i = 1, \cdots, k.$$

Since $\mathscr{B}_i \subset \mathscr{B} \cap \mathscr{R}_i \subset \mathscr{B} \cap \bar{\mathscr{R}}_i$, it follows that (7.3) is true.

Proof. Part 2. To verify (7.4), it is enough to show that the subspaces $\mathscr{B} \cap \bar{\mathscr{R}}_i$ are independent. For then,

$$\dim (\mathscr{B} \cap \bar{\mathscr{R}}_i) = 1, \qquad i = 1, \cdots, k,$$

and so

$$\mathscr{B} \cap \mathscr{R}_i = \mathscr{B} \cap \bar{\mathscr{R}}_i, \qquad i = 1, \cdots, k. \tag{7.6}$$

Assuming (7.6) is true, let $\bar{\mathscr{R}}_i = \hat{\mathscr{R}}_i \oplus \mathscr{R}_i$ and choose C_i, by Lemma 3.1, such that

$$(A + BC_i)\hat{\mathscr{R}}_i \subset \mathscr{R}_i, \qquad C_i r = Cr, \quad r \in \mathscr{R}_i, \quad i = 1, \cdots, k.$$

Then $C_i \in \mathbf{C}(\mathscr{R}_i) \cap \mathbf{C}(\bar{\mathscr{R}}_i)$, so that

$$\begin{aligned} \mathscr{R}_i &= \{A + BC_i | \mathscr{B} \cap \mathscr{R}_i\} \\ &= \{A + BC_i | \mathscr{B} \cap \bar{\mathscr{R}}_i\} \\ &= \bar{\mathscr{R}}_i \end{aligned}$$

which proves (7.4).

We proceed to show that the $\mathscr{B} \cap \bar{\mathscr{R}}_i$ are independent. Write

$$\bar{\mathscr{R}}_i^* = \sum_{j \neq i} \bar{\mathscr{R}}_j.$$

It is even true that

$$\mathscr{B} \cap \bar{\mathscr{R}}_i \cap \bar{\mathscr{R}}_i^* = 0, \qquad i = 1, \cdots, k. \tag{7.7}$$

On the contrary, suppose (7.7) fails for, say, $i = 1$. If $\dim(\mathscr{B} \cap \bar{\mathscr{R}}_1) = 1$, then

$$\mathscr{B} \cap \bar{\mathscr{R}}_1 \subset \bar{\mathscr{R}}_1^*. \tag{7.8}$$

If $\dim(\mathscr{B} \cap \bar{\mathscr{R}}_1) \geqq 2$, and

$$\mathscr{B} \cap \bar{\mathscr{R}}_i \not\subset \bar{\mathscr{R}}_i^*, \qquad i = 1, \cdots, k, \tag{7.9}$$

then

$$\dim\left[\sum_{i=1}^{\mu+1} \mathscr{B} \cap \bar{\mathscr{R}}_i\right] \geqq \dim\left[\sum_{i=1}^{\mu} \mathscr{B} \cap \bar{\mathscr{R}}_i\right] + 1,$$

$\mu = 1, \cdots, k - 1$, that is,

$$\dim\left[\sum_{i=1}^{2} \mathscr{B} \cap \bar{\mathscr{R}}_i\right] \geqq 3;$$

and by induction

$$\dim\left[\sum_{i=1}^{k} \mathscr{B} \cap \bar{\mathscr{R}}_i\right] \geqq k + 1,$$

a contradiction. Thus (7.9) is false; combining this result with (7.8) there follows

$$\mathscr{B} \cap \bar{\mathscr{R}}_\alpha \subset \bar{\mathscr{R}}_\alpha^* \tag{7.10}$$

for some $\alpha \in (1, \cdots, k)$. It will be shown below that there exists $\tilde{C}_\alpha$ such that

$$(A + B\tilde{C}_\alpha)\bar{\mathscr{R}}_\alpha \subset \bar{\mathscr{R}}_\alpha; \qquad (A + B\tilde{C}_\alpha)\bar{\mathscr{R}}_\alpha^* \subset \bar{\mathscr{R}}_\alpha^*. \tag{7.11}$$

Assuming (7.11) is true, we have

$$\bar{\mathscr{R}}_\alpha = \{A + B\tilde{C}_\alpha | \mathscr{B} \cap \bar{\mathscr{R}}_\alpha\} \subset \{A + B\tilde{C}_\alpha | \bar{\mathscr{R}}_\alpha^*\} \subset \bar{\mathscr{R}}_\alpha^* \subset \mathscr{V}_\alpha,$$

and therefore (7.2) fails for $i = \alpha$. With this contradiction, (7.7) is established. It remains to verify the existence of $\tilde{C}_\alpha$. For this we need the following result.

LEMMA 7.1. *Let $\mathscr{V}$, $\mathscr{W}$ be arbitrary. There exists C such that*

$$(A + BC)\mathscr{V} \subset \mathscr{V}; \qquad (A + BC)\mathscr{W} \subset \mathscr{W}$$

if and only if

$$A\mathscr{V} \subset \mathscr{B} + \mathscr{V};$$

$$A\mathscr{W} \subset \mathscr{B} + \mathscr{W};$$

$$A(\mathscr{V} \cap \mathscr{W}) \subset \mathscr{B} + \mathscr{V} \cap \mathscr{W}.$$

Proof. Necessity is obvious. For sufficiency, write

$$\mathscr{V} + \mathscr{W} = \hat{\mathscr{V}} \oplus (\mathscr{V} \cap \mathscr{W}) \oplus \hat{\mathscr{W}},$$

where $\hat{\mathscr{V}}' \subset \mathscr{V}'$; $\hat{\mathscr{W}}' \subset \mathscr{W}'$. By the construction of Lemma 3.2, C can be chosen such that

$$(A + BC)(\mathscr{V}' \cap \mathscr{W}') \subset \mathscr{V}' \cap \mathscr{W}';$$

$$(A + BC)\hat{\mathscr{V}}' \subset \mathscr{V}';$$

$$(A + BC)\hat{\mathscr{W}}' \subset \mathscr{W}'.$$

This completes the proof of the lemma.

Consider now $\bar{\mathscr{R}}_\alpha, \bar{\mathscr{R}}_\alpha^*$. Clearly $A\bar{\mathscr{R}}_\alpha \subset \mathscr{B} + \bar{\mathscr{R}}_\alpha$; $A\bar{\mathscr{R}}_\alpha^* \subset \mathscr{B} + \bar{\mathscr{R}}_\alpha^*$. By (7.3) $\mathscr{B} = \mathscr{B} \cap \bar{\mathscr{R}}_\alpha + \mathscr{B} \cap \bar{\mathscr{R}}_\alpha^*$, and so

$$\begin{aligned} A(\bar{\mathscr{R}}_\alpha \cap \bar{\mathscr{R}}_\alpha^*) &\subset (\mathscr{B} + \bar{\mathscr{R}}_\alpha) \cap (\mathscr{B} + \bar{\mathscr{R}}_\alpha^*) \\ &= \mathscr{B} + (\mathscr{B} + \bar{\mathscr{R}}_\alpha) \cap \bar{\mathscr{R}}_\alpha^* \\ &= \mathscr{B} + (\mathscr{B} \cap \bar{\mathscr{R}}_\alpha^* + \bar{\mathscr{R}}_\alpha) \cap \bar{\mathscr{R}}_\alpha^* \\ &= \mathscr{B} + \bar{\mathscr{R}}_\alpha \cap \bar{\mathscr{R}}_\alpha^*. \end{aligned}$$

By applying Lemma 7.1, the existence of $\tilde{C}_\alpha$ is finally established.

Proof. Part 3. We now prove that (7.2) and (7.3) are sufficient conditions for existence of a solution. Let $\bar{\mathscr{V}}_i$ be the maximal subspace such that

$$A\bar{\mathscr{V}}_i \subset \mathscr{B} + \bar{\mathscr{V}}_i, \qquad \bar{\mathscr{V}}_i \subset \bigcap_{j \neq i} \mathscr{K}_j, \qquad i = 1, \cdots, k. \tag{7.12}$$

It is enough to check that the $\bar{\mathscr{V}}_i$ are *compatible*, in the sense that there exists C such that

$$(A + BC)\bar{\mathscr{V}}_i \subset \bar{\mathscr{V}}_i, \qquad i = 1, \cdots, k.$$

We show first that the subspaces

$$\bar{\mathscr{V}}_i^* = \sum_{j \neq i} \bar{\mathscr{V}}_j$$

are compatible. From (7.12) there follows

$$\begin{aligned} A\bar{\mathscr{V}}_i^* &\subset \mathscr{B} + \bar{\mathscr{V}}_i^* \\ &= \mathscr{B} \cap \bar{\mathscr{V}}_i + \bar{\mathscr{V}}_i^* \qquad \text{(by (7.3))} \\ &= \mathscr{B}_i + \bar{\mathscr{V}}_i^*, \qquad i = 1, \cdots, k, \end{aligned}$$

where $\mathscr{B}_i = \mathscr{B} \cap \bar{\mathscr{V}}_i$. By Lemma 3.2, there exist B_i with $\{B_i\} = \mathscr{B}_i$, and C_i, such that

$$(A + B_iC_i)\bar{\mathscr{V}}_i^* \subset \bar{\mathscr{V}}_i^*, \qquad i = 1, \cdots, k.$$

Choosing a basis $\{v_1, \cdots, v_\mu\}$ for $\bar{\mathscr{V}}_1 + \cdots + \bar{\mathscr{V}}_k$, we define C such that

$$BCv_\nu = \sum_{i=1}^{k} B_iC_iv_\nu, \qquad \nu = 1, \cdots, \mu.$$

Then

$$
\begin{aligned}
(A + BC)\bar{\mathscr{V}}_i^* &= \left(A + B_iC_i + \sum_{j\neq i} B_jC_j\right)\bar{\mathscr{V}}_i^* \\
&\subset (A + B_iC_i)\bar{\mathscr{V}}_i^* + \sum_{j\neq i} \mathscr{B}_j \\
&\subset \bar{\mathscr{V}}_i^* + \sum_{j\neq i} \bar{\mathscr{V}}_j \\
&= \bar{\mathscr{V}}_i^*, \qquad i = 1, \cdots, k.
\end{aligned} \tag{7.13}
$$

This proves compatibility of the $\bar{\mathscr{V}}_i^*$. Now define

$$\mathscr{V}'_i = \bigcap \bar{\mathscr{V}}_j^*, \qquad i = 1, \cdots, k.$$

Clearly, $\mathscr{V}'_i \supset \bar{\mathscr{V}}_i$, $i = 1, \cdots, k$. By (7.13),

$$(A + BC)\mathscr{V}'_i \subset \mathscr{V}'_i, \qquad i = 1, \cdots, k, \tag{7.14}$$

and, furthermore by the second condition of (7.12),

$$\mathscr{V}'_i \subset \bigcap_{j\neq i} \sum_{\alpha\neq j} \bigcap_{m\neq\alpha} \mathscr{K}_m = \bigcap_{j\neq i} \mathscr{K}_j.\text{[2]} \tag{7.15}$$

By (7.14) and (7.15), the $\mathscr{V}'_i$ satisfy the conditions imposed on the $\bar{\mathscr{V}}_i$ in (7.12). Since the $\bar{\mathscr{V}}_i$ are maximal, there results $\mathscr{V}'_i \subset \bar{\mathscr{V}}_i$, and, therefore, $\mathscr{V}'_i = \bar{\mathscr{V}}_i$, $i = 1, \cdots, k$.

Remark 6. If the conditions of Theorem 7.1 are satisfied, then

$$
\begin{aligned}
\sum_{i=1}^{k} \bar{\mathscr{R}}_i &= \sum_{i=1}^{k} \{A + BC|\mathscr{B} \cap \bar{\mathscr{R}}_i\} = \left\{A + BC\middle|\sum_{i=1}^{k} \mathscr{B} \cap \bar{\mathscr{R}}_i\right\} \\
&= \{A + BC|\mathscr{B}\} = \{A|\mathscr{B}\} = \mathscr{E}.
\end{aligned} \tag{7.16}
$$

We turn now to the problem of pole assignment. In contrast to the situation of § 6, it is no longer possible, in general, to vary the spectrum of $A + BC$ on each $\bar{\mathscr{R}}_i$ independently. The following example shows that certain eigenvalues of $A + BC$ may even be fixed for all admissible C.

Let $\mathscr{E} = \mathscr{E}^3$, $k = 2$ and

$$A = \begin{bmatrix} 1 & 0 & 0 \\ 1 & 1 & 1 \\ 0 & 0 & 1 \end{bmatrix}, \qquad B = \begin{bmatrix} 1 & 0 \\ 0 & 0 \\ 0 & 1 \end{bmatrix},$$

$$\mathscr{K}_1 = \begin{bmatrix} 1 \\ 0 \\ 0 \end{bmatrix}^{\perp}, \qquad \mathscr{K}_2 = \begin{bmatrix} 0 \\ 0 \\ 1 \end{bmatrix}^{\perp}.$$

[2] This identity and its dual, $\sum \cap \sum = \sum$, are readily established by using the (modular) distributive rule for subspaces.

It is easily checked that (5.7)–(5.9) have the (unique) solution

$$\mathscr{R}_1 = \left\{ \begin{bmatrix} 1 \\ 0 \\ 0 \end{bmatrix}, \begin{bmatrix} 0 \\ 1 \\ 0 \end{bmatrix} \right\}, \qquad \mathscr{R}_2 = \left\{ \begin{bmatrix} 0 \\ 1 \\ 0 \end{bmatrix}, \begin{bmatrix} 0 \\ 0 \\ 1 \end{bmatrix} \right\}$$

and that C must have the form

$$C = \begin{bmatrix} c_1 & 0 & 0 \\ 0 & 0 & c_2 \end{bmatrix}$$

with arbitrary c_1, c_2. Then

$$\det(A + BC - \lambda I) = (1 + c_1 - \lambda)(1 - \lambda)(1 + c_2 - \lambda).$$

Observe that the eigenvalue $\lambda = 1$, belonging to the eigenvector $(0, 1, 0)'$ of $A + BC$, is fixed.

To discuss the present case in general, we introduce a suitable decomposition of $\mathscr{E}$. Assume that the problem of (5.7)–(5.9) has a solution $C, \bar{\mathscr{R}}_1, \cdots, \bar{\mathscr{R}}_k$, and let $\mathbf{C}$ denote the class of matrices C for which $(A + BC)\bar{\mathscr{R}}_i \subset \bar{\mathscr{R}}_i$, $i = 1, \cdots, k$. We know that the spaces $\bar{\mathscr{R}}_i$ are the unique solutions: for simplicity of notation, write $\mathscr{R}_i$ for $\bar{\mathscr{R}}_i$. Define

$$\mathscr{E}_0 = \bigcap_{i=1}^{k} \mathscr{R}_i^*, \tag{7.17}$$

and let $\mathscr{E}_i$ be any subspace such that

$$\mathscr{R}_i = \mathscr{E}_i \oplus (\mathscr{R}_i \cap \mathscr{E}_0), \qquad i = 1, \cdots, k. \tag{7.18}$$

In the following, J denotes the set of indices $(1, \cdots, k)$, J_0 the set $(0, 1, \cdots, k)$. In intersections and summations involving $\mathscr{R}$'s, the index ranges over J; in those involving $\mathscr{E}$'s, the index ranges over J_0.

Lemma 7.2. *The subspaces $\mathscr{E}_i$ have the properties*

$$\mathscr{E}_0 \oplus \mathscr{E}_1 \oplus \cdots \oplus \mathscr{E}_k = \mathscr{E}, \tag{7.19}$$

$$(A + BC)\mathscr{E}_i \subset \mathscr{E}_i + \mathscr{E}_0, \qquad i \in J_0, \quad C \in \mathbf{C}. \tag{7.20}$$

Proof. Assertion (7.20) is obvious by the fact that the $\mathscr{R}_i$ are $(A + BC)$-invariant. For (7.19), observe first that

$$\mathscr{R}_i \cap \mathscr{E}_0 = \mathscr{R}_i \cap \bigcap_{j \neq i} \mathscr{R}_j^* \cap \mathscr{R}_i^* = \mathscr{R}_i \cap \mathscr{R}_i^*$$

and so, if $i \in J$,

$$\begin{aligned} \mathscr{E}_i \cap \left(\mathscr{E}_0 + \sum_{\substack{j \neq 0 \\ j \neq i}} \mathscr{E}_j \right) &\subset \mathscr{E}_i \cap (\mathscr{E}_0 + \mathscr{R}_i^*) \\ &= \mathscr{E}_i \cap \mathscr{R}_i^* \\ &= \mathscr{E}_i \cap \mathscr{R}_i \cap \mathscr{R}_i^* \\ &= 0. \end{aligned} \tag{7.21}$$

Now for arbitrary subspaces $\mathscr{S}_i$, $i = 1, 2, 3$, if

$$\mathscr{S}_1 \cap (\mathscr{S}_2 + \mathscr{S}_3) = \mathscr{S}_1 \cap \mathscr{S}_2 + \mathscr{S}_1 \cap \mathscr{S}_3,$$

then

$$\mathscr{S}_2 \cap (\mathscr{S}_1 + \mathscr{S}_3) = \mathscr{S}_1 \cap \mathscr{S}_2 + \mathscr{S}_2 \cap \mathscr{S}_3.$$

Applying this fact and using (7.21) we have

$$\mathscr{E}_1 \cap \left(\mathscr{E}_0 + \sum_{j=2}^{k} \mathscr{E}_j\right) = 0 = \mathscr{E}_1 \cap \mathscr{E}_0 + \mathscr{E}_1 \cap \sum_{j=2}^{k} \mathscr{E}_j,$$

and therefore

$$\mathscr{E}_0 \cap \left(\mathscr{E}_1 + \sum_{j=2}^{k} \mathscr{E}_j\right) = \mathscr{E}_0 \cap \mathscr{E}_1 + \mathscr{E}_0 \cap \sum_{j=2}^{k} \mathscr{E}_j$$

$$= \mathscr{E}_0 \cap \sum_{j=2}^{k} \mathscr{E}_j.$$

Repetition of this argument yields, after $k - 2$ steps,

$$\mathscr{E}_0 \cap \left(\mathscr{E}_1 + \sum_{j=2}^{k} \mathscr{E}_j\right) = \mathscr{E}_0 \cap \mathscr{E}_k = 0. \tag{7.22}$$

Equations (7.21), (7.22) state that the $\mathscr{E}_i$, $i \in J_0$, are independent. Finally, by (7.16),

$$\sum_{i=0}^{k} \mathscr{E}_i = \sum_{i=1}^{k} (\mathscr{E}_i + \mathscr{E}_0) \supset \sum_{i=1}^{k} \mathscr{R}_i = \mathscr{E}.$$

Remark 7. If the $\mathscr{R}_i$ are independent, then $\mathscr{E}_0 = 0$ and $\mathscr{E}_i = \mathscr{R}_i$, $i \in J$.

For $i \in J_0$ let P_i be the projection on $\mathscr{E}_i$ along $\sum_{j \neq i} \mathscr{E}_j$, and now let $C \in \mathbf{C}$ be fixed.

LEMMA 7.3. *Let* $\mathscr{B} \cap \mathscr{R}_i = \{b_i\}$, $i \in J$. *Then*

$$\mathscr{E}_i = \{P_i(A + BC) | \{P_i b_i\}\}, \qquad i \in J. \tag{7.23}$$

Proof. By (7.18) and (7.19), $\mathscr{E}_i = P_i \mathscr{R}_i$. By (7.18) and (7.20),

$$P_i \mathscr{R}_i = P_i \sum_{j=1}^{n} (A + BC)^{j-1} \{b_i\}$$

$$= \sum_{j=1}^{n} [P_i(A + BC)]^{j-1} \{P_i b_i\}$$

$$= \{P_i(A + BC) | \{P_i b_i\}\}.$$

LEMMA 7.4.

$$\mathscr{B} \cap \mathscr{E}_0 = 0. \tag{7.24}$$

Proof. By (7.3) and (7.7),

$$\mathscr{B} \cap \mathscr{E}_0 = \left(\sum_{i=1}^{k} \mathscr{B} \cap \mathscr{R}_i\right) \cap \bigcap_{j=1}^{k} \mathscr{R}_j^*$$

$$= \left(\mathscr{B} \cap \mathscr{R}_1 + \sum_{i=2}^{k} \mathscr{B} \cap \mathscr{R}_i\right) \cap \mathscr{R}_1^* \cap \bigcap_{j=2}^{k} \mathscr{R}_j^*$$

$$= \left(\sum_{i=2}^{k} \mathscr{B} \cap \mathscr{R}_i\right) \cap \bigcap_{j=2}^{k} \mathscr{R}_j^*$$

$$\vdots$$

$$= \mathscr{B} \cap \mathscr{R}_k \cap \mathscr{R}_k^* = 0.$$

This completes the proof of Lemma 7.4.

Next let $C = C_1$ be a fixed member of **C**, let $C_2 \in \mathbf{C}$, and write $D = C_2 - C_1$: thus $A + BC_2 = A + BC_1 + BD$. Now $b_i \in \mathscr{R}_i \subset \mathscr{E}_i + \mathscr{E}_0$ $(i \in J)$; and (7.20) yields $BD\mathscr{E}_i \subset \mathscr{E}_i + \mathscr{E}_0$ $(i \in J_0)$; therefore

$$P_i b_j = 0, \qquad P_i BD\mathscr{E}_j = 0, \qquad i, j \in J, \quad i \neq j. \tag{7.25}$$

Also, using (7.24)

$$BD\mathscr{E}_0 \subset \mathscr{B} \cap \mathscr{E}_0 = 0. \tag{7.26}$$

Write

$$BD = \sum_{j=1}^{k} b_j d_j', \tag{7.27}$$

where, as before, $\{b_j\} = \mathscr{B} \cap \mathscr{R}_j$. Then

$$P_i BD\mathscr{E}_j = P_i b_i d_i' \mathscr{E}_j = 0, \qquad i \neq j, \quad i \in J, \quad j \in J_0. \tag{7.28}$$

We can now compute the spectrum Λ of $A + BC_2$. Define

$$A_i = P_i(A + BC_1), \qquad i \in J_0. \tag{7.29}$$

By (7.26) and (7.29),

$$P_0(A + BC_2) = P_0(A + BC_2)(I - P_0) + A_0 P_0, \tag{7.30}$$

and by (7.28) and (7.29),

$$\begin{aligned} P_i(A + BC_2) &= A_i + P_i BD \sum_{j=0}^{k} P_j \\ &= A_i + P_i b_i d_i' P_i, \qquad i \in J. \end{aligned} \tag{7.31}$$

Suppose $\lambda \in \Lambda$, with corresponding (complex) eigenvector ξ. A brief calculation from (7.30), (7.31) shows that either (i) for some $i \in J$, $P_i\xi \neq 0$ and $(A_i + P_i b_i d_i')P_i\xi = \lambda P_i\xi$, or (ii) $\xi = P_0\xi$ and $A_0\xi = \lambda\xi$. Conversely, if $A_0\xi = \lambda\xi$ for $0 \neq \xi \in \mathscr{E}_0$, or $(A_i + P_i b_i d_i')\xi = \lambda\xi$ for $0 \neq \xi \in \mathscr{E}_i$ and some $i \in J$, then $\lambda \in \Lambda$. Therefore

$$\Lambda = \bigcup_{i=0}^{k} \Lambda_i,$$

where Λ_i, $i \in J_0$, is the spectrum of the restriction of $P_i(A + BC_2)$ to $\mathscr{E}_i$. By (7.30), Λ_0 is independent of the choice of C_2, i.e., is fixed uniquely by the requirement $C \in \mathbf{C}$. On the other hand, for $i \in J$ Lemma 7.3 states that $\mathscr{E}_i$ is the controllability space of the pair $(A_i, P_i b_i)$. Hence, any choice of Λ_i can be realized by appropriate choice of d_i: indeed, for any $w \in \mathscr{E}$ there exists d_i such that

$$d_i'x = \begin{cases} w'x & \text{for } x \in \mathscr{E}_i, \\ 0 & \text{for } x \in \sum_{j \neq i} \mathscr{E}_j. \end{cases}$$

These results are summarized in the following theorem.

Theorem 7.2. *Let the conditions of Theorem* 7.1 *be satisfied. If $C \in \mathbf{C}$, the eigenvalues of $A + BC$ can be partitioned into $k + 1$ disjoint sets*

$$\Lambda_i = \{\lambda_{i1}, \cdots, \lambda_{in_i}\}, \qquad i \in J_0,$$

where

$$n_0 = \dim\left(\bigcap_{j=1}^{k} \bar{\mathscr{R}}_j^*\right),$$

$$n_i = \dim(\bar{\mathscr{R}}_i) - \dim(\bar{\mathscr{R}}_i \cap \bar{\mathscr{R}}_i^*), \qquad i \in J.$$

The set Λ_0 and the integers n_i ($i \in J_0$) are fixed for all $C \in \mathbf{C}$. The sets Λ_i ($i \in J$) can be assigned freely (by suitable choice of $C \in \mathbf{C}$) subject only to the requirement that any λ_{ij} with $\operatorname{Im} \lambda_{ij} \neq 0$ occur in Λ_i in a conjugate pair.

Remark 8. If basis vectors are chosen in the $\mathscr{E}_i$, then the system differential equation can be put in a simple "normal" form. Let

$$z_i = P_i x, \qquad i \in J_0,$$

and

$$\dot{x} = (A + BC_2)x + Bv.$$

Multiplying through by P_i and using (7.30), (7.31), we obtain

$$\begin{aligned} \dot{z}_i &= (A_i + P_i b_i d_i')z_i + P_i Bv, \qquad i \in J, \\ \dot{z}_0 &= P_0(A + BC_2)(z_1 + \cdots + z_k) + A_0 z_0 + P_0 Bv. \end{aligned} \tag{7.32}$$

Let K be an $m \times m (= k \times k)$ matrix such that $BK = [b_1 \cdots b_k]$ and put $v = Kw$, $w \equiv (w_1, \cdots, w_k)'$. Since $b_i \in \mathscr{E}_i \oplus \mathscr{E}_0$, we have

$$b_i = P_i b_i + P_0 b_i \equiv \hat{b}_i + \hat{b}_{i0}.$$

Adopting n_i-dimensional representations of the z_i, etc., we see that (7.32) can be written as

$$\begin{aligned} \dot{z}_i &= (\hat{A}_i + \hat{b}_i \hat{d}_i')z_i + \hat{b}_i w_i, \qquad i \in J, \\ \dot{z}_0 &= \sum_{j \neq 1}^{k} \hat{A}_{0j} z_j + \hat{A}_0 z_0 + \hat{B}_0 w. \end{aligned} \tag{7.33}$$

Equation (7.33) exhibits the system (2.1) as an array of k decoupled subsystems, each completely controllable by an independent scalar input w_i, plus one additional subsystem which is driven by the others and by w. Finally, since $\mathscr{R}_i \cap \mathscr{E}_0 = \mathscr{R}_i \cap \mathscr{R}_i^* \subset \mathscr{N}_i$, it follows by (5.8) and (7.18) that $\mathscr{E}_i + \mathscr{N}_i = \mathscr{E}$, that is, $H_i \mathscr{E}_i = \mathscr{H}_i$.

Remark 9. The decoupled system is acceptable in practice only if the eigenvalues in the fixed set Λ_0 are all stable. It is possible to check for stability of Λ_0 as follows. Recall that $\mathscr{R}_i \subset \mathscr{E}_i + \mathscr{E}_0$ ($i \in J$) and note from (7.20) that $A(\mathscr{E}_i + \mathscr{E}_0) \subset \mathscr{E}_i + \mathscr{E}_0 + \mathscr{B}$ ($i \in J$). Furthermore,

$$\mathscr{E}_i + \mathscr{E}_0 \subset \bigcap_{j \neq i} \mathscr{N}_j, \qquad i \in J.$$

It follows by Theorem 4.3 and the maximality of the $\mathscr{R}_i$ ($=\bar{\mathscr{R}}_i$) that

$$\mathscr{R}_i = \{A + BC | \mathscr{B} \cap (\mathscr{E}_i + \mathscr{E}_0)\}$$

for any C with the property (7.20). That is, (7.20) is both necessary and sufficient that $C \in \mathbf{C}$. Thus, to compute Λ_0 it is necessary only to compute the spectrum of $A + BC_0$ (restricted to $\mathscr{E}_0$) where C_0 is any matrix such that $(A + BC_0)\mathscr{E}_0 \subset \mathscr{E}_0$.

Concluding remark. This article represents a preliminary investigation of the general decoupling problem formulated in § 5. The results for the special cases of § 6 and § 7 suggest the possibility of a complete and detailed geometric theory of linear multivariable control, in which the concept of controllability subspace would play a central role. Specific problems for future study include not only that of § 5 but also the problem of decoupling by adjunction of suitable dynamics (augmentation of the state space), and the problem of sensitivity. As formulated, decoupling represents a "hard" constraint, an all-or-nothing algebraic property. Of course, for applications a quantitative approach via "soft" constraints might also prove rewarding.

It is clear that an adequate qualitative theory of large linear multivariable systems is currently lacking; and equally clear that, with computers, such a theory would find wide application.

REFERENCES

[1] Z. V. REKASIUS, *Decoupling of multivariable systems by means of state variable feedback*, Proc. Third Allerton Conference on Circuit and System Theory, Urbana, Illinois, 1965, pp. 439–447.

[2] P. L. FALB AND W. A. WOLOVICH, *Decoupling in the design and synthesis of multivariable control systems*, IEEE Trans. Automatic Control, AC-12 (1967), pp. 651–659.

[3] E. G. GILBERT, *The decoupling of multivariable systems by state feedback*, this Journal, 7 (1969), pp. 50–63.

[4] W. M. WONHAM, *On pole assignment in multi-input controllable linear systems*, IEEE Trans. Automatic Control, AC-12 (1967), pp. 660–665.

[5] M. HEYMANN, *Pole assignment in multi-input linear systems*, Ibid., AC-13 (1968), pp. 748–749.

System Theory on Group Manifolds and Coset Spaces

R. W. BROCKETT

THE 1960s witnessed important developments in control theory revolving around the use of state space methods. Many of these developments were first reported in some of the papers reprinted in this volume. The principal results were revolutionary because they provided fundamentally new approaches to problems in optimization and estimation which could be solved in no other way. The concept of state space is absolutely central, for instance, to understanding recursive estimation and optimal control of linear systems with white noise inputs and quadratic performance indices.

As the decade of the 1960s gave way to the decade of the 1970s there remained a great deal of excitement concerning the power of the state space point of view, and it is not surprising that there were attempts to extend the state space theory of linear control systems into the nonlinear domain. Several important papers appeared in the early 1970s, and two of these (this paper by Brockett, and [14]) as well as a later entry ([7]) are reprinted in the present volume. The most interesting common thread was the emerging general awareness that differentiable manifolds provided the right state space setting in which to study finite dimensional nonlinear control systems. (See the references in the paper by Brockett.)

The two papers of 1972 appearing in this volume (this paper by Brockett, and the one by Sussmann and Jurdjevic [14] that follows it) were watershed efforts in exploring the essential differences between the linear and nonlinear theories of control. Although it is now widely recognized, it was a somewhat revolutionary observation made in the paper by Brockett that differential geometric methods are central to treating global questions involving strongly nonlinear systems (e.g., in controlling rapid, large-angle slewing maneuvers in spacecraft). As the analytical tools from differential geometry became more refined, this observation became more widely appreciated, and the field of *geometric nonlinear control theory* was born.

The realization that nonlinear systems should be studied using the tools of differential geometry was singularly important because it motivated researchers to think about control systems in a different and fundamentally new way. From the geometric viewpoint, a new and deeper understanding of classical linear control theory emerged in which important classical results were seen as special cases of more general nonlinear results. It also led, very early on, to the discovery that there were important parts of the nonlinear theory for which there were no exact counterparts in the linear setting.

A few words contrasting Brockett's paper with that of Sussmann and Jurdjevic [14] may be in order. Sussmann and Jurdjevic developed their treatment within the very general state space framework of differentiable manifolds. Careful analysis within this very general setting showed how the topology of the reachable sets depended on the (Lie) algebraic structure of equations of motion. The group and coset manifolds treated by Brockett were more specialized but had enough additional structure to facilitate a more wide ranging study of control theoretic issues. While the setting was specialized, all the key features of very general nonlinear systems were present. At the time when the paper appeared, the concept of *controllability* for nonlinear systems was still not completely understood, and in Brockett's paper we find one of the earliest efforts to explain the important distinctions that must be drawn between systems with and without drift—in Brockett's notation:

$$\dot{X}(t) = \left(A + \sum_{i=1}^{m} u_i(t)B_i\right) X(t) \tag{1}$$

and

$$\dot{X}(t) = \sum_{i=1}^{m} u_i(t)B_iX(t) \tag{2}$$

respectively. The symmetric or drift-free case (2) is interesting both in the extent to which it differs from (1) and also in that it admits a very complete characterization of controllability. More specifically, the system (2) is said to be *controllable* if for any two points X_0 and X_1 in the state space and any $T > 0$, there exist piecewise continuous (or merely integrable) control functions $u_1(\cdot), \ldots, u_m(\cdot)$ such that the corresponding trajectory defined by (2) and $X(0) = X_0$ satisfies $X(T) = X_1$. This is a very strong notion of controllability that turns out to be natural for both symmetric systems, of the form (2), and classical constant-coefficient finite dimensional linear systems, but that turns out to be generally too restrictive for systems of the form (1). Indeed, systems with drift (1), where the evolution equation has the uncontrolled component A, are almost never controllable in this sense. (See [14], for the complete details.)

Both Brockett and Sussmann-Jurdjevic [14] have adopted the language and style of differential geometry and Lie theory in developing a theoretical framework for nonlinear control systems. In Brockett's paper, controllability questions are posed in terms of transitive actions of subgroups on orbits. The paper begins by proving a version of Chow's theorem in the setting of matrix Lie algebras. The proof consists of a sequence of steps that compute Lie brackets and rescale time so as to show that for a system of the form (2) it is possible to steer a trajectory in the direction of any Lie bracket of the matrices B_i. From this it is a short step, taken with the help of a basic result of Wei and Norman, to show that (2) is controllable (in the sense of the above definition) precisely on the orbit $\mathcal{G} \cdot X_0$, where $\mathcal{G} = \{\exp \mathcal{A}\}_G$ is the group generated generated by all finite products of matrices of the form e^C, where $C \in \mathcal{A}$ = the linear subspace of $\mathrm{R}^{n\times n}$ spanned by the coefficient matrices $B_1, \ldots, B_m$; X_0 is a given initial condition of (2).

The paper goes on to discuss the controllability of systems having the additional structure of the coefficient matrices B_i taking values in the classical semisimple Lie algebras. This was the first place in which semisimple Lie algebras appeared in the control and systems engineering literature, and the importance of introducing these ideas was that they opened a new window on applications involving controlled physical systems. Brockett does not treat the nonsymmetric case (1) at all extensively, but he considers a special case of (1) in which the vanishing of certain Lie brackets allows us to conclude that the system is controllable in the strong sense of the above definition. This case further specializes to include constant coefficient finite dimensional linear systems as a subcase. *Observability* is also treated only briefly (a page and a half), but Brockett makes the important observation that in the nonlinear case, there is no duality between controllability and observability.

It is interesting to observe that all the controllability results in this paper involve determining the dimension of the linear span of a finite number of matrices which are computed as Lie brackets of the coefficient matrices appearing in (1) and (2). All Lie bracket computations are explicitly prescribed and are finite in number. (The Lie algebras encountered in [14], on the other hand, are typically infinite dimensional.) The observability results involve a similar degree of simplicity and explicitness, and thus the control theory of systems defined on matrix Lie groups and coset spaces is comparable to that of finite dimensional linear systems in terms of computational complexity and level of explicitness. Unfortunately, this degree of completeness is not found in more general nonlinear systems, and it is only when there is a special (e.g., polynomial, [1], [2]) structure present that the theory rests on such explicit, finitely verifiable conditions.

In the process of pointing out new research directions, the paper offered the field a definitive perspective on nonlinear control theory and its applications. After the appearance of this paper by Brockett and [14], geometric nonlinear control theory developed along two somewhat different but complementary lines. One line, which continues to grow and prosper, has been aimed at the development of new geometrically based principles of control. Prominent examples come from the literature on nonholonomic motion planning (e.g., [11]) and *sub-Riemannian geometry* (e.g., [5] and [4]). The other line has carried on the effort to extend important ideas rooted in classical control into the nonlinear domain. The paper [7], which also appears in this volume, was an early report of this work, and the textbooks [8] and [10] summarize the larger history of the effort. The paper's influence on the control of mechanical systems has been especially deep, and it continues to be a primary source in this literature; see, e.g., [13], [9], [12], [6], and [3].

References

[1] J. Baillieul, "The geometry of homogeneous polynomial dynamical systems," *Nonlinear Analysis: Theory, Methods and Applications*, **4**(5):879–900, September 1980.

[2] J. Baillieul, "Controllability and observability of polynomial dynamical systems," *Nonlinear Analysis: Theory, Methods and Applications*, **5**(5):543–552, April 1981.

[3] J. Baillieul, "The geometry of controlled mechanical systems," in *Mathematical Control Theory*, J. Baillieul and J.C. Willems, Eds., Springer-Verlag (New York), 1998.

[4] A.M. Bloch, P.E. Crouch, and T.S. Ratiu, *Sub-Riemannian Optimal Control Problems*, Fields Institute Communications, AMS (Providence, RI), **3**:35–48, 1994.

[5] R.W. Brockett, "Control theory and singular Riemannian geometry," in *New Directions in Applied Mathematics*, Springer-Verlag (New York), pp. 13–27, 1982.

[6] L.E. Faibusovich, "Collective Hamiltonian method in optimal control problems," *Cybernetics+*, **25**(2):230–237, March–April 1989.

[7] R. Hermann and A.J. Krener, "Nonlinear controllability and observability," *IEEE Trans. Automat. Contr.*, **AC-22**:728–740, 1977.

[8] A. Isidori, *Nonlinear control systems*, 3rd ed., Springer-Verlag (New York), 1995.

[9] N.E. Leonard, "Control synthesis and adaptation for an underactuated autonomous underwater vehicle," *IEEE J. Oceanic Eng.*, **20**(3):211–220, July 1995.

[10] H. Nijmeijer and A. van der Schaft, *Nonlinear Dynamical Control Systems*, Springer-Verlag (New York), 1990.

[11] R.M. Murray and S.S. Sastry, "Nonholonomic motion planning: steering using sinusoids," *IEEE Trans. Autom. Control*, **38**(5):700–716, 1993.

[12] V. Ramakrishna, M.V. Salapaka, M. Dahleh, H. Rabitz and A. Peirce, "Controllability of molecular systems," *Phys. Rev. A*, **51**(2):960–966, February 1995.

[13] K. Spindler, "Optimal attitude control of a rigid body," *Appl. Math. Opt.*, **34**(1):79–90, July–August 1996.

[14] H.J. Sussmann and V. Jurdjevic, "Controllability of Nonlinear Systems," *J. Diff. Eqns.*, **12**:95–116, 1972.

J.B.

SYSTEM THEORY ON GROUP MANIFOLDS AND COSET SPACES*

R. W. BROCKETT†

Abstract. The purpose of this paper is to study questions regarding controllability, observability, and realization theory for a particular class of systems for which the state space is a differentiable manifold which is simultaneously a group or, more generally, a coset space. We show that it is possible to give rather explicit expressions for the reachable set and the set of indistinguishable states in the case of autonomous systems. We also establish a type of state space isomorphism theorem. These results parallel, and in part specialize to, results available for the familiar case described by $\dot{x}(t) = Ax(t) + Bu(t)$, $y(t) = Cx(t)$. Our objective is to reduce all questions about the system to questions about Lie algebras generated from the coefficient matrices entering in the description of the system and in that way arrive at conditions which are easily visualized and tested.

1. Introduction. A standard assumption in modern control theory is that the state space is a vector space. This assumption is both valid and natural in many situations, but there is a significant class of problems for which it cannot be made. Typical of these are certain problems which arise in the control of the attitude of a rigid body. The state space in this case is not a vector space. Linearization often destroys the essence of the problem—even if one can work locally—and in any case new and different methods are needed for treating global questions.

In this paper we substitute the following hypothesis for the usual vector space assumptions. We let $\mathscr{F}$ and $\mathscr{C}$ be matrix groups and study

$$\dot{X}(t) = \left(A + \sum_{i=1}^{\nu} u_i(t)B_i\right)X(t), \qquad y(t) = \mathscr{C}X(t), \qquad X \in \mathscr{F},$$

where A and B_i belong to the Lie algebra associated with $\mathscr{F}$, the u_i are the controls, and the notation $\mathscr{C}X(t)$ is to be interpreted as being a coset in $\mathscr{F}$. We also study vector systems of a similar type in that we can view their evolution as occurring in a coset space. The results concern the explicit construction of the reachable set and a characterization of observability which is easily tested. Our main point is that this class of systems is in many ways not more difficult than linear systems of the usual type in $\mathbb{R}^n$.

There is a moderately large literature on the use of Chow's results [1] and related ideas to study controllability, including the work of Hermann, Kucera, Hermes, Haynes and Lobry (see [2]–[6]). This work is relevant here but we are directly interested in controllability only in so far as it contributes to the identification of a framework in which we can study a full range of system theoretic questions, including observability and realization theory. Notice that it is impossible to pass

* Received by the editors August 24, 1971, and in revised form October 27, 1971. Presented at the NSF Regional Conference on Control Theory, held at the University of Maryland Baltimore County, August 23–27, 1971.

† Division of Engineering and Applied Physics, Harvard University, Cambridge, Massachusetts 02138. This work was supported in part by the U.S. Office of Naval Research under the Joint Services Electronics Program by Contract N00014-67-A-0298-0006 and by the National Aeronautics and Space Administration under Grant NGR 22-007-172.

Reprinted with permission from *SIAM Journal on Control*, R. W. Brockett, "System Theory on Group Manifolds and Coset Spaces" Vol. 10, 1972, pp. 265–284.

directly from controllability results to observability results in the present setup because there is no clear notion of duality. The main motivation for this work came from some work on Lie algebraic methods in differential equations (see [7]–[10]) and, above all, from being confronted with certain physical problems where linear theory was simply inadequate.

Some unpublished work [18], [19] by Jurdjevic and Sussmann is related to this paper. In particular they give in [19] an alternate proof of our Theorem 5 and make a serious study of the unsymmetric case (treated only superficially in Theorem 7 here). We also mention a recent paper by Elliott [20].

2. Examples. We postpone the development of the subject long enough to present a few simple examples which will help justify why the assumptions are set up the way they are.

Example 1 (Control systems design). Consider the problem of determining the gain, k, in the system

$$\dot{x}(t) = Ax(t) - k(t)bcx(t)$$

so as to achieve good performance relative to an index of the form

$$\eta = \int_0^\infty x'(t)Mx(t)\,dt, \qquad M = M' \geqq 0.$$

If a particular initial state is chosen and $k(\cdot)$ is selected so as to minimize η, then the performance might be bad relative to some other initial state. In cases where the initial state is not known, it is much more realistic to pick a collection of initial state vectors and to pick k in such a way as to minimize a weighted average of the individual performances. In fact, just to ensure stability it is necessary to average over at least n linearly independent initial states. If exactly n are chosen, then k should be regarded as controlling the evolution of the matrix equation

$$\dot{\Phi}(t) = (A - k(t)bc)\Phi(t), \qquad \Phi(0) = [x_1, x_2, \cdots, x_n].$$

The state space is then the space of nonsingular $n \times n$ matrices, $\mathscr{G}\ell(n)$.

Example 2 (Rigid body control). The orientation of a rigid body relative to some fixed set of axes is described by a 3×3 orthogonal matrix A which satisfies the differential equation

$$\begin{bmatrix} \dot{a}_{11}(t) & \dot{a}_{12}(t) & \dot{a}_{13}(t) \\ \dot{a}_{21}(t) & \dot{a}_{22}(t) & \dot{a}_{23}(t) \\ \dot{a}_{31}(t) & \dot{a}_{32}(t) & \dot{a}_{33}(t) \end{bmatrix} = \begin{bmatrix} 0 & \omega_3(t) & -\omega_2(t) \\ -\omega_3(t) & 0 & \omega_1(t) \\ \omega_2(t) & -\omega_1(t) & 0 \end{bmatrix} \begin{bmatrix} a_{11}(t) & a_{12}(t) & a_{13}(t) \\ a_{21}(t) & a_{22}(t) & a_{23}(t) \\ a_{31}(t) & a_{32}(t) & a_{33}(t) \end{bmatrix}.$$

The ω's themselves are usually controlled via the equations

$$\dot{\omega}_1(t) = [(I_2 - I_3)/I_1]\omega_2(t)\omega_3(t) + n_1(t)/I_1,$$
$$\dot{\omega}_2(t) = [(I_3 - I_1)/I_2]\omega_1(t)\omega_3(t) + n_2(t)/I_2,$$
$$\dot{\omega}_3(t) = [(I_1 - I_2)/I_3]\omega_1(t)\omega_2(t) + n_3(t)/I_3.$$

The state space for the first set of equations is $\mathscr{S}\mathscr{O}(3)$—the set of 3×3 orthogonal matrices. The state space for the second set of equations is $\mathbb{R}^3$—Cartesian 3-space

For our present purpose suppose that the center of mass of the body is fixed and suppose that the observed output of this system is a pencil beam of light generated by a light source which is mounted in the body along a line passing through the center of mass. In this case the output is $\mathscr{C}X(t)$, where $\mathscr{C}$ is a subgroup which corresponds to a rotation about the pencil beam (an undetectable motion).

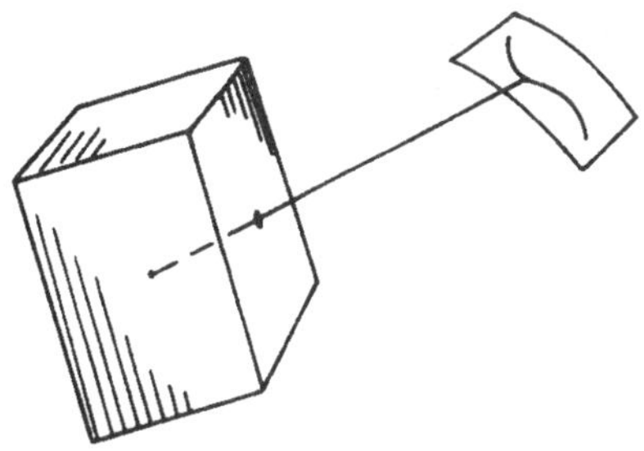

FIG. 1. *Illustration of the observability of a rigid body*

Example 3 (A model for DC to DC conversion). The electrical network shown in Fig. 1 contains switches which are to be manipulated in such a way as to transfer the energy stored on the capacitor 1 to capacitor 2. In order to have a sensible physical model we demand that there be exactly one path through the inductor at all times.

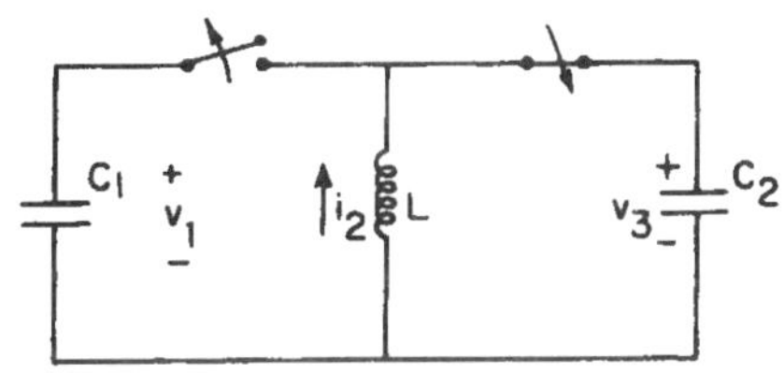

FIG. 2. *An electrical network for which energy is conserved*

The equations of motion are

$$\begin{bmatrix} \dot{x}_1(t) \\ \dot{x}_2(t) \\ \dot{x}_3(t) \end{bmatrix} = \begin{bmatrix} 0 & s_1(t) & 0 \\ -s_1(t) & 0 & s_2(t) \\ 0 & -s_2(t) & 0 \end{bmatrix} \begin{bmatrix} x_1(t) \\ x_2(t) \\ x_3(t) \end{bmatrix},$$

where $x_1 = v_1\sqrt{C_1}$, $x_3 = v_3\sqrt{C_2}$, $x_2 = i_2\sqrt{L}$ and s_1 and s_2 are dependent on the switch positions and take on the values 1 or 0. We have $s_1 = 1$ and $s_2 = 0$ if the switch on the left is closed, and we have $s_1 = 0$, $s_2 = 1$ if the switch on the right is closed.

3. Lie algebras and Lie groups. Let $\mathbb{R}^{n \times n}$ denote the set of real $n \times n$ matrices; $\mathbb{R}^{n \times n}$ is a vector space of dimension n^2. By a Lie algebra $\mathscr{L}$ in $\mathbb{R}^{n \times n}$ we understand a subset of $\mathbb{R}^{n \times n}$ which is a vector space and which has the property that if A and B belong to $\mathscr{L}$, thus so does $[A, B] = AB - BA$. If $\mathscr{L}_1$ and $\mathscr{L}_2$ are Lie algebras in $\mathbb{R}^{n \times n}$, and their intersection $\mathscr{L}_1 \cap \mathscr{L}_2$ is also a Lie algebra, then if A and B belong to $\mathscr{L}_1$ and $\mathscr{L}_2$ and both are algebras, then $[A, B]$ belongs to both $\mathscr{L}_1$ and $\mathscr{L}_2$. The union $\mathscr{L}_1 \cup \mathscr{L}_2$ of two Lie algebras, the sum $\mathscr{L}_1 + \mathscr{L}_2$ of two Lie algebras and the commutator $[\mathscr{L}_1, \mathscr{L}_2]$ of two Lie algebras are not necessarily Lie algebras.

Given an arbitrary subset of $\mathbb{R}^{n \times n}$ we can add additional elements to it so as to imbed it in a Lie algebra. To obtain the smallest Lie algebra which contains a given set $\mathscr{N}$ we first add to $\mathscr{N}$ all linear combinations of elements in $\mathscr{N}$ so as to get a real vector space $\mathscr{N}_1$. Then we commute elements in $\mathscr{N}_1$ to get $\mathscr{N}_2 = \mathscr{N}_1 + [\mathscr{N}_1, \mathscr{N}_1]$; if this is not contained in $\mathscr{N}_1$, then we form $\mathscr{N}_3 = \mathscr{N}_2 + [\mathscr{N}_2, \mathscr{N}_2]$, etc. Clearly this process stops in a finite number of steps since at each stage we increase the dimension of the vector space by at least one and the dimension is upper bounded by n^2. We call this Lie algebra the Lie algebra generated by $\mathscr{N}$ and denote it by $\{\mathscr{N}\}_A$.

If $\mathscr{M}$ is a set of nonsingular matrices in $\mathbb{R}^{n \times n}$, we let $\{\mathscr{M}\}_G$ denote the multiplicative matrix group generated by $\mathscr{M}$, i.e., the smallest group in $\mathbb{R}^{n \times n}$ which contains $\mathscr{M}$ and is closed under multiplication and inversion. If $\mathscr{N}$ is a linear subspace of $\mathbb{R}^{n \times n}$, then the set

$$\mathscr{M} = \{M : M = e^{N_1} e^{N_2} \cdots e^{N_p}, N_i \in \mathscr{N}, p = 0, 1, 2, \cdots\}$$

contains no singular matrices since $\det(\exp N_i) = \exp(\operatorname{tr} N_i) > 0$. Clearly $\mathscr{M}$ is closed under multiplication and inversion and, in our notation,

$$\mathscr{M} = \{\exp \mathscr{N}\}_G.$$

Let $\mathscr{L}$ be a Lie algebra. At each point M in $\{\exp \mathscr{L}\}_G$ there is a one-to-one map ϕ_M from a neighborhood of 0 in $\mathscr{L}$ onto a neighborhood of M in $\{\exp \mathscr{L}\}_G$ which is defined by

$$\phi_M : \mathscr{L} \to \{\exp \mathscr{L}\}_G, \qquad \phi_M(L) = e^L M.$$

This map has a smooth inverse which shows that $\{\exp \mathscr{L}\}_G$ is a locally Euclidean space of dimension equal to the dimension of $\mathscr{L}$. We may check that the maps ϕ_M^{-1} satisfy the conditions for a C^∞-manifold in the sense of [11, p. 97]. Thus we may give $\{\exp \mathscr{L}\}_G$ the structure of a differentiable manifold. This justifies our referring to $\{\exp \mathscr{L}\}_G$ as a *group manifold.*

If $\mathscr{A}$ is a linear subspace of $\mathbb{R}^{n \times n}$ which is not necessarily a Lie algebra, we might inquire as to the relation between $\{\exp \mathscr{A}\}_G$ and $\{\exp \{\mathscr{A}\}_A\}_G$. Clearly the latter contains the former. The following theorem claims that they are identical.

THEOREM 1. *Let $\mathscr{A}_1, \mathscr{A}_2, \cdots, \mathscr{A}_p$ be a collection of linear subspaces of $\mathbb{R}^{n \times n}$. Then*

$$\{\exp \mathscr{A}_1, \exp \mathscr{A}_2, \cdots, \exp \mathscr{A}_p\}_G = \{\exp \{\mathscr{A}_1, \mathscr{A}_2, \cdots, \mathscr{A}_p\}_A\}_G.$$

Before proving Theorem 1 it is appropriate to make a few remarks about its relationship to the controllability literature. (Perhaps a glance at Theorem 5 would help at this point.) In considering the equation

$$\dot{X}(t) = \left(\sum_{i=1}^{m} u_i(t) A_i\right) X(t)$$

as a differential equation in $\mathcal{GL}(n)$ it is clear from the theorem of Frobenius [12] that the solution passing through $X \in \mathcal{GL}(n)$ lies in $\{\exp \{A_i\}_A\}_G X$ because at each point X in $\mathcal{GL}(n)$. The collection $\{A_i X\}_A$ is an involutive distribution which spans the tangent space of $\{\exp \{A_i\}_A\}_G X$ at X. Wei and Norman [9] confirm this fact by giving (locally valid) formulas for the solution of this differential equation in terms of the functions $u_i(\cdot)$ and the structural constants of the Lie algebra generated by the A_i (without pointing out the differential geometric interpretation of the result). On the other hand, if we regard $\dot{X}(t) = (\sum_{i=1}^m u_i(t)A_i)X(t)$ as a control problem, then the natural question is not what manifold contains the solution, but rather what set can be attained from a given point, given freedom over the choice of $(u_1, u_2, \cdots, u_m)$. The results of Chow [1] (see also Hermann [2]) are applicable here. Chow showed under a suitable regularity condition that the set of points reachable for the vector system $\dot{x}(t) = \sum_{i=1}^m u_i(t)f_i[x(t)]$ using piecewise constant controls is the same as the set reachable for

$$\dot{x}(t) = \sum_{i=1}^{n} v_i(t)g_i[x(t)],$$

where $\{g_i(x)\}$ is a basis for the involutive distribution generated by $\{f_i(x)\}$. That is, $\{g_i(x)\}$ spans a vector space which includes $\{f_i(x)\}$ and is closed under the Lie bracket operation

$$[f, g] = \frac{\partial f}{\partial x} g - \frac{\partial g}{\partial x} f.$$

In our case the Lie bracket of $A_i X$ and $A_j X$ is $[A_i, A_j]X$. Thus we see that for the differential equation in question the reachable set includes $\{\exp \{A_i\}_A\}_G$, and the theorem of Frobenius ensures that it includes nothing more.

The proof of Theorem 1 given below could be shortened considerably by the use of these ideas. The reason for preferring the longer proof given here is that it is constructive, it is self-contained (nothing harder than the implicit function theorem is used) and it has the merit of proving a theorem about $n \times n$ matrices using the notation and tools natural to that subject.

Proof. We give a proof which relies on an implicit function theorem which, under suitable hypothesis, ensures the existence of a solution of α equations in $\beta < \alpha$ unknowns. (See [13, pp. 29–30].) We also need the Baker- Hausdorff formula which asserts that

$$e^{At}L e^{-At} = L + [At, L] + \frac{1}{2}[At, [At, L]] + \frac{1}{3!}[At, [At, [At, L]]] + \cdots .$$

Note that the norm of the $(n + 1)$th term in this series is less than $\|L\|2^n\|A\|^n/n!$ so that the series is majorized by the series $\|L\|[1 + 2\|A\| + (2\|A\|)^2/2! + \cdots] = \|L\| \cdot e^{2\|At\|}$ and hence is absolutely and uniformly convergent on $-T \leq t T$ for all T.

Let $\{A_1, A_2, \cdots, A_n\} \subset \mathcal{A}_1 \cup \mathcal{A}_2 \cup \cdots \cup \mathcal{A}_p$ be a basis for $\mathcal{A}_1 + \mathcal{A}_2 + \cdots + \mathcal{A}_p$ and let $\mathcal{L}$ be the Lie algebra generated by $\{A_1, A_2, \cdots, A_r\}$. Assume this algebra is of dimension q. There exists a basis for $\mathcal{L}$ which consists of terms of the form

$$\begin{aligned}
L_1 &= A_1, \\
L_2 &= A_2, \\
&\cdots \\
L_r &= A_r, \\
L_{r+1} &= [A_{k(r+1)}, A_{l(r+1)}], \\
L_{r+2} &= [A_{k(r+2)}, A_{l(r+2)}], \\
&\cdots \\
L_{r+s} &= [A_{k(r+s)}, A_{l(r+s)}], \\
L_{r+s+1} &= [A_{k(r+s+1)}, [A_{l(r+s+1)}, A_{m(r+s+1)}]], \\
&\cdots \\
L_q &= [\cdots [A_{k(q)}, \cdots, A_{m(q)}] \cdots].
\end{aligned}$$

We are quite explicit here because at certain points in our proof it is necessary to regard these expressions as formal expressions as opposed to matrices.

We introduce the following special notation. The operator EXP maps formal expressions into matrices. It is defined on A_i and its commutators (i.e., formal expressions such as A_i, $[A_i, A_j]$, $[A_i, [A_j, A_k]]$, etc.) as follows:

$$\mathrm{EXP}\, A_i t = e^{A_i t},$$

$$\mathrm{EXP}\, [A_i, A_j] t = \begin{cases} e^{A_i\sqrt{t}} e^{A_j\sqrt{t}} e^{-A_i\sqrt{t}} e^{-A_j\sqrt{t}}, & t \geqq 0, \\ e^{A_j\sqrt{|t|}} e^{A_i\sqrt{|t|}} e^{-A_j\sqrt{|t|}} e^{-A_i\sqrt{|t|}}, & t < 0. \end{cases}$$

The definition is completed by recursion. If A and B are formal expressions, then

$$\mathrm{EXP}\, [A, C] t = \begin{cases} (\mathrm{EXP}\, A\sqrt{t})(\mathrm{EXP}\, C\sqrt{t})(\mathrm{EXP}\, A\sqrt{t})^{-1}(\mathrm{EXP}\, C\sqrt{t})^{-1}, & t \geqq 0, \\ (\mathrm{EXP}\, C\sqrt{|t|})(\mathrm{EXP}\, A\sqrt{|t|})(\mathrm{EXP}\, C\sqrt{t})^{-1}(\mathrm{EXP}\, A\sqrt{t})^{-1}, & t < 0. \end{cases}$$

We now show that if B is a formal expression, then

$$\mathrm{EXP}\, Bt = I + Bt + o(t),$$

where $o(t)/t$ goes to zero as t goes to zero. We associate with each formal expression an integer n, called its degree, which is the largest number of times any particular entry of the formal expression is bracketed. To carry out this proof we use an induction on the degree of the expression.

Suppose A and C are two formal expressions of degree n or less. Suppose that we have shown that for any formal expression of degree n or less,

$$\mathrm{EXP}\, At = I + At + o(t).$$

Now in view of the definition of EXP we can expand EXP At and EXP Ct in a convergent power series involving fractional powers of t. Let us say

$$\text{EXP}\, At = I + At + E(t) + Dt^2 + o(t^2),$$

$$\text{EXP}\, Ct = I + C + G(t) + Ft^2 + o(t^2),$$

where E and G involve powers of t between t^1 and t^2 and E and F are the respective coefficients of t^2. The power series expansions for the inverses are then

$$(\text{EXP}\, At)^{-1} = I - At - E(t) + (A^2 - D)t^2 + o(t^2),$$

$$(\text{EXP}\, Ct)^{-1} = I - Ct - G(t) + (C^2 - F)t^2 + o(t^2),$$

as is verified by multiplication with the corresponding expressions for EXP At and EXP Ct respectively. Now using these expressions we have for t nonnegative,

$$\begin{aligned}\text{EXP}\, Bt &= \text{EXP}\, [A, C]t \\ &= (I + A\sqrt{t} + E(t) + Dt^2 + o(t))(I + C\sqrt{t} + G(\sqrt{t}) + Ft + o(t)) \\ &\quad \cdot (I - A\sqrt{t} + E(t) + Dt^2 + o(t))((I - C\sqrt{t} - G(\sqrt{t}) + C^2 - F)t + o(t)) \\ &= I + [A, C]t + Ft + A^2t/2 + A^2t/2 - A^2t + (C^2 - F)t - C^2t + o(t) \\ &= I + [A, C]t + o(t),\end{aligned}$$

and the case $t < 0$ leads to the same result.

Well-known properties of the matrix exponential function let one conclude that for $|t| > 0$, EXP Bt is continuously differentiable. The above argument shows that EXP Bt is differentiable with respect to t in a neighborhood of $t = 0$ and

$$\frac{d}{dt}\text{EXP}\, Bt|_{t=0} = B.$$

Hence we have for the basis elements L_v,

$$\frac{d}{dt}\text{EXP}\, L_v t|_{t=0} = L_v.$$

Now consider a function of $u = (u_1, u_2, \cdots, u_q)$ and $v = (v_1, v_2, \cdots, v_q)$ which maps $\mathbb{R}^q \times \mathbb{R}^q$ into $\mathbb{R}^{n \times n}$ and which is defined by

$$\begin{aligned}F(u, v) = -I &+ (\text{EXP}\, L_1 u_1)(\text{EXP}\, L_2 u_2) \cdots (\text{EXP}\, L_q u_q) \\ &\cdot \exp(-L_1 v_1 - L_2 v_2 \cdots - L_q v_q).\end{aligned}$$

Clearly $F(0, 0) = 0$. Now the linear approximation of F at $(u, v) = (0, 0)$ is given by

$$\begin{aligned}F_{(u,v)}(u, v)|_{(0,0)}(\delta u, \delta v) &= L_1 \delta u_1 + L_2 \delta u_2 + \cdots + L_q \delta u_q - L_1 \delta v_1 \\ &\quad - L_2 \delta v_2 - \cdots - L_q \delta v_q\end{aligned}$$

so that the range space of $F_{(u,v)}(u, v)|_{(0,0)}(u, 0)$ is the q-dimensional subspace of

$\mathbb{R}^{n\times n}$ spanned by $\{L_i\}$. Now $F(u, v) + I$ is a finite product of exponentials which we write as

$$F(u, v) + I = \exp(A_{i_1}u_{i_1}^{p_1})\exp(A_{i_2}u_{i_2}^{p_2})\cdots\exp(A_{i_\nu}u_{i_\nu}^{p_\nu})$$
$$\cdot\exp(-L_1v_1 - L_2v_2 - \cdots - L_qv_q).$$

Since the Baker-Hausdorff formula lets one write

$$e^{A_it}A_k e^{-A_it} = A_k + [A_i, A_k]t + \cdots,$$

we see that $e^{A_it}A_k e^{-A_it}$ belongs to the Lie algebra generated by the A's. Moreover, it is continuous with respect to t and at $t = 0$ takes on the value A_k. Using this result repeatedly we see that for each $\{u_{i_k}\}$ we can find R_{i_k} in $\mathscr{L}$ such that

$$\exp(A_{i_1}u_{i_1}^{p_1})\exp(A_{i_2}u_{i_2}^{p_2})\cdots\exp(A_{i_k}u_{i_k}^{p_k})A_{i_k}\exp(A_{i_{k+1}}u_{i_{k+1}}^{p_{k+1}})\cdots\exp(A_{i_\nu}u_{i_\nu}^{p_\nu})$$
$$\cdot\exp(L_1v_1 + L_2v_2 + \cdots + L_qv_q)$$
$$= (F(u, v) + I)R_{i_k}(u_{i_1}, u_{i_2}, \cdots, u_{i_\nu})$$

simply by pushing A_k past the exponentials one at a time. Clearly $R_{i_k}(0, 0, \cdots) = A_{i_k}$. Thus we see that for a and b small,

$$F_{(u,v)}(u, v)|_{(a,b)}(\delta u, \delta v) = (F(a, b) + I)\left(\sum_{i=1}^{q} S_i(a, b)\delta u_i + H_i(a, b)\delta v_i\right)$$

for some $S_i(a, b)$ and $H_i(a, b)$ in $\mathscr{L}$. Since $S_i(0, 0) = H_i(0, 0) = L_i$ and since S_i and H_i depend continuously on their arguments, this establishes that the Jacobian of the map $F: \mathbb{R}^q \times \mathbb{R}^q \to \mathbb{R}^{n\times n}$ must have rank q in a neighborhood of $(0, 0)$; and hence, by the implicit function theorem cited earlier, there exists an $\varepsilon > 0$ and a map $\phi: \mathbb{R}^q \to \mathbb{R}^q$ such that if $\|v\| < \varepsilon$, then

$$F(\phi(v), v) = 0.$$

Since $F(u, v) = 0$ implies that

$$\text{EXP } L_1u_1 \text{ EXP } L_2u_2 \cdots \text{EXP } L_qu_q = \exp(L_1v_1 + L_2v_2 + \cdots + L_qv_q)$$

we conclude that there exists $\varepsilon_1 > 0$ such that if $L \in \mathscr{L}$ and $\|L\| < \varepsilon_1$, we can write

$$e^L = \exp(A_{i_1}u_{i_1})\exp(A_{i_2}u_{i_2})\cdots\exp(A_{i_\nu}u_{i_\nu}).$$

Now for any $L \in \mathscr{L}$ it follows that $\|(1/m)L\| < \varepsilon_1$ for some integer m, and thus we can express e^L as $e^{L/m}e^{L/m}\cdots e^{L/m}$. Likewise we can express $e^{L_1}e^{L_2}\cdots e^{L_q}$ in this form.

Let $\mathscr{K}$ and $\mathscr{L}$ be Lie algebras in $\mathbb{R}^{n\times n}$. It can happen that $\{\exp\mathscr{K}\}_G$ is a bounded subset of $\mathbb{R}^{n\times n}$ which is not closed, and it can happen that the closure of $\{\exp\mathscr{K}\}_G$ equals $\{\exp\mathscr{L}\}_G$ with $\mathscr{K} \neq \mathscr{L}$. The skew-line on the torus [11] is an easy example. Also, $\{\exp\mathscr{K}\}_G$ is not necessarily simply connected. Nonetheless, we have the following result which we deduce from Theorem 1 rather then sending the reader to the literature.

COROLLARY 1. *If $\mathscr{K}$ and $\mathscr{L}$ are Lie algebras in $\mathbb{R}^{n\times n}$, then $\{\exp\mathscr{K}\}_G \subset \{\exp\mathscr{L}\}_G$ if and only if $\mathscr{K} \subset \mathscr{L}$, and $\{\exp\mathscr{K}\}_G = \{\exp\mathscr{L}\}_G$ if and only if $\mathscr{K} = \mathscr{L}$.*

Proof. For both statements the sufficiency is obvious. To establish necessity in the first case notice that if $\{\exp \mathscr{K}\}_G \subset \{\exp \mathscr{L}\}_G$, then by Theorem 1,

$$\{\exp \mathscr{K}\}_G = \{\exp \mathscr{K}, \exp \mathscr{L}\}_G = \{\exp \{\mathscr{K}, \mathscr{L}\}_A\}_G.$$

Suppose $\mathscr{L}$ is of dimension n. To obtain a proof by contradiction, suppose that $\mathscr{K}$ is not contained in $\mathscr{L}$. Then $\{\mathscr{K}, \mathscr{L}\}_A$ is of dimension $n + 1$ or greater. Then $\exp \mathscr{L}$ is an n-dimensional manifold and $\{\exp \{\mathscr{K}, \mathscr{L}\}\}$ is not, which contradicts $\{\exp \mathscr{L}\} = \{\exp \{\mathscr{L}, \mathscr{K}\}\}$. To establish necessity in the second case repeat this argument verbatim but with "contained in" replaced by "equals" both verbally and symbolically.

The notation $\mathrm{ad}_A^0 B = B, \mathrm{ad}_A^1 B = [A, B], \mathrm{ad}_A^2 B = [A, [A, B]]$, etc., is standard. If $\mathscr{K}$ and $\mathscr{L}$ are Lie algebras, we use the notation $\{\mathrm{ad}_{\mathscr{L}} \mathscr{K}\}_A$ to denote the Lie algebra generated by $\mathscr{K}$ under commutation with elements of $\mathscr{L}$. That is,

$$\{\mathrm{ad}_{\mathscr{L}} \mathscr{K}\}_A = \{\mathscr{K}, [\mathscr{L}, \mathscr{K}], \cdots, [\mathscr{L}, [\mathscr{L} \cdots [\mathscr{L}, \mathscr{K}] \cdots]] \cdots\}_A.$$

This algebra may also be described as the intersection of all Lie algebras which contain $\mathscr{K}$ and are closed under commutation with $\mathscr{L}$. If $\mathscr{F}$ and $\mathscr{G}$ are groups, we introduce an analogous notation. The smallest group which contains $\mathscr{F}$ and all products of the type GFG^{-1} for G in $\mathscr{G}$ and F in $\mathscr{F}$ will be denoted by $\{\mathrm{AD}_{\mathscr{G}} \mathscr{F}\}_G$. This group may be described as the intersection of all groups which contain $\mathscr{F}$ and are closed under conjugation with elements of $\mathscr{G}$. If $\mathscr{G}$ is $\{\exp \mathscr{L}\}_G$ and $\mathscr{F}$ is $\{\exp \mathscr{K}\}_G$, then clearly $\{\mathrm{AD}_{\mathscr{G}} \mathscr{F}\}_G$ consists of products of terms of the form

$$M = e^{L_1} e^{K_1} e^{-L_1} e^{L_2} e^{K_2} e^{-L_2} \cdots e^{L_m} e^{K_m} e^{-L_m}.$$

THEOREM 2. *Let $\mathscr{K}$ and $\mathscr{L}$ be Lie algebras in $\mathbb{R}^{n \times n}$. Then*

$$\left\{ \bigcup_{M \in \{\exp \mathscr{L}\}_G} M \mathscr{K} M^{-1} \right\}_A = \{\mathrm{ad}_{\mathscr{L}} \mathscr{K}\}_A$$

and

$$\{\mathrm{AD}_{\{\exp \mathscr{L}\}_G} \{\exp \mathscr{K}\}_G\}_G = \{\exp \{\mathrm{ad}_{\mathscr{L}} \mathscr{K}\}_A\}_G.$$

Proof. From the Baker–Hausdorff formula we see at once that if L belongs to $\mathscr{L}$ and K belongs to $\mathscr{K}$, then $e^L K e^{-L}$ belongs to $\{\mathrm{ad}_{\mathscr{L}} \mathscr{K}\}_A$. Thus the right side of the first equality in question contains the left. On the other hand, expressions of the following type belong to the left side:

$$e^{\alpha L} \frac{1}{\alpha} K e^{-\alpha L} - \frac{1}{\alpha} K = [L, K] + o(\alpha),$$

$$e^{\alpha L} \left(\frac{1}{\alpha} [K, L] + o(\alpha) \right) e^{-\alpha L} - \frac{1}{\alpha} [K, L] + o(\alpha) = [L, [L, K]] + o(\alpha), \quad \text{etc.}$$

Since $\mathbb{R}^{n \times n}$ is a finite-dimensional space and since $\{\bigcup_{M \in \{\exp \mathscr{L}\}_G} M \mathscr{K} M^{-1}\}_A$ is a linear subspace, it is closed. Thus $[L, K]$, $[L[L, K]] \cdots$ belong to this set and the first equality is seen to hold.

The second statement is obtained by exponentiating the first. This gives

$$\left\{ \exp \left\{ \bigcup_{M \in \{\exp \mathscr{L}\}_G} M \mathscr{L} M^{-1} \right\}_A \right\}_G = \{\exp \{\mathrm{ad}_{\mathscr{L}} \mathscr{K}\}_A\}_G,$$

but since $e^L e^K e^{-L} = \exp(e^L K e^{-L})$, we see that

$$\left\{\exp\left\{\bigcup_{M\in\{\exp\mathscr{L}\}G} M\mathscr{L}M^{-1}\right\}_A\right\}_G = \{\mathrm{AD}_{\{\exp\mathscr{L}\}G}\{\exp\mathscr{L}\}_G\}_G$$

so the result follows.

The next theorem states a purely group theoretic result which, although easily proved, is stated formally because we need it in our study of observability.

THEOREM 3. *Let $\mathscr{H}$ and $\mathscr{R}$ be subgroups of a group $\mathscr{G}$. Let $\mathscr{P}$ be the subset of $\mathscr{G}$ defined as*

$$\mathscr{P} = \{P: RPR^{-1} \in \mathscr{H}, \text{ all } R \in \mathscr{R}\}.$$

Then $\mathscr{P}$ is a subgroup of $\mathscr{H}$, $\mathscr{R}\mathscr{P}$ is a subgroup of $\mathscr{G}$, and $\mathscr{P}$ is a normal subgroup of $\mathscr{R}\mathscr{P}$. Thus $\mathscr{R} \cap \mathscr{P}$ is a normal subgroup of $\mathscr{R}$ and $\mathscr{R}\mathscr{P}/\mathscr{P}$ is isomorphic to $\mathscr{R}/\mathscr{R} \cap \mathscr{P}$.

Proof. Suppose P_1 and P_2 belong to $\mathscr{P}$. Then for each R in $\mathscr{R}$ there exist $H_1(R)$ and $H_2(R)$ in $\mathscr{H}$ such that $RP_1R^{-1}(RP_2^{-1}R^{-1}) = H_1(R)\cdot[H_2(R_2)]^{-1}$. Since $\mathscr{H}$ is a group, this means $RP_1P_2^{-1}R^{-1}$ belongs to $\mathscr{H}$, and thus that $\mathscr{P}$ is a subgroup of $\mathscr{G}$. Clearly it is a subgroup of $\mathscr{H}$ since the choice $R = I$ is possible. To see that $\mathscr{R}\mathscr{P}$ is a group, note that if R_1 and R_2 belong to $\mathscr{R}$ and P_1 and P_2 belong to $\mathscr{P}$, then

$$R_1P_1(R_2P_2)^{-1} = R_1P_1R_2^{-1}R_2P_2^{-1}R_2^{-1} = R_1R_2^{-1}(R_2P_1R_2^{-1})(R_2P_2^{-1}R_2^{-1}).$$

Since $\mathscr{R}$ is a group and since $\mathscr{P}$ is a group which has the property that if P belongs to $\mathscr{P}$ then so does RPR^{-1} for each R in $\mathscr{R}$, we see that this product belongs to $\mathscr{R}\mathscr{P}$. Clearly $\mathscr{P}$ is a normal subgroup of $\mathscr{R}\mathscr{P}$ since $RP\mathscr{P}P^{-1}R^{-1} = \mathscr{P}$ for each RP in $\mathscr{R}\mathscr{P}$. By the second isomorphism theorem (Rotman [14, p. 26]) $\mathscr{R} \cap \mathscr{P}$ is normal in $\mathscr{R}$ and $\mathscr{R}\mathscr{P}/\mathscr{P} \simeq \mathscr{R}/\mathscr{R} \cap \mathscr{P}$.

We now state and prove a Lie algebraic analogue of this theorem. Algebraic tests for observability will be derived from this result.

THEOREM 4. *Let $\mathscr{H}$ and $\mathscr{L}$ be Lie algebras in $\mathbb{R}^{n\times n}$. Let $\mathscr{P}$ be defined as*

$$\mathscr{P} = \{P: RPR^{-1} \in \{\exp\mathscr{H}\}_G, \text{ all } R \in \{\exp\mathscr{L}\}_G\}.$$

If $\mathscr{K}$ is a Lie algebra in $\mathbb{R}^{n\times n}$, then $\{\exp\mathscr{K}\}_G \subset \mathscr{P}$ if and only if $\{\mathrm{ad}_{\mathscr{L}}\mathscr{K}\}_A \subset \mathscr{H}$. There exists a unique Lie algebra $\mathscr{K}_1$ such that $\{\mathrm{ad}_{\mathscr{L}}\mathscr{K}_1\}_A \subset \mathscr{H}$ and $\mathscr{K}_1$ contains all other Lie algebras having this property.

Proof. Suppose $\{\mathrm{ad}_{\mathscr{L}}\mathscr{K}\}_A \subset \mathscr{H}$. Then for L_i in $\mathscr{L}$ and K_i in $\mathscr{K}$ we see from Theorem 2 that $\{\exp\mathscr{H}\}_G$ contains

$$RPR^{-1} = e^{L_1}e^{L_2}\cdots e^{L_p}e^{K_1}e^{K_2}\cdots e^{K_q}e^{-L_p}\cdots e^{-L_2}e^{-L_1}.$$

By the hypothesis and Corollary 1, $\{\exp\{\mathrm{ad}_{\mathscr{L}}\mathscr{K}\}_A\}_G \subset \{\exp\mathscr{H}\}_G$.

On the other hand, if for all L_i in $\mathscr{L}$ and all K_i in $\mathscr{K}$ we have

$$RPR^{-1} = e^{L_1}e^{L_2}\cdots e^{L_p}e^{K_1}e^{K_2}\cdots e^{K_q}e^{-L_p}\cdots e^{-L_2}e^{-L_1} \in \{\exp\mathscr{H}\}_G,$$

then since $\{\exp\mathscr{H}\}_G$ is a group, we see that $\{\mathrm{AD}_{\{\exp\mathscr{L}\}G}\{\exp\mathscr{K}\}_G\}_G \subset \{\exp\mathscr{H}\}_G$, and again from Theorem 2 and Corollary 1 we see that $\{\mathrm{ad}_{\mathscr{L}}\mathscr{K}\}_A \subset \mathscr{H}$.

Finally, notice that if $\{\mathrm{ad}_{\mathscr{L}}\mathscr{K}_1\}_A \subset \mathscr{H}$ and $\{\mathrm{ad}_{\mathscr{L}}\mathscr{K}_2\}_A \subset \mathscr{H}$, then $\{\mathrm{ad}_{\mathscr{L}}(\mathscr{K}_1 + \mathscr{K}_2)\}_A \subset \mathscr{H}$. Thus there is a largest Lie algebra with this property.

4. Controllability on group manifolds. The first question of a system theoretic character which we investigate is that of controllability. Since we want to emphasize global results, we work with the most elementary type of evolution equation appropriate to our present setting, namely,

$$\dot{X}(t) = (A + \sum u_i(t)B_i)X(t).$$

The choice of control affects the direction in which X moves. However, A is a constant over which there is no control. This evolution equation has the property that the change of variable $X \to XP$ for P nonsingular leaves the equation unchanged. This invariance gives the vector field which a given choice of $\{u_i(t)\}$ establishes on $\mathscr{G}\ell(n)$, a particularly simple form.

THEOREM 5. *Consider the linear dynamical system*

$$\dot{X}(t) = \left(\sum_{i=1}^{m} u_i(t)B_i\right)X(t), \qquad X = \text{an } n \times n \text{ matrix}.$$

Given a time $t_a > 0$ and given two nonsingular matrices X_1 and X_2, there exist piecewise continuous controls which steer the state from X_1 at $t = 0$ to X_2 at $t = t_a$ if and only if $X_2X_1^{-1}$ belongs to $\{\exp\{B_i\}_A\}_G$.

Proof. Sufficiency. Theorem 1 asserts that any matrix M in $\{\exp\{B_i\}_A\}_G$ can be written as a finite product, say

$$M = e^{B_{i_1}\alpha_1} e^{B_{i_2}\alpha_2} \cdots e^{B_{i_m}\alpha_m}.$$

Suppose $X_2X_1^{-1} = M$. Divide the interval $0 \leqq t \leqq t_a$ up into m equal intervals $[t_i, t_{i+1})$ whereby $t_i = i \cdot t_a/m$. Let $t_a/m = \beta^{-1}$. On the interval $[0, t_1)$ all controls are zero except the i_mth control, which takes on the value $\alpha_m\beta$. On the interval $[t_1, t_2)$ all controls are zero except the i_{m-1}th, which takes on the value $\alpha_{m-1}\beta$, etc., down to the last interval on which all controls are zero except the i_1st, which takes on the value $\alpha_1\beta$. Since the differential equation is linear and constant on each of the subintervals, the solution is a product of exponentials and the result follows.

Necessity. To show that X_2 cannot be reached from X_1 unless $X_2X_1^{-1}$ is of the form $e^{L_m} e^{L_{m-1}} \cdots e^{L_1}$ we assume the contrary and obtain a contradiction. Suppose that $u_1(\cdot), \cdots, u_m(\cdot)$ is a control which steers the system from X_1 at $t = 0$ to X_2 at $t = t_a$. By Theorem 1 of [9] we know that there exists a sequence of times $t_0, t_1, t_2, \cdots, t_m$ such that on each of the subintervals $[t_i, t_{i+1}]$ the transition matrix of

$$\dot{X}(t) = \left(\sum_{i=1}^{q} u_i(t)B_i\right)x(t)$$

can be written as $e^{H_i(t)}$ for some $H_i(\cdot)$ in $\mathscr{L}$. Thus we can write

$$X(t_a) = e^{L_m} e^{L_{m-1}} \cdots e^{L_1} X_0,$$

which establishes the contradiction.

As an application which emphasizes the ease with which we can study global questions using this theorem we observe the following results relating to the classical groups. Here J is given by

$$J = \begin{bmatrix} 0 & I \\ -I & 0 \end{bmatrix},$$

and a matrix is called symplectic if $\theta' J \theta = J$.

THEOREM 6. *Consider the system of Theorem 5. Given a time $t_a > 0$ and given two nonsingular $n \times n$ matrices X_1 and X_2 with* $\det X_1 X_2 > 0$, *there exists a piecewise continuous control which steers the state from X_1 at $t = 0$ to X_2 at $t = t_a$ if $\{B_i\}_A$:*

(i) *spans $\mathbb{R}^{n \times n}$,*
(ii) *spans the $(n^2 - 1)$-dimensional subspace of $\mathbb{R}^{n \times n}$ consisting of the zero trace matrices and* $\det X_1 = \det X_2$,
(iii) *spans the $(n(n+1)/2)$-dimensional subset of $\mathbb{R}^{n \times n}$ consisting of the set of matrices which satisfy $JA + A'J = 0$ and $X_2 X_1^{-1}$ is symplectic,*
(iv) *spans the $(n(n-1)/2)$-dimensional subset of $\mathbb{R}^{n \times n}$ consisting of all skew symmetric matrices, and $X_2 X_1^{-1}$ is orthogonal.*

Proof. As is well known, any nonsingular matrix can be written as θR with $\theta'\theta = I$ and $R = R' > 0$. Also, real orthogonal matrices with positive determinants and real symmetric positive definite matrices have real logarithms. Moreover, in case (iii) the factors in the polar representation inherit the property of the group itself, which is to say that the θ and R in the polar representation of a symplectic matrix are symplectic. To complete the proof we need only invoke Theorem 5 since the previous remarks justify our writing $X_2 X^{-1} = e^{\Omega} e^{S}$ with $\Omega = -\Omega'$ and $S = S'$ both in the appropriate Lie algebras.

The results of Theorems 5 and 6 are somewhat unsatisfactory in that the A term is absent. The following theorem describes one way in which this can be relaxed.

THEOREM 7. *Consider the linear dynamical system*

$$\dot{X}(t) = \left(A + \sum_{i=1}^{\nu} u_i(t) B_i\right) X(t), \qquad X = \text{an } n \times n \text{ matrix.}$$

Suppose that $[\mathrm{ad}_A^k B_i, B_j] = 0$ for $i, j = 1, 2, \cdots, \nu$ and $k = 0, 1, \cdots, n^2 - 1$. Let $\mathcal{H}$ be the linear subspace of $\mathbb{R}^{n \times n}$ spanned by $\mathrm{ad}_A^k B_i$ for $i = 1, 2, \cdots, \nu$ and $k = 0, 1, \cdots, n^2 - 1$. Then given a time $t_a > 0$ and two $n \times n$ matrices X_1 and X_2 there exist continuous controls which steer the system from the state X_1 at $t = 0$ to the state X_2 at $t = t_a$ if and only if there exists H in $\mathcal{H}$ such that

$$X_2 = e^{At_a} e^{H} X_1.$$

Proof. First of all, notice that

$$\begin{aligned}
\frac{d^k}{dt^k}[e^{At} B_i e^{-At}, B_j]\bigg|_{t=0} &= \frac{d^{k-1}}{dt^{k-1}}[e^{At}[A, B_i] e^{-At}, B_j]\bigg|_{t=0} \\
&= \frac{d^{k-2}}{dt^{k-2}}[e^{At}(\mathrm{ad}_A^2 B_i) e^{-At}, B_j]\bigg|_{t=0} \\
&\cdots\cdot \\
&= [e^{At}(\mathrm{ad}_A^k B_i) e^{-At}, B_j]|_{t=0} \\
&= [\mathrm{ad}_A^k B_i, B_j].
\end{aligned}$$

Thus $[e^{At} B_i e^{-At}, B_j]$ is identically zero if $[\mathrm{ad}_A^k B_i, B_j] = 0$ for $k = 0, 1, 2, \cdots$. However, ad_A is a linear operator from an n^2-dimensional space into itself so that by the Cayley–Hamilton theorem all powers above $n^2 - 1$ are lineary dependent on the first $n^2 - 1$. Thus under the hypothesis of the theorem statement, $[e^{At}B_i e^{-At}, B_j]$ vanishes identically. Also

$$0 = e^{At}B_i e^{-At}B_j - B_j e^{At}B_i e^{-At}$$
$$= e^{A\sigma}(e^{At}B_i e^{-At})B_j e^{-A\sigma} - e^{A\sigma}B_j e^{At}B_i e^{-At} e^{-A\sigma}.$$

Now let $t + \sigma = \beta$ and $\gamma = \sigma$. Thus for all β and γ,

$$0 = [e^{A\beta}B_i e^{-A\beta}, e^{A\gamma}B_j e^{-A\gamma}].$$

For the purpose of solving the differential equation we introduce $Z(t) = e^{-At}X(t)$ and observe that

$$\dot{Z}(t) = \left(\sum_{i=1}^{v} u_i(t) e^{-At}B_i e^{At}\right) Z(t).$$

But recall (see e.g. Martin [15]) that the solution of $\dot{Z}(t) = B(t)Z(t)$ is $\exp \int_0^t B(\sigma)\,d\sigma$ if $[B(t), B(\sigma)]$ vanishes for all t and σ. Thus we can write

$$Z(t) = \exp\left(\int_0^t \sum_{i=1}^{n} u_i(t) e^{-At}B_i e^{At}\,dt\right) Z(0).$$

It is a well-known and frequently used fact (e.g., [16, p. 79]) that the image space of the map taking continuous functions into $\mathbb{R}^p$ according to the rule $x = L(u) = \int_0^{t_1} e^{A\sigma}bu(\sigma)\,d\sigma$, is spanned by the first p derivatives of $e^{At}b$ evaluated at zero. Using this fact here we see that for each H in $\mathscr{H}$ and $t_a > 0$ we have a continuous u defined on $[0, t_a]$ such that

$$Z(t_a) = e^{H}Z(0).$$

Therefore in terms of X we see that we can reach at t_a any X which can be expressed as $e^{At_a}e^{H}x(0)$ with H in $\mathscr{H}$.

As an application of this result we derive a familiar relationship.

Example 4. Consider the system in $\mathbb{R}^n$.

$$\dot{x}(t) = \tilde{A}x(t) + \sum_{i=1}^{m} b_i u_i(t); \qquad x(0) \text{ is given}.$$

Related to this is the matrix system in $\mathbb{R}^{(n+1)\times(n+1)}$:

$$\dot{X}(t) = \begin{bmatrix} \tilde{A} & 0 \\ 0 & 0 \end{bmatrix} X(t) + \sum_{i=1}^{m} u_i(t) \begin{bmatrix} 0 & b_i \\ 0 & 0 \end{bmatrix} X(t).$$

Let A and B_i be the matrices appearing in this expression. In this case $[\mathrm{ad}_A^k B_i, B_j]$ vanishes as required and so the reachable set from $x(0) = I$ is

$$\mathscr{R}(t) = \exp\begin{bmatrix} At & 0 \\ 0 & 0 \end{bmatrix}\{\exp H;\, H \in \mathscr{H}\},$$

where $\mathscr{H}$ is the subspace spanned by $\mathrm{ad}_A^k B_i$. A computation gives

$$\mathrm{Ad}_A^k B_i = \begin{bmatrix} 0 & A^k b_i \\ 0 & 0 \end{bmatrix}$$

so that the reachable set at t is

$$\mathscr{R} = \left\{ X : X = \begin{bmatrix} e^{At} & H \\ 0 & 1 \end{bmatrix}; H \in \operatorname{range}(B, AB, \cdots, A^{n-1}B) \right\},$$

where we have used the fact that $e^{At}\mathscr{K} = \mathscr{K}$ for all t and $\mathscr{K} = \operatorname{range}(B, AB, \cdots, A^{n-1}B)$.

5. Observability. In order to get a theory having a scope comparable to linear theory, it is necessary to treat observability. The choice of an appropriate form of the observational equation is critical for the success of the overall theory. As it turns out, the natural choice is indicated by the second example in § 2.

Let $\mathscr{F}$ be a matrix group and let $\mathscr{C}$ be a subgroup. Consider the system evolving in $\mathscr{F}$,

$$\dot{X}(t) = \left(A + \sum_{i=1}^{\nu} u_i(t)B_i\right)X(t), \qquad y(t) = \mathscr{C}X(t),$$

by which we mean that instead of observing $X(t)$ directly, we observe to what equivalence class $X(t)$ belongs with respect to the equivalence relation in $\mathscr{F}$ defined by $\mathscr{C}$. Thus $y(t)$ takes on values in the coset space $\mathscr{F}/\mathscr{C}$ which is generally not a group manifold (see § 7).

We call two states X_1 and X_2 *distinguishable* if there exists some control which gives rise to different outputs for the two starting states. In general the zero control is not adequate to distinguish between all states which are distinguishable as contrasted with the situation one finds for linear systems.

THEOREM 8. *Let $\mathscr{C}$ be a matrix group and suppose that the set of points $\mathscr{R}$ reachable from the identity for the system*

$$\dot{X}(t) = \left(A + \sum_{i=1}^{\nu} u_i(t)B_i\right)X(t), \qquad y(t) = \mathscr{C}X(t),$$

is a group. Then the set of initial states which are indistinguishable from the identity is given by

$$\mathscr{P} = \{P : RPR^{-1} \in \mathscr{C} \text{ for all } R \in \mathscr{R}\}.$$

$\mathscr{P}$ is a normal subgroup of $\mathscr{P}\mathscr{R}$ and a subgroup of $\mathscr{C}$.

Proof. Suppose that X is a starting state for the given equation which is indistinguishable from the identity. That means that for each R in $\mathscr{R}$ there is $C(R)$ in $\mathscr{C}$ such that

$$C(R)RX = R.$$

Since $\mathscr{R}$ and $\mathscr{C}$ are groups, we can take inverses to get

$$RXR^{-1} \in \mathscr{C}.$$

Thus the set $\mathscr{P}$ is exactly those states indistinguishable from the identity. The remainder of the conclusions come from Theorem 3.

THEOREM 9. *Let $\mathscr{H}$ and $\mathscr{L}$ be Lie algebras in $\mathbb{R}^{n \times n}$, and suppose that all the points reachable from the identity for*

$$\dot{X}(t) = \left(A + \sum_{i=1}^{\nu} u_i(t)B_i\right)X(t), \qquad y(t) = \{\exp \mathscr{H}\}_G X(t).$$

are $\{\exp \mathscr{L}\}_G$. *Then the set of initial states* $\mathscr{P}$ *which are indistinguishable from the identity contains* $\{\exp \mathscr{K}\}_G$ *if and only if* $\{\mathrm{ad}_{\mathscr{L}} \mathscr{K}\}_A \subset \mathscr{H}$. *Therefore a necessary condition for all states to be distinguishable from the identity is that* $\mathscr{H}$ *contain no subalgebra* $\mathscr{K}$ *such that* $\{\mathrm{ad}_{\mathscr{L}} \mathscr{K}\}_A \subset \mathscr{H}$.

Proof. Theorem 8 gives a characterization of $\mathscr{P}$ which permits one to bring to bear Theorem 4. Theorem 4 immediately gives the desired result.

One might be tempted to conclude that if there is no nontrivial algebra $\mathscr{K}$ meeting the requirements of Theorem 9, then all initial states are distinguishable. This is not true because $\mathscr{P}$ can be a discrete subgroup and hence not trivial and yet not expressible as $\{\exp \mathscr{K}\}_G$ for any Lie algebra $\mathscr{K}$. The next example illustrates this.

Example 5. In the numerical integration of the equations of motion of a rigid body one usually avoids Euler angle representation and uses instead quaternion or direction cosine representations. As is well known, the group of unit quaternions covers $\mathscr{SO}(3)$ twice. This causes an ambiguity in going from $\mathscr{SO}(3)$ to the group of unit quaternions. This example illustrates this idea. Consider an equation in the group of unit quaternions $\mathscr{Q}$ which we parametrize in the usual way $(a^2 + b^2 + c^2 + d^2 = 1)$:

$$\frac{d}{dt}\begin{bmatrix} a & -b & -c & -d \\ b & a & -d & c \\ c & d & a & -b \\ d & -c & b & a \end{bmatrix} = \begin{bmatrix} 0 & u_1 & u_2 & u_3 \\ -u_1 & 0 & u_3 & -u_2 \\ -u_2 & -u_3 & 0 & u_1 \\ -u_3 & +u_2 & -u_1 & 0 \end{bmatrix}\begin{bmatrix} a & -b & -c & -d \\ b & a & -d & c \\ c & d & a & -b \\ d & -c & b & a \end{bmatrix},$$

$$y(t) = \mathscr{C}\begin{bmatrix} a & -b & -c & -d \\ b & a & -d & c \\ c & d & a & -b \\ d & -c & b & a \end{bmatrix},$$

where $\mathscr{C}$ is the subgroup given by

$$\mathscr{C} = \{\exp \mathscr{H}\}_G, \qquad \mathscr{H} = \begin{bmatrix} 0 & \alpha & 0 & 0 \\ -\alpha & 0 & 0 & 0 \\ 0 & 0 & 0 & \alpha \\ 0 & 0 & -\alpha & 0 \end{bmatrix}.$$

Now it is true that $\{\exp \mathscr{H}\}_G$ includes I and $-I$, and it is also true that this pair of elements forms a normal subgroup of $\mathscr{Q}$. Thus I as an initial state cannot be distinguished from $-I$. Yet there is no nontrivial Lie algebra $\mathscr{K}$ such that $\{\mathrm{ad}_{\mathscr{L}} \mathscr{K}\}_A \subset \mathscr{H}$.

6. Realization theory. One of the central results in linear system theory is the fact that any two time-invariant, controllable and observable realizations of a given time-invariant input–output map are related to each other in a very simple way. Our purpose here is to establish a similar theorem in this context.

Suppose we have two systems

$$\dot{X}(t) = \left(\sum_{i=1}^{m} u_i(t)B_i\right)X(t), \qquad y(t) = \mathscr{C}X(t),$$

$$\dot{Z}(t) = \left(\sum_{i=1}^{m} u_i(t)G_i\right)Z(t), \qquad y(t) = \mathscr{H}Z(t).$$

We assume that

(i) the systems are observable in the sense that no two initial states give rise to the same response y for all piecewise continuous inputs, and

(ii) there exist one-to-one maps, say $c(\cdot)$ and $h(\cdot)$, both mapping into a set S such that if each system starts at the identity state and if each system receives the same input, $c(\mathscr{C}X(t)) = h(\mathscr{H}Z(t))$ for all future time. A pair of systems meeting these criteria will be said to be *observable realizations of the same input–output map.* We emphasize that $X(t)$ and $Z(t)$ are square matrices but not necessarily of the same dimension.

Suppose we have two observable realizations of the same input–output map. Let $u(\cdot)$ be a nonzero piecewise constant control defined on $[0, 1]$ which when applied to the X system takes the state $X(0) = I$ into the state $X(1) = I$. Then of course it must do the same for the Z system because they are observable realizations of the same input–output map. Thus we see that if $i_1, i_2, \cdots, i_q$ is a collection of integers with $1 \leqq i_k \leqq m$ and if α_i are any real numbers such that

$$e^{\alpha_1 B_{i_1}} e^{\alpha_2 B_{i_2}} \cdots e^{\alpha_q B_{i_q}} = I,$$

then

$$e^{\alpha_1 G_{i_1}} e^{\alpha_2 G_{i_2}} \cdots e^{\alpha_q G_{i_q}} = I.$$

Let $L_1, L_2, \cdots, L_r$ be a set of commutator expressions in $B_1, B_2, \cdots, B_m$ such that $\{L_i\}$ forms a basis for $\{B_i\}_A$. Let $K_1, K_2, \cdots, K_r$ be in an analogous expression obtained by replacing B_1 by G_1, B_2 by G_2, etc. Let S be an arbitrary commutator expression in $B_1, B_2, \cdots, B_m$ and let T be the analogous commutator expression in $G_1, G_2, \cdots, G_m$. Then in the notation of the proof of Theorem 1, there exist differentiable functions $\alpha_i(\rho)$ such that for $|\rho|$ small,

$$\text{EXP}\,\alpha_1(\rho)L_1\,\text{EXP}\,\alpha_2(\rho)L_2 \cdots \text{EXP}\,\alpha_r(\rho)L_r = \text{EXP}\,\rho S$$

and

$$\text{EXP}\,\alpha_1(\rho)K_1\,\text{EXP}\,\alpha_2(\rho)K_2 \cdots \text{EXP}\,\alpha_r(\rho)K_r = \text{EXP}\,\rho T.$$

Since the α_i are differentiable we can write (prime denotes derivative)

$$I + \rho \sum_{i=1}^{m} \alpha_i'(0)L_i + o(\rho^2) = I + \rho S + o(\rho^2)$$

and

$$I + \rho \sum_{i=1}^{m} \alpha_i'(0)K_i + o(\rho^2) = I + \rho T + o(\rho^2).$$

Thus if

$$S = \sum_{i=1}^{r} \gamma_i L_i,$$

then

$$T = \sum_{i=1}^{r} \gamma_i K_i.$$

From this we see that the algebra $\{G_i\}_A$ is generated from $\{B_i\}$ in exactly the same way as the algebra $\{B_i\}_A$ is generated from $\{B_i\}$ and thus that the algebras are isomorphic. We summarize this discussion with a theorem.

THEOREM 10. *Consider the two systems*

$$\dot{X}(t) = \left(\sum_{i=1}^{m} u_i(t) B_i X(t)\right), \qquad y(t) = \mathscr{C} X(t),$$

$$\dot{Z}(t) = \left(\sum_{i=1}^{m} u_i(t) G_i Z(t)\right), \qquad y(t) = \mathscr{H} Z(t),$$

where X and Z are $n \times n$ and $q \times q$ respectively. Suppose that these systems are observable realizations of the same input–output map. Then $\{B_i\}_A$ and $\{G_i\}_A$ are isomorphic as Lie algebras, and moreover if $L_1, L_2, \cdots, L_r$ are commutator expressions in $\{B_i\}$ which form a basis for $\{B_i\}_A$ and if $K_1, K_2, \cdots, K_r$ are the analogous expressions in G_i obtained by replacing B_i by G_i, then $K_1, K_2, \cdots, K_r$ is a basis for $\{G_i\}_A$, and if

$$[L_i, L_j] = \sum_{k=1}^{r} \gamma_{ijk} L_k,$$

then

$$[K_i, K_j] = \sum_{k=1}^{r} \gamma_{ijk} K_k.$$

Of course this does not mean that the reachable sets from I, namely $\{\exp\{B_i\}_A\}_G$ and $\{\exp\{G_i\}_A\}_G$, are isomorphic as groups. For example, the group of unit quaternions and the group of 3×3 orthogonal matrices have isomorphic Lie algebras, yet they are not isomorphic as groups.

7. System theory on coset spaces. In this section we reinterpret our results in a somewhat different way. This interpretation leads to some facts about systems on manifolds which do not admit a group structure. In particular we have in mind the n-sphere $S^n = \{x : x'x = 1, x \in \mathbb{R}^{n+1}\}$ which, as is well known, does not admit a Lie group structure except for the cases $n = 1$ and 3.

Let $M \subset \mathbb{R}^n$ be a manifold. Let $\mathscr{G}$ be a matrix group in $\mathbb{R}^{n \times n}$. We say that $\mathscr{G}$ *acts on* M if for every $x \in M$ and every $G \in \mathscr{G}$, Gx belongs to M. By the *orbit* of $\mathscr{G}$ through x we mean the set of points $\mathscr{G}x = \{y : y = Gx, G \in \mathscr{G}\}$. We say that $\mathscr{G}$ acts *transitively* on M if it acts on M and if for every pair of points x, y in M, there exists G in $\mathscr{G}$ such that $Gx = y$. If $\mathscr{G}$ acts transitively on M, then at any point $x \in M$ there will be a subset $\mathscr{H}_x \subset \mathscr{G}$ such that for each $H \in \mathscr{H}_x$, $Hx = x$. Clearly if $H_1 \in \mathscr{H}_x$ and $H_2 \in \mathscr{H}_x$, then $H_1 H_2 x = H_1 x = x$ and $H^{-1}x = x$ so that $\mathscr{H}_x$ is a subgroup.

We call $\mathscr{H}_x$ the *isotropy group* at x. Notice that if $Gx = y$, then $y = G\mathscr{H}_x x = G\mathscr{H}_x G^{-1} y$, and thus $G\mathscr{H}_x G^{-1}$ is the isotropy group at y—all isotropy groups are conjugate in $\mathscr{G}$. Now suppose M is a manifold for which there actually exists a group $\mathscr{G}$ acting transitively. Pick a point $x \in M$. Define in $\mathscr{G}$ an equivalence relation whereby $G_1 \sim G_2$ if and only if $G_1 = G_2 H_x$ for some $H_x \in \mathscr{H}_x$. There is a one-to-one correspondence between this space of equivalence classes, $\mathscr{G}/\mathscr{H}_x$, and M. In this case we call M a *coset space*.

We study systems in which the state is represented as an n-vector and the evolution is governed by

$$\dot{x}(t) = \left(A + \sum_{i=1}^{v} u_i(t)B_i\right)x(t), \qquad y(t) = \mathscr{C}x(t). \tag{1}$$

By $\mathscr{C}x(t)$ we mean an equivalence class of vectors, x_1 being equivalent to x_2 if and only if $Cx_1 = x_2$ for some C in $\mathscr{C}$.

Let $\mathscr{L}$ be a Lie algebra generated by $\{A, B_i\}$ and let $M \subset \mathbb{R}^n$ be a manifold such that $\{\exp \mathscr{L}\}_G$ acts on it. Then the above equation can be thought of as evolving on the manifold $M \subset \mathbb{R}^n$, for if $x(0) \in M$, then regardless of the control, $x(t) \in M$ for all $t > 0$. If there exists a differentiable manifold $M \subset \mathbb{R}^n$ such that $\{\exp \mathscr{L}\}_G$ acts on M, then we shall say that (1) is *well-posed* on M.

Example 6. Consider the n-sphere, S^n. Let $B_1, B_2, \cdots, B_m$ be $(n+1) \times (n+1)$ skew symmetric matrices. Clearly, the system

$$\dot{x}(t) = \left[\sum_{i=1}^{m} u_i(t)B_i\right]x(t), \qquad y(t) = \mathscr{C}x(t),$$

is well-posed on S^n since $\{\exp \mathscr{L}\}_G$ consists of orthogonal matrices, and orthogonal transformations preserve norm. If we can observe only the first component of x, then we should let $\mathscr{C}$ be the subsets of $\mathscr{SO}(n+1)$ consisting of those matrices which have a 1 in the first column and first row. That is,

$$\mathscr{C} = \begin{bmatrix} 1 & 0 \\ 0 & \mathscr{SO}(n) \end{bmatrix}.$$

With respect to controllability we can say given any two vectors x_1 and x_2 in S^n there exists a piecewise continuous control which steers the system from x_1 to x_2 if and only if $x_2 = Rx_1$ for some R in $\{\exp \mathscr{L}\}_G$, where $\mathscr{L}$ is the Lie algebra generated by $\{A_i\}$. Also, an arbitrary point can be transferred to an arbitrary point if and only if $\{\exp \mathscr{L}\}_G$ acts transitively on S^n.

At the same time we might observe that any x_0 such that $\|x_0\| = 1$ can be transferred to any x_1 such that $\|x_1\| = 1$ if and only if $\{\exp \mathscr{L}\}_G$ acts transitively on S^n. This second point of view is useful because it puts the problem of controllability on S^n in contact with standard results in geometry. In particular a great deal is known about Lie groups which act transivitely on S^n. (See Samelson [17, p. 26].)

As for observability, we note that two initial states x_1 and x_2 in S^n give rise to the same y if and only if for all R in $\{\exp \mathscr{L}\}_G$ there exists $C(R)$ in $\mathscr{C}$ such that $Rx_1 = C(R)Rx_2$, which is to say that $R^{-1}C(R)Rx_2 = x_1$.

We now abstract from this example the essential features and state formally a result which summarizes the development.

THEOREM 11. *Consider the dynamical system* $(x(t) \in \mathbb{R}^n)$

$$\dot{x}(t) = \left(\sum_{i=1}^{q} u_i(t)B_i\right)x(t), \qquad y(t) = \{\exp \mathscr{H}\}_G x(t),$$

which is well-posed on the manifold $M \subset \mathbb{R}^n$. *Let* $\mathscr{L}$ *be the Lie algebra generated by* $\{B_i\}$. *A given state* x_2 *is reachable from* x_1 *if and only if* $x_2 = Nx_1$ *for some* N *in* $\{\exp \mathscr{L}\}_G$. *Let* $\mathscr{P} = \{P : RPR^{-1} \in \{\exp \mathscr{H}\}_G$ *for all* $R \in \exp \mathscr{L}\}$. *Two states* x_1 *and* x_2 *are indistinguishable if and only if* $x_2 = Px_1$ *for some* P *in* $\mathscr{P}$. *In particular, two states* x_1 *and* x_2 *are indistinguishable if* $x_2 = Px_1$ *for* P *in* $\{\exp \mathscr{K}\}_G$ *with* $\mathscr{K}$ *being any Lie algebra such that* $\{\mathrm{ad}_{\mathscr{L}} \mathscr{K}\}_A \subset \mathscr{H}$.

Example 7. Consider the submanifold M of $\mathbb{R}^{n+1}$ consisting of those points whose last coordinate is 1. The evolution equation in M,

$$\frac{d}{dt}\begin{bmatrix} x(t) \\ 1 \end{bmatrix} = \begin{bmatrix} A & Bu(t) \\ 0 & 0 \end{bmatrix}\begin{bmatrix} x \\ 1 \end{bmatrix}, \qquad y(t) = \left(\exp \begin{bmatrix} 0 & \ker C \\ 0 & 0 \end{bmatrix}\right)\begin{bmatrix} x(t) \\ 1 \end{bmatrix}$$

corresponds to the more familiar $\dot{x}(t) = Ax(t) + Bu(t), y(t) = Cx(t)$. Using Theorem 7 we see that for the associated group equation

$$\dot{X}(t) = \begin{bmatrix} A & 0 \\ 0 & 0 \end{bmatrix} + \sum_{i=1}^{m} u_i(t)\begin{bmatrix} 0 & b_i \\ 0 & 0 \end{bmatrix} X(t)$$

the reachable set at time t consists of those matrices which can be written as

$$\mathscr{R} = \left\{X : X = \begin{bmatrix} e^{At} & x \\ 0 & 1 \end{bmatrix};\ x \in \mathrm{range}\,(B, AB, \cdots, A^{n-1}B)\right\}.$$

Thus if $B, AB, \cdots, A^{n-1}B$ spans $\mathbb{R}^n$, then the reachable group acts transitively on M and we have controllability.

As for observability, we note that

$$\exp \begin{pmatrix} 0 & \ker C \\ 0 & 0 \end{pmatrix} = \begin{pmatrix} I & \ker C \\ 0 & 1 \end{pmatrix}.$$

The subalgebras of $\mathscr{H}$ which are closed under commutation with $\mathscr{L}$ correspond to the linear subspaces of $\ker C$ which are invariant under A.

Acknowledgment. Among the many people who made useful suggestions on earlier expositions I want to mention in particular D. Elliott, V. Jurdjevic, H. Sussmann, Jacques Willems and Jan Willems. I also want to thank H. Rosenbrock who some years ago acquainted me with work relating to differential equations and Lie algebras.

REFERENCES

[1] W. L. CHOW, *Über System von linearen partiellen Differentialgleichungen erster Ordnung*, Math. Ann., 117 (1939), pp. 98–105.

[2] R. HERMANN, *On the accessibility problem in control theory*, Proc. International Symposium on Nonlinear Differential Equations and Nonlinear Mechanics, Academic Press, New York, 1963, pp. 325–332.

[3] J. KUČERA, *Solution in large of control problem*: $\dot{x} = (A(1-u) + Bu)x$, Czech. Math. J., 16 (1966), no. 91, pp. 600–623.

[4] J. Kucera, *Solution in large of control problem:* $\dot{x} = (Au + Bv)x$, Czech. Math. J., 17 (1967), no. 92, pp. 91–96.

[5] C. Lobry, *Contrôlabilité des systèmes non linéaries*, this Journal, 8 (1970), pp. 573–605.

[6] G. W. Haynes and H. Hermes, *Nonlinear controllability via Lie theory*, this Journal, 8 (1970), pp. 450–460.

[7] W. Magnus, *On the exponential solution of differential equations for a linear operator*, Comm. Pure Appl. Math., 7 (1954), pp. 649–673.

[8] K. T. Chen, *Decomposition of differential equations*, Math. Ann., 146 (1962), pp. 263–278.

[9] J. Wei and E. Norman, *On global representations of the solutions of linear differential equations as a product of exponentials*, Proc. Amer. Math. Soc., 15 (1964), pp. 327–334.

[10] ———, *Lie algebraic solution of linear differential equations*, J. Mathematical Phys., 4 (1963), pp. 575–581.

[11] I. M. Singer and J. A. Thorpe, *Lecture Notes on Elementary Topology and Geometry*, Scott, Foresman and Co., Glenview, Ill., 1967.

[12] S. Sternberg, *Lectures on Differential Geometry*, Prentice-Hall, Englewood Cliffs, N.J., 1964.

[13] L. Auslander and R. MacKenzie, *Introduction to Differentiable Manifolds*, McGraw-Hill, New York, 1963.

[14] J. J. Rotmann, *The Theory of Groups*, Allyn and Bacon, Boston, 1965.

[15] J. F. P. Martin, *Some results on matrices which commute with their derivatives*, SIAM J. Appl. Math., 15 (1967), pp. 1171–1183.

[16] R. W. Brockett, *Finite Dimensional Linear Systems*, John Wiley, New York, 1970.

[17] H. Samelson, *Topology of Lie groups*, Bull. Amer. Math. Soc., 58 (1952), pp. 2–37.

[18] H. J. Sussmann and V. Jurdjevic, *Controllability of nonlinear systems*, J. Differential Equations, to appear.

[19] V. Jurdjevic and H. J. Sussmann, *Control systems on Lie groups*, J. Differential Equations, to appear.

[20] D. L. Elliott, *A consequence of controllability*, J. Differential Equations, 10 (1971), pp. 364–370.

Controllability of Nonlinear Systems

HECTOR J. SUSSMANN AND VELIMIR JURDJEVIC

IN the late 1950s and early 1960s, the introduction of state-space models in the analysis and design of linear control systems added enormous thrust to the development of systematic methods for the solution of major control problems such as stabilization by means of feedback, asymptotic regulation with minimal cost, disturbance rejection, and so on. Instrumental, in this expansion, were the notions of reachability and observability, introduced by Kalman to characterize interaction between input and state, on one hand, and state and output, on the other. An important feature of linear state-space methods was the possibility of obtaining information, from simple linear-algebraic tests on the model parameters, about either the set of all states that can be reached from a given one (using appropriate choices of inputs) or the set of all states that cannot be distinguished from a given one based on the observed output.

Indeed, in view of the impact that these concepts had in the design of linear control systems, the problem of deriving similar characterizations for "nonlinear" systems soon became the focus of intense research effort. It must be stressed, in this respect, that the term "nonlinear" encompasses not just the consideration of phenomena such as saturation or dead-zone, whose influence on the stability and performance of an otherwise linear system was relatively well understood, but especially the kind of analytic nonlinearities found in the equations describing the motion of a rigid body, of a multi-link robot, of an electrical drive, and so on (typically, products between state variables or between input and state variables).

Pioneering authors in this domain have been Hermann [7], Haynes and Hermes [6], Lobry [11], and Elliott [4], who realized that the appropriate language in which the study of the structural properties of a nonlinear system had to be cast was that of the differential geometry. Fundamental in this respect were the observations—based on an earlier result of Chow [3]—that the most natural geometric object into which the set of states reachable from a given point can be embedded is a differentiable *manifold* (i.e., a "surface," and not a subspace as in the case of linear systems) and the fact that this manifold can be "infinitesimally" characterized in terms of repeated derivatives, at the point of interest, of the vector fields that define the "admissible velocities" at this point (namely, the velocity vectors that can be imposed at a given point of the state state by setting the control input).

More specifically, the theorem of Chow (whose derivation predates by more than 20 years the birth of state-space methods in control theory, but which can easily be cast into its language) considers systems described by differential equations of the form

$$\dot{x} = f_1(x)u_1 + f_2(x)u_2 + \cdots + f_m(x)u_m \tag{1}$$

in which $x \in \mathbb{R}^n$ and $f_1(x), \ldots, f_m(x)$ are analytic complete[1] vector fields. Each of the "controls" $u_1, \ldots, u_m$ can take values in the set $\Omega = \{-1, 0, 1\}$. Then, the theorem asserts that the set of states reachable from a given x° is a *submanifold* M of $\mathbb{R}^n$, precisely the unique connected submanifold which contains x° and has, at any of its points, a tangent space spanned by the vectors of the set $F = \{f_1(x), \ldots, f_m(x)\}$ and by all vectors that can be expressed as (repeated) *Lie brackets* of vectors in F.[2] In other words, the vectors of F and their repeated Lie brackets provide a complete characterization of the set M of states reachable from x°, by determining the tangent space to M at any of its points.

In the context of control theory, the main conceptual content of Chow's theorem is that the (repeated) Lie brackets of the vector fields $f_1(x), \ldots, f_m(x)$ define the set of "directions" in which the state vector can be (infinitesimally) pushed. In the case of a linear system

$$\dot{x} = Ax + b_1u_1 + \cdots + b_mu_m$$

this idea specializes to the familiar observation that the set of states reachable from $x = 0$ is not just the span of the vectors $b_1, \ldots, b_m$, but rather the span of all the vectors of the form $A^k b_i$.

[1]A vector field is complete if for any x° the integral curve of $\dot{x} = f_i(x)$ passing through x° at time $t = 0$ is maximally defined over $(-\infty, +\infty)$.

[2]We recall that the Lie bracket of two vector fields $f(x)$, $g(x)$, denoted $[f, g](x)$, is the vector field

$$[f, g](x) = \frac{\partial g}{\partial x} f(x) - \frac{\partial f}{\partial x} g(x).$$

Unfortunately, however, a basic hypothesis in Chow's result is that each of the inputs $u_1, \ldots, u_m$ take values in the set $\Omega = \{-1, 0, 1\}$, which corresponds—in the control-theoretic framework—to the possibility of reversing the sign of the corresponding component of the velocity vector (or, equivalently, of reversing the sign of the flow of time) and/or turning-off every single component. Thus, this result cannot provide a satisfactory answer in the case of systems modeled by equations of the form

$$\dot{x} = f_0(x) + f_1(x)u_1 + f_2(x)u_2 + \cdots + f_m(x)u_m \tag{2}$$

which, despite the formal similarity with (1), suffer from the basic limitation of having a component, namely $f_0(x)$, of the velocity vector (commonly referred to as the "drift term") which cannot be controlled. Many systems of practical interest in control applications fall into the class of systems modeled by equations of the form (2).

A complete and satisfactory answer to a number of problems arising in the analysis of the geometric structure of the sets of reachable states for a control system of the form (2) (actually, for the more general class of systems modeled by equations of the form

$$\dot{x} = F(x, u) \tag{3}$$

with $F(x, 0) \not\equiv 0$) was provided by Sussmann and Jurdjevic in their 1972 paper. This paper can in full right be considered as the paper that laid the foundations of geometric nonlinear systems theory in a comprehensive and fully mathematically rigorous fashion.

Sussmann and Jurdjevic have observed that, in the analysis of a system of the form (3), it is convenient to look separately at two different concepts of reachability: reachability with fixed end-time and reachability with free end-time. The former is used to define the set $\mathbf{A}(x^\circ, T)$ of all states attainable from x° at time T (i.e., the set of all states x for which there exists some input, defined on the *fixed* interval $[0, T]$, such that the integral curve $x(t)$ of (3) satisfies $x(0) = x^\circ$ and $x(T) = x$), while the latter is used to define the set $\mathbf{A}(x)$ of all states attainable from x° (i.e., the set of all states x for which there exists *some* time interval $[0, T]$ and some input, defined on the interval $[0, T]$, such that the integral curve $x(t)$ of (3) satisfies $x(0) = x^\circ$ and $x(T) = x$). The difference between these two concepts is present of course also in the case of linear systems, but nobody had pointed it out before. The reason was mainly because in the study of linear systems the initial state from which reachability is considered is the equilibrium state $x^\circ = 0$, and in this case the two concepts coincide. But, if $x^\circ \neq 0$, it is readily seen that

$$\mathbf{A}(x^\circ, T) = \exp(At)x^\circ + \mathcal{R}$$

where $\mathcal{R}$ is the (standard) set of states reachable from the origin, while

$$\mathbf{A}(x^\circ) = \bigcup_{0 \le T < \infty} \mathbf{A}(x^\circ, T)$$

In other words, the family of sets $\mathbf{A}(x^\circ, T)$, as T ranges over $[0, \infty)$, partitions the set $\mathbf{A}(x^\circ)$.

Assuming that for each fixed $u \in \Omega$, with Ω a fixed set, the vector field $F(x, u)$ is analytic, Sussmann and Jurdjevic have shown that $\mathbf{A}(x^\circ)$ *is contained* in a manifold M for which a tangent space is determined, at each point, by the vector fields of $\mathcal{F} = \{F(x, u) : u \in \Omega\}$ and their repeated Lie brackets, and that $\mathbf{A}(x^\circ)$ *has a nonempty interior in the topology of* M (i.e., contains an open subset of M). In other words, $\mathbf{A}(x^\circ)$ may not coincide with the whole of M (as it was in the particular situation addressed by Chow's theorem), but M has the least dimension among all manifolds which contain $\mathbf{A}(x^\circ)$. Thus, M can be regarded, in a certain sense, as the "most appropriate" manifold into which the reachable set $\mathbf{A}(x^\circ)$ can be embedded. Likewise, they prove that $\mathbf{A}(x^\circ, T)$ is contained in a *submanifold* M_T of M, of codimension at most 1 in M, and that $\mathbf{A}(x^\circ)$ has a nonempty interior in the topology of M_T. Again, the tangent space to M_T is determined, at each point, by certain vector fields of $\mathcal{F}$ and their repeated Lie brackets.

The concepts, the formalism, and the results contained in this paper soon became a standard reference for all authors working in that broad area of research which became known as "nonlinear control theory". The hypothesis of analyticity of the vector fields was soon removed by Sussmann himself [12] and an alternative proof of the property that $\mathbf{A}(x^\circ)$ has a nonempty interior in M, due to Krener [10], also became available. The results were the point of departure for the work of Sussmann [12] on the theory of minimal realizations for nonlinear systems and the work of Hermann and Krener [8] (also included in this volume) on the theory of nonlinear observability, which these authors were able to develop in a dual-like fashion. The concept of Lie bracket played also an instrumental role in the work of Fliess on the characterization of the dimension of a minimal realization of a given nonlinear input-output map, through the concept of Lie rank of a Hankel matrix, introduced in [5]. The principles laid down in the 1972 paper by Sussman and Jurdjevic were instrumental in addressing many problems of "local" controllability and optimal control, a field of research that has been very active during the 1980s and the 1990s (see, e.g., [1], [14], and [15]).

Finally, it is worth stressing that a specific outcome of the theory of nonlinear controllability developed in this paper was the possibility of characterizing (local) decompositions of nonlinear systems into "controllable/uncontrollable" parts, very similar in nature to the decompositions developed some 10 years earlier by Kalman (see the earlier preamble to [9], also included in this volume). The availability of these decompositions was one of the main thrusts for the development of systematic methods for solutions to problems of disturbance decoupling and noninteracting control, which occurred in the late 1970s and early 1980s (see the preamble to [8], later in this volume).

References

[1] J. Baillieul, "Controllability and observability of polynomial systems," *Nonlin. Anal.*, **5**:543–552, 1981.

[2] R. W. Brockett, "System theory on group manifolds and coset spaces," *SIAM J. Contr.*, **10**:265–284, 1972.

[3] W. L. Chow, "Über Systeme von Linearen Partiellen Differentialgleichungen Ester Ordnung," *Math. Ann.*, **117**:98–105, 1938.

[4] D. L. Elliott, "A consequence of controllability," *J. Diff. Eqs.*, **10**:364–370, 1970.

[5] M. Fliess, "Réalisation locale des systèmes non linéaires, algèbres de Lie

filtrées transitives et séries génératrices non commutatives," *Invent. Math.*, **71**:521–537, 1983.

[6] G. W. HAYNES AND H. HERMES, "Nonlinear controllability via Lie theory," *SIAM J. Contr.*, **8**:450–460, 1970.

[7] R. HERMANN, "On the accessibility problem in control theory," in *Int. Symp. on Nonlinear Differential Equations and Nonlinear Mechanics*, Academic Press, pp. 325–332, 1963.

[8] R. HERMANN AND A. J. KRENER, "Nonlinear controllability and observability," *IEEE Trans. Aut. Contr.*, **AC-22**:728–740, 1977.

[9] R.E. KALMAN, "Mathematical description of linear dynamical systems," *SIAM J. Control*, **1**:152–192, 1963.

[10] A.J. KRENER, "A generalization of Chow's theorem and the bang-bang theorem to nonlinear control systems," *SIAM J. Contr.*, **12**:43–52, 1974.

[11] C. LOBRY, "Contrôlabilité des systèmes non linéaires," *SIAM J. Contr.*, **8**: 573–605, 1970.

[12] H. SUSSMANN, "Orbits of families of vector fields and integrability of distributions," *Trans. Am. Math. Soc.*, **180**:171–188, 1973.

[13] H. SUSSMANN, Existence and uniqueness of minimal realizations of non-linear system, *Math. Syst. Theory*, **10**:263–284, 1977.

[14] H. SUSSMANN, Lie brackets and local controllability, *SIAM J. Contr. Optimiz.*, **21**:686–713, 1983.

[15] H. SUSSMANN, "On a general theorem on local controllability," *SIAM J. Contr. Optimiz.*, **25**:158–194, 1987.

A.I.

Controllability of Nonlinear Systems*

HÉCTOR J. SUSSMANN

Department of Mathematics, University of Chicago, Chicago, Illinois 60637

AND

VELIMIR JURDJEVIC

Department of Mathematics, University of Toronto, Toronto, Canada

Received April 26, 1971

1. INTRODUCTION

In this article we study the controllability of nonlinear systems of the form

$$dx/dt = F(x, u). \tag{$*$}$$

Our objective is to establish criteria in terms of F and its derivatives at a point x which will give qualitative information about the sets attainable from x. The study is based primarily on the work of Chow [4] and Lobry [16], although it is similar in its approach to works by other authors in that it makes systematic use of differential geometry (for instance, see Hermann [8, 9], Haynes and Hermes [6], Brockett [2], etc.).

The state variable x is assumed to take values in an arbitrary real, analytic manifold M, rather than in $\mathbf{R}^n$. We chose this generalization because it creates no essential new difficulties while, on the other hand, it allows for certain applications which are not commonly treated in control theory. For instance, when M is a Lie group, then the present results can be specialized to obtain more detailed controllability criteria. Control problems on Lie groups were first considered by Brockett in [2], and will be treated in a forthcoming paper by the authors.

Most of the recent studies on controllability of nonlinear systems have essentially dealt with symmetric systems, i.e., systems of the form $(*)$ with

* This work was performed while the authors were at Harvard University, Division of Engineering and Applied Physics, Cambridge, Mass. The first author was supported by the U.S. Office of Naval Research under the Joint Electronics Program by Contract N00014-67-A-0298-0006. The second author was supported by the National Aeronautics and Space Administration under Grant NGR 22-007-172.

Reprinted with permission from *Journal of Differential Equations*, Hector J. Sussmann and Velimir Jurdjevic, "Controllability of Nonlinear Systems" Vol. 12, 1972, pp. 95–116.

the property that $F(x, -u) = -F(x, u)$ (Hermann [9], Haynes and Hermes [6], Lobry [16]). As remarked by Lobry in [16], the consideration of symmetric systems often excludes interesting situations arising from mechanics. In these cases the system is of the form

$$dx/dt = A(x) + H(x) \cdot u.$$

A notable exception is the work by Lobry [17]. Lobry stated (and proved for the case of two vector fields in $\mathbf{R}^3$) the result for nonsymmetric systems that appears here as Theorem 3.1.

Our results apply to nonsymmetric systems. We obtain some general information about the geometric structure of the attainable sets showing that they "practically" are submanifolds (see Theorems 4.4 and 4.5 for the precise statements). This information yields a complete answer to the problem of deciding when the sets attainable from a point x have a nonempty interior. The criteria obtained involve purely *algebraic* manipulations of F and its derivatives (of all orders) *at the point* x (see the Remark below).

In particular, our results contain those of Kučera [14]. In this connection we observe that our proofs are of interest even for the case treated by Kučera (see Sussmann [21]).

We have omitted the consideration of nonautonomous systems; they can be treated analogously by the familiar procedure of reduction to an autonomous system (i.e., by considering the state variable to be defined in $M \times \mathbf{R}$).

The organization of the article is as follows: in Section 2 we introduce notations and basic concepts; in addition, we quote some well-known basic results which will be used later. In Section 3 we prove our main results in differential geometric terminology.

In Section 4 we apply these results to control systems. We derive the algebraic criteria mentioned above (Corollaries 4.6 and 4.7), and we prove two "global results": we show that, for a large class of manifolds, accessibility (i.e., the property that, for any given x, the set of points attainable from x has a nonempty interior) implies strong accessibility (i.e., that for any given x and any given fixed positive t, the set of points attainable from x at time t has a nonempty interior). We also show that, for a still larger class, including the Euclidean spaces, controllability implies strong accessibility.

Finally, Section 5 contains examples. We show how our results can be used to derive the classical controllability criteria for the system

$$dx/dt = Ax + Bu.$$

We also derive the results of Kučera and indicate some generalizations.

Remark. An assumption that is made throughout the article is that F is an *analytic* function of x. This guarantees that all the information about the system is actually contained in F and its derivatives (of all orders) at a given point x. The analyticity assumption *cannot* be relaxed without destroying the theory (cf. Example 5.3).

Another assumption that we make is that the trajectories of the system are everywhere defined. As opposed to the previous one, this assumption is not essential (except for the "global" Theorems 4.9 and 4.10). We use it, however, because it considerably simplifies all the proofs.

2. Preliminaries

We shall assume that the reader is familiar with the fundamental notions of differential geometry. All the definitions and basic concepts utilized in this paper can be found in standard books, (for instance [1, 3, 7, 13 and 19]).

The following notations will be used throughout:

- $\mathbf{R}$ the set of real numbers.
- $\mathbf{R}^n$ n-dimensional Euclidean space.
- M_x the tangent space to the manifold M at the point x.
- TM the tangent bundle of the manifold M.
- $V(M)$ the set of all analytic vector fields on the analytic manifold M.

We will regard $V(M)$ as a Lie algebra over the reals. For any X and Y in $V(M)$, we will denote the Lie product by $[X, Y]$ (i.e., $[X, Y] = XY - YX$). All the manifolds will be assumed to be paracompact. Recall that a submanifold of a paracompact manifold is paracompact. Also, a connected paracompact manifold is a countable union of compact sets. These facts imply (cf. Lobry [16, p. 589]):

Lemma 2.1. *Let M be a (paracompact) manifold of dimension n. Let S be a k-dimensional connected submanifold of M. If $k < n$, then the set of points of S has an empty interior in M.*

A subset D of $V(M)$ will be called *involutive* if, whenever X and Y belong to D, then $[X, Y]$ also belongs to D. A *subalgebra* of $V(M)$ is an involutive subspace. Let $D \subset V(M)$. An *integral manifold* of D is a connected submanifold S of M with the property that $S_x = \mathscr{L}(D(x))$ for every $x \in S$, where $D(x) = \{X(x) : X \in D\}$, and where $\mathscr{L}(D(x))$ is the subspace of M_x spanned by $D(x)$. We state the following basic result about integral manifolds:

Lemma 2.2. *Let D be an involutive subset of $V(M)$, and let $x \in M$. Then x is contained in a unique maximal integral manifold of D (here "maximal" means "maximal with respect to inclusion").*

This result is classical if the dimension of $\mathscr{L}(D(x))$ is the same for each $x \in M$ (Chevalley [3]). For a proof in the general case, see Lobry [16].

If $D \subset V(M)$, we denote the smallest subalgebra of $V(M)$ which contains D by $\mathscr{T}(D)$, and the maximal integral manifold of $\mathscr{T}(D)$ through x by $I(D, x)$. Recall that, if X is a vector field on M, then α is an *integral curve* of X if α is a smooth mapping from a closed interval I, $I \subset \mathbf{R}$, into M such that

$$d\alpha(t)/dt = X(\alpha(t)) \qquad \text{for all} \quad t \in I.$$

DEFINITION 2.3. If D is a subset of $V(M)$, then an *integral curve of* D is a mapping α from a real interval $[t, t']$ into M such that there exist $t = t_0 < t_1 < \cdots < t_k = t'$, and elements $X_1, \ldots, X_k$ of D with the property that the restriction of α to $[t_{i-1}, t_i]$ is an integral curve of X_i for each $i = 1, 2, \ldots, k$. We have the following elementary fact:

LEMMA 2.4. *Let* $D \subset V(M)$. *Let* $\alpha : [t_0, t_1] \to M$ *be an integral curve of* D, *and let* $\alpha(t) = x$ *for some* $t \in [t_0, t_1]$. *Then,* $\alpha(s) \in I(D, x)$ *for all* $s \in [t_0, t_1]$.

Proof. It is sufficient to consider the case when α is an integral curve of X, $X \in D$. For each maximal integral manifold S of $\mathscr{T}(D)$, let $J(S)$ be the set of all $s \in [t_0, t_1]$ such that $\alpha(s) \in S$. From the local existence and uniqueness of solutions of ordinary differential equations it follows that, if $s \in J(S)$, then there exists $r > 0$ such that $(s - r, s + r) \cap [t_0, t_1] \subset J(S)$. Thus, $J(S)$ is open relative to $[t_0, t_1]$. Since the maximal integral manifolds of $\mathscr{T}(D)$ are disjoint, we have that, for some maximal integral manifold S, $[t_0, t_1] \subset J(S)$. But $\alpha(t) \in I(D, x)$; therefore, our proof is complete.

Chow's theorem provides a partial converse to the above lemma. If $D \subset V(M)$, then D is *symmetric* if, whenever $X \in D$, $-X$ also belongs to D. We can now state Chow's theorem as follows:

LEMMA 2.5. *Let* $D \subset V(M)$ *be symmetric, and let* $x \in M$. *Then, for every* $y \in I(D, x)$ *there exists an integral curve* $\alpha : [0, T] \to M$ *of* D, *with* $T \geqslant 0$, *such that* $\alpha(0) = x$ *and* $\alpha(T) = y$.

In other words, every point of the maximal integral manifold of $\mathscr{T}(D)$ through x can be reached in positive time by following an integral curve of D having x as its initial point.

DEFINITION 2.6. Let $D \subset V(M)$, and let $x \in M$. If $T \geqslant 0$, then, for any $y \in M$, y is *D-reachable from x at time T* if there exists an integral curve α of D defined on $[0, T]$ such that $\alpha(0) = x$ and $\alpha(T) = y$. The set of all points D-reachable from x at time T is denoted by $L_x(D, T)$. The union of $L_x(D, t)$ for $0 \leqslant t < \infty$ (respectively for $0 \leqslant t \leqslant T$) is denoted by $\mathbf{L}_x(D)$ (respectively, $\mathbf{L}_x(D, T)$).

3. Integrability of Families of Analytic Vector Fields

As an introduction to the general situation we first consider the case when D is a symmetric subset of $V(M)$. Chow's theorem can be utilized to obtain a necessary and sufficient condition for $\mathbf{L}_x(D)$ to have a nonempty interior in M. Let $n = \dim M = \dim \mathscr{T}(D)(x)$. Then $I(D, x)$ is an n-dimensional submanifold of M, and hence is open in M. By Chow's theorem we have that $\mathbf{L}_x(D) = I(D, x)$. We conclude that $\mathbf{L}_x(D)$ is open in M. Conversely (and without invoking the symmetry of D), if $\dim \mathscr{T}(D)(x) < n$, then $I(D, x)$ is a connected submanifold of M of dimension less than n; then from Lemma 2.1 it follows directly that $I(D, x)$ has an empty interior in M. Since $\mathbf{L}_x(D) \subset I(D, x)$, $\mathbf{L}_x(D)$ also has an empty interior. Thus, if D is symmetric, a necessary and sufficient condition for $\mathbf{L}_x(D)$ to have a nonempty interior in M is that $\dim \mathscr{T}(D)(x) = \dim M$. Moreover, this condition is necessary even in the nonsymmetric case (Lobry [16]). We shall show that it is also sufficient. For this purpose we shall assume that the elements of D are complete—recall that a vector field X is *complete* if the integral curves of X are defined for all real t [13, p. 13].

Theorem 3.1. *Let M be an n-dimensional analytic manifold, and let $D \subset V(M)$ be a family of complete vector fields. A necessary and sufficient condition for $\mathbf{L}_x(D)$ to have a nonempty interior in M is that* $\dim \mathscr{T}(D)(x) = n$. *Moreover, if this condition is satisfied, then for each $T > 0$, the interior of $\mathbf{L}_x(D, T)$ is dense in $\mathbf{L}_x(D, T)$ [thus, in particular, $\mathbf{L}_x(D, T)$ has a nonempty interior].*

Proof. We already know that the condition of the theorem is necessary. So we assume that $\dim \mathscr{T}(D)(x) = n$, and we prove the second statement. Clearly, this will imply that $\mathbf{L}_x(D)$ has a nonempty interior in M. Without loss of generality we can assume that D is finite. Let $D = \{X_1, \ldots, X_k\}$. For each $i = 1, 2, \ldots, k$, let $\Phi_i(t, \cdot)$ be the one-parameter group of diffeomorphisms induced by X_i (i.e., $t \to \Phi_i(t, y)$ is the integral curve of X_i which passes through y at $t = 0$; the fact that it is defined for all real t follows from the completeness of X_i). If m is a natural number, $\mathbf{t} = (t_1, \ldots, t_m)$ is an element of $\mathbf{R}^m$, and $\mathbf{i} = (i_1, \ldots, i_m)$ is an m-tuple of natural numbers between 1 and k, then we denote the element $\Phi_{i_1}(t_1, \Phi_{i_2}(t_2, \ldots, \Phi_{i_m}(t_m, x) \cdots))$ by $\Phi_{\mathbf{i}}(\mathbf{t}, x)$. Let $\pm D$ be the family of vector fields obtained from D by adjoining the vector fields $-X_1, \ldots, -X_k$ to D. Then, $\pm D$ is symmetric, and $\dim \mathscr{T}(\pm D)(x) = n$. From Chow's theorem we conclude that $\mathbf{L}_x(\pm D)$ is open in M. Clearly the elements of $\mathbf{L}_x(\pm D)$ are exactly those elements of M which are of the form $\Phi_{\mathbf{i}}(\mathbf{t}, x)$ for some m, some m-tuple $\mathbf{i}$, and some $\mathbf{t} \in \mathbf{R}^m$. For each $\mathbf{i}$, and for each natural number $N > 0$, let $A(\mathbf{i}, N)$ be the set of all points of M of the form $\Phi_{\mathbf{i}}(\mathbf{t}, x)$, where $\|\mathbf{t}\| \leqslant N$ (here $\|\mathbf{t}\| = |t_1| + \cdots + |t_m|$). Since $A(\mathbf{i}, N)$ is the image of the compact set $\{\mathbf{t} : \|\mathbf{t}\| \leqslant N\}$ under the continuous mapping $\mathbf{t} \to \Phi_{\mathbf{i}}(\mathbf{t}, x)$, we have that $A(\mathbf{i}, N)$ is compact. Also, since $\mathbf{L}_x(\pm D)$ is the union of the sets $A(\mathbf{i}, N)$ (taken over m, $\mathbf{i}$ and N), it follows from the category theorem that, for some $\mathbf{i}$ and N, the set $A(\mathbf{i}, N)$ has a nonempty interior in M. For such an $\mathbf{i}$, let $F : R^m \to M$ be defined by $F(\mathbf{t}) = \Phi_{\mathbf{i}}(\mathbf{t}, x)$. Then F is an analytic mapping whose image has a nonempty interior in M. By Sard's theorem (Sternberg [19]), the differential $dF_{\mathbf{t}}$ of F at $\mathbf{t}$ must have rank n for some $\mathbf{t} \in \mathbf{R}^m$. Since $dF_{\mathbf{t}}$ depends analytically on $\mathbf{t}$, it follows that the set $\Omega^{\#} = \{\mathbf{t} : \mathbf{t} \in \mathbf{R}^m, \text{rank } dF_{\mathbf{t}} < n\}$ has an empty interior. Let $\Omega = \mathbf{R}^m - \Omega^{\#}$. Then Ω is open and dense in $\mathbf{R}^m$.

Let $T > 0$, and $y \in \mathbf{L}_x(D, T)$. We now show that y is in the closure of the interior of $\mathbf{L}_x(D, T)$. It is clearly sufficient to assume that $y \in L_x(D, t)$, where $0 \leqslant t < T$ (for each point of $\mathbf{L}_x(D, T)$ is in the closure of $\bigcup \{L_x(D, t) : 0 \leqslant t < T\}$). Let $y = \Phi_{\mathbf{j}}(\mathbf{s}, x)$ where $\mathbf{j} = (j_1, \ldots, j_p)$, $\mathbf{s} = (s_1, \ldots, s_p)$, $s_1 > 0, \ldots, s_p > 0$, and $s_1 + \cdots + s_p = t$. Let

$$U = \Omega \cap \{\mathbf{t} : \|\mathbf{t}\| < T - t\} \cap \{\mathbf{t} : t_1 > 0, \ldots, t_m > 0\}.$$

U is open, and its closure contains the origin $\mathbf{0}$ of $\mathbf{R}^m$. Since $dF_{\mathbf{t}}$ has rank n at each point $\mathbf{t} \in U$, it follows that $F(U)$ is open. Let $V = \{\Phi_{\mathbf{j}}(\mathbf{s}, F(\mathbf{t})) : \mathbf{t} \in U\}$. V is the image of $F(U)$ under the diffeomorphism $z \to \Phi_{\mathbf{j}}(\mathbf{s}, z)$; therefore, V is open in M and, moreover, every element of V is D-reachable from x at time $\|\mathbf{s}\| + \|\mathbf{t}\| = t + \|\mathbf{t}\| < T$ (here we use essentially the fact that $t_1, \ldots, t_m$ are nonnegative). It remains to be shown that y belongs to the closure of V. Let $\{\mathbf{t}_q\}$ be a sequence of elements of U which converges to $\mathbf{0}$. Then

$$\lim \Phi_{\mathbf{j}}(\mathbf{s}, F(\mathbf{t}_q)) = \Phi_{\mathbf{j}}(\mathbf{s}, F(\mathbf{0})) = \Phi_{\mathbf{j}}(\mathbf{s}, x) = y.$$

This completes the proof of the theorem.

We now want to state an analogous theorem for the sets $L_x(D, T)$. For this purpose, we shall introduce a Lie subalgebra $\mathscr{T}_0(D)$ of $\mathscr{T}(D)$ which will be related to these sets in the same way as $\mathscr{T}(D)$ is related to the sets $\mathbf{L}_x(D, T)$. The aim of the following informal remarks is to motivate our definition of $\mathscr{T}_0(D)$. We shall ignore the fact that time has to be positive. Moreover, we shall assume, for simplicity, that D consists of three vector fields X_1, X_2 and X_3. Let Φ_1, Φ_2 and Φ_3 be the corresponding one-parameter groups. It is clear that $\mathscr{T}(D)$ has the following "geometric interpretation". $\mathscr{T}(D)(x)$ is, for each $x \in M$, the set of all limiting directions of curves through x that are entirely contained in $\mathbf{L}_x(D)$. Thus, for instance, if $i = 1, 2, 3$, then all the points in the curve $t \to \Phi_i(t, x)$ are attainable from x (recall that we are forgetting about positivity), and this is reflected in the fact that $X_i(x)$ belongs to $\mathscr{T}(D)(x)$. Similarly, the curves $\alpha_{ij}(t) = \Phi_i(-t, \Phi_j(-t, \Phi_i(t, \Phi_j(t, x))))$ are also contained in $\mathbf{L}_x(D)$. By the well-known geometric interpretation of the Lie bracket (cf. Helgason [7, p. 97]), the limiting direction of α_{ij} is $[X_i, X_j](x)$ (after a reparametrization). Thus, it is clear why $[X_i, X_j]$ belongs to $\mathscr{T}(D)$. Obviously, a similar argument works for the brackets of higher order. The geometrical meaning of $\mathscr{T}(D)$ is now obvious.

If $\mathscr{T}_0(D)$ is going to play the desired role it is clear that $\mathscr{T}_0(D)(x)$ will have to be the set of all limiting directions of curves γ through x such that $\gamma(t)$ is "attainable from x in zero units of time" for all t. Notice that the curves $\alpha_{ij}(t)$ of the preceding paragraph have this property. Indeed, $\alpha(t)$ can be reached from x by "moving forward" in time $2t$ units, and then "backward" another $2t$ units. This shows that the vector fields $[X_i, X_j]$ are reasonable candidates for membership in $\mathscr{T}_0(D)$. A similar argument applies to higher order brackets, such as $[X_i, [X_j, X_k]]$, etc. On the other hand, a vector field such as X_i should *not* be included in $\mathscr{T}_0(D)$ *by definition*, because we do not know whether the points $\Phi_i(t, x)$, $t \neq 0$, can be reached from x in 0 units of time (but, of course, it may happen that some X_i will belong to $\mathscr{T}_0(D)$ anyhow; for instance, we could have $X_1 = [X_2, X_3]$). However, *the vector fields* $X_i - X_j$ *will have to be included*, because $(X_i - X_j)(x)$ is the limiting direction of the curve $t \to \Phi_j(-t, \Phi_i(t, x))$. In other words, *the subspace generated by the differences* $X_i - X_j$ *will have to* be included in $\mathscr{T}_0(D)$. This subspace can also be defined as the set of all linear combinations $\lambda_1 X_1 + \lambda_2 X_2 + \lambda_3 X_3$ such that $\lambda_1 + \lambda_2 + \lambda_3 = 0$ (that all the differences $X_i - X_j$ are linear combinations of this type is trivial; conversely, if $Y = \lambda_1 X_1 + \lambda_2 X_2 + \lambda_3 X_3$ with $\lambda_1 + \lambda_2 + \lambda_3 = 0$, then $Y = \lambda_1 X_1 + \lambda_2 X_2 + (-\lambda_1 - \lambda_2) X_3$, i.e., $Y = \lambda_1(X_1 - X_3) + \lambda_2(X_2 - X_3)$).

We conclude that the reasonable candidates for membership in $\mathscr{T}_0(D)$ are: (i) all the brackets $[X_i, X_j]$, $[X_i, [X_j, X_k]]$, etc., and (ii) all the sums $\lambda_1 X_1 + \lambda_2 X_2 + \lambda_3 X_3$, where $\sum \lambda_i = 0$. Notice that the subspace generated by (i) is clearly the ***derived algebra*** of $\mathscr{T}(D)$ (by definition, the derived algebra of a Lie algebra L is the subalgebra L' of L generated by all the brackets $[X, Y]$, $X \in L$, $Y \in L$; it is easy to check that L' is in fact an *ideal* of L; cf. Helgason [7, p. 133]).

We now return to our formal development. Let $\mathscr{T}'(D)$ denote the derived algebra of $\mathscr{T}(D)$. Motivated by the previous remarks, we define $\mathscr{T}_0(D)$ to be the set of all sums $X + Y$, where X is a linear combination $\sum_{i=1}^{p} \lambda_i X_i$ with $X_1, \ldots, X_p \in D$ and $\sum \lambda_i = 0$, and where $Y \in \mathscr{T}'(D)$. It is obvious that $\mathscr{T}_0(D)$ is an ideal of $\mathscr{T}(D)$. One shows easily that $\mathscr{T}(D)$ is the set of all vector fields of the form $\sum_{i=1}^{p} \lambda_i X_i + Y$, where $X_1, \ldots, X_p$ belong to D, Y belongs to $\mathscr{T}'(D)$, and $\lambda_1, \ldots, \lambda_p$ are reals (but $\lambda_1 + \cdots + \lambda_p$ need not be zero). From this it follows immediately that $\mathscr{T}_0(D)$ is a subspace of $\mathscr{T}(D)$ of codimension zero or one. The codimension will be zero if and only if some $X \in D$ belongs to $\mathscr{T}_0(D)$ (in which case every $X \in D$ will belong to $\mathscr{T}_0(D)$). Similarly, for each $x \in M$, if $k = \dim \mathscr{T}(D)(x)$, then the dimension of $\mathscr{T}_0(D)(x)$ will either be k or $k - 1$.

We shall also be interested in associating to each $D \subset V(M)$, a set D^* of vector fields in the manifold $M \times \mathbf{R}$. Recall that the tangent space to $M \times \mathbf{R}$ at a point (x, r) $(x \in M, r \in \mathbf{R})$ is identified, in a natural way, to the direct sum $M_x \oplus \mathbf{R}_r$. If $x \in V(M)$, $Y \in V(\mathbf{R})$, we define the vector field $X \oplus Y \in V(M \times \mathbf{R})$ by

$$(X \oplus Y)(x, r) = (X(x), Y(r)).$$

The set D^* is defined to be the set of all vector fields $X \oplus \partial/\partial t$, where $X \in D$, and where $\partial/\partial t$ is the "canonical" vector field on $\mathbf{R}$ (i.e. $(\partial/\partial t)f = f'$). Using the identity $[X \oplus X', Y \oplus Y'] = [X, Y] \oplus [X', Y']$, one shows easily that $\mathscr{T}'(D^*)$ is the set of all vector fields of the form $X \oplus \mathbf{0}$, where $X \in \mathscr{T}'(D)$ and $\mathbf{0}$ is the zero vector field. Therefore, $\mathscr{T}(D^*)$ is the set of vector fields of the form

$$\sum_{i=1}^{p} \lambda_i \left(X_i \oplus \frac{\partial}{\partial t}\right) + Y \oplus \mathbf{0},$$

where $X_1, \ldots, X_p$ belong to D, $Y \in \mathscr{T}'(D)$, and $\lambda_1, \ldots, \lambda_p$ are scalars.

THEOREM 3.2. *Let M be an analytic n-dimensional manifold, and let D be a family of complete analytic vector fields on M. Let $x \in M$, and let $T > 0$. Then $L_x(D, T)$ has a nonempty interior in M if and only if $\dim \mathscr{T}_0(D)(x) = n$. Moreover, in this case, the interior of $L_x(D, T)$ is dense in $L_x(D, T)$.*

Proof. The main idea in this proof is to modify our problem so that we can "keep track" of the time elapsed while we move along an integral curve of D. We shall then apply Theorem 3.1 to the modified system. We shall work in the manifold $M \times \mathbf{R}$. As in the preceding paragraphs, we let the family D^* of vector fields on $M \times \mathbf{R}$ be defined by $D^* = \{X \oplus (\partial/\partial t) : X \in D\}$. It is clear that there is a one-to-one correspondence between integral curves α of D such that $\alpha(0) = x$, and integral curves β of D^* such that $\beta(0) = (x, 0)$. This correspondence is given by assigning to each curve α the curve $t \to (\alpha(t), t)$. It follows that $y \in L_x(D, T)$ if and only if $(y, T) \in \mathbf{L}_{(x,0)}(D^*, T)$. We show that $L_x(D, T)$ has a nonempty interior in M if and only if $\mathbf{L}_{(x,0)}(D^*)$ has a nonempty interior in $M \times \mathbf{R}$. Assume that $L_x(D, T)$ has a nonempty interior in M, and let V be a nonempty open set such that $V \subset L_x(D, T)$. Let $X \in D$, and let Φ be the one-parameter group of diffeomorphisms of M generated by X. Consider the mapping $F : V \times \mathbf{R} \to M \times \mathbf{R}$ defined by $F(v, t) = (\Phi(t, v), T + t)$. It is immediate that the differential of F has rank $n + 1$ everywhere. Therefore F maps open sets onto open sets. Since $F(V \times (0, \infty)) \subset \mathbf{L}_{(x,0)}(D^*)$, we conclude that $\mathbf{L}_{(x,0)}(D^*)$ has a nonempty interior in $M \times \mathbf{R}$.

To prove the converse, assume that $\mathbf{L}_{(x,0)}(D^*)$ has a nonempty interior in $M \times \mathbf{R}$. By Theorem 3.1, for each t with $0 < t < T$, $\mathbf{L}_{(x,0)}(D^*, t)$ has a nonempty interior in $M \times \mathbf{R}$. Let V be a nonempty open subset of M, and let W be a nonempty open subset of $\mathbf{R}$ such that $V \times W \subset \mathbf{L}_{(x,0)}(D^*, t)$. Let $s \in W$. Since $V \times \{s\} \subset \mathbf{L}_{(x,0)}(D^*, t)$, we conclude that $V \subset L_x(D, s)$. Let $X \in D$, and let Φ be the corresponding one-parameter group on M. Denote the mapping $y \to \Phi(T - s, y)$ by G. Then $G(V)$ is open. Since $G(V)$ is contained in $L_x(D, T)$, it follows that $L_x(D, T)$ has a nonempty interior.

We conclude from Theorem 3.1 that $L_x(D, T)$ has a nonempty interior if and only if $\dim \mathscr{T}(D^*)(x, 0) = n + 1$. To complete the proof of the first part of our statement, we must show that this last condition holds if and only if $\dim \mathscr{T}_0(D)(x) = n$. We recall, from the remarks preceding this proof, the fact that every $X^* \in \mathscr{T}(D^*)$ can be expressed as

$$X^* = \sum_{i=1}^{p} \lambda_i \left(X_i \oplus \frac{\partial}{\partial t}\right) + Y \oplus \mathbf{0} \qquad (\#)$$

where $X_1, \ldots, X_p$ belong to D and $Y \in \mathscr{T}'(D)$. Now assume that

$$\dim \mathscr{T}(D^*)(x, 0) = n + 1.$$

Let $v \in M_x$. Then $(v, 0)$ must belong to $\mathscr{T}(D^*)(x, 0)$, so that $(v, 0) = X^*(x, 0)$, where $X^* \in \mathscr{T}(D^*)$. Then formula (#) holds for suitable λ_i, X_i, Y. Therefore,

$$v = \left(\sum \lambda_i X_i + Y\right)(x),$$

and

$$\mathbf{0} = \sum \lambda_i \frac{\partial}{\partial t}(0).$$

The last equality implies that $\sum \lambda_i = 0$, so that the vector field $\sum \lambda_i X_i + Y$ belongs to $\mathscr{T}_0(D)$. Thus $v \in \mathscr{T}_0(D)(x)$. We have shown that $M_x \subset \mathscr{T}_0(D)(x)$. Therefore the dimension of $\mathscr{T}_0(D)(x)$ is n. Conversely, let $\dim \mathscr{T}_0(D)(x) = n$. Let $v \in M_x$. Then $v \in \mathscr{T}_0(D)(x)$, so that

$$v = \left(\sum \lambda_i X_i + Y\right)(x),$$

where the X_i belong to D, $Y \in \mathscr{T}'(D)$ and $\sum \lambda_i = 0$. Therefore,

$$\begin{aligned}(v, 0) &= \left(\left(\sum \lambda_i X_i + Y\right) \oplus \left(\sum \lambda_i\right) \frac{\partial}{\partial t}\right)(x, 0) \\ &= \left(\sum \lambda_i \left(X_i \oplus \frac{\partial}{\partial t}\right) + Y \oplus \mathbf{0}\right)(x, 0).\end{aligned}$$

This shows that $(v, 0)$ belongs to $\mathscr{T}(D^*)(x, 0)$. Pick an $X \in D$. Then $(X \oplus \partial/\partial t)(x, 0)$ belongs to $D^*(x, 0)$ by definition, and $(X \oplus \mathbf{0})(x, 0)$ belongs to $\mathscr{T}(D^*)(x, 0)$ by the previous remarks. Therefore $(\mathbf{0}, \partial/\partial t(0))$ belongs to $\mathscr{T}(D^*)(x, 0)$. We have thus shown that $\mathscr{T}(D^*)(x, 0)$ contains all the vectors $(v, 0)$, $v \in M_x$, and also the vector $(\mathbf{0}, \partial/\partial t(0))$. Therefore $\mathscr{T}(D^*)(x, 0) = (M \times \mathbf{R})_{(x,0)}$, so that $\dim \mathscr{T}(D^*)(x, 0) = n + 1$ as stated.

We now prove the second part of the theorem. As we remarked earlier, there is no loss of generality in assuming that D is finite. Let $y \in L_x(D, T)$. Using the notations of the proof of Theorem 3.1, let $y = \Phi_{\mathbf{i}}(\mathbf{t}, x)$, where $\mathbf{i} = (i_1, ..., i_m)$, and where $\mathbf{t} \in \mathbf{R}^m$ is such that $t_i > 0$ for $i = 1, ..., m$ and $\| \mathbf{t} \| = T$. Let $\{s_k\} \subset (0, t_m)$ be such that $\lim_{k \to \infty} s_k = 0$. Since our condition for $L_x(D, T)$ to have a nonempty interior is independent of T, we conclude that $L_x(D, t)$ has a nonempty interior for all $t > 0$. In particular, for each $k > 0$, there exists x_k which belongs to the interior of $L_x(D, s_k)$. Let $\mathbf{t}_k = (t_1, ..., t_{m-1}, t_m - s_k)$, and let $y_k = \Phi_{\mathbf{i}}(\mathbf{t}_k, x_k)$. For each $k > 0$, y_k belongs to $L_x(D, T)$; since $\Phi_{\mathbf{i}}$ is a diffeomorphism, y_k is in the interior of $L_x(D, T)$. Also, $x_k \to x$ as $k \to \infty$ because D is finite and $s_k \to 0$. Since $\Phi_{\mathbf{i}}$ is continuous in both variables, and since $\mathbf{t}_k \to \mathbf{t}$, we have that $y_k \to y$, and our theorem is proved.

The results is the previous theorems can be utilized to obtain information about the sets $\mathbf{L}_x(D, T)$ and $L_x(D, T)$, even when $\dim \mathscr{T}(D)(x) < n$.

THEOREM 3.3. *Let $D \subset V(M)$ be a family of complete vector fields. Then, for each $T > 0$, the set $\mathbf{L}_x(D, T)$ is contained in $I(D, x)$. Moreover, in the topology of $I(D, x)$, the interior of $\mathbf{L}_x(D, T)$ is dense on $\mathbf{L}_x(D, T)$. $L_x(D, T)$ has a nonempty interior in $I(D, x)$ if and only if* $\dim \mathscr{T}_0(D(x)) = \dim \mathscr{T}(D)(x)$ and, *in this case, the interior of $L_x(D, T)$ is dense in $L_x(D, T)$.*

Proof. If $X \in \mathscr{T}(D)$, then X is tangent to $I(D, x)$. Thus there is a well-defined restriction $X^\#$ of X to $I(D, x)$. We denote the set of all such restrictions of elements of D by $D^\#$. Since $[X, Y]^\# = [X^\#, Y^\#]$, it follows that $\mathscr{T}(D)^\# = \mathscr{T}(D^\#)$. Analogously, we have that $\mathscr{T}_0(D)^\# = \mathscr{T}_0(D^\#)$. If we now apply the previous theorems to the family $D^\#$ of vector fields in $I(D, x)$, we get all the conclusions of the theorem.

COROLLARY 3.4. *Let S be a maximal integral manifold of $\mathscr{T}(D)$. Then the dimension of $\mathscr{T}_0(D)(x)$ is the same for all $x \in S$.*

Proof. If $\dim \mathscr{T}(D)(x) = k$ then, for each $x \in S$, the dimension of $\mathscr{T}_0(D)(x)$ is either k or $k - 1$. We show that, if $\dim \mathscr{T}_0(D)(x) = k - 1$ for some $x \in S$, then $\dim \mathscr{T}_0(D)(y) = k - 1$ for *all* $y \in S$. Let Ω be a nonempty, open (relative to S) subset of $\mathbf{L}_x(D)$ (this is possible by Theorem 3.3). We first show that, if $y \in \Omega$, then $\dim \mathscr{T}_0(D)(y) = k - 1$. If this were not the case, then necessarily $\dim \mathscr{T}_0(D)(y) = k$. Then $L_y(D, t)$ would have a nonempty interior in S for all $t > 0$. This would imply that $L_x(D, t)$ has a nonempty interior in S. But by our assumption this is impossible. Thus, $\dim \mathscr{T}_0(D)(y) = k - 1$ for all $y \in \Omega$. Since S is connected, and Ω is open in S, we have that $\dim \mathscr{T}_0(D)(y) = k - 1$ for all $y \in S$: therefore, our statement is proved.

We now proceed to study the case when $\dim \mathscr{T}_0(D)(x) = \dim \mathscr{T}(D)(x) - 1$. We begin by proving some preliminary lemmas.

LEMMA 3.5. *Let $D \subset V(M)$ be a family of complete vector fields. If $X \in D$, let $\{\Phi_t\}$ be the one-parameter group generated by X. Then, for every $x \in M$, and every $t \in \mathbf{R}$ the differential $d\Phi_t$ maps $\mathscr{T}_0(D)(x)$ onto $\mathscr{T}_0(D)(\Phi_t(x))$.*

Proof. We first show that for every $y \in M$ there is an $r > 0$ such that, if $v \in \mathscr{T}_0(D)(y)$, then $d\Phi_t(v) \in \mathscr{T}_0(D)(\Phi_t(y))$ for all t with $|t| < r$. It is sufficient to show that for every $y \in M$ and every $v \in \mathscr{T}_0(D)(y)$ there exists an $r > 0$ such that $d\Phi_t(v) \in \mathscr{T}_0(D)(\Phi_t(y))$ for all t with $|t| < r$. Let $y \in M$, and let $v \in \mathscr{T}_0(D)(y)$. If $v = Y(y)$ for some $Y \in \mathscr{T}_0(D)$, then an easy computation shows that there exists a neighborhood of $t = 0$ such that

$$d\Phi_t(v) = \sum_{i=0}^{\infty} \frac{(-1)^i}{i!} [X^{(i)}, Y](\Phi_t(y))t^i$$

for all t in this neighborhood, where $[X^{(0)}, Y] = Y$, and $[X^{(n)}, Y] = [X, [X^{(n-1)}, Y]]$ for $n = 1, 2, \ldots$. Since each term of the above series belongs to $\mathscr{T}_0(D)(\Phi_t(y))$, we have that $d\Phi_t(v) \in \mathscr{T}_0(D)(\Phi_t(y))$ for t sufficiently small. Also, for such t we have that $d\Phi_t(\mathscr{T}_0(D)(y)) = \mathscr{T}_0(D)(\Phi_t(y))$; this is so because $d\Phi_t$ is one-to-one, and $\dim \mathscr{T}_0(D)(y) = \dim \mathscr{T}_0(D)(\Phi_t(y))$ (Corollary 3.4). It follows easily that the set of all t such that $d\Phi_t(\mathscr{T}_0(D)(x)) = \mathscr{T}_0(D)(\Phi_t(x))$ is both open and closed. If $v \in \mathscr{T}_0(D)(x)$, we can conclude that $\Phi_t(v) \in \mathscr{T}_0(D)(\Phi_t(v))$ for all t. This completes our proof.

As we remarked earlier, $\mathscr{T}_0(D)$ is a subalgebra of $\mathscr{T}(D)$. We will denote the maximal integral manifold of $\mathscr{T}_0(D)$ through x by $I_0(D, x)$. If $X \in D$ then, by the previous lemma, $\Phi_t(I_0(D, x))$ is a maximal integral manifold of $\mathscr{T}_0(D)$.

LEMMA 3.6. *Let $D \subset V(M)$ be a family of complete vector fields. Let X and Y be elements of D, and let $\{\Phi_t\}$ and $\{\Psi_t\}$ be their corresponding one-parameter groups. If S is a maximal integral manifold of $\mathscr{T}_0(D)$ then, for any $t \in \mathbf{R}$, $\Phi_t(S) = \Psi_t(S)$.*

Proof. Let X, Y, Φ_t, Ψ_t and S satisfy the conditions of the lemma. Let P be the maximal integral manifold of $\mathscr{T}(D)$ which contains S. If $\dim P = \dim S$, then $S = P$, and $\Phi_t(S) = S = \Psi_t(S)$. Assume that $\dim S = k = \dim(P) - 1$. We first show that there is an $r > 0$ such that

$\Phi_t(S) = \Psi_t(S)$ whenever $|t| < r$. Let $x \in S$. The mapping $(s, t) \to \Phi_t(s)$ has rank $k + 1$ at $(x, 0)$. Let Ω be a neighborhood of x in S, and let $\delta > 0$ be such that this mapping, restricted to $\Omega \times (-\delta, \delta)$, is a diffeomorphism onto an open subset Ω^* of P. If $y \in \Omega^*$, let $s(y)$ and $f(y)$ be such that $\Phi_{f(y)}(s(y)) = y$. Clearly, f is analytic in Ω^*, and $f(y) = 0$ if and only if $y \in \Omega$. Moreover, $Xf \equiv 1$ in Ω^*. For every t such that $|t| < \delta$, the set $\Phi_t(\Omega)$ is an integral manifold of $\mathscr{T}_0(D)$. The vector field $Y - X$ is tangent to $\Phi_t(\Omega)$ and, since f is constant on $\Phi_t(\Omega)$, it follows that $Yf \equiv Xf$ on $\Phi_t(\Omega)$. Since Ω^* is the union of the sets $\Phi_t(\Omega)$ over $-\delta < t < \delta$, we conclude that $Yf \equiv Xf \equiv 1$ on Ω^*. Let $r > 0$ be such that the curve $t \to \Phi_{-t}(\Psi_t(x))$, defined on $-r < t < r$, is contained in Ω^*. In addition, let $r < \delta$. Let $g(t) = f(\Phi_{-t}(\psi_t(x))$. Then g is analytic in $(-r, r)$, and moreover $g(t) = f(\Psi_t(x)) - t$. We have that $g' = (Yf)(\Psi_t(x)) - 1 \equiv 0$ and, since $g(0) = 0$ it follows that $g \equiv 0$ on $(-r, r)$. But this means that $\Phi_{-t}(\Psi_t(x)) \in \Omega$ for all $t \in (-r, r)$. Hence, if $|t| < r$, the manifold $\Phi_{-t}(\Psi_t(S))$ intersects S. Since $\Phi_{-t}(\Psi_t(S))$ is a maximal integral manifold of $\mathscr{T}_0(D)$, it follows that $\Phi_{-t}(\Psi_t(S)) = S$, and that $\Psi_t(S) = \Phi_t(S)$. Let A be the set of all t such that $\Phi_\tau(S) = \Psi_\tau(S)$ for all τ in a neighborhood of t. Then A is obviously open, and we have shown that $0 \in A$. It follows easily from the preceding argument that A is closed. Therefore $\Phi_t(S) = \Psi_t(S)$ for all real t, and our proof is complete.

According to the above lemma, if $D \subset V(M)$ and if $x \in M$, then the manifold $\Phi_t(I_0(D, x))$ depends only on t, and not on the particular choice of X. We shall denote this manifold by $I_0{}^t(D, x)$. It is clear that $I_0{}^t(D, x)$ could be defined as the maximal integral manifold of $\mathscr{T}_0(D)$ passing through y, where y is an arbitrary point of $L_x(D, t)$.

Finally, we prove a factorization property of maps that will be utilized several times.

Lemma 3.7. *Let E be a locally convex vector space, let $K \subset E$, and let C be a convex dense subset of K. Let $F : K \to I(D, x)$ be a continuous mapping such that $F(C)$ is contained in a maximal integral manifold S of $\mathscr{T}_0(D)$. Then $F(K)$ is contained in S, and F, as a mapping from K into S, is continuous.*

Proof. If $\dim S = \dim I(D, x)$, then $S = I(D, x)$, and the conclusion follows trivially. Therefore, we shall assume that $\dim S = \dim I(D, x) - 1$.

Let $k \in K$, let $X \in D$, and let $\{\Phi_t\}$ be the one-parameter group induced by X. Then, as in the proof of Lemma 3.6, we can find a neighborhood Ω of $F(k)$ in $I_0(D, F(k))$, and a positive number δ, such that the mapping $(s, t) \to \Phi_t(s)$ is a diffeomorphism of $\Omega \times (-\delta, \delta)$ onto an open subset Ω^* of $I(D, x)$. Let U be an open convex neighborhood of k such that $F(U \cap K) \subset \Omega^*$. For each $t \in (-\delta, \delta)$, the set $\Phi_t(\Omega)$ is an integral manifold of $\mathscr{T}_0(D)$; therefore, if

$\Phi_t(\Omega)$ intersects S, then $\Phi_t(\Omega)$ is contained and open in S. Let $A = \{t : |t| < \delta, \Phi_t(\Omega) \subset S\}$. It follows that $S \cap \Omega^\#$ is the union of the sets $\Phi_t(\Omega)$, $t \in A$. These sets are mutually disjoint and, since S is separable, it follows that A is at most countable. Let $y \to (s(y), f(y))$ be the inverse in $\Omega^\#$ of the map $(s, t) \to \Phi_t(s)$. Then the function g defined in $U \cap K$ by $g(m) = f(F(m))$ is continuous. Since $F(U \cap C) \subset S \cap \Omega^\#$, we conclude that $g(m) \in A$ for all $m \in U \cap C$. But A is at most countable, and $U \cap C$ is convex; therefore g is constant on $U \cap C$. Since $U \cap C$ is dense in $U \cap K$, we have that g is constant on $U \cap K$. Obviously $g(k) = 0$, and therefore $g(m) = 0$ for all $m \in U \cap K$; thus $F(m) \in \Omega$. This shows that Ω contains a point of S; hence $\Omega \subset S$, and $F(k) \in S$. This proves the first part of the lemma.

To prove the second part, let $\{k_n\} \subset K$ converge to k. Since F is continuous, $F(k_n) \to F(k)$. For large n, $s(F(k_n))$ is defined. Since s is continuous, $s(F(k_n))$ converges to $s(F(k))$ in S. But $g(k_n) = 0$, and therefore $s(F(k_n)) = F(k_n)$. Similarly, $s(F(k)) = F(k)$. We have thus shown that $F(k_n)$ converges to $F(k)$ in S, and our proof is complete.

Remark 3.8. If is clear that the preceding lemma is valid under weaker assumptions about C and K. For instance, it is sufficient to assume that, for every $k \in K$ and for every neighborhood U of k, there exists a neighborhood V of k such that $V \subset U$ and $V \cap C$ is connected.

We now state and prove the theorem towards which we have been aiming.

THEOREM 3.9. *Let $D \subset V(M)$ be a set of complete vector fields, and let $x \in M$. Then, for each $T > 0$, $L_x(D, T) \subset I_0{}^t(D, x)$ and, moreover, the interior of $L_x(D, T)$ (relative to $I_0{}^t(D, x)$) is dense in $L_x(D, T)$ (and is, in particular, nonempty).*

Proof. If $\dim \mathscr{T}_0(D)(x) = \dim \mathscr{T}(D)(x)$, then we have from Corollary 3.4 that $\mathscr{T}_0(D)(y) = \mathscr{T}(D)(y)$ for all $y \in I(D, x)$. Therefore, $I_0(D, x) = I(D, x) = I_0{}^t(D, x)$ and our conclusion follows from Theorem 3.3. Assume that $\dim \mathscr{T}_0(D)(x) = k = \dim \mathscr{T}(D)(x) - 1$. It is clear from Lemma 3.6 that, if α is an integral curve of D such that $\alpha(0) = x$, then $\alpha(T) \in I_0{}^T(D, x)$; hence, $L_x(D, T) \subset I_0{}^T(D, x)$.

We now show that, if $y \in L_x(D, T)$, then y is the limit of points which belong to the interior of $L_x(D, T)$. Let $D = \{X_1, ..., X_k\}$ and let $y = \Phi_{\mathrm{i}}(\mathbf{T}, x)$, where $\|\mathbf{T}\| = T$, and $T_i > 0$ for $i = 1, 2, ..., m$ (the notations here are the same as in the proof of Theorem 3.1). Let $\mathrm{j} = (j_1, ..., j_s)$ be an s-tuple of integers between 1 and k such that the rank of $\mathbf{t} \to \Phi_{\mathrm{j}}(\mathbf{t}, x)$ is equal to $\dim \mathscr{T}(D)(x)$ for all t in an open dense subset Ω of $\mathbf{R}^s$. Let $\Omega' = \{\mathbf{t} : \mathbf{t} \in \mathbf{R}^s, t_i > 0 \text{ for } i = 1, ..., s\} \cap \Omega$. Let $\{\mathbf{t}_p\} \subset \Omega$ be a sequence that converges to 0, and let $\mathbf{T}_p = (T_1, ..., T_{m-1}, T_m - \|\mathbf{t}_p\|)$. We can assume that $\|\mathbf{t}_p\| < T_m$ for all $p > 0$. If we let $y_p = \Phi_{\mathrm{i}}(\mathbf{T}_p, \Phi_{\mathrm{j}}(\mathbf{t}_p, x))$, then $y_p \in L_x(D, T)$. We next show that y_p is in the interior of $L_x(D, T)$ relative to $I_0^T(D, x)$. Since the mapping $z \to \Phi_i(T_p, z)$ is a diffeomorphism from $I_0^{\|t_p\|}(D, x)$ onto $I_0^T(D, x)$, it suffices to show that $\Phi_j(t_p, x)$ is in the interior of $L_x(D, \| t_p \|)$. Let $V_p = \{t : t \in \mathbf{R}^s, t_1 > 0, \ldots, t_s > 0, \| t \| = \| t_p \|\}$. Clearly, if $t \in V_p$, then $\Phi_j(t, x) \in L_x(D, \| t_p \|)$. Let $F_p : V_p \to I_0^{\|t_p\|}(D, x)$ be defined by $F_p(t) = \Phi_j(t, x)$. We show that F_p is analytic. Since F_p is analytic as a map from V_p into $I(D, x)$, it suffices to show that it is continuous. But this follows from the previous lemma, because V_p is convex. The rank of $t \to \Phi_j(t, x)$ is equal to $\dim \mathscr{T}(D)(x)$ at $t = t_p$. Since V_p is a submanifold of $\mathbf{R}^s$ of codimension 1, it follows that the rank of F_p at t_p is equal to the dimension of $I_0^{\|t_p\|}(D, x)$. Thus, $F_p(V_p)$ contains a neighborhood of $F_p(t_p)$ in $I_0^{\|t_p\|}(D, x)$. It follows that $\Phi_j(t_p, x)$ is in the interior of $L_x(D, \| t_p \|)$. By the previous remark we conclude that y_p is interior to $L_x(D, T)$ in $I_0^T(D, x)$. There remains to be shown that y_p converges to y in $I_0^T(D, x)$. The mapping $(t, s) \to \Phi_i(t, \Phi_j(s, x))$ is continuous as a map from $\mathbf{R}^m \times \mathbf{R}^s$ into $I(D, x)$. The set $V = \{(t, s) : t_i > 0,\ s_j > 0,\ i = 1, \ldots, m,\ j = 1, \ldots, s,\ \| t \| + \| s \| = T\}$ is convex, and is mapped into $I_0^T(D, x)$. Therefore, the previous lemma is applicable, and we conclude that $y_p \to y$ in $I_0^T(D, x)$. This proves our theorem.

4. Applications to Control Systems

We shall consider systems of the form

$$dx(t)/dt = F(x(t), u(t))$$

defined on an analytic manifold M. The functions u belong to a class $\mathscr{U}$ of "admissible controls". We make the following assumptions about $\mathscr{U}$ and the system function F:

(i) *The elements of $\mathscr{U}$ are piecewise continuous functions defined in* $[0, \infty)$, *having values in a locally path connected set Ω, $\Omega \subset \mathbf{R}^m$* ($\Omega$ is locally path connected if, for every $w \in \Omega$ and every neighborhood V of ω, there exists a neighborhood U of ω such that $U \subset V$, and $U \cap \Omega$ is path connected). In addition, we assume that *$\mathscr{U}$ contains all the piecewise constant functions with values in Ω*, and that *every element of $\mathscr{U}$ has finite one-sided limits at each point of discontinuity.* $\mathscr{U}$ is endowed with the topology of uniform convergence on compact intervals.

(ii) *$F : M \times \Omega \to TM$ is jointly continuously differentiable. For each $u \in \Omega$, $F(\cdot, u)$ is a complete analytic vector field on M.* We know that for each $x \in M$, $u \in \mathscr{U}$, the differential equation

$$dx(t)/dt = F(x(t), u(t)), \qquad x(0) = x, \tag{1}$$

has a solution defined for all $t \in [0, \delta)$, where $\delta > 0$. We denote such a solution by $\Pi(x, u, \cdot)$, and *we assume that* $\Pi(x, u, t)$ *is defined for all* $t \in [0, \infty)$.

For the above-defined control system we now state the basic controllability concepts. We say that $y \in M$ is *attainable from* $x \in M$ at time t ($t \geqslant 0$), if there exists $u \in \mathscr{U}$ such that $\Pi(x, u, t) = y$. For each $x \in M$, we let $A(x, t)$ denote the set of all points attainable from x at time t. If $0 \leqslant t < \infty$, we define $\mathbf{A}(x, t) = \bigcup_{s \leqslant t} A(x, s)$ and $\mathbf{A}(x) = \bigcup_{t \geqslant 0} \mathbf{A}(x, t)$. We say that the system is *controllable from* x if $\mathbf{A}(x) = M$, and that it is *controllable* if it is controllable from every $x \in M$. We say that the system has *the accessibility property from* x if $\mathbf{A}(x)$ has a nonempty interior, and that it has *the accessibility property* if it has the accessibility property from every $x \in M$. Finally, we shall say that the system has *the strong accessibility property from* x if $A(x, t)$ has a nonempty interior for some $t > 0$, and that it has *the strong accessibility property* if it has the strong accessibility property from x for every $x \in M$.

For $\omega \in \Omega$, let $X_\omega = F(\cdot, \omega)$; from assumption (ii) it follows that X_ω is a complete analytic vector field on M. Throughout the remaining part of this article we let $D = \{X_\omega : \omega \in \Omega\}$.

LEMMA 4.1. *For each* $x \in M$, $\mathbf{A}(x)$ *is contained in* $I(D, x)$.

The proof is identical to that of Lemma 2.4, and will therefore be omitted.

Remark 4.2. It is easy to see that the control system defined by restricting F to $I(D, x)$ satisfies the same assumptions as the original system. Since it can be readily verified that the map $u \to \Pi(x, u, t)$ is continuous as a map from $\mathscr{U}$ into M, it follows that *this map is also continuous as a map from* $\mathscr{U}$ *into* $I(D, x)$.

We now want to obtain a result for $A(x, t)$ which is similar to that of Lemma 4.1. It is here that the assumption about Ω will be utilized. Let $\mathscr{P}$ be the class of piecewise constant Ω-valued functions defined on $[0, \infty)$. Clearly, $\mathscr{P}$ is dense in $\mathscr{U}$. Moreover, the local connectedness of Ω implies that the condition of Remark 3.8 is satisfied (this can be easily verified, and we omit the proof). Thus, we can apply Lemma 3.7, with $C = \mathscr{P}$ and $K = \mathscr{U}$, to obtain the following result:

LEMMA 4.3. *Let* $x \in M$. *For each* $t \geqslant 0$, $A(x, t) \subset I_0{}^t(D, x)$.

Proof. Since $\mathscr{U}$ contains $\mathscr{P}$, we have that $L_x(D, t) \subset A(x, t)$. Let $G : \mathscr{U} \to I(D, x)$ be defined by $G(u) = \Pi(x, u, t)$. We have that $G(\mathscr{P}) = L_x(D, t)$, and by Theorem 3.9, $G(\mathscr{P}) \subset I_0{}^t(D, x)$. Now our conclusion follows immediately from Lemma 3.7, and the proof is complete.

The above lemmas combined with the theorems of the preceding section yields the following results;

Theorem 4.4. *Let $x \in M$. Then $\mathbf{A}(x) \subset I(D, x)$. Moreover, for every $T > 0$, the interior of $\mathbf{A}(x, T)$ relative to $I(D, x)$ is dense in $\mathbf{A}(x, T)$ (and, in particular, is nonempty).*

Proof. The first part is just the statement of Lemma 4.1. To prove the second part, we can assume that $I(D, x) = M$ (if not, replace the original system by its restriction to $I(D, x)$, cf. Remark 4.2). Since $\mathbf{L}_x(D, T)$ is dense in $\mathbf{A}(x, T)$, our conclusion follows immediately from Theorem 3.1.

Theorem 4.5. *Let $x \in M$. Then, for each $t > 0$, $A(x, t) \subset I_0^t(D, x)$ and, moreover, the interior of $A(x, t)$ relative to $I_0^t(D, x)$ is dense in $A(x, t)$ (and, in particular, is nonempty).*

Proof. The first part is just the statement of Lemma 4.3. To prove the second part, we apply Lemma 3.7 to the function G of Lemma 4.3, and we get that G is continuous as a map into $I_0^t(D, x)$; therefore, $L_x(D, t)$ is dense in $A(x, t)$ relative to $I_0^t(D, x)$. Our conclusion now follows immediately from Theorem 3.9, and the proof is complete.

The following two controllability criteria follow immediately from Theorems 4.4 and 4.5, and from Lemma 2.1:

Corollary 4.6. *The system has the accessibility property from x if and only if $\dim \mathscr{T}(D)(x) = \dim M$. In this case $\mathbf{A}(x, T)$ has a nonempty interior for every $T > 0$.*

Corollary 4.7. *The system has the strong accessibility property from x if and only if $\dim \mathscr{T}_0(D)(x) = \dim M$. In this case $A(x, T)$ has a nonempty interior for every $T > 0$.*

The preceding results can be utilized to derive relationships between accessibility and strong accessibility. Even though the latter property seems much stronger than the former, we show that, for a very large class of manifolds (including the spheres S^n for $n > 1$, and all compact semisimple Lie groups, but not $\mathbf{R}^n$), it is in fact implied by it. On the other hand, for a still larger class of manifolds (including $\mathbf{R}^n$) controllability (which trivially implies accessibility), is sufficient to guarantee strong accessibility (the fact that controllability implies that $\dim \mathscr{T}(D^*)(x) = n + 1$ for all x was proved by Elliott in [5]).

Consider a system on a connected n-dimensional analytic manifold M, having the accessibility property but not having the strong accessibility property. Let D be the family of associated vector fields. By Corollary 4.6, $\dim \mathscr{T}(D)(x) = n$ for all $x \in M$. By Corollary 3.4 the number $\dim \mathscr{T}_0(D)(x)$ is independent of x. Since this number is either n or $n - 1$, Corollary 4.7

implies that $\dim \mathcal{T}_0(D)(x) = n - 1$ for all $x \in M$. Choose a fixed $X \in D$, and use Φ_t to denote the one-parameter group generated by X (i.e., for every $y \in M$, the integral curve of X that passes through y at $t = 0$ is the curve $t \to \Phi_t(y)$). Define a mapping F from the manifold $S \times \mathbf{R}$ into M by

$$F(s, t) = \Phi_t(s).$$

One shows easily that F is a local diffeomorphism *onto* M. Moreover, $S \times \mathbf{R}$ is connected. In fact, we have (see [18] for the definition of a covering projection):

LEMMA 4.8. *The map F is a covering projection.*

Before we prove Lemma 4.8, we show how the results mentioned above follow from it.

THEOREM 4.9. *Let M be a manifold whose universal covering space is compact. Then every system having the accessibility property has the strong accessibility property.*

Proof. If the universal covering space of M is compact, then every covering space of M is compact. If it were possible to have a system on M having the accessibility property but not the strong accessibility property, we could define, for such a system, S and F as above. It would follow that $S \times \mathbf{R}$ is compact, which is clearly a contradiction.

Remark. If $n > 1$, the sphere S^n is simply connected (and compact). Therefore Theorem 4.9 applies. Also, if M is a connected compact semisimple Lie group (for instance $SO(n)$, if $n > 2$), the universal covering group of M is also compact [7, p. 123] and, therefore, Theorem 4.9 applies in this case as well.

THEOREM 4.10. *Let M be a manifold whose fundamental group has no elements of infinite order. Then every controllable system on M has the strong accessibility property.*

Proof. A controllable system obviously has the accessibility property. Assume it does not have the strong accessibility property. Define S and F as before. We show that F is one-to-one. Otherwise, there would exist s_0, $s_0' \in S$ and a $T \neq 0$ such that $F(s_0', T) = \Phi_T(s_0') = F(s_0, 0) = s_0$. Therefore $\Phi_T(S) = S$. Define $H : S \times \mathbf{R} \to S \times \mathbf{R}$ by $H(s, t) = (\Phi_T(s), t - T)$. Then H is well defined, because $\Phi_T(S) = S$, and is a homeomorphism. Moreover, if $(s, t) \in S \times \mathbf{R}$,

$$F(H(s, t)) = \Phi_{t-T}(\Phi_T(s)) = \Phi_t(s) = F(s, t).$$

Therefore H is a covering transformation [18, Chap. 2]. Moreover, if s is a point of S and if t belongs to $\mathbf{R}$, then $H^m(s, t) = (\Phi_{mT}(s), t - mT)$, so that $H^m(s, t) \neq (s, t)$ for every m. Let γ be a curve in $S \times \mathbf{R}$ which joins (s, t) to $H(s, t)$. Let π denote the fundamental group of M (with base point $F(s, t)$). Then the composite δ of γ and F represents an element of π, and it is easy to see that δ has infinite order. This is a contradiction. Therefore F must be one-to-one. On the other hand, the points that are attainable from S_0 must belong to S_t $(= \Phi_t(S))$ for some nonnegative t (cf. Theorem 4.5). Therefore, the points in S_{-t} are not attainable, if $t > 0$. Thus, the system is not controllable, and we have reached a contradiction.

Remark. Theorem 4.10 applies, in particular, to any simply connected manifold, such as $\mathbf{R}^n$.

Proof of Lemma 4.8. We must show that every point of M has a neighborhood that is evenly covered by F. Let $m \in M$. Since F is a local diffeomorphism onto, there exist $s \in S$, $t \in \mathbf{R}$, $\epsilon > 0$ and a connected neighborhood U of s in S such that $F(s, t) = m$ and that the restriction of F to $U \times (t - \epsilon, t + \epsilon)$ is a diffeomorphism onto an open subset V of M. We claim that V is evenly covered. Let $A = \{\tau : \Phi_\tau(S) = S\}$. For each $\tau \in A$, let $U_\tau = \Phi_\tau(U)$. Since $\Phi_\tau : S \to S$ is a diffeomorphism, it follows that U_τ is open in S and connected for each $\tau \in A$. We first show that, if $0 < |\tau - \eta| < 2\epsilon$, $\tau \in A$, $\eta \in A$, then U_τ and U_η are disjoint. Assume they are not. Then $\Phi_T(U_\tau)$ and $\Phi_T(U_\eta)$ are not disjoint, for any T. Choose T such that both $T + \tau$ and $T + \eta$ belong to $(t - \epsilon, t + \epsilon)$. Let $u = \Phi_{T+\tau}(u_1) = \Phi_{T+\eta}(u_2)$ be a common element, where u_1 and u_2 belong to U. Then the points $(u_1, T + \tau)$ and $(u_2, T + \eta)$ belong to $U \times (t - \epsilon, t + \epsilon)$. Since the restriction of F to this set is one-to-one, it follows that $\tau = \eta$, which is a contradiction. For each $\tau \in A$, let $W_\tau = U_\tau \times (t - \tau - \epsilon, t - \tau + \epsilon)$. We shall conclude our proof that V is evenly covered by showing:

(a) the sets W_τ are open, connected and pairwise disjoint,

(b) for each $\tau \in A$, F maps W_τ diffeomorphically onto V, and

(c) the inverse image of V under F is the union of the sets W_τ.

The first two assertions of (a) are obvious. If τ and η belong to A, and $\tau \neq \eta$, then either $|\tau - \eta| < 2\epsilon$ or $|\tau - \eta| \geqslant 2\epsilon$. In the first case W_τ and W_η must be disjoint, because U_τ and U_η are disjoint. In the second case, W_τ and W_η are also disjoint, because the intervals $(t - \tau - \epsilon, t - \tau + \epsilon)$ and $(t - \eta - \epsilon, t - \eta + \epsilon)$ cannot have a point in common.

To prove (b), take $\tau \in A$. Define $G : U \times (t - \epsilon, t + \epsilon) \to W_\tau$ by $G(u, \sigma) = (\Phi_\tau(u), \sigma - \tau)$. Clearly, G is a diffeomorphism from $U \times (t - \epsilon, t + \epsilon)$ onto W_τ. Moreover, if $u \in U$, $t - \epsilon < \sigma < t + \epsilon$, then $F(G(u, \sigma)) = \Phi_{\sigma-\tau}(\Phi_\tau(u)) =$

$\Phi_\sigma(u) = F(u, \sigma)$. Since the restriction of F to $U \times (t - \epsilon, t + \epsilon)$ is a diffeomorphism onto V, the same must be true for the restriction of F to W_τ.

Finally, we prove (c). Let $u \in S$, $\sigma \in \mathbf{R}$ be such that $F(u, \sigma) \in V$. Then there exist $u' \in U$, $\sigma' \in (t - \epsilon, t + \epsilon)$ such that $F(u', \sigma') = F(u, \sigma)$. Therefore $u = \Phi_{\sigma' - \sigma}(u')$. This implies, in particular, that $\tau = \sigma' - \sigma$ belongs to A, and that $u \in U_\tau$. Moreover, since $t - \epsilon < \sigma' < t + \epsilon$, it follows that $t - \tau - \epsilon < \sigma < t - \tau + \epsilon$. Therefore $(u, \sigma) \in W_\tau$.

The proof of Lemma 4.8 is now complete.

5. Examples

Example 5.1. Let $M = \mathbf{R}^n$, $\Omega = \mathbf{R}^m$, and let $F : M \times \Omega \to TM$ be defined by $F(x, u) = Ax + Bu$, where A and B are, respectively, $n \times n$ and $n \times m$ real matrices. We have that $D = \{A(\cdot) + Bu : u \in \mathbf{R}^m\}$. Let b_i denote the i-th column of B. Then, as shown by Lobry [16], $\mathscr{T}(D)(x)$ contains the vectors

$$Ax \pm b_i, \pm Ab_i, ..., \pm A^{n-1}b_i, \qquad i = 1, ..., m.$$

It is not difficult to see that the above set of vectors forms a system of generators for $\mathscr{T}(D)(x)$. From Corollary 4.6 we get that $\mathbf{A}(0, t)$ has a nonempty interior in $\mathbf{R}^n$ if and only if $\{\pm b_i, \pm Ab_i, ..., \pm A^{n-1}b_i, i = 1, 2, ..., m\}$ has rank n; equivalently, $\mathbf{A}(0, t)$ *has a nonempty interior in* $\mathbf{R}^n$ *if and only if* $\operatorname{rank}[B, AB, ..., A^{n-1}B] = n$.

Since, obviously, $\mathscr{T}_0(D)(0) = \mathscr{T}(D)(0)$, we conclude that $A(0, t)$ has a nonempty interior whenever $\mathbf{A}(0, t)$ does. The above statements, along with the fact that $A(0, t)$ and $\mathbf{A}(0, t)$ are linear subspaces of $\mathbf{R}^n$, imply that, if $\operatorname{rank}[B, AB, ..., A^{n-1}B] = n$, then for each $t > 0$ $A(0, t) = \mathbf{A}(0, t) = \mathbf{A}(0) = \mathbf{R}^n$ (Kalman [12]). Thus, in this example, the accessibility property is equivalent to controllability. This is, of course, not true in general.

Example 5.2. Let $M = \mathbf{R}^n$, $\Omega = \{u \in \mathbf{R}^m : 0 \leqslant u_i \leqslant 1, i = 1, ..., m\}$, and let $F(x, u) = (A_0 + \sum_{i=1}^m A_i u_i)x$ for all $(x, u) \in \mathbf{R}^n \times \Omega$, where $A_0, ..., A_m$ are $n \times n$ real matrices. Then D is the set of all vector fields X_u where $X_u(x) = (A_0 + \sum_{i=1}^m u_i A_i)x$. The set $\mathbf{M}^n$ of all $n \times n$ real matrices is a Lie algebra, with the bracket defined by $[P, Q] = PQ - QP$. To each matrix P there corresponds a vector field $V(P)$ defined by $V(P)(x) = Px$. It is easy to check that $V([Q, P]) = [V(P), V(Q)]$. Using this fact, the spaces $\mathscr{T}(D)(x)$ and $\mathscr{T}_0(D)(x)$ can be readily computed:

$$\mathscr{T}(D)(x) = \{Px : P \in \mathbf{L}\},$$

and

$$\mathscr{T}_0(D)(x) = \{Px : P \in L\},$$

where $\mathbf{L}$ is the Lie algebra spanned by $A_0, \ldots, A_m$, and L is the ideal of $\mathbf{L}$ spanned by $A_1, \ldots, A_m$. We remark that for this example the theory of Section 4 is valid *even if $\mathscr{U}$ is the set of all bounded and measurable Ω-valued functions*. This is so because the only properties of the class of admissible controls that were utilized in Section 4 were: (a) that the class of piecewise constant controls is dense in $\mathscr{U}$ (in the topology of uniform convergence), and (b) that, if $\{u_\alpha\}$ are elements of $\mathscr{U}$ that converge uniformly to u, then $\Pi(u_\alpha, x, t)$ converges to $\Pi(u, x, t)$.

In our example, both (a) and (b) remain valid if the topology of uniform convergence is replaced by that of *weak* convergence. This is easy to verify, and we shall not do it here (see Kučera [14]). Moreover, the set of Ω-valued measurable functions defined in $[0, T]$ is weakly compact. It follows that the sets $\mathbf{A}(x, T)$, $A(x, T)$ are compact for each $T > 0$. Denote their interiors (relative to $I(D, x)$ and $I_0^T(D, x)$, respectively) by int $\mathbf{A}(x, T)$, int $A(x, T)$. It follows that $\mathbf{A}(x, T)$ is the closure of int $\mathbf{A}(x, T)$, and that $A(x, T)$ is the closure of int $A(x, T)$. Therefore, our results contain those of Kučera (in this connection, see also Sussmann [21]).

Remark. The result of the preceding example is a particular case of a more general situation. Let G be a Lie group, and let M be an analytic manifold on which G acts analytically to the left. Then there is a homomorphism λ from the Lie algebra of G into $V(M)$, defined by

$$\lambda(X)(m) = (d/dt)[\exp(tX) \cdot m],$$

the derivative being evaluated at $t = 0$. If $X_0, \ldots, X_k$ belong to the Lie algebra of G, we can consider the control problem

$$\frac{dx}{dt} = X_0'(x) + \sum_{i=1}^{k} u_i X_i'(x),$$

where $X_i' = \lambda(X_i)$. Example 5.2 results by letting $G = GL(n, \mathbf{R})$ and $M = \mathbf{R}^n$.

EXAMPLE 5.3. This example shows that the analyticity assumptions are essential. Consider the following two systems defined in the (x, y) plane:

$$(S_1) \quad \dot{x} = f_1(x, y, u), \qquad \dot{y} = g_1(x, y, u),$$

and

$$(S_2) \quad \dot{x} = f_2(x, y, u), \qquad \dot{y} = g_2(x, y, u).$$

Let $f_1 \equiv f_2 \equiv 1$, $g_1 \equiv 0$, and $g_2(x, y, u) = \varphi(x)$ where φ is a C^∞ function which vanishes for $-\infty < x < 1$, and which is equal to 1 for $x > 2$. It is clear that for (S_1) the set $\mathbf{A}((0, 0))$ is the half line $\{(x, y) : y = 0, x \geqslant 0\}$ while, for (S_2), $\mathbf{A}((0, 0))$ has a nonempty interior. However, both systems are identical in a neighborhood of (0, 0).

Acknowledgments

We are grateful to Professor R. W. Brockett for his encouragement and advice. Also, we wish to thank Dr. E. H. Cattani, Dr. L. Marino and an anonymous referee for helpful suggestions.

References

1. R. L. Bishop and R. J. Crittenden, "Geometry of Manifolds," Academic Press, New York, 1964.
2. R. W. Brockett, System theory on group manifolds and coset spaces, *SIAM J. Control*, to appear.
3. C. Chevalley, "Theory of Lie Groups," Princeton University Press, Princeton, NJ, 1946.
4. W. L. Chow, Über Systeme von linearen partiellen Differential-gleichungen erster Ordnung, *Math. Ann.* **117** (1939), 98–105.
5. D. L. Elliott, A consequence of controllability, *J. Diff. Eqs.* **10** (1971), 364–370.
6. G. W. Haynes and H. Hermes, Nonlinear controllability via Lie theory, *SIAM J. Control* **8** (1970), 450–460.
7. S. Helgason, "Differential Geometry and Symmetric Spaces," Academic Press, New York, 1962.
8. R. Hermann, E. Cartan's geometric theory of partial differential equations, *Advances in Math.* **1** (1965), 265–315.
9. R. Hermann, On the accessibility problem in control theory, *in* "International Symposium on Nonlinear Differential Equations and Nonlinear Mechanics," pp. 325–332, Academic Press, New York, 1963.
10. R. Hermann, The differential geometry of foliations II, *J. Math. Mech.* **11** (1962), 305–315.
11. V. Jurdjevic, Abstract control systems: controllability and observability, *SIAM J. Control* **8** (1970), 424–439.
12. R. E. Kalman, Y. C. Ho, and K. S. Narendra, Controllability of linear dynamical systems, *Contrib. Diff. Eqs.* **1** (1963), 189–213.
13. S. Kobayashi and K. Nomizu, "Foundations of Differential Geometry," Vol. I, Interscience, New York, 1963.
14. J. Kučera, Solution in large of control problem: $\dot{x} = (A(1 - u) + Bu)x$, *Czech. Math. J.* **16** (91) (1966), 600–623.
15. J. Kučera, Solution in large of control problem: $\dot{x} = (Au + Bv)x$, *Czech. Math. J.* **17** (92) (1967), 91–96.
16. C. Lobry, Contrôlabilité des systèmes non linéaires, *SIAM J. Control* **8** (1970), 573–605.

17. C. Lobry, Une propriété de l'ensemble des états accessibles d'un système guidable, *C. R. Acad. Sci. Paris* **272** Series A (1971), 153–156.
18. E. H. Spanier, "Algebraic Topology," McGraw-Hill, New York, 1966.
19. S. Sternberg, "Lectures on Differential Geometry," Prentice Hall, Englewood Cliffs, NJ, 1964.
20. H. J. Sussmann, The bang-bang problem for certain control systems in $GL(n, \mathbf{R})$, *SIAM J. Control*, to appear.
21. H. J. Sussmann, The control problem $\dot{x} = (A(1-u) + Bu)x$: a comment on an article by J. Kučera, *Czech. Math. J.*, to appear.

Dissipative Dynamical Systems—Part I: General Theory

JAN C. WILLEMS

EARLIER papers and discussions in this volume explained how the works of Popov [7], Yakubovich [13], and Kalman [4] led to the formulation of the Positive Real Lemma, through which passivity, a network theory concept, became useful for feedback control. In his input/output analysis, Zames [14] used passivity as a defining property of positive operators on extended inner product spaces and pursued a nonlinear analogy of Bode's gain-phase characteristics. In this analogy, passivity is viewed as a "phase" property behind both Popov and Circle Criteria. In Kalman's analysis of the inverse optimal Linear-Quadratic problem [5], a passivity property was shown to be necessary for optimality.

In the mid-1960s, Jan Willems was among a select group of young researchers seeking to develop a general theory to unify newly discovered feedback properties. His theory of dissipative dynamical systems [12] was a major step in this direction. Only the general theory part of this two-part paper is reprinted here. The second part of the paper demonstrates the ability of this theory to unify previous stability and optimality results for linear systems. While most of the readers immediately saw the significance of the second part of the paper, the impact of the first part has been growing over a longer period.

Jan Willems defines dissipativity as an input/output property but, in contrast to Zames, does not avoid the state-space and Lyapunov functions. Instead, he generalizes Lyapunov functions with his new concept of a storage function $S(x)$. For an input/output pair u, y he introduces the notion of supply rate $w(u, y)$ and defines a system to be dissipative if the increment of storage, $S(x_1) - S(x_0)$, is not larger than the integral of the supply rate along any state-space trajectory $x(t)$ from $x(t_0) = x_0$ to $x(t_1) = x_1$. A significant result of the paper is that input/output knowledge does not determine the storage function uniquely. Rather, there are two variational problems that can be posed, whose solutions determine two storage functions, the so-called *available storage* and *required supply*, and these two storage functions underbound and overbound any storage function that can be associated with the system. In the linear case, storage functions are associated with solutions of linear matrix inequalities. The extremal available storage and required supply functions correspond to extremal solutions of these inequalities which are frequently also stabilizing or antistabilizing solutions of Riccati equations.

Reviewing the results of Popov [8] and Zames [14] for feedback systems consisting of two nonlinear blocks, Willems points out that the small gain stability conditions are established with supply rate $w = u^T u - y^T y$, while $w = u^T y$ is the supply rate used to define passivity.

The work of Willems stimulated Hill and Moylan [3] to derive a nonlinear analog of the PR Lemma and to quantify the "excess" or "shortage" of passivity using the supply rate $w = u^T y - \nu u^T u - \rho y^T y$ where ν and ρ are constants that can be negative or positive. In feedback interconnections, ν and ρ for individual blocks can be changed by feedforward and feedback loop transformations. These results were instrumental for feedback passivation developed a decade later by Kokotović and Sussmann [6], Byrnes et al. [2], and for nonlinear feedback design by Sepulchre et al. [9].

The concept of dissipativity was also crucial in recent extensions of linear robust control theory to nonlinear systems, as summarized by van der Schaft [11]. It also arose in the dynamic game approach to the disturbance attenuation problem by Başar and Bernhard [1], where the optimal (min-max) feedback control $u(x)$ guarantees that the designed system is dissipative with respect to the disturbance input. For $z^T := [y^T, \; u^T]$, the supply rate is of the form $w = \gamma^2 |d|^2 - |z|^2$, and the L_2-gain of the closed-loop system is $\frac{\|z\|_2}{\|d\|_3} \le \gamma$.

As shown by Sontag and Wang [10], input-to-state stability, another currently popular concept, is directly related to dissipativity.

REFERENCES

[1] T. BAŞAR AND P. BERNHARD, *$\mathcal{H}_\infty$ Optimal Control and Related Minimax Design Problems*, Birkhäuser (Boston), second edition, 1995.

[2] C.I. BYRNES, A. ISIDORI, AND J.C. WILLEMS, "Passivity, feedback equivalence, and global stabilization of minimum phase systems," *IEEE Trans. Aut. Contr.*, **AC-36**:1228–1240, 1991.

[3] D. HILL AND P. MOYLAN, "The stability of nonlinear dissipative systems," *IEEE Trans. Aut. Contr.*, **AC-21**(5):708–711, 1976.

[4] R.E. KALMAN, "Lyapunov functions for the problem of Lur'e in automatic control," *Proceedings of the National Academy of Sciences of the United States of America*, **49**:201–205, 1963.

[5] R.E. KALMAN, "When is a linear control system optimal?" *Transactions of the ASME, Series D, J. Basic engineering*, **86**:1–10, 1964.

[6] P.V. KOKOTOVIĆ AND H.J. SUSSMANN, "A positive real condition for global stabilization of nonlinear systems," *Syst. Contr. Lett.*, **19**:177–185, 1989.

[7] V.M. POPOV, "Absolute stability of nonlinear systems of automatic control," *Automation and Remote Control*, **22**:857–875, 1962. Translated from *Avtomatika i Telemekhanika*, **22**:961–979, 1961.

[8] V.M. POPOV, "Hyperstability and optimality of automatic systems with several control functions," *Rev. Roumaine Sci. Tech. Electrotechn. et Energ.*, **9**:629–690, 1964.

[9] R. SEPULCHRE, M. JANKOVIĆ, AND P. KOKOTOVIĆ, *Constructive Nonlinear Control*. Springer-Verlag (New York), 1997.

[10] E.D. SONTAG AND Y. WANG, "On characterizations of the input-to-state-stability property," *Syst. Contr. Lett.*, **24**:351–359, 1995.

[11] A.J. VAN DER SCHAFT, *L_2-Gain and Passivity Techniques in Nonlinear Control*, Springer-Verlag (New York), 1996.

[12] J.C. WILLEMS, "Dissipative dynamical systems Part I: General theory; Part II: Linear systems with quadratic supply rates," *Archive for Rational Mechanics and Analysis*, **45**:321–393, 1972.

[13] V.A. YAKUBOVICH, "The solution of certain matrix inequalities in automatic control theory," *Doklady Akademii Nauk*, **143**:1304–1307, 1962.

[14] G. ZAMES, "On the input-output stability of time-varying nonlinear feedback systems-Parts I and II," *IEEE Trans. Aut. Contr.*, **AC-11**:228–238 & 465–476, 1966.

P.V.K.

Dissipative Dynamical Systems
Part I: General Theory

JAN C. WILLEMS

Communicated by C. TRUESDELL

Contents

Abstract

The first part of this two-part paper presents a general theory of dissipative dynamical systems. The mathematical model used is a state space model and dissipativeness is defined in terms of an inequality involving the storage function and the supply function. It is shown that the storage function satisfies an *a priori* inequality: it is bounded from below by the available storage and from above by the required supply. The available storage is the amount of internal storage which may be recovered from the system and the required supply is the amount of supply which has to be delivered to the system in order to transfer it from the state of minimum storage to a given state. These functions are themselves possible storage functions, *i.e.*, they satisfy the dissipation inequality. Moreover, since the class of possible storage functions forms a convex set, there is thus a continuum of possible storage functions ranging from its lower bound, the available storage, to its upper bound, the required supply. The paper then considers interconnected systems. It is shown that dissipative systems which are interconnected via a neutral interconnection constraint define a new dissipative dynamical system and that the sum of the storage functions of the individual subsystems is a storage function for the interconnected system. The stability of dissipative systems is then investigated

Reprinted with permission from *Archive for Rational Mechanics and Analysis*, Jan C. Willems, "Dissipative Dynamical Systems—Part I: General Theory" Vol. 45, 1972, pp. 321–351.

and it is shown that a point in the state space where the storage function attains a local minimum defines a stable equilibrium and that the storage function is a Lyapunov function for this equilibrium. These results are then applied to several examples. These concepts and results will be applied to linear dynamical systems with quadratic supply rates in the second part of this paper.

1. Introduction

Dissipative systems are of particular interest in engineering and physics. The dissipation hypothesis, which distinguishes such systems from general dynamical systems, results in a fundamental constraint on their dynamic behavior. Typical examples of dissipative systems are electrical networks in which part of the electrical energy is dissipated in the resistors in the form of heat, viscoelastic systems in which viscous friction is responsible for a similar loss in energy, and thermodynamic systems for which the second law postulates a form of dissipation leading to an increase in entropy.

In the first part of this paper we hope to provide an axiomatic foundation for a general theory of dissipative systems. In the course of doing this we examine the concepts of an internal storage function and of a dissipation function.

There will be an obvious search for generality in the theoretical discussion of the first part of this paper. This stems from a belief that in studying specialized classes of dynamical systems it is important to keep the axioms separated. Such a procedure has more than just an aesthetic appeal: it allows one to pinpoint clearly what is a consequence of what.

My interest in dissipative systems stems from their implications on the stability of control systems. One of the main results in stability theory states that a feedback system consisting of a passive dynamical system in both the forward and the feedback loop is itself passive and thus stable. Moreover, the sum of the stored "energies" in the forward loop and in the feedback loop is a Lyapunov function for the closed loop system. The existence of a stored energy function is rather simple to eastablish since it is equivalent to the passivity assumption. It was in computing this stored energy function that we encountered some difficulties. It became clear that there is no uniqueness of the stored energy function, rather that there is a range of possible stored energy functions for a system with a prescribed input/output behavior.

In this paper these concepts are studied in detail and generalized. The terminology *dissipative* will be used as a generalization of the concept of passivity and *storage function* as a generalization of the concept of stored energy or entropy.

One of the main results obtained in this paper is that the storage function is as a rule not uniquely defined by the input/output behavior. It is shown that the storage function associated with a dissipative dynamical system satisfies an *a priori* inequality: it is bounded from below by the available storage and from above by the required supply. Moreover, and possibly more important, there is a continuum of possible storage functions between these upper and lower bounds.

This situation has important consequences. To give but one example, consider the familiar area of linear viscoelasticity. This is a typical example of a situation where the internal physical mechanism which is responsible for a stress/strain relationship is admittedly not completely understood. For many applications, one is, however, satisfied with an input/output description in terms of a relaxation function which may be obtained experimentally. Such an input/output description has, in fact, become the starting point of a general approach to the description of materials with memory. Nevertheless, the literature insists on postulating the knowledge of an internal energy function. It should be realized that this destroys some of the advantages of working with an input/output description since this knowledge of an internal energy function cannot be obtained from the relaxation function but requires additional information about the physical process. (In the present example one may often circumvent this difficulty by determining the heat production as well as the stress/strain relation, but this problem remains very fundamental in the context of thermodynamic systems where it is unclear what is being dissipated while the entropy increases.)

There are several methods for further reducing the number of possible storage functions. One rather obvious method is to consider a system as an interconnection of dissipative subsystems. Another possibility is by assuming additional qualitative internal properties for the system. A typical example is by *postulating* internal symmetry conditions as the Onsager-Casimir reciprocal relations. These will be examined in the second part of the paper.

We shall use the state space formalism for representing systems with memory. This feature is felt to be essential and the absence of the state space formalism in continuum mechanics and thermodynamics is somewhat disturbing. It is indeed customary in these areas to assume that the functionals appearing in the constitutive equations of materials with memory may depend on the entire past history (see for example [1] and [2]). This approach, however, does not recognize the idea of "equivalent histories": two histories are said to be *equivalent* if they bring the system into the same state and are thus indistinguishable under future experiments. Hence, one should constrain *a priori* the constitutive relations of any internal function as, for example, the internal energy or the entropy to take on the same value for equivalent (but not necessarily identical) histories. The state space formalism is the natural way for incorporating this constraint. There has, in fact, been some recent work by ONAT [3, 4] which deals with the construction of state space models for continuum systems.

We consider this paper as a contribution to mathematical system theory. The methods employed are those which have grown out of the modern developments of control theory; some of the auxiliary results, particularly in the second part of the paper, are drawn from network synthesis and optimal control theory. The implications of the results obtained and the methods used ought to be of interest to physicists, in particular those concerned with continuum mechanics and thermodynamics. We have tried to make the paper self-contained by being as explicit as possible whenever known results are being used.

2. Dynamical Systems

A dynamical system is viewed as an abstract mathematical object which maps *inputs* (causes, excitations) into *outputs* (effects, responses) via a set of intermediate variables, the *state*, which summarizes the influence of past inputs. The following

22*

lengthy definition is concerned with continuous systems (the time-interval of definition is the real line). In order to avoid unnecessary complications mainly of a notational nature, we will restrict ourselves to stationary (*i.e.*, time-invariant, nonaging) systems. The time-varying case is briefly discussed in Section 6.

Definition 1. A (continuous stationary) dynamical system Σ is defined through the sets $U, \mathcal{U}, Y, \mathcal{Y}, X$ and the maps ϕ and r. These satisfy the following axioms:

(i) $\mathcal{U}$ is called the *input space* and consists of a class of U-valued functions on* R. The set U is called the set of *input values*. The space $\mathcal{U}$ is assumed to be closed under the shift operator, *i.e.*, if $u \in \mathcal{U}$ then the function u_T defined by $u_T(t)=u(t+T)$ also belongs to $\mathcal{U}$ for any $T \in R$;

(ii) $\mathcal{Y}$ is called the *output space* and consists of a class of Y-valued functions on R. The set Y is called the set of *output values*. The space $\mathcal{Y}$ is assumed to be closed under the shift operator, *i.e.*, if $y \in \mathcal{Y}$ then the function y_T defined by $y_T(t)=y(t+T)$ belongs to $\mathcal{Y}$ for any $T \in R$;

(iii) X is an abstract set called the *state space*;

(iv) ϕ is called the *state transition function* and is a map from $R_2^+ \times X \times \mathcal{U}$ into X. It obeys the following axioms:

- (iv)$_a$ (*consistency*): $\phi(t_0, t_0, x_0, u)=x_0$ for all $t_0 \in R$, $x_0 \in X$, and $u \in \mathcal{U}$;
- (iv)$_b$ (*determinism*): $\phi(t_1, t_0, x_0, u_1)=\phi(t_1, t_0, x_0, u_2)$ for all $(t_1, t_0) \in R_2^+$, $x_0 \in X$, and $u_1, u_2 \in \mathcal{U}$ satisfying $u_1(t)=u_2(t)$ for $t_0 \leqq t \leqq t_1$;
- (iv)$_c$ (*semi-group property*): $\phi(t_2, t_0, x_0, u)=\phi(t_2, t_1, \phi(t_1, t_0, x_0, u), u)$ for all $t_0 \leqq t_1 \leqq t_2$, $x_0 \in X$, and $u \in \mathcal{U}$;
- (iv)$_d$ (*stationarity*): $\phi(t_1+T, t_0+T, x_0, u_T)=\phi(t_1, t_0, x_0, u)$ for all $(t_1, t_0) \in R_2^+$, $T \in R$, $x_0 \in X$, and $u, u_T \in \mathcal{U}$ related by $u_T(t)=u(t+T)$ for all $t \in R$;

(v) r is called the *read-out function* and is a map from $X \times U$ into Y;

(vi) the Y-valued function $r(\phi(t, t_0, x_0, u), u(t))$ defined for $t \geqq t_0$ is, for all $x_0 \in X$, $t_0 \in R$ and $u \in \mathcal{U}$, the restriction to $[t_0, \infty)$ of a function $y \in \mathcal{Y}$. This means that there exists an element $y \in \mathcal{Y}$ such that $y(t)=r(\phi(t, t_0, x_0, u), u(t))$ for $t \geqq t_0$.

A dynamical system thus generates outputs from inputs as follows: the system starts off in some initial state x_0 at time t_0 and an input u is applied to it. Then the state at time t_1 is given by $\phi(t_1, t_0, x_0, u)$. The output resulting from this experiment is given by $y(t)=r(\phi(t, t_0, x_0, u), u(t))$ and is defined for $t \geqq t_0$. It is important (for applications to systems described by partial differential equations for example) to realize that state transitions, and thus outputs, need only be defined in the forward time direction.

We call $\phi(t_1, t_0, x_0, u)$ "*the state at time t_1 reached from the initial state x_0 at time t_0 by applying the input u to the dynamical system Σ*" and $r(x, u)$ "*the output*

---

* We are using the following notation: R=the real numbers; R^n=n-dimensional Euclidean space; R^+=the nonnegative real numbers; R_2^+=the causal triangular sector of R_2 defined by $R_2^+=\{(t_2, t_1) \in R_2 \mid t_2 \geqq t_1\}$; R^e=the extended real number system$=\{-\infty\} \cup R \cup \{+\infty\}$.

due to the presence of state x and the input-value u". We will denote the function $r(\phi(t, t_0, x_0, u), u(t))$ defined for $t \geqq t_0$ unambiguously by $y(t_0, x_0, u)$.

Definition 1 is precise and yet very general. By a suitable choice of the state space, the state transition function, and the read-out function, it includes all common deterministic models used in classical physics, in circuit theory, and control theory.

The axiom of determinism is the crucial one. It expresses at the same time a fundamental property of the state and an important restriction on the class of systems which qualify for dynamical systems in the above sense. It states that the initial state summarizes the effect of past inputs in the sense that for future responses it does not matter how the system was brought into this state; it also implies that the state and thus the output before some time are not influenced by the values of the input after that time. We are hence in effect restricting our attention to systems in which future inputs do not affect past and present outputs. The idea is simple: since all experimental evidence indicates that physical systems indeed satisfy this property of causality, we require this to be preserved in the model.

It should be emphasized that the read-out function is required to be a memoryless map in the sense that the output only depends on the *present* value of the state and the input. All dynamical effects (*i.e.*, those phenomena involving memory) are required to be taken care of by the state.

The above definition is commonly used in mathematical system theory (see, for instance, references [5, 6]). Although physicists have been groping for a similar concept for a long time, it is only for systems in which the input space consists of only one element (*i.e.*, the autonomous dynamical systems of classical mechanics) that such mathematical structures have been introduced in a formal way. In the framework of Definition 1 the state at every moment completely describes the present situation. It is, however, impossible to deduce *a priori*, in physical terms, what will be the state. This, indeed, is a very difficult problem even for relatively simple systems, and it appears to be the cause for much of the reluctance of introducing this concept in physics. The approach which has been taken for describing materials with memory is to allow the outputs to be a function of the whole past history of the input. This is particularly prominent in the pioneering work of TRUESDELL, COLEMAN, and NOLL [1, 2]. Another approach is that of ONAT [3, 4] where the state is constructed in terms of observables. These two extreme points of view are particular cases of Definition 1, but we see no compelling reason to adhere to either of them. The first approach does not recognize the idea of equivalent histories, and the second approach will lead to difficulties when we consider isolated systems for example.

In view of this dichotomy, it would appear to be useful to allow some time discussing these state space concepts further. Let us take the point of view that all the information the experimenter may obtain about a system is a table of input functions in $\mathcal{U}$ versus the corresponding output functions in $\mathcal{Y}$. The so-called *problem of realization* is to define a state space X and the functions ϕ and r in such a way that the resulting dynamical system in state space form generates the given input/output pairs by a suitable choice for the initial state in each tabulated experiment. This problem has attracted a great deal of attention in the literature. Both the questions, "Does a state space realization exist?" and "What are the maps ϕ and r?", have been examined. For the first question we mention the work of ZADEH [7] and for the second question the work by YOULA [8] and, especially, KALMAN [5], among others. The existence question essentially only requires a determinism postulate on the input/output pairs. The construction of ϕ and r is understandably much more intricate but has been satisfactorily resolved for large classes of systems. In particular, there exists a very elegant solution to this problem for linear systems with a finite number of degrees of freedom. This material is considered to be of prime importance and can be found in a number of recent texts (*e.g.*, [10]).

We now consider an important particular case of this realization problem. Assume that F is a given map from $\mathscr{U}$ into $\mathscr{Y}$ satisfying the postulate of determinism which states that inputs u_1, $u_2 \in \mathscr{U}$ satisfying $u_1(t)=u_2(t)$ for $t \leqq t_0$ yield outputs $y_1=Fu_1$ and $y_2=Fu_2$ which similarly satisfy $y_1(t)=y_2(t)$ for $t \leqq t_0$. Assume in addition that this map is stationary *i.e.*, two inputs u_1, $u_2 \in \mathscr{U}$ related by $u_1(t)=u_2(t+T)$ yield outputs $y_1=Fu_1$ and $y_2=Fu_2$ which are similarly related by $y_1(t)=y_2(t+T)$. The question is to realize F by a dynamical system in state space form. The solution to this problem is by no means unique. One possibility is to consider the function $f\colon R^+ \to U$ defined by $f(s)=u(t-s)$ for $s \geqq 0$ as the state at time t resulting from the input u. It is clear how the state transition function and the read-out function may be defined from here [11]. This state space realization is of course completely inefficient: in trying to store sufficient information about the past inputs, we decided to store the whole past input. The most efficient and natural state space realization of F is the one obtained by considering as the state at time t the equivalence class of those inputs up to time t which yield the same output after time t regardless of how the input is continued after time t. More specifically, in this realization we start with the space of functions $f\colon R^+ \to U$ satisfying $f(s)=u(-s)$, $s \geqq 0$, for some $u \in \mathscr{U}$. We then group these functions into equivalence classes by letting $f_1 \sim f_2$ if $y_1=Fu_1$, and $y_2=Fu_2$ satisfy $y_1(t)=y_2(t)$ for $t \geqq 0$ whenever $u_1(-t)=f_1(t)$, $u_2(-t)=f_2(t)$, and $u_1(t)=u_2(t)$ for $t \geqq 0$. The latter realization is sometimes called a "*minimal* realization" and plays a central role in control theory [5, 10]. A similar idea has been proposed by ONAT [3, 4] in a restricted context.

The point of view taken in this paper is that the state space realization is given, *i.e.*, it has been inferred from previous considerations what the state space is. We do not demand minimality since, in our opinion, there is no compelling reason for doing so: minimality is very much a function of the class of experiments and observations which are allowed, is sensitive to modelling, and is not necessarily a good physical assumption. Neither do we adhere to the idea that the state is the whole past input since this point of view leads to nonsensical situations. Consider for example an electrical *RLC* network which has a given set of charges on the C's and fluxes through the L's. Does it make sense to allow the stored energy of such a system to depend on exactly how these charges and fluxes came about? The whole question of what the state space of a physical system is requires much consideration. In this paper we have taken the easy way out by assuming that this has already been decided.

3. Dissipative Dynamical Systems

In this section the concepts, which will be the basis for the further developments, are introduced. Assume that a dynamical system Σ is given together with a real-valued function w defined on $U \times Y$. This function will be called the *supply rate*. We assume that for any $(t_1, t_0) \in R_2^+$, $u \in U$, and $y \in Y$, the function $w(t) = w(u(t), y(t))$ satisfies* $\int_{t_0}^{t_1} |w(t)|\, dt < \infty$, i.e., w is locally integrable.

Definition 2. A dynamical system Σ with supply rate w is said to be *dissipative* if there exists a nonnegative function S: $X \to R^+$, called the *storage function*, such that for all $(t_1, t_0) \in R_2^+$, $x_0 \in X$, and $u \in U$,

$$S(x_0) + \int_{t_0}^{t_1} w(t)\, dt \geqq S(x_1)$$

where $x_1 = \phi(t_1, t_0, x_0, u)$ and $w(t) = w(u(t), y(t))$, with $y = y(t_0, x_0, u)$.

The above inequality will be called the *dissipation inequality*. Note that $\oint w(t)\, dt \geqq 0$ with $\oint$ indicating that the dynamical system is taken from a particular initial state to the same terminal state along some path in state space. This condition is in itself inadequate as a definition for dissipativeness but dynamical systems which are dissipative in such cyclic motions only are of independent interest.

The approach taken here proceeds from the knowledge, from physical considerations, that the dynamical system is dissipative and thus that the storage function exists. The fact that this storage function is "defined" via an inequality requires further analysis. Central in this analysis is the question: "In how far is S defined by the dissipation inequality?" (The question is not so much "Does a storage function exist?" but rather "What can it be?")

A crucial role will be played in the sequel by a quantity termed the *available storage:* it is the maximum amount of storage which may at any time have been extracted from a dynamical system. The notion of available storage is a generalization of the concept of "available energy" [11, 12, 13] studied in control theory and of "recoverable work" encountered in the theory of viscoelasticity [14, 15].

Definition 3. The *available storage*, S_a, of a dynamical system Σ with supply rate is the function from X into R^e defined by

$$S_a(x) = \sup_{\substack{x \to \\ t_1 \geqq 0}} - \int_0^{t_1} w(t)\, dt$$

where the notation $x \to$ denotes the supremum over all motions starting in state x at time 0 and where the supremum is taken over all $u \in \mathcal{U}$.

The available storage is an essential function in determining whether or not a system is dissipative. This is shown in the following theorem:

* The shorthand notation $w(t)$ for $w(u(t), y(t))$ will be used whenever it is obvious from the context what x_0, t_0, and u are.

Theorem 1. *The available storage, S_a, is finite for all $x \in X$ if and only if Σ is dissipative. Moreover, $0 \leqq S_a \leqq S$ for dissipative dynamical systems and S_a is itself a possible storage function.*

Proof. Assume first that $S_a < \infty$: it will be shown that Σ is then dissipative. It suffices therefore to show that S_a is a possible storage function. Notice that $S_a \geqq 0$ since $S_a(x)$ is the supremum over a set of numbers which contains the zero element ($t_1 = 0$). Consider now the quantity $S_a(x_0) + \int_{t_0}^{t_1} w(t)\,dt$. We have to show that this quantity is not less than $S_a(x_1)$ whenever w is evaluated along a trajectory generated by an input u which transfers the state from x_0 at t_0 to x_1 at t_1. The proof of this is quite simple although writing out the details is somewhat laborious. The idea is the following: in extracting the available storage from Σ when it is in state x_0 we could first take Σ along the path generated by u, thus transferring Σ to x_1, and then extract the available storage with Σ in state x_1. This combined process is clearly a suboptimal procedure for extracting the storage originally present with Σ in state x_0. Formalizing this idea immediately leads to the desired dissipation inequality for S_a. Assume next that Σ is dissipative. Then

$$S(x_0) + \int_{t_0}^{t_1} w(t)\,dt \geqq S(x_1) \geqq 0$$

which shows that

$$S(x_0) \geqq \sup_{\substack{x_0 \to \\ t_1 \geqq 0}} - \int_0^{t_1} w(t)\,dt = S_a(x_0).$$

Hence $S_a < \infty$ as claimed. This ends the proof of Theorem 1. ||

Theorem 1 gives a method which in theory may be used to verify whether or not a dynamical system is dissipative and this procedure does not require knowledge of the storage functions. In this sense it is an input/output test. Note that the theorem only states that the available storage may be the storage function. Usually it will not be the actual storage function. In fact, under certain additional assumptions (*e.g.*, the Onsager-Casimir reciprocal relations) it may be shown that it will *not* be the actual storage function. This fact should be kept in mind when interpreting the results of [12, 14, 15]. A dynamical system which has the available storage as its actual storage function has the interesting (and unusual) property that all of its internal storage is available to the outside via its external terminals.

It is convenient to introduce at this point the concept of reachability. This notion is related to controllability and plays a central role in mathematical systems theory.

Definition 4. The state space of the dynamical system Σ is said to be *reachable* from x_{-1} if for any $x \in X$ there exists a $t_{-1} \leqq 0$ and $u \in \mathscr{U}$ such that

$$x = \phi(0, t_{-1}, x_{-1}, u).$$

It is said to be *controllable* to x_1 if for any $x \in X$ there exists a $t_1 \geqq 0$ and a $u \in \mathscr{U}$ such that $x_1 = \phi(t_1, 0, x, u)$.

Theorem 1 emphasizes what happens when the system starts off in a particular state. One may similarly examine what happens when the system ends up in a particular state. We will therefore introduce the concept of required supply. This is done by letting the system start in a given state and by bringing it to its present state in the most efficient manner, *i.e.*, by using no more supply from the outside than is absolutely necessary. The notion of required supply has been introduced in [11]. Although one could choose any point in state space as the initial state, it is most logical to assume that the system starts in a state of minimum storage.

Assumption. It will be assumed that there exists a point $x^* \in X$ such that $S(x^*) = \min_{x \in X} S(x)$ and that the storage function S has been normalized to $S(x^*) = 0$.

Definition 5. The required supply, S_r, of a dissipative dynamical system Σ with supply rate w is the function from X into R^e defined by

$$S_r(x) = \inf_{\substack{x^* \to x \\ t_{-1} \leqq 0}} \int_{t_{-1}}^{0} w(t)\,dt$$

where the notation $\inf_{\substack{x^* \to x \\ t_{-1} \leqq 0}}$ denotes* the infimum over all $u \in \mathcal{U}$ and $t_{-1} \leqq 0$ such that $x = \phi(0, t_{-1}, x^*, u)$.

Theorem 2. (i) *Assume that the state space of Σ is reachable from x_{-1}. Then Σ is dissipative if and only if there exists a constant K such that*

$$\inf_{\substack{x_{-1} \to x \\ t_{-1} \leqq 0}} \int_{t_{-1}}^{0} w(t)\,dt \geqq K \quad \textit{for all } x \in X.$$

Moreover,

$$S_a(x_{-1}) + \inf_{\substack{x_{-1} \to x \\ t_{-1} \leqq 0}} \int_{t_{-1}}^{0} w(t)\,dt$$

is a possible storage function.

(ii) *Let Σ be a dissipative dynamical system and assume that $S(x^*) = 0$. Then $S_r(x^*) = 0$ and $0 \leqq S_a \leqq S \leqq S_r$. Moreover, if the state space Σ is reachable from x^* then $S_r < \infty$ and the required supply S_r is a possible storage function.*

Proof. (i) By reachability and Theorem 1 we see that Σ is dissipative if and only if $S_a(x_{-1}) < \infty$. Any $K \leqq -S_a(x_{-1})$ will thus yield the inequality in part (i) of the theorem statement. It remains to be shown that

$$S_a(x_{-1}) + \inf_{\substack{x_{-1} \to x \\ t_{-1} \leqq 0}} \int_{t_{-1}}^{0} w(t)\,dt$$

is a possible storage function. This function is clearly nonnegative. To prove that it satisfies the dissipation inequality, consider the following idea: in taking the system from x_{-1} to x_1 at t_1, we can first take it to x_0 at t_0 while minimizing the supply and then take it from x_0 at t_0 to x_1 at t_1 along the path for which we are to

* This notation, along with the similar one introduced in Definition 3, will be used throughout.

demonstrate the dissipation inequality. This results in a suboptimal policy for taking the system to x_1 and the formalization of this procedure leads to the desired dissipation inequality.

(ii) That $S_r(x^*)=0$ is obvious. Moreover, any $u\in\mathscr{U}$ resulting in a transfer from x^* at t_{-1} to x at 0 satisfies $S(x)\leqq\int_{t_{-1}}^{0} w(t)\,dt$ by the dissipation inequality. The inequality $S_r(x)\geqq S(x)$ follows by taking the infimum at the right-hand side. Assume now that the state space of Σ is reachable. Then clearly $S_r\leqq\infty$. It remains to be shown that S_r is a possible storage function. This, however, follows from (i). ‖

It is an immediate consequence of the normalization $S(x^*)=0$ that for a dissipative system any motion starting in x^* at t_0 satisfies $\int_{t_0}^{t_1} w(t)\,dt\geqq 0$ for all $u\in\mathscr{U}$ and $t_1\geqq t_0$. Thus the net supply flow is into the system. This idea has been proposed [16, 17, 18, 19] as a definition of passivity. It has the advantage of being an input/output concept which does not involve introduction of state space notions. However implicit in this approach is the fact that one knows the state of minimum internal storage.

Note that the required supply is in general a function of S and x^*. Usually, however, the point of minimum storage is a unique *a priori* known equilibrium point which may thus be shown to be independent of S and this ambiguity does not arise.

Remarks. 1. Under the assumptions of reachability from x_{-1} and controllability to x_1 we always have the following inequalities for a dissipative system:

$$S(x_1)+\sup_{\substack{x\to x_1\\ t_1\geqq 0}} -\int_0^{t_1} w(t)\,dt\leqq S(x)\leqq S(x_{-1})+\inf_{\substack{x_{-1}\to x\\ t_{-1}\leqq 0}}\int_{t_{-1}}^{0} w(t)\,dt.$$

Note however that the lower bound on S thus obtained is itself in general not a possible storage function because it need not be nonnegative.

2. Often a state space model of a dynamical system is constructed on the basis of an input/output description. Particularly important realizations are the minimal realization mentioned earlier and the realization in which the state is the whole past history. It is quite simple to associate a storage function with these realizations when one has determined a storage function on a particular state space X. For example, defining $S(u_{(-\infty,0)})=S(x(0))$ leads to a storage function on a state space which keeps track of the whole past input history. The available storage function of these realizations will in fact agree on that part of the state space which is reachable along some past history. Assuming that for t sufficiently small every element of $\mathscr{U}$ is equal to a fixed constant u^* (typically the zero element of some vector space) such that $w(u^*, y^*)=0$ and that $\int_{-\infty}^{0} w(u(t), y(t))\,dt$ exists and is nonnegative (thus the state at "$t=-\infty$" is assumed to be the state of minimal storage), then we may actually also evaluate the required supply for the realization in which the state keeps track of the whole past history. This does not require any infimization and is simply equal to $\int_{-\infty}^{0} w(u(t), y(t))\,dt$. It may in principle

be different for every history. Moreover, the dissipation inequality holds with equality for this storage function. (This fact does not conflict with Theorem 4 since this realization will never be controllable.)

If one works with the minimal realization then one may associate a storage function by defining $S(\tilde{x}_{\min})=S(x)$ where x is a state in the equivalence class $\tilde{x}_{\min}$. After elimination of the non-reachable states, one thus divides the state space X into equivalence classes and defines the storage to be the storage of an arbitrary element in this class. The available storage functions of these realizations again agrees on that part of the state space which is reachable along some past history. The required storage may now take on more values in X than in $X_{\min}$. An interesting consequence of the above reasoning is that the notion of available storage is defined purely as an input/output concept for states which are reachable. Thus, taking equivalence classes as the state or the whole past history as the state leads to the same value for the available storage function. This reemphasizes the importance of Theorem 1 as an input/output test for dissipativeness. There is an interesting paper by DAY [33] which has used the concept of available storage (or "useful work" as it is called in [33]) in setting up an axiomatic theory of thermodynamics. Although the technical details are quite different, the ideas exploited in that paper appear to be very much along the lines of those on which Theorem 1 is based.

To summarize the above results, we have shown that the storage function of a dissipative dynamical system satisfies the *a priori* inequality $S_a \leqq S \leqq S_r$, *i.e.*, a dissipative system can supply to the outside only a fraction of what it has stored and can store only a fraction of what has been supplied to it. The available storage always satisfies the dissipation inequality, as does the required supply for systems with a state space which is reachable from a point of minimum storage. (This show that the above inequality is the best of its type.) Of course not every function bounded by this *a priori* inequality will be a possible storage function. It appears to be difficult to state other general properties of the set of possible storage functions. One interesting property is its convexity:

Theorem 3. *The set of possible storage functions of a dissipative dynamical system forms a convex set. Hence* $\alpha S_a+(1-\alpha) S_r$, $0 \leqq \alpha \leqq 1$, *is a possible storage function for a dissipative dynamical system whose state space is reachable from* x^*.

Proof. This theorem is an immediate consequence of the dissipation inequality. ||

The ultimate test for a theory of dissipative systems is whether or not there exists a (possibly idealized) "physical" system which realizes the input/output exchange process and which has the desired storage function. Such a synthesis program based on interconnecting ideal elements may in fact be carried out for linear systems with a finite number of degrees of freedom and quadratic supply functions. Some results in this direction will be indicated in Part II.

We now proceed with a few remarks regarding the evaluation of the available storage and the required supply:

(i) If the state of minimum storage x^* is an equilibrium point corresponding to the constant input $u^* \in \mathscr{U}$ (*i.e.*, $\phi(t,0,x^*,u^*)=x^*$ for all $t \geqq 0$) and if $w(u^*,y^*)=0$,

then

$$S_r(x) = \lim_{t_{-1}\to-\infty} \inf_{x^*\to x} \int_{t_{-1}}^{0} w(t)\,dt;$$

(ii) If for all $x \in X$ there exists a $u \in \mathcal{U}$ such that $w(u, y) \leqq 0$ (*i.e.*, the external termination may always be adjusted so that the supply flows out of the system), then

$$S_a(x) = \lim_{t_1\to\infty} \sup_{x\to} \int_{0}^{t_1} w(t)\,dt;$$

(iii) The concept of required supply assumes that there exists a point $x^* \in X$ such that $S(x^*) = \min_{x^*\in X} S(x)$. There need however not be a point of minimum storage. One may then define $S_r(x)$ by considering a sequence of states $\{x_n\}$ with $\lim_{n\to\infty} S(x_n) = \inf_{x\in X} S(x)$ and define

$$S_r(x) = \lim_{n\to\infty} S_{r,n}(x) \quad \text{where} \quad S_{r,n}(x) = \inf_{\substack{x_n\to x\\ t_{-1}\leqq 0}} \int_{t_{-1}}^{0} w(t)\,dt.$$

We now show how to treat conservative systems as particular cases of dissipative dynamical systems.

Definition 6. A dissipative dynamical system Σ with supply rate w and storage function S is said to be *lossless* if for all $(t_1, t_0) \in R_2^+$, $x_0 \in X$, and $u \in \mathcal{U}$

$$S(x_0) + \int_{t_0}^{t_1} w(t)\,dt = S(x_1)$$

where $x_1 = \phi(t_1, t_0, x_0, u)$.

The following theorem is immediate:

Theorem 4. *Let Σ be a lossless dissipative dynamical system and assume that $S(x^*) = 0$. If the state space is reachable from x^* and controllable to x^*, then $S_a = S_r$ and thus the storage function is unique and given by*

$$S(x) = \int_{t_{-1}}^{0} w(t)\,dt$$

with any $t_{-1} \leqq 0$ and $u \in \mathcal{U}$ such that $x = \phi(0, t_{-1}, x^*, u)$, or

$$S(x) = -\int_{0}^{t_1} w(t)\,dt$$

with any $t_1 \geqq 0$ and $u \in \mathcal{U}$ such that $x^* = \phi(t_1, 0, x, u)$.

The condition $S_a = S_r$ which implies uniqueness of the storage function is in itself not sufficient to imply losslessness. We could call such systems *quasi-lossless* since they may be transferred between states without dissipation provided; however, this transfer is executed optimally. An arbitrary transfer instead is expected to involve dissipation.

An interesting property of dissipative dynamical systems is the following:

(i) For dissipative dynamical systems with $x(t_0) = x^*$, $\int_{t_0}^{t_1} w(t)\,k(t)\,dt \geqq 0$ for all bounded functions k with $k(t) \geqq 0$ and $\dot{k}(t) \leqq 0$;

(ii) For lossless dynamical systems with $x(t_0)=x^*$ and $\phi(t_1, t_0, x^*, u)=x^*$, $\int_{t_0}^{t_1} w(t)\, k(t)\, dt \geqq 0$ for all bounded functions k with $\dot{k}(t) \leqq 0$.

These inequalities formalize the idea that for a dissipative system with no initial storage the supply flows into the system before part of it is recovered whereas in addition all of it gets recovered in a lossless system. These expressions generalize similar inequalities obtained in [20, 21, 22].

We conclude this section with a discussion of the concept of a dissipation function.

Definition 7. A real-valued function d: $X \times U \rightarrow R$ is said to be the ***dissipation rate*** of a dissipative dynamical system Σ with supply rate w and storage function S if for all $(t_1, t_0) \in R_2^+$, $x_0 \in X$, and $u \in \mathscr{U}$

$$S(x_0) + \int_{t_0}^{t_1} \big(w(t) + d(t)\big)\, dt = S(x_1)$$

where $x_1 = \phi(t_1, t_0, x_0, u)$.

It is clear that d being nonnegative implies dissipativeness. Moreover, since the dynamical system Σ is lossless with respect to the new supply rate $(w+d)$ it follows that the dissipation rate d uniquely determines the storage function S provided the appropriate reachability and controllability conditions are satisfied. The converse, *i.e.*, that dissipativeness implies the existence of a nonnegative dissipation rate is also the case under some technical smoothness conditions. The set of dissipation rate functions for a given dissipative system forms a convex set.

Remarks. Note that if $S(\phi(t, 0, x, u))$ is differentiable at $t=0$ for all $x \in X$ and $u \in \mathscr{U}$, then the dissipation inequality is equivalent to

$$\dot{S}(x, u) \leqq w\big(r(x, u), u\big)$$

for all $x \in X$ and $u \in U$ where $\dot{S}$ denotes $\left.\frac{d}{dt} S(\phi(t, 0, x, u))\right|_{t=0}$. This definition is more standard but slightly less general than the one proposed here. The dissipation function d is then given by

$$d = \dot{S} - w.$$

4. Interconnected Systems

The main result obtained in the previous section yields an *a priori* bound on the storage function of a dissipative dynamical system. Moreover these bounds themselves define possible storage functions and the storage function is thus uniquely determined by the dissipation inequality if and only if the required supply equals the available storage. This situation is the exception and as a rule there are consequently many possible storage functions. If we consider the implications of this result to physical systems which dissipate energy or to thermodynamic systems, then we conclude that experiments on a physical system will usually only give bounds on the stored energy function or on the entropy function. This result is unexpected in the sense that in classical physical systems this ambiguity does not arise: we thus expect that the additional structural assumptions implicit in such systems will greatly reduce the number of possible storage functions and often render it unique.

In this section we examine one such possibility: it will be shown that by considering a given dissipative system as an interconnection of dissipative subsystems the number of possible storage functions is greatly reduced. Other qualitative assumptions (linearity, reciprocity, *etc.*) on the system will be investigated in Part II.

The idea of an interconnected system is actually quite simple, albeit somewhat difficult to formalize. We start with a collection of dynamical systems $\{\Sigma_\alpha\}$ where α ranges over some given index set A. For simplicity we will assume that A is a finite set. More general interconnections involve the introduction of a measure on A which would take us somewhat astray. Each Σ_α is determined, as in Definition 1, by a septuplet $\{U_\alpha, \mathscr{U}_\alpha, Y_\alpha, \mathscr{Y}_\alpha, X_\alpha, \phi_\alpha, r_\alpha\}$. We assume that the inputs and outputs of each dynamical system Σ_α are divided into two groups, *i.e.*,

$$U_\alpha = U_\alpha^e \times U_\alpha^i, \quad \mathscr{U}_\alpha = \mathscr{U}_\alpha^e \times \mathscr{U}_\alpha^i, \quad Y_\alpha = Y_\alpha^e \times Y_\alpha^i, \quad \text{and} \quad \mathscr{Y}_\alpha = \mathscr{Y}_\alpha^e \times \mathscr{Y}_\alpha^i$$

when the superscripts e and i stand for the adjectives *external* and *interconnecting*.

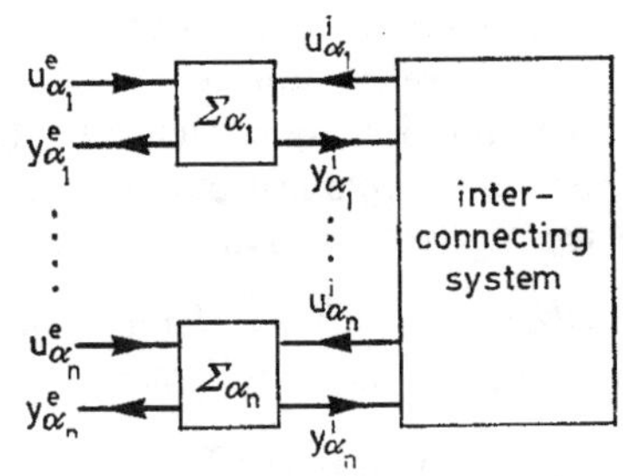

Fig. 1. Illustrating the concept of an interconnected system

Next we introduce the notion of an *interconnecting function* which is simply a function* $f\colon \prod_{\alpha\in A} (U_\alpha^i \times Y_\alpha^i) \to V$ where V is some vector space, and of the *interconnection constraint* which states that $f(\prod_{\alpha\in A} (U_\alpha^i \times Y_\alpha^i)) = 0$: it is thus a relation between the instantaneous values of the inputs and the outputs. The idea of an interconnected system is illustrated in Figure 1 and indicates that the external inputs u_α^e are given but that the interconnecting inputs u_α^i are to be determined implicitly. More precisely, given any $u_\alpha^e \in \mathscr{U}_\alpha^e$ and $x_\alpha \in X_\alpha$, $\alpha \in A$, we may attempt to solve the implicit equations

$$f(\prod_{\alpha\in A} (u_\alpha^i(t) \times r_\alpha^i(\phi_\alpha(t, t_0, x_\alpha(u_\alpha^e, u_\alpha^i)), (u_\alpha^e(t), u_\alpha^i(t))))) = 0, \qquad t \geqq t_0,$$

for some set of the functions $u_\alpha^i \in \mathscr{U}_\alpha$. This equation is of course not necessarily uniquely solvable and this fact needs to be assumed explicitly. Notice that only

* The notation $\prod_{\alpha\in A}$ stands for the Cartesian set product.

the values of $u_\alpha^i(t)$ for $t \geqq 0$ enter in the equations. Hence we can only hope these equations determine u_α^i on the half line $t \geqq t_0$. However it may be expected that under reasonable assumptions there will be a map from the set of states and external input values into the set of internal inputs and outputs which determines a solution to these equations.

We will thus assume *explicitly* that the interconnected system, denoted by $\prod_{\alpha \in A} \Sigma_\alpha / f$, is well-posed in the sense that it defines unique functions ϕ and r and

$$\psi_\alpha^i: \prod_{\alpha \in A} X_\alpha \times \prod_{\alpha \in A} U_\alpha^e \to U_\alpha^i, \qquad \alpha \in A,$$

such that:

(i) the septuplet $\{\prod_{\alpha \in A} U_\alpha^e, \prod_{\alpha \in A} \mathscr{U}_\alpha^e, \prod_{\alpha \in A} Y_\alpha^e, \prod_{\alpha \in A} \mathscr{Y}_\alpha^e, \prod_{\alpha \in A} X_\alpha, \phi, r\}$ defines a dynamical system in the sense of Definition 1;

(ii) the function $u_\alpha^i(t) = \psi_\alpha^i(\phi(t, t_0, \prod_{\alpha \in A} x_\alpha, \prod_{\alpha \in A} u_\alpha^e), \prod_{\alpha \in A} u_\alpha^e(t))$ defined for $t \geqq t_0$ is the restriction to $[t_0, \infty)$ of a function in $\mathscr{U}_\alpha^i$;

(iii) $\phi(t, t_0, \prod_{\alpha \in A} x_\alpha, \prod_{\alpha \in A} u_\alpha^e) = \prod_{\alpha \in A} \phi_\alpha(t, t_0, x_\alpha, (u_\alpha^e, u_\alpha^i))$ with u_α^i as in (ii);

(iv) $r(\prod_{\alpha \in A} x_\alpha, \prod_{\alpha \in A} u_\alpha^e) = \prod_{\alpha \in A} r_\alpha^e(x_\alpha, (u_\alpha^e, \psi_\alpha^i(\prod_{\alpha \in A} x_\alpha, \prod_{\alpha \in A} u_\alpha^e)))$;

(v) $f(\prod_{\alpha \in A} (\psi_\alpha^i(\prod_{\alpha \in A} x_\alpha, \prod_{\alpha \in A} u_\alpha^e), r_\alpha^i(x_\alpha, (u_\alpha^e, \psi_\alpha^i(\prod_{\alpha \in A} x_\alpha, \prod_{\alpha \in A} u_\alpha^e)))) = 0.$

It is easy to verify that the above conditions formalize the intuitive conditions one expects a well-posed interconnected dynamical system to satisfy. Examples of interconnected systems will be given in Section 7. Note that although the interconnected system may have many state space realizations we are insisting on using the one with state space the Cartesian product of the state spaces of the individual subsystems. This is indeed a natural thing to do since the interconnection itself introduces no memory.

We now introduce the concept of dissipation in this framework. Assume therefore that each dynamical system Σ_α has associated with it an *external supply rate*, w_α^e, defined on $U_\alpha^e \times Y_\alpha^e$ and an *interconnecting supply rate*, w_α^i, defined on $U_\alpha^i \times Y_\alpha^i$.

Definition 8. Consider the dynamical systems Σ_α with interconnecting supply rates w_α^i. Then the interconnection defined by the interconnection constraint $f(\prod_{\alpha \in A} (u_\alpha^i \times y_\alpha^i)) = 0$ is said to be *neutral* if all u_i^α and y_i^α satisfying this equality yield $\sum_{\alpha \in A} w_\alpha^i(u_\alpha^i, y_\alpha^i) = 0$.

In terms of Figure 1, an interconnection is thus said to be neutral if the interconnecting system itself is lossless with respect to the supply rate $\sum_{\alpha \in A} w_\alpha^i$. Thus the mere interconnection does not introduce any new supply or dissipation. One thus expects the dissipativeness of the interconnected system to be a consequence of the dissipativeness of the individual subsystems. That this is indeed the case is shown in the following theorem:

Theorem 5. *Let Σ_α, $\alpha \in A$, be a collection of dissipative dynamical systems with supply rates $w_\alpha = w_\alpha^e + w_\alpha^i$ and storage functions S_α. Let f be a neutral interconnection* *constraint. Then the interconnected system $\Sigma = \prod_{\alpha \in A} \Sigma_\alpha | f$ is itself dissipative with respect to the supply rate $w = \sum_{\alpha \in A} w_\alpha^e$ and $S = \sum_{\alpha \in A} S_\alpha$ is a storage function for Σ.*

Proof. Summing both sides of the inequality

$$S_\alpha(x_\alpha(t_0)) + \int_{t_0}^{t_1} (w_\alpha^e(t) + w_\alpha^i(t))\, dt \geqq S_\alpha(x_\alpha(t_1))$$

over $\alpha \in A$ and using the assumption $\sum_{\alpha \in A} w_\alpha^i(t) = 0$ leads to the desired dissipation inequality for Σ. ‖

The above theorem is intuitively obvious. Note however that by considering only storage functions which are additive in the sense that $S(\prod_{\alpha \in A} x_\alpha) = \sum_{\alpha \in A} S_\alpha(x_\alpha)$ one obtains only part of the admissible storage functions for Σ. It is easy to see that since the interconnection introduces additional constraints on the inputs $u_\alpha = (u_\alpha^e, u_\alpha^i)$, we always have the inequality $S_a \leqq \sum_{\alpha \in A} S_{\alpha,a} \leqq \sum_{\alpha \in A} S_{\alpha,r} \leqq S_r$, with equality holding exceptionally. Thus one obtains a unique additive storage function for the interconnected system if and only if $S_{\alpha,a} = S_{\alpha,r}$ for each $\alpha \in A$.

In many physical systems encountered in practice, *e.g.*, in lumped electrical networks or in continuum systems, one may postulate *a priori* that the system is an interconnection of dissipative systems and use this qualitative property to describe the system in terms of "local" states, *i.e.*, to take the state space $X = \prod_{\alpha \in A} X_\alpha$ and furthermore to require that the storage function be of the type $S(\prod_{\alpha \in A} x_\alpha) = \sum_\alpha S_\alpha(x_\alpha)$. This natural requirement on the storage function of a dissipative dynamical system which consists of a family of dissipative systems interconnected by means of a neutral interconnection serves to reduce greatly the number of possible storage functions. This requirement leads to a *unique* storage function whenever it is possible to regard Σ as the interconnection of lossless systems with memory (capacitors and inductors, elastic systems) and a dissipative system without memory (resistors, friction elements). The lossless part possesses a unique storage function by Theorem 4 (under the additional hypothesis of reachability and controllability) whereas the dissipative part does not contribute to the storage since its state space is the empty set. The storage of the original system is thus given by the storage in the lossless subsystem and is consequently unique.

In concluding this section we remark that the above method of considering interconnected systems is implicit in most treatments of dissipative systems. It is based on a qualitative assumption on the system (the idea of "simple" materials) and sometimes it results in the uniqueness of the storage function. This is however by no means always the case, and typical examples of areas where this nonuniqueness remains are linear viscoelasticity and the modern treatments of materials with memory in continuum mechanics and thermodynamics where the nonuniqueness of the storage function at the elementary particle level remains. In other words, one has to make more assumptions (or, equivalently, obtain more knowledge about the physics) in order to derive the storage function (internal energy, entropy, *etc.*) from the constitutive equations.

5. Stability

In this section we examine the stability of dissipative systems. As is to be expected, only some technical conditions are required in order for dissipativeness to imply stability of an equilibrium at a local minimum of the storage function.

We shall be concerned with the stability of an equilibrium state, and in order to make this study meaningful we need to isolate the system from its environment. Moreover, since stability is concerned with convergence the concept of a distance function on the state space needs to be introduced. Assume therefore that the following assumptions hold:

(i) The system is *isolated, i.e.*, the input space consists of one element only. In order to preserve stationarity we assume that this element is the constant function $u(t)=u^*$;

(ii) $x^* \in X$ is an *equilibrium* point, *i.e.*, $\phi(t, t_0, x^*, u^*)=x^*$ for all $t \geqq t_0$;

(iii) X is a subset of a normed space and $\| \ \|$ denotes its norm;

(iv) $\phi(t, t_0, x_0, u^*)$ is continuous in t for $t \geqq t_0$;

(v) $w(u^*, r(x, u^*)) \leqq 0$ for all $x \in X$ in a neighborhood of x^*.

The following stability definition is a standard one in the context of Lyapunov stability theory [23]:

Definition 9. The equilibrium point x^* of Σ is said to be *stable* if given $\varepsilon>0$ there exists a $\delta(\varepsilon)>0$ such that $\|x_0-x^*\| \leqq \delta$ implies that

$$\|\phi(t, t_0, x_0, u^*)-x^*\| \leqq \varepsilon \qquad \text{for all } t \geqq t_0.$$

A very useful method for proving stability is by means of Lyapunov functions. The notion of a Lyapunov function is introduced in the following definition. It is a slight variation of the usual definition:

Definition 10. A real-valued function V defined on the state space X of Σ is said to be a *Lyapunov function* in the neighborhood of the equilibrium point x^* if

(i) V is *continuous* at x^*;

(ii) V attains a strong *local minimum* at x^*, *i.e.*, there exists a continuous function α: $R^+ \to R^+$ with $\alpha(\sigma)>0$ for $\sigma>0$ such that $V(x)-V(x^*) \geqq \alpha(\|x-x^*\|)$ for all $x \in X$ in a neighborhood of x^*;

(iii) V is *monotone nonincreasing* along solutions in the neighborhood of x^*, *i.e.*, the real-valued function $V(\phi(t, t_0, x_0, u^*))$ is monotone nonincreasing at $t=t_0$ for all x_0 in a neighborhood of x^*.

It is a standard exercise in (ε, δ)-manipulations to show that an equilibrium point $x^* \in X$ is stable if there exists a Lyapunov function in the neighborhood of x^*. This leads to the following theorem:

Theorem 6. *An equilibrium point $x^* \in X$ of a dissipative dynamical system Σ is stable if the storage function S is continuous and attains a strong local minimum at x^*. Moreover S is a Lyapunov function in the neighborhood of x^*.*

Proof. It suffices to show that $S(\phi(t, t_0, x_0, u^*))$ is monotone non-increasing at $t=t_0$ if $\|x_0-x^*\|$ is sufficiently small. By the dissipation inequality $S(\phi(t, t_0, x_0, u^*))$ is indeed monotone nonincreasing for all $t \geqq t_0$ and for all x^*. ‖

Note that S attains a strong local minimum at x^* if S_a does—consequently this condition may usually be verified without explicit knowledge of S. Note also that the fact that x^* is an equilibrium point itself follows if (i), (iii), (iv), (v) are satisfied and if S is continuous and attains a strong local minimum at x^*.

The consideration of the storage function is an extremely useful tool in stability investigations and by properly choosing the supply rates one may indeed obtain an interpretation for most of the existing stability criteria. In constructing a storage function it is natural to proceed to the evaluation of either the available storage or the required supply. These however lead to variational problems and it is only in exceptional circumstances that one may solve such problems, particularly if the dynamical system Σ is nonlinear. The concept of interconnected systems becomes in fact very useful in this context: it allows one to construct storage functions which correspond to neither the available storage nor the required supply, and which may be constructed by solving variational problems for—presumably less involved—subsystems of the original dynamical system Σ. This procedure will be illustrated in Section 7.2.

One may refine the basic result of Theorem 6 in several directions. Some of these are briefly discussed below:

(i) roughly speaking local minima of the storage function define stable equilibria and vice versa;

(ii) under appropriate additional hypotheses one may conclude that all trajectories actually approach the point of minimum storage. These additional hypotheses require the system to be strongly dissipative in the sense that no trajectory (other than the equilibrium) is free of dissipation. This strong form of dissipation is studied in [11]. We note here that one will usually not obtain $\dot{S}$ to be negative definite but merely semi-definite. The so-called invariance principles [24, 25] are thus very useful in establishing asymptotic stability in this context;

(iii) local maxima of the storage function will define an unstable equilibrium if all trajectories in its neighborhood involve some dissipation;

(iv) if $w(u^*, y)=0$ for all $\mathscr{Y} \in Y$, and if the system is lossless, then local minima *and* maxima of the storage function define stable equilibra.

6. Nonstationary Dynamical Systems

All of the above theory and results have been based on the hypothesis that the dynamical system under consideration is stationary. This stationarity has been postulated on two distinct levels:

(i) it has been assumed that the dynamical system Σ is itself stationary, *i.e.*, the constitutive equations defined by the maps ϕ and r are invariant under shifts of the time axis;

(ii) the storage functions have been assumed to be time-invariant, *i.e.*, the function S did not involve an explicit time dependence.

Although it may not seem so at first glance, assumptions (i) and (ii) are separate since (i) mainly refers to input/output stationarity whereas (ii) supplements this with internal stationarity. There are important types of physical systems (*e.g.*, rotating electrical machines) which are externally stationary but internally time-varying.

We view stationarity postulates as an *a priori* qualitative assumption imposed on the mathematical model of the dynamical systems under consideration. In this section we will indicate the modifications required to extend the above definitions to time-varying systems. Once the conceptual framework is appropriately expanded, one may indeed generalize the results to the time-varying case without difficulty.

1. The following definition generalizes Definition 1. In contrast with most similar definitions which have appeared in the literature we allow for the state space itself to be time-varying.

Definition 1′. A (continuous) *dynamical system*, Σ, is defined through the sets $\{U_t, \mathscr{U}, Y_t, \mathscr{Y}, X_t\}$, $t \in R$, and the maps ϕ_{t_1, t_0}, $(t_1, t_0) \in R_2^+$, and r_t, $t \in R$. These satisfy the following axioms:

(i) $\mathscr{U}$ is called the *input space* and consists of a class of functions $u(t)$, $t \in R$, taking their values at time t in the set of *input values* U_t;

(ii) $\mathscr{Y}$ is called the *output space* and consists of a class of functions $y(t)$, $t \in R$, taking their values at time t in the set of *output values* Y_t;

(iii) X_t is called the *state space* at time $t \in R$;

(iv) ϕ_{t, t_0} is called the *state transition function* and maps $X_{t_0} \times \mathscr{U}$ into X_{t_1}. It satisfies the analogous axioms of $(\text{iv})_a$, $(\text{iv})_b$, and $(\text{iv})_c$ of Definition 1;

(v) r_t is called the *read-out function* and is a map from $X_t \times U_t$ into Y_t;

(vi) the function $r_t(\phi(t, t_0, x_0, u), u(t))$ defined for $t \geqq t_0$ is the restriction to $[t_0, \infty)$ of an element of $\mathscr{Y}$.

The solution of the problem of state space realization in terms of equivalence classes goes through unchanged.

2. A (time-varying) dynamical system with supply rate at time t $w_t: U_t \times Y_t \to R$ is said to be *dissipative* if there exists a nonnegative function $S_t: X_t \to R^+$, called the *storage function*, such that

$$S_{t_0}(x_0) + \int_{t_0}^{t_1} w_t(t)\, dt \geqq S_{t_1}(x_1).$$

The available storage is defined by

$$S_{t_0, a}(x) = \sup_{\substack{x \to \\ t_1 \geqq t_0}} - \int_{t_0}^{t_1} w_t(t)\, dt,$$

whereas the definition of required supply necessitates again the notion of a point of minimal storage. Assume then that $x_t^* \in X_t$ minimizes $S_t(x)$ over $x \in X_t$ and as-

23*

sume in addition that $S_t(x_t^*)=0$ (this postulate now involves more than simply adjusting an additive constant). The *required supply* then becomes

$$S_{t_0,r}(x)=\inf_{\substack{x^*(t_1)\to x\\ t_{-1}\leq t_0}} \int_{t_{-1}}^{t_0} w_t(t)\,dt.$$

The results of Theorems 1 and 2 follow with the obvious modifications in notation. The available storage and the required supply are thus bounds on the storage functions and are themselves possible storage functions.

7. Applications

In this section we shall present a series of applications which serve to illustrate the previous theoretical developments.

7.1. Systems with a Finite Number of Degrees of Freedom

Consider the dynamical system described by the set of first order ordinary differential equations

$$\dot{\boldsymbol{x}}=\boldsymbol{f}(\boldsymbol{x},\boldsymbol{u}), \qquad \boldsymbol{y}=\boldsymbol{g}(\boldsymbol{x},\boldsymbol{u})$$

and assume that the supply function is given by

$$w=\langle \boldsymbol{u},\boldsymbol{y}\rangle=\boldsymbol{u}'\boldsymbol{y} \quad \text{(prime denotes transposition).}$$

Here, $\boldsymbol{x}\in R^n$, $\boldsymbol{u}$, $\boldsymbol{y}\in R^m$, and it is assumed that $\boldsymbol{f}$ and $\boldsymbol{g}$ are Lipschitz continuous in $\boldsymbol{x}$ and $\boldsymbol{u}$ jointly. It is well known that this implies that the above differential equation has a unique solution for any $\boldsymbol{x}(t_0)\in R^n$ and any locally square integrable $\boldsymbol{u}(t)$. Moreover the resulting functions $\boldsymbol{x}(t)$ and $\boldsymbol{y}(t)$ are themselves also locally square integrable.

The above differential equation thus describes a dynamical system in the sense of Definition 1 with $U=Y=R^m$, $X=R^n$, and $\mathscr{U}=\mathscr{Y}$ the locally square integrable R^m-valued functions defined on R. The differential equation itself defines the state transition map ϕ whereas the relation $\boldsymbol{y}=\boldsymbol{g}(\boldsymbol{x},\boldsymbol{u})$ describes the read-out function r. Note also that the supply function is locally integrable for $u\in\mathscr{U}$ and $y\in\mathscr{Y}$.

The problem at hand is (i) to determine conditions on $\boldsymbol{f}$ and $\boldsymbol{g}$ which make the dynamical system under consideration dissipative with respect to the given supply function and (ii) to discover the possible storage functions. If we restrict ourselves to sufficiently smooth storage functions then we are asking to find those functions $S: R^n\to R^+$ satisfying

$$\frac{d}{dt}S(\boldsymbol{x})=\nabla_{\boldsymbol{x}}S(\boldsymbol{x})\cdot\boldsymbol{f}(\boldsymbol{x},\boldsymbol{u})\leq\langle \boldsymbol{u},\boldsymbol{y}\rangle=\langle \boldsymbol{u},\boldsymbol{g}(\boldsymbol{x},\boldsymbol{u})\rangle$$

for all $\boldsymbol{x}\in R^n$ and $\boldsymbol{u}\in R^m$. BROCKETT [26] has in fact proposed this as a definition of passivity. This equivalent statement hardly solves the problem. The question of dissipativeness is by Theorem 1 equivalent to whether or not

$$\inf_{u\in\mathscr{U}_0}\int_0^\infty \langle \boldsymbol{u},\boldsymbol{g}(\boldsymbol{x},\boldsymbol{u})\rangle\,dt,$$

subject to the constraints $\dot{x}=f(x, u)$; $x(0)=x_0$, is finite for all $x_0 \in R^n$. The value of this infimum (which is seen to be nonpositive by taking $u=0$) yields the negative of the available storage function. This variational problem and the analogous one involved in the computation of the required supply are standard problems in optimal control and these techniques will be used in Part II to obtain some specific answers to the above questions. At the level of generality posed here it is impossible to obtain necessary and sufficient conditions on f and g for dissipativeness, but some interesting special cases offer a great deal of further insight:

(i) Consider the particular case (corresponding to elastic systems and to capacitive networks):

$$\dot{x}=u; \quad y=g(x).$$

In this case it is convenient to derive the conditions for dissipativeness directly from the dissipation inequality. Restricting ourselves again to sufficiently smooth storage functions (the available storage will in fact be smooth as a result of the assumption on f and g made earlier), we see that the dissipation inequality demands that there exists an S: $R^n \to R^+$ such that

$$\nabla_x S(x) \cdot u \leqq \langle u, g(x) \rangle$$

for all $u \in R^m$ and $x \in R^n$. This is the case if and only if the function $g'(x)$ is the gradient of a nonnegative function. It is well known that this requires $\frac{\partial g_i(x)}{\partial x_j} = \frac{\partial g_j(x)}{\partial x_i}$. This condition may be obtained in a different manner by noticing that $\int_0^t w(t)\, dt = \int_{x_0}^{x_1} g'(x)\, dx$. The integral on the right is bounded from below for a given x_0 and x_1 only if it is path independent which in turn requires $g'(x)$ to be the gradient of a real-valued function.

The necessary and sufficient conditions for dissipativeness may thus be expressed in terms of $g(x)$ by:

(i) $\frac{\partial g_i(x)}{\partial x_j} = \frac{\partial g_j(x)}{\partial x_i}$;

(ii) the path integral $P(x) = \int_{x^*}^{x} g'(x)\, dx$ is bounded from below.

Here, x^* is arbitrary and the function P differs from S only by an additive constant. It thus follows that the system is dissipative if and only if it is lossless. The storage function is thus unique and plays the role of a potential function since it determines the dynamical equations by

$$\dot{x}=u; \quad y=\nabla_x' S(x).$$

Note also that in this case one obtains reciprocity (condition (i)) as a result of dissipativeness. This is by no means a general property of dissipative systems however.

(ii) If we add "resistive" terms to the equations of motion studied in (i) then we obtain the dynamical system

$$\dot{x}=u; \quad y=g_1(x)+g_2(u).$$

If we assume (without loss of generality) that $g_2(\mathbf{0})=\mathbf{0}$ and concentrate again on sufficiently smooth storage functions then the dissipation inequality demands that there exists an $S: R^n \to R^+$ such that

$$\nabla_x S(x) \cdot u \leqq \langle g_1(x)+g_2(u), u \rangle \quad \text{for all } x \in R^n \text{ and } u \in R^m.$$

This inequality is satisfied if and only if

$$\nabla_x S(x)=g_1'(x) \quad \text{for all } x \in R^n \quad \textit{and} \quad \langle u, g_2(u) \rangle \geqq 0 \quad \text{for all } u \in R^m.$$

The conditions for dissipativeness are then those obtained in (i) augmented by the additional requirement $\langle u, g_2(u) \rangle \geqq 0$. The storage function is again unique (although the system need *not* be lossless) and is up to an additive constant given by the path integral $\int_{x^*}^{x} g_1'(x)\,dx$. The dissipation function (also unique) is given by $d(x, u)=\langle u, g_2(u) \rangle$. Notice that as a consequence of dissipativeness, we obtain reciprocity of the "elastic" terms $(g_1(x))$ but *not* of the "resistive" terms $(g_2(u))$.

The system studied here may be considered as the interconnection of the three systems:

$$\Sigma_1: \dot{x}_1=u_1; \quad y_1=g_1(x_1) \quad \text{with} \quad w_1=\langle u_1, y_1 \rangle,$$

$$\Sigma_2: y_2=g_2(u_2) \quad \text{with} \quad w_2=\langle u_2, y_2 \rangle,$$

$$\Sigma_3: y_3=y_4=-u; \quad y=u_3+u_4 \quad \text{with} \quad w_3=\langle u_3, y_3 \rangle+\langle u_4, y_4 \rangle+\langle u, y \rangle$$

with the neutral interconnection constraint

$$y_3=-u_1; \quad y_4=-u_2; \quad u_3=y_1; \quad u_4=y_2.$$

The storage function (but not its uniqueness) follows directly from here. The variables with subscripts represent the interconnecting inputs, outputs, and supply rates.

(iii) Consider the system described by the equations (u_2 is scalar-valued):

$$\dot{x}_1=u_1; \quad y_1=g_1(x_1, x_2)$$

$$\dot{x}_2=f_2(x_1, x_2, u_2); \quad y_2=g_2(x_1, x_2, u_2).$$

A simple calculation shows that dissipativeness implies that

$$\nabla_{x_1} S(x_1, x_2)=g_1'(x_1, x_2)$$

and

$$\nabla_{x_2} S(x_1, x_2) \cdot f_2(x_1, x_2, u_2) \leqq u_2\, g_2(x_1, x_2, u_2).$$

for all x_1, x_2, and u_2. Thus only part of the dependence of the storage function on the state vector is determined by the dynamical equations.

The above dynamical system is a particular case of the one studied by COLEMAN [27] and GURTIN [28] (see [2], Chapter 3). These authors obtained in fact very similar results. It should be realized however that for dynamical systems described in this amount of generality one needs a lot more information about the physics of the situation in order to obtain a unique storage function.

7.2. *Stability of Feedback Systems*

Consider the dynamical systems Σ_1 and Σ_2 and assume that $U_1=U_2=Y_1=Y_2$ are inner product spaces. Assume now that Σ_1 and Σ_2 are interconnected via the constraint $u_2=y_1$ and $u_1=-y_2$. This results in the feedback system shown in Figure 2.

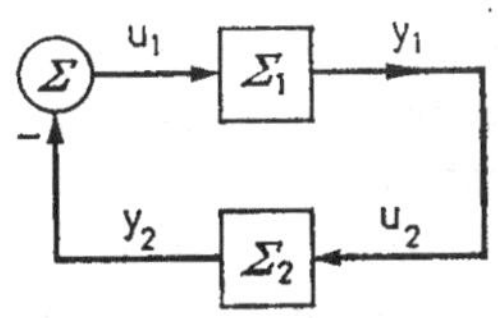

Fig. 2. Feedback system

The theory of dissipative systems discussed above offers a powerful method for investigating the stability of this feedback system. Assume that we associate the supply rate $w_1(u_1, y_1)$ with Σ_1 and the supply rate $w_2(u_2, y_2)$ with Σ_2. If w_1 and w_2 are such that $w_1(u, y)+w_2(y, -u)=0$ for all u and y, then the feedback system may be considered as an interconnected system with the neutral interconnection constraint: $u_2=y_1$, $u_1=-y_2$. Thus in order to prove stability it then suffices to show that Σ_1 is dissipative with respect to w_1 and that Σ_2 is dissipative with respect to w_2 (or, equivalently, with respect to αw_2 for some $\alpha>0$).

It may be verified that essentially all of the frequency-domain stability criteria which have recently appeared in the literature [17, 18] are based on this principle. Particularly important choices of the supply rates are $w_1=\|u_1\|^2-\|y_1\|^2$, $w_2=\|u_2\|^2-\|y_2\|^2$; $w_1=\langle u_1, y_1\rangle$, $w_2=\langle u_2, y_2\rangle$; and $w_1=\langle u_1+ay_1, u_1+by_1\rangle$, $w_2=-ab\left\langle u_2-\frac{1}{a}y_2, u_2-\frac{1}{b}y_2\right\rangle$. The stability theorems resulting from these choices of the supply rates are known as the small loop gain theorem, the positive operator theorem, and the conic operator theorem. The interpretation of these stability principles in terms of dissipative systems gives further insight in these results and unifies the existing conditions.

As an example, consider the autonomous dynamical system described by the set of first order ordinary differential equations:

$$\Sigma: \dot{\boldsymbol{x}}=\boldsymbol{A}\boldsymbol{x}-\boldsymbol{B}\boldsymbol{f}(\boldsymbol{C}\boldsymbol{x})$$

where $\boldsymbol{x}\in R^n$, $\boldsymbol{f}: R^m\to R^m$, and $\boldsymbol{A}$, $\boldsymbol{B}$, and $\boldsymbol{C}$ are matrices of appropriate dimensions. We assume again that $\boldsymbol{f}$ is Lipschitz continuous. Let $\boldsymbol{f}(\boldsymbol{0})=\boldsymbol{0}$; then the trajectory $\boldsymbol{x}(t)=\boldsymbol{0}$ is a solution to this differential equation and the stability properties of this solution have been the subject of a number of recent papers in the control theory literature. Particularly the construction of Lyapunov functions is a matter of great practical importance. The best known result in this area is the so-called

Popov criterion [29, 30] which answered a long-standing question known as the "Lur'e problem". We will reproduce this result for a representative case using the theory of dissipative systems. In doing so we obtain a whole class of Lyapunov functions in a systematic manner and extend the results recently obtained in [11].

We begin with viewing this dynamical system as the feedback interconnection of two dynamical systems, namely:

$$\Sigma_1:\ \dot{\boldsymbol{x}}_1 = \boldsymbol{A}\boldsymbol{x}_1 + \boldsymbol{B}\boldsymbol{u}_1; \quad \boldsymbol{y}_1 = \boldsymbol{C}\boldsymbol{A}\boldsymbol{x}_1 + \boldsymbol{C}\boldsymbol{Q}\boldsymbol{x}_1 + \boldsymbol{C}\boldsymbol{B}\boldsymbol{u}_1$$

and

$$\Sigma_2:\ \dot{\boldsymbol{x}}_2 = -\boldsymbol{Q}\boldsymbol{x}_2 + \boldsymbol{u}_2; \quad \boldsymbol{y}_2 = \boldsymbol{f}(\boldsymbol{x}_2)$$

with the interconnection constraint $\boldsymbol{u}_1 = -\boldsymbol{y}_2$; $\boldsymbol{u}_2 = \boldsymbol{y}_1$. The matrix $\boldsymbol{Q}$ is an arbitrary $(n \times n)$ matrix which features in the conditions for stability. It is clear that the above interconnection constraint defines a neutral interconnection with respect to the supply function $w_1 = \langle \boldsymbol{u}_1, \boldsymbol{y}_1 \rangle$ and $w_2 = \langle \boldsymbol{u}_2, \boldsymbol{y}_2 \rangle$. This interconnection leads to a "closed" system since we only have interconnecting, but no external, inputs and outputs. It is a simple matter involving only algebraic manipulations to show that the interconnection of Σ_1 and Σ_2 via the given interconnection constraint yield $\boldsymbol{x}_1(t) = \boldsymbol{x}(t)$ and $\boldsymbol{x}_2(t) = \boldsymbol{C}\boldsymbol{x}(t)$ provided the initial conditions are chosen as $\boldsymbol{x}_1(0) = \boldsymbol{x}(0)$ and $\boldsymbol{x}_2(0) = \boldsymbol{C}\boldsymbol{x}(0)$. The philosophy behind this equivalence is shown in Figure 3.

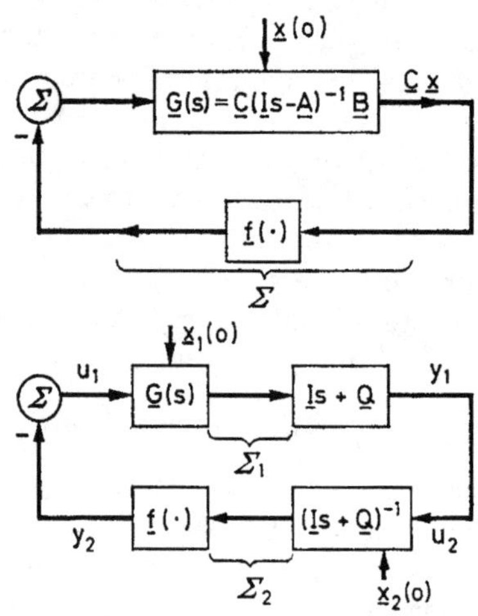

Fig. 3. Illustrating the interconnected system studied in Section 7.2

We now postulate the conditions for stability. These are:*

(i) $\{\boldsymbol{A}, \boldsymbol{B}, \boldsymbol{C}\}$ is a minimal realization of $G(s) = \boldsymbol{C}(\boldsymbol{I}s - \boldsymbol{A})^{-1}\boldsymbol{B}$

* Re $\lambda\{\boldsymbol{A}\}$ denotes the real part of an arbitrary eigenvalue of $\boldsymbol{A}$, the matrix inequality $\boldsymbol{P} \geqq \boldsymbol{0}$ indicates that the Hermitian matrix $\boldsymbol{P}$ is nonnegative definite, and "minimal realization" is system theory jargon which will be explained in detail in Part II. Positive real functions have been studied, particularly in the context of electrical network synthesis, and will be discussed in Part II.

(ii) $(Is+Q)G(s)$ is a positive real function of s

(iii) f is the gradient of a nonnegative function, *i.e.*,

$$\frac{\partial f_i(\sigma)}{\partial \sigma_j}=\frac{\partial f_j(\sigma)}{\partial \sigma_i} \quad \text{for all } \sigma \in R^m,$$

and the path integral $\int_0^z f'(\sigma)\,d\sigma \geqq 0$ for all $z \in R^m$

(iv) $f'(\sigma)\,Q\sigma \geqq 0$ for all $\sigma \in R^m$.

This stability claim will be verified with the aid of a suitable Lyapunov function. The idea behind the above conditions is that they make both Σ_1 and Σ_2 into dissipative systems. In fact, conditions (i) and (ii) ensure that Σ_1 is dissipative with respect to $w_1=\langle u_1, y_1\rangle$. The available storage $x'Q_a x$ and the required supply $x'Q_r x$ are positive definite quadratic forms. These functions and the other possible storage functions which are quadratic will be the subject of study in Part II of this paper. In order to verify that Σ_2 is dissipative with respect to $w_2=\langle u_2, y_2\rangle$, we shall use conditions (iii) and (iv). Consider thus

$$\eta=-\inf\int_{T_1}^{T_2}\langle u_2, y_2\rangle\,dt$$

subject to $\dot{x}_2=-Qx_2+u_2$; $y_2=f(x_2)$. We may eliminate u_2 and y_2 from this integral in terms of x_2. This yields

$$\eta=-\inf\left\{\int_{x_2(T_1)}^{x_2(T_2)} f'(x_2)\,dx_2+\int_{T_1}^{T_2} f'(x_2(t))\,Qx_2(t)\,dt\right\}.$$

Note that the first integral is independent of path by (iii) and the second one has an integrand which is always nonnegative by (iv). This last integral may be made arbitrarily small in the evaluation of the available storage and the required supply. The dynamical system Σ_2 is thus dissipative and its storage function is uniquely given by

$$S_2(x_2)=\int_0^{x_2} f'(\sigma)\,d\sigma \geqq 0.$$

The interconnected system is hence dissipative and has $S_1(x_1)+S_2(x_2)$ as an admissible storage function. Restricted to initial conditions $x_2(0)=Cx_1(0)$, this statement implies that the function $x'Q_s x+\int_0^{Cx} f'(\sigma)\,d\sigma$ is nonincreasing along solutions of Σ whenever $Q_s=Q_s'$ defines a quadratic storage function of Σ_1. Since $Q_s=Q_s'\geqq Q_a=Q_a'\geqq \varepsilon I$ for some $\varepsilon \geqq 0$ it follows that this Lyapunov function establishes the stability of the solution $x(t)=0$. By strengthening condition (ii) to include $\operatorname{Re}\lambda\{A\}<0$ we may in fact show, using this Lyapunov function, that all solutions approach their equilibrium solution $x=0$ as $t\to\infty$.

7.3. *Electrical Networks*

Consider an electrical network with n external ports and with a number of internal nodes and branches. We shall denote the external voltages and currents by the n-vectors V and I, respectively. Assume that the network contains resistors, capacitors, inductors, and lossless memoryless elements (*e.g.*, transformers, gyrators, *etc.*) which need not be specified further. Let n_R, n_C, and n_L denote the number of resistors, capacitors, and inductors, and denote the voltage across the elements and the current into these elements by the n_R-vectors V_R, I_R, the n_C-vectors V_C, I_C, and the n_L-vectors V_L, I_L respectively. We take the sign conventions shown in Figure 4.

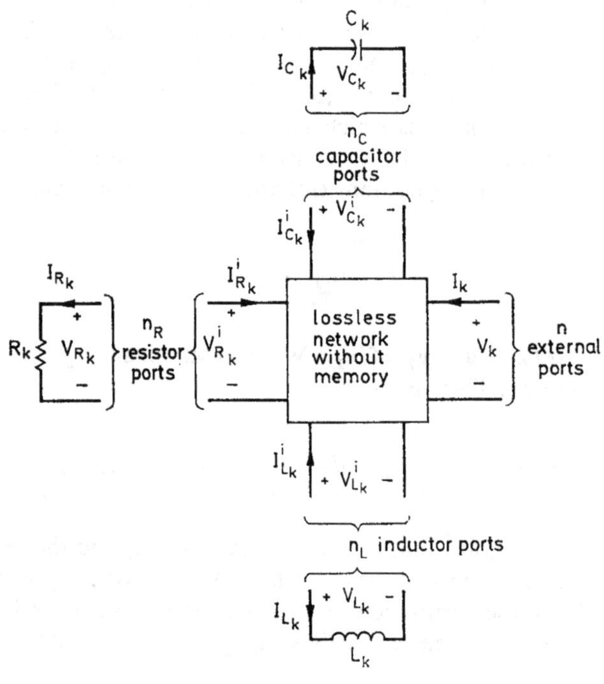

Fig. 4. The electrical n-port considered in Section 7.3

We now turn to the question of what should be considered inputs and outputs. This is a somewhat annoying question since what are the most convenient variables to work with depends on the network under consideration. In fact it has recently become apparent that the so-called scattering representation [31] (input$=v+\rho i$; output$=v-\rho i$, $\rho>0$) is by and large the most convenient model to consider. We shall consider here a somewhat simpler case and assume that [34]

(i) V is the input and I is the output;

(ii) the characteristic of the kth resistor is given by

$$V_{R_k}=R_k(I_{R_k})I_{R_k} \quad \text{with} \quad R_k \geqq 0.$$

This leads to the relation $\boldsymbol{V}_R=\boldsymbol{R}(\boldsymbol{I}_R)\boldsymbol{I}_R$.

(iii) the characteristic of the kth capacitor is given by

$$I_{C_k}=C_k(V_{C_k})\frac{dV_{C_k}}{dt} \quad \text{with} \quad C_k \geqq \varepsilon > 0.$$

This leads to the relation $\boldsymbol{I}_C=\boldsymbol{C}(\boldsymbol{V}_C)\dfrac{d\boldsymbol{V}_C}{dt}$.

(iv) the characteristic of the kth inductor is given by

$$V_{L_k}=L_k(I_{L_k})\frac{dI_{L_k}}{dt} \quad \text{with} \quad L_k \geqq \varepsilon > 0.$$

This leads to the relation $\boldsymbol{V}_L=\boldsymbol{L}(\boldsymbol{I}_L)\dfrac{d\boldsymbol{I}_L}{dt}$.

(v) the part of the network which neither involves dissipation nor memory defines an instantaneous relation from the voltages across the external ports, $\boldsymbol{V}$, the currents into the resistor ports, $\boldsymbol{I}_R^i$, the voltages across the capacitor ports, $\boldsymbol{V}_C^i$, and the currents into the inductor ports, $\boldsymbol{I}_L^i$, into the currents into the network at the external ports, $\boldsymbol{I}$, the voltages across the resistor ports, $\boldsymbol{V}_R^i$, the currents into the capacitor ports, $\boldsymbol{V}_C^i$, and the voltages across the inductor parts, $\boldsymbol{V}_L^i$. It is also assumed that $\langle \boldsymbol{V}, \boldsymbol{I}\rangle + \langle \boldsymbol{I}_R^i, \boldsymbol{V}_R^i\rangle + \langle \boldsymbol{V}_C^i, \boldsymbol{I}_C^i\rangle + \langle \boldsymbol{I}_L^i, \boldsymbol{V}_L^i\rangle = 0$.

The interconnected network may thus be considered a dynamical system with input $\boldsymbol{V}$, output $\boldsymbol{I}$, and state $\begin{bmatrix} \boldsymbol{V}_C \\ \hdashline \boldsymbol{I}_L \end{bmatrix}$. It is a neutral interconnection of dissipative systems with interconnection constraint $\boldsymbol{V}_R=\boldsymbol{V}_R^i$, $\boldsymbol{V}_C=\boldsymbol{V}_C^i$, $\boldsymbol{V}_L=\boldsymbol{V}_L^i$, $\boldsymbol{I}_R=-\boldsymbol{I}_R^i$, $\boldsymbol{I}_C=-\boldsymbol{I}_C^i$, and $\boldsymbol{I}_L=-\boldsymbol{I}_L^i$. The external supply rate is $\langle \boldsymbol{V}, \boldsymbol{I}\rangle$ and the internal supply rates are $\langle \boldsymbol{I}_R, \boldsymbol{V}_R\rangle$, $\langle \boldsymbol{V}_C, \boldsymbol{I}_C\rangle$, and $\langle \boldsymbol{I}_L, \boldsymbol{V}_L\rangle$. The stored energy in the capacitors and inductors is uniquely defined by

$$E(\boldsymbol{V}_C, \boldsymbol{I}_L)=\sum_1^{n_C} E_k(V_k)+\sum_1^{n_L} E_l(I_l)$$

with

$$E_k(V_{C_k})=\int_0^{V_{C_k}} vC_k(v)\,dv \quad \text{and} \quad E(I_{L_k})=\int_0^{I_{L_k}} iL_k(i)\,di.$$

Thus in standard electrical networks no ambiguity in the stored energy function arises. This is an immediate consequence of the fact that we are able to consider these systems as an interconnection of very simple subsystems in which the elements with memory are lossless. The individual dynamical subsystems involving memory are in fact described by first order scalar differential equations.

7.4. Thermodynamics

Consider a thermodynamic system at a uniform temperature. Assume that the mathematical model used for describing this system is in the form of a dynamical system in the sense of Definition 1 and that the outputs to this dynamical system Σ contain (possibly among other things) w, the rate of work done on the system, q, the rate of heat delivered to the system, and q/T where T denotes the temperature of the system. We assume that every admissible input and every initial state yield functions w, q, and q/T which are locally integrable.

The first and the second laws of thermodynamics may then be formulated by stating that a thermodynamic system is dissipative and lossless with respect to the supply rate $(w+q)$ and dissipative with respect to the supply rate $-\frac{q}{T}$. In terms of our definitions this implies the existence of two nonnegative functions E and $-S$ defined on the state space X of Σ such that every motion with $x_1=\phi(t_1, t_0, x_0, u)$ yields

$$E(x_0)+\int_{t_0}^{t_1}(w(t)+q(t))\,dt=E(x_1) \qquad \text{(Conservation of Energy)}$$

$$S(x_0)+\int_{t_0}^{t_1}\frac{q(t)}{T(t)}\,dt\leqq S(x_1) \qquad \text{(Clausius' inequality).*}$$

The function E is called the *internal energy* and S is called the *entropy*.

It follows from the results obtained earlier that E is uniquely defined once the equations of the dynamical system are given but that in general there will be many possible entropy functions. Two particular possibilities, S_a and S_r, may be computed *a priori* via the variational problems

$$S_a(x)=-\sup_{\substack{x\to\\ t_1\geqq 0}}\int_0^{t_1}\frac{q(t)}{T(t)}\,dt$$

and

$$S_r(x)=-\inf_{\substack{x^*\to x\\ t_{-1}\leqq 0}}-\int_{t_{-1}}^{0}\frac{q(t)}{T(t)}\,dt$$

where x^* is a point of maximal entropy normalized to $S(x^*)=0$. We may also conclude that, whatever the actual entropy may be, it satisfies the *a priori* inequality $S_r\leqq S\leqq S_a\leqq 0$. For reversible thermodynamic systems, *i.e.*, when Σ is

* DAY [35] has recently written a paper in which he shows how to replace this axiom by one involving the heat delivered and absorbed and the maximum and minimum temperature attained.

lossless with respect to $-\frac{q}{T}$ as well, the entropy is given unambiguously by $S=S_a=S_r$ provided the state space of Σ is reachable from x^* and controllable to x^*.

It should be emphasized that this ambiguity in the entropy function for irreversible thermodynamic systems is fundamental: the dynamical equations do not provide enough information to define the entropy uniquely. This difficulty has long been advertised by MEIXNER [32].

8. Conclusions

In the first part of this paper we have attempted to outline a general theory of dissipative dynamical systems. The mathematical model employed is a so-called state space model in which the map which generates outputs from inputs is viewed as the composition of a state transition map and a memoryless read-out function. This type of model is standard in control theory and dynamic estimation theory and it is argued that this model offers conceptual advantages for describing general physical systems with memory.

The definition of a dissipative dynamical system postulates the existence of a storage function which satisfies a dissipation inequality involving a given function called the supply rate. In many applications one knows from physical considerations that a storage function exists but it is often a difficult task to determine it. It is then shown that this difficulty is genuine and that the dynamical equations are insufficient to specify the storage function uniquely. However, the storage function satisfies an *a priori* bound, *i.e.*, it is bounded from below by the available storage and from above by the required supply. The available storage is the amount of internal storage which may be recovered from the system and the required supply is the amount of supply which has to be delivered to the system in order to transfer it from a state of minimum storage to a given state. Both these functions are themselves possible storage functions and their evaluation may be posed as variational problems.

These ideas were then applied to interconnected systems and it was established that for interconnected systems with interconnections which instantaneously redistribute the supply (the so-called neutral interconnections), the sum of the storage functions of the individual subsystems is a possible storage function for the interconnected system.

The stability properties of dissipative systems were then investigated and it was shown that states for which the storage function attains a local minimum are locally stable and that the storage function is a suitable Lyapunov function.

Part II of this paper will be devoted to an examination of linear systems with quadratic supply rates.

This research was supported in part by the National Science Foundation under Grant No. GK-25781 and in part by the U.K. Science Research Council. This paper was prepared while the author was a Senior Visiting Fellow at the Department of Applied Mathematics and Theoretical Physics of the University of Cambridge, Cambridge, England.

References

1. Truesdell, C., Elements of Continuum Mechanics. Berlin-Heidelberg-New York: Springer 1966.
2. Truesdell, C., Rational Thermodynamics. New York: McGraw-Hill 1969.
3. Onat, E. T., The Notion of State and Its Implications in Thermodynamics of Inelastic Solids, pp. 292–314 in: Proc. 1966 IUTAM Symposium, H. Parkus & L. I. Sedov, Ed. Berlin-Heidelberg-New York: Springer 1968.
4. Onat, E. T., Representation of Inelastic Mechanical Behavior by Means of State Variables, pp. 213–224 in: Proc. 1968 IUTAM Symposium, B. A. Boley, Ed. Berlin-Heidelberg-New York: Springer 1970.
5. Kalman, R. E., P. L. Falb, & M. A. Arbib, Topics in Mathematical System Theory. New York: McGraw-Hill 1969.
6. Desoer, C. A., Notes for a Second Course on Linear Systems. New York: Nostrand Reinhold 1970.
7. Zadeh, L. A., The Concepts of System, Aggregate, and State in System Theory, in: System Theory, L. A. Zadeh & E. Polak, Ed. New York: McGraw-Hill 1969.
8. Youla, D. C., The synthesis of linear dynamical systems from prescribed weighting patterns. J. SIAM Appl. Math. **14**, No. 3, 527–549 (1966).
9. Balakrishnan, A. V., Foundations of the state-space theory of continuous systems I. J. Computer and System Sci. **1**, 91–116 (1967).
10. Brockett, R. W., Finite Dimentional Linear Systesm. New York: Wiley 1970.
11. Willems, J. C., The construction of Lyapunov functions for input-output stable systems. SIAM J. Control **9**, No. 1, 105–134 (1971).
12. Baker, R. A., & A. R. Bergen, Lyapunov stability and Lyapunov functions of infinite dimensional systems. IEEE Trans. Automatic Control AC-**14**, 325–334 (1969).
13. Estrada, R. F., & C. A. Desoer, Passivity and stability of systems with a state representation. Int. J. Control **13**, No. 1, 1–26 (1971).
14. Breuer, S., & E. T. Onat, On recoverable work in linear visocelasticity. ZAMP **15**, No. 1, 12–21 (1964).
15. Breuer, S., & E. T. Onat, On the determination of free energy in linear viscoelastic solids. ZAMP **15**, No. 1, 184–191 (1964).
16. Youla, D. C., L. J. Castriota, & H. J. Carlin, Bounded real scattering matrices and the foundations of linear passive network theory. IRE Trans. on Circuit Theory CT-**6**, 102–124 (1959).
17. Zames, G., On the input-output stability of time-varying nonlinear feedback systems. Part I: Conditions derived using concepts of loop gain, conicity, and positivity; Part II: Conditions involving circles in the frequency plane and sector nonlinearities. IEEE Trans. Automatic Control AC-**11**, 228–238, 465–476 (1966).
18. Willems, J. C., The Analysis of Feedback Systems. Cambridge, Mass.: The M.I.T. Press 1971.
19. Meixner, J., On the theory of linear passive systems. Arch. Rational Mech. Anal. **17**, 278–296 (1964).
20. Carlin, H. J., Network theory without circuit elements. Proc. IEEE **55**, No. 4, 482–497 (1967).
21. Gruber, M., & J. L. Willems, On a Generalization of the Circle Criterion. Proc. of the Fourth Annual Allerton Conf. on Circuit and System Theory, pp. 827–835, 1966.
22. Freedman, M., & G. Zames, Logarithmic variation criteria for the stability of systems with time-varying gains. SIAM J. on Control **6**, 487–507 (1968).
23. Willems, J. L., Stability Theory of Dynamical Systems. Nelson, 1970.
24. Yoshizawa, T., The Stability Theory by Liapunov's Second Method. Math. Soc. Japan, 1966.
25. La Salle, J. P., An Invariance Principle in the Theory of Stability, pp. 277–286, in: Differential Equations and Dynamical Systems, J. K. Hale & J. P. La Salle, Ed. New York: Academic Press 1967.
26. Brockett, R. W., Path Integrals, Liapunov Functions and Quadratic Minimization. Proc. 4th Annual Allerton Conf. on Circuit and System Theory, Monticello, Ill., pp. 685–698, 1966.

27. COLEMAN, B. D., Thermodynamics of materials with memory. Arch. Rational Mech. Anal. **17**, 1–46 (1964).
28. GURTIN, M., On the thermodynamics of materials with memory. Arch. Rational Mech. Anal. **28**, 40–50 (1968).
29. POPOV, V. M., Hyperstability and optimality of automatic systems with several control functions. Rev. Roumaine Sci. Tech. Electrotechn. et Energ. **9**, No. 4, 629–690 (1964).
30. KALMAN, R. E., Lyapunov functions for the problem of Lur'e in automatic control. Proc. Nat. Acad. Sci. U.S.A. **49**, 201–205 (1963).
31. NEWCOMB, R. W., Linear Multiport Synthesis. New York: McGraw-Hill 1966.
32. MEIXNER, J., On the Foundation of Thermodynamics of Processes, pp. 37–47 in: A Critical Review of Thermodynamics, E. B. STUART, B. GAL-OR & A. J. BRAINARD, Ed. Mono Book Corp. 1970.
33. DAY, W. A., Thermodynamics based on a work axiom. Arch. Rational Mech. Anal. **31**, 1–34 (1968).
34. BRAYTON, R. K., & J. K. MOSER, A theory of nonlinear networks. Quart. Appl. Math. **22**, 1–33, 81–104 (1964).
35. DAY, W. A., A theory of thermodynamics for materials with memory. Arch. Rational Mech. Anal. **34**, 86–96 (1969).

Decision and Control Sciences Group
Electronic Systems Laboratory
Department of Electrical Engineering
Massachusetts Institute of Technology

(Received January 19, 1972)

On Self Tuning Regulators

K. J. ÅSTRÖM AND B. WITTENMARK

IN 1973 adaptive control was in one sense an old subject. The 1950s had seen a vigorous development, where virtually all the basic concepts were born. At the beginning of the 1960s the fundamental theoretical tools and formulations were laid out: Bellman, [3], had shown how Dynamic Programming was a natural tool for decision making under uncertainty and for showing how to employ optimally the increasing process knowledge that observations provided. Feldbaum, [5], had with *Dual Control* pinpointed the basic conflict between applying control and probing the system to secure more information (for better future control). Kalman, [8], had outlined the precursor to the self-tuning regulator: least-squares identification of a difference equation model of the process coupled with a regulator that was optimal (in some sense) for the current model estimate. Self-oscillating adaptive systems (SOAS) and Model Reference Adaptive Control (MRAC) had been devised and also tested in various applications. An excellent description of the status of adaptive control around this time can be found in [2].

It can perhaps be said that enthusiasm for adaptive control declined from the mid-1960s. There were a number of setbacks: flight tests of aircaraft using adaptive control to handle varying dynamics had not proven to be successful and model reference control had convergence difficulties when used for more than gain adjustment. Optimal (dual) control proved to be impossibly complex for use in practice. A number of papers appeared around 1970 describing approximations to dual control, but they merely emphasized the complexity of the issues involved.

It is in this light that the impact of this paper by Åström-Wittenmark should be seen. Like many important results, the paper's impact comes from the simplicity of the main idea it put forward. The approach is to combine two methods well known at the time: recursive estimation of the parameters of a linear discrete-time model of the system

$$\begin{aligned} y(t) + a_1 y(t-1) + \cdots + a_{n_a} y(t-n_a) \\ = b_1 u(t-n_k) + \cdots + b_{n_b} u(t-n_k-n_b+1) \end{aligned} \quad (1)$$

and application of dead-beat control using the estimated model. When the delay, n_k, is 1, the dead-beat control is

$$\begin{aligned} u(t) = \frac{1}{b_1}(\hat{a}_1 y(t) + \cdots + \hat{a}_{n_a} y(t-n_a+1) \\ -\hat{b}_2 u(t-1) - \cdots - \hat{b}_{n_b} u(t-n_b+1) \end{aligned} \quad (2)$$

A useful reparametrization of the system that permits direct estimation of the regulator parameters when $n_k > 1$ is described in the paper.

The main technical contribution of the paper is the following: *(under certain conditions) if the self-tuning regulator converges, it converges to the minimum variance controller.* The surprising aspect of this result is that it also applies when (1) is subject to colored noise, in which case the true dead-beat controller gives bad control, and the least-squares parameter estimates are biased. These two errors cancel each other, so to speak.

This result can be understood by observing that the parameter estimates are driven by the correlation between the control error and the past inputs and outputs. Theorem 5.1 in the paper shows that if the algorithm converges, then the control error becomes uncorrelated with all past data, so that minimum variance control is obtained.

Convergence itself is not claimed or proven in the paper, but examples and approximate analysis establish that the regulator in fact "often" converges to the optimal regulator. It later turned out that there are cases where convergence does not occur [9].

The paper also describes important extensions, such as feedforward, nonminimum phase systems, and nonlinear systems—problems that would keep the community busy for the next few decades.

The paper by Åström and Wittenmark had an immediate impact. Literally thousands of papers on self-tuning regulation, both theoretical and applied, appeared in the next decade. On the theoretical front, the paper left open the questions of convergence and stability and this inspired much subsequent research. A pioneering paper by Monopoli [10], although not complete, pointed the way, but the first proof of convergence of deterministic, continuous-time, adaptive systems systems, due to Feuer and Morse [6], did not appear until 1978; convergence was achieved by an ingenious but complex controller, *not* of the certainty equivalence type, that employed *nonlinear damping* to dominate an error (arising in all adaptive controllers) due to the rate of change of parameter error. A second proof (in a deterministic, discrete-time context), closer in spirit to the Åström-Wittenmark paper and resulting in a much simpler controller, was provided in [7], also included in this volume. The approach in this paper (also included in this volume) used certainty equivalence, as does Åström-Wittenmark, and achieved convergence by slowing down the identifier (by using error normalization, also proposed in [4]) thus reducing the error due to rate of change of the parameter error. Further details on this, and a brief discussion on subsequent work on this topic, can be found later in this volume, in the preamble to [7].

At the applied end of the spectrum, the paper (by Åström and Wittenmark) inspired, almost uniquely, many industrial applications. Providers of control equipment developed many hardware and software products that implement variants of the self-tuning regulator. Some of these developments are described in [1]. The lasting influence of the paper is perhaps best judged by the fact that today there are many thousands of control loops in practical use that have been designed using the self-tuning concept.

One may wonder why the impact of the self-tuning regulator was so massive and immediate. I believe there are several reasons: its simplicity, an almost "magical" robustness property (described in the paper), some early successful industrial applications, and the newly coined term *self-tuning*, all combined to attract a considerable amount of attention. The self-tuning regulator revitalized the field of adaptive control that had lost, in the early 1970s, some of its earlier lustre.

References

[1] K.J. Åström and B. Wittenmark, *Adaptive Control*, Addison-Wesley (Reading, MA), 2nd edition, 1995.

[2] K.J. Åström, "Adaptive control around 1960," *IEEE Control Systems (Magazine)*, **AC-16**(3):44–49, 1996.

[3] R. Bellman, *Adaptive Control Processes – A Guided Tour*, Princeton Univ. Press (Princeton, NJ), 1961.

[4] B. Egardt, *Stability of Model Reference Adaptive and Self-Tuning Regulators*, Lund Inst. Techn. Tech. Report (Lund, Sweden), 1978.

[5] A.A. Feldbaum, "Dual control theory I," *Avtomatica i Telemekhanika*, **21**(9):1240–1249, 1960. (English translation: *Automat. Remote Contr.*, **21**(9):874–880, 1961.)

[6] A. Feuer and A. S. Morse, "Adaptive control of single-input single-output linear systems," *IEEE Trans. Automat. Contr.*, **AC-23**(4):557–569, 1978.

[7] G. C. Goodwin, P.J. Ramadge, and P. E. Caines, "Discrete time multivariable adaptive control," *IEEE Trans. Automat. Contr.*, **AC-25**(3):449–456, 1980.

[8] R. E. Kalman, "Design of a self-optimizing control system," *Trans. ASME, J. Basic Engineering*, **80**:468–478, 1958.

[9] L. Ljung, "On positive real transfer functions and the convergence of some recursive schemes," *IEEE Trans. Automat. Contr.*, **AC-22**:539–551, 1977.

[10] R.V. Monopoli, "Model reference adaptive control with an augmented error," *IEEE Trans. Automat. Contr.*, **AC-19**:474–484, 1974.

L.L.

On Self Tuning Regulators*

Sur les Régulateurs Auto-Syntonisants

Über selbsteinstellende Regler

О самонастраивающихся регуляторах

K. J. ÅSTRÖM and B. WITTENMARK

Control laws obtained by combining a least squares parameter estimator and a minimum variance strategy based on the estimated parameters have asymptotically optimal performance.

Summary—The problem of controlling a system with constant but unknown parameters is considered. The analysis is restricted to discrete time single-input single-output systems. An algorithm obtained by combining a least squares estimator with a minimum variance regulator computed from the estimated model is analysed. The main results are two theorems which characterize the closed loop system obtained under the assumption that the parameter estimates converge. The first theorem states that certain covariances of the output and certain cross-covariances of the control variable and the output will vanish under weak assumptions on the system to be controlled. In the second theorem it is assumed that the system to be controlled is a general linear stochastic *n*th order system. It is shown that if the parameter estimates converge the control law obtained is in fact the minimum variance control law that could be computed if the parameters of the system were known. This is somewhat surprising since the least squares estimate is biased. Some practical implications of the results are discussed. In particular it is shown that the algorithm can be feasibly implemented on a small process computer.

1. INTRODUCTION

IT HAS been shown in several cases that linear stochastic control theory can be used successfully to design regulators for the steady state control of industrial processes. See Ref. [1]. To use this theory it is necessary to have mathematical models of the system dynamics and of the disturbances. In practice it is thus necessary to go through the steps of plant experiments, parameter estimation, computation of control strategies and implementation. This procedure can be quite time consuming in particular if the computations are made off-line. It might also be necessary to repeat the procedure if the system dynamics or the characteristics of the disturbances are changing as is often the case for industrial processes.

* Received 2 March 1972; revised 12 September 1972. The original version of this paper was presented at the 5th IFAC Congress which was held in Paris, France during June 1972. It was recommended for publication in revised form by Associate Editor A. Sage.

From a practical point of view it is thus meaningful to consider the control of systems with constant but unknown parameters. Optimal control problems for such systems can be formulated and solved using non-linear stochastic control theory. The solutions obtained are extremely impractical since even very simple problems will require computations far exceeding the capabilities of todays computers. For systems with constant but unknown parameters it thus seems reasonable to look for strategies that will converge to the optimal strategies that could be derived if the system characteristics were known. Such algorithms will be called *self-tuning* or *self-adjusting* strategies. The word adaptive is not used since adaptive, although never rigorously defined, usually implies that the characteristics of the process are changing. The problem to be discussed is thus simpler than the adaptive problem in the sense that the system to be controlled is assumed to have constant parameters.

The purpose of the paper is to analyse one class of self-adjusting regulators. The analysis is restricted to single-input single-output systems. It is assumed that the disturbances can be characterized as filtered white noise. The criterion considered is the minimization of the variance of the output. The algorithms analysed are those obtained on the basis of a separation of identification and control. To obtain a simple algorithm the identification is simply a least squares parameter estimator.

The main result is a characterization of the closed loop systems obtained when the algorithm is applied to a general class of systems. It is shown in Theorem 5.1 that if the parameter estimates converge the closed loop loop system obtained will be such that certain covariances of the inputs and the outputs of the closed loop system are zero. This is shown under weak assumptions on the system to be

Reprinted with permission from *Automatica*, K. J. Åström and B. Wittenmark, "On Self Tuning Regulators" Vol. 9, 1973, pp. 185–199.

controlled. Moreover if it is assumed that the system to be controlled is a sampled finite dimensional linear stochastic system with a time delay in the control signal it is demonstrated in Theorem 5.2 that, if the parameter estimates converge, the corresponding regulator will actually converge to the minimum variance regulator. This is true, in spite of the fact that the least squares estimate is biased.

The major assumptions are that the system is minimum phase, that the time delay is known and that a bound can be given to the order of the system. The first two assumptions can be removed at the price of a more complicated algorithm.

The paper is organized as follows: sections 2 and 3 provide background and a motivation. The algorithm is given in section 4. Control strategies for systems with known parameters are given in section 2. Least squares parameter estimation is briefly reviewed in section 3. Some aspects on the notion of identifiability is also given in section 3. The algorithm presented in section 4 is obtained simply by fitting the parameters of a least squares structure as was described in section 3 and computing the corresponding minimum variance control strategy as was described in section 2. The possible difficulty with non-identifiability due to the feedback is avoided by fixing one parameter.

The main result is given as two theorems in section 5. We have not yet been able to prove that the algorithm converges in general. In section 6 it is, however, shown that a modified version of the algorithm converges for a first order system. The convergence properties of the algorithm are further illustrated by the examples in section 7. Some practical aspects of the algorithm as well as some problems which remain to be solved are given in section 8. In particular it is shown that the algorithm is easily implemented on a minicomputer.

2. MINIMUM VARIANCE CONTROL

This section gives the minimum variance strategy for a system with known parameters. Consider a system described by

$$y(t)+a_1y(t-1)+\dots+a_ny(t-n)=b_1u(t-k-1)$$

$$+\dots+b_nu(t-k-n)+\lambda[e(t)+c_1e(t-1)$$

$$+\dots+c_ne(t-n)],$$

$$t=0,\ \pm1,\ \pm2,\dots \tag{2.1}$$

where u is the control variable, y is the output and $\{e(t),\ t=0, \pm1, \pm2, \dots\}$ is a sequence of independent normal (0, 1) random variables. If the forward shift operator q, defined by

$$qy(t)=y(t+1)$$

and the polynomials

$$A(z)=z^n+a_1z^{n-1}+\dots+a_n$$

$$B(z)=b_1z^{n-1}+\dots+b_n,\ b_1\neq0$$

$$C(z)=z^n+c_1z^{n-1}+\dots+c_n$$

are introduced, the equation (2.1) describing the system can be written in the following compact form:

$$A(q)y(t)=B(q)u(t-k)+\lambda C(q)e(t). \tag{2.2}$$

It is well known that (2.1) or (2.2) is a canonical representation of a sampled finite dimensional single-input single-output dynamical system with time delays in the output and disturbances that are gaussian random processes with rational spectral densities.

The model (2.1) also admits a time delay τ in the system input which need not be a multiple of the sampling interval. The number k corresponds to the integral part of τ/h, where h is the sampling interval.

Let the criterion be

$$V_1=Ey^2(t) \tag{2.3}$$

or

$$V_2=E\frac{1}{N}\sum_1^N y^2(t). \tag{2.4}$$

The optimal strategy is then

$$u(t)=-\frac{q^kG(q)}{B(q)F(q)}y(t) \tag{2.5}$$

where F and G are polynomials

$$F(z)=z^k+f_1z^{k-1}+\dots+f_k \tag{2.6}$$

$$G(z)=g_0z^{n-1}+g_1z^{n-2}+\dots+g_{n-1} \tag{2.7}$$

determined from the identity

$$q^kC(q)=A(q)F(q)+G(q). \tag{2.8}$$

Proofs of these statements are given in [2]. The following conditions are necessary:

—The polynomial B has all zeroes inside the unit circle. Thus the system (2.1) is minimum phase.

—The polynomial C has all zeroes inside the unit circle.

These conditions are discussed at length in [1]. Let it suffice to mention here that if the system (2.1) is non-minimum phase the control strategy (2.5) will still be a minimum variance strategy. This strategy will, however, be so sensitive that the slightest variation in the parameters will create an unstable closed loop system. Suboptimal strategies which are less sensitive to parameter variations are also well known. This paper will, however, be limited to minimum phase systems.

3. PARAMETER ESTIMATION

For a system described by (2.1) it is thus straight forward to obtain the minimum variance regulator, if the parameters of the model are known. If the parameters are not known it might be a possibility to try to determine the parameters of (2.1) using some identification scheme and then use the control law (2.5) with the true parameters substituted by their estimates. A suitable identification algorithm is the maximum likelihood method which will give unbiased estimates of the coefficients of the A, B and C polynomials. The maximum likelihood estimates of the parameters of (2.1) are, however, strongly non-linear functions of the inputs and the outputs. Since finite dimensional sufficient statistics are not known it is not possible to compute the maximum likelihood estimate of the parameters of (2.1) recursively as the process develops. Simpler identification schemes are therefore considered.

The least squares structure

The problem of determining the parameters of the model (2.1) is significantly simplified if it is assumed that $c_i=0$ for $i=1, 2, \ldots, n$. The model is then given by

$$A(q)y(t)=B(q)u(t-k)+\lambda e(t+n). \tag{3.1}$$

The parameters of this model can be determined simply by the least squares method [3]. The model (3.1) is therefore referred to as a *least squares model.*

The least squares estimate has several attractive properties. It can easily be evaluated recursively. The estimator can be modified to take different model structures, e.g. known parameters, into account. The least squares estimates will converge to the true parameters e.g. under the following conditions.

- —The output $\{y(t)\}$ is actually generated from a model (3.1).
- —The residuals $\{e(t)\}$ are independent.
- —The input is persistently exciting, see Ref. [3].
- —The input sequence $\{u(t)\}$ is independent of the disturbance sequence $\{e(t)\}$.

These conditions are important. If the residuals are correlated the least squares estimate will be biased. If the input sequence $\{u(t)\}$ depends on $\{e(t)\}$ it may not be possible to determine all parameters.

When the inputs are generated by a feedback they are correlated with the disturbances and it is not obvious that all the parameters of the model can be determined. Neither is it obvious that the input generated in this way is persistently exciting of sufficiently high order. A simple example illustrates the point.

Example 3.1

Consider the first order model

$$y(t)+ay(t-1)=bu(t-1)+e(t). \tag{3.2}$$

Assume that a linear regulator with constant gain

$$u(t)=\alpha y(t) \tag{3.3}$$

is used. If the parameters a and b are known the gain $\alpha=a/b$ would obviously correspond to a minimum variance regulator. If the parameters are not known the gain $\alpha=\hat{a}/\hat{b}$ where $\hat{a}$ and $\hat{b}$ are the least squares estimates of a and b could be attempted. The least squares parameter estimates are determined in such a way that the loss function

$$V(a, b)=\sum_1^N[y(t+1)+ay(t)-bu(t)]^2 \tag{3.4}$$

is minimal with respect to a and b. If the feedback control (3.3) is used the inputs and outputs are linearly related through

$$u(t)-\alpha y(t)=0. \tag{3.5}$$

Multiply (3.5) by $-\gamma$ and add to the expression within brackets in (3.4).

Hence

$$V(a, b)=\sum_1^N[y(t+1)+(a+\alpha\gamma)y(t)-(b+\gamma)u(t)]^2$$

$$=V(a+\alpha\gamma, b+\gamma).$$

The loss function will thus have the same value for all estimates a and b on a linear manifold. Thus the two parameters a and b of the model (3.2) are not identifiable when the feedback (3.3) is used.

The simple example shows that it is in general not possible to estimate all the parameters of the model (3.1) when the input is generated by a feedback. Notice, however, that all parameters can be estimated if the control law is changed. In the particular example it is possible to estimate both parameters of the model, if the control law (3.3) is replaced by

$$u(t)=\alpha y(t-1)$$

or

$$u(t)=\alpha_1 y(t)+\alpha_2 y(t-1)$$

or if a time varying gain is used.

4. THE ALGORITHM

In order to control a system with constant but unknown parameters the following procedure could be attempted. At each step of time determine the parameters of the system (3.1) using least squares estimation based on all previously observed inputs and outputs as was described in section 3. Then determine a control law by calculating the minimum variance strategy for the model obtained. To compute the control law the identity (2.8) must be resolved in each step. The problem of computing the minimum variance regulator is simplified if it is observed that by using the identity (2.8) the system (3.1) can be written as

$$\begin{aligned} y(t+k+1)+\alpha_1 y(t)+\ldots+\alpha_m(t-m+1) \\ =\beta_0[u(t)+\beta_1 u(t-1)+\ldots+\beta_l u(t-l)] \\ +\varepsilon(t+k+1) \end{aligned} \tag{4.1}$$

where $m=n$ and $l=n+k-1$. The coefficients α_i and β_i are computed from the parameters a_i and b_i in (3.1) using the identity (2.8). The disturbance $\varepsilon(t)$ is a moving average of order k of the driving noise $e(t)$.

The minimum variance strategy is then simply

$$\begin{aligned} u(t)=\frac{1}{\beta_0}[\alpha_1 y(t)+\ldots+\alpha_m y(t-m+1)] \\ -\beta_1 u(t-1)-\ldots-\beta_l u(t-l). \end{aligned} \tag{4.2}$$

In order to obtain simple computation of the control strategy it could thus be attempted to use the model structure (4.1) which also admits least squares estimation. The trade-off for the simple calculation of the control law is that k more parameters have to be estimated.

As was shown in Example 3.1 all parameters of the model (4.1) can not necessarily be determined from input–output observations if the input is generated by a feedback (4.2) with constant parameters. In order to avoid a possible difficulty it is therefore assumed that the parameter β_0 is given. It will be shown in section 6 that the choice of β_0 is not crucial.

Summing up, the algorithm can be described as follows.

Step 1 parameter estimation

At the sampling interval t determine the parameters $\alpha_1,\ldots\alpha_m,\ \beta_1,\ldots,\beta_l$ of the model

$$\begin{aligned} y(t)+\alpha_1 y(t-k-1)+\ldots+\alpha_m y(t-k-m) \\ =\beta_0[u(t-k-1)+\beta_1 u(t-k-2) \\ +\ldots+\beta_l u(t-k-l-1)]+\varepsilon(t) \end{aligned} \tag{4.1}$$

using least squares estimation based on all data available at time t, i.e.

$$\sum_{k=0}^{t}\varepsilon^2(k)$$

minimum. The parameter β_0 is assumed known.

Step 2 control

At each sampling interval determine the control variable from

$$\begin{aligned} u(t)=\frac{1}{\beta_0}[\alpha_1 y(t)+\ldots+\alpha_m y(t-m+1)] \\ -\beta_1 u(t-1)-\ldots-\beta_l u(t-l) \end{aligned} \tag{4.2}$$

where the parameters α_i and β_i are those obtained in Step 1.

The control law (4.2) corresponds to

$$\begin{aligned} u(t)=\frac{\alpha_1+\alpha_2 q^{-1}+\ldots+\alpha_m q^{-m+1}}{\beta_0[1+\beta_1 q^{-1}+\ldots+\beta_l q^{-l}]}y(t) \\ =\frac{q^{l-m+1}\mathscr{A}(q)}{\mathscr{B}(q)}y(t). \end{aligned} \tag{4.3}$$

Since the least squares estimate can be computed recursively the algorithm requires only moderate computations.

It should be emphasized that the algorithm is not optimal in the sense that it minimizes the criterion (2.3), or the criterion (2.4). It fails to minimize (2.3) because it is not taken into account that the parameter estimates are inaccurate and it fails to minimize (2.4) because it is not dual in FELDBAUM's sense [4]. These matters are discussed in [2]. It will however, be shown in section 5 that the algorithm has nice asymptotic properties.

The idea of obtaining algorithms by a combination of least squares identification and control is old. An early reference is KALMAN [5]. The particular algorithm used here is in essence the same as the one presented by PETERKA [6]. A similar algorithm where the uncertainties of the parameters are also considered is given in WIESLANDER–WITTENMARK [7].

5. MAIN RESULTS

The properties of the algorithm given in the previous section will now be analysed. We have

Theorem 5.1

Assume that the parameter estimates $\alpha_i(t)$, $i=1,\ldots,m$, $\beta_i(t)$, $i=1,\ldots l$ converge as $t\to\infty$ and that the closed loop system is such that the output is *ergodic* (in the second moments). Then the closed loop system has the properties

$$Ey(t+\tau)y(t)=r_y(\tau)=0 \quad \tau=k+1,\ldots k+m \tag{5.1}$$

$$Ey(t+\tau)u(t)=r_{yu}(\tau)=0 \quad \tau=k+1,\ldots,k+l+1. \tag{5.2}$$

Proof

The least squares estimates of the parameters $\alpha_1, \alpha_2, \ldots, \alpha_m, \beta_1, \beta_2, \ldots, \beta_l$ is given by the equations

$$\left[\begin{array}{cccc|ccc}
\Sigma y(t)^2 & \Sigma y(t)y(t-1) & \ldots & \Sigma y(t)y(t-m+1) & -\beta_0\Sigma y(t)u & \ldots & -\beta_0\Sigma y(t)u(t-l) \\
\Sigma y(t)y(t-1) & & & \Sigma y(t-1)y(t-m+1) & \vdots & & \vdots \\
\vdots & & & \vdots & & & \\
\Sigma y(t)y(t-m+1) & & & \Sigma y^2(t-m+1) & -\beta_0\Sigma y(t-m+1)u(t-1) & \ldots & -\beta_0\Sigma y(t-m+1)u(t-l) \\
\hline
-\beta_0\Sigma y(t)u(t-1) & & & & \beta_0^2\Sigma u^2(t-1) & \ldots & \beta_0^2\Sigma u(t-1)u(t-l) \\
\vdots & & & & & & \vdots \\
-\beta_0\Sigma y(t)u(t-l) & \ldots & & & & \ldots & \beta_0^2\Sigma u^2(t-l)
\end{array}\right].$$

$$\begin{bmatrix} \alpha_1 \\ \alpha_2 \\ \vdots \\ \alpha_m \\ \beta_1 \\ \vdots \\ \beta_l \end{bmatrix} = \begin{bmatrix} -\Sigma y(t+k+1)y(t)+\beta_0\Sigma u(t)y(t) \\ \Sigma y(t+k+1)y(t-1)+\beta_0\Sigma u(t)y(t-1) \\ \vdots \\ -\Sigma y(t+k+1)y(t-m+1)+\beta_0\Sigma u(t)y(t-m+1) \\ -\beta_0\Sigma y(t+k+1)u(t-1)-\beta_0^2\Sigma u(t)u(t-1) \\ \vdots \\ \beta_0\Sigma y(t+k+1)u(t-l)-\beta_0^2\Sigma u(t)u(t-l) \end{bmatrix} \tag{5.3}$$

where the sums are taken over N_0 values. See Ref. [3].

Assume that the parameters converge. For sufficiently large N_0 the coefficients of control law (4.2) will then converge to constant values. Introduction of (4.2) into (5.3) gives

$$\begin{aligned}
&\Sigma y(t+k+1)y(t)=0 \\
&\Sigma y(t+k+1)y(t-1)=0 \\
&\quad\vdots \\
&\Sigma y(t+k+1)y(t-m+1)=0 \\
&\Sigma y(t+k+1)u(t-1)=0 \\
&\quad\vdots \\
&\Sigma y(t+k+1)u(t-l)=0.
\end{aligned}$$

Using the control law (4.2) it also follows that

$$\Sigma y(t+k+1)u(t)=0.$$

Under the ergodicity assumption the sums can furthermore be replaced by mathematical expectations and the theorem is proven.

Remark 1

Notice that the assumptions on the system to be controlled are very weak. In particular it is not necessary to assume that the system is governed by an equation like (2.1) or (3.1).

Remark 2

It is sufficient for ergodicity that the system to be controlled is governed by a difference equation of finite order, e.g. like (2.1), and that the closed loop system obtained by introducing the feedback law into (2.1) gives a stable closed loop system.

Remark 3

The self tuning algorithm can be compared with a PI-regulator. If the state variables of a deterministic system with a PI-regulator converge to steady state values, the control error must be zero irrespective of the properties of the system. Analogously theorem 5.1 implies that if the parameter estimates of the self tuning algorithm converge the covariances (5.1) and (5.2) are zero.

Remark 4

Theorem 5.1 holds even if the algorithm is modified in such a way that the parameter estimation (Step 1) is not done in every step.

If it is assumed that the system to be controlled is governed by an equation like (2.1) it is possible to show that the conditions (5.1) and (5.2) in essence imply that the self tuning regulator will converge to a minimum variance regulator. We have

Theorem 5.2

Let the system to be controlled be governed by the equation (2.1). Assume that the self tuning algorithm is used with $m=n$ and $l=n+k-1$. If the parameter estimates converge to values such that the corresponding polynomials $\mathscr{A}$ and $\mathscr{B}$ have no common factors, then the corresponding regulator (4.2) will converge to a minimum variance regulator.

Proof

Assume that the least squares parameter estimates converge. The regulator is then given by (4.3) i.e.

$$u(t)=\frac{q^{l-m+1}\mathscr{A}(q)}{\mathscr{B}(q)}y(t)=\frac{q^k\mathscr{A}(q)}{\mathscr{B}(q)}$$

where the coefficients of $\mathscr{A}$ and $\mathscr{B}$ are constant. Since the system to be controlled is governed by (2.1) the closed loop system becomes

$$[A(q)\mathscr{B}(q)-B(q)\mathscr{A}(q)]y(t)=\lambda C(q)\mathscr{B}(q)e(t). \tag{5.4}$$

The closed loop system is of order $r=n+l$. Introduce the stochastic process $\{v(t)\}$ defined by

$$v(t)=\lambda\frac{q^lC(q)}{A(q)\mathscr{B}(q)-B(q)\mathscr{A}(q)}e(t). \tag{5.5}$$

Then

$$y(t)=q^{-l}\mathscr{B}(q)v(t) \tag{5.6}$$

and

$$u(t)=q^{-m+1}\mathscr{A}(q)v(t). \tag{5.7}$$

Multiplying (5.6) and (5.7) by $y(t+\tau)$ and taking mathematical expectations gives

$$r_y(\tau)=r_{yv}(\tau)+\beta_1r_{yv}(\tau+1)\ldots+\beta_lr_{yv}(\tau+l) \tag{5.8}$$

$$r_{yu}(\tau)=\alpha_1r_{yv}(\tau)+\alpha_2r_{yv}(\tau+1)$$

$$+\ldots+\alpha_mr_{yv}(\tau+m-1). \tag{5.9}$$

Furthermore it follows from Theorem 5.1, equations (5.1) and (5.2), that the left member of (5.8) vanishes for $\tau=k+1,\ldots,k+m$ and that the left member of (5.9) vanishes for $\tau=k+1,\ldots,k+l+1$. We thus obtain the following equation for $r_{yv}(\tau)$.

$$\begin{bmatrix} 1 & \beta_1 & \ldots & & \beta_l & 0 & \ldots & 0 \\ 0 & 1 & \beta_1 & & & \cdot & & \cdot \\ \cdot & & \cdot & & & \cdot & & \cdot \\ \cdot & & & \cdot & & & \cdot & \cdot \\ \cdot & & & & \cdot & & & 0 \\ 0 & \ldots & 0 & 1 & \beta_1 & & & \beta_l \\ 0 & \alpha_1 & \ldots & & \alpha_m & 0 & \ldots & 0 \\ \cdot & & \cdot & & & \cdot & & \cdot \\ \cdot & & & \cdot & & & \cdot & \cdot \\ & & & & & & & 0 \\ 0 & \ldots & & & \alpha_1 & & \ldots & \alpha_m \end{bmatrix}\begin{bmatrix} r_{yv}(k+1) \\ r_{yv}(k+2) \\ \cdot \\ \cdot \\ \cdot \\ \\ \\ \\ \\ \\ r_{yv}(k+l+m) \end{bmatrix}=\begin{bmatrix} 0 \\ 0 \\ \cdot \\ \cdot \\ \cdot \\ \\ \\ \\ \\ \\ 0 \end{bmatrix}. \tag{5.10}$$

Since the polynomials $\mathscr{A}$ and $\mathscr{B}$ have no common factor it follows from an elementary result in the theory of equations [8, p. 145] that the $(l+m) \times (l+m)$-matrix of the left member of (5.10) is non-singular.

Hence

$$r_{yv}(\tau)=0 \qquad \tau=k+1, \dots, k+l+m. \tag{5.11}$$

Since v is the output of a dynamical system of order $r=n+l=m+l$ driven by white noise it follows from (5.11) and the Yule–Walker equation that

$$r_{yv}(\tau)=0 \qquad \tau \geqslant k+1. \tag{5.12}$$

The equation (5.8) then implies

$$r_y(\tau)=0 \qquad \tau \geqslant k+1. \tag{5.13}$$

The output of the closed loop system is thus a moving average of white noise of order k. Denote this moving average by

$$y(t)=q^{-k}F(q)e(t) \tag{5.14}$$

where F is a polynomial of degree k. It follows from (5.4) and (5.14) that

$$q^k C = FA - \frac{BF\mathscr{A}}{\mathscr{B}}.$$

Since $q^k C$ and FA are polynomials $BF\mathscr{A}/\mathscr{B}$ must also be a polynomial.

Hence

$$q^k C = FA + G \tag{5.15}$$

where

$$G = -\frac{BF\mathscr{A}}{\mathscr{B}} \tag{5.16}$$

is of degree $n-1$ and F of degree k. A comparison with (2.8) shows, however, that (5.15) is the identity which is used to derive the minimum variance strategy. It thus follows from (2.5) that the minimum variance strategy is given by

$$u(t) = -\frac{q^k G}{BF} y(t).$$

The equation (5.16) then implies that

$$-\frac{q^k G}{BF} = \frac{q^k \mathscr{A}}{\mathscr{B}} \tag{5.17}$$

and it has thus been shown that (4.2) is a minimum variance strategy.

Remark

The conditions $m=n$ and $l=n+k-1$ mean that there are precisely the number of parameters in the model that are required in order to obtain the correct regulator for the process (2.1). Theorem 5.2 still holds if there are more parameters in the regulator in the following cases.

(i) Theorem 5.2 still holds if $m=n$ and $l \geqslant n+k-1$. In this case the order of the system is $r=n+l$ and since $m=n$ the equation (5.10) implies (5.11)–(5.13) and the equation (5.16) is changed to

$$G = -\frac{q^{l-k-m+1} BF\mathscr{A}}{\mathscr{B}}. \tag{5.16'}$$

The rest of the proof remains unchanged.

(ii) If $m \geqslant n$ and $l=n+k-1$ the theorem will also hold. The closed loop system is of order $r \leq m+l$ but the equation (5.10) will still imply (5.11) and (5.16) is changed to (5.16′). The rest of the proof remains unchanged.

(iii) If $m>n$ and $l>n+k-1$ the theorem does not hold because $\mathscr{A}$ and $\mathscr{B}$ must have a common factor if the parameter estimates converge. It can, however, be shown that if the algorithm is modified in such a way that common factors of $\mathscr{A}$ and $\mathscr{B}$ are eliminated before the control signal is computed, Theorem 5.2 will still hold for the modified algorithm.

6. CONVERGENCE OF THE ALGORITHM

It would be highly desirable to have general results giving conditions for convergence of the parameter estimates. Since the system (2.1) with the regulator (4.2) and the least squares estimator is described by a set of nonlinear time dependent stochastic difference equations the problem of a general convergence proof is difficult. So far we have not been able to obtain a general result. It has, however, been verified by extensive numerical simulations that the algorithm does in fact converge in many cases. The numerical simulations as well as analysis of simple examples have given insight into some of the conditions that must be imposed in order to ensure that the algorithm will converge.

A significant simplification of the analysis is obtained if the algorithm is modified in such a way that the parameter estimates are kept constant over long periods of time. To be specific a simple example is considered.

Example 6.1

Let the system be described by

$$y(t)+ay(t-1)=bu(t-1)+e(t)+ce(t-1) \quad |c|<1. \tag{6.1}$$

Assume that the control law

$$u(t)=\alpha_n y(t) \tag{6.2}$$

is used in the time interval $t_n<t<t_{n+1}$ where the parameter α_n is determined by fitting the least squares model

$$y(t+1)+\alpha y(t)=u(t)+\varepsilon(t+1) \tag{6.3}$$

to the data $\{u(t), y(t), t=t_{n-1}, \ldots, t_n-1)\}$.

The least squares estimate is given by

$$\alpha_n=-\frac{\sum_{t=t_{n-1}}^{t_n-2} y(t)[y(t+1)-u(t)]}{\sum_{t=t_{n-1}}^{t_n-2} y^2(t)}$$

$$=\alpha_{n-1}-\frac{\sum_{t=t_{n-1}}^{t_n-2} y(t+1)y(t)}{\sum_{t=t_{n-1}}^{t_n-2} y^2(t)} \tag{6.4}$$

where the last equality follows from (6.2). Assume that $t_n-t_{n-1}\to\infty$ and that

$$|a-b\alpha_{n-1}|<1 \tag{6.5}$$

which means that the closed loop system used during the time interval $t_{n-1}<t<t_n$ is stable then

$$\alpha_n=\alpha_{n-1}-\frac{r_y(1)}{r_y(0)} \tag{6.6}$$

where $r_y(\tau)$ is the covariance function of the stochastic process $\{y(t)\}$ defined by

$$y(t)+(a-b\alpha_{n-1})y(t-1)=e(t)+ce(t-1). \tag{6.7}$$

Straightforward algebraic manipulations now give

$$\alpha_n=\alpha_{n-1}-\frac{(c-a+b\alpha_{n-1})(1-ac+bc\alpha_{n-1})}{1+c^2-2ac+2bc\alpha_{n-1}}. \tag{6.8}$$

The problem is thus reduced to the analysis of the nonlinear differerence equation given by (6.8).

Introduce

$$x_n=\alpha_n-\frac{a-c}{b} \tag{6.9}$$

the equation (6.8) then becomes

$$x_{n+1}=g(x_n)=(1-b)x_n+\frac{b^2cx_n^2}{1-c^2+2bcx_n}. \tag{6.10}$$

The point $x=0$ is a fixed point of the mapping g which corresponds to the optimal value of gain of the feedback loop, i.e. $\alpha=(a-c)/b$. The problem is thus to determine if this fixed point is stable. Since the closed loop system is assumed to be stable it is sufficient to consider

$$\frac{c-1}{b}<x<\frac{c+1}{b}. \tag{6.11}$$

Three cases have to be investigated

1. $c=0$
2. $c>0$ or $c<0$, $0<b\leqslant 1$
3. $c>0$ or $c<0$, $1<b<2$.

For all cases $g(x)\approx(1-b)x$ if x is small. This implies that solutions close to 0 converge to $x=0$ if $0<b<2$.

Case 1

The equation (6.10) reduces to

$$x_{n+1}=(1-b)x_n$$

and the fixed point $x=0$ is stable if $|1-b|<1$ i.e. $0<b<2$.

Case 2

The pincipal behavior of $g(x)$ when $c>0$ and $0<b\leqslant 1$ is shown in Fig. 1. It is straightforward to verify that all initial values in the stability region $(c-1)/b<x<(c+1)/b$ will give solutions which converge to zero.

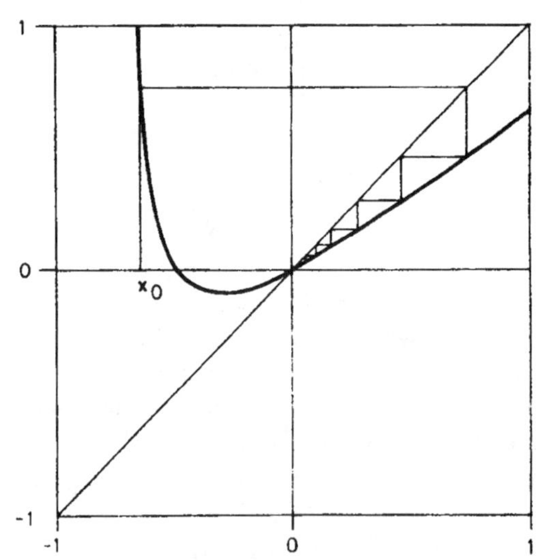

FIG. 1. Graph of the function g when $0<b\leqslant 1$. The figure is drawn with the parameter values $a=-0{\cdot}5$, $b=0{\cdot}5$ and $c=0{\cdot}7$.

Case 3

The function $g(x)$ for this case is shown in Fig. 2. It is not obvious that x will converge to zero, because there might exist a "limit cycle".

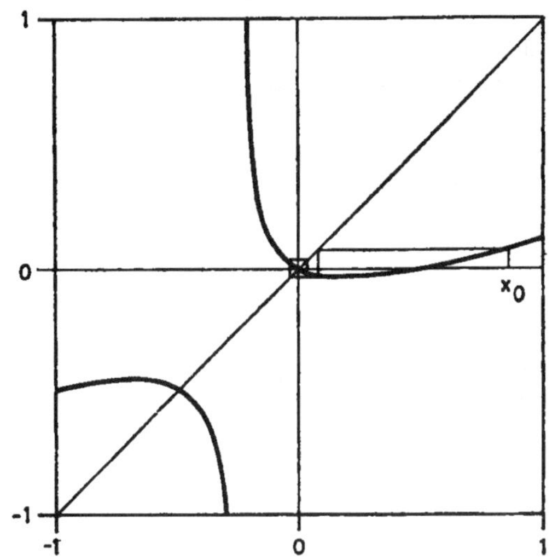

FIG. 2. Graph of the function g when $1<b<2$. The figure is drawn with the parameter values $a=-0{\cdot}5$, $b=1{\cdot}5$ and $c=0{\cdot}7$.

If $c>0$ and starting with $x_o>0$ then if $g(x_o)<0$ it can be shown that after two iterations the new value x_2 will satisfy $0<x_2<x_o$ i.e.

$$0<g(g(x))<x \qquad \text{if } x>0, g(x)<0.$$

If $(c-1)/b<x<0$ then $g(x)$ will be positive and can be taken as a new initial value for which the condition above will be satisfied. If $c<0$ then it can be shown that $g(g(x))>x$ if $x<0$ and $g(x)>0$.

Summary

From the analysis above we can conclude that $x=0$ is a stable fixed point if

$$-1<c<1$$

and

$$0<b<2. \tag{6.12}$$

The example shows that under the condition (6.12) the version of the self-tuning algorithm where the parameters of the control law are kept constant over long intervals will in fact converge. In the analysis above $\beta_o=1$ was chosen. If $\beta_o\neq 1$ then the condition (6.12) is replaced by

$$0<b/\beta_o<2 \tag{6.12'}$$

or

$$0.5b<\beta_o<\infty. \tag{6.13}$$

The condition (6.13) implies that it is necessary to pick the parameter β_o in a correct manner. The algorithm will always converge if β_o is greater than b. Under-estimation may be serious and the value $\beta_o<0{\cdot}5b$ gives an unstable algorithm.

The analysis presented in the simple example can be extended to give stability conditions for the modified algorithm in more complex cases. The analysis required is tedious.

7. SIMULATIONS

The results in section 5 are given under the assumption that the least squares estimator really converges, but yet we have not been able to give general conditions for convergence. But simulation of numerous examples have shown that the algorithm has nice convergence properties.

This section presents a number of simulated examples which illustrate the properties of the self-tuning algorithm.

Example 7.1

Let the system be

$$y(t)+ay(t-1)=bu(t-1)+e(t)+ce(t-1) \tag{7.1}$$

with $a=-0{\cdot}5$, $b=3$ and $c=0{\cdot}7$. The minimum variance regulator for the system is

$$u(t)=\frac{a-c}{b}y(t)=-0{\cdot}4y(t). \tag{7.2}$$

A regulator with this structure can be obtained by using the self-tuning algorithm based on the model

$$y(t+1)+\alpha y(t)=\beta_o u(t)+\varepsilon(t+1). \tag{7.3}$$

Figure 3 shows for the case $\beta_o=1$ how the parameter estimate converges to the value $\alpha=-0{\cdot}4$ which corresponds to the minimum variance strategy (7.2).

In Figure 4 is shown the expected variance of the output if the current value of α should be used for all future steps of time. Notice that the algorithm has practically adjusted over 50 steps.

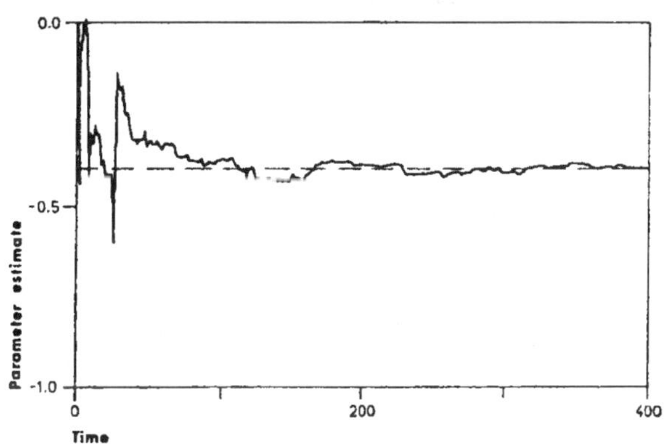

FIG. 3. Parameter estimate $a(t)$ obtained when the self tuning algorithm based on the model (7.3) is applied to the system given by (7.1). The minimum variance regulator corresponds to $a=-0{\cdot}4$ and is indicated by the dashed line.

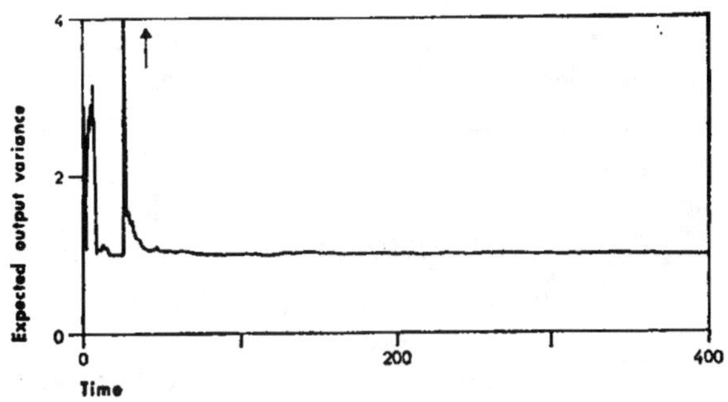

FIG. 4. Expected variance of the output of Example 7.1 if the control law obtained at time t is kept constant for all future times. Notice that the estimate at time $t=26$ would give an unstable system.

The analysis of Example 6.1 shows that, since $b>2$, and $\beta_o=1$ the modified self-tuning algorithm obtained when the parameters of the controller are kept constant over long intervals is unstable. The simulation in Example 7.1 shows that at least in the special case a conservative estimate of the convergence region is obtained by analysing the modified algorithm. If the value of b is increased further it has been shown that the algorithm is unstable. Unstable realizations have been found for $b=5$. In such cases it is of course easy to obtain a stable algorithm by increasing β_o. This requires, however, a knowledge of the magnitude of b.

The system of Example 7.1 is very simple. For instance, if no control is used the variance will still be reasonably small. The next example is more realistic in this aspect.

Example 7.2

Consider the system

$$y(t)-1.9y(t-1)+0.9y(t-2)=u(t-2)$$
$$+u(t-3)+e(t)-0.5e(t-1). \tag{7.4}$$

If no control is used the variance of the output is infinite. Also notice that $B(z)=z-1$. The assumption that B has all zeroes inside the unit circle is thus violated. The minimum variance strategy for the system is

$$u(t)=-1.76y(t)+1.26y(t-1)-0.4u(t-1)$$
$$+1.4u(t-2). \tag{7.5}$$

A regulator with this structure is obtained by using the self-tuning algorithm with the model

$$y(t+2)+\alpha_1 y(t)+\alpha_2 y(t-1)$$
$$=u(t)+\beta_1 u(t-1)+\beta_2 u(t-2)+\varepsilon(t+2). \tag{7.6}$$

The convergence of the parameters is shown in Fig. 5. Figure 6 shows the accumulated losses when

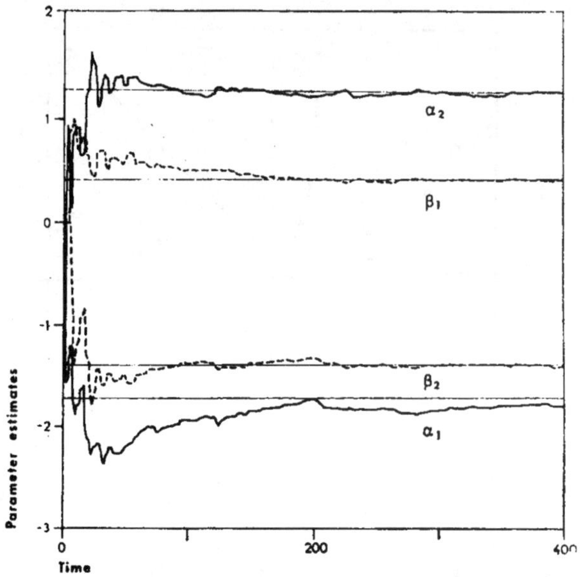

FIG. 5. Parameter estimates α_1, α_2, β_1 and β_2 obtained when the self tuning algorithm based on (7.6) is applied to the system given by (7.4). The thin lines indicate the parameter values of the minimum variance strategy.

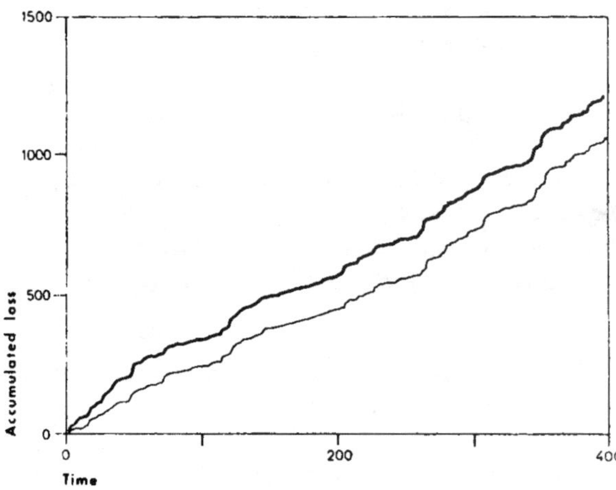

FIG. 6. Accumulated loss

$$\sum_{s=1}^{t} y^2(s)$$

for a simulation of the system (7.4) when using the self tuning algorithm (thick line) and when using the optimal minimum variance regulator (7.5) (thin line).

using the self-tuning algorithm and when using the optimal minimum variance regulator (7.5).

In both examples above, the models in the self-tuning algorithm have had enough parameters so it could converge to the optimal minimum variance regulator. The next example shows what happens when the regulator has not enough parameters.

Example 7.3

Consider the system

$$y(t)-1.60y(t-1)+1.61y(t-2)-0.776y(t-3)$$
$$=1.2u(t-1)-0.95u(t-2)+0.2u(t-3)+e(t)$$
$$+0.1e(t-1)+0.25e(t-2)+0.87e(t-3). \tag{7.7}$$

The polynomial $A(z)$ has two complex zeroes near the unit circle ($+0{\cdot}4 \pm 0{\cdot}9i$) and one real zero equal to 0·8.

If a self-tuning regulator is determined based on a model with $m=3$ and $l=2$ it will converge to the minimum variance regulator as expected. Figure 7(a) shows a small sample of the output together with the sample covariance of the output, $\hat{r}_y(\tau)$.

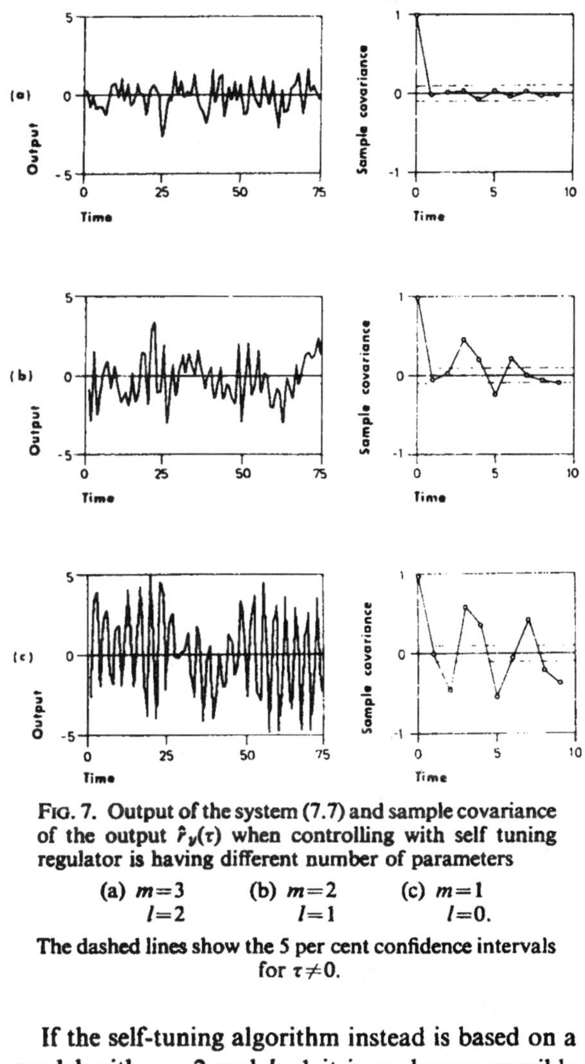

FIG. 7. Output of the system (7.7) and sample covariance of the output $\hat{r}_y(\tau)$ when controlling with self tuning regulator is having different number of parameters

(a) $m=3$, $l=2$ (b) $m=2$, $l=1$ (c) $m=1$, $l=0$.

The dashed lines show the 5 per cent confidence intervals for $\tau \neq 0$.

If the self-tuning algorithm instead is based on a model with $m=2$ and $l=1$ it is no longer possible to obtain the minimum variance regulator for the system since there are not parameters enough in the self-tuning regulator. Theorem 5.1 indicates, however, that if the self-tuning regulator converges, its parameters will be such that the covariances $r_y(1)$, $r_y(2)$, $r_{yu}(1)$ and $r_{yu}(2)$ are all zero. The simulation shows that the algorithm does in fact converge with $\beta_o=1{\cdot}0$. The covariance function of the output is shown in Fig. 7(b). It is seen that the sample covariances $\hat{r}_y(1)$ and $\hat{r}_y(2)$ are within the 5 per cent confidence interval while $\hat{r}_y(3)$ is not as would be expected from Theorem 5.1.

If a self-tuning algorithm is designed based on a model with $m=1$, $l=0$ then Theorem 5.1 indicates that $r_y(1)$ should vanish. Again the simulation shows that the algorithm does in fact converge and that the sample covariance $\hat{r}_y(1)$ does not differ significantly from zero. See Fig. 7(c).

When using regulators of lower order than the optimal minimum variance regulator, the parameters in the controller will not converge to values which for the given structure gives minimum variance of the output. In Table 1 is shown the variance of the output for the system above when using different regulators.

The loss when using the self-adjusting regulator is obtained through simulations. The optimal regulator is found by minimizing $r_y(0)$ with respect to the parameters in the controller.

TABLE 1

		Loss $\frac{1}{N}\sum_{t=1}^{N} y^2(t)$	
m	l	Self-adjusting	Optimal
3	2	1·0	1·0
2	1	2·5	1·9
1	0	4·8	3·4

The previous examples are all designed to illustrate various properties of the algorithm. The following example is a summary of a feasibility study which indicates the practicality of the algorithm for application to basis weight control of a paper machine.

Example 7.4

The applicability of minimum variance strategies to basis weight control on a paper machine was demonstrated in [9]. In this application the control loop is a feedback from a wet basis weight signal to thick stock flow. The models used in [9] were obtained by estimating the parameters of (2.1) using the maximum likelihood method. In one particular case the following model was obtained.

$$y(t)=\frac{b_1q^{-1}+b_2q^{-2}}{1+a_1q^{-1}+a_2q^{-2}}u(t-2)+v(t) \qquad (7.8)$$

where the output y is basis weight in g/m^2 and the control variable is thick stock flow in g/m^2. The disturbance $\{v(t)\}$ was a drifting stochastic process which could be modelled as

$$v(t)=\lambda\frac{1+c_1q^{-1}+c_2q^{-2}}{(1+a_1q^{-1}+a_2q^{-2})(1-q^{-1})}e(t) \qquad (7.9)$$

where $\{e(t)\}$ is white noise. The sampling interval was 36 sec and the numerical values of the parameters obtained through identification were as follows

$$a_1 = -1{\cdot}283$$

$$a_2 = 0{\cdot}495$$

$$b_1 = 2{\cdot}307$$

$$b_2 = -2{\cdot}025$$

$$c_1 = -1{\cdot}438$$

$$c_2 = 0{\cdot}550$$

$$\lambda = 0{\cdot}382.$$

To investigate the feasability of the self-tuning algorithm for basis weight control, the algorithm was simulated using the model (7.8) where the disturbance v was the actual realization obtained from measurements on the paper machine. The parameters of the regulator were chosen as $k=1$, $l=3$, $m=4$ and $\beta_0=2{\cdot}5$ and the initial estimates were set to zero. The algorithm is thus tuning 7 parameters.

The results of the simulation are shown in Figs. 8–10. Figure 8 compares the output obtained when using the self-tuning algorithm with the result obtained when using the minimum variance regulator computed from the process model (7.8) with the disturbance given by (7.9). The reference value was 70 g/m^2. In the worst case the self-tuning regulator gives a control error which is about 1 g/m^2 greater than the minimum variance regulator. This happens only at two sampling intervals.

After about 75 sampling intervals (45 min) the output of the system is very close to the output obtained with the minimum variance regulator.

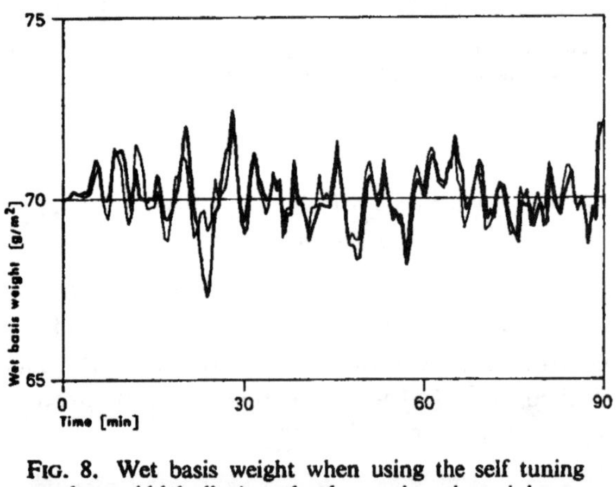

FIG. 8. Wet basis weight when using the self tuning regulator (thick line) and when using the minimum variance regulator based on maximum likelihood identification (thin line). The reference value for the controller was 70g/m².

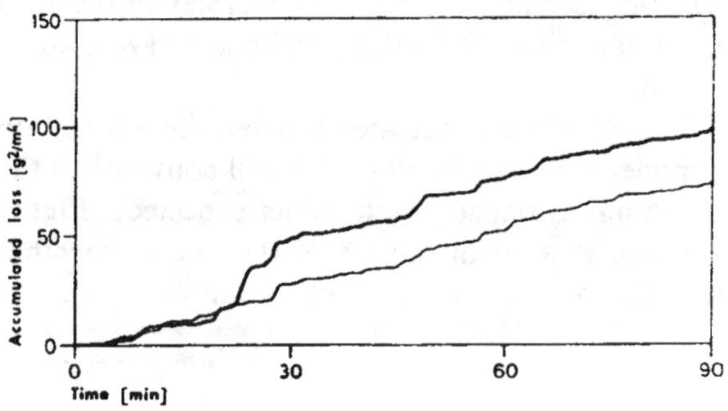

FIG. 9. Accumulated loss for Example 7.4 when using the self tuning regulator (thick line) and when using the minimum variance regulator (thin line).

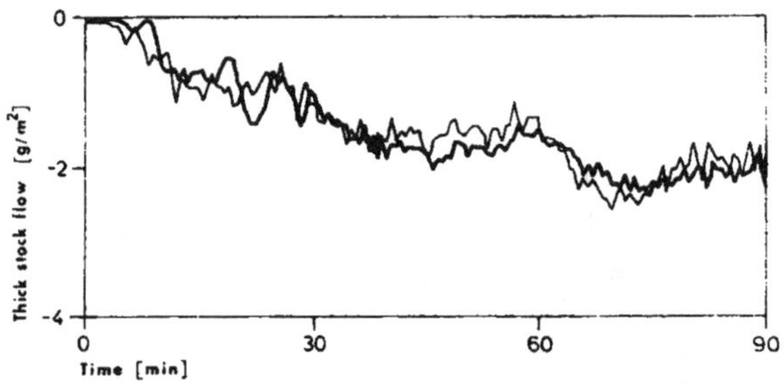

FIG. 10. The control signal in g/m² for Example 7.4 when using the self tuning regulator (thick line) and when using the minimum variance regulator (thin line).

Figure 9 compares the accumulated losses

$$V(t) = \sum_{n=0}^{t} y^2(n)$$

obtained with the minimum variance regulator and the self-tuning regulator. Notice that in the time interval (21, 24) minutes there is a rapid increase in the accumulated loss of the self-tuning regulator of about 17 units. The largest control error during this interval is 2·7 g/m^2 while the largest error of the minimum variance regulator is 1 g/m^2. The accumulated losses over the last hour is 60 units for the self-tuning regulator and 59 units for the minimum variance regulator.

The control signal generated by the self-tuning algorithm is compared with that of the minimum variance regulator in Fig. 10. There are differences in the generated control signals. The minimum variance regulator generates an output which has more rapid variations than the output of the self-tuning regulator.

The parameter estimates obtained have not converged in 100 sampling intervals. In spite of this the regulator obtained will have good performance as has just been illustrated. The example thus indicates that the self-tuning algorithm could be feasible as a basis weight regulator.

8. PRACTICAL ASPECTS

A few practical aspects on the algorithm given in section 4 are presented in this section which also covers some possible extensions of the results

A priori knowledge

The only parameters that must be known à *priori* are k, l, m and β_0. If the algorithm converges it is easy to find out if the à *priori* guesses of the parameters are correct simply by analyzing the sample covariance of the output. Compare Example 7.3. The parameter β_0 should be an estimate of the corresponding parameter of the system to be controlled. The choice of β_0 is not critical as was shown in the Examples 6.1 and 7.1. In the special cases studied in the examples an under-estimate led to a diverging algorithm while an over-estimate was safe.

Implementation on process computers

It is our belief that the self-tuning algorithm can be conveniently used in process control applications. There are many possibilities. The algorithm can be used as a tool to tune regulators when they are installed. It can be installed among the systems programs and cycled through different control loops repetitively to ensure that the regulators are always properly tuned. For critical loops where the parameters are changing it is also possible to use a dedicated version which allows slowly drifting parameters.

A general self-tuning algorithm requires about 40 FORTRAN statements. When compiled using the standard PDP 15/30 FORTRAN compiler the code consists of 450 memory locations. The number of memory locations required to store the data is $(l-1+m)^2+3(l-1+m)+2k+4$. Execution times on a typical process computer (PDP 15) without floating point hardware are given in the table below. The major part of the computing is to update the least squares estimate.

Number of parameters $l+m$	Execution time ms
1	5
3	16
5	34
8	69

Improved convergence rates

The results of this paper only shows that if the parameters converge the regulator obtained will tend to a minimum variance regulator. Nothing is said about convergence rates, which of course is of great interest from practical as well as theoretical points of view. There are in fact many algorithms that have the correct asymptotic properties. Apart from the algorithm given in section 4 we have the algorithm which minimizes (2.4). But that algorithm is impossible to use due to the computational requirements. It is of interest to investigate if other possible algorithms have better convergence rates than the algorithm of section 4. No complete answer to this problem is yet known. A few possibilities will be indicated. It could be attempted to take into account that the parameter estimates are uncertain. See Refs. [2, 7 and 10]. The least squares identifier can be improved upon by introducing exponential weighting of past data. This has in some cases shown to be advantageous in simulations. Algorithms of this type have in simulations been shown to handle slowly drifting parameters.

Another possibility is to assume that the parameters are Wiener processes, which also can be incorporated fairly easily [2, 7]. It has been verified by simulation that the region of convergence can be improved by introducing a bound on the control signal.

Feed forward

In many industrial applications the control can be improved considerable if feed forward is used. The self tuning regulators in this paper can include feed forward control by changing the process model (4.1) to

$$
\begin{aligned}
y(t+k+1)&+\alpha_1 y(t)+\ldots+\alpha_m y(t-m+1)\\
&=\beta_0[u(t)+\beta_1 u(t-1)+\ldots+\beta_l u(t-l)]+\gamma_1 s(t)\\
&\quad+\ldots+\gamma_p s(t-p+1)+\varepsilon(t+k+1) \qquad (8.1)
\end{aligned}
$$

where $s(t)$ is a known disturbance.

The parameters α_i, β_i and γ_i can be identified as before and the control law (4.2) will be changed to

$$
\begin{aligned}
u(t)=\frac{1}{\beta_0}[&\alpha_1 y(t)+\ldots+\alpha_m y(t-m+1)-\gamma_1 s(t)\\
&-\ldots-\gamma_p s(t-p+1)]-\beta_1 u(t-1)\\
&-\ldots-\beta_l u(t-l). \qquad (8.2)
\end{aligned}
$$

Nonminium phase systems

Difficulties have been found by a straightforward application of the algorithm to nonminimum phase systems, i.e. systems where the polynomial B has zeroes outside the unit circle.

Several ways to get around the difficulty have been found. By using a model with $B(z)=\beta_0$ it has in many cases been possible to obtain stable algorithms at the sacrifice of variance.

It is well-known that the minimum variance regulators are extremely sensitive to parameter variations for nonminimum phase systems [1]. This is usually overcome by using suboptimal strategies which are less sensitive [1]. The same idea can be used for the self-tuning algorithms as well. The drawback is that the computations increase because the polynomials F and G of an identity similar to (2.8) must be determined at each step of the iteration. An alternative is to solve a Riccati-equation at each step.

Multivariable and nonlinear systems

It is possible to construct algorithms that are similar to the one described in section 4 for multivariable and nonlinear systems as long as a model structure which is linear in the parameters [3, p. 131] is chosen. For multivariable systems the structure given in equation (3.2) of Ref. [3] can thus be attempted. Analyses of the properties of the algorithm obtained when applied to a multivariable or a nonlinear system are not yet done.

9. CONCLUSIONS

The paper has been concerned with control of systems with constant but unknown parameters. The analysis has been limited to single-input single-output systems with disturbances in terms of filtered white noise. A control algorithm based on least squares parameter estimation and a minimum variance regulator computed from the estimated parameters has been analysed. Assuming that the parameter estimates converge the closed loop system has been analysed. A characterization of the closed loop system has been given under weak assumption on the system to be controlled. Under stronger assumptions on the system to be controlled it has been shown that the regulator obtained will actually converge to the minimum variance regulator if the estimates converge.

Since the closed loop system is characterized as a nonlinear stochastic system it is very difficult to give general conditions that guarantee that the estimates converge. The convergence has only been treated for simple examples and under further assumptions as in section 6. But simulations of numerous examples indicate that the algorithm has nice convergence properties.

The simplicity of the algrorithm in combination with its asymptotic properties indicate that it can be useful for industrial process control. The feasibility has also been demonstrated by experiments on real processes in the paper and mining industries.

10. REFERENCES

[1] K. J. Åström: *Introduction to Stochastic Control Theory*. Academic Press, New York (1970).

[2] K. J. Åström and B. Wittenmark: Problems of identification and control. *J. Math. Analysis Applic.* **34**, 90–113 (1971).

[3] K. J. Åström and P. Eykhoff: System identification—A survey. *Automatica* **7**, 123–162 (1971).

[4] A. A. Feldbaum: Dual control theory I–IV. *Aut. Remote Control* **21**, 874–880, 1033–1039 (1961); **22**, 1–12, 109–121 (1962).

[5] R. E. Kalman: Design of a self optimizing control System. *Trans. ASME* **80**, 468–478 (1958).

[6] V. Peterka: Adaptive Digital Regulation of Noisy Systems, 2nd Prague IFAC Symposium on Identification and Process Parameter Estimation (1970).

[7] J. Wieslander and B. Wittenmark: An approach to adaptive control using real time identification. *Automatica* **7**, 211–217 (1971).

[8] L. E. Dickson: *First Course in the Theory of Equations*. Wiley, New York (1922).

[9] K. J. Åström: Computer control of a paper machine—An application of linear stochastic control theory. *IBM J. Res. Development* **11**, 389–405 (1967).

[10] J. B. Farison *et al.*: Identification and control of linear discrete systems. *IEEE Trans. Aut. Control* **12**, 438–442 (1967).

Résumé—On considère le problème du contrôle d'un système avec paramètres constants mais inconnus. L'analyse se limite aux systèmes discrets à entrée unique et sortie unique. Un algorithme obtenu en combinant un estimateur à carrés minimum avec un régulateur à variance minimum calculée du modèle estimé, est analysé. Les résultats principaux sont deux théorèmes qui caractérisent le système à boucle fermée obtenu en supposant que les estimations des paramètres convergent. Le premier théorème dit que certaines covariances de la sortie et certaines covariances transverses de la variable de contrôle et de la sortie disparaitront avec des suppositions faibles du système contrôlé. Dans le second théorème il est supposé que le système à contrôler est un système général linéaire stochastique du nème ordre. Il est montré que si les estimations des paramètres convergent, la loi de contrôle obtenue est en fait la loi de contrôle de variance minimale qui pourrait être calculée si les paramètres du système étaient connus. Ceci est quelque peu surprenant car l'estimation des carrés minimum est partiale. On discute certaines des implications pratiques des résultats. Il est montré en particulier qu'il est possible d'appliquer l'algorithme à un petit ordinateur.

Zusammenfassung—Betrachtet wird das Problem der Steuerung eines Systems mit konstanten, aber unbekannten Parametern. Die Analyse wird auf Systeme mit Diskretzeit und einem Eingang bzw. einem Ausgang beschränkt. Ein durch Kombination einer Schätzeinrichtung und der Methode der kleinsten Quadrate mit einer aus dem Schätzmodell berechneten Regeleinrichtung erhaltener Algorithmus wird analysiert. Die Hauptergebnisse sind zwei Theoreme, die unter der Annahme, daß die Parameterschätzungen konvergieren, den erhaltenen geschlossenen Regelkreis charakterisieren. Das erste Theorem konstatiert, daß bestimmte Kovarianzen des Ausganges und bestimmte Kreuz-Kovarianzen der Steuervariablen und des Ausgangs unter schwachen Annahmen über das zu regelnde System verschwinden. Im zweiten Theorem wird angenommen, daß das zu regelnde System ein allgemeines lineares stochastisches System n-ter Ordnung ist. Gezeigt wird, daß bei Konvergenz der Parameterschätzung des erhaltenen Steuergesetzes in der Tat das Steuergesetz bei minimaler Varianz ist, das berechnet werden kann, wenn die Parameter des Systems bekannt waren. Das ist etwa überraschend, weil die Schätzung nach den kleinsten Quadraten angesteuert wird. Einige praktische Folgerungen aus den Ergebnissen werden diskutiert. Speziell wird gezeigt, daß der Algorithmus auf einem kleinen Prozeßrechner leicht verwirklicht werden kann.

Резюме—Рассматривается проблема регулирования систем при помощи постоянных но неизвестных параметров. Анализ ограничивается системами дискретного времени с одним вводом и с одним выводом. Анализируется алгоритм, полученный путем комбинирования оценок наименьших квадратов с регулировкой минимальных колебаний, вычисленных из расчетной модели. Основными результатами являются две теоремы, характеризующие систему замкнутого контура, полученного в предположении, что оценки параметров сходятся. Первая теорема утверждает, что некоторая ковариантность вывода и некоторая пересекающая ковариантность контрольных переменных вывода исчезнут при небольших допущениях в регулируемой системе. Во второй теореме предполагается, что контролируемая система является общей линейной стохастической системой порядка n:th. Показано, что, если оценки параметров сходятся, то полученный закон регулировкя ивляется фактически законом регулировки минимального колебания, которое могло бы быть вычислено при известных параметрах системы. Это до некоторой степени неожиданно, так как оценка наименьших квадратов явияется смещенной. Обсуждаются некоторые практические выводы из результатов. В частности, показано, что этот алгоритм может быть применен на небольшой счетно-вычислительной машине для продессов.

Nonlinear Controllability and Observability

ROBERT HERMANN AND ARTHUR J. KRENER

AT the time of its publication in 1977, this paper represented, for the analysis of the "observability" properties of a nonlinear system, the same kind of breakthrough that the earlier 1972 paper by Sussmann and Jurdjevic (also included in this volume) had represented for the analysis of the reachability properties of those systems. In particular, the paper developed the appropriate language and mathematical tools for the analysis of certain properties of "local" observability, which lend themselves to an appealing algebraic test as well as the possibility of finding decompositions of the system into "observable/unobservable" parts.

In an abstract system-theoretic setup, a system is considered to be "observable" if in its state-space it is not possible to find pairs of indistinguishable states, i.e., pairs of initial states that induce two identical input-output maps. This notion yields a very simple characterization in the case of linear systems, where the output is the sum of a function of the input (the zero-state response) and a function of the initial state (the zero-input response), which in turn is a linear function of its argument. In fact, in this case, two states are distinguishable if and only if the associated zero-input responses are different, i.e., if their difference produces a zero-input response that is not identically zero. In a nonlinear system, on the other hand, two states are distinguishable if there is at least one input function (not necessarily the zero-input) for which the output responses are different, and this input may in principle depend on the particular pair considered. In other words, the study of observability in a nonlinear system must explicitly take into account the way in which inputs affect the response, and this makes the analysis substantially more involved.

Pioneering authors in the study of observability for nonlinear systems were Brockett [2], d'Alessandro *et al.* [3], Sussmann [11], [12], and Sontag [10]. In particular, Sussmann, generalizing the work of Brockett, [2], on Lie groups, showed that under appropriate hypotheses the relation of indistinguishability between states is a closed regular equivalence relation, which can be used to define a reduced system (with state defined on the associated "quotient" manifold) having the same input-output behavior as the original system but no pairs of indistinguishable states. Sontag studied the observability of polynomial discrete-time nonlinear systems, and introduced the notion of "observation space."

The approach of Hermann and Krener to the analysis of the observability properties of nonlinear systems was similar to the one that had proved very successful a few years earlier in the analysis of the controllability properties, namely the approach of differential geometry. In particular, considering systems described by equations of the form

$$\dot{x} = F(x, u) \qquad y = h(x) \tag{1}$$

in which $x \in \mathbb{R}^n$, $F(x, u)$ is an analytic vector field for every $u \in \Omega$, with Ω a fixed set, and $h(x)$ is an analytic $\mathbb{R}^p$-valued map, they realized that the appropriate geometric object to look at was the subspace (of differential forms) spanned by the differentials of the p components of the vector $h(x)$ and their (repeated) Lie derivatives[1] along the vector fields of the set $\mathcal{F} = \{F(x, u) : u \in \Omega\}$.

In particular they proved that if, at a given point x°, this subspace has dimension equal to that of the state-space, then all states in a neighborhood of this point are distinguishable form x°. In other words, the computation of (repeated) Lie derivatives of the differentials of the elements of $h(x)$ along the vector fields of $\mathcal{F}$ yields a simple test for observability, very much in the same way the computation of the (repeated) Lie brackets of the vector fields of $\mathcal{F}$ yields a simple test for reachability. This test amounts to checking the rank of a matrix having n columns, whose rows consist of the differentials of the elements of $h(x)$ as well as their (repeated) derivatives along the vector fields of $\mathcal{F}$ and appears as a natural extension of the familiar test which determines the observability of a linear

[1]The differential of a real-valued function $\lambda(x)$ is the one-form

$$d\lambda(x) = \left(\frac{\partial \lambda}{\partial x_1} \cdots \frac{\partial \lambda}{\partial x_n}\right)$$

and the Lie derivative of a one-form $\omega(x)$ along a vector field $f(x)$, denoted $L_f\omega(x)$ is the one-form

$$L_f\omega(x) = f^{\mathrm{T}}(x)\left(\frac{\partial \omega^{\mathrm{T}}}{\partial x}\right)^{\mathrm{T}} + \omega(x)\frac{\partial f}{\partial x}$$

system

$$\dot{x} = Ax + Bu \qquad y = Cx$$

by looking at the rank of a matrix whose rows are the rows of C (the differentials of the elements of Cx) and rows of the form CA^k (the differentials of the kth Lie derivative of Cx along the vector field Ax).

In case the subspace in question does not have dimension n at x°, but its dimension is constant, say d, then Hermann and Krener also pointed out that, in a neighborhood of x°, it is possible to find d real-valued functions $\varphi_1(x), \ldots, \varphi_d(x)$, whose differentials span that subspace, and these can be used to define, locally near x°, a reduced (d-dimensional) system having the same input-output behavior as the original one, but no pairs of indistinguishable states. This observation had a significant fallout in subsequent years, as it became one of the main ingredients for understanding the solution of the problems of decoupling and non-interacting control via state-feedback, thus paving the way to the systematic development of geometric methods for nonlinear feedback design.

In particular, this observation and the results of [6], suggested to Krener and to the author of this preamble (in the course of the joint work which led to the paper [5]) that the set of functions $\varphi_1(x), \ldots, \varphi_d(x)$, augmented with a complementary set of $n-d$ functions having independent differentials, could be used for a local change of coordinates which highlights a decomposition into observable/unobservable parts. For example, in a single-input single-output system modeled by equations of the form

$$\dot{x} = f(x) + g(x)u \qquad y = h(x) \tag{2}$$

the change of coordinates in question yields the decomposition

$$\begin{aligned} \dot{x}_1 &= f_1(x_1, x_2) + g_1(x_1, x_2)u \\ \dot{x}_2 &= f_2(x_2) + g_2(x_2)u \\ y &= h_2(x_2) \end{aligned}$$

where x_1 is of dimension d and x_2 has dimension $n-d$, and where it is clear that, from an input-output point of view, it is only the subsystem formed by the second and third equation (i.e., the "x_2-subsystem") that matters (extensive discussions of these decompostions can be found in [4]).

Now, it is immediately clear from this that if the state equation of (2) is perturbed by an additive term of the form $p(x)w$ in which w represents an extra input channel and, in the new coordinates, the vector field $p(x)$ transforms into a vector of the form

$$\begin{pmatrix} p_1(x_1, x_2) \\ 0 \end{pmatrix}$$

then the input w has no effect on the output y (whatever the initial state might be). In other words, the input w is decoupled from the output y. From this to the next stage, i.e., understanding how to use feedback so as to enforce decompositions of this kind, the step was short and the geometric theory of nonlinear feedback design was born.

Another relevant offspring of Hermann and Krener's theory of nonlinear observability, which appeared a few years later, was the identification of conditions under which there exists a local change of coordinates $z = \Phi(x)$ transforming a system of the form

$$\dot{x} = f(x) \qquad y = h(x) \tag{3}$$

to a system of the form

$$\dot{z} = Az + \phi(y) \qquad y = Cz$$

with (A, C) an observable pair (see [1] and [7]). The importance of this result is that for a system having this property it is very easy to construct a state observer, in the form

$$\dot{\xi} = (A - GC)\xi - \phi(y) - Gy$$

which yields an observation error $e = \Phi(x) - \xi$ satisfying $\dot{e} = (A - GC)e$. The kind of "observability" property thus characterized was used in many applications and has led, for instance, to the development of the nonlinear adaptive controllers studied by Marino and Tomei [9] and by Krstic et al. [8].

References

[1] D. Bestle and M. Zeitz, "Canonical form observer design for nonlinear time-variable systems," *Int. J. Contr.*, **38**:419–431, 1983.

[2] R.W. Brockett, "System theory on group manifolds and coset spaces," *SIAM J. Contr.*, **10**:265–284, 1972.

[3] P. D'Alessandro, A. Isidori, and A. Ruberti, "Realization and structure theory of bilinear dynamical systems," *SIAM J. Contr.*, **12**:517–535, 1974.

[4] A. Isidori, *Nonlinear Control Systems: An Introduction*, 3rd ed., Springer-Verlag (New York), 1995.

[5] A. Isidori, A.J. Krener, C. Gori Giorgi, and S. Monaco, "Nonlinear decoupling via feedback: a differential geometric approach," *IEEE Trans. Aut. Contr.*, **AC-26**:331–345, 1981.

[6] A.J. Krener, "A decomposition theory for differentiable systems," *SIAM J. Contr. Optimiz.*, **15**:289–297, 1977.

[7] A.J. Krener and A. Isidori, "Linearization by output injection and nonlinear observers," *Syst. Contr. Lett.*, **3**:47–52, 1983.

[8] M. Krstic, I. Kanellakopoulos, and P. Kokotovic, *Nonlinear Adaptive Control Design*, Wiley (New York), 1995.

[9] R. Marino and P. Tomei, "Global adaptive output feedback control of nonlinear systems," *IEEE Trans. Aut. Contr.*, **AC-38**:17–48, 1993.

[10] E. Sontag, *Polynomial Response Maps*, Springer-Verlag (New York), 1979.

[11] H. Sussmann, "Existence and uniqueness of minimal realizations of nonlinear system," *Math. Syst. Theory*, **10**:263–284, 1977.

[12] H. Sussmann, "Single input observability of continuous time systems," *Math. Syst. Theory*, **12**:371–393, 1979.

A.I.

Nonlinear Controllability and Observability

ROBERT HERMANN AND ARTHUR J. KRENER, MEMBER, IEEE

Abstract—**The properties of controllability, observability, and the theory of minimal realization for linear systems are well-understood and have been very useful in analyzing such systems. This paper deals with analogous questions for nonlinear systems.**

I. Introduction

FREQUENTLY, control systems of the following form are used to model the behavior of physical, biological, or social systems,

$$\Sigma: \quad \begin{aligned} \dot{x} &= f(x,u) \\ y &= g(x) \end{aligned} \qquad (1)$$

where $u \in \Omega$, a subset of $\boldsymbol{R}^l$, $x \in M$, a C^∞ connected manifold of dimension m, $y \in \boldsymbol{R}^n$ and f and g are C^∞ functions.

The control variable u represents the externally applied controls or exogenous inputs to the system and the output variable y represents the observable parameters of the system. The state variable x may or may not be directly measurable and is used to represent the memory of the system. The past history of Σ affects its future evolution only through information conveyed by this variable.

The state space M may be deficient in one of two ways. It may be too small to adequately represent the full variety of memory states, i.e., it may fail to distinguish between real states where some control exictes different observable effects. If this is the case, then the mathematical system Σ fails to adequately model the real system and hence, must be revised.

On the other hand, the state space may be too large. The system may not be controllable, i.e., if Σ is known to be in a given state $x^0 \in M$ at some time there may be other states $x \in M$ where the system cannot possibly get to or have come from using the given set of inputs. Or there may be distinct states which are indistinguishable from an input-output point of view, i.e., if the same input is applied in either of these states then the same output results. These problems can be caused by ignoring possible real inputs or real observables, in which case these must be added to the model. If, after all the significant input and output variables have been incorporated into Σ, the state space M is still too large, then one would like to apply a systematic technique to reduce M and still preserve the input-output structure of the model.

Using the ideas of controllability and observability, in the early 1960's Kalman and others carried out this program for linear systems. The similar questions for nonlinear systems were not effectively treated until the early 1970's. Based on the work of Chow [5], Hermann [9], Haynes–Hermes [8], and Brockett [9] and working independently, Lobry [21], [22], Sussman–Jurdjevic [25], and Krener [19], [20] developed the nonlinear analog of linear controllability in terms of the Lie algebra $\mathcal{F}$ of vector fields on M generated by the vector fields $f(\cdot,u)$ corresponding to constant controls $u \in \Omega$.

It was shown that if the dimension of $\mathcal{F}$ is constant or if the system Σ is analytic, then there exists a unique maximal submanifold M' of M through x^0 which carries all the trajectories of Σ passing through x^0 such that any point on this submanifold can be reached from x^0 going forward and backward along the trajectories of the system. In particular if the dimension of $\mathcal{F}(x^0)$ is m then $M = M'$ and so the system is "controllable" in some sense, to be made precise in Section II. If it is less than m then Σ can be restricted to M' where it is "controllable." This is one half of reducing the state space M.

For linear systems

$$\begin{aligned} \dot{x} &= Ax + Bu \\ y &= Cx \end{aligned} \qquad (1)$$

where $u \in \boldsymbol{R}^l$, $x \in \boldsymbol{R}^m$, $y \in \boldsymbol{R}^n$, this reduces to the well-known criterion that the rank of the matrix

$$(B : AB : \cdots : A^{m-1}B)$$

be m.

The other half of the program of reducing the state space was supplied by Sussmann [28] for analytic or symmetric systems. This generalized the work of Brockett [3] on Lie groups. Sussmann noted that indistinguishability is an equivalence relation on M and showed that for analytic "controllable" or symmetric "controllable" systems, indistinguishability is a closed regular equivalence relation so that the quotient is another manifold. He also showed that the quotient inherits a system which has the same input–output behavior as Σ but is "controllable" and observable, i.e., has no indistinguishable states.

In this paper we take a different approach to nonlinear observability, related to that of previous authors [16]–[18] but which is more in the spirit of the approach to nonlinear controllability described above. In particular we develop results involving observability which are analogous

Manuscript received December 16, 1976; revised June 7, 1977. R. Hermann's work was supported in part by a National Research Council Senior Postdoctoral Research Associateship at Ames Research Center (NASA), and in part by the National Science Foundation under Grant MCS75-07993. A. J. Krener's work was supported in part by the National Science Foundation under Grant MPS75-05248. Paper recommended by J. Davis, Chairman of the Stability, Nonlinear, and Distributed Systems Committee.

R. Hermann is with the Department of Mathematics, Rutgers University, New Brunswick, NJ 08903.

A. J. Krener is with the Department of Mathematics, University of California, Davis, CA 95616.

Reprinted from *IEEE Transactions on Automatic Control*, Vol. AC-22, October 1977, pp. 728–740.

to the controllability results depending on the dimension of $\mathcal{F}(x)$. The relevant objects in this study are $\mathcal{G}$, the smallest linear space of functions on M which contains the observations $g_1(x),\cdots,g_n(x)$ and which is closed with respect to Lie differentiation by $\mathcal{F}$, and the differentials of $\mathcal{G}$ denoted by $d\mathcal{G}$. If the dimension of $d\mathcal{G}$ is constant over M then indistinguishability need not be regular but there is a related regular equivalence relation which we can use to factor M. On the quotient there exists a system with the same input-output behavior as Σ but which is "observable" in the sense that neighboring points are distinguishable. In particular if the dimension of $d\mathcal{G}(x)$ is always m then Σ has this property. For linear systems this reduces to the well-known criterion that the matrix

$$\begin{bmatrix} C \\ \hline CA \\ \hline \cdot \\ \cdot \\ \cdot \\ \hline CA^{m-1} \end{bmatrix}$$

be of rank m.

In order to bring out the "duality" between "controllability" and "observability" (which is, mathematically, just the duality between vector fields and differential forms), we first review in Section II the known facts concerning nonlinear controllability and then in Section III we discuss our approach to nonlinear observability. Finally Section IV deals with the question of minimality for nonlinear systems.

II. Nonlinear Controllability

We consider the system Σ described in Section I. Recall a C^∞ (analytic) *m-dimensional manifold* M is a Hausdorff topological space with a C^∞ (analytic) structure, i.e., a countable cover of coordinate charts (U^α, x_α) where

1) U^α is an open set in M,

2) $x_\alpha = \text{col}(x_{\alpha 1},\cdots,x_{\alpha m}): U^\alpha \to \boldsymbol{R}^m$ is a homeomorphism onto its range, and

3) if (U^α, x_α) and (U^β, x_β) are two such coordinate charts then the change of coordinates $x_\beta \circ x_\alpha^{-1}: x_\alpha(U^\alpha \cap U^\beta) \to x_\beta(U^\alpha \cap U^\beta)$ is C^∞ (analytic).

For further details we refer the reader to Hermann [10] or Boothby [2]. A non-Hausdorff manifold is a non-Hausdorff topological space with such a structure. If M is not Hausdorff then solutions of ordinary differential equations on M need not be unique.

Throughout this section we simplify notation by assuming M admits globally defined coordinates $x = \text{col}(x_1,\cdots,x_m)$. This allows us to identify the points of M with their coordinate representations and to describe control systems in the familiar fashion (1). However when dealing with nonlinear observability in the next section we shall be forced to consider several coordinate charts.

It is with no loss of generality that we assume $y \in \boldsymbol{R}^n$. If $y \in N$, a C^∞ manifold, then by the Whitney Imbedding Theorem, N can be imbedded in $\boldsymbol{R}^n$ for some n which then can be taken as the range of g.

The assumption of infinite differentiability for M, f, and g is not essential but is only invoked to avoid counting the degree of differentiability needed in a particular argument. Occasionally we consider *analytic systems* where these are assumed to be an analytic manifold and analytic mappings. Nonautonomous systems are handled in the standard fashion by assuming time is one of the state variables.

We also assume the system is complete, i.e., for every bounded measurable control $u(t)$ and every $x^0 \in M$ there exists a solution of the differential equation $\dot{x} = f(x(t), u(t))$ satisfying $x(t^0) = x^0$ and $x(t) \in M$ for all $t \in \boldsymbol{R}$. We use the notation $(u(t),[t^0,t^1])$ to denote functions defined on $[t^0,t^1]$.

Given a subset $U \subseteq M$, x^1 is U-accessible from x^0 (denoted by $x^1 A_U x^0$) if there exists a bounded measurable control $(u(t),[t^0,t^1])$ satisfying $u(t) \in \Omega$ for $t \in [t^0,t^1]$ such that the corresponding solution $(x(t),[t^0,t^1])$ of the differential equation (1) satisfies $x(t^0) = x^0$, $x(t^1) = x^1$ and $x(t) \in U$ for all $t \in [t^0,t^1]$. M-accessibility and A_M are simply referred to as *accessibility* and A. Given any relation R on M we use the notation

$$R(x^0) = \{x^1 \in M : x^1 R x^0\}.$$

For example $A(x^0)$ is the set of *points accessible from* x^0. The system Σ is said to be *controllable at* x^0 if $A(x^0) = M$ and *controllable* if $A(x) = M$ for every $x \in M$.

If Σ is controllable at x^0 it still may be necessary to travel a considerable distance or for a long time to reach points near to x^0. As a result this type of controllability is not always of use and so we introduce a local version of this concept. Σ is *locally controllable at* x^0 if for every neighborhood U of x^0, $A_U(x^0)$ is also a neighborhood of x^0; Σ is *locally controllable* if it is locally controllable at every $x \in M$. (This is called local–local controllability by Haynes and Hermes [8].)

Accessibility is a reflexive and transitive relation but for nonlinear systems it need not be symmetric. For this reason we need a weaker relation. Given an open set $U \subseteq M$ there is a unique smallest equivalence relation on U which contains all U-accessible pairs. We call this relation *weak U-accessibility* and denote it by WA_U. It is easy to see that $x' WA_U x''$ iff there exists $x^0,\cdots,x^k$ such that $x^0 = x'$, $x^k = x''$ and either $x^i A_U x^{i-1}$ or $x^{i-1} A_U x^i$ for $i = 1,\cdots,k$. Weak M-accessibility and WA_M are simply referred to as weak accessibility and WA. Σ is *weakly controllable at* x^0 if $WA(x^0) = M$ in which case $WA(x) = M$ for all x and so Σ is *weakly controllable*.

Notice that weak controllability is a global concept and does not reflect the behavior of Σ restricted to a neighborhood of x^0. So again we introduce a local concept. Σ is *locally weakly controllable at* x^0 if for every neighborhood U of x^0, $WA_U(x^0)$ is a neighborhood of x^0. Σ is *locally weakly controllable* if it is locally weakly controllable at every $x \in M$. Clearly (local) controllability implies (local) weak controllability and it is not hard to show using the transitivity of (weak) accessibility and the connectivity of M that local (weak) controllability implies (weak) controllability, i.e., we have the following implications:

$$\begin{array}{ccc} \Sigma \text{ locally controllable} & \Rightarrow & \Sigma \text{ controllable} \\ \Downarrow & & \Downarrow \\ \Sigma \text{ locally weakly controllable} & \Rightarrow & \Sigma \text{ weakly controllable.} \end{array}$$

In general there are no other implications but for autonomous linear systems it can be shown that all four concepts are equivalent.

The following result gives an intuitive interpretation of local weak controllability. Loosely put it shows that Σ is locally weakly controllable iff one needs local coordinates of dimension m to distinguish the trajectories of Σ from any initial point.

Theorem 2.1: Σ is locally weakly controllable iff for every $x \in M$ and every neighborhood U of x the interior of $A_U(x) \neq \varnothing$.

Proof: Suppose Σ is weakly controllable. Given any $x^0 \in M$ and any neighborhood U of x^0 we can choose $u^1 \in \Omega$ such that $f^1(x) = f(x, u^1)$ is not zero at x^0 (assuming $m > 0$, if $m = 0$ then the result is immediate). Let $s \to \gamma_s^1(x)$ denote the flow of f^1, i.e., the family of solutions of the differential equation

$$\frac{d}{ds}\gamma_s^1(x) = f^1\left(\gamma_s^1(x)\right)$$

satisfying the initial conditions

$$\gamma_0^1(x) = x.$$

For some $\epsilon > 0$ the set $V^1 = \{\gamma_s^1(x^0) : 0 < s < \epsilon\}$ is a submanifold of U of dimension 1.

Suppose inductively V^{j-1} is a $j-1$ dimensional submanifold of U defined by

$$V^{j-1} = \Big\{ \gamma_{s_{j-1}}^{j-1} \circ \cdots \circ \gamma_{s_1}^1(x^0) : (s_1, \cdots, s_{j-1}) \text{ in some open subset of the positive orthant of } \boldsymbol{R}^{j-1} \Big\}$$

where $\gamma_{s_i}^i(x)$ is the flow of $f^i(x) = f(x, u^i)$ for some $u^i \in \Omega$. Clearly $V^{j-1} \subseteq A_U(x^0)$. If $j \leq m$ we construct V^j by choosing a $u^j \in \Omega$ and $x^{j-1} \in V^{j-1}$. This is always possible for if not then every trajectory of Σ starting on V^{j-1} would remain on V^{j-1} for a while. This contradicts the local weak controllability of Σ.

It follows that we can choose an open subset of the positive orthant of $\boldsymbol{R}^j$ such that the map $(s_1, \cdots, s_j) \mapsto \gamma_{s_j}^j \circ \cdots \circ \gamma_{s_1}^1(x^0)$ is an imbedding of the subset into U. Call the range V^j. We can continue until $j = m$, V^m is a nonempty open subset of $A_U(x^0)$ and so the interior of $A_U(x^0) \neq \varnothing$.

As for the converse suppose the interior of $A_U(x^0) \neq \varnothing$, we choose a control $(u(t), [t^0, t^1])$ such that the corresponding trajectory $(x(t), [t^0, t^1])$ satisfies $x(t^0) = x^0$, $x(t^1) = x^1 \in$ interior of $A_U(x^0)$ and $x(t) \in U$ for $t \in [t^0, t^1]$. Let $\gamma_t(x, t^0)$ denote the time dependent vector field $f_t(x) = f(x, u(t))$, i.e.,

$$\frac{d}{dt}\gamma_t(x, t^0) = f_t\left(\gamma_t(x, t^0)\right)$$

$$\gamma_{t^0}(x, t^0) = x.$$

Then $\gamma_{t^0}(\cdot, t^1)$ is a diffeomorphism of a neighborhood V of x^1 onto a neighborhood of x^0. Moreover we can choose $V \subseteq A_U(x^0)$ sufficiently small so that $\gamma_{t^0}(V, t^1) \subseteq WA_U(x^0)$. □

The advantage of local weak controllability over the other forms of controllability discussed above is that it lends itself to a simple algebraic test. First we introduce some additional mathematical concepts.

The set of all C^∞ vector fields on M is an infinite dimensional real vector space denoted by $\mathcal{X}(M)$ and also a Lie algebra under the multiplication defined by the Jacobi bracket $[h_1, h_2]$ given by

$$[h_1, h_2](x) = \frac{\partial h_2}{\partial x}(x) h_1(x) - \frac{\partial h_1}{\partial x}(x) h_2(x)$$

where h_1, h_2 and $[h_1, h_2] \in \mathcal{X}(M)$. Elements of $\mathcal{X}(M)$ are represented by column m-vector valued functions of x. For any fixed $h_1 \in \mathcal{X}(M)$ the real linear transformation from $\mathcal{X}(M)$ into itself which sends $h_2 \mapsto [h_1, h_2]$ is called Lie differentiation with respect to h_1 and is denoted by L_{h_1}.

Each constant control $u \in \Omega$ defines a vector field $f(x, u) \in \mathcal{X}(M)$, we denote by $\mathcal{F}^0$ the subset of all these vector fields. Let $\mathcal{F}$ denote the smallest subalgebra of $\mathcal{X}(M)$ which contains $\mathcal{F}^0$. A typical element of $\mathcal{F}$ is a finite linear combination of elements of the form

$$\left[f^1\left[f^2\left[\cdots\left[f^{k-1}, f^k\right]\cdots\right]\right]\right]$$

where $f^i(x) = f(x, u^i)$ for some constant $u^i \in \Omega$. We denote by $\mathcal{F}(x)$ the space of tangent vectors spanned by the vector fields of $\mathcal{F}$ at x. Σ is said to satisfy the *controllability rank condition at* x^0 if the dimension of $\mathcal{F}(x^0)$ is m; Σ satisfies the *controllability rank condition* if this is true for every $x \in M$.

Theorem 2.2: If Σ satisfies the controllability rank condition at x^0 then Σ is locally weakly controllable at x^0.

Proof: The proof is very similar to Theorem 2.1. We start by choosing a neighborhood of U of x^0 small enough so that Σ satisfies the controllability rank condition at every $x \in U$. We construct a sequence of submanifolds as before but this time there is a different reason why one can always choose $u^j \in \Omega$ and $x^{j-1} \in V^{j-1}$ such that $f^j(x) = f(x, u^j)$ is not tangent to V^{j-1} at x^{j-1}. If this is not possible then $\mathcal{F}$ restricted to V^{j-1} is a subalgebra of $\mathcal{X}(V^{j-1})$ which implies the dimension of $\mathcal{F}(x) \leq j-1 < m$ on $V^{j-1} \subseteq U$ which is a contradiction.

It follows that for every neighborhood U of x^0, the interior of $A_U(x^0)$ is not empty. The second half of the proof of Theorem 2.1 implies $WA_U(x^0)$ is a neighborhood of x^0. □

From the above we see that if Σ satisfies the controllability rank condition then it is locally weakly controllable. The converse is almost true as we shall see later on in Theorems 2.5 and 2.6.

Suppose the trajectories of Σ are required to satisfy the initial condition

$$x(t^0)=x^0,$$

then Σ defines a map from inputs to outputs as follows. Each admissible input $(u(t),[t^0,t^1])$ gives rise to a solution $(x(t),[t^0,t^1])$ of $\dot{x}=f(x,u(t))$ satisfying the initial condition. This, in turn, defines an output $(y(t),[t^0,t^1])$ by $y(t)=g(x(t))$. We denote this map by

$$\Sigma_{x^0}:(u(t),[t^0,t^1])\mapsto(y(t),[t^0,t^1])$$

and call it the *input–output map of* Σ *at* x^0. Given a map

$$\beta:(u(t),[t^0,t^1])\mapsto(y(t),[t^0,t^1])$$

from inputs to outputs, the pair (Σ,x^0) is said to be a realization of β if $\Sigma_{x^0}=\beta$.

Now suppose Σ is neither locally weakly controllable nor weakly controllable. Given the input–output map Σ_{x0} we would like to find another realization (Σ',z^0) of this map which is weakly controllable in some sense. The obvious way to proceed is to find a submanifold M' of M which contains x^0 and all the trajectories of Σ passing through x^0 then let Σ' be the restriction of Σ to M' and $z^0=x^0$. If M' is chosen small enough then hopefully Σ' will be weakly controllable in some sense.

To carry out this program we introduce some mathematical tools. A connected submanifold M' of M is an *integral submanifold* of $\mathfrak{F}$ if at each $x\in M'$ the tangent space to M' at x is contained in $\mathfrak{F}(x)$. M' is a *maximal integral submanifold* of $\mathfrak{F}$ if it is not properly contained in any other integral submanifold of $\mathfrak{F}$. There are two important cases when $\mathfrak{F}$ has maximal integral submanifolds.

Frobenius Theorem [*10*]*:* If the dimension of $\mathfrak{F}(x)=k$ for every $x\in M$, then there exists a partition of M into maximal integral submanifolds of $\mathfrak{F}$ all of dimension k.

Hermann–Nagano Theorem [*11*], [*12*], [*14*], [*23*]*:* If the system is analytic then there exists a partition of M into maximal integral submanifolds of $\mathfrak{F}$ of varying dimensions. The dimension of $\mathfrak{F}(x)$ can vary but it will be constant on each submanifold of the partition and equal to the dimension of that submanifold.

The relationship of these two theorems with controllability is given by the following.

Chow Theorem [*5*]*:* If either of the above is satisfied and M' is the maximal integral submanifold of $\mathfrak{F}$ containing x^0 then $M'=WA(x^0)$.

This leads to the following.

Theorem 2.3: Suppose (Σ,x^0) is a realization of an input–output map such that either

1) $\mathfrak{F}(x)$ is of constant dimension, or

2) Σ is analytic,

then there is a locally weakly controllable realization (Σ',x^0) of the same input–output map on the maximal integral submanifold M' of $\mathfrak{F}$ containing x^0. In fact, Σ' satisfies the controllability rank condition.

For completeness we mention the case when Σ does not satisfy either of the hypotheses of Theorem 2.3. It has been shown by Sussmann [29] that for any C^∞ system, $WA(x^0)$ can be given the structure of a C^∞ manifold, in general it will not be an integral submanifold of $\mathfrak{F}$.

Theorem 2.4 (Sussmann [*29*]*):* Given any realization (Σ,x^0), there exists a weakly controllable realization (Σ',x^0) of the same input–output map on $M'=WA(x^0)$.

Notice that unlike Theorem 2.3 the above does not guarantee that Σ' is locally weakly controllable or satisfies the controllability rank condition. Locally weakly controllable systems "almost" satisfy the controllability rank condition.

Theorem 2.5: If Σ is locally weakly controllable then the controllability rank condition is satisfied generically, i.e., on an open dense subset of M.

Proof: For any system the controllability rank condition is satisfied on an open subset of M, possibly empty. To see that it is dense for locally weakly controllable systems suppose there exists an open subset U of M where the dimension of $\mathfrak{F}(x)<m$. Without loss of generality we can assume $\dim\mathfrak{F}(x)=k<m$ for all $x\in U$. For some $x^0\in U$, let U' denote the maximal integral submanifold of $\mathfrak{F}(x)$ in U given by the Frobenius Theorem, then $A_U(x^0)\subseteq U'$ and so using Theorem 2.1 we see Σ is not locally weakly controllable at x^0. □

For analytic systems we can strengthen the above, in fact, weak controllability, local weak controllability, and the controllability rank condition are equivalent.

Theorem 2.6: If Σ is analytic then Σ is weakly controllable iff it is locally weakly controllable iff the controllability rank condition is satisfied.

Proof: We have already shown that for C^∞ systems the controllability rank condition implies local weak controllability which implies weak controllability. The reverse implications follow for analytic systems if we show that x^0WAx^1 implies that the dimension of $\mathfrak{F}(x^0)$ and $\mathfrak{F}(x^1)$ are the same. For then, if Σ is weakly controllable, the dimension of $\mathfrak{F}(x)$ must be constant and hence equal to m by the Frobenius and Chow Theorems.

To show that x^0WAx^1 implies the dimension of $\mathfrak{F}(x^0)$ and $\mathfrak{F}(x^1)$ are the same, it suffices to consider the case $x^1=\gamma(x^0)$ where γ is the family of solutions of the vector field $f(x)=f(x,u)$ for some constant $u\in\Omega$ and $x>0$. The map $(\partial/\partial x)\gamma_{-s}(x^1)$ is a linear isomorphism from the tangent space at x^1 to the tangent space at x^0. For any $h\in\mathfrak{X}(M)$, the Campbell–Baker–Hausdorff formula allows us to expand $(\partial/\partial x)\gamma_{-s}(x^1)h(x^1)$ in a convergent series

$$\frac{\partial}{\partial x}\gamma_{-s}(x^1)h(x^1)=\sum_{k=0}^{\infty}(L_f)^k h(x^0)\frac{s^k}{k!}.$$

In particular, if $h\in\mathfrak{F}$ then the right side of the above is a vector in $\mathfrak{F}(x^0)$ so $(\partial/\partial x)\gamma_{-s}(x^1)$ carries $\mathfrak{F}(x^1)$ into $\mathfrak{F}(x^0)$. Therefore, the dimension of $\mathfrak{F}(x^1)\leqslant$ dimension of $\mathfrak{F}(x^0)$. Reversing the argument shows that the inverse map $(\partial/\partial x)\gamma_s(x^0)$ carries $\mathfrak{F}(x^0)$ into $\mathfrak{F}(x^1)$ and the result follows. □

Example 2.7: Consider the linear system (2). In this

case $\mathcal{F}^0=\{Ax+Bu:u\in\Omega\}$, so the Lie algebra is generated by the vector fields $\{Ax,B_{*1},\cdots,B_{*l}\}$ where B_{*j} denotes the jth column of B considered as a constant vector field. Computing brackets yields

$$[Ax,B_{*j}]=-AB_{*j} \qquad [B_{*j},B_{*k}]=0$$
$$[Ax[Ax,B_{*j}]]=A^2B_{*j} \qquad [B_{*j}[Ax,B_{*k}]]=0$$

and so on. The Cayley–Hamilton Theorem implies that $\mathcal{F}$ is spanned by the linear vector fields Ax and the constant vector fields A^iB_{*j} where $i=0,\cdots,m-1$ and $j=1,\cdots,l$.

This system is analytic so the Hermann–Nagano Theorem guarantees the existence of maximal integral submanifolds of $\mathcal{F}$ through each $x\in\mathbf{R}^m$. In particular if we let M' denote the maximal integral submanifold through $x=0$ then M' must contain the integral curves of all linear combinations of the constant vector fields $\{A^iB_{*j}\}$ starting at 0 and hence must include the linear subspace spanned by these vectors. In fact M' is precisely this subspace because at each $x\in M'$ the tangent space to M' contains Ax also.

Notice that in this context the controllability rank condition reduces to the well-known linear controllability condition that

$$\operatorname{rank}(B:AB:\cdots:A^{m-1}B)=m.$$

The controllability rank condition only implies local weak controllability but for linear systems it also implies controllability (see [4] for details).

If the controllability rank condition fails and the rank of the above matrix is $m'<m$, then by restricting the system to $M'=\mathbf{R}^{m'}$, the maximal integral submanifold of $\mathcal{F}$ through 0, we obtain a controllable linear system. Notice that in this case the dimensions of the maximal integral manifolds of $\mathcal{F}$ can vary, for if $Ax\notin\operatorname{span}\{A^iB_{*j}\}$ then the maximal integral manifold of $\mathcal{F}$ through x will be of dimension $>m'$. It cannot be a linear subspace of $\mathbf{R}^m$ because it does not contain 0 which must lie in a maximal integral manifold of dimension m'.

Example 2.8: Consider the linear system (2) but where $A(t)$, $B(t)$, and $C(t)$ are C^∞ matrix valued functions of time. We adjoin another state variable $x_0=t$ and rewrite the system equations

$$\begin{aligned}\dot{x}_0&=1\\ \dot{x}&=A(x_0)x+B(x_0)u\\ y_0&=x_0\\ y&=C(x_0)x.\end{aligned}\tag{3}$$

The structure of $\mathcal{F}$ is similar to that for autonomous systems, for example,

$$\left[\begin{pmatrix}1\\ A(x_0)x\end{pmatrix},\begin{pmatrix}0\\ B_{*j}(x_0)\end{pmatrix}\right]=\begin{bmatrix}0\\ \dfrac{\partial}{\partial x_0}B_{*j}(x_0)-A(x_0)B_{*j}(x_0)\end{bmatrix}$$

$$\left[\begin{pmatrix}0\\ B_{*j}(x_0)\end{pmatrix},\begin{pmatrix}0\\ B_{*k}(x_0)\end{pmatrix}\right]=\begin{pmatrix}0\\0\end{pmatrix}.$$

From this it can be shown that the controllability rank condition is equivalent to the familiar requirement that

$$\operatorname{rank}(B(t):\Delta_cB(t):\Delta_c^2B(t):\cdots)=m$$

for every $t\in\mathbf{R}$ where

$$\Delta_cB(t)=\frac{d}{dt}B(t)-A(t)B(t),$$
$$\Delta_c^rB(t)=\left(\frac{d}{dt}-A(t)\right)\Delta_c^{r-1}B(t).$$

See [4], [6], or [15].

Example 2.9: Consider the bilinear system

$$\dot{x}=Ax+\sum_{i=1}^{l}u_iB^ix,$$

where $u\in\Omega=\mathbf{R}^l$ and M is either $\mathbf{R}^m$ or an m-dimensional subgroup of $Gl(h,\mathbf{R})$, the group of invertible $h\times h$ matrices.

The Jacobi brackets of linear vector fields Dx, Ex is seen to be

$$[Dx,Ex]=[D,E]x$$

where

$$[D,E]=ED-DE$$

is the commutator of the matrices D and E. $\mathcal{F}$ is isomorphic to a Lie subalgebra of $gl(m\mathbf{R})$ if $M=\mathbf{R}^m$ and $gl(h,\mathbf{R})$ if M is a subgroup of $Gl(h,\mathbf{R})$. ($gl(m,\mathbf{R})$ is the Lie algebra of all $m\times m$ matrices with commutation as the bracket). Let F denote the group

$$F=\{\exp X:X\in\mathcal{F}\}.$$

The maximal integral manifold M' of $\mathcal{F}$ through x_0 is just the orbit of x^0 under F

$$M'=Fx^0$$

and $\mathcal{F}$ satisfies the controllability rank condition iff F acts transitively on M.

III. Nonlinear Observability

We consider the system Σ as described in Section I and the input–output map of the pair (Σ,x^0) as described in Section II. A pair of points x^0 and x^1 are indistinguishable (denoted x^0Ix^1) if (Σ,x^0) and (Σ,x^1) realize the same input–output map, i.e., for every admissible input $(u(t),[t^0,t^1])$

$$\Sigma_{x^0}(u(t),[t^0,t^1])=\Sigma_{x^1}(u(t),[t^0,t^1]).$$

Indistinguishability I is an equivalence relation on M. Σ is said to *be observable at* x^0 if $I(x^0)=\{x^0\}$ and Σ is *observable* if $I(x)=\{x\}$ for every $x\in M$.

Notice that the observability of Σ does not imply that every input distinguishes points of M. If, however, the output is the sum of a function of the initial state and a function of the input, as it is for linear systems, then it is easy to see that if any input distinguishes between two initial states then every input does.

Notice also that observability is a global concept; it might be necessary to travel a considerable distance or for a long time to distinguish between points of M. Therefore we introduce a local concept which is stronger than observability. Let U be a subset of M and $x^0, x^1 \in U$. We say x^0 is *U-indistinguishable* from x^1 $(x^0 I_U x^1)$ if for every control $(u(t),[t^0,t^1])$, whose trajectories $(x^0(t),[t^0,t^1])$ and $(x^1(t),[t^0,t^1])$ from x^0 and x^1 both lie in U, fails to distinguish between x^0 and x^1, i.e., if $x^0(t)\in U$ and $x^1(t)\in U$ for $t\in[t^0,t^1]$, then

$$\Sigma_{x^0}(u(t),[t^0,t^1]) = \Sigma_{x^1}(u(t),[t^0,t^1]).$$

U-indistinguishability is not, in general, an equivalence relation on U for it fails to be transitive. This is related to the fact that Σ restricted to U is not necessarily complete. (See Sussmann [27] for a fuller discussion of this point.) However, we can still define Σ to be *locally observable at* x^0 if for every open neighborhood U of x^0, $I_U(x^0)=\{x^0\}$, and Σ is *locally observable* if it is so at every $x\in M$.

On the other hand one can weaken the concept of observability; in practice it may suffice to be able to distinguish x^0 from its neighbors. Therefore we define Σ to be *weakly observable at* x^0 if there exists a neighborhood U of x^0 such that $I(x^0)\cap U=\{x^0\}$ and Σ is *weakly observable* if it is so at every $x\in M$.

Notice once again that it may be necessary to travel considerably far from U to distinguish points of U, so we make a last definition, Σ is *locally weakly observable at* x^0 if there exists an open neighborhood U of x^0 such that for every open neighborhood V of x^0 contained in U, $I_V(x^0) =\{x^0\}$ and is *locally weakly observable* if it is so at every $x\in M$. Intuitively, Σ is locally weakly observable if one can instantaneously distinguish each point from its neighbors.

It can easily be seen that the relationships between the various forms of observability parallel that of controllability, i.e.,

$$\begin{array}{ccc} \Sigma \text{ locally observable} & \Rightarrow & \Sigma \text{ observable} \\ \Downarrow & & \Downarrow \\ \Sigma \text{ locally weakly observable} & \Rightarrow & \Sigma \text{ weakly observable.} \end{array}$$

In general there are no other implications but for autonomous linear systems it can be shown that all four are equivalent. The advantage of local weak observability over the other concepts is that it lends itself to a simple algebraic test. To describe it we need some additional tools.

Let $C^\infty(M)$ denote the infinite dimensional real vector space of all C^∞ real valued functions on M. Elements of $\mathcal{X}(M)$ act as linear operators on $C^\infty(M)$ by Lie differentiation. If $h\in\mathcal{X}(M)$ and $\varphi\in C^\infty(M)$ then $L_h(\varphi)\in C^\infty(M)$ is given by

$$L_h(\varphi)(x)=\frac{\partial\varphi}{\partial x}(x)h(x).$$

The gradient $d\varphi=\partial\varphi/\partial x=(\partial\varphi/\partial x_1,\cdots,\partial\varphi/\partial x_m)$ is a row vector valued function.

We denote by $\mathcal{G}^0$ the subset of $C^\infty(M)$ consisting of $g_1,\cdots,g_n$, and by $\mathcal{G}$ the smallest linear subspace of $C^\infty(M)$ containing $\mathcal{G}^0$ which is closed with respect to Lie differentiation by elements of $\mathcal{F}^0$. An element of $\mathcal{G}$ is a finite linear combination of functions of the form

$$L_{f^1}(\cdots(L_{f^k}(g_i))\cdots)$$

where $f^j(x)=f(x,u^j)$ for some constant $u^j\in\Omega$.

If $h_1,h_2\in\mathcal{X}(M)$ and $\varphi\in C^\infty(M)$ then

$$L_{h_1}(L_{h_2}(\varphi))-L_{h_2}(L_{h_1}(\varphi))=L_{[h_1,h_2]}(\varphi).$$

From this it follows that $\mathcal{G}$ is closed under Lie differentiation by elements of $\mathcal{F}$ also.

Let $\mathcal{X}^*(M)$ denote the real linear space of one forms on M, i.e., all finite $C^\infty(M)$ linear combinations of gradients of elements of $C^\infty(M)$. These are represented by row m-vector valued functions of x. The pairing between one forms and vector fields denoted by $\langle\omega,h\rangle\in C^\infty(M)$ is just multiplication of $1\times m$ and $m\times 1$ matrix valued functions of x.

We define a subset of $\mathcal{X}^*(M)$ by $d\mathcal{G}^0=\{d\varphi:\varphi\in\mathcal{G}^0\}$ and a subspace $d\mathcal{G}=\{d\varphi:\varphi\in\mathcal{G}\}$. Just as vector fields act on functions and other vector fields by Lie differentiation, they act on one forms according to the definition

$$L_h(\omega)(x)=\left(\frac{\partial\omega^*}{\partial x}(s)h(x)\right)^*+\omega(x)\frac{\partial h}{\partial x}(x)$$

where $\omega\in\mathcal{X}^*(M)$, $h\in\mathcal{X}(M)$ and $*$ denotes transpose. The three kinds of Lie differentiation are related by the following Liebnitz-type formula

$$L_{h_1}\langle\omega,h_2\rangle=\langle L_{h_1}\omega,h_2\rangle+\langle\omega,[h_1,h_2]\rangle.$$

If $\omega=d\varphi$, then L_h and d commute

$$L_h(d\varphi)=d(L_h(\varphi)).$$

From this it follows that $d\mathcal{G}$ is the smallest linear space of one forms containing $d\mathcal{G}^0$ which is closed with respect to Lie differentiation by elements of $\mathcal{F}^0$ (or $\mathcal{F}$). Elements of $d\mathcal{G}$ are finite linear combinations of one forms of the form

$$d(L_{f^1}(\cdots(L_{f^k}(g_i))\cdots))=L_{f^1}(\cdots(L_{f^k}(dg_i))\cdots)$$

where $f^j(x)=f(x,u^j)$ for some constant $u^j\in\Omega$. As before we denote by $d\mathcal{G}(x)$ the space of vectors obtained by evaluating the elements of $d\mathcal{G}$ at x. The space $d\mathcal{G}(x^0)$ determines the local weak observability of Σ at x^0. Σ is said to satisfy the *observability rank condition at* x^0 if the dimension of $d\mathcal{G}(x^0)$ is m, Σ satisfies the *observability rank condition* if this is true for every $x\in M$.

Theorem 3.1: If Σ satisfies the observability rank condition at x^0 then Σ is locally weakly observable at x^0.

The proof depends on the following.

Lemma 3.2: Let V be any open subset of M. If $x^0,x^1\in V$ and $x^0I_Vx^1$ then $\varphi(x^0)=\varphi(x^1)$ for every $\varphi\in\mathcal{G}$.

Proof: If $x^0 I_V x^1$ then for any $k \geq 0$, any constant controls $u^1, \cdots, u^k \in \Omega$, small $s_1, \cdots, s_k \geq 0$ and g_i, $i = 1, \cdots, n$, we have

$$g_i\left(\gamma^k_{s_k} \circ \cdots \circ \gamma^2_{s_2} \circ \gamma^1_{s_1}(x^0)\right) = g_i\left(\gamma^k_{s_k} \circ \cdots \circ \gamma^2_{s_2} \circ \gamma^1_{s_1}(x^1)\right).$$

Here $\gamma^i_{s_i}(x)$ denotes the flow of $f^i(x) = f(x, u^i)$. Differentiating with respect to $s_k, \cdots, s_1$, at 0 yields

$$L_{f^1}\left(\cdots\left(L_{f^k}(g_i)\right)\cdots\right)(x^0) = L_{f^1}\left(\cdots\left(L_{f^k}(g_i)\right)\cdots\right)(x^1).$$

$\mathcal{G}$ is spanned by functions of this form so the lemma follows. □

Proof of Theorem 3.1: If the dimension of $d\mathcal{G}(x^0) = m$ then there exists m functions $\varphi_1, \cdots, \varphi_m \in \mathcal{G}$ such that $d\varphi(x^0), \cdots, d\varphi_m(x^0)$ are linearly independent. Define a map

$$\Phi: x \mapsto \operatorname{col}\left(\varphi_1(x), \cdots, \varphi_m(x)\right).$$

The Jacobian of Φ at x^0 is nonsingular so Φ restricted to some open neighborhood U of x^0 is one to one. If $V \subseteq U$ is an open neighborhood of x^0 then Lemma 3.2 implies $I_V(x^0) = \{x^0\}$ so Σ is locally weakly observable at x^0. □

From this we see that if Σ satisfies the observability rank condition then Σ is locally weakly observable. The converse is almost true as we shall see later on in Theorems 3.11 and 3.12.

Suppose (Σ, x^0) is a realization of an input–output map which is not observable in any of the above senses. We turn our attention to finding such a realization (Σ', z^0) of the same input–output map. To understand the difficulties involved consider the following.

Example 3.3: Let $u \in \Omega = \boldsymbol{R}$, $x \in M = \boldsymbol{R}$, $y \in \boldsymbol{R}^2$, $x(0) = x^0 = 0$, and

$$\dot{x} = u \quad y_1 = \cos x \quad y_2 = \sin x.$$

Clearly the system satisfies both controllability and observability rank conditions. Therefore, it is locally weakly controllable and locally weakly observable. It is not observable because x^0 and $x^k = x^0 + 2k\pi$ are indistinguishable for any x^0 and any integer k.

To obtain an observable system with the same input–output behavior as the original we must identify x^0 and x^k, that is, define a system Σ' on $M' = S^1$, the unit circle by $\theta(0) = \theta^0 = 0$ and

$$\dot{\theta} = u \quad y_1 = \cos\theta \quad y_2 = \sin\theta.$$

Note that (Σ, x^0) and (Σ', θ^0) realize the same input–output map.

This example seems to imply that one can obtain an observable realization from one that it is not by "factoring" by the relation I of indistinguishability, the new system lives on the state space $M' = M/I$. However, given an equivalence relation R on M it is not always true that M/R with the quotient topology is Hausdorff and admits a C^∞ structure in such a way that the canonical projection $\pi: M \to M/R$ is a *submersion*, i.e., a C^∞ map of

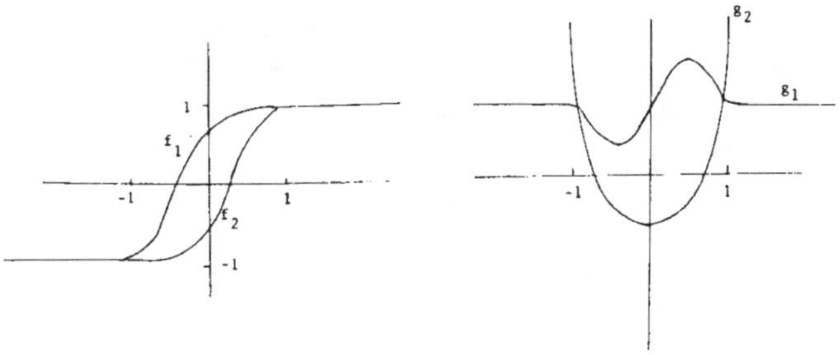

Fig. 1. Dynamics and observers of Example 3.4.

maximal rank. A necessary and sufficient condition for the quotient topology on M/R to be Hausdorff is that R be a *closed equivalence relation*, that is, the graph of R be a closed subset of $M \times M$. An equivalence relation R which admits a C^∞ structure on M/R compatible with the projection is called *regular*. A necessary and sufficient condition for regularity is that the graph of R be a regularly imbedded submanifold of $M \times M$ and that the map $(x^0, x^1) \mapsto x^0$ from the graph of R onto M be a submersion [23].

It is not hard to see using the continuity of solutions of differential equations with respect to initial conditions that for any C^∞ system Σ the relation I is closed. However, the following example due to Sussmann shows that it need not be regular even when the controllability and observability rank conditions are satisfied.

Example 3.4: Let $u \in \Omega = \{(u_1, u_2): u_i \geq 0\}$ $x \in M = \boldsymbol{R}$, $y \in \boldsymbol{R}^2$ and

$$\dot{x} = u_1 f_1(x) + u_2 f_2(x)$$

$$y_1 = g_1(x), \quad y_2 = g_2(x).$$

We choose $f_i, g_i: \boldsymbol{R} \to \boldsymbol{R}$ to be C^∞ functions with the following graphs. See Fig. 1.

Since f_1 and f_2 have no common zeros the system satisfies the controllability rank condition and is locally weakly controllable. Since dg_1 and dg_2 have no common zeros the system satisfies the observability rank condition and is locally weakly observable.

From the graphs of g_1 and g_2 we see the only possible pairs of indistinguishable points are x and $-x$ where $|x| \geq 1$. Since $f_i(x) = 1$ if $x \geq 1$ and $f_i(x) = -1$ if $x \leq -1$ it can be seen that these pairs are indistinguishable.

If we quotient the state space $M = \boldsymbol{R}$ by the equivalence relation of indistinguishability, the result is clearly not a manifold for it looks like a circle with a ray attached. At this point one can ask what additional assumptions on Σ are needed in order to insure that I be a closed and regular equivalence relation. Sussmann has proved the following.

Theorem 3.5 [27]: If (Σ, x^0) is a weakly controllable realization and either

a) Σ is analytic or

b) Σ is symmetric (i.e., $\forall u \in \Omega\ \exists v \in \Omega$ such that $f(x,u) = -f(x,v)\ \forall x \in M$), then I is a closed and regular equivalence relation. Moreover, there exists a system Σ' on $M' = M/I$ such that $(\Sigma', I(x^0))$ is an observable and weakly controllable realization of the same input–output map. If Σ is analytic, then Σ' is also locally observable.

In some sense Theorem 2.3b is the dual of Theorem 3.5a and Theorem 2.4 is the dual of Theorem 3.5b. What we would like to discuss now is the dual of Theorem 2.3a. Proceeding analogously we might expect that I is a regular equivalence relation if the dimension of $d\mathcal{G}(x)$ is constant over M but Example 3.4 shows this not to be the case. Just as we have used the relation WA in addition to A when studying nonlinear controllability so must we introduce another relation for nonlinear observability.

We call this new relation strong indistinguishability, x^0 and x^1 are *strongly indistinguishable* (denoted by x^0SIx^1) if there exists a continuous curve α: $[0,1]\to M$ such that $\alpha(0)=x^0$, $\alpha(1)=x^1$ and $x^0I\alpha(s)$ for all $s\in[0,1]$. Clearly SI is an equivalence relation, x^0SIx^1 implies x^0Ix^1, and Σ weakly observable at x^0 implies $SI(x^0)=\{x^0\}$.

As we shall demonstrate in a moment, if the dimension of $d\mathcal{G}(x)$ is constant over M then SI is a regular equivalence relation and the quotient M', possibly non-Hausdorff, inherits a locally weakly observable system Σ' which realizes the same input–output map as Σ. Before proving this we must introduce some more machinery. Let $\mathcal{K}(x)$ denote the space of all tangent vectors at x which annihilate every element of $d\mathcal{G}(x)$

$$\mathcal{K}(x)=\{\tau\in T_xM:\langle d\varphi(x),\tau\rangle=0,\qquad \forall\varphi\in\mathcal{G}\}.$$

If the dimension of $d\mathcal{G}(x)$ is constant, say k, over M then the dimension of $\mathcal{K}(x)$ is constant, $m-k$. The importance of $\mathcal{K}(x)$ is explained by the corollaries to the following lemma.

Lemma 3.6: Suppose the dimension of $d\mathcal{G}(x)$ is k for every $x\in M$. Let $f_t(x)$ be a time-dependent vector field on M such that $f_t(\cdot)\in\mathcal{F}$ for every t [for example, if $u(t)$ is an admissible control and $f_t(x)=f(x,u(t))$]. Let $\gamma_t(x^0,t^0)$ be the flow of f_t, i.e.,

$$\frac{d}{dt}\gamma_t(x^0,t^0)=f_t(\gamma_t(x^0,t^0))$$
$$\gamma_0(x^0,t^0)=x^0.$$

Then

$$d\mathcal{G}(\gamma_t(x^0,t^0))\frac{\partial}{\partial x}\gamma_t(x^0,t^0)=d\mathcal{G}(x^0).$$

Proof: Since $\gamma_{t+s}(x^0,t^0)=\gamma_s(\gamma_t(x^0,t^0),t)$, it suffices to consider only small $|t-t^0|$. Choose $d\varphi_1,\cdots,d\varphi_k\in d\mathcal{G}$ which are linearly independent on a neighborhood U of x^0. A straightforward calculation yields

$$\frac{d}{dt}\left(d\varphi_i(\gamma_t(x^0,t^0))\frac{\partial\gamma_t}{\partial x}(x^0,t^0)\right)$$
$$=L_{f_t}(d\varphi_i)(\gamma_t(x^0,t^0))\frac{\partial\gamma_t}{\partial x}(x^0,t^0).$$

On the other hand we can choose functions $\lambda_{ij}(t)$ such that

$$d\varphi_i(\gamma_t(x^0,t^0))\frac{\partial\gamma_t}{\partial x}(x^0,t^0)=\sum_{j=1}^{m}\lambda_{ij}(t)dx_j$$

for $i=1,\cdots,k$. Moreover if $f\in\mathcal{F}$ and $d\varphi\in d\mathcal{G}$ then $L_f(d\varphi)\in d\mathcal{G}$ so there exists functions $\mu_{ir}(t)$ such that

$$L_{f_t}(d\varphi_i)(\gamma_t(x^0,t^0))=\sum_{r=1}^{k}\mu_{ir}(t)d\varphi_r(\gamma_t(x^0,t^0))$$

for $i=1,\cdots,k$.

Combining these three equations we obtain

$$\frac{d}{dt}\left(\sum_{j=1}^{m}\lambda_{ij}(t)\right)=\sum_{r=1}^{k}\sum_{j=1}^{m}\mu_{ir}(t)\lambda_{rj}(t)$$

for $i=1,\cdots,k$. This is a linear homogeneous differential equation so there exist invertible linear transformations $\Lambda(t):\boldsymbol{R}^k\to\boldsymbol{R}^k$ such that

$$(\lambda_{ij}(t))=\Lambda(t)(\lambda_{ij}(t^0)).$$

The lemma follows since $d\mathcal{G}(x^0)$ is spanned by

$$d\varphi_i(x^0)=\sum_{j=1}^{m}\lambda_{ij}(t^0)dx_j$$

for $i=1,\cdots,k$ and for small $|t-t^0|$, $d\mathcal{G}(\gamma_t(x^0,t^0))\partial/\partial x\gamma_t(x^0,t^0)$ is spanned by

$$d\varphi_i(\gamma_t(x^0,t^0))\frac{\partial\gamma_t}{\partial x}(x^0,t^0)=\sum_{j}\lambda_{ij}(t)dx_j$$

for $i=1,\cdots,k$. □

Remark: The above depends heavily on the fact that the dimension of $d\mathcal{G}(x)$ is constant so that $d\mathcal{G}$ is "locally finitely generated." For similar results see Hermann [10].

A curve $\alpha:[0,1]\to M$ is *piecewise* C^∞ if it is C^∞ at all but finitely many points of $[0,1]$ and left (right) limits of α and all its derivatives exist at every point of $(0,1]$ ($[0,1)$).

We define another equivalence relation on M, x^0Hx^1 if there exists a continuous and piecewise C^∞ curve $\alpha:[0,1]\to M$ such that $\alpha(0)=x^0$, $\alpha(1)=x^1$ and

$$\frac{d}{ds}\alpha(s)\in\mathcal{K}(\alpha(s)).$$

Corollary 3.7: Assuming the dimension of $d\mathcal{G}(x)$ is constant, if x^0Hx^1 then x^0SIx^1.

Proof: Let $(u(t),[t^0,t^1])$ be any admissible control with flow $\gamma_t(x,t^0)$. The ith component of the output at time t when the system is started at $\alpha(s)$ at time t^0 is

$$y_i(t)=g_i(\gamma_t(\alpha(s),t^0)).$$

The derivative with respect to s is given by

$$\frac{d}{ds}y_i(t)=\left\langle dg_i(\gamma_t(\alpha(s),t^0))\frac{\partial\gamma_t}{\partial x}(\alpha(s),t^0),\frac{d\alpha}{ds}(s)\right\rangle$$

which is identically zero by the last lemma.

This next result shows that $\mathcal{F}$ is a collection of symmetry vector fields (in the sense of Sussmann [26]) for the relation H.

Corollary 3.8: Assume that the dimension of $d\mathcal{G}(x)$ is constant. Let $f \in \mathcal{F}$ with flow $\gamma_t(x)$. If $x^0 H x^1$ then $\gamma_t(x^0) H \gamma_t(x^1)$ for all $t \in \boldsymbol{R}$.

Proof: Let $\alpha:[0,1] \to M$ be a continuous and piecewise C^∞ curve such that $\alpha(0)=x^0$, $\alpha(1)=x^1$ and $d/ds\,\alpha(s) \in \mathcal{K}(\alpha(s))$. Define a curve, $\beta:[0,1]\to M$ by $\beta(s) = \gamma_t(\alpha(s))$. Clearly β is continuous, piecewise C^∞, and $\beta(0)=\gamma_t(x^0)\,\beta(1)=\gamma_t(x^1)$. Moreover for $\varphi \in \mathcal{G}$

$$\langle d\varphi(\beta(s)), \frac{d}{ds}\beta(s)\rangle = \frac{d}{ds}\varphi(\beta(s)) = \frac{d}{ds}\varphi(\gamma_t(\alpha(s)))$$

$$= \langle d\varphi(\gamma_t(\alpha(s)))\frac{\partial \gamma_t}{\partial x}(\alpha(s)), \frac{d}{ds}\alpha(s)\rangle$$

$$=0.$$

So $d/ds\beta(s) \in \mathcal{K}(\beta(s))$. □

Theorem 3.9: Suppose the dimension of $d\mathcal{G}(x)$ is k, then SI is a regular equivalence relation and there exists a locally weakly observable system Σ' on the k-dimensional non-Hausdorff manifold $M' = M/SI$ which has the same input–output properties as Σ. More precisely if $\pi: M \to M'$ is the canonical projection then (Σ, x^0) and $(\Sigma', \pi(x^0))$ realize the same input–output map for every $x^0 \in M$. If Σ is (locally)(weakly) controllable then so is Σ'. If Σ satisfies the controllability rank condition then so does Σ' and moreover M' is Hausdorff.

Proof: An outline goes like this.

From Corollary 3.7 we see that H equivalence implies SI equivalence. We first show that H is a regular equivalence relation and that we can define a system Σ' on $M' = M/H$ with the same input–output properties as Σ. We then note that Σ' satisfies the observability rank condition, hence Σ' is locally weakly observable from which it follows that $H = SI$.

Remark: By Corollary 3.7 $x^0 H x^1$ implies $x^0 SI x^1$ which by definition implies $x^0 I x^1$ which by Lemma 3.2 implies $\varphi(x^0)=\varphi(x^1)$ for every $\varphi \in \mathcal{G}$.

Given any $x^0 \in M$ there exists $d\varphi_1, \cdots, d\varphi_k \in d\mathcal{G}$ which are linearly independent at x^0. After reordering the x coordinates we can suppose that $d\varphi_1, \cdots, d\varphi_k$, $dx_{k+1}, \cdots, dx_m$ are linearly independent at x^0. Define $z_i(x) = \varphi_i(x) - \varphi_i(x^0)$, $i=1,\cdots,k$, $z_i(x) = x_i - x_i^0$, $i=k+1,\cdots,m$ and

$$U = \{x : |z_i(x)| < \epsilon\}.$$

If ϵ is chosen sufficiently small, then (U,x) is a coordinate chart around x^0.

Claim If $x^1, x^2 \in H(U)$ then $x^1 H x^2$ iff $\varphi_i(x^1) = \varphi_i(x^2)$ for $i=1,\cdots,k$. The only if part follows immediately from the above remark. As for the converse suppose first that $x^1, x^2 \in U$ and $\varphi_i(x^1)=\varphi_i(x^2)$ for $i=1,\cdots,k$. Then $z_i(x^1) = z_i(x^2)$ *for* $i=1,\cdots,k$. Clearly $\partial/\partial z_{k+1}, \cdots, \partial/\partial z_m \in \mathcal{K}(x)$ for every $x \in U$ and these vector fields can be used to get from x^1 to x^2 within the z-cube U so $x^1 H x^2$. More generally if $x^1, x^2 \in H(U)$ and $\varphi_i(x^1)=\varphi_i(x^2)$ for $i = 1,\cdots,k$ then there exists $p^j \in U$ such that $x^j H p^j$ and $\varphi_i(p^j)=\varphi_i(x^j)$ for $i=1,\cdots,k$ and $j=1,2$. From before we know $p^1 H p^2$ which by transitivity implies $x^1 H x^2$.

We next put a C^∞ structure on $M' = M/H$. Notice that $H(U)$ is open and $\pi^{-1}(\pi(H(U))) = H(U)$ so $V = \pi(H(U))$ is open in the quotient topology on M'. The functions of $\mathcal{G}$ are constant on H-equivalence classes so every $\varphi \in \mathcal{G}$ can be pulled down to M', i.e., there exists a continuous $\varphi': M' \to \boldsymbol{R}$ such that $\varphi' \circ \pi = \varphi$.

In particular we define $z_i' = \varphi_i' - \varphi_i(x^0)$ for $i=1,\cdots,k$ and $z' = (z_1', \cdots, z_k')$. The claim implies that the map $z': V \to \boldsymbol{R}^k$ is one to one. Moreover it is open because if W is an open subset of V then

$$z'(W) = \pi_k \circ z \circ (\pi^{-1}(W) \cap U)$$

where $\pi_k: \boldsymbol{R}^m \to \boldsymbol{R}^k$ is the projection on the first k factors. Therefore $z': V \to \boldsymbol{R}^k$ is a homeomorphism into and (V,z') a coordinate chart. In the coordinates (U,z) and (V,z') on M and M', the projection π is just π_k and clearly a submersion.

M is covered by a countable number of charts (U,z) and so M' is covered by the corresponding (V,x'). Verifying that changes of coordinates on M' are C^∞ reduces to checking that every $\varphi': M' \to \boldsymbol{R}$ is C^∞ on (V,z'). But if $\varphi \in \mathcal{G}$ then on (U,z)

$$d\varphi = \Sigma \lambda_i(z)\, dz_i$$

so $\varphi(z) = \varphi(z_1,\cdots,z_k)$. Clearly $\varphi'(z') = \varphi(z_1',\cdots,z_k')$ is C^∞. This shows H is a regular equivalence relation on M.

Next we define Σ' on M' locally. In the coordinate system (U,z) on M, the dynamics of Σ are given by

$$\dot z = \frac{\partial z}{\partial x}(x(z))\, f(x(z),u).$$

In particular for $i=1,\cdots,k$

$$\dot z_i = \frac{\partial \varphi_i}{\partial x}(x(z))\, f(x(z),u) = L_{f(\cdot,u)}(\varphi_i)(x(z)).$$

The right side is an element of $\mathcal{G}$ and hence pulls down to a function $f_i'(z',u)$ on V'. We define the dynamics of Σ' on (V,z') by

$$\dot z' = f'(z',u).$$

Clearly if $z(t)$ is a curve in U generated by the dynamics of Σ under the control $u(t)$ satisfying $z(t^0)=z^0$ and if $z'(t)$ is the curve in V generated by the dynamics of Σ' under the control $u(t)$ satisfying $z'(t^0) = \pi(z^0)$ then for small $|t-t^0|$

$$z'(t) = \pi(z(t)).$$

From this it follows that if Σ is (locally) (weakly) controllable so is Σ'.

Moreover each $g_i \in \mathcal{G}$ and hence pulls down to a function g_i' on M' which we use to define the output of Σ' on (V,Z') as

$$y = g'(z').$$

The outputs of Σ from z^0 under $u(t)$ and Σ' from $\pi(z^0)$ under $u(t)$ are the same because $g = g' \circ \pi$ and so

$$g(z(t)) = g' \circ \pi(z(t)) = g'(z'(t)).$$

Notice M' is of dimension k.

We turn now to the relationship between $\mathcal{F}$ and $\mathcal{G}$ of Σ and $\mathcal{F}'$ and $\mathcal{G}'$ of Σ'. Note that

$$f'(\pi(x), u) = \frac{\partial \pi}{\partial x}(x) f(x, u).$$

It follows from this and the definition of $\mathcal{F}$ and $\mathcal{F}'$ that $f' \in \mathcal{F}'$ iff there exists an $f \in \mathcal{F}$ such that

$$f'(\pi(x)) = \frac{\partial \pi}{\partial x}(x) f(x).$$

In particular

$$\mathcal{F}'(\pi(x)) = \frac{\partial \pi}{\partial x}(x) \mathcal{F}(x).$$

and since π is a submersion this shows that if Σ satisfies the controllability rank condition so does Σ'.

Furthermore

$$g' \circ \pi = g$$

and so

$$L_f g_i(x) = L_{f'}(g_i')(\pi(x)).$$

Repeated applications of this formula show that $\varphi' \in \mathcal{G}'$ iff there exists on $\varphi \in \mathcal{G}$ such that

$$\varphi'(\pi(x)) = \varphi(x).$$

In particular the coordinate functions $z_i' = \varphi_i'$ are in $\mathcal{G}'$ so Σ' satisfies the observability rank condition and therefore is locally weakly observable.

This also implies that H and SI are the same relation. We have already seen that $x^0 H x^1$ implies $x^0 SI x^1$. Suppose the converse does not hold; there exists $x^0, x^1 \in M$ such that $x^0 SI x^1$ but $x^0 H x^1$. Let $\alpha(s)$ be the arc of Σ-indistinguishable points joining x^0, x^1, since Σ and Σ' have the same input-output properties, $\pi(\alpha(s))$ is an arc of Σ'-indistinguishable points joining $\pi(x^0)$ and $\pi(x^1)$. Moreover this arc is not constant because $\pi(x^0) \neq \pi(x^1)$. This contradicts the local weak observability of Σ'.

If Σ satisfies the controllability rank condition then M' can be seen to be Hausdorff. Given $x^0, x^1 \in M$ first suppose there exists a $\varphi \in \mathcal{G}$ such that $\varphi(x^0) \neq \varphi(x^1)$. Define disjoint open sets

$$U^0 = \{x \in M : |\varphi(x) - \varphi(x^0)| < |\varphi(x^1) - \varphi(x^0)|/2\}$$
$$U^1 = \{x \in M : |\varphi(x) - \varphi(x^1)| > |\varphi(x^1) - \varphi(x^0)|/2\}.$$

Since φ is constant on H equivalence classes

$$\pi^{-1}(\pi(U^i)) = U^i$$

and so $\pi(U^0)$ and $\pi(U^1)$ are disjoint open neighborhoods of $\pi(x^0)$ and $\pi(x^1)$.

On the other hand suppose $\varphi(x^0) = \varphi(x^1)$ for every $\varphi \in \mathcal{G}$. Using the controllability rank condition we choose $f_1, \cdots, f_m \in \mathcal{F}$ which are linearly independent at x^0. We also choose $d\varphi_1, \cdots, d\varphi_k \in d\mathcal{G}$ which are linearly independent at x^0. Then the $k \times m$ matrix with i–jth element

$$\langle d\varphi_i(x), f_j(x) \rangle = L_{f_j}(\varphi_i)(x)$$

is of rank k at x^0. Without loss of generality we can assume the first k columns of this matrix are linearly independent at x^0. The elements of this matrix are all in $\mathcal{G}$ so the first k columns are also linearly independent at x^1 and therefore $d\varphi_1, \cdots, d\varphi_k$ are linearly independent at x^1.

As before we can construct a coordinate cube (U^0, z) around x^0 using $\varphi_1, \cdots, \varphi_k, x_{k+1}, \cdots, x_m$. In a similar fashion using the same ϵ and $\varphi_1, \cdots, \varphi_k$ but perhaps different $x_{k+1}, \cdots, x_m$ we construct a cube (U^1, z) around x^1. Given $\delta > 0$ define $m-k$ slices S_δ^0 and S_δ^1 in U^0 and U^1 by

$$S_\delta^j = \{x \in U^j : z_i(x) = 0, \qquad i = 1, \cdots, k \quad \text{and} \quad |z_i(x)| < \delta, \qquad i = k+1, \cdots, m\}.$$

Let $\gamma_s^i(x)$ denote the flow of $f^i(x)$ for $i = 1, \cdots, k$ and C_δ^k denote the cube of side 2δ around 0 in $\boldsymbol{R}^k$. Define maps

$$\beta_j : C_\delta^k \times S_\delta^j \to M$$

by

$$\beta_j(s_1, \cdots, s_n, x) = \gamma_{s_k}^k \circ \cdots \gamma_{s_1}^1(x)$$

where $x \in S_\delta^j$ and $|s_i| < \delta$ for $i = 1, \cdots, k$. Of course S_δ^j is a $m-k$ dimensional cube so we can view these maps as

$$\beta^j : C_\delta^m \to M$$

where C_δ^m is an m-cube of side 2δ. More precisely if $(s_1, \cdots, s_m) \in C_\delta^m$ then

$$\beta^j(s_1, \cdots, s_m) = \beta^j(s_1, \cdots, s_k, x)$$

where $x \in S_\delta^j$ with $z_i(x) = 0$ for $i = 1, \cdots, k$ and $z_i(x) = s_i$ for $i = k+1, \cdots, m$. For sufficiently small δ these maps are diffeomorphisms onto open neighborhoods W^j of x^j contained in U^j. In fact $(W^j, s_1, \cdots, s_m)$ are coordinate charts at x^j. If δ is chosen sufficiently small then we can assume that for every $x \in W^1$ and for every $(s_1, \cdots, s_n) \in C_\delta^k$ we have

$$\gamma_{s_1}^1 \circ \cdots \circ \gamma_{s_k}^k(x) \in U^1.$$

Now suppose there exists $p^0 \in W^0$, $p^1 \in W^1$ such that $p^0 H p^1$. In particular suppose $p^0 = \gamma_{s_k}^k \circ \cdots \circ \gamma_{s_1}^1(q^0)$ where $q^0 \in S_\delta^0$. Let

$$q^1 = \gamma_{-s_1}^1 \circ \cdots \circ \gamma_{-s_k}^k(p^1).$$

then by assumption $q^1 \in U^1$ and by Corollary 3.8, $q^0 H q^1$. Because $q^0 \in S_\delta^0$, $x^0 H q^0$ so $x^0 H q^1$. This implies that $\varphi_i(x^0) = \varphi_i(q^1)$ for $i = 1, \cdots, k$ and so by assumption $\varphi_i(q^1) = \varphi_i(x^1)$ for $i = 1, \cdots, k$. By our previous claim, $q^1 H x^1$ and so $x^0 H x^1$.

We have just shown that if $\pi(W^0)\cap\pi(W^1)\neq\emptyset$ then $\pi(x^0)=\pi(x^1)$. It follows that if $\pi(x^0)\neq\pi(x^1)$ then these are disjoint open neighborhoods of $\pi(x^0)$ and $\pi(x^1)$. □

The following simple example shows that if Σ fails to satisfy the controllability rank condition then M' need not be Hausdorff.

Example 3.10: Let $M=\mathbf{R}^2\setminus\{(x_1,0):x_1\leqslant 0\}\, y\in\mathbf{R}^1$ and

$$\dot{x}=0 \quad y=x_1.$$

The dimension of $d\mathcal{G}(x)$ is one for all x so we can form the quotient $M'=M/SI$. However this is a non-Hausdorff manifold because $\pi(0,1)$ and $\pi(0,-1)$ are distinct points without a pair of disjoint neighborhoods. Notice that factoring M by the closure of the relation SI is not a way around this problem for the closure of SI is not a regular relation.

Theorem 3.5a and b and Theorem 3.9 exhibit three situations where realizations can be made observable in some sense. If Σ is analytic then Σ' is locally observable, if Σ is symmetric then Σ' is observable and if $d\mathcal{G}(x)$ of Σ is of constant dimension then Σ' satisfies the observability rank condition and hence is locally weakly observable. The converse of this last remark is "almost" true.

Theorem 3.11: If Σ is locally weakly observable then the observability rank condition is satisfied generically.

Proof: For any system the observability rank condition is satisfied on an open subset of M, possibly empty. Suppose there exists an open subset U of M where the dimension of $d\mathcal{G}(x)<m$. Without loss of generality we can assume $\dim d\mathcal{G}(x)=k<m$ for $x\in U$. Choose $x^0\in U$, an open set V such that $x^0\in V\subseteq U$ and a C^∞ function $\varphi: M\to\mathbf{R}$ such that $\varphi(x)=1$ for all $x\in V$, $\varphi(x)\neq 0$ for all $x\in U$ and $\varphi(x)=0$ for all $x\in M\setminus U$. Consider the system Σ' defined by

$$\dot{x}=\varphi(x)f(x,u)$$
$$y=g(x)$$
$$x(0)=x^0.$$

The state space of Σ' is U and Σ' is complete. It is easy to see $\mathcal{F}'(x)=\mathcal{F}(x)$ for all $x\in U$. From this it follows that $d\mathcal{G}'(x)=d\mathcal{G}(x)$ for all $s\in U$, and so $\dim d\mathcal{G}(x)=k$ for all $x\in U$. We can apply Theorem 3.9 to Σ' on U so Σ' is not locally weakly observable. Since Σ and Σ' agree on V, Σ is not locally weakly observable either. □

Recall that for analytic systems, weak controllability, local weak controllability, and the controllability rank condition are equivalent. With regard to observability an analogous result holds for analytic systems which are weakly controllable.

Theorem 3.12: If Σ is a weakly controllable analytic system then Σ is weakly observable iff it is locally weakly observable iff the observability rank condition is satisfied.

Proof: It suffices to show that weak observability implies the observability rank condition. By Theorem 2.6 Σ satisfies the controllability rank condition, we proceed in a similar fashion to the proof of that theorem to show the dimension of $d\mathcal{G}(x)$ is constant.

Let $x^1=\gamma_s(x^0)$ where γ denotes the flow of $f\in\mathcal{F}$. The adjoint of $(\partial/\partial x)\gamma_s(x^0)$ carries one forms at x^1 to one forms at x^0 according to the rule

$$\omega(x^1)\mapsto\omega(x^1)\frac{\partial}{\partial x}\gamma_s(x^0),$$

for $\omega\in\mathcal{X}^*(M)$.

This map also has a Campbell–Baker–Hausdorff expansion

$$\omega(x^1)\frac{\partial}{\partial x}\gamma_s(x^0)=\sum_{k=0}^{\infty}(L_f)^k\omega(x^0)\frac{s^k}{k!}.$$

In particular, if $\omega\in H$ then the adjoint of $(\partial/\partial x)\gamma_s(x^0)$ carries $d\mathcal{G}(x^1)$ into $d\mathcal{G}(x^0)$ so the dimension of $d\mathcal{G}(x^1)$ is less than or equal to the dimension $d\mathcal{G}(x^0)$. A similar argument using the adjoint of $(\partial/\partial x)\gamma_{-s}(x^1)$ shows the reverse inequality.

Therefore, if x^0WAx^1 then the dimensions of $d\mathcal{G}(x^0)$ and $d\mathcal{G}(x^1)$ are the same. Since Σ is weakly controllable, this implies that the dimension of $d\mathcal{G}(x)$ is constant.

Suppose $\dim d\mathcal{G}(x)=k$ then we apply Theorem 3.9 and identify strongly indistinguishable points. But Σ weakly observable implies there are no strongly indistinguishable points so k must equal m. □

To relate the results of this section to the well-known linear theory, we consider the following.

Example 3.11: Consider the linear system

$$\dot{x}=Ax+Bu$$
$$y=Cx.$$

In Example 2.5 we saw that $\mathcal{F}$ was spanned by $\{Ax, A^rB_{*j}: r=0,\cdots,m-1,\ j=1,\cdots,l\}$ where B_{*j} denotes the jth column of B. Let C_{i*} denote the ith row of C then for $r\geqslant 0$, $\rho\geqslant 0$

$$L_{Ax}C_{i*}yA^rx=C_{i*}A^{r+1}x$$
$$L_{A^rB_{*j}}C_{i*}A^\rho x=C_{i*}A^{r+\rho}B_{*j}$$
$$L_{Ax}C_{i*}A^\rho B_{*j}=L_{A^rB_{*j}}C_{i*}A^\rho B_{*j}=0.$$

Therefore by the Cayley–Hamilton Theorem $\mathcal{G}$ is spanned by

$$\{C_{i*}A^rx, C_{i*}A^rB_{*j}: i=1,\cdots,m, j=1,\cdots,l \text{ and } r=0,\cdots,m-1\}$$

and $d\mathcal{G}(x)$ is spanned by

$$\{C_{i*}A^r: i=1,\cdots,m,\ r=0,\cdots,m-1\}.$$

Clearly $d\mathcal{G}(x)$ is of constant dimension. The observability rank condition reduces to the well-known linear observability condition that

$$\text{rank}\begin{bmatrix} C \\ CA \\ \cdot \\ \cdot \\ \cdot \\ CA^{m-1} \end{bmatrix} = m.$$

If the observability rank condition fails then we define $\mathcal{K}(x)$ to be the constant linear space of column vectors orthogonal to $d\mathcal{G}(x)$. Two points x^0 and x^1 are H equivalent (or SI equivalent) if $x^1 - x^0 \in \mathcal{K}(x)$ because

$$\alpha(s) = x^0 + s(x^1 - x^0)$$

is a curve tangent to $\mathcal{K}(x)$ from x^0 to x^1. Factoring M by $\mathcal{K}(x)$ results in a locally weakly observable system which because of linearity is also locally observable ($SI = I$ for linear systems).

Example 3.12: Suppose the linear system is nonautonomous as in Example 2.6. As in formula 2.1 we add time as a state variable which we can observe directly. Then direct calculation shows that the observability rank condition is equivalent to

$$\text{rank}\begin{bmatrix} C(t) \\ \cdot \\ \cdot \\ \cdot \\ \Delta_o C(t) \\ \cdot \\ \cdot \\ \cdot \\ \Delta^2{}_* C(t) \cdot \\ \cdot \\ \cdot \\ \cdot \end{bmatrix} = m$$

for every $t \in \boldsymbol{R}$ where

$$\Delta_o C(t) = \frac{d}{dt} C(t) + C(t) A(t),$$

$$\Delta_o^r C(t) = \frac{d}{dt} \Delta_o^{r-1} C(t) + \Delta_o^{r-1} C(t) A(t).$$

See [3] or [5].

Example 3.13: Consider the bilinear system of Example 2.8

$$\dot{x} = Ax + \sum_{i=1}^{l} u_i B^i x.$$

If $M = \boldsymbol{R}^m$ (or a subgroup of $Gl(h, \boldsymbol{R})$), let C be a $n \times m$ ($n \times h$) matrix and

$$y = Cx,$$

then

$$\mathcal{G} = \{ C_{ix} Dx : i = 1, \cdots, n;\ Dx \in \mathcal{F} \}$$

$$d\mathcal{G} = \{ C_{ix} D : i = 1, \cdots, n;\ Dx \in \mathcal{F} \}.$$

See [3] for further details.

IV. Minimality

A linear system is said to be *minimal* if it is controllable and observable. As is well known, two minimal linear systems initialized at 0 which realize the same input–output map differ only by a linear diffeomorphism of the state spaces, [1], [4]. A nonlinear system which is observable, weakly controllable and either analytic or symmetric is called *minimal* by Sussmann [29]. He has shown that two minimal nonlinear systems which realize the same input–output map from their respective initial states differ only by a diffeomorphism of the state spaces.

A nonlinear system Σ is *locally weakly minimal* if it is locally weakly controllable and locally weakly observable. Two locally weakly minimal realizations of a given input–output map need not be diffeomorphic as is seen by Example 3.3, but the following theorem shows they must be of the same state dimension which is minimal over all possible realizations.

Let Σ, Σ' be two nonlinear systems with control set $\Omega = \Omega' \subseteq \boldsymbol{R}^l$ states spaces M and M' of dimension m and m' and output with values in $\boldsymbol{R}^n$ given by

$$\Sigma : \dot{x} = f(x, u)$$
$$y = g(x)$$
$$\Sigma' : \dot{z} = f'(z, u)$$
$$y = g'(z).$$

Theorem 4.1: Suppose (Σ, x^0) and (Σ', z^0) realize the same input–output map. If Σ' is locally weakly minimal then $m \geqslant m'$.

Proof: Without loss of generality we can assume that Σ' satisfies the controllability and observability rank condition at z^0. (If not, by Theorems 2.5 and 3.11 we can find a control $(u(t), [t^0, t^1])$ and a corresponding Σ' trajectory $(z(t), [t^0, t^1])$ with $z(t^0) = z^0$, $z(t^1) = z^1$ such that these conditions are satisfied at z^1. Moreover if x^1 is the endpoint of the corresponding Σ trajectory then (Σ, x^1) and (Σ', z^1) realize the same input–output map.)

Since Σ' satisfies the observability rank condition at z^0 we can find a neighborhood V of z^0 and functions $\varphi_1', \cdots, \varphi_{m'}' \in \mathcal{G}'$ such that the map

$$\Phi' = \text{col}(\varphi_1', \cdots, \varphi_m') : V \to \boldsymbol{R}^{m'}$$

is a diffeomorphism into.

Because Σ' satisfies the controllability rank condition, as in Theorems 2.1 and 2.2 we can find controls $u^1, \cdots, u^m$ and corresponding vector fields $f'^i(z) = f'(z, u^i)$ with flows $\gamma'^i_s(z)$ such that the map

$$\Psi'(s_1, \cdots, s_m) = \gamma'^m_{s_m} \circ \cdots \gamma'^1_{s_1}(z^0)$$

is a diffeomorphism from some open subset S of the positive orthant of $\boldsymbol{R}^{m'}$ into M'.

Let φ_i, Φ, f^i, γ^i, Ψ be the corresponding objects of Σ. Since (Σ, x^0) and (Σ', z^0) realize the same input–output map $\Phi \circ \Psi = \Phi' \circ \Psi'$ and so

$$S \subseteq R^{m'} \xrightarrow{\Psi} M \xrightarrow{\Phi} R^{m'}$$

is a diffeomorphism. But clearly this is only possible if $m \geqslant m'$.

Corollary 4.2: If (Σ, x^0) and (Σ', z^0) are locally weakly minimal realizations of the same input–output map then their state spaces are of the same dimension.

References

[1] B. D. O. Anderson and J. B. Moore, *Linear Optimal Control.* Englewood Cliffs, NJ: Prentice-Hall, 1971.

[2] W. M. Boothby, *An Introduction to Differential Manifolds and Riemannian Geometry.* New York: Academic, 1975.

[3] R. W. Brockett, "System theory on group manifolds and coset spaces," *SIAM J. Contr.*, vol. 10, pp. 265–284, 1972.

[4] ——*Finite Dimensional Linear Systems.* New York: Wiley, 1970.

[5] W. L. Chow, "Uber Systeme von Linearen Partiellen Differentialgleichungen erster Ordnung," *Math. Ann.*, vol. 117, pp. 98–105, 1939.

[6] H. D'Angelo, *Linear Time-Varying Systems.* Boston, MA: Allyn and Bacon, 1970.

[7] E. W. Griffith and K. S. P. Kumar, "On the observability of nonlinear systems, I," *J. Math. Anal. Appl.*, vol. 35, pp. 135–147, 1971.

[8] G. W. Haynes and H. Hermes, "Non-linear controllability via Lie theory," *SIAM J. Contr.*, vol. 8, pp. 450–460, 1970.

[9] R. Hermann, "On the accessibility problem in control theory," in *Int. Symp. on Nonlinear Differential Equations and Nonlinear Mechanics.* New York: Academic Press, 1963, pp. 325–332.

[10] ——*Differential Geometry and the Calculus of Variations.* New York: Academic, 1968.

[11] ——"The differential geometry of foliations, II," *J. of Math. and Mech.*, vol. 11, pp. 303–316, 1962.

[12] ——"Some differential geometric aspects of the Lagrange variational problem," *J. Math.*, vol. 111, no. 6, pp. 634–673, 1962.

[13] ——"Existence in the large of parallelism homomorphisms," *Trans. Amer. Math. Soc.*, vol. 108, pp. 170–183, 1963.

[14] ——"Cartan connections and the equivalence problem for geometric structures," *Contribution to Differential Equations*, vol. 3, 1964, pp. 199–248.

[15] H. Hermes and J. P. Lasalle, *Functional Analysis and Time Optimal Control.* New York: Academic, 1969.

[16] Y. M.-L. Kostyukovskii, "Observability of nonlinear controlled systems," *Automat. Remote Contr.*, vol. 9, pp. 1384–1396, 1968.

[17] ——"Simple conditions of observability of nonlinear controlled systems," *Automat. Remote Contr.*, vol. 10, pp. 1575–1584, 1968.

[18] S. R. Kou, D. L. Elliot, and T. J. Tarn, "Observability of nonlinear systems," *Inform. Contr.*, vol. 22, pp. 89–99, 1973.

[19] A. J. Krener, "A generalization of the Pontryagin maximal principle and the bang-bang principle," Ph.D. dissertation, University of California, Berkeley, CA, 1971.

[20] ——"A generalization of Chow's theorem and the bang-bang theorem to nonlinear control problems," *SIAM J. Contr.*, vol. 12, pp. 43–52, 1974.

[21] C. Lobry, "Controllabilité des systèmes non linéaires," *SIAM J. Contr.*, vol. 8, pp. 573–605, 1970.

[22] ——"Quelques aspects qualitatifs de la théorie de la commande," Thèse de doctorat d'état, Université de Grenoble, Grenoble, France, 1972.

[23] T. Nagano, "Linear differential systems with singularities and an application to transitive Lie algebras," *J. Math. Soc. Japan*, vol. 18, pp. 398–404, 1966.

[24] J. P. Serre, *Lie Groups and Lie Algebras.* New York: Benjamin, 1965.

[25] H. J. Sussmann and V. J. Jurdjevic, "Controllability of nonlinear systems," *J. Differential Equations*, vol. 12, pp. 95–116, 1972.

[26] H. J. Sussmann, "Minimal realizations of nonlinear systems," in *Geometric Methods in Systems Theory*, Mayne and Brockett, Ed., Dordrecht, The Netherlands: D. Riedel, 1973.

[27] ——"A generalization of the closed subgroup theorem to quotients of arbitrary manifolds," *J. Differential Geometry*, vol. 10, pp. 151–166, 1975.

[28] ——"Observable realizations of finite dimensional nonlinear autonomous systems," *Math. Systems Theory*, to be published.

[29] ——"Existence and uniqueness of minimal realizations of nonlinear systems I: Initialized systems," *Math. Systems Theory.*

Analysis of Recursive Stochastic Algorithms

LENNART LJUNG

INTEREST in identification algorithms and their use in adaptive control commenced very early, the former having its roots in the literature on time series analysis and the latter in dynamic programming which was immediately perceived, on its introduction in the 1950s, to be the natural tool for decision making under uncertainty. Adaptive control required *recursive* identification, stimulating a substantial activity in this topic. An obvious starting point for analyzing these recursive algorithms was the older literature on stochastic approximation [5], [8]. However, in recursive identification, unlike the models considered in the traditional stochastic approximation literature, the regression (observation) vector $\phi(t)$ is allowed to depend on all previous values of the parameter estimate, preventing use of the older techniques. It was obviously important to answer questions about the convergence of such identification algorithms: for example,

- under what circumstances does the algorithm converge?
- what is its rate of convergence?
- what meaning can be attached to the point to which convergence occurred?

These questions had started to present themselves in analyses conducted several years previously, such as references 1–5 of this paper by Ljung. Among other things, these references had shown that convergence need not necessarily occur, and also that the convergence property could be associated with the stability properties of a deterministic *ordinary differential equation* despite the fact that discrete-time identification algorithms were being considered.

The paper by Ljung starts with a very general form of recursive algorithm:

$$x(t) = x(t-1) + \gamma(t)Q(t; x(t-1), \phi(t)) \tag{1}$$

In this equation, $x(\cdot)$ is a sequence of vectors, of dimension n, constituting parameter estimates. It is hypothesised that in the background there rests a system described by a "true" parameter, and one would hope that the parameter estimates converge to the true parameter. In Equation (1), Q is another n-vector, most commonly not dependent explicitly on time, and depending on the current parameter estimate and a so-called observation vector, $\phi(t)$ which is drawn from the "true" system, but may be noisy; entries of $\phi(\cdot)$ may be noise-contaminated inputs and outputs of the true system so that it is *not* necessarily a sequence of independent, identically distributed random variables commonly demanded by stochastic approximation. Lastly, $\gamma(\cdot)$ is a sequence of positive scalars. Since there is interest in parameter convergence, and since in general Q varies randomly, it is clear that parameter convergence can only occur if

$$\gamma(t) \to 0 \quad \text{as} \quad t \to \infty \tag{2}$$

This is the type of assumption that became very familiar in the literature on stochastic approximation. It carries with it the great advantage that analysis of a fundamentally stochastic system can be replaced by analysis of a deterministic system, and there is some degree of comfort in a theorem which says that parameters converge to a deterministic limit. On the other hand, there is also a great disadvantage. With γ going to zero, less and less attention is paid to the "noisy" observations. This means that should the plant whose parameters are being estimated, and from whom observations are derived, not have constant parameter values, but have parameter values that change in time (perhaps slowly) the ability to track those varying parameters vanishes as γ goes to zero. In effect, only strictly time-invariant plants can be identified.

It is important to understand at a broad level the basis of the argument which allows the analysis to be done by consideration of an ordinary differential equation. There is a general appeal to averaging theory ideas. More precisely, suppose that in Equation (1) $\gamma(t)$ is very small over a long interval. Then over this interval, $x(\cdot)$ does not change much, and presumably ϕ takes a large number of values with some statistical regularity properties. These properties may well be x dependent, especially if a controller is being implemented to generate plant inputs, and the parameters in that controller are computed using the estimate x. In this circumstance, the sequences $\phi(t)$ and $x(t)$ will certainly not be independent.

Considering $x(t)$ as approximately a constant over a long interval, and assuming that $\phi(t)$ takes a range of values during this interval, one agrees to replace say, s, successive updates of Equation (1) by an update involving a single gain

$$\sum_{t+1}^{t+s} \gamma(k) \tag{3}$$

and to replace Q by its average value conditioned on $x(t-1)$. The result is still then a difference equation. However, one can regard it as an Euler approximation of an ordinary differential equation, and the ordinary differential equation comes out to be something like Equation (4) below, where $f(x)$ is the expected value of Q conditioned on x.

$$\frac{d}{d\tau}x^D(t) = f(x^D(\tau)) \tag{4}$$

The above paragraph sets out a heuristic justification for the assertion that the behavior of the recursive algorithm can be linked to the behavior of the deterministic ordinary differential equation. Of course, the paper sets out a series of assumptions, under which formal theorems can be stated.

Broadly speaking, these theorems are of the type: if the differential equation has an invariant set with an associated domain of attraction, and if the recursive equation gives a parameter estimate which from time to time lies in the domain of attraction, then the recursive equation will have a solution converging to the invariant set.

If the invariant set is a point, then the unknown parameter vector can be considered as being learnt (identified). However, if it is not a point, the unknown parameter cannot be identified, although some of its properties can be learnt. The second group of results focuses on the question of learning the whole parameter vector.

A third important result of the paper relates the trajectories of the deterministic ordinary differential equation and the stochastic recursive equation, putting a bound on the probability that these can never be far apart. Thus the trajectories of the differential equation illustrate how the parameter values will change on a transient basis—at least for a typical run.

The value of the paper for applications is that a whole collection of important situations are captured by the general theorems of the paper. Important applications include equation error identification and self-tuning regulators. Stochastic approximation turns out to be a special case of the theory.

Readers familiar with equation identification will probably be aware that in the presence of noise, bias in the parameter estimates is to be expected. It is a measure of the power of the theory of the paper that in this case, where parameters may converge but not to the correct values, one nevertheless has a hard result.

Where did the paper lead to? As might be expected, a series of incremental advances and a widening of the example classes were able to be found. Secondly, the paper flagged the great importance of averaging ideas for recursive identification and for adaptive control. The averaging derives from the fact that the time scale associated with the time-constants of all but the identification process, and the time constant of the identification process itself, are very different. This separation of time scales found application in more than one book dealing with adaptive control; for example, [1] and [6]. Systematic use was made of the the techniques of the paper to analyze and classify recursive identification algorithms (including the extended Kalman filter) in [4]. A simpler approach to averaging analysis that preserves its full power is presented in [7]. The same family of problems, using different machinery, is addressed in [3] and [2].

The paper also gave results underpinning later work addressing the very difficult issue of the behavior of algorithms in which the adaptive gain sequence does not go to zero. There had been earlier analysis of such algorithms, essentially for FIR systems (this was particularly driven by the school of Widrow, and many results can be found in [9]), but these earlier results fell completely short of addressing situations where there was closed-loop control and the systems were not finite impulse response. Such situations are being addressed in isolated papers, but there seems as yet no textbook distillation embodying the key ideas.

References

[1] B.D.O. Anderson et al., *Stability of Adaptive Systems*, MIT Press (Cambridge, MA), 1986.

[2] A. Benveniste, M. Metivier, and P. Priouret, *Adaptive Algorithms and Stochastic Approximation*, Springer-Verlag (New York), 1990.

[3] H.J. Kushner and D.S. Clark, *Stochastic Approximation Methods for Contrained and Unconstrained Systems*, Springer-Verlag (New York), 1978.

[4] L. Lyung and T. Söderström, *Theory and Practice of Recursive Identification*, MIT Press (Cambridge, MA), 1983.

[5] H. Robbins and S. Monro, "A stochastic approximation method," *Ann Math. Stat.*, **22**:400‘407, 1951.

[6] S. Sastry and M. Bodson, *Adaptive Control, Stability, Convergence, and Robustness*, Prentice Hall (Englewood Cliffs, NJ), 1989.

[7] V. Solo and X. Kong, *Adaptive Signal Processing Algorithms: Stability and Performance*, Prentice Hall (Englewood Cliffs, NJ), Information and System Science Series, 1995.

[8] M.T.W. Wasan, *Stochastic Approximation*, Cambridge University Press (New York), 1969.

[9] B. Widrow and S. Stearns, *Adaptive Signal Processing*, Prentice Hall (Englewood Cliffs, NJ), 1985.

B.D.O.A.

Analysis of Recursive Stochastic Algorithms

LENNART LJUNG, MEMBER, IEEE

Abstract—Recursive algorithms where random observations enter are studied in a fairly general framework. An important feature is that the observations may depend on previous "outputs" of the algorithm. The considered class of algorithms contains, e.g., stochastic approximation algorithms, recursive identification algorithms, and algorithms for adaptive control of linear systems.

It is shown how a deterministic differential equation can be associated with the algorithm. Problems like convergence with probability one, possible convergence points and asymptotic behavior of the algorithm can all be studied in terms of this differential equation. Theorems stating the precise relationships between the differential equation and the algorithm are given as well as examples of applications of the results to problems in identification and adaptive control.

Manuscript received February 23, 1976; revised March 2, 1977. Paper recommended by R. Monopoli, Chairman of the Adaptive, Learning Systems, Pattern Recognition Committee. A major part of this work has been supported by the Swedish Board for Technical Development under Contract 733546.

The author is with the Department of Electrical Engineering, Linköping University, Linköping, Sweden.

I. Introduction

RECURSIVE algorithms, where stochastic observations enter are common in many fields. In the control and estimation literature such algorithms are widely discussed, e.g., in connection with adaptive control, (adaptive) filtering and on-line identification. The convergence analysis of the algorithms is not seldom difficult. As a rule, special techniques for analysis are used for each type of application and often the convergence properties have to be studied only by simulation.

In this paper a general approach to the analysis of the asymptotic behavior of recursive algorithms is described. In effect, the convergence analysis is reduced to stability analysis of a deterministic, ordinary differential equation. This technique is believed to be a fairly general tool and to have a wide applicability. Applications to various prob-

Reprinted from *IEEE Transactions on Automatic Control*, Vol. AC-22, August 1977, pp. 551–575.

lems have been published in [1]–[4] and some theory was presented in [5].

The objective of the present paper is to give a comprehensive presentation of formal results and useful techniques for the convergence analysis, as well as to illustrate with several examples how the techniques can be applied.

In Section II a general recursive algorithm is described and discussed. A heuristic treatment of the convergence problem is given in Section III and this leads to the basic ideas of the present approach. Section IV contains a discussion of the conditions which are imposed on the algorithm in order to prove the formal results. These theorems are given in Sections V and VI. The theorems suggest certain techniques for the convergence analysis, and these aspects are treated in Section VII. Several examples of how the theorems may be used, some of them reviewing previous applications are given in Section VIII.

II. The Algorithm

A general recursive algorithm can be written

$$x(t)=x(t-1)+\gamma(t)Q(t;x(t-1),\varphi(t)), \tag{1}$$

where $x(\cdot)$ is a sequence of n-dimensional column vectors, which are the objects of our interest. We shall refer to $x(\cdot)$ as "the *estimates*," and they could, e.g., be the current estimates of some unknown parameter vector. They could, however, also be parameters that determine a feedback law of an adaptive controller, etc., and we shall be precise about the character of $x(\cdot)$ only in the examples below. The sequence $\gamma(\cdot)$ is assumed throughout the paper to be a sequence of positive scalars. The m-dimensional vector $\varphi(t)$ is an *observation* obtained at time t, and these are the objects that cause $x(t-1)$ to be updated to take new information into account. (The notion "observation" does not have to be taken literally. The variable φ may very well be the result of certain treatment of actual measurements.) The observations are in general functions of the previous estimates $x(\cdot)$ and of a sequence of random vectors $e(\cdot)$. This means that the observation is a random variable, which may be affected by previous estimates. This is the case, e.g., for adaptive systems, when the input signal is determined on the basis of previous estimates. If the experiment designer has some test signal at his disposal, this may be included in $e(\cdot)$.

The function $Q(\cdot;\cdot,\cdot)$ from $R\times R^n\times R^m$ into R^n is a deterministic function with some regularity conditions to be specified below. This function, together with the choice of the scalar "gain" sequence $\gamma(\cdot)$ determine entirely the algorithm.

We shall not work with completely general dependence of $\varphi(t)$ on $x(\cdot)$, but the following structure for the generation of $\varphi(\cdot)$ will be used:

$$\varphi(t)=A(x(t-1))\varphi(t-1)+B(x(t-1))e(t). \tag{2}$$

Here $A(\cdot)$ and $B(\cdot)$ are $m|m$ and $m|r$ matrix functions.

Remark: It is perhaps more natural to think of an observation $\tilde{\varphi}(t)$ as the (lower dimensional) output of a dynamical system like (2), $\tilde{\varphi}(t)=C(x(t-1))\varphi(t)$. However, this case is naturally subsumed in the present one, since $\varphi(t)$ may enter in Q only as the combination $\tilde{\varphi}(t)$.

The assumption (2) seems to be appropriate for many applications. The same results as those below can be obtained also for nonlinear dynamics.

$$\varphi(t)=g(t;\varphi(t-1),x(t-1),e(t)) \tag{3}$$

and the proofs for this case are given in [7].

Throughout this paper it is assumed that the estimates are desired to converge to some "true" or "optimal" value(s). Since $Q(t,x(t-1),\varphi(t))$ is a random variable, with, in general, nonzero variance, convergence can take place only if the noise is rejected by paying less and less attention to the noisy observations, i.e., by letting

$$\gamma(t)\to 0 \quad \text{as } t\to\infty. \tag{4}$$

In tracking problems, when a time-varying parameter is to be tracked using algorithm (1), this condition is not feasible. Then $\gamma(t)$ usually tends to some small, nonzero value, the size of which depends on what is known about the variability of the tracked parameters and about the noise characteristics. This case is not treated here, but it is reasonable to assume that analysis under the condition (4) also will have some relevance for the case of tracking slowly varying parameters.

Suppose we have a linear, stochastic, discrete-time system, governed by a linear output feedback law, which at time t is determined by $x(t-1)$. Then the behavior of this overall system can be described by (2), with $\varphi(t)$ consisting of lagged inputs and outputs. Therefore the algorithms (1) and (2) can be understood as archetypical for adaptive control of a linear system. Indeed this setup is useful for analysis of certain adaptive controllers, as will be exemplified below, but the basic algorithm (1), (2) also covers several other cases of interest.

III. Heuristic Analysis

The algorithm (1), (2) is fairly complex to analyze, being a time-variant, stochastic, nonlinear difference equation. Notice also that the correction $x(t)-x(t-1)$ depends in general via $\varphi(t)$ implicitly on all old $x(s)$. Therefore, while (1) certainly is recursive from the user's point of view, it is not so for analysis purposes.

In this section we shall illustrate heuristically how a differential equation can be associated with (1), and how it seems reasonable that asymptotic properties of (1) may be studied in terms of this differential equation. The formal analysis and results follow in the next two sections.

Consider

$$x(t)=x(t-1)+\gamma(t)Q(x(t-1),\varphi(t)), \tag{5}$$

where for simplicity we let Q be time independent. As remarked before, $\varphi(t)$ depends on all previous estimates:

$$\varphi(t)=\sum_{j=1}^{t}\left(\prod_{i=j+1}^{t}A(x(i-1))\right)B(x(j-1))e(j). \quad (6)$$

Now, if (2) is exponentially stable, then the first terms in (6) will be very small and, for some M,

$$\varphi(t)\approx\sum_{j=t-M}^{t}\left(\prod_{i=j+1}^{t}A(x(i-1))\right)B(x(j-1))e(j).$$

Moreover, it follows from (5) and (4) that the difference $x(t)-x(t-1)$ becomes smaller as t increases. Therefore, for sufficiently large t, we have $x(k)\approx x(t)$; $t>k>t-2M$. Hence,

$$\varphi(k)\approx\sum_{j=k-M}^{k}\left[A(x(t))\right]^{k-j}B(x(t))e(j)$$

$$\approx\sum_{j=1}^{k}\left[A(x(t))\right]^{k-j}B(x(t))e(j)\triangleq\bar{\varphi}(k;x(t)) \quad (7)$$

for $t>k>t-M$. Furthermore,

$$Q(x(k-1),\varphi(k))\approx Q(x(t),\bar{\varphi}(k;x(t)))=f(x(t))+w(k) \quad (8)$$

where

$$f(x)=EQ(x,\bar{\varphi}(k;x))$$

and hence $w(k)$ is a random variable with zero mean.

Using (8), we can approximately evaluate

$$x(t+s)=x(t)+\sum_{k=t+1}^{t+s}\gamma(k)Q(x(k-1),\varphi(k))$$

$$\approx x(t)+f(x(t))\sum_{k=t+1}^{t+s}\gamma(k)+\sum_{k=t+1}^{t+s}\gamma(k)w(k)$$

$$\approx x(t)+f(x(t))\sum_{t+1}^{t+s}\gamma(k), \quad (9)$$

where the last step should follow since the last term is a zero mean random variable which is dominated by the second term. Expression (9) suggests that the sequence of estimates more or less follows the difference equation

$$x^D(\tau+\Delta\tau)=x^D(\tau)+\Delta\tau f(x^D(\tau)) \quad (10)$$

where $\Delta\tau$ corresponds to

$$\sum_{t+1}^{t+s}\gamma(k).$$

It is useful to interpret (10) as a way of solving the differential equation ($\Delta\tau$ small),

$$\frac{d}{d\tau}x^D(\tau)=f(x^D(\tau)), \quad (11)$$

where the (fictitious) time τ relates to the original time t in (5) by

$$\tau_t=\sum_{k=1}^{t}\gamma(k). \quad (12)$$

We therefore have some reason to believe that the sequence of estimates $x(\cdot)$ asymptotically should follow the trajectories $x^D(\cdot)$ of (11).

(We could also have related (9) to the difference equation

$$x^D(t)=x^D(t-1)+\gamma(t)f(x^D(t-1)), \quad (13)$$

but the differential equation is easier to handle since it is time-invariant.)

It now seems reasonable that asymptotic properties of algorithm (1), (2) may be studied in terms of the differential equation (11). This heuristic discussion is perhaps not very convincing, but along a similar path, though with far more technical labor, formal results to this effect can be proven. These are given below.

IV. Assumptions on the Algorithm

In order to prove the formal results, certain regularity conditions on the functions Q, A, and B and on the driving "noise" term e, have to be introduced. Some of these are fairly technical, but it is believed that none is very restrictive. Several sets of assumptions are possible, and we shall give a few. In particular, there is a possibility to treat the sequence $e(\cdot)$ either in a stochastic or in a deterministic framework.

We shall start by giving a formal definition of $\bar{\varphi}$ used in the previous section. Let

$$D_S=\{x|A(x)\ \text{has all eigenvalues strictly inside the unit circle}\}.$$

Then for each $x\in D_S$, there exists a $\lambda=\lambda(x)$, such that

$$|A(x)^k|<C\lambda(x)^k;\qquad \lambda(x)<1. \quad (14)$$

Take $\bar{x}\in D_S$ and define the random variables $\bar{\varphi}(t,\bar{x})$ and $v(t,\lambda,c)$, $\lambda<1$, by

$$\bar{\varphi}(t,\bar{x})=A(\bar{x})\bar{\varphi}(t-1,\bar{x})+B(\bar{x})e(t);\qquad \bar{\varphi}(0,\bar{x})=0 \quad (15)$$

$$v(t,\lambda,c)=\lambda v(t-1,\lambda,c)+c|e(t)|;\qquad v(0,\lambda,c)=0. \quad (16)$$

Let D_R be an open, connected subset of D_S. The regularity conditions will be assumed to be valid in D_R. Now, the first set of assumptions is the following.

A.1: $e(\cdot)$ is a sequence of independent random variables (not necessarily stationary or with zero means).

A.2: $|e(t)|<C$ with probability one (w.p.1) all t.

A.3: The function $Q(t,x,\varphi)$ is continuously differentiable w.r.t x and φ for $x\in D_R$. The derivatives are, for fixed x and φ, bounded in t.

A.4: The matrix functions $A(\cdot)$ and $B(\cdot)$ are Lipschitz continuous in D_R.

A.5: $\lim_{t\to\infty} EQ(t,\bar{x},\bar{\varphi}(t,\bar{x}))$ exists for $\bar{x}\in D_R$ and is denoted by $f(\bar{x})$. The expectation is over $e(\cdot)$.

A.6: $\Sigma_1^\infty \gamma(t) = \infty$

A.7: $\Sigma_1^\infty \gamma(t)^p < \infty$ for some p.

A.8: $\gamma(\cdot)$ is a decreasing sequence.

A.9: $\lim_{t\to\infty} \sup[1/\gamma(t) - 1/\gamma(t-1)] < \infty$.

These conditions will be referred to as "assumptions A." A.1 introduces the stochastic structure into the setup. While A.2 certainly is most reasonable for all practical purposes, it is somewhat unattractive from a theoretical point of view, since it excludes, e.g., the common Gaussian models for noise. Below (assumptions B) are given conditions which allow more general noise. Conditions A.3 and A.4 are reasonable regularity properties, and A.5 is the basic assumption that makes it possible to associate (1), (2) with a differential equation. Condition A.6 makes it possible for the algorithm (1) to move the estimate to the desired limit, regardless of the initial value, and it is thus obviously necessary. A.7 gives a condition on how fast $\gamma(t)$ must tend to zero. This is considerably less restrictive than the usually given condition

$$\sum_1^\infty \gamma^2(t) < \infty. \tag{17}$$

Conditions A.8 and A.9 are motivated by technical arguments in the proofs, but they have so far not appeared to be restrictive. For example, it is easy to see that the sequence $\gamma(t) = Ct^{-\alpha}$ satisfies A.6–A.9 for $0 < \alpha \leq 1$.

If we would like to alleviate A.2, further regularity conditions on Q are required. This gives us our second set of assumptions. ($\mathfrak{B}(\bar{x},\rho)$ denotes a ρ-neighborhood of $\bar{x}$, i.e., $\mathfrak{B}(\bar{x},\rho) = \{x \mid |\bar{x}-x| < \rho\}$.)

B.1: $e(\cdot)$ is a sequence of independent random variables (not necessarily stationary or with zero means).

B.2: $E|e(t)|^p$ exists and is bounded in t for each $p > 1$.

B.3: The function $Q(t,x,\varphi)$ is Lipschitz continuous in x and φ: $|Q(t,x_1,\varphi_1) - Q(t,x_2,\varphi_2)| < \mathcal{K}_1(x,\varphi,\rho,v)\{|x_1 - x_2| + |\varphi_1 - \varphi_2|\}$ for $x_i \in \mathfrak{B}(x,\rho)$ for some $\rho = \rho(x) > 0$ where $x \in D_R$; $\varphi_i \in \mathfrak{B}(\varphi, v)$, $v \geq 0$.

B.4: $|\mathcal{K}_1(x,\varphi_1,\rho,v_1) - \mathcal{K}_1(x,\varphi_2,\rho,v_2)| \leq \mathcal{K}_2(x,\varphi,\rho,v,w) \cdot \{|\varphi_1 - \varphi_2| + |v_1 - v_2|\}$ for $\varphi_i \in \mathfrak{B}(\varphi,w)$ and $v_i \in \mathfrak{B}(v,w)$.

B.5: $A(\cdot)$ and $B(\cdot)$ are Lipschitz continuous in D_R.

B.6: $\lim_{t\to\infty} EQ(t,\bar{x},\bar{\varphi}(t,\bar{x}))$ exists for $\bar{x}\in D_R$ and is denoted by $f(\bar{x})$. The expectation is over $e(\cdot)$.

B.7: For $x \in D_R$, the random variables $Q(t,x,\bar{\varphi}(t,x))$, $\mathcal{K}_1(x,\bar{\varphi}(t,x), \rho(x), v(t,\lambda,c))$ and $\mathcal{K}_2(x,\bar{\varphi}(t,x), \rho(x), v(t,\lambda,c), v(t,\lambda,c))$ have bounded p-moments for all $p > 1$, and all $\lambda < 1$, $c < \infty$. Here $\bar{\varphi}(\cdot,x)$ and $v(\cdot,\lambda,c)$ are the random variables defined by (15) and (16).

B.8: $\Sigma_1^\infty \gamma(t) = \infty$.

B.9: $\Sigma_1^\infty \gamma(t)^p < \infty$ for some p.

B.10: $\gamma(\cdot)$ is a decreasing sequence.

B.11: $\lim_{t\to\infty} \sup[1/\gamma(t) - 1/\gamma(t-1)] < \infty$.

Conditions B.4, B.3, and B.7 admittedly look somewhat complex, but they are as a rule easy to check in a given situation, especially since $Q(t,x,\varphi)$ is a simple function of x and φ in most applications. The conditions B.3 and B.4 effectively require that $Q(t,x,\varphi)$ is twice continuously differentiable and B.7 implies that Q and its derivatives must not increase too rapidly with φ and v.

In these two cases the algorithm (1), (2) is treated directly in a stochastic framework, due to assumption A.1 = B.1. In certain cases it may not be suitable to treat $e(\cdot)$ in (2) as a sequence of random variables. Naturally the algorithm (1), (2) still makes sense, even if $e(\cdot)$ is a given, deterministic sequence. Convergence of (1) will than depend, among other things, on the properties of this sequence $e(\cdot)$. In such a case we may work with the following assumptions. Let $\mathcal{K}_1$ be defined as in B.3 and let $\bar{\varphi}(\cdot,\bar{x})$ and $v(\cdot,\lambda,c)$ be given by (15), (16). Introduce the quantities $z(\cdot,\bar{x})$ and $k_v(\cdot,\bar{x},\lambda,c)$ by

$$z(t,\bar{x}) = z(t-1,\bar{x}) + \gamma(t)\left[Q\left(t,\bar{x},\bar{\varphi}(t,\bar{x})\right) - z(t-1,\bar{x})\right]$$
$$z(0,\bar{x}) = 0 \tag{18a}$$

$$\begin{aligned} k_v(t,\bar{x},\lambda,c) &= k_v(t-1,\bar{x},\lambda,c) + \gamma(t) \\ &\quad \cdot\left[\mathcal{K}_1\left(\bar{x},\bar{\varphi}(t,\bar{x}),\rho(\bar{x}),v(t,\lambda,c)\right)\right. \\ &\quad \left.\cdot\left(1 + v(t,\lambda,c)\right) - k_v(t-1,\bar{x},\lambda,c)\right]; \end{aligned}$$
$$k_v(0,\bar{x},\lambda,c) = 0. \tag{18b}$$

Notice that for the common choice $\gamma(t) = 1/t$, (18a) implies that

$$z(t,\bar{x}) = \frac{1}{t}\sum_{k=1}^{t} Q\left(k;\bar{x},\bar{\varphi}(k,\bar{x})\right)$$

and analogously for $k_v(t,\bar{x},\lambda,c)$. The assumptions then are as follows.

C.1: The function $Q(t,x,\varphi)$ is Lipschitz continuous in x and φ: $|Q(t,x_1,\varphi_1) - Q(t,x_2,\varphi_2)| < \mathcal{K}_1(x,\varphi,\rho,v)\{|x_1 - x_2| + |\varphi_1 - \varphi_2|\}$ for $x_i \in \mathfrak{B}(x,\rho)$ for some $\rho = \rho(x) > 0$ where $x \in D_R$; $\varphi_i \in \mathfrak{B}(\varphi,v)$, $v \geq 0$.

C.2: The matrix functions $A(\cdot)$ and $B(\cdot)$ are Lipschitz continuous in D_R.

C.3: $z(t,\bar{x})$ as defined by (18a) converges for all $\bar{x} \in D_R$ as $t\to\infty$. Denote the limit by $f(\bar{x})$.

C.4: $k_v(t,\bar{x},\lambda,c)$ defined by (18b) converges to a finite limit as $t\to\infty$ for all $\bar{x}\in D_R$, $\lambda < 1$ and $c < \infty$.

C.5: $\Sigma_1^\infty \gamma(t) = \infty$.

C.6: $\gamma(t) \to 0$ as $t\to\infty$.

C.7: If the matrices A and B in (2) do not depend on x, then Lipschitz continuity w.r.t to φ is not required in C.1 and we may take $c = 0$ in C.4. That is, we may take $v \equiv 0$ in C.1 and C.4.

When these assumptions are used, no stochastic framework has to be introduced. The statements about the behavior of $x(\cdot)$ to be given below are true as long as $e(\cdot)$ is such that C.3 and C.4 hold. If a stochastic framework *is* imposed and C.3, C.4 hold with probability one, then the statements about $x(\cdot)$ will be true w.p.1. This is, essentially, the approach taken in [5] and [8], which also contain a detailed study of algorithms like (18) (esp. [8, ch. 4]). There several different sets of conditions implying convergence of (18) are given. In fact, conditions B imply that C.3 and C.4 hold w.p.1. It may in this context be remarked that there is actually a tradeoff between condition A.7=B.9 and conditions B.2 and B.7. The largest p for which B.2 and B.7 need to hold is twice the p for which A.7 holds. Therefore, if $\gamma(\cdot)$ *is* subject to (17), then only the fourth moments of e, Q, and $\mathcal{K}_i$ have to be bounded. This is discussed further in [5] and [8], and we shall not pursue it here.

V. Main Theorems

The function $f(\cdot)$ defined in A.5, B.6, or C.3 is the basic object of interest. As the heuristic discussion in Section III indicated the differential equation

$$\frac{d}{d\tau}x^D(\tau)=f(x^D(\tau)) \tag{19}$$

will be relevant for the asymptotic behavior of the algorithm (1), (2). The exact relationships between (19) and (1), (2) are given in three theorems. The first one concerns convergence of (1).

Theorem 1: Consider the algorithm (1), (2) subject to assumptions A, B or C. Let $\bar{D}$ be a compact subset of D_R such that the trajectories of (19) that start in $\bar{D}$ remain in a closed subset $\bar{D}_R$ of D_R for $\tau>0$. Assume that

1) there is a random variable C such that

$$x(t)\in\bar{D} \text{ and } |\varphi(t)|<C \quad \text{infinitely often (i.o.)w.p.1} \tag{20}$$

2) the differential equation (19) has an invariant set D_c with domain of attraction $D_A\supset\bar{D}$. (21)

Then $x(t)\to D_c$ with probability one as $t\to\infty$. □

Remarks: By (20) is meant that there exists with probability one a subsequence t_k, possibly depending on the realization ω, such that $x(t_k)\in\bar{D}$ and $|\varphi(t_k)|<C(\omega)$, $k=1,2,\cdots$. This condition, which we may call the "boundedness condition," is further discussed in Section VI.

An invariant set of a differential equation is a set such that the trajectories remain in there for $-\infty<\tau<\infty$. The domain of attraction of an invariant set D_c consists of those points from which the trajectories converge into D_c as τ tends to infinity. It is obviously an open set. See, e.g., [9]. An interesting special case is when the invariant set D_c is just a stationary point of (19) say x^*, with $f(x^*)=0$. Then the theorem proves convergence of $x(t)$ to x^*. By $x(t)\to D_c$ is meant that

$$\inf_{x\in D_c}|x(t)-x|\to 0.$$

The phrase "w.p.1" naturally does not apply in the case with assumptions C. In order to verify the stability condition (21) analytically, usually the Lyapunov theory has to be applied. Theorem 1 can be given a formulation, which does not refer to any differential equation, but directly relates to a Lyapunov function associated with $f(x)$. In that way also the assumption about the trajectories of (19) remaining in $\bar{D}_R$ and boundedness of $\bar{D}$ can be avoided.

Corollary 1: Consider the algorithm (1), (2) subject to assumptions A, B or C. Let $\bar{D}$ be a closed subset of D_R (possibly $\bar{D}=D_R=R^n$) such that (20) holds. Assume that there exists a twice differentiable function $V(x)$ in D_R, such that

$$V'(x)f(x)\leqslant 0 \qquad x\in D_R. \tag{22}$$

Then either

$$x(t)\to D_c=\{x|x\in D_R \text{ and } V'(x)f(x)=0\} \quad \text{w.p.1 as } t\to\infty$$

or

$$\{x(t)\} \text{ has a cluster point on the boundary of } D_R. \quad □$$

Our second theorem concerns the set of possible convergence points. It can be used to prove failure of convergence by showing that the "desired" or "true" parameter value does not belong to this set.

Theorem 2: Consider algorithm (1), (2) subject to assumptions A or B. Suppose that $x^*\in D_R$ has the property that

$$P(x(t)\to\mathfrak{B}(x^*,\rho))>0 \qquad \text{for all } \rho>0. \tag{23}$$[1]

Furthermore, suppose that

$Q(t,x^*,\bar{\varphi}(t,x^*))$ has a covariance matrix bounded from below by a strictly positive definite matrix, (24)

and that

$EQ(t,x,\bar{\varphi}(t,x))$ is continuously differentiable with respect to x in a neighborhood of x^* and the derivatives converge uniformly in this neighborhood as t tends to infinity.

Then

$$f(x^*)=0 \tag{25a}$$

and

$$H(x^*)=\frac{d}{dx}f(x)\Big|_{x=x^*} \text{ has all eigenvalues in the LHP } (\mathrm{Re}\,z\leqslant 0). \quad □ \tag{25b}$$

The matrix $H(x^*)$ defines, of course, the linear differential equation obtained from (19) by linearization around x^*. Therefore this theorem essentially states that the algorithm can converge only to stable stationary points of the differential equation (19).

If $f(x)=-(d/dx)V(x)$, which might be the case if the algorithm is based on criterion-minimization, then $V(x)$ can be chosen as a Lyapunov function for the differential equation (19). Since $(d/d\tau)V(x(\tau))=-|f(x(\tau))|^2$, we see

[1] $P(A)$ = The probability of the event A.

that the stationary points of (19), together with the point $\{\infty\}$, form an invariant set with global domain of attraction. Moreover, if the stationary points are isolated, it follows from Theorem 2 that only stable ones, i.e., local minima, are possible convergence points. It also follows from Theorem 1 that the estimates cannot oscillate between different minima. Collecting all this we obtain a corollary to Theorems 1 and 2.

Corollary 2: Suppose that $D_R = R^n$, that $f(x) = -(d/dx)V(x)$ and that $V(x)$ has isolated stationary points. Assume that $|\varphi(t)| < C$ i.o. w.p.1. Then, w.p.1, $x(t)$ tends either to a local minimum of $V(x)$ (i.e., $V''(x)$ positive semidefinite) or to infinity as t tends to infinity.

Finally, our third theorem relates the trajectories of the differential equation (19) to the paths of the algorithm (1), (2). The result is formulated as follows. Let $x(t)$, $t = t_0, \cdots$, be generated by (1), (2). The values can be plotted with the sample numbers t as the abscissa. It is also possible to introduce as before a fictitious time τ by

$$\tau_t = \sum_{k=1}^{t} \gamma(k). \tag{26}$$

Suppose that the estimates $x(t)$ are plotted against this time τ. See Fig. 1(a). Let $x^D(\tau, \tau_{t_0}, x(t_0))$ be the solution of (19) with initial value $x(t_0)$ at time τ_{t_0}. Also plot this solution in the same diagram, as in Fig. 1(b). Let I be a set of integers. The probability that *all* points $x(t)$, $t \in I$, simultaneously are within a certain distance ϵ from the trajectory is estimated in the following theorem.

Theorem 3: Consider algorithm (1), (2) under assumptions A or B. Assume that $f(x)$ is continuously differentiable, and that (20) holds. Assume that the solutions to (19) with initial conditions in $\bar{D}$ are exponentially stable, and let I be a set of integers, such that $\inf|\tau_i - \tau_j| = \delta_0 > 0$ where $i \neq j$ and $i, j \in I$. Then for any $p \geq 1$ there exist constants K, ϵ_0 and T_0 that depend on p, $\bar{D}$, and δ_0, such that for $\epsilon < \epsilon_0$ and $t_0 > T_0$.

$$P\left\{ \sup_{\substack{t \in I \\ t > t_0}} |x(t) - x^D(\tau_t; \tau_{t_0}, x(t_0))| > \epsilon \right\} \leq \frac{K}{\epsilon^{4p}} \sum_{j=t_0}^{N} \gamma(j)^p \tag{27}$$

where $N = \sup i$; $i \in I$, which may be ∞.

Remark: In the proof of Theorem 3 it is assumed that the exponential stability of the solution $x^D(\tau, \tau_{t_0}; x(t_0))$ is ensured by a quadratic Lyapunov function for the (linear and time-varying) variational equation around this solution, cf., e.g., [10].

Although the proof of Theorem 3 provides an estimate of K from given constants, we do not intend to use (27) to obtain numerical bounds for the probability. The point of the theorem is that a connection between the differential equation (19) and the algorithm (1), (2) is established. In particular, we notice that, due to A.7 there is a p such that the RHS of (27) becomes arbitrarily small when t_0 increases, and ϵ is fixed. This means that the estimates stay close to the corresponding trajectory with higher and higher probability as t_0 increases. Another way of interpreting (27) is that the gain sequence $\gamma(\cdot)$ can be scaled so that $x(\cdot)$ stays arbitrarily close to $x^D(\cdot)$, with an arbitrary high degree of probability.

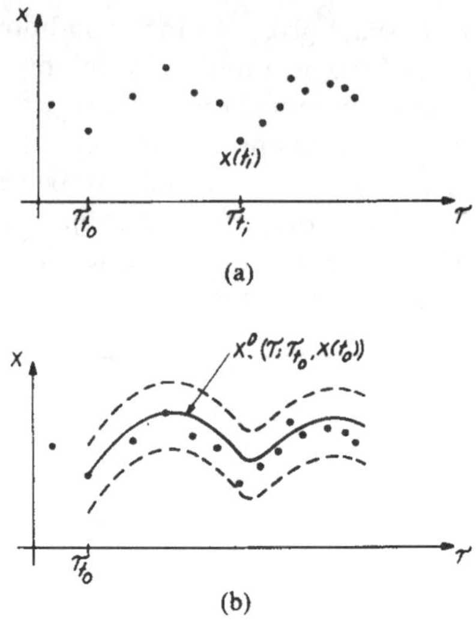

Fig. 1. (a) and (b) To illustrate Theorem 3.

A result that is related to Theorem 3 in the case $A(x) \equiv 0$ is given in [35]. Interesting connections between stochastic-approximation type algorithms and a corresponding differential equation have also recently been made by Kushner [36], [37], using weak convergence theory.

The proof of Theorems 1, 2, and 3 are long and technical. They are given in Appendices I, II, and III, respectively. The idea of the proofs of Theorems 1 and 3 follow the discussion in Section III. However, a considerable amount of technicalities are required to rigorously justify the "approximatively equal" signs.

In Appendix V extensions of the theorems, e.g., to continuous time algorithms and to cases when the limit in A.5 or B.6 does not exist are also commented upon.

VI. The Boundedness Condition

In this section we shall discuss condition (20). The reason why it is required is twofold. First, obviously $x(t)$ must be inside D_R (with $\varphi(t)$ not too large to prevent an immediate jump) for the differential equation to be valid at all, and also inside D_A to get "caught" by a trajectory converging to D_c. Second, and perhaps less obviously, even if $D_R = D_A = R^n$ it may happen that $x(t)$ tends to infinity. The reason for divergence is that if $Q(t, x, \varphi)$ increases rapidly with $|x|$ it may happen that the correction $\gamma(t)Q(t, x(t-1), \varphi(t))$ always is too large even though $\gamma(t)$ tends to zero. Another reason is that the variance of the "noise" $Q(t, x, \varphi) - f(x)$ may increase so fast with $|x|$ that a "random walk" effect becomes predominately close to infinity.

From a practical point of view, the question of boundedness of the estimates may seem uninteresting, since no implementation of (1) will allow that $x(t)$ tends to infinity. It will be kept bounded either by deliberate measures or due to, e.g., overflow in the computer. Now the measures to keep $x(t)$ in a bounded area may not be completely arbitrary to obtain convergence.

A feature that can be used when $D_R = D_A = R^n$, is to introduce a saturation in $Q(\cdot;\cdot,\cdot)$ so that $|Q(t;x,\varphi)| < \mathcal{K}$. This is further discussed in [7].

Another possibility of preventing $x(t)$ from tending to infinity is to project $x(t)$ into a bounded area if $|x(t)|$ is too large or if $x(t)$ does not belong to a desired area, say D_S. In fact, if $A(x)$ is a known function of x, it is common, and often even necessary to test if $x(t) \in D_S$ and project it into D_S otherwise. We then have an algorithm of the following type:

$$x(t) = \left[x(t-1) + \gamma(t)Q(t, x(t-1), \varphi(t))\right]_{D_1, D_2} \tag{28}$$

$$\varphi(t) = \begin{cases} A(x(t-1))\varphi(t-1) + B(x(t-1))e(t) \\ \qquad\qquad \text{if } x(t-1) \in D_1 \\ \text{a value in a given compact subset of } R^m \\ \qquad\qquad \text{if } x(t-1) \notin D_1 \end{cases} \tag{29}$$

where $D_1 \supset D_2$, and

$$[z]_{D_1, D_2} = \begin{cases} z & \text{if } z \in D_1 \\ \text{some value in } D_2 & \text{if } z \notin D_1. \end{cases}$$

It should be clear that D_1, D_2 cannot be chosen arbitrarily. Loosely speaking, the trajectories of (19) that start in D_2 must not leave the area D_1. Otherwise there may be an undesired cluster point on the boundary of D_1. This may be formalized as follows.

Theorem 4: Consider the algorithm (28), (29) subject to assumptions A, B or C. Let $D_1 \subset D_R$ be an open bounded set containing the compact set D_2. Let $\tilde{D} = D_1 \backslash D_2$ (D_1 "minus" D_2). Assume that $D_2 \subset D_A$, with D_A defined as in Theorem 1. Suppose that there exists a twice differentiable function $U(x) > 0$, defined in a neighborhood of $\tilde{D}$ with properties

$$\sup_{x \in \tilde{D}} U'(x)f(x) < 0, \tag{30}$$

$$\begin{aligned} U(x) &> C_1 \qquad \text{for } x \notin D_1 \\ U(x) &< C_2 < C_1 \qquad \text{for } x \in D_2. \end{aligned} \tag{31}$$

Then Theorem 1 holds without assumption (20).

The proof of Theorem 4 is given in Appendix IV.

Assumption (30) clearly makes $U(\cdot)$ a Lyapunov function in $\tilde{D}$, while (31) formalizes the intuitive notion of trajectories from D_2 never leaving D_1. We may remark that (30), (31) hold, e.g., if the trajectories of (19) do not intersect the boundary of D_1 "outwards" and D_2 is sufficiently close to D_1.

VII. How to Use the Theorems

The intuitive content of the theorems of Section V is that the algorithm (1), (2),

$$x(t) = x(t-1) + \gamma(t)Q(t, x(t-1), \varphi(t)) \tag{32a}$$

$$\varphi(t) = A(x(t-1))\varphi(t-1) + B(x(t-1))e(t) \tag{32b}$$

can be studied and analyzed in terms of the differential equation

$$\frac{d}{d\tau}x^D(\tau) = f(x^D(\tau)), \tag{33}$$

where

$$f(x) = \lim_{t \to \infty} EQ(t, x, \bar{\varphi}(t, x)). \tag{34}$$

The precise statements about the relations between (32) and (33) of Theorems 1-3 may be summarized in a somewhat looser language as follows.

a) $x(t)$ can converge only to stable stationary points of (33).

b) If $x(\cdot)$ belongs to the domain of attraction of a stable stationary point x^* of (33) i.o. w.p.1, then $x(t)$ converges w.p.1 to x^* as t tends to infinity.

c) The trajectories of (33) are "the asymptotic paths" of the estimates $x(\cdot)$, generated by (32).

These statements are fairly attractive intuitively, and they suggest certain unified techniques to analyze recursive algorithms. We shall illustrate this below, but let us here point out some aspects.

By the result a) the possible convergence points of (32) may be determined and studied. That a possible convergence point must be a zero of (34) is fairly obvious and it may be derived without reference to any differential equation. However, the observation that among these stationary points only stable ones are candidates for being limit points of (32) is a most important complement and it is probably less obvious without the present interpretation in terms of the differential equation. Perhaps the main use of result a) is to prove failure of convergence. It may be remarked that usually an algorithm is constructed so that the desired limit indeed is a stationary point. Consequently the possible lack of convergence is then due to the unstable character of the stationary point, so it is the complement (25b) that is the key result for proving divergence.

Result b) is the result by which convergence can be proved. In many cases it is not easy to find a proper Lyapunov function to prove global stability of (33), and sometimes the RHS of (33) is quite complex. For certain algorithms, though, in particular those arising from criteria-minimization, it is possible to do this analytically, and some examples will be given below.

While analytic treatment of (33) may be difficult, it is always possible to solve it numerically when the dimension of x is not too large. In that way insight can be gained into the global stability properties of the differential equation, the stationary points and their character. In view of result c) the trajectories thus obtained are also relevant for the asymptotic behavior of the algorithm. Therefore, numerical solution of (33) is a valuable complement to simulation of (32). Due to the time scaling (26) in the differential equation, this reveals more rapidly the asymptotic properties and the stationary points of the algorithm. Since the estimates change more and more slowly, due to (4), it is seldom not difficult to decide from simulations only whether the estimates have settled around a limit value or are just converging slowly. In addition, it might be difficult to tell from a simulation if a certain effect is an inherent feature of the algorithm or just depends on random influence. Numerical solution of (33) may resolve such questions.

The regularity conditions A, B, or C are usually not very difficult to verify as demonstrated below. Notice in particular that the observation φ may have a "dummy" character, since it may have to be extended to fit into the assumptions (2) and A.1. For example, if the sequence $\{e(\cdot)\}$ in (2) is a stationary process with rational spectral power density, then it can be modeled as the output of a stable, linear filter with independent random variables as input. Hence, by extending the vector φ and adjoining this filter to (2), assumption A.1 will hold. Even if this may lead to a vector φ with a large dimension, the complexity of the algorithm and of the analysis is not affected. What matters is the calculation of

$$EQ\left(t,\bar{x},\bar{\varphi}(t,\bar{x})\right)$$

and this can often be done without explicit expressions for $A(\cdot), B(\cdot), \varphi(\cdot)$, and even for $Q(\cdot,\cdot,\cdot)$.

In the next section we shall apply the method to a few examples and illustrate how the techniques of the items above may be used.

VIII. Examples

Example 1—Stochastic Approximation Algorithms

Consider the problem of solving

$$E_\varphi Q(x,\varphi)=0 \tag{35}$$

for x. Here "E_φ" denotes the expectation with respect to φ, while the vector x is considered as a fixed parameter. Quantities $Q(x,\varphi(t))$, $t=1,2,\cdots$, are available for any x, where the distribution of the random vector $\varphi(\cdot)$ does not depend on x. Robbins–Monro [11] proved that under certain assumptions the scheme ("the Robbins–Monro scheme")

$$x(t)=x(t-1)+\gamma(t)Q\left(x(t-1),\varphi(t)\right) \tag{36}$$

gives a sequence of estimates that converges to the (a) solution of (35) in the mean square sense. Convergence w.p.1 of (36) has then been studied in several papers, e.g., [12]–[14], and the theorems of these studies do not differ very much conceptually from Theorem 1. The functions used in the convergence theorems of, e.g., [12] or [13] can be interpreted as Lyapunov functions for the differential equation (33) and condition (20) of Theorem 1 is ensured by further conditions on this function, rather than by the more practically oriented Theorem 4, see, e.g., [12, condition A], or [13, condition B, p. 184]. It can also be remarked that the condition

$$E(x-x^*)^TQ(x,\varphi)<0 \tag{37}$$

frequently used in Tsypkin's work, see, e.g., [15], [16], clearly can be understood as a stability condition for (33) with $V(x)=\|x-x^*\|^2$ as the Lyapunov function. Tsypkin's condition

$$EQ^T(x,\varphi)Q(x,\varphi)\leqslant C(1+\|x\|^2) \tag{38}$$

is then a variant of the "boundedness condition."

The convergence results thus obtained are, however, essentially restricted to the case $\varphi(\cdot)$ being independent random variables ($A(\cdot)=0$) and $\gamma(\cdot)$ satisfying (17) which is quite restrictive for control and estimation application. These conditions are inherently tied to the use of martingale theory in the proofs and cannot easily be dispensed with. Our Theorem 1 when applied to (36) is thus more general in that $\varphi(\cdot)$ may be dependent (generated as white noise through a linear filter) and $\gamma(\cdot)$ has only to satisfy A.7. This is satisfied, e.g., for $\gamma(t)=Ct^{-\alpha}$ $0<\alpha\leqslant 1$, while (17) admits only $1/2<\alpha\leqslant 1$. Notice that slowly decreasing gain sequences may be of interest in practice to achieve fast convergence of the sequence of estimates. We must, however, admit that we in return require more regularity of Q and of $e(\cdot)$. On the other hand, nonsmoothness of the involved functions is seldom a problem in applications, and we believe that our version of the convergence theorem is more widely applicable.

In addition, Theorems 2 and 3 are important results for convergence analysis, and we are not aware of similar previous results for the Robbins–Monro scheme.

In many applications it is of interest to minimize a function $E_\varphi J(x,\varphi)=P(x)$ with respect to x. If the derivative of J with respect to x can be calculated, the stationary points of $P(x)$ can be found as solutions of

$$E_\varphi\left[\frac{\partial}{\partial x}J(x,\varphi)\right]=0.$$

This is a problem that can be solved using the Robbins–Monro scheme and then Corollary 2 of Theorems 1 and 2 is quite useful.

If the derivative of J cannot be calculated it seems natural to replace it with some difference approximation. This was suggested by Kiefer and Wolfowitz [17] and their procedure has also been used for various control and estimation problems. Kushner has in several recent papers discussed interesting variants of this procedure, see e.g., [18], [19]. Our theorems are not directly applicable to the Kiefer–Wolfowitz scheme as they stand, since condition A.3 (or B.3) is not valid. The reason is that the function Q in this case increases to infinity with t. For the case of additive noise to the function to be minimized, however, it can readily be shown that Theorems 1–4 hold anyway. Details are given in [5] and [7].

Stochastic approximation algorithms have been applied to a broad variety of problems in control theory, see e.g., Tsypkin [15], [16], Fu [20], and Saridis *et al.* [21]. The approach is known as "learning systems," and in this framework estimation and identification problems, adaptive control, supervised and unsupervised pattern recognition, etc. can be treated.

An approach that is related to stochastic approximation

is suggested by Aizerman *et al.* [13]. Their "Potential Function Method" can be applied to various problems in machine learning.

Therefore the Robbins–Monro scheme appears in various disguises in many control and estimation algorithms, and consequently the described techniques can be applied to these. A particular example is given below. □

Example 2–An Automatic Classifier

A classifer receives scalar valued signals $\varphi(t)$ which may belong to either of two *a priori* unknown classes A and B. The classifier must find a classification rule, i.e., a number $c(t)$ such that $\varphi(t)$ is classified as A if $\varphi(t) \leqslant c(t)$ and B otherwise. The number $c(t)$ can, e.g., be determined as follows:

$$c(t) = (x^A(t) + x^B(t))/2$$

where

$$x^A(t) = \begin{cases} x^A(t-1) + \gamma(t)[\varphi(t) - x^A(t-1)], \\ \qquad \text{if } \varphi(t) \text{ is classified as } A \\ x^A(t-1), \qquad \text{otherwise.} \end{cases} \tag{39}$$

$x^B(t)$ is defined analogously. Clearly, $x^A(t)$ is the mean value of the outcomes classified as A. This scheme is discussed by Tsypkin [22] and Braverman [23].

Let $\varphi(t)$ have the distribution shown in Fig. 2 consisting of two triangular distributions. The probability of outcomes in the left triangle is λ. We assume that $\varphi(\cdot)$ is a sequence of independent random variables. Clearly, it is desirable that the classification rule, the number $c(t)$, should converge to some value between -1 and $+1$. Introduce

$$x(t) = \begin{bmatrix} x^A(t) \\ x^B(t) \end{bmatrix}.$$

Then (39) can be written

$$x(t) = x(t-1) + \gamma(t) Q(x(t-1), \varphi(t)) \tag{40}$$

where

$$Q(x,\varphi) = \begin{bmatrix} Q^A(x^A, x^B, \varphi) \\ Q^B(x^A, x^B, \varphi) \end{bmatrix}$$

and

$$Q^A(x^A, x^B, \varphi) = \begin{cases} \varphi - x^A, & \text{if } \varphi < \frac{1}{2}(x^A + x^B) - \delta \\ 0, & \text{if } \varphi > \frac{1}{2}(x^A + x^B) + \delta \end{cases}$$

and Q^B analogously

where the values for $\frac{1}{2}(x^A + x^B) - \delta \leqslant \varphi \leqslant \frac{1}{2}(x^A + x^B) + \delta$ are such that Q^A is a continuously differentiable function of φ and x. Here δ is some small positive number.

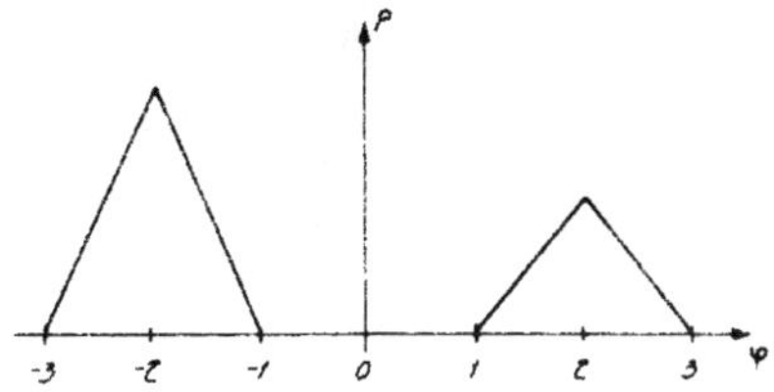

Fig. 2. Probability density function of the random variable to be classified by the automatic classifier.

Clearly, the algorithm (40) together with the observation equation

$$\varphi(t) = e(t) \tag{41}$$

is a simple case of (32). Since $\varphi(\cdot) = e(\cdot)$ is bounded we may use assumptions A. Obviously A.1 to A.5 are satisfied, and let us assume that $\gamma(\cdot)$ is such that A.6–A.9 hold. (Here A.3 holds in virtue of our somewhat artificial modification of Q^A; but this example will illustrate that a heuristic use of the present convergence results will reveal important features of the algorithm.)

$E_\varphi Q(x,\varphi) = f(x)$ is readily computed as follows. For a given x the corresponding classification point is $c(x) = (x^A + x^B)/2$. $f^A(x)$ is then the mean value of the distribution left of the point $c(x)$, minus x^A. $f^B(x)$ is found correspondingly. The algebraic expression for $f(x)$ as a function of x and λ is lengthy and is omitted.

We first note that by construction, the estimates are confined to the area $\bar{D}$: $3 > x^B \geqslant x^A > -3$. Therefore condition (20) of Theorem 1 is trivially satisfied. Analytical treatment of the differential equation $\dot{x} = f(x)$ is not easy, but its trajectories can easily be determined by numerical solution and they are shown in Fig. 3 for two choices of λ. For the case $\lambda = 0.5$, [Fig. 3(a)] there is convincing evidence that the point $x^* = (-2, 2)$ is a stable stationary point with global domain of attraction. Therefore, for $\lambda = 0.5$ it follows from Theorem 1 that $x(t) \to x^*$ w.p.1 as $t \to \infty$, which gives a correct classification rule $c^* = 0$. The case $\lambda = 0.99$ [Fig. 3(b)] corresponds to a common situation where errors that occur rather seldom (1 percent), "outliers," shall be detected. In this case there are two stable stationary points of the differential equation. $x^* = (-2, 2)$ and $x^{**} = (-2.3, -1.4)$. There is obviously a nonzero probability that $x(t)$ belongs to the domain of attraction of x^{**} i.o. Therefore Theorem 1 shows that for $\lambda = 0.99$, and for any starting value $x(0)$ there is a nonzero probability (that depends on $x(0)$) that $x(t) \to x^{**}$ as $t \to \infty$. This gives an asymptotic classification rule $c^{**} = -1.8$, that classifies 39 percent of the "correct values" as outliers. For this case simulations of the classifier are shown in Fig. 4. In fact, the simulation leading to the undesired value c^{**} appeared only after several (257) attempts and from simulations only it might have been tempting to conclude general convergence to c^*.

In this example it is cumbersome to find a suitable Lyapunov function for the stability problem. However, as seen in Fig. 3 numerical solution of the differential equation yields sufficient insight into the stability properties. Such detailed information can naturally be obtained only if the dimensionality of the problem is small.

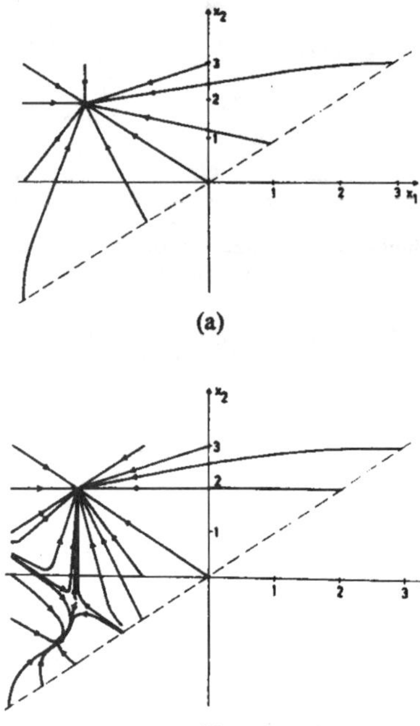

Fig. 3. Trajectories for the ODE that is associated with self-learning classifier (40). (a) $\lambda=0.5$. (b) $\lambda=0.99$.

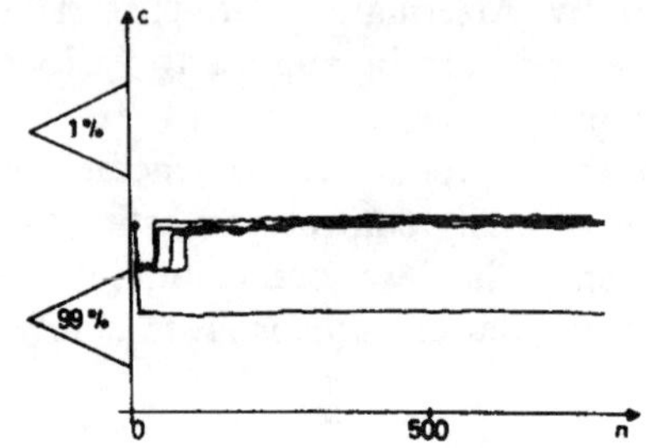

Fig. 4. Simulations of the classifier (40) for the case $\lambda=0.99$.

Example 3—Equation Error Identification Methods

A common way of modeling dynamic systems is as a vector difference equation (VDE),

$$y(t)+A_1y(t-1)+\cdots+A_ny(t-n) = B_1u(t-1)+\cdots+B_nu(t-n), \quad (42)$$

where $y(t)$ and $u(t)$ are column vectors and A_i and B_i are matrices of appropriate dimensions. Introduce

$$\theta=(A_1\cdots A_nB_1\cdots B_n)^T$$
$$\psi(t)=\left(-y(t-1)^T\cdots-y(t-n)^Tu(t-1)\cdots u(t-n)^T\right)^T. \quad (43)$$

Then (42) can be written

$$y(t)=\theta^T\psi(t). \quad (44)$$

We may remark that (44) also covers several other interesting estimation problems, not necessarily related to system identification.

Usually, the true system cannot be described exactly in the form (44). Suppose that it can be described as

$$y(t)=\theta_0^T\psi(t)+v(t) \quad (45)$$

where $v(\cdot)$ is a disturbance that can be modeled as

$$v(t)=D(q^{-1})e_1(t). \quad (46)$$

Here $D(q^{-1})$ is a matrix with rational functions of the backward shift operator q^{-1} as entries and $e_1(\cdot)$ is a stationary sequence of independent random vectors with finite moments. It is assumed that the denominator polynomials in $D(z)$ (z replacing q^{-1}) have all roots outside the unit circle, i.e., $D(q^{-1})$ is an exponentially stable filter.

Even if an exact description of the system is impossible, a θ can be determined that gives a model (44) which describes the recorded data as well as possible. Often θ is determined by minimizing a criterion based on the equation error

$$\|y(t)-\theta^T\psi(t)\|^2. \quad (47)$$

Several algorithms based on the idea of somehow minimizing (47) have been suggested in the literature, see e.g., [24] and also [25] for a comprehensive treatment. The probably best known method of this type is the least squares algorithms, see, e.g., [24]. Then the sum

$$\sum_{t=1}^{N}\|y(t)-\theta^T\psi(t)\|^2 \quad (48)$$

is minimized w.r.t θ to obtain the estimate $\theta(N)$ based on measurement up to time N. An important and well known feature of this method is that the sequence of estimates can be obtained recursively as

$$\theta(t)=\theta(t-1)+\gamma(t)\mathcal{K}(t)\left[y(t)-\theta(t-1)^T\psi(t)\right]^T \quad (49a)$$
$$\mathcal{K}(t)=R^{-1}(t-1)\psi(t)\Big/\left[1+\gamma(t)\left(\psi(t)^TR^{-1}(t-1)\psi(t)-1\right)\right] = R^{-1}(t)\psi(t) \quad (49b)$$
$$R(t)=R(t-1)+\gamma(t)\left[\psi(t)\psi(t)^T-R(t-1)\right] \quad (49c)$$

(usually (49c) is written in terms of $R^{-1}(t)$, which makes it of "Riccati type"). For the minimization of (48) $\gamma(t)$ has to be taken as $1/t$. Other sequences $\gamma(\cdot)$ correspond to criteria where old measurements are discounted, which often is relevant in practice.

Let us assume that the input to the process is determined as

$$u(t)=F(q^{-1})e_2(t)+H\left(q^{-1};\theta(t-1)\right)y(t) \quad (50)$$

where $F(q^{-1})$ and $H(q^{-1},\theta)$ are matrices with rational functions of the backward shift operator q^{-1} as entries. Let $e_2(\cdot)$ be a stationary sequence of random vectors with finite moments, that are mutually independent and also independent of $e_1(\cdot)$. Moreover, $H(q^{-1},\theta)$ is a causal operator that allows output feedback terms in the input.

This feedback law may depend on the current parameter estimate as is further discussed in Example 5.

It is clear that the rational filters in (46) and (50) can be represented in a state space form,

$$z_v(t+1)=A_v z_v(t)+B_v e_1(t+1); \qquad v(t)=(I\ 0\cdots 0)z_v(t) \tag{51}$$

$$z_u(t+1)=A_u(\theta(t))z_u(t)+z_u+B_{ue}e_2(t+1)+B_{uy}y(t+1);$$
$$u(t)=(I\ 0\cdots 0)z_u(t), \tag{52}$$

where $z_v(\cdot)$ and $z_u(\cdot)$ are the corresponding state vectors of appropriate dimensions. We may now form the "observation vector,"

$$\varphi(t)=\left[y(t)^T,\psi(t)^T,z_v(t)^T,z_u(t)^T\right]^T \tag{53}$$

which obeys

$$\varphi(t)=A_\varphi(\theta(t-1))\varphi(t-1)+B_\varphi\begin{pmatrix}e_1(t)\\ e_2(t)\end{pmatrix} \tag{54}$$

where the matrix $A_\varphi(\cdot)$ is formed from (45), (43), (51), (52) in an obvious manner. Its eigenvalues are the poles of the filters $D(q^{-1})$, $F(q^{-1})$ and of the closed loop system which is obtained for (42) with a constant feedback (50) using $\theta(t-1)$. There are also a number of eigenvalues in the origin, arising from the shifting of the vector $\psi(t)$. Notice that $A_\varphi(\theta)$ depends on θ only since the feedback filter $H(q^{-1};\theta)$ does. Let us take

$$x(t)=\left(\theta(t)^T \operatorname{col}^T R(t)\right)^T. \tag{55}$$

Then (49) takes the form

$$x(t)=x(t-1)+\gamma(t)Q(t;x(t-1);\varphi(t)) \tag{56}$$

with an obvious definition of $Q(t;x,\varphi)$ from (49). Therefore the algorithm (49) together with (54) is of the general form (32). Let us check if assumptions B of Section IV are satisfied. Conditions B.1 and B.2 are satisfied due to our assumptions. By straightforward calculations it is readily shown that B.3 is satisfied in the open area $D_R=\{x|R>0\}$, [cf., (55)] e.g., with

$$\mathcal{K}_1(x,\varphi,\rho,v)=(|\theta|+\rho)(1+|\varphi|+v)^2/(1-\rho|R^{-1}|)^2 \tag{57}$$

for $\rho=\rho(x)<1/|R^{-1}|$. Then B.4 will be satisfied with

$$\mathcal{K}_2(x,\varphi,\rho,v,w)=(|\theta|+\rho)(|\varphi|+2w+v)/(1-\rho|R^{-1}|)^2. \tag{58}$$

Condition B.5 is satisfied if the matrix $H(q^{-1};\theta)$ is Lipschitz continuous in θ. For condition B.6 we define

$$\bar\varphi(t,\bar x)=A_\varphi(\bar\theta)\bar\varphi(t-1,\bar x)+B_\varphi\begin{pmatrix}e_1(t)\\ e_2(t)\end{pmatrix}$$
$$\left(\bar x=(\bar\theta^T\operatorname{col}^T\bar R)^T\right).$$

Since $e_i(\cdot)$ are stationary, $\bar\varphi(t,\bar x)$ will approach stationarity exponentially, for all x, such that $\bar\theta$ makes the closed-loop system stable. Therefore the limits

$$f(\bar\theta)=\lim_{t\to\infty} E\bar\psi(t,\bar x)\left[\bar y(t,\bar x)-\bar\theta^T\bar\psi(t,\bar x)\right]^T \tag{59a}$$

$$G(\bar\theta)=\lim_{t\to\infty} E\bar\psi(t,\bar x)\bar\psi(t,\bar x)^T \tag{59b}$$

are well defined where $\bar y(t,\bar x)$ and $\bar\psi(t,\bar x)$ are the corresponding parts of $\bar\varphi(t,\bar x)$, and

$$\lim_{t\to\infty} EQ(t;\bar x,\bar\varphi(t,\bar x))=\begin{bmatrix}\bar R^{-1}f(\bar\theta)\\ \operatorname{col}\left(G(\bar\theta)-\bar R\right)\end{bmatrix} \tag{60}$$

so B.6 is satisfied. Moreover, from (57) and (58) it follows that B.7 holds, since all moments of $\bar\varphi(t,x)$ and $v(t,\lambda,c)$ exist. Conditions B.8–B.11 about the sequence $\gamma(\cdot)$ are assumed to be satisfied.

The conclusion therefore is that the differential equation

$$\frac{d}{d\tau}\theta(\tau)=R^{-1}(\tau)f(\theta(\tau)) \tag{61a}$$

$$\frac{d}{d\tau}R(\tau)=G(\theta(\tau))-R(\tau) \tag{61b}$$

can be associated with the algorithm (49). In the remaining part of this example, we shall assume that the feedback matrix H *does not depend on* θ (i.e., there is no adaptive feedback), that the matrix $F(z)$ has full rank *a.e.* z and that $e_i(\cdot)$ are full rank processes. (Adaptive feedback is further discussed in Example 5.) This means that the matrix $A_\varphi(\cdot)$ does not depend on θ, $\bar\varphi(t,\bar x)=\varphi(t)$, and $\bar y(t,\bar x)=y(t)$, so the values in (59) are directly defined in terms of input–output covariances. In particular, the matrix G is independent of θ; $G(\theta)=G$.

Introduce

$$r=E\varphi(t)v(t)^T \tag{62}$$

and we have, using (45),

$$f(\theta)=G\cdot(\theta_0-\theta)+r. \tag{63}$$

Hence, (61) can be rewritten as

$$\frac{d}{d\tau}\theta(\tau)=R^{-1}(\tau)G\left[(\theta_0+G^{-1}r)-\theta(\tau)\right] \tag{64a}$$

$$\frac{d}{d\tau}R(\tau)=G-R(\tau). \tag{64b}$$

With

$$\tilde\theta(\tau)=\theta(\tau)-(\theta_0+G^{-1}r)$$

and

$$V(\tilde\theta,R)=\tilde\theta^TR\tilde\theta$$

we have

$$\frac{d}{d\tau}V(\tilde{\theta}(\tau),R(\tau)) = -2\tilde{\theta}^T G\tilde{\theta} + \tilde{\theta}^T(G-R(\tau))\tilde{\theta}$$
$$= -\tilde{\theta}^T(\tau)(G+R(\tau))\tilde{\theta}(\tau) < 0 \quad (65)$$

so that V is a Lyapunov function for (64) or (61) that assures that the stationary point

$$\theta^* = \theta_0 + G^{-1}r \quad (66)$$

has a domain of attraction equal to D_R. Therefore condition (21) of Theorem 1 is satisfied with $D_A = D_R$. To check condition (20) we note first that the assumption on full rank and finite moments of $e_i(\cdot)$ implies that $CI > G > \delta I$ for some $\delta > 0$, $C < \infty$. Therefore also $(\delta/2)I < R(t) < 2CI$ and $|\varphi(t)| < 2C$ i.o. w.p.1. We also note that (49a) can for large $\theta(t-1)$ be written

$$\theta(t) \approx (I - R^{-1}(t)\psi(t)\psi(t)^T)\theta(t-1)$$

which shows that $\theta(t)$ can, w.p.1, not tend to infinity. Hence $x(t)$ belongs to a compact subset of D_R i.o. w.p.1 and condition (20) is satisfied. Theorem 1 now implies that

$$\theta(t) \to \theta^* \qquad \text{w.p.1 as } t \to \infty.$$

In particular, we see that the least squares estimate is consistent only if $r = 0$, which essentially is the same as requiring that $v(\cdot)$ is a sequence of uncorrelated random variables, and that the current $u(t)$ is uncorrelated with future $v(s)$, $s > t$.

Other variants of equation error methods are treated analogously.

These facts are, of course, well known, [24], [25], at least for the case $\gamma(t) = 1/t$, but one reason for this example is that the analysis extends into less trivial problems. □

Example 4–Equation Error Methods–Assumptions C

The model (44) and the criterion (48) appear in several contexts, like curve fitting, etc., where a stochastic framework often is not imposed. Let $\gamma(t) = 1/t$ and denote

$$\psi(t)y(t)^T = \eta(t) \quad (67a)$$

$$\psi(t)\psi(t)^T = \Lambda(t). \quad (67b)$$

Then the algorithm (49) can be written

$$\Theta(t) = \Theta(t-1) + \frac{1}{t}R^{-1}(t)\left[\eta(t) - \Lambda(t)\Theta(t-1)\right] \quad (68a)$$

$$R(t) = R(t-1) + \frac{1}{t}\left[\Lambda(t) - R(t-1)\right]. \quad (68b)$$

We shall in this example illustrate how assumptions C can be applied to the algorithm (68) to infer convergence directly from the properties of the sequences $\{\eta(t)\}$ and $\{\Lambda(t)\}$. Take in this example the observation to be

$$\varphi(t) = \begin{pmatrix} \eta(t) \\ \operatorname{col}\Lambda(t) \end{pmatrix} \quad (69)$$

and in (32b) $A(x) \equiv 0$, $B(x) \equiv I$. The estimate is as before given by (55). The condition C.1 has already been verified and $\mathcal{K}_1$ can this time be taken as

$$\mathcal{K}_1(x,\varphi,\rho,v) = 1 + |x| + \rho + |\Lambda| + v. \quad (70)$$

Condition C.2 is trivially satisfied. The variable $z(t,\bar{x})$ defined by (18a) is given by

$$z(t,\bar{x}) = \begin{bmatrix} \frac{1}{t}\bar{R}^{-1}\sum_{s=1}^{t}\left[\eta(s) - \Lambda(s)\bar{\Theta}\right] \\ \operatorname{col}\frac{1}{t}\sum_{s=1}^{t}\left[\Lambda(s) - \bar{R}\right] \end{bmatrix}$$

$$= \begin{bmatrix} \bar{R}^{-1}\left\{\left[\frac{1}{t}\sum_{s=1}^{t}\eta(s)\right] - \left[\frac{1}{t}\sum_{s=1}^{t}\Lambda(s)\right]\bar{\Theta}\right\} \\ \operatorname{col}\left[\frac{1}{t}\sum_{s=1}^{t}\Lambda(s)\right] - \operatorname{col}\bar{R} \end{bmatrix} \quad (71)$$

where

$$\bar{x} = \begin{pmatrix} \bar{\Theta} \\ \operatorname{col}\bar{R} \end{pmatrix}.$$

Therefore, if the sequences $\{\eta(s)\}$ and $\{\Lambda(s)\}$ are such that

$$\frac{1}{t}\sum_{s=1}^{t}\eta(s) \to \bar{\eta} \quad (72a)$$

and

$$\frac{1}{t}\sum_{s=1}^{t}\Lambda(s) \to \bar{\Lambda}, \quad (72b)$$

then C.3 holds with

$$f(\bar{x}) = \begin{bmatrix} \bar{R}^{-1}(\bar{\eta} - \bar{\Lambda}\bar{\Theta}) \\ \operatorname{col}\bar{\Lambda} - \bar{R} \end{bmatrix}. \quad (73)$$

Since A and B do not depend on x (see C.7) condition C.4 is satisfied if

$$\frac{1}{t}\sum_{s=1}^{t}|\Lambda(s)| \quad \text{converges} \quad (74)$$

which is implied by (72b) if $\Lambda(s) = \psi(s)\psi(s)^T$. The boundedness condition of Theorem 1 has been verified in

Example 3. The differential equation

$$\frac{d}{d\tau}\Theta(\tau)=R^{-1}(\tau)\left[\bar{\eta}-\bar{\Lambda}\Theta(\tau)\right] \tag{75a}$$

$$\frac{d}{d\tau}R(\tau)=\bar{\Lambda}-R(\tau) \tag{75b}$$

is as shown in Example 3 globally asymptotically stable with stationary point

$$\Theta^*=\bar{\Lambda}^{-1}\bar{\eta} \tag{76}$$

if $\bar{\Lambda}$ is positive definite. Hence Theorem 1 implies that $\Theta(t)$ given by (68) converges to Θ^* defined in (76) for any sequences $\{\Lambda(s)\}$ and $\{\eta(s)\}$ such that (72) holds with a positive definite $\bar{\Lambda}$. This is by no means surprising since (68) in fact only is a way of recursively writing

$$\Theta(t)=\left[\sum_{s=1}^{t}\Lambda(s)\right]^{-1}\sum_{s=1}^{t}\eta(s)$$

using the matrix inversion lemma. However, it illustrates that our method of analysis does not bring in "unnecessary" assumptions, even though it is concerned with a more general algorithm.

Moreover, for a stochastic approximation version of (68),

$$\Theta(t)=\Theta(t-1)+\frac{1}{t}\left[\eta(t)-\Lambda(t)\Theta(t-1)\right] \tag{77}$$

we have exactly the same convergence result under the condition (72), since the associated differential equation then is

$$\frac{d}{d\tau}\Theta(\tau)=\bar{\eta}-\bar{\Lambda}\Theta(\tau). \tag{78}$$

In this case no explicit expression for $\Theta(t)$ is available to infer this result directly. In fact, the algorithm (77) has been studied in [26], [27] using fairly elaborate methods, since it was found that the usual stochastic approximation convergence results could not be applied due to the correlation in the sequences $\{\Lambda(s)\}$ and $\{\eta(s)\}$. ■

These four examples have all been for the case where $A(\cdot)$ in (32b) actually does not depend on x. The convergence part in this case can, at least under further assumptions, not seldom be treated by more conventional statistical methods. When $A(\cdot)$ does depend on x, conventional approaches become much more difficult, and in fact, also in the proof of Theorem 1, a major burden is to keep control over the coupled stability questions in (32a) and (32b). The inclusion of x-dependent A-matrices becomes necessary for more elaborate recursive identification methods in which the observed data are processed through filters that are formed from current parameter estimates. This is the case, e.g., for the extended Kalman filter, the extended least squares method, output error methods, like Landau's scheme [28] etc. Since a companion paper [29] is devoted to the last two schemes we shall not go into any detail here. Further analysis of some recursive identification schemes using the present approach can also be found in [1], [30], and [31].

It was remarked above that the structure (32) can be understood to be patterned after adaptive control of linear systems. We shall therefore conclude with such an example.

Example 5—Self-Tuning Regulators

We shall in this example discuss an application to the self-tuning regulator, described in [32]; see also [2] and [4]. This regulator is based on least squares identification, (49), and the output feedback law is determined from the current parameter estimates. Usually the feedback law is chosen to be a minimum variance regulator, [32], but here it could be a general linear regulator as in (50), where perhaps in most cases F is zero.

In this case the matrix $A_\varphi(\theta)$ in (54) does depend on θ since the feedback term does. However, the point now is that, in contrast to conventional analysis of the least squares algorithm, most of what was said in Example 3 still holds. Up to (61) the development was quite general. This differential equation is valid also in the case of adaptive feedback, although G and r now are functions of θ. If $v(\cdot)$ is a sequence of independent random variables, then $r=0$ and (64) and (65) hold. (Equation (65) does not hold if $r\neq 0$ depends on θ.) We therefore still find that the points defined by

$$D_c=\{\theta\,|\,f(\theta)=0\}$$

form an invariant set with global domain of attraction. Clearly $\theta_0\in D_c$, and whether D_c contains more points depends on the choice of feedback law and model order. There is a further complication before Theorem 1 can be applied. In this case the area D_S is unknown, i.e., the area of such θ that inserted in a constant feedback law (50) makes the closed-loop system stable ($A_\varphi(\theta)$ in (54) has all eigenvalues inside the unit circle). Therefore we cannot guarantee stability by projecting θ into D_S. Hence condition (20) of Theorem 1 has to be verified by other considerations, e.g., by showing that the overall system has a certain stability property as in [33]. But when this is shown, Theorem 1 proves convergence of $\theta(t)$ into D_c w.p.1. Let us repeat that this holds for the case of arbitrary feedback law, but under the assumption that $v(\cdot)$ is white noise. For general noise $v(\cdot)$ the convergence analysis is more cumbersome, but it can be performed in certain special cases, [2], [4], [29]. We refer also to these papers for more details on how Theorems 1, 2, and 3 can be used in the analysis of self-tuning regulators. Numerical solution of the associated differential equation has turned out to be a valuable tool here, and it has been used in [34] as well as in the references above.

IX. Conclusions

Recursive algorithms like (1) have been analyzed in various contexts. However, we would like to stress again that when φ in (1) is generated as in (2), the analysis becomes more difficult. The reason is that (1) no longer is recursive in x for analysis purposes: the whole history of $x(\cdot)$ enters in each step of (1). Moreover, the coupled stability problems between (1) and (2) are intricate. But the structure (1), (2) is nonetheless common in estimation and control problems; a typical example is adaptive control of linear stochastic systems. The analysis of this case is also known to be usually very difficult.

With the present approach we are able to give a general treatment of (1), (2) under assumptions that do not appear to be restrictive. The examples indicate that the theorems may be applied to rather diverse problems, and perhaps the technique also may serve as a basis for a unified approach to the analysis of adaptive controllers. In addition, an extension is obtained for the conventional convergence results in the simple case where $A(\cdot)$ in (2) is independent of x. We may remark that the analysis is restricted to the asymptotic behavior, convergence, possible convergence points etc. of the algorithm. Two related algorithms which are associated with the same differential equation may differ noticably in transient behavior and convergence rate.

In the described theory, we would like to stress the intuitive content of the theorems and the methodology of analysis as outlined in Section VII. It is no doubt important to appreciate the exact formulations of the theorems and to know the exact conditions under which they are valid. But it is perhaps equally rewarding to use the properly defined differential equation as a general instrument for analysis in a more heuristic fashion. This may be exemplified in Theorem 3, which has a fairly technical formulation and is probably more valuable as a "moral support" for studying the trajectories of the differential equation, than in its literal sense.

Appendices

The proofs of Theorems 1–4 are given in Appendices I–IV. They are slightly compressed versions of the proofs in [7], in that some calculations of technical nature are omitted. Proofs for a simple special case of the algorithm (1), (2) are also given in [6]. These are naturally less technical than the present ones, and reveal perhaps more clearly the underlying ideas. The basic path of the proofs of Theorems 1 and 3 follows the heuristic outline of Section III.

Some notational conventions in the proofs should be noted. "C" will denote any constant, that need not be the same in different parts of the proof. Important dependencies of the constants will be given as arguments, while indexed constants are "global" throughout the proof. "$\mathcal{B}(\bar{x},\rho)$" denotes as before an open ρ-neighborhood of $\bar{x}$. Realizations in the sample space Ω will be denoted by "ω." The abbreviation "i.o." denotes as before "infinitely often."

Appendix I
Proof of Theorem 1

Outline

In order to prove Theorem 1, we shall first show that the estimates provided by the algorithm locally and asymptotically follow the trajectories of the associated differential equation. This will be done in Lemma 1 under assumptions C and the proof of this follows the intuitive outline of Section III. After that the local behavior of the algorithm is thus established, this is used to prove that all cluster points of the sequence $\{x(t)\}$ must belong to D_c. This is done by means of a Lyapunov function, the existence of which is inferred from the stability condition (21). A possible clusterpoint outside the set D_c would yield a decreasing value of the Lyapunov function along the corresponding trajectory of the differential equation. Since Lemma 1 proves that we follow this trajectory asymptotically, a contradiction is obtained. The value of the Lyapunov function is decreasing outside D_c and it is not possible to return a given point outside D_c infinitely often. This is the intuitive path of proof. The many technicalities tend to obscure the simple idea, and the proof will be structured as much as possible to enhance the basic ideas.

Finally it is proved in Lemma 2 that assumptions A or B imply assumptions C for the case $\gamma(n)=1/n$. An outline of the general proof will also be given, but the details of this are omitted, and the reader is referred to [7] for them.

Lemma 1: Consider the algorithm (1), (2) under assumptions C. Let $\bar{x}\in D_R$ and define the number $m(n,\Delta\tau)$ such that

$$\sum_{n}^{m(n,\Delta\tau)} \gamma(t)\rightarrow\Delta\tau \quad \text{as } n\rightarrow\infty. \tag{I.1}$$

Assume that

$$x(n)\in\mathcal{B}(\bar{x},\rho) \tag{I.2a}$$

where

$$\rho=\rho(\bar{x}) \quad \text{is sufficiently small} \tag{I.2b}$$

and that

$$|\varphi(n)|\leqslant C_{\varphi}. \tag{I.3}$$

Then, there exists a value $\Delta\tau_0=\Delta\tau_0(\bar{x},\rho)$ and a number $N_0=N_0(\bar{x},\rho)$ such that for $\Delta\tau<\Delta\tau_0$ and $n>N_0$

$$x(m(n,\Delta\tau))=x(n)+\Delta\tau f(\bar{x})+q_1(n,m,\bar{x})+q_2(n,m,\bar{x}) \tag{I.4}$$

where $m=m(n,\Delta\tau)$ and

$$q_1(n,m,\bar{x})\rightarrow 0 \qquad \text{as } n\rightarrow\infty \tag{I.5}$$

and

$$|q_2(n,m,\bar{x})| \leq \Delta\tau \cdot C_1|x(n)-\bar{x}| + C_2\Delta\tau^2 \quad \text{(I.6)}$$

where C_1 and C_2 depend on x and ρ, but not on $n > N_0$ or $\Delta\tau$.

Proof of Lemma 1: The proof is structured into 4 steps. In *Step 1* an explicit expression for $x(j)$ where

$$n \leq j \leq m(n,\Delta\tau) \quad \text{(I.7)}$$

is derived. This expression shows that if certain terms are small, then $x(j)$ is close to what would be obtained if $Q(t,x(t-1),\ \varphi(t))$ is replaced by $Q(t,\bar{x},\bar{\varphi}(t,\bar{x}))$ [with $\bar{\varphi}(t,\bar{x})$ defined by (15)]. As a first step to show that these terms indeed are small an expression for $\varphi(t)-\bar{\varphi}(t,\bar{x})$ is derived in *Step 2*. This expression is used in *Step 3* to prove that assumptions C.3 and C.4 imply that the terms are small. A complication so far has been that in order to prove these things it must be assumed that $x(s)$ remains in a small neighborhood of $\bar{x}$ for $n \leq s \leq j$. In *Step 4* it is proved that $x(s)$ actually will remain in this neighborhood up to $s = m(n,\Delta\tau)$ if $\Delta\tau$ is chosen sufficiently small (and dependent only on $\bar{x}$) and n sufficiently large. This will conclude the proof.

Introduce $j = j(n)$ such that

$$n \leq j \leq m(n,\Delta\tau) \quad \text{(I.8)}$$

and

$$x(k) \in \mathfrak{B}(\bar{x},2\rho), \qquad k = n, n+1, \cdots, j-1. \quad \text{(I.9)}$$

Step 1—An Expression for x(j)

Directly from the algorithm (1) the following expression is obtained:

$$\begin{aligned} x(j) &= x(n) + \sum_{s=n+1}^{j} \gamma(s) Q(s;x(s-1),\varphi(s)) \\ &= x(n) + \sum_{n+1}^{j} \gamma(s) Q(s;\bar{x},\bar{\varphi}(s;\bar{x})) \\ &\quad + \sum_{n+1}^{j} \gamma(s)\left[Q(s,x(s-1),\varphi(s)) - Q(s;\bar{x},\bar{\varphi}(s;\bar{x}))\right] \end{aligned} \quad \text{(I.10)}$$

where $\bar{\varphi}(s;\bar{x})$ is defined by (15). End Step 1.

The first goal is to show that the last term in (I.10) is "small," and in order to do this we first consider $\varphi(s) - \bar{\varphi}(s;\bar{x})$.

Step 2—An Estimate for $|\varphi(t)-\bar{\varphi}(t,\bar{x})|$

We claim that

$$\prod_{k=n}^{t} A(x(k)) \leq C\left[\frac{1+\lambda(\bar{x})}{2}\right]^{t-n} \triangleq C\tilde{\lambda}(\bar{x})^{t-n} \quad \text{(I.11)}$$

if

$x(k) \in \mathfrak{B}(\bar{x},2\rho); \qquad n \leq k \leq t$ for sufficiently small

$$\rho = \rho(\bar{x}) \quad \text{(I.12)}$$

where $\lambda(\bar{x})$ is defined by (14).

The idea obviously is that since the matrix $A(\bar{x})$ is exponentially stable, exponential stability of the time-varying difference equation (2) given by $A(x(k))$ will be guaranteed if $x(k)$ varies in a sufficiently small neighborhood of $\bar{x}$. The formal proof consists of straightforward calculations given in [7], but omitted here. The "sufficiently small" ρ mentioned in (I.2b) refers to the fact that ρ should be so small that (I.12) implies (I.11).

From (2) it follows that

$$\varphi(t) = \left[\prod_{k=n}^{t} A(x(k))\right]\varphi(n) + \sum_{j=n}^{t}\left[\prod_{k=j}^{t} A(x(k))\right] B(j)e(j) \quad \text{(I.13)}$$

which together with (I.11) and (I.3) implies that

$$|\varphi(t)| \leq C \cdot C_{\varphi}\tilde{\lambda}^{t-n} + \sum_{j=n}^{t} C\tilde{\lambda}^{t-j}|B(j)||e(j)| \quad \text{(I.14)}$$

if (I.12) holds. Since the variable

$$\tilde{\varphi}(t) \triangleq \varphi(t) - \bar{\varphi}(t;\bar{x})$$

obeys

$$\begin{aligned} \tilde{\varphi}(t+1) &= A(\bar{x})\tilde{\varphi}(t) + \left[A(x(t)) - A(\bar{x})\right]\varphi(t) \\ &\quad + \left[B(x(t)) - B(x)\right]e(t+1) \end{aligned} \quad \text{(I.15)}$$

we have that

$$\begin{aligned} \tilde{\varphi}(t) &= A(\bar{x})^{t-n}\tilde{\varphi}(n) \\ &\quad + \sum_{j=n}^{t-1} A(\bar{x})^{t-s}\left[\{A(x(j)) - A(\bar{x})\}\varphi(j)\right. \\ &\quad \left. + \{B(x(j)) - B(\bar{x})\}e(j+1)\right]. \end{aligned} \quad \text{(I.16)}$$

If K_A and K_B denote the Lipschitz constants of $A(\cdot)$ and $B(\cdot)$, respectively (see assumption C.2), then from (I.16) and (14) it follows that

$$\begin{aligned} |\tilde{\varphi}(t)| &\leq \bar{C} \cdot \lambda(\bar{x})^{t-n}\{|\varphi(n)| + |\bar{\varphi}(n,\bar{x})|\} \\ &\quad + \bar{c}\sum_{j=n}^{t} \lambda(\bar{x})^{t-j}\left[K_A|x(j)-\bar{x}||\varphi(j)|\right. \\ &\quad \left. + K_B|x(j)-\bar{x}||e(j+1)|\right]. \end{aligned} \quad \text{(I.17)}$$

Clearly, with $v(n,\lambda,c)$ defined by (16),

$$\bar{c}\{|\varphi(n)| + |\bar{\varphi}(n,\bar{x})|\} \leq v(n,\lambda(\bar{x}),c) \quad \text{(I.18)}$$

for some constant c.

Now, inserting (I.14) and (I.18) in (I.17) gives

$$|\bar{\varphi}(t)| \leq \lambda(\bar{x})^{t-n} v(n,\lambda(\bar{x}),c) + c\cdot \max(K_A, K_B)$$
$$\cdot \max_{n\leq j\leq t-1} |x(j)-\bar{x}| \cdot \left\{ \sum_{j=n}^{t-1} \lambda(\bar{x})^{t-j} \tilde{\lambda}^{j-n} \cdot C_\varphi \right.$$
$$+ \sum_{j=n}^{t-1} \sum_{k=n}^{j} \lambda(\bar{x})^{t-j} \tilde{\lambda}^{j-k} |e(k)|$$
$$\left. + \sum_{j=n}^{t-1} \lambda(\bar{x})^{t-j} |e(j+1)| \right\}.$$

Sum first over j in the double sum and introduce c' and λ such that

$$|k\tilde{\lambda}^k + \lambda^k| < c'\lambda^k \qquad \lambda < 1. \tag{I.19}$$

Then

$$|\varphi(t) - \bar{\varphi}(t,\bar{x})| \leq \lambda(\bar{x})^{t-n} v(n,\lambda(\bar{x}),c) + \max_{n\leq j\leq t-1} |x(j)-x| \cdot v(t,\lambda,c). \tag{I.20}$$

In particular,

$$\lambda(\bar{x})^{t-n} v(n,\lambda(\bar{x}),c) \leq v(t,\lambda(\bar{x}),c)$$

and therefore also

$$|\varphi(t) - \bar{\varphi}(t,\bar{x})| \leq v(t,\lambda,c) \tag{I.21}$$

where λ can be taken as $(3+\lambda(x))/4$ and c is a constant that is obtained from c_φ, $\bar{c}$ in (14), c in (I.11), c' in (I.19), K_A, K_B, and $\max|B(x)|(x \in \mathfrak{B}(\bar{x},\rho))$. However, since assumption C.4 shall hold for any finite c, there is no point in tracing the expression for c through the calculations.

We summarize this step as follows: Choose 2ρ in (I.9) so small that (I.12) implies (I.11) for this ρ. Then (I.20) and (I.21) hold for $n \leq t \leq j$, where j is defined by (I.8), (I.9).

Note that in case the matrices A and B in (2) do not depend on x, then $\varphi(t) = \bar{\varphi}(t,\bar{x})$ and we may take $v(t,\lambda,c) \equiv 0$ in all calculations to come. This justifies the remark C.7. End Step 2.

We can now return to the expression (I.10) for $x(j)$. From assumption C.1 and (I.21) it follows that

$$|Q(s,x(s-1),\varphi(s)) - Q(s,\bar{x},\bar{\varphi}(s;\bar{x}))|$$
$$\leq \mathcal{K}_1(\bar{x},\bar{\varphi}(s,\bar{x}),2\rho,v(s,\lambda,c))$$
$$\cdot [|x(s-1)-\bar{x}| + |\varphi(s) - \bar{\varphi}(s;\bar{x})|]$$
$$\leq \max_{n\leq k\leq s-1} |x(k)-\bar{x}| \cdot \mathcal{K}_1(\bar{x},\bar{\varphi}(s,\bar{x}),2\rho,v(s,\lambda,c))$$
$$\cdot\{1+v(x,\lambda,c)\} + \mathcal{K}_1(\bar{x},\bar{\varphi}(s,\bar{x}),2\rho,v(s,\lambda,c))$$
$$\cdot \lambda(\bar{x})^{s-n} v(n,\lambda,c) \tag{I.22}$$

where the last inequality follows from (I.20).

We thus find that

$$x(j) = x(n) + \sum_{s=n}^{j} \gamma(s) f(\bar{x}) + R_1(n,j) + R_2(n,j) \tag{I.23}$$

where

$$R_1(n,j) = \sum_{s=n}^{j} \gamma(s)[Q(s,\bar{x},\bar{\varphi}(s;\bar{x})) - f(\bar{x})] \tag{I.24}$$

and

$$|R_2(n,j)| \leq \max_{n\leq k\leq j-1} |x(k)-\bar{x}| \cdot q_3(n,j) + q_4(n,j) \tag{I.25}$$

where

$$q_3(n,j) = \sum_{s=n}^{j} \gamma(s) \mathcal{K}_1(\bar{x},\bar{\varphi}(s;\bar{x}),2\rho,v(s,\lambda,c)) \cdot \{1+v(s,\lambda,c)\} \tag{I.26}$$

and

$$q_4(n,j) = \max_{n\leq s\leq j-1} \{\mathcal{K}_1(\bar{x},\bar{\varphi}(s;\bar{x}),2\rho,v(s,\lambda,c)) \cdot v(s,\lambda,c)\cdot\gamma(s)\} \cdot \sum_{s=n}^{j} \lambda(\bar{x})^{s-n}. \tag{I.27}$$

Step 3a—$\max_{n\leq j\leq m(n,\Delta\tau)} |R_1(n,j)| \to 0$ *as* $n\to\infty$

Let the maximum be attained for $j = j^*(n)$.

Solving (18a) we obtain

$$z(j^*(n),\bar{x}) = z(n,\bar{x}) + \sum_{k=n+1}^{j^*(n)} \gamma(k) \cdot [Q(k,\bar{x},\bar{\varphi}(k,\bar{x})) - z(k-1,\bar{x})].$$

Now, let n tend to infinity. By assumption C.3

$$\lim_{n\to\infty} z(n,\bar{x}) = \lim_{n\to\infty} z(j^*(n),\bar{x}) = f(\bar{x})$$

and hence

$$\lim_{n\to\infty} \sum_{k=n+1}^{j^*(n)} \gamma(k)[Q(k,\bar{x},\bar{\varphi}(k,x)) - z(k-1,\bar{x})]$$
$$= \lim_{n\to\infty} \sum_{k=n+1}^{j^*(n)} \gamma(k)[Q(k,\bar{x},\bar{\varphi}(k,\bar{x})) - f(\bar{x})]$$
$$= \lim_{n\to\infty} R_1(n,j^*(n)) = 0$$

End Step 3a.

Step 3b:

There exists an N_1 such that

$$|q_3(n,j(n))| \leq 2\Delta\tau \bar{k}_v \qquad \text{for } n > N_1.$$

where $\bar{k}_v$ is the limit of $k_v(t,\bar{x},\lambda,c)$ defined in C.4. This step is completely analogous to Step 3a, using the definition (18b) and assumption C.4.

End Step 3b.

Step 3c—$q_4(n,j(n)) \to 0$ *as* $n\to\infty$

From (18b)

$$\gamma(s)\cdot v(s,\lambda,c)\cdot \mathcal{K}_1(\bar{x},\bar{\varphi}(s,\bar{x}),2\rho,v(s,\lambda,c)) = k_v(s)-k_v(s-1)+\gamma(s)k_v(s-1).$$

The RHS tends to zero according to C.4 and C.6 and the claim is proved. End Step 3c.

Expressions (I.23)–(I.27) together with Step 3 would complete the proof of Lemma 1, if $j(n)$ were equal to $m(n,\Delta\tau)$, that is if $x(k)$ remains in $\mathcal{B}(\bar{x},2\rho)$ for all k up to $m(n,\Delta\tau)$. This is proven in the final step.

Step 4

There exists a $\Delta\tau_0=\Delta\tau_0(\bar{x},\rho)$ and a $N_0=N_0(\bar{x},\rho)$ such that for $\Delta\tau<\Delta\tau_0$ and $n>N_0$, the expression (I.9) will hold for $j=m(n,\Delta\tau)$.

By the definition of $j(n)$

$$x(j-1)\in\mathcal{B}(\bar{x},2\rho) \tag{I.28}$$

and

$$\max_{n\leqslant k\leqslant j-1}|x(k)-\bar{x}|<2\rho. \tag{I.29}$$

We shall show that (I.28) and (I.29) imply that also

$$x(j)\in\mathcal{B}(\bar{x},2\rho) \tag{I.39}$$

for sufficiently large n and small $\Delta\tau$. By induction it then follows that

$$x(k)\in\mathcal{B}(\bar{x},2\rho) \qquad k=n,\cdots,m(n,\Delta\tau). \tag{I.31}$$

We have

$$|x(j)-\bar{x}|\leqslant|x(j)-x(n)|+|x(n)-\bar{x}|$$
$$\leqslant|f(\bar{x})|\cdot\sum_{s=n}^{j}\gamma(s)+|R_1(n,j)|+|R_2(n,j)|+\rho \tag{I.32}$$

using (I.23) and (I.2). From (I.24)–(I.27) using Step 3b and (I.29) it follows that

$$\begin{aligned}|x(j)-\bar{x}|&\leqslant|f(\bar{x})|\cdot\Delta\tau+|R_1(n,j)|\\&\quad+4\rho\cdot\Delta\tau\cdot\bar{k}_v+|q_4(n,j)|+\rho\\&\leqslant|f(\bar{x})|\cdot\Delta\tau+\max_{n\leqslant j\leqslant m(n,\Delta\tau)}|R_1(n,j)|\\&\quad+4\rho\cdot\Delta\tau\cdot k_v+|q_4(n,m(n,\Delta\tau))|+\rho.\end{aligned} \tag{I.33}$$

Now, choose

$$\Delta\tau<\Delta\tau_0=\min\left\{\frac{\rho}{4|f(\bar{x})|},\frac{1}{16\bar{k}_v}\right\} \tag{I.34}$$

and then n so large that

$$q_5(n,m(n,\Delta\tau))\triangleq\max_{n\leqslant j\leqslant m(n,\Delta\tau)}|R_1(n,j)| +|q_4(n,m(n,\Delta\tau))|<\rho/2 \tag{I.35}$$

which is possible according to Steps 3a–3c.

It now follows from (I.33)–(I.35) that

$$|x(j)-\bar{x}|<2\rho$$

and hence that (I.30) and (I.31) hold. End Step 4.

From (I.33)–(I.35) it also follows that

$$|x(j)-\bar{x}|\leqslant|f(\bar{x})|\Delta\tau+q_5(n,m(n,\Delta\tau)) +4\bar{\rho}\Delta\tau\cdot\bar{k}_v+|x(n)-\bar{x}|. \tag{I.36}$$

Since the RHS does not depend on j, this inequality holds for all $n\leqslant j\leqslant m(n,\Delta\tau)$ for sufficiently small $\Delta\tau$ and large n.

Therefore, from (I.23)–(I.27) and (I.36)

$$x(m(n,\Delta\tau))=x(n)+\Delta\tau f(x)+q_1(n,m,\bar{x})+q_2(m,n,\bar{x}) \tag{I.37}$$

where

$$|q_1(n,m,\bar{x})|\leqslant|R_1(n,m)|+q_4(n,m)+q_5(n,m) +2\bar{k}_v\Delta\tau+|f(\bar{x})|\left(\Delta\tau-\sum_{k=n}^{m}\gamma(k)\right) \tag{I.38}$$

and

$$|q_2(n,m,\bar{x})|\leqslant2\bar{k}_v\Delta\tau\left[|f(\bar{x})|\cdot\Delta\tau+4p\Delta\tau\cdot\bar{k}_v+|x(n)-\bar{x}|\right] \tag{I.39}$$

for $\Delta\tau<\Delta\tau_0$ and $n>N_0$.

The expression (I.37) with (I.38) and (I.39) is the assertion of the lemma since $q_1(n,m(n,\Delta\tau)\bar{x})$ lends to zero according to Steps 3a–3c and (I.1) [remember that q_5 is defined in (I.35)].

End proof of Lemma 1.

We now proceed with the proof of the main theorem.

Proof of Theorem 1: We consider in the following a sequence ("realization") $\{e(t)\}$ such that assumptions C.3 and C.4 hold and such that condition (20) of Theorem 1 holds.

It follows from the converse stability theorems (see e.g., Krasovskij [39] or Hahn [40]) that the stability assumption (21) of Theorem 1 for $D_c=\{x^*\}$ implies the existence of a function V with properties

1) $V(x)$ is infinitely differentiable;
2) $0\leqslant V(x)<1$ for $x\in D_A$ and $V(x)=0\Leftrightarrow x=x^*$;
3) $(d/d\tau)V(x(\tau))=V'(x)f(x)\leqslant0$ for $x\in D_A$ and equality holds only for $x=x^*$.

[For the case with general invariant set D_c, see Zubov [41] for the proper theorems.]

For convenience in formulations below, we shall let $D_c=\{x^*\}$. An outline of the rest of the proof is as follows:

Step 1: A convergent subsequence $x(n_k)$ tending to $\bar{x}$ is considered. Then $x(n_k)$ is close to $\bar{x}$ infinitely often, and according to Lemma 1, $x(m(n_k,\Delta\tau))$ will approximately be $x(n_k)+\Delta\tau f(\bar{x})$. This means that $V(x(m(n_k,\Delta\tau)))$ is strictly less than $V(x(n_k))$ if $\bar{x} \neq x^*$. The formal proof is somewhat lengthy and involves several elaborate choices of constants. The result is, however, intuitively clear. The proof of Step 1 extends to (I.45).

Step 2: From the above result it is quite clear that x^* must be a cluster point to $x(n)$, since $V(x(n))$ has a tendency to decrease everywhere in D_A except for $x = x^*$.

Step 3: If there is another cluster point to $\{x(n)\}$ than x^*, say $\hat{x}$, the sequence must move from x^* to $\hat{x}$ infinitely many times. But then $V(x(n))$ is increasing, which contradicts the result of Step 1. Hence, only one cluster point exists and convergence follows.

Step 1: Any cluster point of $\{x(t)\}$ outside D_c yields strictly decreasing values of V.

From assumption (20) of Theorem 1 there exists a subsequence, such that

$$x(\bar{n}_k) \in \bar{D} \text{ and } |\varphi(\bar{n}_k)| < C. \tag{I.40}$$

Since $\bar{D}$ is compact, there exists at least one cluster point to $x(\cdot)$ in $\bar{D}$. Let the cluster point be denoted by $\bar{x}$ and let n_k be a subsequence of $\bar{n}_k$ such that

$$x(n_k) \to \bar{x} \qquad \text{as } k \to \infty. \tag{I.41}$$

Consequently, for arbitrarily small $\epsilon > 0$,

$$|x(n_k) - \bar{x}| < \epsilon \qquad k > K_0(\epsilon). \tag{I.42}$$

Consider now

$$V[x(m(n_k, \Delta\tau))] - V[x(n_k)],$$

where m is defined as in Lemma 1. Denote $n_k = k'$ and $m(n_k, \Delta\tau) = k''$, and use the mean value theorem. This gives

$$\begin{aligned} V[x(k'')] - V[x(k')] &= V'(\xi)[x(k'') - x(k')] \\ &= V'(\bar{x})[x(k'') - x(k')] + [\xi - \bar{x}]^T \\ &\quad \cdot V''(\xi')[x(k'') - x(k')] \end{aligned} \tag{I.43}$$

where ξ and ξ' belong to $\mathfrak{B}(\bar{x}, \epsilon + \Delta\tau)$.

Now take $\epsilon < \rho(\bar{x})$, and we can in view of (I.42) apply Lemma 1 to $x(n_k)$, which gives

$$x(k'') - x(k') = \Delta\tau f(\bar{x}) + q_1(k', k'', \bar{x}) + q_2(k', k'', \bar{x})$$

where q_i are subject to (I.5), (I.6).

Insert this into (I.43)

$$V[x(k'')] - V[x(k')] = \Delta\tau V'(\bar{x}) f(\bar{x}) + R_3(\Delta\tau, n_k \bar{x}) \tag{I.44}$$

where

$$\begin{aligned} R_3(\Delta\tau, n_k, \bar{x}) = (\xi - x)^T V''(\xi')(x(k'') - x(k')) \\ + V'(\bar{x})\{q_1 + q_2\}. \end{aligned}$$

Now suppose that the cluster point $\bar{x}$ is different from the desired convergence point x^*. Then $V'(\bar{x}) f(\bar{x}) = -\delta$, $\delta > 0$.

Then (I.44) implies that by a proper choice of ϵ, $\Delta\tau$, and K (see [7] for a detailed account), we have

$$V(x(m(n_k, \Delta\tau))) < V(\bar{x}) - \Delta\tau\delta/64 \qquad k > K. \tag{I.45}$$

End Step 1.

This means that if $\bar{x}$ is a cluster point different from x^* the sequence $x(n)$ will i.o. be interior to $\tilde{D} = \{x | x \in \bar{D}_R$ and $V(x) \leq V(\bar{x}) - \Delta\tau\delta/64\}$. The conclusion that $x \in \bar{D}_R$ follows from our assumption that trajectories that start in $\bar{D}_R$ remain in there. This is required in order to apply Lemma 1 to new points in $\tilde{D}$. The set $\tilde{D}$ is compact. Consequently, another cluster point must exist that yields a smaller value of V. Moreover, since $x(t) \in \mathfrak{B}(\bar{x}, 2\rho)$ $n_k \leq t \leq m(n_k, \Delta\tau)$ we have from (I.14)

$$|\varphi(k'')| \leq C\tilde{\lambda}^{k''-k'}|\varphi(k')| + C|B| \sum_{k=k'}^{k''} \tilde{\lambda}^{k''-k'}|e(k+1)| \tag{I.46}$$

and hence $|\varphi(m(n_k, \Delta\tau))| < C$ i.o. so the argument can be repeated, again applying Lemma 1 to this new cluster point.

Step 2: Suppose (I.45) holds for any subsequence $\{x(n_k)\}$ that converges to a point different from x^*. Then

$$\liminf_{t \to \infty} V(x(t)) = 0. \tag{I.47}$$

Consider $\inf V(x)$ taken over all cluster points in D_A. Let this value be U. Since the set of cluster points in D_A is compact, there exists a cluster point $\hat{x}$, such that $V(\hat{x}) = U$. If now $U > 0$, $V'(\hat{x}) f(\hat{x})$ will be strictly negative ($= -\hat{\delta}$) and from (I.45) $V(x(k))$ takes a value less than $U - \hat{\delta}\Delta\tau/64$ i.o., which contradicts U being the infinum. Hence, $U = 0$, which means that x^* is a cluster point. End Step 2. □

Step 3: From (I.45) and (I.47) it follows that

$$\limsup_{t \to \infty} V(x(t)) = 0. \tag{I.48}$$

Let $\rho^* = \rho(x^*)$ be the region for which (2) is exponentially stable for $x(k) \in \mathfrak{B}(x^*, \rho^*)$, as in (I.11), (I.12).

Suppose that

$$\limsup_{n \to \infty} V(x(n)) = \bar{W} > 0. \tag{I.49}$$

Take $W < \bar{W}$ such that

$$\{x | V(x) \leq W\} \subset B(x^*, \rho^*)$$

and consider the interval $I = [W/3, 2W/3]$. (We may of course choose any subinterval; there is no particular reason to divide it up in thirds.)

Since x^* is a cluster point and since $V(x(n))$ is supposed to have a subsequence tending to W, this interval I is crossed "upwards" and "downwards" infinitely many times by $V(x(n))$.

We shall now proceed to show that (I.49) would imply that there must be a subsequence of $V(x(n))$ that belongs to I, by proving that in $\mathfrak{B}(x^*, \rho^*)$ the "step size" $x(n+1) - x(n)$ tends to zero.

First, let $x(\tilde{n}_k)$ be a subsequence tending to x^*, such that $|\varphi(\tilde{n}_k)| < C$. (The existence of such a sequence follows from Step 2 and the stability argument (I.46), using the fact that $\rho = \rho(\bar{x})$ is bounded from below by a positive constant in $\bar{D}$.)

For $t > \tilde{n}_k$, but such that $x(t)$ "remains" in $B(x^*, \rho^*)$ we have, (I.46)

$$|\varphi(t)| \leqslant C\tilde{\lambda}^{t-\tilde{n}_k}|\varphi(\tilde{n}_k)| + C|B|\sum_{k=n_k}^{t}\tilde{\lambda}^{t-k}|e(k+1)| \leqslant v(t,\tilde{\lambda},c)$$

with $v(t,\tilde{\lambda},c)$ defined by (16).

Hence,

$$\begin{aligned}|\gamma(t)Q(t,x(t),\varphi(t))| &\leqslant \gamma(t)|Q(t,x^*,0)| \\ &\quad + \gamma(t)\mathcal{K}_1(x^*,0,\rho^*,v(t,\tilde{\lambda},c))(|x(t)-x^*| + |\varphi(t)|) \\ &\leqslant \gamma(t)|Q(t,x^*,0)| + \gamma(t)\mathcal{K}_1(x^*,0,\rho^*,v(t,\tilde{\lambda},c)) \\ &\quad \cdot(\rho^* + v(t,\tilde{\lambda},c))\end{aligned} \tag{I.50}$$

where the first inequality follows from assumption C.1.

It follows from Step 3a of the proof of Lemma 1 that the first term of the RHS of (I.50) tends to zero and from Step 3c that the second one tends to zero. Consequently, inside $B(x^*,\rho^*)$ the step size tends to zero, and hence there will be a subsequence of $V(x(n))$ entirely in the interval I. Consider now a special, convergent sequence of "upcrossings," that is a subsequence of this.

Let the sequence n_k' be defined such that

$$V(x(n_k'-1)) < W/3 \tag{I.51}$$

$$V(x(n_k')) \geqslant W/3 \tag{I.52}$$

$$V(x(n_k'+s_k)) > 2W/3 \tag{I.53a}$$

where

$$s_k \text{ is the first } s \text{ for which } V(x(n_k+s)) \notin I \tag{I.53b}$$

$$x(n_k') \to \bar{x} \qquad \text{as } k \to \infty. \tag{I.54}$$

From (I.51) and (I.52) it is clear that $V(\bar{x}) = W/3$ and let $V'(\bar{x})f(\bar{x}) = -\tilde{\delta}$. From (I.45) we now have that

$$V(x(m(n_k',\Delta\tau))) < W/3 - \tilde{\delta}\Delta\tau/64. \tag{I.55}$$

This means that $V(x(n_k'+s_k')) \notin I$ where $s_k' = m(n_k',\Delta\tau) - n_k'$.

Suppose now that there is a $s_k < s_k'$ such that $V(x(n_k' + s_k)) > 2W/3$. However, V is continuous and $x(n_k'+s)$, $(s < m(n_k',\Delta\tau) - n_k')$ will belong to an arbitrarily small neighborhood of $x(n_k')$ for sufficiently small $\Delta\tau$ according to Lemma 1. Therefore (I.55) implies a contradiction to the existence of a subsequence n_k' with properties (I.51)–(I.54). Hence, no interval I may exist, W must be zero and (I.48) follows. End Step 3.

Equation (I.48) implies according to the properties of $V(\cdot)$ that $x_n \to x^*$. This concludes the proof of Theorem 1.

End proof of Theorem 1.

We shall now proceed to prove that assumptions A or B imply assumptions C.3 and C.4. This will be done only for the case $\gamma(t) = 1/t$, which anyway should be the most common one. We may also remark that since the correction term $Q(\cdot,\cdot,\cdot)$ may be time-varying, this case also includes cases where

$$\gamma(t) = \frac{c(t)}{t}; c(t) \to c \qquad \text{as } t \to \infty \tag{I.56}$$

be redefining the factors of the product

$$\gamma(t)Q(t,x,\varphi).$$

Lemma 2. Suppose the algorithm (1), (2) is subject to assumptions A or B. Assume that $\gamma(t) = 1/t$. Then C.3 holds w.p.1 as well as Steps 3b and 3c of the proof of Lemma 1. If, in addition

$$E\mathcal{K}_1(\bar{x},\bar{\varphi}(t,\bar{x}),\rho(\bar{x}),v(t,\lambda,c))(1+v(t,\lambda,c))$$

converges at $t\to\infty$, then also C.4 holds w.p.1.

Proof: For $\gamma(t) = 1/t$ we have from (18a)

$$z(t,\bar{x}) = \frac{1}{t}\sum_{k=1}^{t} Q(k;\bar{x},\bar{\varphi}(k,\bar{x})). \tag{I.57}$$

We shall apply the following ergodicity result of Cramér and Leadbetter [38, pp. 94–96] to (I.57).

Let $f(k)$ be a sequence of random variables with zero means and covariances

$$|Ef(k)f(s)| \leqslant \frac{k^p + s^p}{1+|k-s|^q}; \qquad 0 \leqslant 2p < q < 1. \tag{I.58}$$

Then

$$\frac{1}{t}\sum_{k=1}^{t} f(k) \to 0 \qquad \text{w.p.1 as } t\to\infty \qquad \blacksquare$$

(The proof in [38] is given for the continuous time case, but it goes through without changes for the discrete time case.)

In order to calculate the covariances, introduce for $k > s$ the variable

$$\bar{\varphi}_s^0(k,\bar{x})$$

by

$$\bar{\varphi}_s^0(k+1,\bar{x}) = A(\bar{x})\bar{\varphi}_s^0(k,\bar{x}) + B(\bar{x})e(k+1); \quad \bar{\varphi}_s^0(s,\bar{x}) = 0.$$

This variable is independent of $\bar{\varphi}(s,\bar{x})$ according to assumption A.1 = B.1, and

$$|\bar{\varphi}(k,\bar{x}) - \varphi_s^0(k,\bar{x})| \leqslant \bar{c}\cdot\lambda(\bar{x})^{k-s}|\bar{\varphi}(s,\bar{x})| \cdot(\leqslant v(k,\lambda(\bar{x}),\bar{c})).$$

Hence, using assumption A.3 or B.3

$$|Q(k,\bar{x},\bar{\varphi}(k,\bar{x})) - Q(k,\bar{x},\bar{\varphi}_s^0(k,\bar{x}))| \leqslant \mathcal{K}_1(\bar{x},\bar{\varphi}(k,\bar{x}),0,v(k,\lambda(\bar{x}),\bar{c}))\bar{c}\lambda(\bar{x})^{k-s}|\bar{\varphi}(s,\bar{x})|.$$

Moreover, from A.2, A.3 (according to which $\mathcal{K}_1$ and $\bar{\varphi}(s,\bar{x})$ are bounded) or B.7,

$$E|Q(k,\bar{x},\bar{\varphi}(k,\bar{x}))-Q(k,\bar{x},\bar{\varphi}_s^0(k,\bar{x}))|^2 < c\cdot\lambda(\bar{x})^{k-s}. \tag{I.59a}$$

Now

$$\begin{aligned}
&|\operatorname{cov}\{Q(k,\bar{x},\bar{\varphi}(k,\bar{x})),Q(s,\bar{x},\bar{\varphi}(s,\bar{x}))\}| \\
&\quad=|\operatorname{cov}\{Q(k,\bar{x},\bar{\varphi}_s^0(k,\bar{x}))+[Q(k,\bar{x},\bar{\varphi}(k,\bar{x})) \\
&\qquad -Q(k,\bar{x},\bar{\varphi}_s^0(k,\bar{x}))],Q(s,\bar{x},\bar{\varphi}(s,x))\}| \\
&\quad=|0+\operatorname{cov}\{[Q(k,\bar{x},\bar{\varphi}(k,\bar{x}))-Q(k,\bar{x},\bar{\varphi}_s^0(k,\bar{x}))], \\
&\qquad Q(s,\bar{x},\bar{\varphi}(s,x))\}| \leq c\cdot\lambda(\bar{x})^{k-s},
\end{aligned} \tag{I.59b}$$

where the second equality follows since $\bar{\varphi}_s^0(k,\bar{x})$ and $\bar{\varphi}(s,\bar{x})$ are independent, and the last inequality follows from Schwarz' inequality, (I.59a), and assumption B.7.

Hence, (I.58) holds and

$$\frac{1}{t}\sum_{k=1}^{t}[Q(k,\bar{x},\bar{\varphi}(k,\bar{x}))-EQ(k,\bar{x},\bar{\varphi}(k,\bar{x}))]\to 0 \quad \text{w.p.1 as } t\to\infty$$

which according to assumption A.5 = B.6 implies that

$$\frac{1}{t}\sum_{k=1}^{t}Q(k,\bar{x},\bar{\varphi}(k,\bar{x}))\to f(\bar{x}) \qquad \text{w.p.1 as } t\to\infty. \tag{I.60}$$

Notice that (I.60) holds for any given $\bar{x}$ w.p.1. For a given realization ω outside a null set it does not immediately follow that it holds for all $x\in D_R$. To conclude that, we first note that (I.60) will hold w.p.1 simultaneously over a dense denumerable, subset of D_R. (The union of a denumerable number of null sets is a nullset.) Hence, for a given realization outside a null set (I.60) holds for all $\bar{x}$ in a dense subset of D_R. But since $Q(\cdot,\cdot,\cdot)$ and $f(\cdot)$ are continuous in x, we may extend this set to D_R itself.

The proof of the second part of Lemma 2 is analogous. (Under assumptions A it is trivial.)

End proof of Lemma 2.

The proof for general sequences $\{\gamma(n)\}$, subject to A.8, A.9 is achieved by first proving that

$$E\left|\sum_{k=n}^{m(n,\Delta\tau)}\gamma(k)[Q(k,\bar{x},\bar{\varphi}(k,\bar{x})) - EQ(k,\bar{x},\bar{\varphi}(k,\bar{x}))]\right|^{2p} < C\cdot\gamma(n)^p \tag{I.61}$$

and then proving convergence by the Borel–Cantelli Lemma and assumption A.7.

We shall need the estimate (I.61) in Appendix III, and let us therefore prove (I.61) for $p=1$. The case with general p is proven analogously, though with more technical labor. The details are given in [7]. Denote for short

$$Q_k=Q(k,\bar{x},\bar{\varphi}(k,\bar{x}))$$

and assume that Q_k is a scalar and that $EQ_k=0$. Then

$$\begin{aligned}
E\left(\sum_{k=n}^{m}\gamma(k)Q_k\right)^2 &= \sum_{k=n}^{m}\sum_{s=n}^{m}\gamma(k)\gamma(s)EQ_kQ_s \\
&\leq (n)^2C\sum_{k=n}^{m}\sum_{s=n}^{m}\lambda^{|k-s|}\leq C\gamma(n)^2(m-n)
\end{aligned} \tag{I.62}$$

where the first inequality follows from assumption A.8 and (I.59b). It remains now only to prove that

$$(m(n,\Delta\tau)-n)\leq c/\gamma(n). \tag{I.63}$$

Since $\gamma(\cdot)$ is decreasing

$$2\Delta\tau\geq\sum_{n}^{m}\gamma(k)\geq(m-n)\gamma(m)$$

or

$$(m-n)\leq c/\gamma(m). \tag{I.64}$$

Moreover, from assumption A.9

$$\frac{1}{\gamma(n+1)}-\frac{1}{\gamma(n)}<c$$

we have

$$\gamma(n+1)>\gamma(n)(1-c\gamma(n+1))$$

or, upon repetition

$$\gamma(m)>\gamma(n)\prod_{n+1}^{m}(1-c\gamma(j))$$

$$\sim\gamma(n)\exp\left(-c\cdot\sum_{n+1}^{m}\gamma(j)\right)\sim c'\gamma(n).$$

This, together with (I.64) implies (I.63) and the proof of (I.61) is complete for $p=1$.

Proof of Corollary to Theorem 1: The function $V(x)$ with the property (22), whose existence is assumed in the corollary plays the role of the Lyapunov function throughout the proof of Theorem 1. Since, in the formulation of the corollary there is no guarantee that $x(m(n,\Delta\tau))$ will belong to D_R even if $x(n)$ does, the possibility of a cluster point on the boundary of D_R cannot be excluded. Moreover, if $\bar{D}$ is unbounded the existence of a cluster point in $\bar{D}$ is not guaranteed. But if no cluster point exists, then the sequence $\{x(n)\}$ will tend to infinity, which in that case will be a boundary point of D_R.

End proof of corollary.

Appendix II
Proof of Theorem 2

To prove (25a) we first assume that $f(x^*)\neq 0$.

We first note that according to Lemma 2, assumptions A or B imply that (I.4)–(I.6) hold w.p.1.

Let $\Delta\tau(x^*)=\Delta\tau^*$ be "the sufficiently small" $\Delta\tau$ as defined in Lemma 1.

Take

$$\rho^* < \Delta\tau^*|f(x^*)|/4 \qquad \text{(II.1)}$$

and let $\Omega^*=\{\omega|x(t)\to\mathfrak{B}(x^*,\rho^*)\}$ with $P(\Omega^*)=P^*>0$.

If $x(t)$ converges to $\mathfrak{B}(x^*,\rho^*)$, it is in particular inside $\mathfrak{B}(x^*,2\rho^*)$ infinitely often. Therefore, there is a cluster point $\bar{x}$ inside $\mathfrak{B}(x^*,2\rho^*)$. Also, inside $\mathfrak{B}(x^*,2\rho^*)$ the observation equation (2) will be stable for sufficiently small ρ^*, according to Step 2 of the proof of Lemma 1.

Hence, we can select a subsequence $\{x(n_k)\}$ tending to $\bar{x}$, such that $|\varphi(n_k)|<C$, i.e., (I.3) holds. Then Lemma 1 implies that

$$x(m(n_k,\Delta\tau^*))\approx x(n_k)+f(x^*)\cdot\Delta\tau^*$$

asymptotically as $k\to\infty$ for $\omega\in\Omega^*$ except on a null set. This means according to (II.1) that $x(m(n_k,\Delta\tau^*))$ is outside $\mathfrak{B}(x^*,2\rho^*)$, and the assumed convergence is contradicted.

To illustrate the basic idea of the proof of (25b), consider the special case

$$Q(t,x,\varphi(t+1))=Ax+e(t+1)$$

where A is an $n|n$ matrix and $e(\cdot)$ is a sequence of independent random variables with zero mean values. Suppose that A has an eigenvalue λ with $\text{Re}\lambda>0$, and let L be a corresponding left eigenvector. Let $\tau(n)=Lx(n)$ and $\epsilon(n)=Le(n)$. The condition on $\text{cov}\,Q$ implies that $\{\epsilon(n)\}$ is not identically zero. Then the algorithm (1) can be written

$$\tau(n+1)=\tau(n)+\gamma(n+1)\big[\lambda\tau(n)+\epsilon(n+1)\big]$$

and

$$\tau(m)=\Gamma(n,m)\cdot\left\{\tau(n)+\sum_{n+1}^{m}\tilde{\beta}_k^m\epsilon(k)\right\}$$

where

$$\Gamma(n,m)=\prod_{n+1}^{m}(1+\lambda\gamma(k))\sim\exp\left\{\lambda\sum_{n+1}^{m}\gamma(k)\right\} \qquad \text{(II.2)}$$

and

$$\tilde{\beta}_k^m=\gamma(k)\prod_{n+1}^{k}(1+\lambda\gamma(j))^{-1}.$$

Since $\tau(n)$ and the sum of random variables are independent and $\Gamma(n,m)$ tends to infinity as m increases, it follows that $\tau(m)$ will, with probability one, not tend to zero as m tends to infinity. Hence, $x(m)$ will not converge to 0 $(=x^*)$ with nonzero probability.

The general case is proven by linearization around x^* and the additional terms are taken care of by appropriate approximations. Like in the proof of Lemma 1, this leads to several technicalities, but the basic idea remains the same as above.

To keep the Appendices within reasonable size, these technicalities will not be given here. The full proof can be found in [7]. Proofs for related algorithms are given in [6] and [1].

Appendix III
Proof of Theorem 3

The idea of this proof is to apply Lemma 1 to obtain local estimates of how much $x(t)$ differs from the corresponding trajectory, and then linking such estimates together making use of the stability property.

Heuristic Outline of How the Estimates are linked Together

The idea of how the local estimates are extended to global ones can be geometrically expressed as follows, cf. Fig. 5.

Assume that the estimate at time τ_k is in the interval A. The trajectories that start in A belong at time $\tau_k+\Delta\tau_k$ to the interval B which is smaller than A since the trajectory is stable. Now, the estimates obtained by the algorithm differ from the trajectories with a small quantity according to Lemma 1. Denote this distance by C. During the time interval $\Delta\tau_k$, the estimates have not diverged from the nominal trajectory if $A\leqslant B+2C$.

To achieve this, A and $\Delta\tau_k$ must be chosen with care. The interval $\Delta\tau_k$ must be large enough to let the trajectories converge sufficiently, and small enough to limit second-order effects and the noise influence.

The formal proof will be developed in 5 steps. In Step 1 the details of the application of Lemma 1 are given. Step 2 deals with the implications of the stability assumption. In Step 3 an interval for $\Delta\tau_k$ is selected, corresponding to $\Delta\tau_k$ not being too small nor too large as described in the heuristic outline. In Step 4 it is shown that the estimates stay with the given ϵ-region as depicted in Fig. 5 if certain stochastic variables are less than a given value. Step 5 calculates the probability that they are less than this value.

Step 1—Application of Lemma 1:

Order the set of indices $I=\{n_i\}$ such that

$$n_1<n_2<\cdots<n_k<n_{k+1}<\cdots.$$

Denote $\Delta\tau_k=\tau_{n_{k+1}}-\tau_{n_k}$. Then by taking $\bar{x}=x(n_k)$ in Lemma 1 we obtain

$$x(n_{k+1})=x(n_k)+\Delta\tau_k f[x(n_k)] + q_1(n_k,n_{k+1},\bar{x})+q_2(n_k,n_{k+1},\bar{x}) \qquad \text{(III.1)}$$

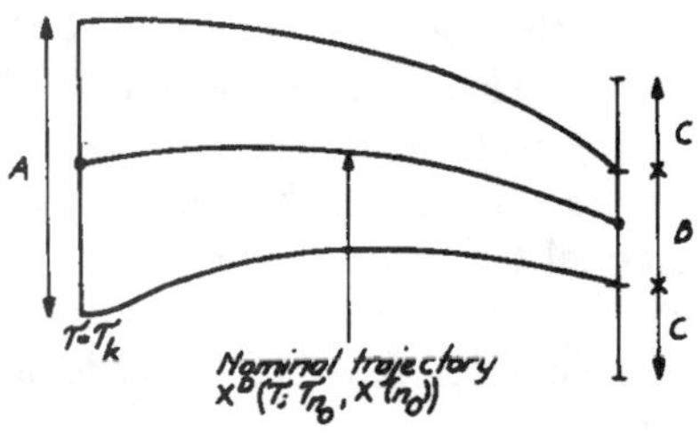

Fig. 5. To illustrate the idea of the proof of Theorem 3.

where $q_2(n_k, n_{k+1}, \bar{x})$ is given by (I.6) and (I.39). Since $x_{n_k} = x$,

$$|q_2(n, \Delta\tau_k; \bar{x})| \leq C_2 \Delta\tau_k^2. \tag{III.2}$$

Moreover $q_1(n_k, n_{k+1}, \bar{x})$ is given by (I.38).

From Lemma 1 we also know that (III.1), (III.2) are valid only if $\Delta\tau_k$ is sufficiently small and n_k so large that

$$|q_1(n_k, n_{k+1}, x(n_k))| < \rho/2. \tag{III.3}$$

We first note that the constant C_2 in (III.2) can be taken to be globally valid in $\bar{D}$, according to the expression (I.39) and since $|f(x)|$ and k_v are bounded in $\bar{D}$. Moreover, there is a common lower bound to $\Delta\tau_0(x)$ for $x \in \bar{D}$, which can be realized as follows.

The radius 2ρ, which is so small that (I.12) implies (I.11) depends on $\bar{x}$ and is a measure of how fast $A(x)$ changes in a neighborhood of $\bar{x}$. Since $A(x)$ is Lipschitz continuous in $\bar{D}$ and $\bar{D}$ is compact, the radius $\rho(\bar{x})$ will have a positive lower bound as $\bar{x}$ varies over $\bar{D}$. Denote this by $\bar{\rho}$. Let [cf., (I.34)]

$$\overline{\Delta\tau}_0 = \inf_{x \in D} \left(\frac{\bar{\rho}}{|f(\bar{x})|}, \frac{1}{16\bar{k}_v} \right) \tag{III.4}$$

which is strictly positive, since $|f(\bar{x})|$ and $\bar{k}_v$ are bounded in $\bar{D}$.

Then (III.1), (III.2) is valid for $\Delta\tau_k < \overline{\Delta\tau}_0$ if (III.3) holds.

By solving the differential equation (19) from τ to $\tau + \Delta\tau$ with $\bar{x}$ as the initial condition we have

$$|x^D(\tau + \Delta\tau; \tau, \bar{x}) - (\bar{x} + \Delta\tau \cdot f(\bar{x}))| \leq L\Delta\tau^2 \tag{III.5}$$

where L can be taken globally in $\bar{D}$. Introduce the following abbreviated notation

$$x_i^D(j) = x^D(\tau_{n_j}; \tau_{n_i}, x(n_i)). \tag{III.6}$$

Equations (III.1), (III.2), and (III.5) now imply in the notation of (III.6)

$$|x(n_{k+1}) - x_k^D(k+1)| \leq (c_2 + L)\Delta\tau^2 + q_1(n_{k+1}, n_k, x(n_k)) \tag{III.7}$$

if (III.3) holds. Introduce

$$M = C_2 + L. \tag{III.8}$$

The estimate (III.7) with the expression (I.38) for q_1 is the basis for the rest of the proof.

End Step 1.

Step 2—Stability of the Trajectories

According to the assumptions of the theorem there exists a function $V(\Delta x, \tau)$ that is quadratic in Δx and such that

$$\frac{d}{d\tau} V(\Delta x, \tau) < -C|\Delta x|^2 \qquad C > 0 \tag{III.9}$$

along solutions of the variational equation of (19). Since we assume $V(\Delta x, \tau)$ to be quadratic in Δx it is no additional loss of generality to assume that

$$V(\Delta x, \tau) = |\Delta x|^2$$

since it is always possible to make a (time-dependent) change of metric.

Then (III.9) implies that

$$|x^D(\tau + \Delta\tau; \tau, x + \Delta x) - x^D(\tau + \Delta\tau; \tau, x)| \leq (1 - \lambda\Delta\tau)|\Delta x| \tag{III.10}$$

for some $1 > \lambda > 0$.

End Step 2.

Step 3—Selection of an Interval for $\Delta\tau_k$

To obtain the upper and lower bounds for $\Delta\tau_k$ discussed under the heuristic outline, let

$$\epsilon_0 = \min\left(\frac{4\delta_0 M}{\lambda}, \frac{\overline{\Delta\tau}_0 M 4}{3\lambda}, \frac{4}{\lambda}\sqrt{\rho M} \right). \tag{III.11}$$

Choose $\epsilon < \epsilon_0$. Then automatically

$$\lambda\epsilon/8M < \Delta\tau_k \qquad \text{all } k \tag{III.12}$$

since $\Delta\tau_k > \delta_0$ according to the assumptions of the theorem. An upper bound for $\Delta\tau_k$ can be obtained by possibly extending the set I, cf., beginning of Step 1. Do this so that

$$\frac{\lambda\epsilon}{4M} < \Delta\tau_k < \frac{3\lambda\epsilon}{4M}. \tag{III.13}$$

The resulting set I may depend on ϵ. According to (III.11), automatically $\Delta\tau_k < \overline{\Delta\tau}_0$.

End Step 3.

Step 4—The Estimates Remain in an ϵ-Neighborhood of the Corresponding Trajectory if $q_1(n_{k+1}, n_k, x(n_k))$ is Sufficiently Small, all $n_k \in I$

Assume that

$$|q_1(n_{k+1}, n_k, x(n_k))| < r(\epsilon) \tag{III.14a}$$

where

$$r(\epsilon) = \frac{\lambda^2 \epsilon^2 \cdot 3}{M \cdot 16}. \tag{III.14b}$$

According to (III.11),

$$r(\epsilon) < \delta/2. \tag{III.14c}$$

Assume also that

$$|x_0^D(k) - x(n_k)| \leq \epsilon. \tag{III.15}$$

Since $\Delta\tau_k$ now is assumed to be less than $\overline{\Delta\tau}_0$ and since (III.14) is assumed to hold, Lemma 1 can be applied, giving according to (III.7) and (III.14) [notation as in (III.6)]

$$|x(n_{k+1})-x_k^D(k+1)|\leq M\cdot\Delta\tau_k^2+r(\epsilon). \quad \text{(III.16)}$$

Now

$$\begin{aligned}|x_0^D(k+1)-x(n_{k+1})|&\leq|x_0^D(k+1)-x_k^D(k+1)|\\&\quad+|x_k^D(k+1)-x(n_{k+1})|\\&\leq(1-\lambda\Delta\tau_k)|x_0^D(k)-x(k)|\\&\quad+M\cdot\Delta\tau_k^2+r(\epsilon)\\&\leq(1-\lambda\Delta\tau_k)\epsilon+M\cdot\Delta\tau_k^2+\frac{\lambda^2\epsilon^2\cdot 3}{M\cdot 16}\\&=\epsilon+M\left[\left(\Delta\tau_k-\frac{\lambda\epsilon}{4M}\right)\left(\Delta\tau_k-\frac{3\lambda\epsilon}{4M}\right)\right]\leq\epsilon \quad \text{(III.17)}\end{aligned}$$

where the second inequality follows from (III.10) and (III.16), the third from (III.15) and (III.14b), and the last one from (III.13).

The conclusion is that if $x(n_k)$ lies in an ϵ-neighborhood of the trajectory, so will $x(n_{k+1})$ if (III.14) holds. If (III.14) holds for all $n_k\in I$, then

$$\sup_{n_k\in I}|x(n_k)-x_0^D(k)|\leq\epsilon. \quad \text{(III.18)}$$

End Step 4.

It remains now only to estimate the probability that (III.14) holds for all $n_k\in I$.

Step 5—The Probability That (III.14) Holds

The expression for $q_1(n_{k+1},n_k,x(n_k))$ may according to (I.38) and (I.22) be decomposed as follows

$$\begin{aligned}|q_1(m,n,\bar{x})|\leq&\left|\sum_{k=n}^{m}\gamma(k)\right.\\&\left.\cdot\left[Q(k,\bar{x},\bar{\varphi}(k,\bar{x}))-EQ(k,\bar{x},\bar{\varphi}(k,\bar{x}))\right]\right|\\&+\sum_{k=n}^{m}\gamma(k)\mathcal{K}_1(\bar{x},\bar{\varphi}(k,\bar{x}),2\bar{p},v(k,\lambda,c))\\&\cdot\lambda^{k-n}v(n,\lambda,c)\\&+\left|\Delta\tau f(\bar{x})-\sum_{k=n}^{m}\gamma(k)EQ(k,\bar{x},\bar{\varphi}(k,\bar{x}))\right|. \quad \text{(III.19)}\end{aligned}$$

The last term in this expression is deterministic and tends to zero according to assumption A.5 = B.6 and according to (I.1) ($m=m(n,\Delta\tau)$). Let N_1 be such that this last term is less than $r(\epsilon)/2$ for $n_k>N_1$.

For the first term of (III.19) the estimate of its $2p$-absolute moment, (I.61) holds. The second term of (III.19) has a mean that is bounded by

$$C\cdot\sum_{n}^{m}\gamma(k)\lambda^{k-n}\leq\frac{c}{1-\lambda}\gamma(n)$$

according to assumptions B.7 (or A.3) and B.10. Analogously to the first term of (III.19) its $2p$-absolute moment around the mean is bounded by $C\cdot\gamma(n)^p$. (Much better estimates are possible, but uninteresting.) Denote the two first terms of (III.19) by $q_1^A(m,n,\bar{x})$ and the last one by $q_1^B(m,n,\bar{x})$. Then

$$q_1(m,n,\bar{x})=q_1^A(m,n,\bar{x})+q_1^B(m,n,\bar{x}) \quad \text{(III.20)}$$

where

$$E|q_1^A(n_{k+1},n_k,x(n_k))|^{2p}\leq C\cdot\gamma(n_k)^p \quad \text{(III.21)}$$

and

$$|q_1^B(n_{k+1},n_k,x(n_k))|<r(\epsilon)/2, \qquad n_k>N_1. \quad \text{(III.22)}$$

Chebyshev's inequality gives, using (III.21)

$$P\left(|q_1^A(n_{k+1},n_k,x(n_k))|>r(\epsilon)/2\right)\leq\left(\frac{2}{r(\epsilon)}\right)^{2p}\cdot C\cdot\gamma(n_k)^p. \quad \text{(III.23)}$$

The probability that (III.14) does not hold for some $n_k>N_1$, $n_k\in I$ is thus bounded by

$$\left(\frac{2}{r(\epsilon)}\right)^{2p}\cdot C\cdot\sum_{n_k=N_1}^{N}\gamma(n_k)^p\leq\left(\frac{2}{r(\epsilon)}\right)^{2p}\cdot C\cdot\sum_{j=N_1}^{N}\gamma(j)^p. \quad \text{(III.24)}$$

Combining the conclusion of Step 4 with this result gives the assertion of the theorem. The number T_0 mentioned in the theorem equals N_1 defined by (III.22).

Remark: The inequality in (III.24) is somewhat wasteful. For example, if $\gamma(t)=1/t$, then with

$$\Delta\tau_k=\sum_{t=n_k}^{n_{k+1}}\frac{1}{t}$$

implies that $n_{k+1}\approx n_k(1+\Delta\tau_k)$ for small $\Delta\tau_k$ and taking $p=1$ in (III.24) an upper bound for the LHS of (III.24) is given by

$$\left(\frac{2}{r(\epsilon)}\right)^2\cdot C\cdot\sum_{n_k=N_1}^{N}\frac{1}{N_1(1+\overline{\Delta\tau}_0)^k}\leq\frac{1}{N_1}\left(\frac{2}{r(\epsilon)}\right)^2\cdot C\cdot\frac{4M}{\lambda\epsilon}\leq\bar{c}/\epsilon^5. \quad \text{(III.25)}$$

End proof of Theorem 3.

Appendix IV
Proof of Theorem 4

In virtue of the projection we know that $x(t)$ belongs to a compact area i.o. that is part of D_R. We could therefore apply Theorem 1 directly, apart from the fact that the projection algorithm (28), (29) differs from the algorithm (1), (2) treated in Theorem 1.

It therefore suffices to show that the "projection" takes place at most a finite number of times w.p.1. After the last time $x(t)$ is forced into D_2 the projection algorithm coincides with the basic algorithm (1), (2) and the proof of Theorem 1 is valid.

If indeed, the estimate $x(t)$ were outside D_1 infinitely often, then it would have to pass from D_2 to outside D_1 i.o., i.e., to a higher value of $U(x(t))$ (see (31)) in spite of the force trying to decrease U according to (30). In Step 3 of the proof of Theorem 1 it was proved that this is impossible, and hence the projection facility in (28) is used only a finite number of times. Also, the estimates cannot remain in $\tilde{D}$ from a certain time on, since condition (30) shows that (using Step 2 of the proof of Theorem 1) they will be forced into D_2.

Appendix V
On Some Extensions

Continuous Time Algorithms: It should be clear that the theorems and proofs given here are valid also for continuous time algorithms, with proper and straightforward modifications.

The Case when Assumption A.5 = B.6 does not hold: The basic assumption, defining the RHS of the differential equation is A.5 = B.6.

Let $EQ(t,\bar{x},\bar{\varphi}(t,\bar{x}))$ be denoted by $f(\bar{x},t)$. In the proof it is only used that

$$\sum_{k=n}^{m(n,\Delta\tau)} \gamma(k) f(\bar{x},k) \to \Delta\tau f(\bar{x}) \quad \text{as } n\to\infty \qquad \text{(V.1)}$$

which, for the case $\gamma(k)=1/k$ is equivalent to

$$\frac{1}{t}\sum_{k=1}^{t} f(\bar{x},k)\to f(\bar{x}) \quad \text{as } t\to\infty. \qquad \text{(V.2)}$$

Notice that (V.1) or (V.2) very well may hold even if A.5 = B.6 does not hold. It is consequently unnecessarily restrictive to assume convergence of $f(\bar{x},t)$ as $t\to\infty$. Moreover, if (V.1) does not hold, it is sufficient to require the existence of a twice differentiable function $V(s)$ [which plays the role of Lyapunov function for the in this case undefined differential equation] such that

$$V'(\bar{x})\left[\sum_{k=n}^{m(n,\Delta\tau)} \gamma(k) f(\bar{x},k)\right] < -\delta(\bar{x})\cdot\Delta\tau \qquad \text{(V.3)}$$

for all sufficiently large n, where $\delta(\bar{x})>0$ for $\bar{x}\notin D_c$.

More Complex Generation of the Observations: In some cases, like for the extended Kalman filter, which is presently being analyzed using the methods of this paper, a more complex mechanism replaces (2):

$$\varphi(t)=A(z(t-1))\varphi(t-1)+B(z(t-1))e(t) \qquad \text{(V.4a)}$$

$$z(t)=h(z(t-1),x(t)). \qquad \text{(V.4b)}$$

Equation (V.4) if of course just a special case of the general, nonlinear dynamics (3). To treat this case let $\bar{z}(\bar{x})$ denote the limit of the recursion

$$z(t,\bar{x})=h(z(t-1,\bar{x}),\bar{x}), \qquad t\to\infty \qquad \text{(V.5)}$$

and define $\bar{\varphi}(t,\bar{x})$ through

$$\bar{\varphi}(t,\bar{x})=A(\bar{z}(\bar{x}))\bar{\varphi}(t-1,\bar{x})+B(\bar{z}(\bar{x}))e(t). \qquad \text{(V.6)}$$

The theorems will then hold in the same formulations as before, if the following three conditions on (V.4b) hold:

$$|z(t)-z(t,\bar{x})|\leq C\cdot \max_{r\leq k\leq t}|\bar{x}-x(k)| \qquad \text{if } z(r)=z(r,\bar{x}) \qquad \text{(V.7)}$$

$$|\bar{z}(\bar{x})-z(t,\bar{x})|\leq c\mu^t\cdot|\bar{z}(\bar{x})-z(0,\bar{x})|; \quad \mu<1 \qquad \text{(V.8)}$$

and

$$|z(t)|+|\varphi(t)|<c \text{ i.o.} \qquad \text{(V.9)}$$

To prove this we note that assumption A.4, (V.7) and (V.8) imply that

$$|A(z(j))-A(\bar{z})|\leq c \max_{n\leq k\leq j}|x(k)-\bar{x}| + c\mu^{j-n}|\bar{z}(\bar{x})-z(n)|.$$

If we use this estimate in (I.16) together with (V.9) we arrive at the same expression as in (I.17) and the proof continues without further changes.

Acknowledgment

I would like to express my sincere gratitude to Profs. K. J. Åström, H. J. Kushner and Ya. Z. Tsypkin for important discussions on the subject of this paper. I have also benefitted from the referees' competent suggestions on how to improve the readability of the Appendices.

References

[1] L. Ljung, T. Söderstrom, and I. Gustavsson, "Counterexamples to general convergence of a commonly used recursive identification method," *IEEE Trans. Automat. Contr.*, vol. AC-20, pp. 643–652, Oct. 1975.

[2] K. J. Åström, U. Borisson, L. Ljung, and B. Wittenmark, "Theory and applications of adaptive regulators based on recursive identification," in *Proc. 6th IFAC Congress*, Boston, MA, 1975. An extended version of this paper will appear in *Automatica*, Sept. 1977.

[3] L. Ljung and S. Lindahl, "Convergence properties of a method for state estimation in power systems," *Int. J. Contr.*, vol. 22, pp. 113–118, July 1975.

[4] L. Ljung and B. Wittenmark, "Analysis of a class of adaptive regulators," in *Proc. IFAC Symp. on Stochastic Control*, Budapest, Sept. 1974.

[5] L. Ljung, "Convergence of recursive, stochastic algorithms," in *Proc. IFAC Symp. on Stochastic Control*, Budapest, Sept. 1974.

[6] —— "Strong convergence of a stochastic approximation algorithm," *Annals of Statistics*, to be published; also available as Rep. LiTH-ISY-I-0126, Linköping University, Linköping, Sweden.

[7] —— "Theorems for the asymptotic analysis of recursive, stochastic algorithms," Dep. Automat. Contr., Lund Institute of Technology, Lund, Sweden, Rep. 7522.

[8] —— "Convergence of recursive stochastic algorithms," Division of Automatic Control, Lund Institute of Technology, Lund, Sweden, Rep. 7403, Feb. 1974.

[9] W. Hahn, *Stability of Motion*. Berlin, Germany: Springer-Verlag, 1967.

[10] R. W. Brockett, *Finite Dimensional Linear Systems*. New York: Wiley, 1970.

[11] H. Robbins and S. Monro, "A stochastic approximation method," *Ann. Math. Stat.*, vol. 22, pp. 400–407, 1951.

[12] J. Blum, "Multidimensional stochastic approximation methods," *Ann. Math. Stat.*, vol. 25, pp. 737–744, 1954.

[13] M. A. Aizermann, E. M. Braverman, and L. I. Rozonoer, "Metod Potentsialnykh funktsij v teorii obuchenija mashin" (The method of potential functions in the theory of machine learning), *Izd. Nauka*, Moscow, 1970 (in Russian).

[14] M. T. Wasan, *Stochastic Approximation*. Cambridge University Press, 1969.

[15] Ya. Z. Tsypkin, *Adaption and Learning in Automatic Systems*. New York: Academic, 1971.

[16] ——, *Foundations of the Theory of Learning Systems*. New York: Academic, 1973.

[17] J. Kiefer and J. Wolfowitz, "Stochastic estimation of the maximum of a regression function," *Ann. Math. Stat.*, vol. 23, pp. 462–466, 1952.

[18] H. J. Kushner, "Stochastic approximation algorithms for the local optimization of functions with non-unique stationary points," *IEEE Trans. Automat. Contr.*, vol. AC-17, pp. 646–655, 1972.

[19] H. J. Kushner and T. Gavin, "Stochastic approximation type methods for constrained systems: Algorithms and numerical results," *IEEE Trans. Automat. Contr.*, vol. AC-19, pp. 349–358, 1974.

[20] K. S. Fu, "Learning system theory," in *System Theory*, L. A. Zadeh and E. Polak, Eds. New York: McGraw-Hill, 1969, pp. 425–466.

[21] G. N. Saridis, Z. J. Nikolic, and K. S. Fu, "Stochastic approximation algorithms for systems identification, estimation and decomposition of mixtures," *IEEE Trans. Syst. Sci. Cybern.*, vol. SSC-5, pp. 8–15, 1969.

[22] Ya. Z. Tsypkin, "Self-learning—What is it?," *IEEE Trans. Automat. Contr.*, vol. AC-13, pp. 608–612, 1968.

[23] E. M. Braverman, "The method of potential functions in the problem of training machines to recognize patterns without a teacher," *Automat. Remote Contr.*, vol. 27, pp. 1748–1770, 1966.

[24] K. J. Åström and P. Eykhoff, "System identification—A survey," *Automatica*, vol. 7, pp. 123–164, 1971.

[25] J. M. Mendel, *Discrete Techniques of Parameter Estimation*. New York: Marcel Dekker, 1973.

[26] L. L. Scharf and D. C. Farden, "Optimum and adaptive array processing in frequence-wavenumber space," in *Proc. IEEE Conf. on Decision and Control*, Phoenix, AZ, 1974, pp. 604–609.

[27] D. C. Farden, "Stochastic approximation with correlated data," Dep. Elect. Eng., Colorado State University, Fort Collins, CO, ONR Tech. Rep. 11.

[28] I. D. Landau, "Unbiased recursive identification using model reference adaptive techniques," *IEEE Trans. Automat. Contr.*, vol. AC-21, pp. 194–202, Apr. 1976.

[29] L. Ljung, "On positive real transfer functions and the convergence of some recursive schemes," this issue, pp. 539–551.

[30] T. Söderström, L. Ljung, and I. Gustavsson, "A comparative study of recursive identification methods," Automatic Control, Lund Institute of Technology, Lund, Sweden, Rep. 7427, 1974.

[31] L. Ljung, "Convergence of an adaptive filter algorithm," *Int. J. Contr.*, 1977, to be published.

[32] K. J. Åström and B. Wittenmark, "On self-tuning regulators," *Automatica*, vol. 9, pp. 185–199, 1973.

[33] L. Ljung and B. Wittenmark, "On a stabilizing property of adaptive regulators," in *Preprints 4th IFAC Symp. Identification*, Part 3, Tbilisi, USSR, Paper 19.1, pp. 380–389, 1976.

[34] K. J. Åström and B. Wittenmark, "Analysis of a self-tuning regulator for nonminimum phase systems," in *Proc. IFAC Symp. on Stochastic Control*, Budapest, 1974.

[35] D. P. Derevitskii and A. L. Fradkov, "Two models for analyzing the dynamics of adaptation algorithms," *Automat. and Remote Contr.*, vol. 35, pp. 59–67, 1974.

[36] H. J. Kushner, "General convergence results for stochastic approximations via weak convergence theory," Brown University, Providence, RI, LCDS Tech. Rep. 76-1, 1976.

[37] ——, "Convergence of recursive adaptive and identification procedures via weak convergence theory," *IEEE Trans. Automat. Contr.*, to be published.

[38] H. Cramér and M. R. Leadbetter, "*Stationary and Related Stochastic Processes*. New York: Wiley, 1967.

[39] N. N. Krasovskij, *Stability of Motion*. Stanford, CA: Stanford University Press, 1963.

[40] W. Hahn, *Stability of Motion*. Berlin, Germany: Springer-Verlag, 1967.

[41] V. I. Zubov, *Methods of A. M. Lyapunov and their Applications*. Groningen, Holland: Noordhoff, 1964.

Lennart Ljung (S'74–M'75), for a photograph and biography see page 551 of this issue.

Discrete-Time Multivariable Adaptive Control

GRAHAM C. GOODWIN, PETER J. RAMADGE, AND PETER E. CAINES

ADAPTIVE control, so great is its appeal, has been studied for almost forty years. Its early history is one of diverse, but interesting, heuristic endeavors. One of the first problems studied, and one that has formed the focus of much subsequent research, is that of *adaptive model reference control* in which the controller employs an estimate of the unknown plant to cause the output y to track $y^* = Hr$, where r is the reference, H a prespecified reference model, and y^* the desired output. Another consistent theme is *certainty equivalence*, regarding (and using) the estimated plant as if it were the true plant; the motivation is obvious . . . if the estimated parameters converge to the true parameters, the certainty equivalent controller converges to a satisfactory (optimal) controller for the true plant. However, adaptive controllers can perform satisfactorily even if, as is often the case, the estimated parameters do *not* converge to their true values; this was dramatically shown in the paper by Åström and Wittenmark [1], also included in this volume. Early use of Lyapunov theory to establish stability was restricted to minimum phase plants with relative degree one. Plants with a relative degree higher than one posed a major problem [17] that was not satisfactorily resolved until the late 1970s; indeed, the paper by Goodwin, Ramadge, and Caines was a major contribution to the resolution of this formidable problem and a substantial stimulus to subsequent research.

Suppose the system to be controlled is described by

$$Ay = Bu$$

where A and B are polynomials of degree n and m, respectively, in the shift operator q, and A is monic; for continuous-time systems q is replaced by the derivative operator d/dt. The system may be re-parameterized (as shown in [1] and [15], the latter in 'state-space language') to obtain

$$Ey = Ry + Su$$

where $E = E_1E_2$ is monic and Hurwitz, and E_1 and E_2 have degrees n and m, respectively. Hence (ignoring exponentially decaying terms)

$$y = Ry_E + Su_E = \phi^T\theta$$

where $y_E := (1/E)y$, $u_E := (1/E)u$, θ is a vector of the coefficients of the polynomials R and S, and the regressor vector ϕ has components $q^i y_E$, $i = 0, \ldots, n-1$, and $q^i u_E$, $i = 0, \ldots, n$. The estimation equation $y = \phi^T\theta$ may be used in many different ways to provide an estimate $\hat{\theta}$ of the unknown parameter θ. Suppose the desired closed-loop transfer function is $H = (N/E)$ so that the desired output is $y^* = Hr$, and that $b_m = s_n = 1$ is known. To determine the control, the system equation may be written in the form

$$E_2y = Ry_{E_1} + Su_{E_1} = \psi^T\theta$$

where $\psi := E_2\phi$ (the vector ψ has components $q^i y_{E_1}, i = 0, \ldots, n-1$, and $q^i u_{E_1}$, $i = 0, \ldots, n$). The degree of E_1 is n so that $\psi^T\theta$ may be written as $\eta^T\theta + u$. If θ is known (and the system is minimum phase), the control u may be chosen (as $u = E_2y^* - \eta^T\theta$) to cause the output to satisfy

$$E_2y = \eta^T\theta + u = E_2y^*$$

so that the *tracking error* $y - y^*$ decays to zero (and all signals remain bounded). If θ is unknown, a tempting strategy is to employ the certainty equivalent control $u = E_2y^* - \eta^T\hat{\theta}$ with the result that $E_2y^* = \psi^T\hat{\theta}$, and the output is now given by

$$E_2y = E_2y^* + \psi^T\tilde{\theta}$$

The tracking error $e_T := y - y^*$ satisfies

$$e_T = e + \hat{e_T}$$

where $\hat{y} := \phi^T\hat{\theta}$ is the estimated output, $e := y - \hat{y} = \phi^T\tilde{\theta}$ is the prediction error (Monopoli's augmented error), $\tilde{\theta} := \theta - \hat{\theta}$ is the parameter error, and $\hat{e_T} := (\hat{y} - y^*)$ is an estimate of the tracking error e_T. Since

$$\hat{e}_T = [(1/E_2)\psi^T]\hat{\theta} - (1/E_2)[\psi^T\hat{\theta}]$$

it is also known as the 'swapping error'; $\hat{e}_T$ is zero if $\hat{\theta}$ is constant, but otherwise depends on the rate of change of $\hat{\theta}$ (in the continuous-time case it is proportional to $(d/dt)\hat{\theta}$).

Monopoli [15] proposed use of the certainty equivalent control law in conjuction with a simple gradient estimator $(d/dt)\hat{\theta} = \phi e$; however, as was later pointed out, convergence of the tracking error to zero was not established except for the case of unity relative degree. The first proof of convergence (for continuous-time systems) was achieved in [4] by modifying the *controller*; nonlinear damping was added to ensure convergence of $\hat{e}_T$ to zero; this enabled convergence of e_T (and boundedness of all signals) to be proven. Because of the complexity of the controller, and of the accompanying analysis, the paper [4] did not have the impact it deserved. The second and much simpler proof (for discrete-time systems), presented in this paper by Goodwin, Ramadge, and Caines, achieved convergence by modifying the *estimator*; the rate of change of $\hat{\theta}$ was reduced (thereby reducing $\hat{e}_T$) by the introduction of error normalization (replacing e in the estimator algorithm by $\tilde{e} := e/[1 + |\phi|^2]^{1/2}$). Error normalization was independently proposed in [3].

The relevant result is Lemma 3.1 of the paper by Goodwin, Ramadge, and Caines, which essentially states:

(a) *Suppose the* estimator *is such that the normalized prediction error* $\tilde{e} = e/[1 + |\phi|^2]^{1/2}$ *lies in* ℓ_2.

(b) *Suppose the* controller *is such that the regressor vector* ϕ *satisfies the growth condition*

$$|\phi(t)|^2 \leq c[1 + \max\{|e(\tau)|^2 \mid \tau \in [0, t]\}$$

for all $t \geq 0$.
Then:

(i) $|\phi(t)|$ *is uniformly bounded, and*

(ii) $e(t) \to 0$ as $t \to \infty$.

The result is simple to state and prove. It provided a simple means for establishing convergence (and boundedness of all signals) for a wide range of adaptive controllers and contributed to an explosive re-awakening of interest in adaptive control. An important feature of the result is its modularity: condition (a) on the estimator can be established independently of (b), i.e., *independently of the control law*. The result was presented in an electrifying seminar at Yale University (New Haven, Connecticut) in late 1978 and was rapidly extended to the deterministic continuous-time case in [16] and [21]. All three papers appeared in the same issue of the *IEEE Transactions on Automatic Control*; ironically, the paper by Goodwin, Ramadge, and Caines was the only one that did not appear as a regular paper. A parameter estimation perspective of the continuous-time results was given in [5], providing an analogous modular decomposition of the conditions for convergence. Stochastic versions of the paper by Goodwin, Ramadge, and Caines and of [5] appeared, respectively, in [7] and [6]. Research on convergence and stability has continued to this day. For example, the earlier nonlinear damping approach of [4] was extended in [11] to deal with 'true' output nonlinearities for which certainty equivalent adaptive controllers are inadequate, and backstepping [13] was developed as a systematic tool for designing adaptive controllers for linear and nonlinear systems with high relative degree. Furthermore, switching was introduced to enforce convergence when structural properties (e.g., relative degree) are unknown [18], and modifications were made in the basic algorithm to ensure robustness [19], [9], [12]. Research results of this period were rapidly consolidated in texts such as [8], [20], [14], [2], [22], and [10].

The paper by Goodwin, Ramadge and Caines is an important milestone in the evolution of adaptive control. It contributed much to the richness of a subject that has progressed far and that now appears poised for further significant advances.

REFERENCES

[1] K. J. ÅSTRÖM AND B. WITTENMARK, "On self tuning regulators," *Automatica*, **9**:185–199, 1973.

[2] K. J. ÅSTRÖM AND B. WITTENMARK, *Adaptive Control*, Addison-Wesley (Reading, MA), 1989.

[3] B. EGARDT, *Stability of Adaptive Controllers*, Springer-Verlag (New York), 1979.

[4] A. FEUER AND S. MORSE, "Adaptive control of single input, single output linear systems," *IEEE Trans. Automat. Contr.*, **AC-23**(4):557–569, 1978.

[5] G. C. GOODWIN AND D. Q. MAYNE, "A parameter estimation perspective of continuous time adaptive control," *Automatica*, 23:57–70, 1987.

[6] G. C. GOODWIN AND D. Q. MAYNE, "Continuous-time stochastic model reference adaptive control," *IEEE Trans. Automat. Contr.*, **AC-36**(11): 1254–1263, November 1991.

[7] G. C. GOODWIN, P. J. RAMADGE, AND P. E. CAINES, "Discrete time stochastic adaptive control," *SIAM J. Contr. Optimiz.*, **19**(6):829–853, 1981.

[8] G. C. GOODWIN AND K. S. SIN, *Adaptive Filtering, Prediction and Control*, Prentice Hall (Englewood Cliffs, NJ), 1984.

[9] P. A. IOANNOU AND P. KOKOTOVIČ, *Adaptive Systems with reduced models*, Lecture Notes in Control and Information Sciences, Vol. 47, Springer-Verlag (New York), 1983.

[10] P. A. IOANNOU AND J. SUN, *Robust Adaptive Control*, Prentice Hall (Englewood Cliffs, NJ), 1989.

[11] I. KANELLAKOPOULOS, P. V. KOKOTOVIĆ, AND A. S. MORSE, "Adaptive output-feedback control of systems with output nonlinearities," *IEEE Trans. Automat. Contr.*, **AC-37**(11):1666–1682, 1992.

[12] G. KREISELLMEIER AND B. D. O. ANDERSON, "Robust model reference adaptive control," *IEEE Trans. Automat. Contr.*, **AC-31**(2): 127–132, February, 1986.

[13] M. KRSTIĆ, I. KANELLAKOPOULIS, AND P. V. KOKOTOVIĆ, *Nonlinear and Adaptive Control Design*, John Wiley, New York, 1995.

[14] P. R. KUMAR AND P. P. VARAIYA, *Stochastic Systems: Estimation, Identification and Adaptive Control*, Plenum Press (New York), 1986.

[15] R. V. MONOPOLI, "Model reference adaptive control with an augmented error signal," *IEEE Trans. Automat. Contr.*, **AC-19**:474–482, 1974.

[16] A. S. MORSE, "Global stability of parameter-adaptive control systems," *IEEE Trans. Automat. Contr.*, **AC-25**(3):433–439, 1980.

[17] A. S. MORSE, "Overcoming the obstacle of high relative degree," *Journal of the Society for Instrument and Control Engineers*, **34**:629–636, 1995.

[18] A. S. MORSE, D. Q. MAYNE, AND G. C. GOODWIN, "Applications of Hysteresis Switching in Parameter Adaptive Control," *IEEE Trans. Automat. Contr.*, **AC-37**(11):1343–1354, 1992.

[19] S. M. NAIK, P. R. KUMAR, AND B. E. YDSTIE, "Robust Continuous-Time Adaptive Control by Parameter Projection," *IEEE Trans. Automat. Contr.*, **AC-37**(2):182–197, 1992.

[20] K. S. NARENDRA, *Adaptive and Learning Systems—Theory and Applications*, Plenum Press (New York), 1986.

[21] K. S. NARENDRA, Y.-H. LIN, AND L. S. VALAVANI, "Stable adaptive controller design, Part II: Proof of stability", *IEEE Trans. Automat. Contr.*, **AC-25**(3):440–448, 1980.

[22] S. SASTRY AND M. BODSON, *Adaptive Control: Stability, Convergence and Robustness*, Prentice Hall (Englewood Cliff, NJ), 1989.

D.Q.M. & L.L.

Short Papers

Discrete-Time Multivariable Adaptive Control

GRAHAM C. GOODWIN, MEMBER, IEEE, PETER J. RAMADGE, AND PETER E. CAINES, MEMBER, IEEE

Abstract—This paper establishes global convergence for a class of adaptive control algorithms applied to discrete-time multiinput multioutput deterministic linear systems. It is shown that the algorithms will ensure that the system inputs and outputs remain bounded for all time and that the output tracking error converges to zero.

I. Introduction

A long-standing problem in control theory has been the question of the existence of simple, globally convergent adaptive control algorithms. By this we mean algorithms which, for all initial system and algorithm states, cause the outputs of a given linear system to asymptotically track a desired output sequence, and achieve this with a bounded-input sequence.

There is a considerable amount of literature on continuous-time deterministic adaptive control algorithms [1]. However, it is only recently that global stability and convergence of these algorithms has been studied under general assumptions. Much interest was generated by the innovative configuration proposed by Monopoli [2] whereby the feedback gains were directly estimated and an augmented error signal and auxiliary input signals were introduced to avoid the use of pure differentiators in the algorithm. Unfortunately, as pointed out in [3] the arguments given in [2] concerning stability are incomplete. New proofs for related algorithms have recently appeared [4], [5]. In [4] Narendra and Valavani treat the case where the difference in orders between the numerator and denominator of the system transfer function (relative degree) is less than or equal to two. In [5], Feuer and Morse propose a solution for general linear systems without constraints on the relative degree. The algorithms in [5] use the augmented error concept and auxiliary inputs as in [2]. The Feuer and Morse result seems to be the most general to date for single-input single-output continuous-time systems. However, these results are technically involved and cannot be directly applied to the discrete-time case.

There has also been interest in discrete-time adaptive control for both the deterministic and stochastic case. This area has particular relevance in view of the increasing use of digital technology in control applications [6], [7].

Ljung [8], [9] has proposed a general technique for analyzing convergence of discrete-time stochastic adaptive algorithms. However, in this analysis a question which is yet to be resolved concerns the boundedness of the system variables. For one particular algorithm [10], it has been argued in [11] that the algorithm possess the property that the sample mean-square output is bounded whenever the sample mean-square of the noise is bounded. However, the general question of stability remains unanswered for stochastic adaptive algorithms.

The study of discrete-time deterministic algorithms is of independent interest but also provides insight into stability questions in the stochastic case [12], [15]. Recent work by Ionescu and Monopoli [13] has been concerned with the extension of the results in [2] to the discrete-time case. As for the continuous case, the augmented error method is used.

In this paper we present new results related to discrete-time deterministic adaptive control. Our approach differs from previous work in several major respects although certain aspects of our approach are inspired by the work of Feuer and Morse [5].

The analysis presented here does not rely upon the use of augmented errors or auxiliary inputs. Moreover, the algorithms have a very simple structure and are applicable to multiple-input multiple-output systems with rather general assumptions.

The paper presents a general method of analysis for discrete-time deterministic adaptive control algorithms. The method is illustrated by establishing global convergence for three simple algorithms. For clarity of presentation, we shall first treat a simple single-input single-output algorithm in detail. The results will then be extended to other single-input single-output algorithms including those based on recursive least squares. Finally, the extension to multiple-input multiple-output systems will be presented.

Since the results in this paper were presented a number of other authors [16]–[18] have presented related results for discrete-time deterministic adaptive control algorithms.

II. Problem Statement

In this paper we shall be concerned with the adaptive control of linear time-invariant finite-dimensional systems having the following state space representation:

$$x(t+1)=Ax(t)+Bu(t); \qquad x(0)=x_0 \tag{2.1}$$

$$y(t)=Cx(t) \tag{2.2}$$

where $x(t)$, $u(t)$, $y(t)$ are the $n\times 1$ state vector, $r\times 1$ input vector, and $m\times 1$ output vector, respectively.

A standard result is that the system (2.1), (2.2) can also be represented in matrix fraction, or ARMA, form as

$$A(q^{-1})y(t)=\begin{bmatrix} q^{-d_{11}}B_{11}(q^{-1}) & \cdots & q^{-d_{1r}}B_{1r}(q^{-1}) \\ \vdots & & \vdots \\ q^{-d_{m1}}B_{m1}(q^{-1}) & \cdots & q^{-d_{mr}}B_{mr}(q^{-1}) \end{bmatrix}u(t) \tag{2.3}$$

with appropriate initial conditions. In (2.3), $A(q^{-1})$, $B_{ij}(q^{-1})$ ($i=1,\cdots,m$; $j=1,\cdots,r$) denote scalar polynomials in the unit delay operator q^{-1} and the factors $q^{-d_{ij}}$ represent pure time delays.

Note that it is not assumed that the system (2.1), (2.2) is completely controllable or completely observable, nor is it assumed that (2.3) is irreducible. The system will be required, however, to satisfy the conditions of Lemma 3.2.

It is assumed that the coefficients in the matrixes A, B, C in (2.1), (2.2) are unknown and that the state $x(t)$ is not directly measurable. A feedback control law is to be designed to stabilize the system and to cause the output, $\{y(t)\}$, to track a given reference sequence $\{y^*(t)\}$.

Manuscript received November 30, 1978; revised June 11, 1979 and November 26, 1979. Paper recommended by K. S. Narendra, Past Chairman of the Adaptive, Learning Systems, Pattern Recognition Committee. This work was supported in part by the Australian Research Grants Committee and the Joint Services Electronics Program under Contract N00014-75-C-0648. The work of G. C. Goodwin was supported in part by a Fulbright Grant and the Division of Applied Sciences, Harvard University, Cambridge, MA.

G. C. Goodwin is with the Department of Electrical Engineering, University of Newcastle, Newcastle, N.S.W., Australia.

P. J. Ramadge is with the Department of Electrical Engineering, University of Toronto, Toronto, Ont., Canada.

P. E. Caines is with the Division of Applied Sciences, Harvard University, Cambridge, MA 02138.

Reprinted from *IEEE Transactions on Automatic Control*, Vol. AC-25, 1980, pp. 449–456.

Specifically, we require $y(t)$ and $u(t)$ to be bounded uniformly in t, and

$$\lim_{t\to\infty} y_i(t)-y_i^*(t)=0 \qquad i=1,\cdots,m. \tag{2.4}$$

III. Key Technical Lemmas

Our analysis of discrete-time multivariable adaptive control algorithms will be based on the following technical results.

Lemma 3.1: If

$$\lim_{t\to\infty} \frac{s(t)^2}{b_1(t)+b_2(t)\sigma(t)^T\sigma(t)}=0 \tag{3.1}$$

where $\{b_1(t)\}$, $\{b_2(t)\}$, *and* $\{s(t)\}$ *are real scalar sequences and* $\{\sigma(t)\}$ *is a real p-vector sequence; then subject to*

1) uniform boundedness condition

$$0<b_1(t)<K<\infty \quad \text{and} \quad 0<b_2(t)<K<\infty \tag{3.2}$$

for all $t>0$ *and*

2) linear boundedness condition

$$\|\sigma(t)\|<C_1+C_2 \max_{0<\tau<t} |s(\tau)| \tag{3.3}$$

where $0<C_1<\infty$, $0<C_2<\infty$, *it follows that*

$$\lim_{t\to\infty} s(t)=0 \tag{3.4}$$

and $\{\|\sigma(t)\|\}$ *is bounded.*

Proof: If $\{s(t)\}$ is a bounded sequence, then by (3.3) $\{\|\sigma(t)\|\}$ is a bounded sequence. Then by (3.2) and (3.1) it follows that

$$\lim_{t\to\infty} s(t)=0.$$

Now assume that $\{s(t)\}$ is unbounded. It follows that there exists a subsequence $\{t_n\}$ such that

$$\lim_{t_n\to\infty} |s(t_n)|=\infty$$

and

$$|s(t)|<|s(t_n)| \qquad \text{for } t<t_n. \tag{3.5}$$

Now along the subsequence $\{t_n\}$

$$\left|\frac{s(t_n)}{\left[b_1(t_n)+b_2(t_n)\sigma(t_n)^T\sigma(t_n)\right]^{1/2}}\right| > \frac{|s(t_n)|}{\left[K+K\|\sigma(t_n)\|^2\right]^{1/2}} \quad \text{using (3.2)}$$

$$> \frac{|s(t_n)|}{K^{1/2}+K^{1/2}\|\sigma(t_n)\|}$$

$$> \frac{|s(t_n)|}{K^{1/2}+K^{1/2}[C_1+C_2|s(t_n)|]} \quad \text{using (3.3) and (3.5).}$$

Hence,

$$\lim_{t_n\to\infty}\left|\frac{s(t_n)}{\left[b_1(t_n)+b_2(t_n)\sigma(t_n)^T\sigma(t_n)\right]^{1/2}}\right| > \frac{1}{K^{1/2}C_2}>0$$

but this contradicts (3.1) and hence the assumption that $\{s(t)\}$ is unbounded is false and the result follows. □

In order to use this lemma in proving global convergence of adaptive control algorithms it will be necessary to verify (3.1) (with $s(t)$ interpreted as the tracking error) and to check that assumptions (3.2) and (3.3) are satisfied.

The next lemma will be used to verify that the linear boundedness condition (3.3) is satisfied by an important class of linear time-invariant systems. This class corresponds to those linear time-invariant systems for which the control objective (2.4) can be achieved with a bounded-input sequence and for which the tracking error can be reduced to zero if the system parameters are known.

Lemma 3.2: For the system (2.3) *with* $r=m$, *and subject to*

$$\det\begin{bmatrix} z^{d_{11}-d_1}B_{11}(z) & \cdots & z^{d_{1m}-d_1}B_{1m}(z) \\ \vdots & & \vdots \\ z^{d_{m1}-d_m}B_{m1}(z) & \cdots & z^{d_{mm}-d_m}B_{mm}(z) \end{bmatrix} \neq 0 \tag{3.6}$$

for $|z|<1$ *where*

$$d_i=\min_{1<j<m} d_{ij} \qquad i=1,\cdots,m,$$

if

$$\max_{\substack{0<t<T\\1<i<m}} |y_i(t+d_i)|=m_2, \tag{3.7}$$

then there exist constants m_3, m_4 *which are independent of* T *with* $0<m_3<\infty$, $0<m_4<\infty$ *such that*

$$|u_i(t)|<m_2m_4+m_3 \qquad 0<t<T,\ i=1,\cdots,m.$$

Proof: The result is standard and simply follows from the fact that (3.6) ensures that the system has a stable inverse. □

In the remainder of the paper these results will be used to prove global convergence of a number of adaptive control algorithms. Sections IV–VII will be concerned with adaptive control of single-input single-output systems. Sections VIII and IX will extend these results to the multiple-input multiple-output case.

IV. Single-Input Single-Output Systems

It is well known that for the single-input single-output (SISO) case, the system output of (2.1), (2.2) can be described by

$$A(q^{-1})y(t)=q^{-d}B(q^{-1})u(t) \tag{4.1}$$

where $\{u(t)\}$, $\{y(t)\}$ denote the input and output sequences, respectively, and $A(q^{-1})$, $B(q^{-1})$ are polynomial functions of the unit delay operator q^{-1},

$$A(q^{-1})=1+a_1q^{-1}+\cdots+a_nq^{-n}$$

$$B(q^{-1})=b_0+b_1q^{-1}+\cdots+b_mq^{-m}; \qquad b_0\neq 0.$$

d represents the system time delay. The initial conditions of (2.1) are replaced by initial values of $y(t)$, $0>t>-n$, and $u(t)$, $-d>t>-d-m$.

The following assumptions will be made about the system.

Assumption Set 4:

a) d is known.

b) An upper bound for n and m is known.

c) $B(z)$ has all zeros strictly outside the closed unit disk. (This is necessary to ensure that the control objective can be achieved with a bounded-input sequence.)

We note that, by successive substitution, (4.1) can be rewritten as

$$y(t+d)=\alpha(q^{-1})y(t)+\beta(q^{-1})u(t) \tag{4.2}$$

where

$$\alpha(q^{-1})=\alpha_0+\alpha_1q^{-1}+\cdots+\alpha_{n-1}q^{-n+1}$$

$$\beta(q^{-1})=\beta_0+\beta_1q^{-1}+\cdots+\beta_{m+d-1}q^{-m-d+1}; \qquad \beta_0\neq 0.$$

As previously stated, the control objective is to achieve

$$\lim_{t\to\infty}[y(t)-y^*(t)]=0 \tag{4.3}$$

where $\{y^*(t)\}$ is a reference sequence. It is assumed that $\{y^*(t)\}$ is known *a priori* and that

$$|y^*(t)|<m_1<\infty \qquad \text{for all } t. \tag{4.4}$$

V. SISO PROJECTION ALGORITHM I

Let θ_0 be the vector of system parameters (dimension $p=n+m+d$),

$$\theta_0^T=(\alpha_0,\cdots,\alpha_{n-1},\beta_0,\cdots,\beta_{m+d-1}). \tag{5.1}$$

Then (4.2) can be written

$$y(t+d)=\varphi(t)^T\theta_0 \tag{5.2}$$

where

$$\varphi(t)^T=(y(t),\cdots,y(t-n+1),u(t),\cdots,u(t-m-d+1)). \tag{5.3}$$

Now define the output tracking error as

$$\begin{aligned} e(t+d)&=y(t+d)-y^*(t+d)\\ &=\varphi(t)^T\theta_0-y^*(t+d). \end{aligned} \tag{5.4}$$

By choosing $\{u(t)\}$ to satisfy

$$\varphi(t)^T\theta_0=y^*(t+d) \tag{5.5}$$

it is evident that the tracking error is identically zero. However, since θ_0 is unknown, we replace (5.5) by the following adaptive algorithm:

$$\begin{aligned}\hat\theta(t)=\hat\theta(t-1)+a(t)\varphi(t-d)[1+\varphi(t-d)^T\varphi(t-d)]^{-1}\\ \cdot\left[y(t)-\varphi(t-d)^T\hat\theta(t-1)\right]\end{aligned} \tag{5.6}$$

$$\varphi(t)^T\hat\theta(t)=y^*(t+d) \tag{5.7}$$

where $\hat\theta(t)$ is a p-vector of reals depending on an initial vector $\hat\theta(0)$ and on $y(\tau)$, $0<\tau<t$, $u(\tau)$, $0<\tau<t-d$ via (5.6), and where the gain constant $a(t)$ is computed as follows:

$$\begin{aligned} a(t)&=1 \quad \text{if } [(n+1)\text{th component of right-hand side of (5.6) evaluated using } a(t)=1]\neq 0;\\ &=\gamma \quad \text{otherwise where } \gamma \text{ is a constant in the interval } (\epsilon,2-\epsilon),\ \gamma\neq 1 \text{ and } 0<\epsilon<1. \end{aligned} \tag{5.8}$$

This choice of gain constant prevents the computed coefficient of $u(t)$ in (5.7) being zero. We also remark that the purpose of the coefficient 1 in the term $[1+\phi(t-d)^T\phi(t-d)]^{-1}$ of (5.6) is solely to avoid division by zero. Any positive constant could be used in place of the 1.

Apart from the above modification, the algorithm (5.6) is an orthogonal projection of $\hat\theta(t-1)$ onto the hypersurface $y(t)-\varphi(t-d)^T\theta=0$.

In the analysis of this algorithm, we will first show that the Euclidean norm of the vector $\tilde\theta(t)=\hat\theta(t)-\theta_0$ is a nonincreasing function along the trajectories of the algorithm. This leads to a characterization of the limiting behavior of the algorithm which will allow us to use Lemma 3.1 to establish global convergence.

Lemma 5.1: Along the solutions of (5.6), (5.7),

$$\|\tilde\theta(t)\|^2-\|\tilde\theta(t-1)\|^2<0, \qquad t>0 \tag{5.9}$$

and

$$\lim_{t\to\infty}\frac{\varphi(t)^T\tilde\theta(t)}{[1+\varphi(t)^T\varphi(t)]^{1/2}}=0. \tag{5.10}$$

Proof: Using the definition of $\tilde\theta(t)$, (5.6) may be rewritten as

$$\begin{aligned}\tilde\theta(t)=\tilde\theta(t-1)-a(t)\varphi(t-d)[1+\varphi(t-d)^T\varphi(t-d)]^{-1}\\ \cdot\varphi(t-d)^T\tilde\theta(t-1).\end{aligned} \tag{5.11}$$

Hence,

$$\begin{aligned}\|\tilde\theta(t)\|^2-\|\tilde\theta(t-1)\|^2=a(t)\left[-2+a(t)\frac{\varphi(t-d)^T\varphi(t-d)}{[1+\varphi(t-d)^T\varphi(t-d)]}\right]\\ \cdot\frac{\tilde\theta(t-1)^T\varphi(t-d)\varphi(t-d)^T\tilde\theta(t-1)}{[1+\varphi(t-d)^T\varphi(t-d)]}<0\end{aligned} \tag{5.12}$$

for all values of $\varphi(t-d)$ provided $0<a(t)<2$. This is satisfied by definition (5.8). Then, since $\|\tilde\theta(t)\|^2$ is a bounded nonincreasing function it converges. Setting

$$\epsilon(t)=-\varphi(t-d)^T\tilde\theta(t-1) \tag{5.13}$$

[and noting that

$$a(t)\left[-2+a(t)\frac{\varphi(t-d)^T\varphi(t-d)}{[1+\varphi(t-d)^T\varphi(t-d)]}\right]$$

is bounded away from zero, with $a(t)$ defined as in (5.8)] we conclude, from (5.12), that

$$\lim_{t\to\infty}\frac{\epsilon^2(t)}{[1+\varphi(t-d)^T\varphi(t-d)]}=0$$

and hence

$$\lim_{t\to\infty}\frac{\epsilon(t)}{[1+\varphi(t-d)^T\varphi(t-d)]^{1/2}}=0. \tag{5.14}$$

Now using (5.13) and (5.11) it follows that

$$\epsilon(t)=-\varphi(t-d)^T\tilde\theta(t-d)-\sum_{i=1}^{d-1}a(t-i)\cdot\frac{\varphi(t-d)^T\varphi(t-d-i)}{[1+\varphi(t-d-i)^T\varphi(t-d-i)]}\epsilon(t-i).$$

Then using (5.4) and (5.7) we have that

$$e(t)=-\varphi(t-d)^T\tilde\theta(t-d). \tag{5.15}$$

Hence,

$$\begin{aligned}\frac{\epsilon(t)}{[1+\varphi(t-d)^T\varphi(t-d)]^{1/2}}=\frac{e(t)}{[1+\varphi(t-d)^T\varphi(t-d)]^{1/2}}\\ -\sum_{i=1}^{d-1}a(t-i)\frac{\varphi(t-d)^T}{[1+\varphi(t-d)^T\varphi(t-d)]^{1/2}}\\ \cdot\frac{\varphi(t-d-i)}{[1+\varphi(t-d-i)^T\varphi(t-d-i)]^{1/2}}\\ \cdot\frac{\epsilon(t-i)}{[1+\varphi(t-d-i)^T\varphi(t-d-i)]^{1/2}}.\end{aligned} \tag{5.16}$$

Now by the Cauchy–Schwarz inequality and the fact that $|a(t)|<2$

$$\begin{aligned}0<\left|\frac{a(t-i)\varphi(t-d)^T}{[1+\varphi(t-d)^T\varphi(t-d)]^{1/2}}\frac{\varphi(t-d-i)}{[1+\varphi(t-d-i)^T\varphi(t-d-i)]^{1/2}}\right.\\ \left.\cdot\frac{\epsilon(t-i)}{[1+\varphi(t-d-i)\varphi(t-d-i)]^{1/2}}\right|\\ <\left|\frac{2\epsilon(t-i)}{[1+\varphi(t-d-i)^T\varphi(t-d-i)]^{1/2}}\right|.\end{aligned}$$

Then using (5.14) it follows that

$$\begin{aligned}\lim_{t\to\infty}\left|a(t-i)\frac{\varphi(t-d)^T}{[1+\varphi(t-d)^T\varphi(t-d)]^{1/2}}\right.\\ \cdot\frac{\varphi(t-d-i)}{[1+\varphi(t-d-i)^T\varphi(t-d-i)]^{1/2}}\\ \left.\cdot\frac{\epsilon(t-i)}{[1+\varphi(t-d-i)^T\varphi(t-d-i)]^{1/2}}\right|=0 \quad \text{for } i=1,2,\cdots,d-1.\end{aligned} \tag{5.17}$$

Hence, using (5.16), (5.17), and (5.14)

$$\lim_{t\to\infty} \frac{e(t)}{[1+\varphi(t-d)^T\varphi(t-d)]^{1/2}}=0. \quad (5.18)$$

This establishes (5.10). □

Note that we do not prove, or claim, that $\hat{\theta}(t)$ converges to θ_0. However, the weaker condition (5.10) will be sufficient to establish convergence of the tracking error to zero and boundedness of the system inputs and outputs. These are the prime properties of concern in adaptive control.

Theorem 5.1: Subject to Assumptions 4a)–c); if the algorithm (5.6), (5.7) is applied to the system (2.1), (2.2) ($r=m=1$), then $\{y(t)\}$ and $\{u(t)\}$ are bounded and

$$\lim_{t\to\infty} [y(t)-y^*(t)]=0. \quad (5.19)$$

Proof: Lemma 5.1 ensures that condition (3.1) of Lemma 3.1 is satisfied, with $s(t)=e(t)$, the tracking error, and $\sigma(t)=\varphi(t-d)$ the vector defined by (5.3). Also $b_1(t)=1$, and $b_2(t)=1$. It follows that the uniform boundedness condition (3.2) is satisfied.

Assumption 4c) and Lemma 3.2 ensure that

$$|u(k-d)|<m_3+m_4 \max_{1<\tau<t} |y(\tau)| \qquad \text{for all } 1<k<t.$$

Therefore, using (5.3)

$$\|\varphi(t-d)\|<p\left\{m_3+[\max(1,m_4)] \max_{1<\tau<t} |y(\tau)|\right\}$$

but

$$|e(t)|>|y(t)|-|y^*(t)|>|y(t)|-m_1.$$

Hence,

$$\begin{aligned}\|\varphi(t-d)\|&<p\left\{m_3+[\max(1,m_4)] \max_{1<\tau<t} [|e(\tau)|+m_1]\right\}\\ &=C_1+C_2 \max_{1<\tau<t} |e(\tau)|; \qquad 0<C_1<\infty,\ 0<C_2<\infty\end{aligned}$$

and it follows that the linear boundedness condition (3.3) is also satisfied.

The result now follows by Lemma 3.1 and by noting that boundedness of $\{\|\varphi(t)\|\}$ ensures boundedness of $\{|y(t)|\}$ and $\{|u(t)|\}$. □

VI. SISO Projection Algorithm II

In this section we present an algorithm differing from that of Section V in that the control law is estimated directly. This approach is adopted in [5], and essentially involves the factorization of β_0 from (4.2). A related procedure is used in the self-tuning regulator [10] where it is assumed that the value of β_0 is known.

An advantage of the algorithm is that the precautions required in Section V to avoid division by zero in the calculation of the input are no longer necessary. However, a disadvantage is that additional information is required; specifically, we need to know the sign of β_0 and an upper bound for its magnitude.

Factoring β_0 from (4.2) yields

$$\begin{aligned}y(t+d)=\beta_0(\alpha_0' y(t)+\cdots+\alpha_{n-1}' y(t-n+1)+u(t)\\ +\beta_1' u(t-1)\cdots+\beta_{m+d-1}' u(t-m-d+1)). \quad (6.1)\end{aligned}$$

Let

$$\begin{aligned}e(t+d)&=y(t+d)-y^*(t+d)\\ &=\beta_0\Big(u(t)+\alpha_0' y(t)\cdots\alpha_{n-1}' y(t-n+1)+\beta_1' u(t-1)\\ &\quad\cdots+\beta_{m+d-1}' u(t-m-d+1)-\frac{1}{\beta_0}y^*(t+d)\Big)\\ &=\beta_0\big(u(t)-\varphi(t)^T\theta_0'\big) \quad (6.2)\end{aligned}$$

where

$$\begin{aligned}\varphi(t)^T=&(-y(t)\cdots-y(t-n+1),-u(t-1)\cdots\\ &-u(t-m-d+1),y^*(t+d))\end{aligned}$$

$$\theta_0'^T=\left(\alpha_0',\cdots,\alpha_{n-1}',\beta_1',\cdots,\beta_{n+d-1}',\frac{1}{\beta_0}\right).$$

It is evident that the tracking error can be made identically zero by choosing $\{u(t)\}$ such that

$$u(t)=\varphi(t)^T\theta_0'. \quad (6.3)$$

However, since θ_0' is unknown, the control law will be recursively estimated. The following adaptive algorithm will be considered:

$$\hat{\theta}(t)=\hat{\theta}(t-d)-\frac{1}{\hat{\beta}_0}\varphi(t-d)[1+\varphi(t-d)^T\varphi(t-d)]^{-1}e(t) \quad (6.4)$$

$$u(t)=\varphi(t)^T\hat{\theta}(t) \quad (6.5)$$

where $\hat{\beta}_0$ is a fixed constant and $\hat{\theta}(t)$ is a p-vector of reals depending on d initial values $\hat{\theta}(0),\cdots,\hat{\theta}(d-1)$ and on $y(\tau)$, $0<\tau<t$, $u(\tau)$, $0<\tau<t-d-1$ via (6.4). Note that (6.4) is actually d separate recursions interlaced. (It has recently been pointed out [18] that it is also possible to analyze a single recursion without interlacing using a different technique but the same general principals.)

The analysis of projection algorithm II has much in common with the analysis for projection algorithm I. We will therefore merely state the analogs of Lemma 5.1 and Theorem 5.1 for the algorithm (6.4), (6.5).

Lemma 6.1: Define

$$\tilde{\theta}(t)=\hat{\theta}(t)-\theta_0'. \quad (6.6)$$

□

Then $\|\tilde{\theta}(t+d)\|^2-\|\tilde{\theta}(t)\|^2<0$ *along with the solutions of (6.4) and (6.5) and*

$$\lim_{t\to\infty} \frac{\varphi(t)^T\tilde{\theta}(t)}{[1+\varphi(t)^T\varphi(t)]^{1/2}}=0 \quad (6.7)$$

provided

$$0<\frac{\beta_0}{\hat{\beta}_0}<2.$$ □

Lemma 6.1 is used to prove Theorem 6.1 in the same manner that Lemma 5.1 is used to establish Theorem 5.1. We obtain the following theorem in this way.

Theorem 6.1: Subject to Assumptions 4a)–c) and for $0<\beta_0/\hat{\beta}_0<2$; if the algorithm (6.4), (6.5) is applied to the system (2.1), (2.2), then $\{y(t)\}$ and $\{u(t)\}$ are bounded and

$$\lim_{t\to\infty} [y(t)-y^*(t)]=0. \quad (6.6)$$

We note that the condition $0<\beta_0/\hat{\beta}_0<2$ has been previously conjectured [9], [10] in regard to stochastic self-tuning regulators using least squares. The condition can always be satisfied if the sign of β_0 and an upper bound for the magnitude of β_0 are known.

VII. Adaptive Control Using Recursive Least Squares

The wide-spread use of recursive least squares in parameter estimation indicates that it may find application in the adaptive control context. We treat the unit delay case of algorithm I with the projection (5.6) replaced by recursive least squares.

The adaptive control algorithm then becomes

$$\hat{\theta}(t)=\hat{\theta}(t-1)+\frac{a(t)P(t-2)\varphi(t-1)}{[1+a(t)\varphi(t-1)^TP(t-2)\varphi(t-1)]}\left[y(t)-\varphi(t-1)^T\hat{\theta}(t-1)\right] \quad (7.1)$$

$$P(t)=\left[I-\frac{P(t-1)\varphi(t)\varphi(t)^Ta(t+1)}{1+\varphi(t)^TP(t-1)\varphi(t)a(t+1)}\right]P(t-1) \tag{7.2}$$

$$\varphi(t)^T\hat{\theta}(t)=y^*(t+1) \tag{7.3}$$

where $P(t)$ is a $p\times p$ matrix and the recursion (7.2) is assumed to be initialized with $P(-1)$ equal to any positive definite matrix.

The scalar $a(t)$ in (7.1), (7.2) plays the same role as in Section V and is required only to avoid the nongeneric possibility of division by zero in (7.3) when evaluating $u(t)$. Hence, $a(t)=1$ will almost always work and for $a(t)=1$ we observe that (7.1) and (7.2) are the standard recursive least squares algorithm. The sequence $\{a(t)\}$ may be chosen as in (5.8).

Lemma 7.1: Along with the solutions of (7.1), (7.2), (7.3) the function $V(t)=\tilde{\theta}(t)^TP(t-1)^{-1}\tilde{\theta}(t)$ is a bounded, nonnegative, nonincreasing function and

$$\lim_{t\to\infty}\frac{\varphi(t-1)^T\tilde{\theta}(t-1)}{[1+a(t)\varphi(t-1)^TP(t-2)\varphi(t-1)]^{1/2}}=0 \tag{7.4}$$

where $\tilde{\theta}(t)=\hat{\theta}(t)-\theta_0$.

Proof: From (7.1), (5.2),

$$\tilde{\theta}(t)=\tilde{\theta}(t-1)-\frac{a(t)P(t-2)\varphi(t-1)\varphi(t-1)^T\tilde{\theta}(t-1)}{[1+a(t)\varphi(t-1)^TP(t-2)\varphi(t-1)]}. \tag{7.5}$$

Then using (7.2),

$$\tilde{\theta}(t)=P(t-1)P(t-2)^{-1}\tilde{\theta}(t-1).$$

Thus,

$$P(t-1)^{-1}\tilde{\theta}(t)=P(t-2)^{-1}\tilde{\theta}(t-1). \tag{7.6}$$

Now defining $V(t)$ as $\tilde{\theta}(t-1)^TP(t-1)\tilde{\theta}(t-1)$ we have

$$V(t)-V(t-1)=\tilde{\theta}(t)^TP(t-1)^{-1}\tilde{\theta}(t)-\tilde{\theta}(t-1)^TP(t-2)^{-1}\tilde{\theta}(t-1).$$

Using (7.6)

$$\begin{aligned}V(t)-V(t-1)&=\left[\tilde{\theta}(t)-\tilde{\theta}(t-1)\right]^TP(t-2)^{-1}\tilde{\theta}(t-1)\\&=-a(t)\frac{\tilde{\theta}(t-1)^T\varphi(t-1)\varphi(t-1)^T\tilde{\theta}(t-1)}{[1+a(t)\varphi(t-1)^TP(t-2)\varphi(t-1)]}\end{aligned} \tag{7.7}$$

where we have used (7.5). It is clear from (7.7) that $V(t)$ is a bounded, nonnegative, nonincreasing function and hence converges.

Thus, from (7.7), and since $a(t)$ is bounded away from zero,

$$\lim_{t\to\infty}\frac{\tilde{\theta}(t-1)^T\varphi(t-1)\varphi(t-1)^T\tilde{\theta}(t-1)}{[1+a(t)\varphi(t-1)^TP(t-2)\varphi(t-1)]}=0.$$

Hence,

$$\lim_{t\to\infty}\frac{e(t)}{[1+a(t)\varphi(t-1)^TP(t-2)\varphi(t-1)]^{1/2}}=0 \tag{7.8}$$

where

$$e(t)=-\varphi(t-1)^T\tilde{\theta}(t-1)=y(t)-y^*(t). \tag{7.9}$$

□

Theorem 7.1: Subject to Assumptions 4a)–c) if the algorithm (7.1), (7.2), (7.3) is applied to the system (2.1), (2.2) ($r=m=1$), then $\{y(t)\}$, $\{u(t)\}$ are bounded and

$$\lim_{t\to\infty}[y(t)-y^*(t)]=0. \tag{7.10}$$

Proof: From Lemma 7.1

$$\lim_{t\to\infty}\frac{e(t)}{[1+a(t)\varphi(t-1)^TP(t-2)\varphi(t-1)]^{1/2}}=0. \tag{7.11}$$

Now

$$\frac{|e(t)|^2}{[1+a(t)\varphi(t-1)^TP(t-2)\varphi(t-1)]}>\frac{|e(t)|^2}{[1+2\|\varphi(t-1)\|^2(\lambda_{\max}[P(t-2)])]}. \tag{7.12}$$

Hence, from (7.11) and (7.12)

$$\lim_{t\to\infty}\frac{|e(t)|^2}{[1+2(\lambda_{\max}[P(t-2)])\|\varphi(t-1)\|^2]}=0. \tag{7.13}$$

This will be recognized as being condition (3.1) with $s(t)=e(t)$, $b_1(t)=1$, and $b_2(t)=2(\lambda_{\max}[P(t-2)])$.

To establish the uniform boundedness condition (3.2) we proceed as follows. From (7.2) and the matrix inversion lemma,

$$P(t)^{-1}=P(t-1)^{-1}+a(t)\varphi(t)\varphi(t)^T.$$

Hence,

$$\begin{aligned}x^TP(t)^{-1}x&>x^TP(t-1)x\\&>\lambda_{\min}[P(t-1)^{-1}]\|x\|^2\qquad\text{for each } x\in\mathbf{R}^p.\end{aligned} \tag{7.14}$$

Now choose x as the eigenvector corresponding to the minimum eigenvalue of $[P(t)^{-1}]$.

Then from (7.14)

$$\lambda_{\min}[P(t)^{-1}]>\lambda_{\min}[P(t-1)^{-1}].$$

So $\lambda_{\min}[P(t)^{-1}]$ is a nondecreasing function bounded below by $\lambda_{\min}[P(-1)^{-1}]=K^{-1}>0$.

Hence from (7.13), $0<b_2(t)<2K$. This establishes condition (3.2).

The proof now proceeds as for Theorem 5.1. □

VIII. Multiple-Input Multiple-Output Systems

For the case $m=r>1$, the system (2.1), (2.2) can be represented in the form

$$\begin{bmatrix}A_1(q^{-1}) & & 0\\ & \ddots & \\ 0 & & A_m(q^{-1})\end{bmatrix}y(t)=\begin{bmatrix}q^{-d_{11}}B_{11}(q^{-1}) & \cdots & q^{-d_{1m}}B_{1m}(q^{-1})\\ \vdots & & \vdots\\ q^{-d_{m1}}B_{m1}(q^{-1}) & \cdots & q^{-d_{mm}}B_{mm}(q^{-1})\end{bmatrix}u(t) \tag{8.1}$$

where $A_k(q^{-1})$ and $B_{kl}(q^{-1})$ $1<k<m$, $1<l<m$ are scalar polynomials in the unit delay operator q^{-1} with nonzero constant coefficient.

Using the m identities $1=A_i(q^{-1})F_i(q^{-1})+q^{-d_i}G_i(q^{-1})$ where

$$F_i(q^{-1})=1+f_1^iq^{-1}\cdots+f_{d_i-1}^iq^{-d_i+1}$$

and

$$d_i=\min_{1<j<m}\{d_{ij}\},\qquad i=1,\cdots,m,$$

(8.1) can be written

$$\begin{bmatrix}y_1(t+d_1)\\ \vdots\\ y_m(t+d_m)\end{bmatrix}=\begin{bmatrix}\alpha_1(q^{-1}) & & 0\\ \vdots & & \vdots\\ 0 & & \alpha_m(q^{-1})\end{bmatrix}y(t)+\begin{bmatrix}\beta_{11}(q^{-1}) & \cdots & \beta_{1m}(q^{-1})\\ \vdots & & \vdots\\ \beta_{m1}(q^{-1}) & \cdots & \beta_{mm}(q^{-1})\end{bmatrix}u(t) \tag{8.2}$$

where

$$\alpha_i(q^{-1}) = G_i(q^{-1})$$

and

$$\beta_{ij}(q^{-1}) = F_i(q^{-1}) B_{ij}(q^{-1}) q^{d_i - d_{ij}}. \tag{8.3}$$

It can be seen that (8.2) consists of a set of multiple-input single-output (MISO) systems having a common input vector. The following assumptions will be made about the system.

Assumption Set 8:

a) $d_1, \cdots, d_m$ are known.

b) An upper bound for the order of each polynomial in (8.2) is known.

c) The system (8.1) satisfies condition

$$\det \begin{bmatrix} z^{d_{11}-d_1} B_{11}(z) & \cdots & z^{d_{1m}-d_1} B_{1m}(z) \\ z^{d_{m1}-d_m} B_{m1}(z) & \cdots & z^{d_{mm}-d_m} B_{mm}(z) \end{bmatrix} \neq 0 \qquad \text{for } |z| \leq 1.$$

Condition 8c) deserves comment.

First, for any output component y_i, $1 \leq i \leq m$, there exists at least one polynomial $q^{-d_{ij}} B_{ij}(q^{-1})$, $1 \leq j \leq m$, for which the power of q^{-1} associated with its (nonzero) leading coefficient is d_i. For each such polynomial the associated input $u_j(t)$ appears in $y_i(t+d_i)$ with the least possible delay d_i. Evaluated at $z=0$ condition 8c) requires the matrix of these leading coefficients to be nonsingular.

Second, the genericity of condition 8c) [for the model (8.1)] depends upon the initial parameterization of the system from which (8.1) is computed. This dependence is currently under investigation.

The control objective, as before, is to achieve

$$\lim_{t \to \infty} [y_i(t) - y_i^*(t)] = 0 \qquad i = 1, \cdots, m$$

where $y_i^*(t)$ is a reference sequence. It is assumed that each $\{y_i^*(t)\}$ is known *a priori* and that $|y_i^*(t)| \leq m_1 < \infty$ for all t, $i = 1, \cdots, m$.

IX. MIMO Adaptive Control

This section will be concerned with the multivariable versions of the adaptive control algorithms introduced in Sections V and VI. The multivariable version of the algorithm of Section VII also follows analogously.

A. MIMO Projection Algorithm I

Let θ_0^i be the vector of parameters in $\alpha_i(q^{-1})$ and $\beta_{i1}(q^{-1}) \cdots \beta_{im}(q^{-1})$. Then (8.2) may be written in the form

$$y_i(t+d_i) = \varphi_i(t)^T \theta_0^i, \qquad 1 \leq i \leq m \tag{9.1}$$

where

$$\varphi_i(t)^T = \left(y_i(t) \cdots y_i(t-n_i+1), u(t)^T, \cdots, u(t-m_i-d_i+1)^T\right).$$

Define

$$\begin{aligned} e_i(t+d_i) &= y_i(t+d_i) - y_i^*(t+d_i) \\ &= \varphi_i(t)^T \theta_0^i - y_i^*(t+d_i). \end{aligned} \tag{9.2}$$

It is evident that the tracking error may be made identically zero if it is possible to choose the vector $u(t)$ to satisfy

$$\phi_i(t)^T \theta_0^i = y_i^*(t+d_i), \qquad 1 \leq i \leq m. \tag{9.3}$$

Obviously (9.3) is a set of simultaneous equations in $u(t)$. Now the matrix multiplying $u(t)$ is nonsingular since in (8.3) $\det(\text{diag} F_i(z)) = 1$ at $z=0$ and Assumption 8c) holds. Hence a unique solution $u(t)$ of (9.3) exists at the true parameter value $\dot{\theta}$.

Consider the following adaptive algorithm:

$$\begin{aligned} \hat{\theta}_i(t) = \hat{\theta}_i(t-1) + a(t)\varphi_i(t-d_i)\left[1 + \varphi_i(t-d_i)^T \varphi_i(t-d_i)\right]^{-1} \\ \cdot \left(y_i(t) - \varphi_i(t-d_i)^T \hat{\theta}_i(t-1)\right) \end{aligned} \tag{9.4}$$

$$\varphi_i(t)^T \hat{\theta}_i(t) = y_i^*(t+d_i), \qquad 1 \leq i \leq m \tag{9.5}$$

where $\hat{\theta}_i(t)$ is a p_i $(= n_i + m(m_i + d_i))$ vector of reals depending upon an initial vector $\hat{\theta}_i(0)$ and $y_i(\tau)$, $0 < \tau \leq t$, $u(\tau)$, $0 < \tau \leq t - d_i$ via (9.4).

Clearly, it is critical to ensure that a solution to (9.5) exists for all t. This is guaranteed if the matrix of coefficients of $u(t)$ in (9.5) is nonsingular and this is ensured by the following procedure.

1) At $t=0$ the arbitrary initial value $\hat{\theta}(0)$ of the parameter estimate is chosen so that Assumption Set 8 is satisfied. Hence, the analogous equations to (9.3) evaluated at $\phi(0)$, $y(d_i)$, $1 \leq i \leq m$, $\hat{\theta}(0)$ are solvable for $u(0)$.

2) For $t \geq 1$ the procedure of Lemma 9.1 below guarantees the solvability of the algorithm equations for $u(t)$.

Lemma 9.1: In order that the matrix of coefficients of $u(t)$ in (9.5) is nonsingular for all $t \geq 1$ it is sufficient for $a(t)$ in (9.4) to be chosen as follows:

$$\epsilon < a(t) < 2 - \epsilon$$

where $0 < \epsilon < 1$ and $a(t)^{-1}$ is not an eigenvalue of $-R^{-1}(t-1)V(t)$ with

$$R(t-1) \triangleq [r_1, \cdots, r_m] \tag{9.6}$$

and

$$V(t) \triangleq [v_1, \cdots, v_m]. \tag{9.7}$$

r_i in (9.6) is the vector of coefficients of $u(t)$ in $\hat{\theta}_i(t-1)$, that is,

$$r_i = S_i \hat{\theta}_i(t-1) \tag{9.8}$$

when

$$S_i = [0_{m \times n_i} \quad I_{m \times n} \quad 0_{m \times (m_i + d_i - 1)}]. \tag{9.9}$$

v_i in (9.6) is the vector of changes in the coefficient of $u(t)$, that is,

$$\begin{aligned} v_i = S_i\Big[\varphi_i(t-d_i)\left(1 + \varphi_i(t-d_i)^T \varphi_i(t-d_i)\right)^{-1} \\ \cdot \left(y_i(t) - \varphi_i(t-d_i)^T \theta_i(t-1)\right)\Big]. \end{aligned} \tag{9.10}$$

Proof: The proof will be by induction and we first observe that from (9.4) and (9.6)–(9.10)

$$R(t) = R(t-1) + a(t)V(t). \tag{9.11}$$

Then:

i) $R(0)$ is nonsingular by the initial choice of $\hat{\theta}_i(0)$, $i = 1, \cdots, m$.

ii) Assume $R(t-1)$ is nonsingular. Then from (9.11), using $a(t) \neq 0$,

$$\begin{aligned} \det R(t) &= [\det R(t-1)][\det(I + a(t)R(t-1)^{-1}V(t))] \\ &= [\det R(t-1)](a(t))^m \left[\det\left(\frac{1}{a(t)}\right)I + R(t-1)^{-1}V(t)\right] \\ &= 0 \qquad \text{if and only if } \frac{1}{a(t)} \text{ is an eigenvalue of} \\ &\quad -R(t-1)^{-1}V(t). \end{aligned}$$

But the definition of $a(t)$ ensures $a(t)^{-1}$ is not an eigenvalue of $-R(t-1)^{-1}V(t)$, hence $R(t)$ is nonsingular. However, by i) $R(0)$ is nonsingular and it follows by induction $R(t)$, $t \geq 0$ is nonsingular. □

We note that the above choice of $a(t)$ has been included for technical completeness and that $a(t) = 1$ will almost always work since it is a nongeneric occurrence for 1 to be an eigenvalue of $-R(t-1)^{-1}V(t)$. Also since $-R(t-1)^{-1}V(t)$ has only a finite number of eigenvalues it is always possible to find an $a(t)$ to satisfy the lemma by computation of the eigenvalues of $R(t-1)^{-1}V(t)$.

Theorem 9.1: Subject to Assumptions 8a)–c) if the algorithm (9.4), (9.5) is applied to the system (2.1), (2.2) with $r = m$, then $\{y(t)\}$ and $\{u(t)\}$ are bounded and

$$\lim_{t \to \infty} |y_i(t) - y_i^*(t)| = 0; \qquad 1 \leq i \leq m.$$

Proof: Using Lemma (5.1) for each i, we have

$$\lim_{t \to \infty} \frac{e_i(t)}{\left[1 + \varphi_i(t-d_i)^T \varphi_i(t-d_i)\right]^{1/2}} = 0.$$

The proof now follows that of Theorem 5.1, except in the case that the vector $y(t)$ is unbounded. In this case there exists a subsequence $\{t_n\}$ such that

$$\lim_{t_n\to\infty} \|y(t_n)\| = \infty$$

and

$$|y_i(t+d_i)| < |y_{j(t_n)}(t_n+d_{j(t_n)})| \qquad \text{for some } 1<j(t_n)<m$$

$$\text{and for all } i \; 1<i<m \text{ and for all } t<t_n.$$

It then follows by Lemma 3.2 that there exist constants $0<C_1<\infty$ and $0<C_2<\infty$ such that

$$\|\varphi_i(t_n)\| < C_1 + C_2|y_{j(t_n)}(t_n+d_{j(t_n)})|, \qquad 1<i<m.$$

Since m is finite, there exists a further subsequence $\{t_{n'}\}$ of the subsequence $\{t_n\}$ such that

$$\|\varphi_i(t_{n'})\| < C_1 + C_2|y_i(t_{n'}+d_i)| \qquad \text{for at least one } i,\ 1<i<m$$

and $\{y_i(t_{n'}+d_i)\}$ is unbounded. The remainder of the proof then follows that of Theorem 5.1 where we note that

$$|e_i(t)| = |y_i(t) - y_i^*(t)|.$$

□

B. MIMO Projection Algorithm II

From (8.2), (8.3) the fact that $F_i(z)=1$ for $z=0$, $i=1,\cdots,m$ and by Assumption 8c) we can factor out the nonsingular matrix Γ_0 $(=[\beta_{ij}(0)])$ giving

$$\begin{bmatrix} y_1(t+d_1) \\ \vdots \\ y_m(t+d_m) \end{bmatrix} = \Gamma_0\{u(t) + C(q^{-1})y(t) + D(q^{-1})u(t-1)\} \qquad (9.12)$$

where

$$C(q^{-1}) = \Gamma_0^{-1}\begin{bmatrix} \alpha_1(q^{-1}) & & 0 \\ & \ddots & \\ 0 & & \alpha_m(q^{-1}) \end{bmatrix}$$

and

$$D(q^{-1}) = \Gamma_0^{-1}\left\{\begin{bmatrix} \beta_{11}(q^{-1}) & \cdots & \beta_{1m}(q^{-1}) \\ \vdots & & \vdots \\ \beta_{m1}(q^{-1}) & \cdots & \beta_{mm}(q^{-1}) \end{bmatrix} - \Gamma_0\right\}.$$

Define

$$\begin{bmatrix} e_1(t+d_1) \\ \vdots \\ e_m(t+d_m) \end{bmatrix} = \begin{bmatrix} y_1(t+d_1) \\ \vdots \\ y_m(t+d_m) \end{bmatrix} - \begin{bmatrix} y_1^*(t+d_1) \\ \vdots \\ y_m^*(t+d_m) \end{bmatrix}$$

$$= \Gamma_0\left[u(t) + C(q^{-1})y(t) + D(q^{-1})u(t-1) - \Gamma_0^{-1}\begin{bmatrix} y_1^*(t+d_1) \\ \vdots \\ y_m^*(t+d_m) \end{bmatrix}\right]$$

$$= \Gamma_0\left(u(t) - \theta_0'^T\varphi(t)\right) \qquad (9.13)$$

where $\theta_0'^T$ is an $m\times n'$ matrix whose ith row contains the parameters from the ith rows of $C(q^{-1})$, $D(q^{-1})$, and $E=\Gamma_0^{-1}$. $\varphi(t)$ is an $n'\times 1$ vector containing the appropriate delayed versions of $y(t)$, $u(t-1)$, and $y^*(t)$:

$$\varphi(t)^T = \left(-y(t)^T, -y(t-1)^T, \cdots, -u(t-1)^T, -u(t-2)^T, \cdots, y_1^*(t+d_1), \cdots, y_m^*(t+d_m)\right).$$

Analogously to Section VI we introduce the following adaptive control algorithm,

$$\hat{\theta}(t+d)^T = \hat{\theta}(t)^T - P\begin{bmatrix} e_1(t+d_1) \\ \vdots \\ e_m(t+d_m) \end{bmatrix}[1+\varphi(t)^T\varphi(t)]^{-1}\varphi(t)^T \qquad (9.14)$$

$$u(t) = \hat{\theta}(t)^T\varphi(t) \qquad (9.15)$$

where $d=\max(d_1,\cdots,d_m)$. $\hat{\theta}(t)^T$ is an $m\times n'$ matrix of reals depending on d initial matrices $\hat{\theta}(i)^T$, $1<i<d$ and past data from the system. P is a matrix of constants specified *a priori*. As in Section VI, (9.14) represents d interlaced recursions.

Once the recursions (9.14), (9.15) are initialized in a manner that leads to a unique solution for $u(0)$ it is sufficient to ensure that all successive recursions lead to equations solvable for $u(t)$, $t>1$. In analogy to Lemma 9.1 we have the following.

Lemma 9.2: Define $\tilde{\theta}(t+d)^T = \hat{\theta}(t+d)^T - \theta_0'^T$ *and let* $K=P\Gamma_0$. *If* K^T+K-K^TK *is positive definite, then, along the trajectories of (9.14), (9.15):*

a) $\operatorname{trace}\left[\tilde{\theta}(t+d)^T\tilde{\theta}(t+d)\right] - \operatorname{trace}\left[\tilde{\theta}(t)^T\tilde{\theta}(t)\right] < 0.$

b) $$\lim_{t\to\infty}\frac{e_i(t+d_i)}{[1+\varphi(t)^T\varphi(t)]^{1/2}} = 0, \qquad 1<i<m.$$

Proof:

a) We can rewrite (9.13) using (9.15) as

$$\begin{bmatrix} e_1(t+d_1) \\ \vdots \\ e_m(t+d_m) \end{bmatrix} = \Gamma_0\tilde{\theta}(t)^T\varphi(t).$$

Hence, from (9.14)

$$\tilde{\theta}(t+d)^T = \tilde{\theta}(t)^T - K\tilde{\theta}(t)^T\varphi(t)[1+\varphi(t)^T\varphi(t)]^{-1}\varphi(t)^T \qquad (9.18)$$

and

$$\operatorname{trace}\left(\tilde{\theta}(t+d)^T\tilde{\theta}(t+d)\right) - \operatorname{trace}\left(\tilde{\theta}(t)^T\tilde{\theta}(t)\right)$$

$$= -\operatorname{trace}\left[\left(K^T+K-K^TK\frac{\varphi(t)^T\varphi(t)}{[1+\varphi(t)^T\varphi(t)]}\right)\cdot\left(\tilde{\theta}(t)^T\varphi(t)[1+\varphi(t)^T\varphi(t)]^{-1}\varphi(t)^T\tilde{\theta}(t)\right)\right]$$

$$<0 \qquad \text{if } K^T+K-K^TK \text{ is positive definite.}$$

b) As in the proof of Lemma (7.1) it follows that

$$\lim_{t\to\infty}\operatorname{trace}\left(K^T+K-K^TK\frac{\varphi(t)^T\varphi(t)}{[1+\varphi(t)^T\varphi(t)]}\right)\cdot\left(\tilde{\theta}(t)^T\varphi(t)[1+\varphi(t)^T\varphi(t)]^{-1}\varphi(t)^T\tilde{\theta}(t)\right) = 0.$$

Since K^T+K-K^TK is positive definite then

$$\lim_{t\to\infty}\left[\tilde{\theta}(t)^T\varphi(t)[1+\varphi(t)^T\varphi(t)]^{-1}\varphi(t)^T\tilde{\theta}(t)\right] = 0$$

or

$$\lim_{t\to\infty}\Gamma_0^{-1}\begin{bmatrix} e_1(t+d_1) \\ \vdots \\ e_m(t+d_m) \end{bmatrix}[1+\varphi(t)^T\varphi(t)]^{-1}[e_1(t+d_1)\cdots e_m(t+d_m)]\Gamma_0^{-1} = 0.$$

This implies that (9.17) holds. □

Using Lemma 9.2 and following the proof of Theorem 9.1 we have the following.

Theorem 9.2: Subject to Assumptions 8a)–c), and K^T+K-K^TK *positive definite, if the algorithm (9.14), (9.15) is applied to the system (2.1), (2.2) with* $r=m$, *then the vectors* $y(t)$ *and* $u(t)$ *are bounded and*

$$\lim_{t\to\infty}|y_i(t)-y_i^*(t)| = 0, \qquad 1<i<m.$$

□

X. Nonlinear Systems

Although the analysis in the paper has been carried out for deterministic linear systems, it is clear that it could be readily extended to certain classes of nonlinear systems of known form. The essential points are the form of (5.2) or (6.2), and the linear bound condition (3.3). The latter point would indicate that systems with cone bounded nonlinearities would satisfy the conditions.

XI. Conclusion

The paper has analyzed a general class of discrete-time adaptive control algorithms and has shown that, under suitable conditions, they will be globally convergent. The algorithms have a very simple structure and are applicable to both single-input single-output and multiple-input multiple-output systems with arbitrary time delays provided only that a stable control law exists to achieve zero tracking error. The results resolve a long standing question in adaptive control regarding the existence of simple, globally convergent adaptive algorithms.

References

[1] I. D. Landau, "A survey of model reference adaptive techniques—Theory and applications," *Automatica*, vol. 10, pp. 353–379, 1974.

[2] R. V. Monopoli, "Model reference adaptive control with an augmented error signal," *IEEE Trans. Automat. Contr.*, vol. AC-19, pp. 474–485, Oct. 1974.

[3] A. Feuer, B. R. Barmish, and A. S. Morse, "An unstable dynamical system associated with model reference adaptive control," *IEEE Trans. Automat. Contr.*, vol. AC-23, pp. 499–500, June 1978.

[4] K. S. Narendra and L. S. Valavani, "Stable adaptive controller designs—Direct control," *IEEE Trans. Automat. Contr.*, vol. AC-23, pp. 570–583, Aug. 1978.

[5] A. Feuer and S. Morse, "Adaptive control of single-input single-output linear systems," *IEEE Trans. Automat. Contr.*, vol. AC-23, pp. 557–570, Aug. 1978.

[6] G. A. Dumont and P. R. Bélanger, "Self-tuning control of a titanium dioxide kiln," *IEEE Trans. Automat. Contr.*, vol. AC-23, pp. 532–538, Aug. 1978.

[7] K. J. Åström, U. Borisson, L. Ljung, and B. Wittenmark, "Theory and applications of self tuning regulators," *Automatica*, 19, pp. 457–476, 1977.

[8] L. Ljung, "Analysis of recursive stochastic algorithms," *IEEE Trans. Automat. Contr.*, vol. AC-22, pp. 551–575, Aug. 1977.

[9] ——, "On positive real transfer functions and the convergence of some recursive schemes," *IEEE Trans. Automat. Contr.*, vol. AC-22, pp. 539–551, Aug. 1977.

[10] K. J. Åström and B. Wittenmark, "On self-tuning regulators," *Automatica*, vol. 9, pp. 195–199, 1973.

[11] L. Ljung and B. Wittenmark, "On a stabilizing property of adaptive regulators," *Preprint, IFAC Symp. on Identification*, Tbilisi, U.S.S.R., 1976.

[12] B. Egardt, "A unified approach to model reference adaptive systems and self tuning regulators," Dep. Automat. Contr., Lund Inst. Technol. Tech. Rep., Dec. 1977.

[13] T. Ionescu and R. V. Monopoli, "Discrete model reference adaptive control with an augmented error signal," *Automatica*, vol. 13, pp. 507–517, Sept. 1977.

[14] J. L. Willems, *Stability Theory of Dynamical Systems*. New York: 1970.

[15] G. C. Goodwin, P. J. Ramadge, and P. E. Caines, "Discrete time stochastic adaptive control," *SIAM J. Contr. Optimiz.*, to be published.

[16] A. S. Morse, "Global stability of parameter adaptive control systems," Yale Univ., S & IS Rep. 7902R, Mar. 1979.

[17] K. S. Narendra and Y-H. Lin, "Stable discrete adaptive control," Yale Univ., S & IS Rep. 7901, Mar. 1979.

[18] B. Egardt, "Stability of model reference adaptive and self tuning regulators," Dep. Automat. Contr., Lund Inst. Technol., Tech. Rep., Dec. 1978.

Feedback and Optimal Sensitivity: Model Reference Transformations, Multiplicative Seminorms, and Approximate Inverses

G. ZAMES

THIS paper is credited with the introduction of the subject of H^∞ control theory, a topic at the center of control research during the 1980s and 1990s.

The symbol H^∞ refers to the Hardy space of bounded analytic functions. For continuous-time systems, this corresponds to the space of stable, proper transfer functions (analytic and bounded in the right half-plane). The H^∞ norm of a stable transfer function is its maximum modulus, that is, the peak value on the Bode magnitude plot. Such peak values for closed-loop transfer functions had always been an important measure of performance and robustness in control. In fact, it defines the induced $\mathcal{L}_2$-norm, and its importance in control comes from this connection.

The paper was motivated mainly by three considerations:

1. The classical approach to feedback design for single input-single output plants, in the style of Bode and others, takes place in the frequency domain: The open-loop frequency response is shaped by the addition of a compensator, such as a lead or lag compensator. Zames wondered whether such controllers can be induced by some precise frequency-domain optimality criterion. If so, then multivariable extensions might follow readily, whereas classical methods had not extended satisfactorily.
2. In the H^2 (equivalently, the LQG, or Wiener-Hopf) design formulation, the exogenous inputs (e.g., disturbances or sensor noise) are modeled as stationary signals with known stochastic characteristics. Zames suggested the need to be able to handle a different set of exogenous signals. He reasoned that a natural choice was a weighted ball in the space of finite energy time functions. This leads to an H^∞-type minimization problem. One can think of the H^2 approach as relatively optimistic. White noise scatters its energy over all frequencies. Zames's scheme is instead a worst case approach. The disturbance energy is allowed to be concentrated at the worst possible frequency.
3. Often, plant uncertainty is modeled in the frequency domain. Zames argued that robustness with respect to state models is artificial in many situations, such as where high-frequency dynamics are neglected in obtaining a lower order model.

The paper has two main contributions. First, the formulation of the problem of minimizing the H^∞ norm of the weighted closed-loop sensitivity function for a nominal plant; and second, a study of the ability of feedback (as in the case of the feedback amplifier, discussed in the paper of Black elsewhere in this volume) to reduce uncertainty. Of the two, the first sparked the most research. The literature on H^∞ control is vast (the reference list includes twenty-two books, a sampling of merely the *books* treating the subject in various settings and at various levels), so a survey of the entire landscape is not possible.

Research in the 1980s was devoted mainly to developing a variety of solutions to the pure mathematical problem of H^∞ optimization. A beautiful collaboration arose between control theorists and operator theorists, and the H^∞ sessions at the conferences of the day were charged with excitement. Surprisingly, it turned out that the best way to solve H^∞ optimization problems was to convert them to state-space form. In fact, a breakthrough occurred in 1988 with Doyle and Glover's state-space solution (the formulas for the controller were announced in [11], the full treatment with proofs appearing later in [7]). Their solution involved two Riccati equations, just as for the LQG problem, but in addition a remarkable coupling condition.

The 1990s saw a number of interesting developments. The two-Riccati solution highlights the fact that the H^∞ problem can be posed as a two-person zero-sum differential game: the control signal as the minimizing player versus the disturbance signal as the maximizing player. Başar and Bernhard [1] developed H^∞ theory in this way and extended it to time varying, finite horizon, and nonlinear problems. A more recent trend is the formulation of the H^∞ criterion as a linear matrix inequality, e.g., [3], leading to a convex optimization problem solvable numerically by interior-point methods.

Since one of Zames's goals began with classical control, it is fitting that H^∞ control has filtered down from research journals into undergraduate textbooks, e.g., [2]. In an undergraduate course, the treatment usually involves three problems. In the *nominal performance* problem, a suitable closed-loop transfer function, such as the weighted sensitivity function, is optimized in H^∞ norm for a nominal plant. The closed-loop shape is altered via the weighting function, and this supersedes the older approach of open-loop shaping. In *stability robustness*, a perturbation model of the plant is assumed and a controller is sought that maximizes the size of the perturbation while achieving closed-loop stability. Both of these problems are amenable to analytical solution. The third problem is *performance robustness*, where closed-loop performance and stability are guaranteed for all plant perturbations not greater in size than a prespecified bound. This problem has remained intractable, though it initiated a great deal of study, such as that by Doyle [6] on the structured singular value.

The importance of Zames's paper is that it breaks a trend, a trend away from stochastic control as a way of dealing with uncertainty, and it initiates the trend to give robustness the place in control that it deserves.

References

[1] T. Başar and P. Bernhard, *H^∞-Optimal Control and Related Minimax Design Problems: A Dynamic Game Approach*, 2nd ed., Birkhäuser (Boston), 1995.

[2] P. Bélanger, *Control Engineering: A Modern Approach*, Saunders College Publishing (Philadelphia, PA), 1995.

[3] S. Boyd, L. El Ghaoui, E. Feron, and V. Balakrishnan, *Linear Matrix Inequalities in System and Control Theory*, SIAM (Cherry Hill, NJ), 1994.

[4] T. Chen and B.A. Francis, *Optimal Sampled-Data Control Systems*, Springer-Verlag (New York), 1995.

[5] R.F. Curtain and H. Zwart, *An Introduction to Infinite-Dimensional Linear Systems Theory*, Springer-Verlag (New York), 1995.

[6] J.C. Doyle, "Analysis of feedback systems with structured uncertainty," *IEE Proceedings*, Part D, **133**:45–56, 1982.

[7] J.C. Doyle, K. Glover, P.P. Khargonekar, and B.A. Francis, "State-space solutions to standard H^2 and H^∞ control problems," *IEEE Trans. Auto. Contr.*, **AC-34**:831–847, 1989.

[8] J.C. Doyle, B.A. Francis and A. Tannenbaum, *Feedback Control Theory*, Macmillan (New York), 1992.

[9] A. Feintuch, *Robust Control Theory in Hilbert Space*, Springer-Verlag (New York), 1998.

[10] B.A. Francis, *A Course in H^∞ Control Theory*, Springer-Verlag (New York), 1987.

[11] K. Glover and J.C. Doyle, "State-space formulae for all stabilizing controllers that satisfy an H^∞ norm bound and relations to risk sensitivity," *Syst. Contr. Lett.*, **11**:167–172, 1988.

[12] M. Green and D.J.N. Limebeer, *Linear Robust Control*, Prentice Hall (Englewood Cliffs, NJ), 1995.

[13] B. Hassibi, A.H. Sayed and T. Kailath, *Indefinite-Quadratic Estimation and Control: A Unified Approach to H^2 and H^∞ Theories*, SIAM (Cherry Hill, NJ), 1999.

[14] J.W. Helton and O. Merino, *Classical Control Using H^∞ Methods*, SIAM (Philadelphia, PA), 1998.

[15] R.A. Hyde, *H^∞ Aerospace Control Design: A VSTOL Flight Application*, Springer-Verlag (New York), 1995.

[16] H. Kimura, *Chain-Scattering Approach to H^∞ Control*, Birkhäuser (Boston), 1997.

[17] J. Maciejowski, *Multivariable Feedback Design*, Addison-Wesley (Reading, MA), 1989.

[18] D.C. McFarlane and K. Glover, *Robust Controller Design Using Normalized Coprime Factor Plant Descriptions*, Springer-Verlag (New York), 1990.

[19] M. Morari and E. Zafiriou, *Robust Process Control*, Prentice Hall (Englewood Cliffs, NJ), 1989.

[20] D. Mustafa and K. Glover, *Minimum Entropy H^∞ Control*, Springer-Verlag (New York), 1990.

[21] M.A. Peters and P.A. Iglesias, *Minimum Entropy Control for Time-Varying Systems*, Birkhäuser (Boston), 1997.

[22] S. Skogestad and I. Postlethwaite, *Multivariable Feedback Control: Analysis and Design*, Wiley (New York), 1996.

[23] A.A. Stoorvogel, *The H^∞ Control Problem: A State-Space Approach*, Prentice Hall (Englewood Cliffs, NJ), 1992.

[24] M. Vidyasagar, *Control System Synthesis: A Factorization Approach*, MIT Press (Cambridge, MA), 1985.

[25] K. Zhou with J.C. Doyle, and K. Glover, *Robust and Optimal Control*, Prentice Hall (Englewood Cliffs, NJ), 1996.

B.A.F.

Feedback and Optimal Sensitivity: Model Reference Transformations, Multiplicative Seminorms, and Approximate Inverses

G. ZAMES, FELLOW, IEEE

Abstract—In this paper, the problem of sensitivity reduction by feedback is formulated as an optimization problem and separated from the problem of stabilization. Stable feedback schemes obtainable from a given plant are parameterized. Salient properties of sensitivity reducing schemes are derived, and it is shown that plant uncertainty reduces the ability of feedback to reduce sensitivity.

The theory is developed for input–output systems in a general setting of Banach algebras, and then specialized to a class of multivariable, time-invariant systems characterized by $n \times n$ matrices of H^∞ frequency response functions, either with or without zeros in the right half-plane.

The approach is based on the use of a *weighted seminorm* on the algebra of operators to measure sensitivity, and on the concept of an *approximate inverse*. Approximate invertibility of the plant is shown to be a necessary and sufficient condition for sensitivity reduction. An indicator of approximate invertibility, called a *measure of singularity*, is introduced.

The measure of singularity of a linear time-invariant plant is shown to be determined by the location of its right half-plane zeros. In the absence of plant uncertainty, the sensitivity to output disturbances can be reduced to an optimal value approaching the singularity measure. In particular, if there are no right half-plane zeros, sensitivity can be made arbitrarily small.

The feedback schemes used in the optimization of sensitivity resemble the lead–lag networks of classical control design. Some of their properties, and methods of constructing them in special cases are presented.

I. Introduction

In THIS paper we shall be concerned with the effects of feedback on uncertainty, where uncertainty occurs either in the form of an additive disturbance d at the output of a linear plant P (Fig. 1), or an additive perturbation in P representing "plant uncertainty." We shall approach this subject from the point of view of classical sensitivity theory, with the difference that feedbacks will not only reduce but actually optimize sensitivity in an appropriate sense.

The theory will be developed at two levels of generality. At the higher level, a framework will be sought in which the essence of the classical ideas can be captured. To this end, systems will be represented by mappings belonging to a normed algebra. The object here is to obtain general answers to such questions as: how does the usefulness of feedback depend on plant invertibility? are there measures of sensitivity or plant uncertainty that are natural for optimization? how does plant uncertainty affect the possibility of designing a feedback scheme to reduce plant uncertainty?

Manuscript received October 8, 1979; revised December 4, 1980 and December 17, 1980. Paper recommended by A. Z. Manitius, Past Chairman of the Optimal Systems Committee. An earlier version of this paper [23] was presented the 17th Allerton Conference, October 1979.

The author is with the Department of Electrical Engineering, McGill University, Montreal, P.Q., Canada.

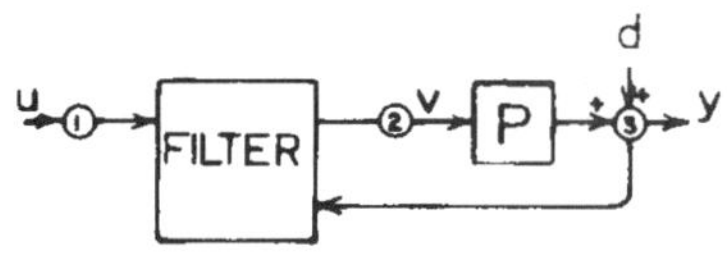

Fig. 1.

At a more practical level, the theory will be illustrated by simple examples involving single variable and multivariable frequency responses. The questions here are: can the classical "lead–lag" controllers be derived from an optimization problem? How do RHP (right half-plane) zeros restrict sensitivity? in multivariable systems without RHP zeros, can sensitivity be made arbitrarily small, and if so how?

A. Motivation

A few observations might serve to motivate this reexamination of feedback theory.

One way of attenuating disturbances is to introduce a filter of the WHK (Wiener–Hopf–Kalman) type in the feedback path. Despite the unquestioned success of the WHK and state-space approaches, the classical methods, which rely on lead–lag "compensators" to reduce sensitivity, have continued to dominate many areas of design. On and off, there have been attempts to develop analogous methods for multivariable systems. However, the classical techniques have been difficult to pin down in a mathematical theory, partly because the purpose of compensation has not been clearly stated. One of our objectives is to formulate the compensation problem as the solution to a well defined optimization problem.

Another motivating factor is the gradual realization that classical theory is not just an old-fashioned way of doing WHK, but is concerned with a different category of mathematical problems. In a typical WHK problem, the quadratic norm of the response to a disturbance d is minimized by a projection method (see Sections III′-A′ and IV-C); in a deterministic version, the power spectrum

Reprinted from *IEEE Transactions on Automatic Control,* Vol. AC-26, April 1981, pp. 301–320.

$|\hat{d}(j\omega)|$ is a *single, known* vector in, e.g., the space $L_2(-\infty,\infty)$; in stochastic versions, d belongs to a *single* random process of *known* covariance properties. However, there are many practical problems in which $|\hat{d}(j\omega)|$ is unknown but belongs to a prescribed set, or d belongs to a class of random processes whose covariances are uncertain but belong to a prescribed set. For example, in audio design, d is often one of a set of narrow-band signals in the 20–20K Hz interval, as opposed to a single, wide-band signal in the same interval. Problems involving such more general disturbance sets are not tractable by WHK or projection techniques. In a feedback context, they are now usually handled by empirical methods resembling those of classical sensitivity. One objective here is to find a systematic approach to problems involving such sets of disturbances.

Another observation is that many problems of plant uncertainty can be stated easily in the classical theory, e.g., in terms of a tolerance-band on a frequency response as in [2], but are difficult to express in a linear-quadratic-state-space framework. One reason for this is that frequency-response descriptions and, more generally, input–output descriptions preserve the operations of system addition and multiplication, whereas state-space descriptions do not. Another reason is that the quadratic norm is hard to estimate for system products (see Sections III′-A′ and IV-B1), whereas the induced norm (or "gain") that is implicit in the classical theory is easier to estimate. We would like to exploit these advantages in the study of plant uncertainty.

Finally, sensitivity theory is one of the few tools available for the study of organization structure: feedback versus open-loop, aggregated versus disaggregated, etc. For example, feedback reduces complexity of identification roughly for the same reason that it reduces sensitivity [12], [13]. However, it is hard to draw definitive conclusions about the effects of organization without some notion of optimality, and such a notion is missing in the old theory.

B. *Weighted Seminorms and Approximate Inverses*

One way of defining the optimal sensitivity of a feedback system, and of addressing some of the issues mentioned in Section I-A, is in terms of an induced norm of the sensitivity operator. However, it will be shown in Section III′-B′ that the primary norm of an operator in a normed algebra is useless for this purpose. Perhaps that is why operator norm optimization has not been pursued extensively in the past.

Instead, we shall introduce an auxiliary "weighted" seminorm, which retains some of the multiplicative properties of the induced norm, but is amenable to optimization. Plant uncertainty will be described in terms of belonging to a sphere in the weighted seminorm.

Approximate invertibility of the plant is one of the features which distinguishes control from, say, communication problems. We shall define the concept of an approximate inverse under a weighted seminorm, and show that sensitivity reduction is possible if there is such an inverse.

C. *Background*

Many of the ideas in this paper are foreshadowed in the classical theory [1], [2] of single-input single-output convolution systems, especially as presented by Horowitz [2], who derived various limits on sensitivity imposed by the plant, and stressed the need to consider plant uncertainty in design. The author posed the feedback problem in a normed algebra of operators on a Banach space, and introduced [4], [5] perturbation formulas of the type

$$(I-P)^{-1}-(I-P_0)^{-1}=(I-P)^{-1}(P-P_0)(I-P_0)^{-1} \tag{1.1}$$

which were used to show that high-gain feedback reduces the sensitivity of linear amplifiers to large nonlinear perturbations [3]–[5]. Desoer studied a related problem in [6], and recently [7] has obtained results for the case of P and P_0 both nonlinear (also see footnote 8). Perkins and Cruz [8] used perturbation formulas similar to (1.1) to calculate the sensitivity of linear multivariable systems. Porter [9] posed various sensitivity problems in Hilbert space, and in a paper with Desantis [10] obtained circle type conditions for sensitivity reduction. Willems [11] has stressed the Banach algebraic aspects of feedback theory.

In [1]–[10], the disturbance is either a fixed vector, or lies in some band of frequencies, and sensitivity is measured in terms of an output norm, as opposed to an induced operator norm. The approach of using weighted operator norms, and relating optimal sensitivity to weighted invertibility via a fractional transformation was used in [12], but has since been reworked and expanded.

D. *Two Problems*

We shall be concerned with the system of Fig. 1. Here, P is a given *plant* with a single (possibly multivariable) input v accessible to control, and an output y to which a disturbance d, not accessible to control, has been added. The plant input v is generated by a filter whose only inputs consist of observations on the plant output y and a reference input u. Two types of problems will be considered.

Problem 1—Disturbance Attenuation: This problem will be the subject of Sections V–VII. Suppose that $u=0$. The input–output behavior of the system between the nodes (2, 3) can be modeled by the flowgraph of Fig. 2, which consists of the plant P and a single additional operator F in the feedback path. The disturbance d is uncertain in the sense that it can be any one of a set of disturbances. Initially (through Section VI) P is assumed to be known exactly, but later (Section VII) to be uncertain. We would like to *characterize the feedback operators F which attenuate the response y to d in some appropriately optimal sense, and examine the effects of uncertainty about P on disturbance attenuation.*

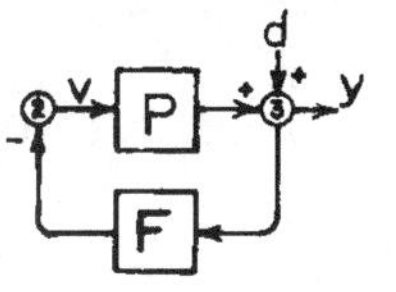

Fig. 2.

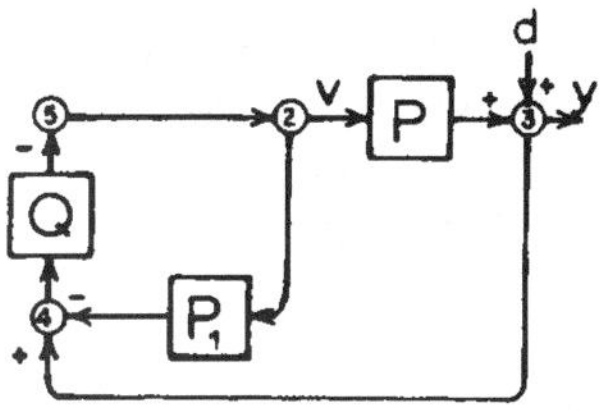

(a)

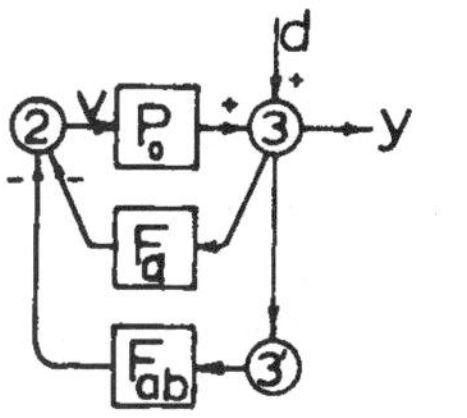

(b)

Fig. 3.

Problem 2—Plant Uncertainty Attenuation: (Problem 2 is the subject of Section VIII.) Suppose that $d=0$, and the plant P is uncertain to the extent that it can be any one of a "ball" of possible plants centered around some nominal value P_1. If the filter is linear, the behavior of the system between nodes (1,2,3) can be modeled by the flowgraph of Fig. 4. The filter can be characterized by a pair of operators (U, F). We would like to *find operators (U, F) which shrink the ball of uncertainty but leave the nominal plant invariant; to find bounds on the optimal shrinkage and to look at its dependence on plant uncertainty.*

E. Outline of the Paper

See Synopsis following Appendixes.

II. Spaces and Algebras of Systems

The purpose of this section is to specify the meaning which will be attached to the terms "frequency-response" and "linear system," and to summarize their properties for later use.

A feature of the input–output approach is that systems can be added, multiplied by other systems or by scalars, and the sums or products obtained are still systems, i.e., they form an *algebra*. Frequently, it will be assumed that the largest amplification produced by a system can be measured by a norm, typically the maximum frequency response amplitude over some region of analyticity; under this assumption the algebra of systems becomes a *normed algebra*. Normed algebras provide the natural setting for the study of system interconnections such as feedback. Their elementary properties will be used freely here, and

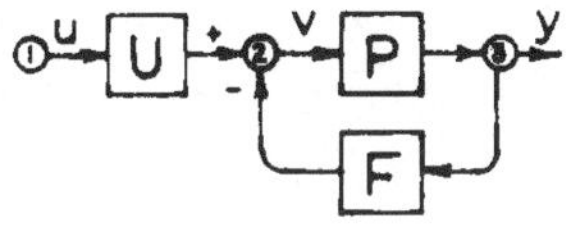

Fig. 4.

may be found in such texts as Naimark [19]. Occasionally, it will be assumed that a normed algebra is a *Banach algebra*, i.e., has the property that every convergent sequence of elements of the algebra has a limit in the algebra.

It will be assumed that all linear spaces and algebras are over the real field.

A. Algebras of Frequency Response Systems

The frequency response of a stable, causal, linear time-invariant system is a function analytic in the right-half of the complex plane. An accepted setting for such functions involves the H^p Hardy spaces [15], which we shall employ with some modifications to accommodate unstable systems.

The algebra H_e^∞ consists of functions $\hat{p}(\cdot)$ of a complex variable $s=\sigma+j\omega$, each of which is analytic in some open half-plane $\mathrm{Re}(s)>\sigma_p$ possibly depending on $\hat{p}(\cdot)$, and is bounded there, i.e., $\hat{p}(s)\leqslant$const. for $\mathrm{Re}(s)>\sigma_p$. The functions in H_e^∞ will be referred to as *causal* frequency responses. If $\hat{p}$ is in H_e^∞, then the domain of definition of $\hat{p}$ can be extended by analytic continuation to a unique, maximal, open half-plane of analyticity $\mathrm{Re}(s)>\sigma_{pm}$, where $\sigma_{pm}\leqslant\sigma_p$. In general, $\hat{p}$ need not be bounded on this maximal open RHP, but if it is, then it can further be extended to the boundary by the limit $\hat{p}(\sigma_{pm}+j\omega \triangleq \lim_{\sigma\to\sigma_{pm}}\hat{p}(\sigma+j\omega)$, which exists for almost all ω provided $\sigma+j\omega$ approaches the boundary nontangentially from the right. Assume that all functions in H_e^∞ have been so extended.

The algebra H^∞ (of *stable causal* frequency responses) consists of functions $\hat{p}$ of H_e^∞ for which $\sigma_p\leqslant 0$, i.e., the region of bounded analyticity includes the RHP. The norm $\|\hat{p}\|=\sup\{|\hat{p}(s)|: \mathrm{Re}(s)>0\}$ is defined on H^∞, making H^∞ a normed and, indeed, Banach algebra.

A *strictly proper* function in H_e^∞ satisfies the condition $\hat{p}(s)\to 0$ as $|s|\to\infty$ in $\mathrm{Re}(s)\geqslant\sigma_p$. The symbols H_{e0}^∞ and H_0^∞ will denote the algebras of strictly proper frequency responses in H_e^∞ and H^∞, respectively. By a straightforward application of the maximum modulus principle, the normed algebra H_0^∞ of strictly proper stable frequency responses has the property that $\|\hat{p}\|=\mathrm{ess\,sup}\{|\hat{p}(j\omega)|: \omega$ real$\}$ for any $\hat{p}$ in H_0^∞, i.e., the norm can be computed from $j\omega$-axis measurements.

Spaces of Inputs and Outputs: For any integer $1\leqslant q<\infty$, the linear space H_e^q consists of functions $\hat{u}(\cdot)$ of a complex variable, each $\hat{u}$ being analytic in some open half-plane $\mathrm{Re}(s)>\sigma_u$ in which the restriction of $\hat{u}$ to any vertical line is in L^q and $\int_{-\infty}^{\infty}|\hat{u}(\sigma+j\omega)|^q\,d\omega\leqslant$const. for all $\sigma>\sigma_u$. Again, the domain of definition of each $\hat{u}$ in H_e^q is extended by analytic continuation to a maximal open RHP of analyticity and then, if $\hat{u}$ is L^q-bounded in this RHP, to its boundary by a nontangential limit. The space H^q consists of functions $\hat{u}$ in H_e^q for which $\sigma_u \leq 0$, and is a Banach space under the norm $\|\hat{u}\| = \sup_{\sigma>0}\{\int_{-\infty}^{\infty}|\hat{u}(\sigma + j\omega)|^q d\omega\}^{1/q}$.

Inputs and outputs will belong either to H_e^∞ or one of the H_e^q spaces, $1 \leq q < \infty$. In the special case $q=2$ it follows from the Paley–Wiener theory that every function of H^2 is a Laplace transform of a time function in $L^2(0, \infty)$ and vice versa. In the general case of H^q, some functions can be viewed as transforms of time functions, and the others as frequency functions that do not appear in physical applications and can be disregarded.

Frequency Response Operators: Let q be any integer, $1 \leq q \leq \infty$, which will be held fixed. For any causal frequency response $\hat{p}(\cdot)$ in H_e^∞ an operator P: $H_e^q \to H_e^q$ is defined by the multiplication $\widehat{Pu}(s) = \hat{p}(s)\hat{u}(s)$. P will be called a causal frequency response operator, and the algebra of all such operators will be denoted by the bordered capital $\mathbb{H}_e^\infty$. Similarly, for each of the algebras of frequency responses H^∞, H_{e0}^∞, and H_0^∞, an algebra of operators mapping H_e^q into itself is defined and denoted by the corresponding bordered capital, i.e., $\mathbb{H}^\infty$, $\mathbb{H}_{e0}^\infty$, or $\mathbb{H}_0^\infty$. In the case of normed frequency response algebras, the corresponding operator algebras are similarly normed, e.g., $\|P\|_{\mathbb{H}^\infty} = \|\hat{p}\|_{H^\infty}$.

The stable operators in $\mathbb{H}^\infty$ map H^q, which is a proper subspace of H_e^q, into H^q, and are in fact completely determined by their behavior on H^q. They can therefore be represented by their restrictions of the form P: $H^q \to H^q$. In sections devoted entirely to stable systems, we shall concentrate on operators of the form P: $H^q \to H^q$ without distinguishing them as restrictions of operators on H_e^q.

B. More General Algebras of Systems

We would like to take an axiomatic approach to the problem of sensitivity reduction by feedback, i.e., to single out the relevant properties of linear systems and postulate them as axioms. For example, the related properties of causality, realizability, and strong or strict causality have definitions [see, e.g., [14] and [11]) reflecting the fact that the response to a sudden input to a physical system can not anticipate the input, and cannot occur instantaneously. These properties of physical systems preclude the pathological phenomena associated with instantaneous response around a feedback loop, and ensure that the feedback operator $(I+P)^{-1}$ is well defined. However, these details are not relevant here. The only items of interest are that causal systems form an algebra of mappings, and that strictly causal systems form a subalgebra whose salient feature is the existence of the inverse $(I+P)^{-1}$ for all of its members, i.e., a "radical." Accordingly we postulate the following.

$\mathfrak{X}$ is a linear space whose elements will be called *inputs* or *outputs*. $\mathbf{A}$ is a linear algebra of linear mappings P: $\mathfrak{X} \to \mathfrak{X}$ with identity I, whose elements will be called *causal operators*. $\mathbf{A}_s$ is a radical of $\mathbf{A}$, i.e.,[1] a proper nontrivial subalgebra of $\mathbf{A}$ with the property that for any P in $\mathbf{A}_s$, the inverse $(I+P)^{-1}$ exists in $\mathbf{A}$, and for any F in $\mathbf{A}$ the products PF and FP are in $\mathbf{A}_s$. The elements of $\mathbf{A}_s$ will be called *strictly causal operators*. (The concept of an algebra of "realizable" systems was introduced by the author in [14]; the related notions of strong causality of Willems [11], and later strict causality of Porter, Saeks, and Desantis are compared in a paper of Feintuch [20].)

An example of the space $\mathfrak{X}$ of inputs and the algebra $\mathbf{A}$ of causal operators is provided by the space H_e^q of (transforms of) inputs and the algebra $\mathbb{H}_e^\infty$ of causal frequency response operators. In this context the algebra $\mathbb{H}_{e0}^\infty$ of strictly proper frequency response operators is an example of an algebra $\mathbf{A}_s$ of strictly causal operators. Henceforth, we shall refer to the strictly proper operators as strictly causal.

In the case of stability, we shall need the fact that a stable input–output system produces a finite amplification of inputs that can be measured by a suitable norm, and that stable systems form an algebra (see, e.g. [5]). Accordingly, we postulate the following.

$\mathfrak{B}$ is a Banach subspace of $\mathfrak{X}$ whose elements will be called *bounded* inputs or outputs [for example, H^2 or $L_2(0, \infty)$]. $\mathbf{B}$ is a normed subalgebra of $\mathbf{A}$ containing the identity I, whose elements will be called *stable* causal operators, under the following assumption: the norm of any P in $\mathbf{B}$ is the $\mathfrak{B}$-induced norm, that is, $\|P\| \triangleq \sup\{\|Pu\|/\|u\|$: u in $\mathfrak{B}$, $u \neq 0\}$, the sup being finite.[2]

If $\mathbb{H}_e^\infty$ is taken as an example of $\mathbf{A}$, then $\mathbb{H}^\infty$ is an example of $\mathbf{B}$.

$\mathbf{B}_s$ is the subalgebra of $\mathbf{B}$ obtained by intersecting $\mathbf{A}_s$ and $\mathbf{B}$, consisting of *strictly-causal stable* operators. It should be noted that $\mathbf{B}_s$ is not a radical of $\mathbf{B}$, as P stable does not imply that $(I+P)^{-1}$ is stable.

For the purpose of estimating the effects of small perturbations, it will be assumed that $\mathbf{B}_s$ has the *small-gain property*, i.e., for any P in $\mathbf{B}_s$, if $\|P\| < 1$ then $(I+P)^{-1}$ is in $\mathbf{B}$. If $\mathbf{B}$ is complete, i.e., a Banach space, this assumption is redundant for then the series $I - P + P^2 - \cdots$ converges to the inverse in $\mathbf{B}$ of $(I+P)$. However, we have applications in mind in which completeness of $\mathbf{B}$ is replaced by other assumptions.

A frequency response $\hat{p} \in H^\infty$ which is not strictly proper can not be realized exactly, but can be approximated by a sequence of strictly proper responses of the form $n(s+n)^{-1}\hat{p}(s)$, $n = 1, 2, \cdots$. The sequence $n(s+n)^{-1}$ is an example of an "identity sequence." More generally, identity sequences will be used to construct strictly proper approximations to improper responses, and are defined as follows: an *identity sequence* $\{I_n\}_{n=1}^\infty$ for $\mathbf{B}$ is a sequence of operators in $\mathbf{B}$ with the property that for any F in $\mathbf{B}$ the sequences $\|I_n F - F\|$ and $\|FI_n - F\|$ approach 0 as $n \to \infty$. It is assumed that $\mathbf{B}_s$ contains an identity sequence for $\mathbf{B}$.

The following well-known (cf. Naimark [19, p. 162]) properties of a normed algebra will be crucial in many

[1] The properties of radicals are discussed in Naimark [19, p. 162].

[2] Whenever the norm of x is not identified by a subscript, it should be taken to be the principal norm of the space to which x belongs.

parts of the paper. For convenience they are proved in Appendix I.

Let P and Q be in B.

Proposition 2.1: a) If $(I+PQ)^{-1}$ is in B, then $(I+QP)^{-1}$ is in B, and the formula $P(I+QP)^{-1}=(I+PQ)^{-1}P$ is valid. b) If R is a radical in B and P is in R, then P has no inverse in B. (Strictly causal operators have no inverses in $\mathbf{B}$.) c) If P and $(I+P)^{-1}$ are in B and $\|P\|<1$, then $\|(I+P)^{-1}\|\leq(1-\|P\|)^{-1}$.

III. Feedback Decomposition: Stabilizing and Stabilized Stages

We proceed to derive a decomposition principle to be employed in disturbance attenuation. Suppose that there is no plant uncertainty, and that the plant and feedback are constrained not to be simultaneously unstable. Under these hypotheses, any closed-loop stable feedback design can be decomposed into two stages: a first stage involving plant stabilization (which can be omitted for stable plants); and a second stage, involving a model reference scheme in which only stable elements are used, and which is automatically closed-loop stable. The choice of a stabilizing stage is independent of, and does not prejudice the choice of the second stage. Having established this fact, we shall be free to concentrate on the second stage of the disturbance attenuation problem under the condition that the plant is stable (or has been stabilized), without loss of generality.

Consider the system of Fig. 2. The plant input v, output y, and disturbance d are all in $\mathcal{X}$ and satisfy the equations

$$y=Pv+d \tag{3.1a}$$

$$v=-Fy \tag{3.1b}$$

in which P and F are operators in $\mathbf{A}_s$ [see Remark 3.1c)]. We shall refer to (3.1) as a *feedback scheme* with plant P and feedback F. Since P is strictly causal, the inverse $(I+PF)^{-1}$ exists in $\mathbf{A}$. Therefore, for each d in $\mathcal{X}$, (3.1) have unique solutions for v and y in $\mathcal{X}$, given by the formulas

$$y=(I+PF)^{-1}d \tag{3.2a}$$

$$v=-F(I+PF)^{-1}d. \tag{3.2b}$$

Let K_{32}: $\mathcal{X}\rightarrow\mathcal{X}$ denote the "closed-loop" operator mapping d to v. K_{32} is an operator in $\mathbf{A}$ given by $K_{32}=-F(I+PF)^{-1}$.

The flowgraphs in this paper are simple, and will be approached informally in order to avoid lengthy definitions. Expressions for some of the subsidiary c.l. (closed-loop) operators, which can be found by inspection, will be listed without derivation as needed.

For a system to be physically realizable on an infinite time interval, it is usual to postulate [14] that all c.l. input–output operators must be stable, though "open-loop" operators such as the plant P and feedback F may be unstable. The set of c.l. operators for (3.1) consists of: $K_{22}=(I+FP)^{-1}$, $K_{23}=PK_{22}$, $K_{33}=(I+PF)^{-1}$, and K_{32} specified above. Accordingly, the feedback scheme (3.1) will be called *c.l. stable* if K_{ij} is in $\mathbf{B}$ for $i,j=2$ or 3.

We shall be interested in situations in which P is at or near some nominal value P_1, and the feedback F appears as an operator variable in an optimization problem whose object is to minimize response to d. Unstable operator variables are difficult to handle, and so our first step will be to show that F has an equivalent realization in terms of a stable operator.

A. The Model Reference Transformation

The flowgraph of Fig. 3 is described by the equations

$$y=Pv+d \tag{3.3a}$$

$$v=-Q(y-P_1v) \tag{3.3b}$$

in which y, v, and d are in $\mathcal{X}$, and P, P_1, Q are in $\mathbf{A}_s$. Equation (3.3) will be called a *model reference scheme* with *comparator* Q, as the output of the plant P with disturbance d added is compared to the output of a model P_1 of the plant without disturbance, and the difference actuates Q.

The two sets of equations (3.1) and (3.3) are called *equivalent* iff every input–output triple (d,v,y) in $\mathcal{X}^3$ satisfying (3.1) satisfies (3.3), and vice versa. Their equivalence will be established under the assumption that the equations

$$Q=F(I+P_1F)^{-1} \tag{3.4a}$$

$$F=Q(I-P_1Q)^{-1} \tag{3.4b}$$

hold. If either equation in (3.4) is valid, then so is the other, and

$$(I-P_1Q)=(I+P_1F)^{-1}. \tag{3.5}$$

To derive (3.5), suppose (3.4a) is valid. Therefore,

$$\begin{aligned}I-P_1Q&=I-P_1F(I+P_1F)^{-1}\\&=(I+P_1F)(I+P_1F)^{-1}-P_1F(I+P_1F)^{-1}\\&=(I+P_1F)^{-1}\end{aligned}$$

and (3.5) is true; here the expression $I=(I+P_1F)(I+P_1F)^{-1}$ and the distributive law for multiplication on the right was used. Equations (3.4a) and (3.5) can now be used to give the identities

$$Q(I-P_1Q)^{-1}=F(I+P_1F)^{-1}(I+P_1F)=F,$$

so (3.4b) is true as claimed. The converse proposition is proved similarly.

Assumption: For the present, and until the end of Section VI, assume that $P=P_1$, i.e., there is no plant uncertainty.

Theorem 1:

a) Any closed-loop stable feedback scheme (3.1) with stable plant $P\in\mathbf{B}_s$ and (not necessarily stable) feedback $F\in\mathbf{A}_s$ is equivalent to a model reference scheme whose branches are all stable, i.e., $Q\in\mathbf{B}_s$ and $P_1=P$, where F and Q are related by (3.4). Conversely, any model reference scheme with stable branches is closed-loop stable, and equivalent to a closed-loop stable feedback scheme subject to (3.4).

b) If (3.4) holds, and d and y satisfy either the feedback or model reference equations they satisfy the equation

$$y=(I-P_1Q)d. \tag{3.6}$$

Proof:

a) For any feedback scheme (3.1), if $(d,v,y)\in\mathcal{X}^3$ satisfies (3.1) and F is given by (3.4b), then (d,v,y) satisfies (3.3), and conversely. Therefore, (3.1) is equivalent to (3.3). If the feedback scheme is c.l. stable, then Q must be stable as it equals $-K_{32}$. If, in addition, P is assumed stable, then all branches in (3.3) are stable, as claimed.

Conversely, by a similar argument, any model reference scheme (3.1) is equivalent to a feedback scheme (3.3). Suppose that the branches of (3.3), namely, P_1, Q, and $P=P_1$ are all stable. Then, all the c.l. operators of (3.3), namely, $\{K_{ij}\}_{i,j=2,3,4,5}$, must be stable because they can be expressed in terms of sums and products of the stable operators P, P_1, Q, and I. The last assertion follows from the expressions for the diagonal c.l. operators K_{ii} of (3.3), namely,

$$K_{22}=I-QP_1,\ K_{33}=I-P_1Q,\ K_{44}=K_{55}=I$$

and the fact, easily checked by inspection, that the remaining c.l. operators K_{ij}, $i\neq j$, are products of the K_{ii} by P, P_1, Q, or I. It follows that (3.3) is c.l. stable.

b) If $P=P_1$ and (d,y) satisfies (3.1) or (3.3), and (3.4) holds, then (3.6) is obtained by substitution of (3.5) into (3.2a). Q.E.D.

The operator $(I-P_1Q)$ appearing in (3.6) will reappear as a factor in most expressions for sensitivity. It will be called the *sensitivity operator* and denoted by E. For equivalent schemes $E=(I+P_1F)^{-1}$.

Remarks 3.1:

a) The model-reference scheme has some remarkable features. Unlike most feedback arrangements, it is a realization which cannot be made unstable by any choice of Q, at least for stable plants in the absence of plant uncertainty. Under these assumptions,[3] any allowable feedback law can be realized in the form of an equivalent model reference scheme, with the guarantee that all branches will be stable, and the closed-loop system automatically stable. *The design of Q, whether for small sensitivity or other purposes, can be accomplished without concern for closed-loop stability.*

In engineering applications, model reference schemes are realizable in principle, but may have undesirable features. For example, they may have high sensitivity to errors in the realization of Q. Unstable inner loops, obtained whenever $(I-P_1Q)^{-1}$ is unstable, may present reliability problems. Even then, the fractional transformation remains advantageous from the viewpoint of theory, as potentially unstable feedbacks F are replaced by stable operators Q. In later sections on plant uncertainty, the flowgraph interpretation of the model reference scheme will provide a convenient guide to perturbation analysis. It will also appear that model-reference schemes have a useful plant-invariance property.

b) Implicit in our notion of an allowable feedback is the view that each feedback realization involves a graph, and that although most of the internal details of the realization may be unimportant, closed-loop stability at *all* internal nodes is essential.

c) Theorem 1 holds even if F and Q are in $\mathbf{A}$ but not strictly causal. However, strict causality is a prerequisite for physical realizability, and will therefore have to be assumed in subsequent theorems.

B. Unstable Plants

The assumption in Theorem 1 that the plant P is stable will now be relaxed. Consider a plant $P_0\in\mathbf{A}_s$ with disturbance d at the output, which is unstable but for which there exists a *stabilizing feedback*, i.e., an operator $F_0\in\mathbf{A}_s$ which gives a c.l. stable feedback scheme on being fed back around P_0. The stabilized system can be incorporated in a model reference scheme, by letting P be the stabilized c.l. operator $P_0(I+F_0P_0)^{-1}$ and d be the stabilized disturbance $(I+P_0F_0)^{-1}d_0$, and Q (or F) can be selected as for a stable plant. At this point the question arises: "can F_0 be selected independently of Q (or F), or could the prior choice of F_0 prejudice the class of achievable systems?"

In general, the choices are not independent, even for stable plants, because the application of two unstable feedbacks in succession may give a result different from the application of a single feedback equal to their sum. Consider the following frequency response example in $\mathbf{H}_e^\infty$. Let $\hat{p}(s)=1$ and $\hat{f}_1(s)=\hat{f}_2(s)=s^{-1}$. The application of a single feedback $\hat{f}(s)$ equal to $\hat{f}_1(s)+\hat{f}_2(s)$ gives a c.l. stable feedback scheme, with c.l. responses 1, $(s+2)^{-1}$, and $s(s+2)^{-1}$. However, if the feedback is split into two branches, the c.l. response across either one of these branches is $(s+1)/s(s+2)$, i.e., the system is not c.l. stable. Popular belief notwithstanding, *c.l. stable systems do not form an additive group under feedback if the complete set of c.l. operators is considered.*

However, if feedbacks are constrained to be stable then choices are independent, as the following construction shows.

Let $P_0\in\mathbf{A}_s$ be an unstable plant which can be stabilized by either one of two feedbacks, F_a and F_b in $\mathbf{A}_s$, and label the resulting feedback schemes (a) and (b), respectively. We would like to find an operator $F_{ab}\in\mathbf{A}_s$ which on being fed back around scheme (a), as shown in Fig. 3(a), produces a *two-stage feedback scheme* equivalent to scheme (b). (Observe that the two-stage feedback scheme has extra nodes in the feedback branches to allow for the possibility of noise sources there.)

Proposition 3.2: If F_a and F_b are stable, then the stable feedback $F_{ab}\triangleq F_b-F_a$ makes the two-stage feedback scheme c.l. stable and equivalent to scheme (b).

Proof: The two-stage scheme is obviously equivalent to scheme (b), and is c.l. stable because its c.l. operators consist of: i) the c.l. operators K_{22}, K_{23}, K_{32}, and K_{33} of scheme (a) or scheme (b), which are stable by hypothesis, or ii) sums and products of the operators listed in i), and

[3]If P is unstable, the parameterization of c.l. stable schemes by a single operator Q is obviously still possible. However, some of our other conclusions, concerning existence of a feedback realization with stable elements, structural stability, or decomposition properties, may no longer be valid.

the operators F_a or F_b which are stable by hypotheses. Q.E.D.

The following *decomposition principle* can be obtained immediately from Theorem 1 and Proposition 3.2. Let $P_0 \in \mathbf{A}_s$ be any unstable plant stabilizable by a set of stable feedbacks in $\mathbf{B}_s$. *Any closed-loop stable feedback scheme employing a stable feedback around the plant P_0 is equivalent to a closed-loop stable scheme consisting of: i) a stabilizing feedback $F_0 \in \mathbf{B}_s$ which can be selected arbitrarily, followed by ii) a model reference scheme with stable operators Q and P_1.*

It follows that under our hypotheses,[3] and in particular under the assumption that plant and feedback are not simultaneously unstable, the problem of sensitivity reduction can be decomposed into two independent problems: stabilization followed by desensitization of a stable system. Henceforth, we shall confine ourselves to the second problem.

III′. Approaches to Feedback-Sensitivity Minimization

A′. Quadratic versus Induced Norms

The main properties of feedback cannot be deduced without some notion of uncertainty. Suppose that the disturbance d is uncertain but belongs to some subset $\mathfrak{D}$ of possible disturbances in $\mathfrak{X}$. From (3.6) it is clear that for disturbances to be attenuated, $(I-P_1Q)$ must be small on $\mathfrak{D}$, i.e., Q must act as an approximate inverse of P_1 on $\mathfrak{D}$. The various approaches to the disturbance attenuation problem are differentiated by the way in which uncertainty is described, and this approximate inversion is metricized and calculated.

A typical WHK approach in a deterministic version could be viewed as follows: $\mathfrak{D}$ consists of the set of disturbances d in $L_2(0,\infty)$ possessing a single, fixed, known power spectrum $|\hat{d}(j\omega)|$ in $L_2(-\infty,\infty)$, and the object of design is to find a filter Q that minimizes the quadratic distance $\|d-P_1Qd\|_{L_2}$, where in general Q depends on $|\hat{d}(j\omega)|$. In the stochastic analog of this problem $\mathfrak{D}$ is a random process characterized by probability-covariance functions and metricized by a quadratic norm. This description of uncertainty has certain limitations that we would like to circumvent, namely, the following.

1) The covariance properties of the random process must be known. In practice, they are often unknown elements of prescribed sets.

This is merely a limitation on the class of random processes for which WHK is valid. More serious from the point of view of feedback theory is the following observation.

2) The quadratic norm on plants employed in the WHK method lacks the multiplicative property $\|PQ\|_{L_2} \leqslant \|P\|_{L_2}\|Q\|_{L_2}$, and in general it may be difficult or impossible to estimate the norm of a product PQ from the norms of P and Q. The product norm $\|PQ\|_{L_2}$ may be large even though $\|P\|_{L_2}$ and $\|Q\|_{L_2}$ are small.

Consequently, if plant uncertainty is metricized by the quadratic norm, its propagation through products and inverses is hard to study. This is a serious limitation in feedback problems in which expressions such as $(I+PF)^{-1}$ play a major role.

By contrast, the "M_m" spec which is widely used in classical design measures the maximum frequency response magnitude, and is essentially the induced operator norm $\|P\| \triangleq \sup\{\|Px\|_{L_2}/\|x\|_{L_2}: x \in L_2\}$, which has the multiplicative property, and is therefore convenient to estimate in cascaded systems. By describing plant uncertainty in terms of a sphere of specified radius in a norm having such a multiplicative property, it is possible to obtain a general approach to problems involving disturbances/random processes which are unknown but belong to prescribed sets, as we shall see in Section IV.

Minimization of an induced norm in effect amounts to a minimax solution. Minimax methods do not necessarily represent uncertainty with greater fidelity than quadratic methods. However, the concern here is less with fidelity than with the ability to handle product systems.

B′. Constraints on the Norm of a Sensitivity Operator

It is natural then to try to pose sensitivity reduction problems in terms of the minimization of norm of the sensitivity operator, and to employ a norm having multiplicative properties. The primary norm of a Banach algebra has such properties, but the following propositions show it to be useless for this purpose.

Proposition 3.3: If P and Q are in a Banach algebra $\mathbf{B}$ and $\|I-PQ\|<1$, then PQ has an inverse in $\mathbf{B}$.

Proof: Denote $(I-PQ)$ by E. As $\|E\|<1$, the power series $I-E+E^2-\cdots$ converges to an operator which inverts $(I-E)$. As $(I-E)=PQ$, $(PQ)^{-1}$ exists. Q.E.D.

Proposition 3.3 has occasionally been interpreted as showing that invertibility is necessary for sensitivity reduction. This interpretation is empty. In fact, since strongly causal operators never have inverses in $\mathbf{B}$ [see Proposition 2.1b)], we have the following.

Corollary 3.4: If PQ is in $\mathbf{B}_s$, then $\|I-PQ\| \geqslant 1$.

It is impossible to make the sensitivity operator less than 1 in the original $\mathbf{B}$ norm. In H_0^∞ this simply means that the frequency response of PQ approaches 0 at infinite frequencies and $(I-PQ)$ approaches 1.

An obvious idea at this point is to make $(I-P_1Q)$ small in norm over some finite frequency band, i.e., over an invariant subspace. The next proposition shows that norms over invariant subspaces usually are not useful measures of sensitivity for optimization purposes. Let $\mathfrak{B}_1$ be a subspace of $\mathfrak{B}$, Π a projection operator onto $\mathfrak{B}_1$, and suppose that $\mathfrak{B}_1$ is invariant[4] under $\mathbf{B}$, i.e., $R\Pi=\Pi R\Pi$ for each R in $\mathbf{B}$. Let α denote the norm of $(I-P_1Q)$ restricted to the subspace $\mathfrak{B}_1$ and optimized over all Q, i.e., $\alpha \triangleq \inf_{Q\in\mathbf{B}_s}\sup\{\|(I-P_1Q)\Pi d\|: d\in\mathfrak{B} \text{ and } \|d\|=1\}$.

Proposition 3.5: For any P in $\mathbf{B}_s$, $\alpha=1$ or $\alpha=0$.

The proof is in Appendix I. If $\alpha=0$, sensitivity can be made arbitrarily small over the subspace $\mathfrak{B}_1$. In practice there are special cases involving "minimum phase" systems

[4] More generally, Proposition 3.5 holds if $\mathbf{B}$ is replaced by any of its norm preserving extensions. Note that Π is not necessarily in the algebra $\mathbf{B}$ or causal.

in which solutions that approach $\alpha=0$ may be useful. More typically, this result is achieved at the expense of increasing the sensitivity without bound on complements of $\mathcal{B}_1$; in such cases, the norm of a restriction of $(I-P_1Q)$ is not a candidate for minimization.

Corollary 3.4 and Proposition 3.5 delineate some of the peculiarities of the sensitivity optimization problem. In one form or another these peculiarities were recognized in the classical theory, and are probably the reason why it stopped short of optimization. We shall try to circumvent them by introducing an auxiliary (semi) norm to which they do not apply.

IV. Multiplicative Seminorms and Approximate Inverses

Uncertainty in a disturbance (or plant) in a linear space can be specified in terms of belonging to a ball of disturbances (or plants) centered at some nominal value, and of radius specified in some norm. Such a description of uncertainty may be cruder than a probabilistic description, but is usually more tractable in feedback problems.

One of the axioms of a norm asserts that only the zero element has zero norm. This axiom is often not needed, and with its elimination a norm is replaced by the slightly more general concept of a seminorm.

A ball in any seminorm can be shown to be a convex set.[4] Conversely, any convex set[5] in a linear space generates a seminorm (see Rudin [22, p. 24]) known as the Minkowski functional of that set. In linear spaces convex sets[4] of uncertainty can therefore always be described in terms of seminorms. We shall employ seminorms to obtain a systematic approach to such sets of uncertainty (cf. the objectives outlined in Section I-A).

In the next section, we shall define classes of left and right seminorms. To motivate the definitions, let us find seminorm descriptions for two disturbance sets which can be generated by the interaction of filters and certain "flat" disturbance sets.

Henceforth, W will denote a stable causal operator of unit norm, which will play the role of a weighting filter. For concreteness, W can be thought of as an operator in $\mathbf{H}_0^\infty$, W: $H^\infty \to H^\infty$, with response $\hat{w}(s)=k(s+k)^{-1}$. $\mathcal{D}$ will denote a flat disturbance set (analogous to white noise) consisting of the unit ball in the space of inputs, in this case in H^∞:

$$\mathcal{D}_1=\{d_1\in H^\infty: |\hat{d}_1(j\omega)|\leq 1\}$$

whose elements are frequency functions of unknown but bounded magnitude. Consider two situations.

1) Let $\mathcal{D}$ be the set mapped by W into flat disturbances, i.e., $W\mathcal{D}=\mathcal{D}_1$. $\mathcal{D}$ can be described as the unit ball in the seminorm $\|\cdot\|_\ell$, defined on H^∞ by the equation $\|d\|_\ell \triangleq \|Wd\|$.

2) Let $\mathcal{D}$ be the set into which W maps flat disturbances, i.e., $W\mathcal{D}_1=\mathcal{D}$. $\mathcal{D}$ is also the set

$$\mathcal{D}=\{d\in \text{range}(W): |\hat{d}(j\omega)|\leq|\hat{w}(j\omega)|\}.$$

$\mathcal{D}$ can be described as the unit ball in the seminorm $\|\cdot\|_r$ defined on the range of W (which is a proper subspace of H_0^∞) by the equation $\|d\|_r \triangleq \|W^{-1}d\|$.

There are many engineering problems in which the apriori information about disturbances is in the form of an upper bound to the magnitudes of their possible frequency responses. The seminorm description 2) is natural for such problems, and 1) occurs in inverse problems.

The seminorm $\|\cdot\|_r$ employs an up-weighting, and $\|\cdot\|_\ell$ employs a down-weighting. These two examples generalize into the notion of left and right seminorms, defined as follows.

A. *Seminorms for Inputs and Outputs*[6]

Let $\mathcal{Y}$ be a $\|\cdot\|$-normed linear space. Let $\|\cdot\|_\ell$ be any seminorm defined on all of $\mathcal{Y}$, and $\|\cdot\|_r$ a seminorm defined on some nontrivial subspace $\mathcal{Y}_r$ of $\mathcal{Y}$. The seminorm $\|\cdot\|_r$ is said to *dominate* $\|\cdot\|_\ell$ iff $\|y\|_\ell\leq\|y\|_r$ for all y in $\mathcal{Y}_r$; this dominance is denoted by $\|\cdot\|_\ell\leq\|\cdot\|_r$.

Definition: A *left seminorm* is any seminorm defined on all of $\mathcal{Y}$ with the property that $\|\cdot\|_\ell\leq\|\cdot\|$. A *right seminorm* is any seminorm defined on a nontrivial subspace $\mathcal{Y}_r$ of $\mathcal{Y}$ with the property that $\|\cdot\|\leq\|\cdot\|_r$.

For example, if W is any $\mathbf{H}^\infty$ filter of unit norm, the expressions $\|y\|_\ell \triangleq \|Wy\|$ and $\|y\|_r \triangleq \|W^{-1}y\|$ define left and right seminorms, on H^∞ and the range of W, respectively. The range of W is a subspace of H^∞, proper whenever $\hat{w}(s)$ has zeros in the right half-plane or at ∞.

B. *Weighted Seminorms for Plants*

Definition: A weighted seminorm is any seminorm $\|\cdot\|_w$ on the $\mathbf{B}$ with the property that $\|\cdot\|_W\leq\|\cdot\|$.

The terms "weighted" and "left" are synonymous. We shall use the term "weighted" to distinguish the left seminorms on $\mathbf{B}$ used as measures of plant sensitivity from the others.

A weighted seminorm on $\mathbf{B}$ is *induced* by a pair $(\|\cdot\|_\ell, \|\cdot\|_r)$, where $\|\cdot\|_\ell$ is a left seminorm on the space $\mathcal{B}$ (of outputs), and $\|\cdot\|_r$ is a right seminorm on a subspace $\mathcal{B}_r\subset\mathcal{B}$ (of inputs), iff $\|\cdot\|_W$ is defined for $A\in\mathbf{B}$ by the equation

$$\|A\|_W=\sup\{\|Au\|_\ell/\|u\|_r: u\in\mathcal{B}_r \text{ and } \|u\|_r\neq 0\}.$$

It follows that $\|I\|_W\leq 1$.

In control problems, weightings are often introduced by filters, which act on disturbances either before entering a plant or after leaving it. For example, let W_ℓ and W_r be linear mappings in $\mathcal{X}\times\mathcal{X}$, each of unit $\mathcal{B}$-induced norm. A

[5]Satisfying the following additional assumptions: 1) if x is in the linear space, then αx is in the set for some real α; 2) if y is in the set then so is $-y$.

[6]*Convention*: Whenever x belongs to a space on which several norms are defined, the unsubscripted norm $\|x\|$ denotes the principal norm. Weighted norms will be designated by subscripts.

left seminorm is defined on $\mathfrak{B}$ by the equation $\|y\|_\ell \triangleq \|W_\ell y\|$. Let $\mathfrak{B}_r$ be the range of W_r; if W is 1:1, a right seminorm is defined on $\mathfrak{B}_r$ by the equation $\|u\|_r \triangleq \|W_r^{-1}u\|$. The pair $(\|\cdot\|_r, \|\cdot\|_\ell)$ induces the weighted seminorm $\|A\|_W = \|W_\ell A W_r\|$ on the space $\mathbf{B}$.

Although weightings produced by filters will be emphasized in this paper, they can be produced by other means. For example, a weighted seminorm on $\mathbf{H}_0^\infty$ is given by the supremum over a shifted half-plane,

$$\|P\|_W = \sup\{|\hat{p}(s)|: \operatorname{Re}(s) \geq \alpha\}, \alpha > 0.$$

1) Multiplicative Seminorms—Symmetric Case: In general, weighted seminorms lack the multiplicative property $\|CD\|_W \leq \|C\|_W \|D\|_W$ of algebra norms. In problems involving the attenuation of a single disturbance (or single random process) this need not matter, as multiplications can be avoided. However, in problems involving plant uncertainty, closed-loop perturbations have the product form $(I-PQ)\Delta P$. We shall employ seminorms with weaker multiplicative properties suitable for such products.

Definition: A *symmetric (weighted) seminorm* on the algebra $\mathbf{B}$ is a weighted seminorm $\|\cdot\|_W$ on the space $\mathbf{B}$ which satisfies the multiplicative inequalities

$$\|CD\|_W \leq \|C\|_W \|D\|, \quad \|CD\|_W \leq \|C\| \cdot \|D\|_W. \tag{4.1}$$

Any operator W: $\mathfrak{X} \to \mathfrak{X}$ of unit $\mathfrak{B}$-induced norm which commutes with all operators of $\mathbf{B}$ defines a symmetric (weighted) seminorm by the equation $\|A\|_W \triangleq \|WA\| = \|AW\|$.

Symmetric seminorms have the property that $\|I\|_W = 1$.

2) Multiplicative Seminorms—General Case: In multivariable systems the plant perturbation ΔP always appears on the right of the product $(I-PQ)\Delta P$, and often does not commute with $(I-PQ)$. In such cases a more general class of multiplicative seminorms will be used. If ΔP is strictly causal, then the product $(I-PQ)\Delta P$ lies in a proper subspace of $\mathbf{B}$ (which is a left ideal, although we have no immediate use for this fact). With such products in mind, we make the following definition.

Definition: Let $\mathbf{B}_r$ be any subspace of $\mathbf{B}$, and $\mathbf{B}\cdot\mathbf{B}_r$ denote the space of products $\{CD: C \text{ in } \mathbf{B} \text{ and } D \text{ in } \mathbf{B}_r\}$. A *multiplicative seminorm* on the space of products $\mathbf{B}\cdot\mathbf{B}_r$ is any weighted seminorm on the space $\mathbf{B}\cdot\mathbf{B}_r$ with the following additional property: there is a left seminorm $\|\cdot\|_{W\ell}$ defined on $\mathbf{B}$, a seminorm $\|\cdot\|_{W1}$ defined on $\mathbf{B}_r$, and the inequality $\|CD\|_W \leq \|C\|_{W\ell}\|D\|_{W1}$ holds[7] for all C in $\mathbf{B}$ and D in $\mathbf{B}_r$.

Note that $\|\cdot\|_{W1}$ is not necessarily a right seminorm.

An example of a multiplicative seminorm is obtained as follows. Let $\|\cdot\|_W$ be the seminorm produced by operators W_ℓ and W_r in the example of Section IV-B [preceding Section IV-B1)]; V: $\mathfrak{X} \to \mathfrak{X}$ be any 1:1 map of unit norm; and $\mathbf{B}_r \triangleq \{VD_1: D_1 \in \mathbf{B}\}$. Define the seminorms $\|C\|_{W\ell} \triangleq \|W_\ell C V\|$ for $C \in \mathbf{B}$, and $\|D\|_{W1} \triangleq \|V^{-1}DW_r\|$ for $D \in \mathbf{B}_r$. Then, $\|\cdot\|_W$ has the multiplicative property claimed, as

$$\|CD\|_W = \|W_\ell CDW_r\| \leq \|W_\ell CV\| \cdot \|V^{-1}DW_r\|$$
$$= \|C\|_{W\ell}\|D\|_{W1}.$$

A *symmetric* seminorm can be viewed as a special case of a multiplicative seminorm on the space of products $\mathbf{B}\cdot\mathbf{B}_r$, in which $\mathbf{B}_r = \mathbf{B}$, and the multiplicative inequality holds for each of two pairs of seminorms, namely, $(\|\cdot\|_W, \|\cdot\|)$ and $(\|\cdot\|, \|\cdot\|_W)$.

C. *Approximate Inverses and Singularity Measures*

Many problems of feedback theory, both classical and modern, can be reduced to the construction of an approximate inverse. Let $\|\cdot\|_W$ be a fixed, weighted seminorm on the space $\mathbf{B}$.

Definition: For any operator P in $\mathbf{B}$, an *approximate right inverse* (in $\mathbf{B}_s$) of P is any operator Q in $\mathbf{B}_s$ for which[8] $\|I-PQ\|_W < \|I\|_W$; the *right singularity measure* (in $\mathbf{B}_s$) of P (under $\|\cdot\|$), denoted by $\mu(P)$, is

$$\mu(P) = \inf\{\|I-PQ\|_W: Q \text{ in } \mathbf{B}_s\}.$$

In general, $\mu(P)$ is a number in the interval $0 \leq \mu(P) \leq 1$. The last inequality follows from the observation that $Q=0$ gives $\|I-PQ\|_W = \|I\|_W$.

In all of the following $\mathbf{H}^\infty$ examples, $\|\cdot\|_W$ will be the symmetric weighted norm defined by $\|A\|_W \triangleq \|WA\|$ for $A \in \mathbf{H}^\infty$, where $W \in \mathbf{H}_0^\infty$ is a fixed (strictly causal) operator of unit norm. For example, W can be the "low-pass" frequency response $\hat{w}(s) = k(s+k)^{-1}$, $k>0$. As $\|\cdot\|_W$ is symmetric, it has the multiplicative properties defined in Section IV-B1).

Example 4.1: P_a is a plant in $\mathbf{H}_0^\infty$ with frequency response $\hat{p}_a(s) = \alpha(s+\alpha)^{-1}$, $\alpha>0$. The sequence of operators Q_n in $\mathbf{H}_0^\infty$ with frequency responses $\hat{q}_n(s) = \alpha^{-1}(s+\alpha) \cdot n^2(s+n)^{-2}$, $n=1,2,\cdots$, satisfies the equation

$$\|\hat{w}(1-\hat{p}\hat{q}_n)\|_{H^\infty} = \sup_{\operatorname{Re}(s)\geq 0}\left|\hat{w}(s)\left(1-\frac{n^2}{(s+n)^2}\right)\right|. \tag{4.1}$$

The right-hand side (RHS) of (4.1) approaches 0 as $n \to \infty$. Therefore, the singularity measure in $\mathbf{H}_0^\infty$ of P_a under $\|\cdot\|_W$ is $\mu(P_a) = 0$. The operators Q_n are approximate inverses, and the sequence Q_n is an example of what will be called an inverting sequence.

Example 4.2: P_a is the operator of Example 4.1; P_h is the "nonminimum phase" operator in $\mathbf{H}^\infty$ with frequency response $\hat{p}_h(s) = (\beta-s)(\beta+s)^{-1}$, $\beta>0$; P_1 is the product, $P_1 \triangleq P_h P_a$. For any Q in $\mathbf{H}_0^\infty$, $\hat{w}(s)[1-\hat{p}_1(s)\hat{q}(s)]$ has the value $\hat{w}(\beta)$ at the zero of $\hat{p}_h(s)$. Therefore, $\|I-P_1Q\|_W \geq |\hat{w}(\beta)|$, and we get the lower bound $\mu(P_1) \geq |\hat{w}(\beta)|$. In

[7] The definition can obviously be generalized for the case of perturbations appearing on the left.

[8] Recall that $\|I\|_W \leq 1$, and $\|I\|_W = 1$ if $\|\cdot\|_W$ is symmetric.

Section V-A and Corollary 6.1 it will be established that in fact $\mu(P_1)=|\hat{w}(\beta)|$.

In these examples we have emphasized approximate inverses under a multiplicative seminorm. In passing, it may be worth mentioning that WHK problems can be viewed as approximate inversion problems in which the weighted seminorm of $(I-PQ)$ is obtained by weighting by a fixed vector $d\in L^2(0,\infty)$, to obtain $\|I-PQ\|_W=\|(I-PQ)d\|_{L^2}$. Here $\mu(P)$ is the irreducible error. However, $\|\cdot\|_W$ lacks the multiplicative properties.

No matter which seminorm is used $\mu(\cdot)$ has the following property.

Proposition 4.3: For any P_a and P_b in $\mathbf{B}$, $\mu(P_bP_a)\geqslant\mu(P_b)$.

Proof: If the contrary is assumed to be true, there is a Q in $\mathbf{B}_s$ for which $\|I-P_bP_aQ\|_W<\mu(P_b)$; P_aQ now acts as an approximate inverse for P_b, and there is a contradiction.

V. Sensitivity to Disturbances and Approximate Invertibility

We shall show that approximate invertibility of the plant is a necessary and sufficient condition for the existence of a feedback to attenuate disturbances, and the optimal sensitivity depends on the measure of singularity of the plant.

Consider the feedback scheme, (3.1), and suppose the plant P in $\mathbf{B}$ equals the nominal P_1. Let $\|\cdot\|_r$ be a fixed right seminorm defined on some subspace $\mathfrak{B}_r$ of (inputs), $\|\cdot\|_\ell$ a fixed left seminorm on (outputs) $\mathfrak{B}$, and $\|\cdot\|_W$ the resulting weighted seminorm induced on $\mathbf{B}$. The *sensitivity to disturbances* η_1 is defined by the equation $\eta_1\triangleq\sup\{\|y\|_\ell/\|d\|_r: d \text{ in } \mathbf{B}_r, \|d\|_r\neq 0\}$. From (3.2a) it follows that $\eta_1=\|(I+P_1F)^{-1}\|_W$. Whenever the equivalence equations (3.4) hold, η_1 also equals $\|I-P_1Q\|_W$.

The sensitivity η_0 obtained when $F=0$ will be called the *open-loop sensitivity*. As $\eta_0=\|I\|_W$, $\eta_0\leqslant 1$, and $\eta_0=1$ whenever $\|\cdot\|_W$ is symmetric.

Theorem 2:

a) A necessary and sufficient condition for the existence of a feedback F in $\mathbf{A}_s$ for which (3.1) is closed-loop stable and the sensitivity η_1 is less than the open-loop value η_0, is that P_1 have an approximate (stable) right inverse Q in the algebra[9] $\mathbf{B}_s$.

b) For any $\epsilon>0$, there is a feedback F in $\mathbf{A}_s$ for which (3.1) is closed-loop stable and for which $\eta_1<\mu(P_1)+\epsilon$, but no F for which (3.1) is closed-loop stable and $\eta_1<\mu(P_1)$, where $\mu(P_1)$ is the measure of right singularity in $\mathbf{B}_s$ of P_1 under $\|\cdot\|_W$.

Proof: By Theorem 1, any closed-loop stable feedback scheme (3.1) with F in $\mathbf{A}_s$ is equivalent to a model reference scheme (3.3) with Q in $\mathbf{B}_s$. Let (3.3) be equivalent to (3.1). η_1 can be expressed by $\eta_1=\|I-P_1Q\|_W$. Therefore, a) $\eta_1<\eta_0$ iff Q is an approximate right inverse of P_1. Also, b) by definition of $\mu(P_1)$, there is a Q in $\mathbf{B}_s$ for which $\|I-P_1Q\|_W<\mu(P_1)+\epsilon$, but none for which $\|I-P_1Q\|_W<\mu(P_1)$. The conclusion concerning F follows from the equivalence of (3.1) and (3.3). Q.E.D.

In general, it may be impossible to attain the sensitivity $\mu(P_1)$.

Definition: A sequence $Q_n\in\mathbf{A}$, $n=1,2,\cdots$, will be called *optimal* for P_1 iff the sequence of sensitivities $\|I-P_1Q_n\|_W$ approaches $\mu(P_1)$ as $n\to\infty$. $\mu(P_1)$ will be called the *optimal sensitivity* for P_1.

A sequence of feedbacks $F_n\in\mathbf{A}_s$ will be called optimal iff the equivalent comparators Q_n are optimal.

Remark: If the disturbance d lies in a balanced set $\mathfrak{D}$ (i.e., d in $\mathfrak{D}$ implies $-d$ in $\mathfrak{D}$), it is simple to show that no open-loop control, obtained by letting $Q=0$ and applying an input at node 2, can make $\|y\|_\ell$ less than $\|d\|_\ell$ for all d in $\mathfrak{D}$. It follows from Theorem 2 that, for right invertible plants, optimal sensitivity achievable with feedback is smaller than without; in other words, ability of control schemes to cope with unknown disturbances depends on their configuration. This can be viewed as a continuation of the internal model principle [16] to seminormed disturbances.

A. An Example of Sensitivity Optimization

We would like to show that feedback optimization is feasible under a seminorm that has the multiplicative properties (4.1), unlike the quadratic norm of WHK methods. If weighting is obtained from a filter, the optimal feedback in $\mathbf{H}^\infty$ resembles a classical lead–lag network. We shall try to demonstrate these points by an example. A more comprehensive theory of H^∞ optimization would take too long to present here.

Let P_1 in $\mathbf{H}_0^\infty$ be any plant with a single $\mathrm{Re}(s)>0$ zero at $s=\beta$, subject to the high frequency restriction $|\omega\hat{p}_1(j\omega)|^{-1}\leqslant$const. for $|\omega|>1$. Let $\|\cdot\|_W$ be the weighted seminorm, $\|P\|_W\triangleq\|WP\|$, where $\hat{w}$ in H_0^∞ satisfies the condition $|\omega^\ell\hat{w}(j\omega)|^{-1}\leqslant$const. for some integer ℓ. An optimal sequence of comparators Q_n in $\mathbf{H}_0^\infty$ is sought.

It will be shown that *a sequence with frequency responses*

$$\hat{q}_m(s)=c_mP_b^{-1}(s)\left[1-\hat{w}(\beta)\hat{w}^{-1}(s)\right](s+m)^{-\ell}(s+n_m)^{-2},\quad m,n_m=1,2,\cdots \qquad (5.1)$$

in which: P_b is the operator of Example 4.2; $c_m=m^\ell n_m^2$; and for each m, n_m is a sufficiently large integer, is such an optimal sequence. In fact, $\mu(P_1)=|\hat{w}(\beta)|$, and the sequence of sensitivities $\|I-P_1Q_m\|_W$ approaches $\mu(P_1)$ as $m\to\infty$.

Q_m will be constructed here, but some of the proof details will be postponed until Section VI. P_1 can be factored into the product[10] $P_1=P_aP_b$, in which $\hat{p}_b(s)=(\beta-s)(\beta+s)^{-1}$, and $|\hat{p}_b(j\omega)|=1$; and $\hat{p}_a$ is in H_0^∞, $\hat{p}_a(j\omega)=$

[9]The fact that given $P\in\mathbf{B}$, the closed-loop system is stable iff $F(I+P_1F)^{-1}$ is stable is pointed out by Desoer–Chan [21, Theorem 3], who scrutinize the relationship between open- and closed-loop stability in the context of convolution algebras.

[10]P_a/P_b are often referred to as all-pass/minimum phase or inner/outer factors.

$\hat{p}_1(j\omega)/\hat{p}_b(j\omega)$ for all real ω, and $\hat{p}_a$ has no zeros in $\mathrm{Re}(s)\geqslant 0$. Optimal sequences will be constructed for P_a and P_b separately.

P_b coincides with the operator of Example 4.2. By the reasoning of that example we get the lower bound $\mu(P_1)\geqslant |\hat{w}(\beta)|$; in fact, for the $\hat{p}_b$ factor alone, $\|\hat{w}(1-\hat{p}_b\hat{q})\|_{H^\infty}\geqslant |\hat{w}(\beta)|$ whenever $\hat{q}$ preserves the analyticity of $\hat{w}\hat{p}_b\hat{q}$ in $\mathrm{Re}(s)\geqslant 0$.

We observe next that the function $\hat{q}_b$: $\mathbb{C}\to\mathbb{C}$, $\hat{q}_b \triangleq \hat{p}_b^{-1}(s)[1-\hat{w}(\beta)\hat{w}^{-1}(s)]$ exactly minimizes $\|\hat{w}(1-\hat{p}_b\hat{q}_b)\|$, since $\hat{w}(s)[1-\hat{p}_b(s)\hat{q}_b(s)]$ equals the lower bound $\hat{w}(\beta)$ for all s in $\mathbb{C}$. Also, $\hat{q}_b$ is analytic in $\mathrm{Re}(s)\geqslant 0$, as $[1-\hat{w}(\beta)\hat{w}^{-1}(s)]$ has a zero at β to cancel the pole $\hat{p}_b^{-1}$. We now combine $\hat{q}_b$ with enough high frequency attenuation to obtain an H^∞ sequence:

$$\hat{q}_{bm}(s)=\hat{p}_b^{-1}(s)\left[1-\hat{w}(\beta)\hat{w}^{-1}(s)\right]m^\ell(s+m)^{-\ell}\qquad m=1,2,\cdots.$$

(It will become apparent that the attenuation $m\ell(s+m)^{-\ell}$ will be at a high enough frequency to have arbitrarily little effect on weighted sensitivity.) The factor $P_a(s)$ can be inverted by the sequence in H_0^∞,

$$\hat{q}_{an}(s)=\hat{p}_a^{-1}(s)n^2(s+n)^{-2},\qquad n=1,2,\cdots.$$

The sequence $\hat{q}_m$ of (5.1) is constructed from the product of $\hat{q}_{an}$ and $\hat{q}_{bm}$. The validity of this construction is established in Corollary 6.1.

Remarks: It may be worth looking at a special case of (5.1) to get a better feel for the kind of feedbacks our approach generates. Let $\hat{p}_a(s)=\alpha(s+\alpha)^{-1}$ and $\hat{w}(s)=k(s+k)^{-1}$, $\alpha>0$, $k>0$. Equation (5.1) gives

$$\hat{q}_m(s)=c_m(s+\beta)(s+m)^{-1}(s+n_m)^{-2}. \qquad (5.2)$$

$c_m \triangleq m^2n_m(k+\beta)^{-1}$. This optimal $\hat{q}_m$ consists of the "lead" factor $c(s+\beta)$, and high frequency poles whose purpose is to make $\hat{q}_m(s)$ strictly proper. The feedback law $\hat{f}(s)$ produced by $\hat{q}_m(s)$ [via (3.4b)] is a lead–lag network typical of classical control.

If the high frequency poles are neglected, the sensitivity operator $E \triangleq I-P_1Q_m$ has the frequency response $\hat{e}(j\omega)=\hat{w}(\beta)\hat{w}^{-1}(s)$; $\hat{e}$ is small at highly weighted frequencies and vice versa. If $\hat{e}(j\omega)$ is compared to the value that would be obtained by letting $\hat{q}(s)=1$, it appears that $\hat{q}_m(s)$ has the effect of trading undesirable low-frequency phase lags $\arg\hat{p}(j\omega)$, for undesirably large magnitudes $|\hat{p}(j\omega)\hat{q}(j\omega)|$ at high frequencies, where they matter less, at least if the specified weighting is correct. This, too, is a typical strategy of classical design.

The growth of $|\hat{e}(j\omega)|$ as $\omega\to\infty$ depends on the decay of $|\hat{w}(j\omega)|$; if the former is too high, the choice of the latter was inappropriate, if $\|\hat{w}(j\omega)|$ is bounded from below, then $|\hat{e}(j\omega)|$ is bounded from above. The dependence of the optimal filter on the weighting W is not surprising, as W describes the convex set of disturbances to be attenuated.

Filters optimal in this sense are known to be very sensitive to plant uncertainty, and are practical only where accurate plant models are available.

B. *Unstable Plants*

If P_0 is an unstable plant in $\mathbf{A}_s$ with output disturbance d_0 in $\mathfrak{B}$, and there is a *stabilizing feedback* F_0 with the property that $P_0(I+F_0P_0)^{-1}$ and $(I+P_0F_0)^{-1}$ are in $\mathbb{B}$, then we can let P_1 in Theorem 2 be the stabilized system $P_0(I+F_0P_0)^{-1}$, and d be the stabilized disturbance $(I+P_0F_0)^{-1}d_0$. In that case, (3.6) takes the form

$$y=(I-P_1Q)(I+P_0F_0)^{-1}d_0. \qquad (5.3)$$

The term $(I+P_0F_0)^{-1}$ contributed by stabilization appears in (5.3) as an extra right weighting on $(I-P_1Q)$. Provided $\|\cdot\|_W$ is modified to inlude the extra weighting, Theorem 2 remains valid.

Remark: We have preferred to separate feedback synthesis into two consecutive stages: 1) stabilization and 2) desensitization of (input–output) stable systems. From the point of view of input–output sensitivity theory, the separation is in a certain sense unavoidable, as the perturbations allowed for robust stabilization are radically different from those for desensitization. This point is elaborated in Appendix II.

C. *Optimal Sequences: Symmetric Case*

Usually, optimal filters for sensitivity reduction are not strictly proper and can not be attained, although they can be approached by sequences of filters of increasing bandwidth. The behavior of such sequences is conveniently described in terms of the concept of an "identity sequence," or "approximate identity" drawn from Banach algebras. Here we shall summarize those properties of identity sequences which are employed in our filter construction.

Let $\|\cdot\|_W$ be a fixed, symmetric[11] weighted seminorm on the algebra $\mathbb{B}$. A *weighted identity sequence* (widseq) is any sequence I_m, $m=1,2,\cdots$ in $\mathbb{B}$ with the property that $\|I_m\|=1$ and, for any A in $\mathbb{B}$, $\|A-I_nA\|_W$ and $\|A-AI_n\|_W$ approach 0 as $n\to\infty$. For any P in $\mathbf{A}$, a sequence Q_n, $n=1,2,\cdots$, is a *right weighted inverting sequence* (winvseq) iff PQ_n is a widseq.

Whenever $\|I-PQ_n\|_W\to 0$ as $n\to\infty$, Q_n must be a right winvseq for P; for then, the inequalities

$$\|A-PQ_nA\|_W\leqslant\|I-PQ_n\|_W\|A\|,$$
$$\|A-APQ_n\|_W\leqslant\|A\|\cdot\|I-PQ_n\|_W,$$

obtained using the multiplicative property of symmetric seminorms, have left-hand sides which converge to 0 as $n\to\infty$.

For example, in $\mathbb{H}^\infty$ the sequence of operators with frequency responses $n^r(s+n)^{-r}$, $n=1,2,\cdots$ and $r>0$ any constant integer, is a widseq for the weighted norm of

[11]Without symmetry a distinction between left and right identity sequences, etc., must be made.

Example 4.1. The sequence Q_n of that example is a winvseq for P_a.

It will be shown that in certain cases an optimal sequence of approximate inverses for a product P_bP_a can be obtained from separate sequences for the factors P_a and P_b, and consequently some optimization problems can be decomposed into simpler ones. Let $\|\cdot\|_a \geq \|\cdot\|_b$ be a pair of symmetric, weighted seminorms on the algebra $\mathbf{B}$. Let P_a in $\mathbf{B}_s$ and P_b in $\mathbf{B}$ have singularity measures $\mu_a(P_a)$ and $\mu_b(P_b)$ under $\|\cdot\|_a$ and $\|\cdot\|_b$, respectively, and optimal sequences $\{Q_{am}\}_{m=1}^{\infty}$ in $\mathbf{B}_s$, $\{Q_{bn}\}_{n=1}^{\infty}$ in $\mathbf{B}$, i.e., $\|I-P_iQ_{in}\|_i \to \mu_i(P_i)$ as $n\to\infty$ for $i=a$ or b.

Lemma 5.1: If $\mu_a(P_a)=0$, then $\mu_b(P_bP_a)=\mu_b(P_b)$, and P_bP_a has an optimal sequence of $\|\cdot\|_b$-weighted right approximate inverses in $\mathbf{B}_s$ of the form $Q_{an_m}Q_{bm}$, where for any $m=1,2,\cdots,n_m$ is a sufficiently large integer.

In fact, for any $\epsilon>0$ and $0<\alpha<1$, if we pick integers m and n_m such that $\|I-P_bQ_{bm}\|_b \leq \mu_b(P_b)+\alpha\epsilon$, and $\|I-P_aQ_{an_m}\|_a \leq (1-\alpha)\epsilon\|P_b\|^{-1}\|Q_{bm}\|^{-1}$, then

$$\|I-P_bP_aQ_{an_m}Q_{bm}\|_b \leq \mu_b(P_b)+\epsilon. \tag{5.4}$$

Proof:

a) $\mu_b(P_bP_a)\geq\mu_b(P_b)$ by Proposition 4.3. Let us establish the opposite inequality. For any $m>0$, $n>0$, we have

$$\begin{aligned}\|I-P_bP_aQ_{an}Q_{bm}\|_b &= \|I-P_bQ_{bm}-P_n(P_aQ_{an}-I)Q_{bm}\|_b\\ &\leq \|I-P_bQ_{bm}\|_b+\|P_b\|\cdot\|P_aQ_{an}-I\|_a\|Q_{bm}\|\end{aligned} \tag{5.5}$$

where the triangle inequality, the multiplicative property of symmetric seminorms, and the dominance $\|\cdot\|_b\leq\|\cdot\|_a$ have been used. The integers m and n specified in the hypothesis exist by definition of $\mu_b(P_b)$ and by the hypothesis that $\mu_a(P_a)=0$. (5.4) now follows from (5.5). As ϵ was arbitrary, (5.4) implies that $\mu_b(P_bP_a)\leq\mu_b(P_b)$. Therefore, $\mu_b(P_bP_a)=\mu_b(P_b)$, and $Q_{an_m}Q_m$ is an optimal sequence. Q.E.D.

VI. Multivariable Systems in $\mathbf{H}^{\infty N}$

At present there is much interest in the design of multivariable systems along classical lines, but there is little formal theory. Our next objective is to show that the theory outlined in Sections III–V provides a suitable framework for multivariable design.

Multivariable frequency responses will be viewed as elements of an n-dimensional version of the Hardy space H^{∞}. Our approach will be to decompose the frequency response matrix into a product of nearly invertible and noninvertible parts, and to design a feedback for each part separately. The validity of approaches based on decompositions will be established in Theorem 3. The optimal sensitivity scheme of Section V-A will be obtained as a corollary. Another corollary will validate the common hypothesis that sensitivity can be made arbitrarily small if the plant is nearly invertible. Finally, a lower bound to sensitivity in terms of the location of RHP zeros will be derived.

A. Terminology

N is a fixed integer. H^{2N} is the Banach space consisting of the N-fold product of H^2, on which the norm of any vector $\hat{u}=(\hat{u}_1,\hat{u}_2,\cdots,\hat{u}_N)$ is $\|\hat{u}\| \triangleq (\sum_{i=1}^{N}\|\hat{u}_i\|^2)^{1/2}$. The elements of H^{2N} can be represented by columns of input or output Laplace transforms.

$H^{\infty N}$ is the algebra of $N\times N$ matrices, $\hat{P}=[\hat{p}_{ij}]$, whose elements $\hat{p}_{ij}$ are causal, stable, frequency responses in H^{∞}. For any $\hat{P}$ in $H^{\infty N}$ and s in $\mathcal{C}$, $\sigma[\hat{P}(s)]$ denotes the square root of the magnitude of the largest eigenvalue of $[\hat{P}(s)]^*\hat{P}(s)$, where $*$ denotes the conjugate transpose of a matrix. $H^{\infty N}$ is a Banach algebra under the norm $\|\hat{P}\|=\sup_{\omega}\sigma[\hat{P}(j\omega)]$.

Each frequency response matrix $\hat{P}$ in $H^{\infty N}$ determines an operator P: $H^{2N}\to H^{2N}$, $\widehat{Pu}=\hat{P}\hat{u}$. The algebra of such operators is denoted by $\mathbf{H}^{\infty N}$, and is a Banach algebra under the norm $\|P\|_{\mathbf{H}^{\infty N}}=\|\hat{P}\|_{H^{\infty N}}$ The norm defined in this way coincides with the norm induced on $\mathbf{H}^{\infty N}$ by H^{2N}, i.e., $\sup\{\|Pu\|$: u in H^{2N} and $\|u\|=1\}$.

A frequency response $\hat{P}$ in $H^{\infty N}$ is called *strictly proper* iff each of its components $\hat{p}_{ij}$ is strictly proper; the corresponding operator P in $H^{\infty N}$ is called *strictly causal*. The subalgebras of strictly proper or strictly causal elements in $H^{\infty N}$ or $\mathbf{H}^{\infty N}$ are denoted by $H_0^{\infty N}$ or $\mathbf{H}_0^{\infty N}$, respectively.

Let $\hat{w}$ be a (scalar) weighting function in H_0^{∞}. The weighted seminorm $\|P\|_W \triangleq \|\hat{w}\hat{P}\|_{H^{\infty N}}$ is defined on $\mathbf{H}^{\infty}$. Observe that $\hat{w}$ commutes with all operators in $\mathbf{H}^{\infty N}$, so that $\|\cdot\|_W$ is symmetric, and the multiplicative inequalities (4.1) are satisfied.

B. A Decomposition Theorem

We shall be interested in frequency response matrices whose rate of approach to zero as $s\to\infty$ is comparable to that of a power of $(1+s)$, and employ the following.

Notation: For any integer n, $(1+s)^nH^{\infty N}$ denotes the set of functions $\hat{P}(s)$ with the property that $(1+s)^{-n}\hat{P}(s)$ is in $H^{\infty N}$.

In the rest of Section VI, plants will be decomposed into products P_aP_b of strictly causal and approximately invertible parts under the following *assumption*:

a) P_a is a strictly causal operator in $\mathbf{H}_0^{\infty N}$; and

b) P_b is in $\mathbf{H}^{\infty N}$. The frequency response $\hat{P}_b$ has a *minimal* approximate right inverse $\hat{Q}_b$ in $(1+s)^tH^{\infty N}$; i.e., $\hat{w}\hat{P}_b\hat{Q}_b$ is in $H^{\infty N}$ and $\|\hat{w}(1-\hat{P}_b\hat{Q}_b)\|_{H^{\infty N}}=\mu(P_b)$.

Theorem 3: If i) $\det\hat{P}_a(s)\neq 0$ for $\mathrm{Re}(s)\geq 0$ and ii) there are constants $c>0$, $\rho>0$, and an integer $k>0$ for which the inequality $|s|^k\sigma[\hat{P}_a(s)]\geq c$ is valid in the region $|s|>\rho$, $\mathrm{Re}(s)\geq 0$, then

a) $\mu(P_bP_a)=\mu(P_b)$, and

b) P_bP_a has an optimal sequence of approximate right inverses Q_m in $\mathbf{H}_0^{\infty N}$, with frequency responses

$$\hat{Q}_m(s) \triangleq \hat{P}_a^{-1}(s)\hat{Q}_b(s)\left[m(s+m)^{-1}\right]^t\left[n_m(s+n_m)^{-1}\right]^{k+1}$$

where for any $m=1,2,\cdots,n_m$ is a sufficiently large integer (given explicitly in Lemma 5.1).

Theorem 3 is an application of Lemma 5.1. Q_m will be constructed out of separate optimal sequences for P_a and P_b, both under the symmetric $\|\cdot\|_W$ norm. First, however, a lemma will be proved.

For any integer $r>0$, let J_n^r denote the sequence in $\mathbf{H}^{\infty N}$ with "low-pass" frequency responses

$$\hat{J}_n^r(s)=n^r(s+n)^{-r}\hat{I}, \qquad n=1,2,\cdots.$$

Lemma 6.1: $\|J_n^r\|=1$; J_n^r is a weighted identity sequence,[12] i.e., for any $r>0$, $\|I-J_n^r\|_W\to 0$ as $n\to\infty$.

Proof: First of all, $\|J_n^r\|=\|I\|\sup_\omega|n(j\omega+n)^{-1}|^r=1$. Second, we have

$$\|I-J_n^r\|_W=\sup_\omega|\hat{w}(j\omega)|\cdot\left|1-\left(\frac{n}{j\omega+n}\right)^r\right| \tag{6.1}$$

$$\leqslant 2\sup_{|\omega|\geqslant\delta}|\hat{w}(j\omega)|+\|\hat{w}\|\sup_{|\omega|<\delta}\left|1-\left(\frac{n}{j\omega+n}\right)^r\right| \tag{6.2}$$

$\delta>0$ being any number. Let $\epsilon>0$ be given. As $\hat{w}$ is in H_0^∞, there is a $\delta>0$ for which the first term in (6.2) is less than $\epsilon/2$; for fixed δ there is an integer n for which the second term is less than $\epsilon/2$; i.e., $\lim_{n\to\infty}\|I-J_n^r\|_W=0$. Q.E.D.

Proof of Theorem 3: First, let us show that the sequence Q_{an} with frequency responses

$$\hat{Q}_{an}\triangleq\hat{P}_a^{-1}\hat{J}_n^{k+1}, \qquad n=1,2,\cdots \tag{6.3}$$

is a right weighted inverting sequence in $\mathbf{H}_0^{\infty N}$ for P_a. As $\det(\hat{P}_a(s))\neq 0$, $\hat{P}_a^{-1}(s)$ is analytic in $\mathrm{Re}(s)\geqslant 0$. The inequality $|s|^k\sigma[\hat{P}_a(s)]\geqslant c$ ensures that the functions defined by (6.3) are in $H_0^{\infty N}$. Now, $\|I-P_aQ_{an}\|_W=\|I-J_n^{k+1}\|_W\to 0$ as $n\to\infty$, by Lemma 6.1. Therefore, $\mu(P_a)=0$ and Q_{an} is a right winvseq in $\mathbf{H}_0^{\infty N}$ for P_a.

Next, consider P_b. By hypothesis, $\hat{P}_b$ has a minimal approximate right inverse $\hat{Q}_b$ in $(1+s)^\ell H^{\infty N}$. The frequency responses

$$\hat{Q}_{bm}\triangleq\hat{Q}_b\hat{J}_m^\ell, \qquad m=1,2,\cdots \tag{6.4}$$

lie in $H^{\infty N}$, and determine operators Q_{bm} in $\mathbf{H}^{\infty N}$. Let us show that Q_{bm} is optimal. Certainly, $\|I-P_bQ_{bm}\|_W\geqslant\mu(P_b)$. Now,

$$\|I-P_bQ_{bm}\|_W=\|(I-P_bQ_b)J_m^\ell+I-J_m^\ell\|_W$$

$$\leqslant\|I-P_bQ_b\|_W\|J_m^{\ell+1}\|+\|I-J_m^\ell\|_W$$

by the triangle inequality and the multiplicative property of symmetric seminorms. As $\|I-P_bQ_b\|_W=\mu(P_b)$ by hypothesis, and by Lemma 6.1 $\|J_m^{\ell+1}\|=1$ and $\lim_{m\to\infty}\|I-J_m^\ell\|_W\to 0$, we obtain $\lim_{m\to\infty}\|I-P_bQ_{bm}\|_W\leqslant\mu(P_b)$. As this upper bound coincides with the lower bound obtained above, $\lim_{m\to\infty}\|I-P_bQ_{bm}\|_W=\mu(P_b)$, i.e., Q_{bm} is a right winvseq in $\mathbf{H}^{\infty N}$ for P_b.

[12] For a symmetric weighting, $\|A-J_n^rA\|_W\leqslant\|I-J_n^r\|_W\|A\|$, so $\|A-J_n^rA\|_W\to 0$ whenever $\|I-J_n^r\|_W$ does; similarly for $\|A-AJ_n^r\|_W$. Therefore, J_n^r is a widseq iff $\lim_{n\to\infty}\|I-J_n^r\|_W=0$.

The conclusions of Theorem 3 are now true by Lemma 5.1. Here $\|\cdot\|_a=\|\cdot\|_b=\|\cdot\|_W$, and Q_m is the product $P_a^{-1}J_{n_m}^{k+1}Q_bJ_m^\ell$. Q.E.D.

Corollary 6.1: The conclusion of the example of Section V-A is true.[13] That example is simply a special case of Theorem 3 for $N=k=1$, and $\hat{q}_b=\hat{p}_b^{-1}(1-\hat{w}(\beta)\hat{w}^{-1})$ in $(1+s)^\ell H^\infty$.

C. Sensitivity Reduction in Nearly Invertible Multivariable Systems

Consider the feedback scheme of equations (3.1)–(3.3). Suppose that the plant $P_1(s)$ is in $\mathbf{H}_0^{\infty N}$ and satisfies the restrictions: $\det P_1(s)\neq 0$ for $\mathrm{Re}(s)\geqslant 0$, and $|s|^k\sigma[\hat{P}_1(s)]>c$ in some region $|s|>p$, $\mathrm{Re}(s)\geqslant 0$, where c, p are constants and k is an integer. The sensitivity $\eta\triangleq\|I-P_1Q\|_W$ is defined[14] as in Section V.

Corollary 6.2:[15] The sensitivity $\|I-P_1Q_n\|_W$ can be made smaller than any $\epsilon>0$ by a comparator Q_n in $\mathbf{H}_0^{\infty N}$ with frequency response

$$\hat{Q}_n(s)=\hat{p}_1^{-1}(s)\left[n(s+n)^{-1}\right]^{k+1}$$

n being a (sufficiently large) integer.

Proof: The hypotheses of Theorem 3 are satisfied, with $P_a=P_1$ and $P_b=I$. Therefore, $\mu(P_1)=0$ and $\|I-P_1Q_n\|_W\to 0$ as $n\to\infty$. Q.E.D.

D. Lower Bounds to Sensitivity

It might be expected that the singularity measure of an $H^{\infty N}$ frequency response matrix would be limited by the location of its RHP zeros, and that the optimal sensitivity would be similarly limited. The following theorem shows this to be true.

Let $\|\cdot\|_W$ be any weighted seminorm on $\mathbf{H}^{\infty N}$ of the form $\|P\|_W=\|\hat{w}\hat{P}\|$, $\hat{w}$ being (a scalar function) in H_0^∞. For any plant P in $\mathbf{H}^{\infty N}$, let the *RHP zeros* of P be the points s_i in $\mathrm{Re}(s)\geqslant 0$ $i=1,2,\cdots$, at which $\det[\hat{P}(s_i)]=0$.

Theorem 4:

$$\mu(P)\geqslant\sup\{|\hat{w}(s_i)|: i=1,2,\cdots\}.$$

Proof: For any Q in $\mathbf{H}^{\infty N}$ let $E\triangleq(I-PQ)$. For any RHP zero s_i, let ξ be a unit vector in the nullspace of the matrix $\hat{P}(s_i)\hat{Q}(s_i)$. Let f: $C\to C$ be the function $f(s)=\xi^*\hat{w}(s)\hat{E}(s)\xi$. Since $\hat{P}(s_i)\hat{Q}(s_i)=0$, $f(s_i)=\xi^*\xi\hat{w}(s_i)=\hat{w}(s_i)$. Now $f(s)$ is an H_0^∞ function, and by the maximum modulus principle attains its maximum on the $j\omega$ axis, i.e., $|f(j\omega)|\geqslant|\hat{w}(s_i)|$ for some ω. But for any transformation A in Euclidean N-space, and any unit N-vector ξ, $\sigma(A)\geqslant\xi^*A\xi$. The last assertion is established by the inequalities

[13] For simplicity we have considered inputs in H^2 here, but the conclusion is easily extended to H^P, $1\leqslant p\leqslant\infty$.

[14] Recall that sensitivity of a feedback scheme equals $\|I-P_1Q\|_W$, by Theorem 1.

[15] This corollary is based on results obtained with D. Bensoussan [17].

$$|\xi^* A\xi|^2 \leq |\xi^*|^2 |A\xi|^2 = |A\xi|^2$$
$$= (A^* A\xi)^* \xi \leq |A^* A\xi| \cdot |\xi|$$
$$\leq \sigma^2(A)|\xi|^2 = \sigma^2(A).$$

We employ this fact to obtain

$$\mu(P) \geq \|E\| = \sup_\omega \sigma\left[\hat{w}(j\omega)\hat{E}(j\omega)\right] \geq \sup_\omega |f(j\omega)| \geq |\hat{w}(s_i)|.$$

The theorem follows. Q.E.D.

Remark: Theorem 4 implies that no feedback can produce small sensitivity if a plant zero is present in any heavily weighted part of the right half-plane. For "low-pass" weightings such as $k(s+k)^{-1}$, $k>0$, this means that if sensitivity is to be reduced, the only RHP zeros possibly allowed are those at very high complex frequencies, $|s_i| \gg k$.

VII. Effects of Plant Uncertainty on Disturbance Attenuation

Two opposing tendencies can be found in most feedback systems. On the one hand, to the extent that feedback reduces sensitivity it reduces the need for plant identification. On the other hand, the less information is available about the plant, the less possible it is to select a feedback to reduce sensitivity. The balance between these tendencies establishes a maximum to the amount of tolerable plant uncertainty and, equivalently, a minimum to the amount of identification needed.

It can be argued that the search for such a minimum should be basic to the theory of adaptive systems. Actually, even the existence of such a minimum appears not to have been stated, perhaps because plant uncertainty is so difficult to study in the WHK framework in the absence of the multiplicative properties (4.1), and because there is no notion of optimality in the classical setup.

Here, we would like to take a step in the direction of articulating these issues, by defining the tradeoff between minimal sensitivity and plant uncertainty and deducing its simpler properties. Sensitivity to disturbances will be considered in this section, and to plant uncertainty in the next.

Let $\mathbf{B}_r$ be a subspace of the causal operators $\mathbf{B}_s$. Consider the feedback (3.1) and model reference (3.3) schemes. Suppose that some nominal plant P_1 in $\mathbf{B}_r$ is specified and (3.4) hold, so the two schemes are equivalent. The true plant P in $\mathbf{B}_r$ will differ from P_1 in general. Let K_{33}: $\mathcal{X} \to \mathcal{X}$, $K_{33}(d)=y$ be the operator in $\mathbf{A}$ mapping disturbances into outputs. K_{33} can be expressed in terms of F [using (3.2a)],

$$K_{33} = (I+PF)^{-1}$$

or in terms of P_1 and Q [using 3.4b)],

$$K_{33} = \left[I + PQ(I-P_1Q)^{-1}\right]^{-1}$$
$$= \left[(I-P_1Q+PQ)(I-P_1Q)^{-1}\right]^{-1}$$
$$= (I-P_1Q)\left[I+(P-P_1)Q\right]^{-1}. \quad (7.1)$$

We shall be interested in the way in which closed-loop operators such as K_{33} behave as functions of the open-loop operators P, P_1, Q, and F. In particular, let us define two functions mapping open-loop into closed-loop operators. Let $\mathbb{E}$: $\mathbf{A}_s^3 \to \mathbf{A}$, $\mathbb{E}(P, P_1, Q) = K_{33}$ be the function relating K_{33} to the model reference variables, and $\mathbb{E}_f: \mathbf{A}_s^2 \to \mathbf{A}$, $\mathbb{E}_f(P, F) = K_{33}$ the function for the feedback variables. Any pair (d, y) in $\mathcal{X}^2$ satisfying (3.1) or (3.3) also satisfies the equations

$$y = \mathbb{E}_f(P, F)d = \mathbb{E}(P, P_1, Q)d. \quad (7.2)$$

The c.l. (closed-loop) operators K_{ij}, $i, j = 2, 3$, of the feedback scheme (3.1) were introduced in Section III, and are well defined for the model reference scheme (3.3) under the assumed equivalence. The model reference scheme shown in Fig. 3 has two extra nodes labeled 4 and 5. We shall avoid the lengthly but straightforward calculation of the remaining c.l. operators, and instead assume the following elementary properties of the model reference scheme: define the operators $K_{44} \triangleq [I+(P-P_1)Q]^{-1}$ and $K_{55} \triangleq [I+Q(P-P_1)]^{-1}$ and let K be the sextuplet of operators $\{K_{44}, K_{55}, P, P_1, Q, I\}$; the set $\{K_{ij}\}_{i,j=2}^5$ of all c.l. operators consists of algebraic combinations (i.e., involving sums and products only of) the operators in K; furthermore $K_{32} = -QK_{44}$.

A. Stabilizing Feedbacks

It is well known that any stable plant which is stabilized by feedback is surrounded by a ball of "admissible" perturbations which preserve closed-loop stability. Here, the radius of such a ball will be calculated, and sensitivity will be defined with respect to plant uncertainty within the ball.

Suppose a right seminorm $\|\cdot\|_r$ to be defined on $\mathbf{B}_r$, and recall that $\|\cdot\|_r \geq \|\cdot\|$. The true plant P will be supposed to lie in a *ball of uncertainty of radius* $\delta \geq 0$ *in* $\mathbf{B}_r$ *around the nominal* P_1, $b(P_1, \delta) \triangleq \{P \text{ in } \mathbf{B}_r : \|P-P_1\|_r \leq \delta\}$. Since $\|P-P_1\| \leq \|P-P_1\|_r$, $b(P_1, \delta)$ is a subset of the ball of radius δ in $\mathbf{B}$.

An operator K in $\mathbf{A}$ which depends on $P \in \mathbf{B}_r$ will be called *bounded* over a ball $b(P_1, \delta)$ iff K is in $\mathbf{B}$ and there is a constant $c \geq 0$ with the property that $\|K\| \leq c$ for all P in $b(P_1, \delta)$. A feedback or model reference scheme will be *c.l. (closed-loop) bounded* on $b(P_1, \delta)$ iff all its c.l. operators are bounded on $b(P_1, \delta)$.

Proposition 7.1: The following statements are equivalent on any ball $b(P_1, \delta)$. i) The feedback scheme is closed-loop bounded; ii) Q is in $\mathbf{B}$ and $K_{44} \triangleq [I+(P-P_1)Q]^{-1}$ is bounded; iii) the model reference scheme is closed-loop bounded.

Proof: We shall prove the sequence of implications. i)⇒ii)⇒iii)⇒i). If i) is true, then K_{44} is bounded by definition. Also, K_{32} is bounded by definition, and therefore in $\mathbf{B}$ for all P in $b(P_1, \delta)$. But P_1 is in $b(P_1, \delta)$, and $K_{32} = -Q$ when $P = P_1$; therefore, Q is in $\mathbf{B}$, and ii) is true. If ii) is true then K_{55} is also bounded, as the equations $K_{55} = [I+Q(P-P_1)]^{-1} = I - QK_{44}(P-P_1)$ imply that

$\|K_{55}\| \leq 1+\|Q\| \cdot \|K_{44}\| \cdot \delta$, for any P in $b(P_1, \delta)$. Therefore, and by the assumed property of the K_{ij}, each c.l. operator K_{ij}, $i, j=2, \cdots, 5$, is an algebraic combination of the operators of K, which are bounded on $b(P_1, \delta)$. It follows that each K_{ij} is bounded there, and that (3.3) is c.l. bounded, i.e., iii) is true. If iii) is true then i) is true, as (3.1) and (3.3) are equivalent. Q.E.D.

Again, only stable Q need be considered. The operators F in $\mathbf{A}_s$ and Q in $\mathbf{B}_s$ will be called *stabilizing* for $b(P_1, \delta)$ iff (3.1) and (3.3), respectively, are c.l. bounded on $b(P_1, \delta)$. The set of all Q in $\mathbf{B}_s$ stabilizing for $b(P_1, \delta)$ will be denoted by $\mathbf{B}_s(P_1, \delta)$. For any P_1 and Q in $\mathbf{B}_s$, the set of real points δ for which Q is stabilizing for $b(P_1, \delta)$ will be denoted by $\Delta(P_1, Q)$, and abbreviated to Δ when dependence on (P_1, Q) is not of interest.

The next lemma shows that any stable Q gives a c.l. bounded scheme (3.1), and is therefore stabilizing, for δ small enough. Let P_1 and Q be in $\mathbf{B}_s$.

Proposition 7.2: If δ satisfies $0 \leq \delta < \|Q\|^{-1}$, then δ is in Δ and for any P in $b(P_1, \delta)$ the inequality

$$\|\{I+(P-P_1)Q\}^{-1}\| < (1-\delta\|Q\|)^{-1} \tag{7.3}$$

holds. Δ is a half-open interval, $[0, \delta_1)$.

Proof: Under the hypothesis, $\|(P-P_1)Q\| < 1$. Therefore, the small gain property ensures that $[I+(P-P_1)Q]^{-1}$ is in $\mathbf{B}$, and (7.3) is true by Proposition 2.1c. Δ is an interval beginning at 0, because $\delta \in \Delta$ and $0 \leq \delta' \leq \delta$ implies that $b(P_1, \delta)$ contains $b(P_1, \delta')$, which implies that $\delta' \in \Delta$. Δ is half open by a standard perturbation argument for the openness of resolvent sets which will be omitted, as this property is not important here. Δ contains $[0, \|Q\|^{-1}]$ because, for any δ in $[0, \|Q\|^{-1}]$, (7.3) implies that K_{44} is bounded on $b(P_1, \delta)$, so Q is stabilizing by Proposition 7.1.

B. Sensitivity

Suppose that right and left seminorms are defined on $\mathfrak{B}$ as in Section V, and induce a weighted seminorm $\|\cdot\|_W$ on $\mathbf{B}$. For the plant P in any ball of uncertainty $b(P_1, \delta)$, and any feedback F in $\mathbf{A}_s$ or comparator Q in $\mathbf{B}_s$, the *sensitivity to disturbances under plant uncertainty* of the equivalent schemes (3.1), (3.3) is defined to be

$$\eta(P_1, Q; \delta) = \sup_{P \in b(P_1, \delta)} \sup\{\|y\|_l / \|d\|_r : d \text{ in } \mathfrak{B}_r, \|d\|_r \neq 0\}$$

if Q is (stabilizing) in $\mathbf{B}_s(P_1, \delta)$, and $\eta(P_1, Q; \delta) = \infty$ otherwise. For any stabilizing Q,

$$\eta(P_1, Q; \delta) = \sup_{P \in b(P_1, \delta)} \|\mathbb{E}(P, P_1, Q)\|_W$$

by (7.2). (Recall, also, that $\mathbb{E}(P, P_1, Q) = \mathbb{E}_f(P, F)$.) $\eta(P_1, Q, \delta)$ is the smallest assured sensitivity for P in the ball. We are interested in finding Q to minimize this sensitivity. Accordingly, we define the *minimal sensitivity to plant uncertainty* to be

$$\eta(P_1, \delta) = \inf_{Q \in \mathbf{B}_s} \sup_{P \in b(P_1, \delta)} \|\mathbb{E}(P, P_1, Q)\|_W$$

with the proviso that the sup is replaced by ∞ if Q is not stabilizing. For any fixed nominal plant[16] P_1 in $\mathbf{B}_s$, $\eta(P_1, \delta)$ is a positive-real valued function of $\delta > 0$.

Theorem 5: For any nominal plant P_1 in $\mathbf{B}_s$, the minimal sensitivity to plant uncertainty $\eta(P_1, \delta)$ is a monotone nondecreasing function of δ for $\delta \geq 0$; $\eta(P_1, \delta)$ approaches the singularity measure $\mu(P_1)$ as $\delta \to 0$; and $\eta(P_1, \delta) = \|I\|_W$ for $\delta = \|P_1\|_r$.

Lemma 7.3: For any P_1 and Q in $\mathbf{B}_s$, $\eta(P_1, Q; \delta)$ is a monotone nondecreasing function of $\delta \geq 0$, finite for δ in Δ. Δ contains the interval $[0, \|Q\|^{-1}]$. For any δ in $[0, \|Q\|^{-1}]$, the inequalities

$$\|I-P_1Q\|_W \leq \eta(P_1, Q; \delta) \leq \|I-P_1Q\|_W + \|I-P_1Q\|_W \delta \|Q\| (1-\delta\|Q\|)^{-1} \tag{7.4}$$

and, if $\|\cdot\|_W$ is symmetric, also the inequality

$$\eta(P_1, Q; \delta) \leq \|I-P_1Q\|_W (1-\delta\|Q\|)^{-1} \tag{7.5}$$

are satisfied.

Proof of Lemma 7.3: The symbol $\uparrow$ will denote a monotone nondecreasing function of $\delta \geq 0$. By definition of Δ, Q is stabilizing for any $b(P_1, \delta)$ with δ in Δ, so $\eta(P_1, Q; \delta)$ is finite on Δ. By Proposition 7.2, Δ is an interval containing $[0, \|Q\|^{-1}]$. δ is a $\uparrow$ function on Δ since it is a sup over sets $b(P_1, \delta)$ which are nested and nondecreasing as δ increases.

From (7.1) we get

$$\mathbb{E}(P, P_1, Q) = (I-P_1Q)\{I-(P_1-P)Q[I+(P-P_1)Q]^{-1}\}. \tag{7.6}$$

The triangle inequality and dominance condition $\|\cdot\|_W \leq \|\cdot\|$ give

$$\|\mathbb{E}(P, P_1, Q)\|_W \leq \|I-P_1Q\|_W + \|(I-P_1Q)(P_1-P)Q[I+(P-P_1)Q]^{-1}\|. \tag{7.7}$$

For δ in $[0, \|Q\|^{-1}]$, the upper bound in (7.4) is obtained from (7.7) by (7.3) and the multiplicative property of norms. The lower bound is valid as P_1 is in $b(P_1 \delta)$, and $E(P_1, P_1, Q) = I - P_1 Q$. If $\|\cdot\|_W$ is symmetric, the multiplicative property of *symmetric* seminorms applied to (7.1) gives

$$\|\mathbb{E}(P, P_1, Q)\|_W \leq \|I-P_1Q\|_W \|\{I+(P-P_1)Q\}^{-1}\|$$

and (7.4) follows. Q.E.D.

Proof of Theorem 5: From Lemma 7.3 and the inequality $\|I-P_1Q\|_W \geq \mu(P_1)$, it can be concluded that for any Q in $\mathbf{B}_s$, $\eta(P_1, Q; \delta)$ is a $\uparrow$ function with values in $[\mu(P_1), \infty]$. As $\eta(P_1, \delta) = \inf_{Q \in \mathbf{B}} \eta(P_1, Q; \delta)$ by definition, $\eta(P_1, \delta)$ is the inf of a set of $\uparrow$ functions with values in $[\mu(P_1, \infty]$. Therefore, $\eta(P_1, \delta)$ is $\uparrow$ and $\eta(P_1, \delta) \geq \mu(P_1)$. Since 0 is in

[16] Again, the assumption that plant and feedback are both strictly causal can be relaxed in Theorem 5, Lemma 5.3, etc.

$\mathbf{B}_s$, $\eta(P_1,0;\delta)$ is in the set; and $\eta(P_1,\delta) \leqslant \eta(P_1,0;\delta) = \|I\|_W$, where the last identity is true by (7.1).

Let us show that $\lim_{\delta \to 0} \eta(P_1,\delta) = \mu(P_1)$. By definition of $\mu(P_1)$, for any $\epsilon > 0$ there is a Q in $\mathbf{B}_s$ satisfying $\|I - P_1 Q\|_W \leqslant \mu(P_1) + \epsilon/2$; as the RHS of (7.4) approaches $\|I - P_1 Q\|_W$ as $\delta \to 0$, there is a $\delta > 0$ for which $\eta(P, Q; \delta) \leqslant \mu(P_1) + \epsilon$. Since ϵ was arbitrary, and it has been shown that $\eta(P_1,\delta) \geqslant \mu(P_1)$, the conclusion follows.

Finally, let us show that $\eta(P_1, \|P_1\|_r) = \|I\|_W$. Suppose the contrary to be true. It has been shown that $\eta(P_1, \|P_1\|_r) \leqslant \|I\|_W$, so $\eta(P_1, \|P_1\|_r) \leqslant \|I\|_W$. Therefore, there is a Q in $\mathbf{B}_s$ for which $\|\mathbf{E}(P, P_1, Q)\|_W < \|I\|_W$. Now, 0 is in $b(P_1, \|P_1\|_r)$ and, by (7.1), $\eta(P_1, \|P_1\|_r) \geqslant \|\mathbf{E}(0, P_1, Q)\|_W = \|I\|_W$ which is a contradiction. Q.E.D.

Remark: Theorem 5 implies that whenever feedback reduces sensitivity for the nominal plant P_1, there are points in the interval $[0, \|P_1\|]$ at which $\eta(P_1, \delta)$ increases, i.e., where "the less we know about P, the less able we are to construct a feedback to attenuate disturbances."

VIII. Filtering of Plant Uncertainty: Invariant Schemes

We now turn to the second problem outlined in the introduction, Section I-D. The plant P lies in a ball of uncertainty[17] around some nominal value P_1, and one object of using feedback is to shrink the size of the uncertainty. Of course, uncertainty can be reduced to zero by disconnecting the input (u in Fig. 4) from the system, but then P_1 is also transformed into zero. Clearly, the problem is trivial unless there is a normalization or constraint on the control law that transforms P_1 into a closed-loop system.

We would prefer as far as possible to separate the reduction of uncertainty from the transformation of P_1, and therefore seek a definition of uncertainty which is independent of the eventual closed-loop system.

If P_1 were a real number, uncertainty could be normalized by specifying it as a percentage of the nominal value. This possibility is not open for noninvertible plants. Instead, we shall achieve a normalized definition of uncertainty by employing the device of a plant-invariant scheme, which leaves the nominal plant invariant while shrinking the ball of uncertainty. Such a scheme will be shown always to be realizable in the form of a model reference scheme.

This device will also enable us to separate the design process into two consecutive stages: 1) reduction of uncertainty and 2) transformation of the nominal plant into a nominal closed-loop system (cf. the separation into estimation and control stages in Kalman filtering).

A possible disadvantage to this approach is that the two stages may be dependent, and yield a suboptimal sensitivity. We shall try to get the best of both worlds, and simultaneously formulate normalized and unnormalized

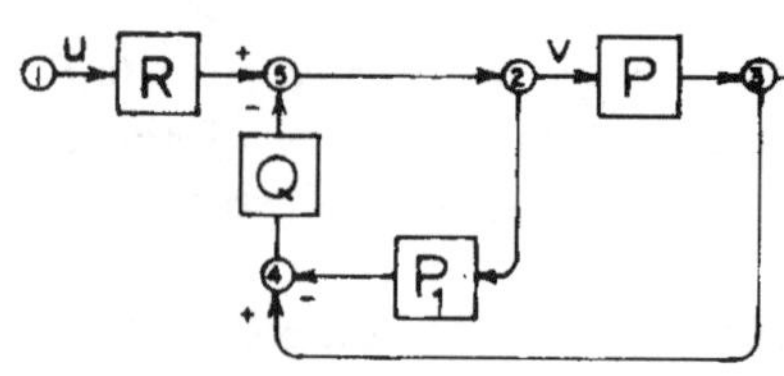

Fig. 5.

versions of the problem, by allowing the plant invariant scheme to be multiplied by a constant R_1, representing a desired nominal control law.

A. Plant Invariant Schemes

Any feedback arrangement of the type shown in Fig. 1, incorporating a plant P with a single accessible input, has an equivalent description in terms of the flowgraph of Fig. 4. This flowgraph is completely specified by the two additional operators U and F, which may be unstable even though the original arrangement is closed-loop stable. Note that Fig. 4 is an enlargement of Fig. 2 by a new branch, U, connected to a new node labeled 1. For simplicity, it will be assumed that $d=0$. The new *feedback scheme* equations are

$$y = Pv \tag{8.1a}$$

$$v = Uu - Fy \tag{8.1b}$$

in which $P \in \mathbf{A}_s$, $F \in \mathbf{A}_s$, and[18] $U \in \mathbf{A}$ are operators, and u, v, and y are in $\mathfrak{X}$. (Equations (8.1) represent an enlargement of (3.1) by the term Uu, under the constraint that $d=0$.) As P is strictly causal, $(I+FP)^{-1}$ exists in $\mathbf{A}$, and for each u in $\mathfrak{X}$, (8.1) have unique solutions for v and y in $\mathfrak{X}$, given by the equations

$$v = (I+FP)^{-1} Uu \tag{8.2a}$$

$$y = P(I+FP)^{-1} Uu. \tag{8.2b}$$

Let K_{12} and K_{13} be the operators mapping $\mathfrak{X}$ into $\mathfrak{X}$ which satisfy, $K_{12}(u) = v$, $K_{13}(u) = y$. By (8.2), $K_{12} = (I+FP)^{-1} U$ and $K_{13} = P(I+FP)^{-1} U$.

The full set of closed-loop operators of (8.1) consists of K_{12}, K_{13}, $K_{11} \triangleq I$, $K_{i1} \triangleq 0$, and the operators K_{ij} defined in Section III, where $i, j = 2, 3$.

A variant of the model-reference transformation of Section III-A will be used. The flowgraph of Fig. 5 is described by the new *model reference scheme* equations

$$y = Pv \tag{8.3a}$$

$$v = Ru - Q(y - P_1 v) \tag{8.3b}$$

in which $P, P_1 \in \mathbf{B}_s$, $R \in \mathbf{A}$, and $u, v, y \in \mathfrak{X}$. For any $u \in \mathfrak{X}$, (8.3) have unique solutions for v and y in $\mathfrak{X}$, as P, P_1, and Q are strictly causal. (Equations (8.3) can be viewed as an enlargement of (3.3) by the term Ru, and subject to the constraint that $d=0$.)

The schemes (8.1) and (8.3) will be called *equivalent* iff

[17] Departures of the true plant from the nominal can be interpreted as uncertain in some applications; in others, they may simply represent known perturbations to be attenuated.

[18] In order to be physically realized, U would have to be approximated by a strictly causal operator. We prefer not to assume that $U \in \mathbf{A}_s$ as this simplifies the presentation. As U is followed by a strictly causal element, and has no feedback around it, there is no loss of generality.

every triple (u, v, y) in $\mathfrak{X}^3$ satisfying (8.1) satisfies (8.3) and vice versa. Equivalence will be established subject to the equations

$$Q=F(I+P_1F)^{-1} \tag{8.4a}$$

$$F=Q(I-P_1Q)^{-1} \tag{8.4b}$$

$$R=(I+FP_1)^{-1}U \tag{8.5a}$$

$$U=(I+FP_1)R. \tag{8.5b}$$

Equation (8.4) is a repetition of (3.4). If either of (8.5) holds then, clearly, both are true.

For equivalent schemes, let $\mathbb{E}(P, P_1, Q)$ be defined as in Section VII; Let $\mathbb{K}$ be the function $\mathbb{K}$: $\mathbf{A}_s^3 x\mathbf{A}\to\mathbf{A}$, $\mathbb{K}(P, P_1, Q, R)=K_{13}$, which maps operators appearing in (8.3) into the c.l. (closed-loop) operator K_{13}. Whenever all operators except P are regarded as fixed, $\mathbb{K}(P, P_1, Q, R)$ will be denoted by $\mathbb{K}(P)$. $\mathbb{K}(P)$ can be expressed in terms of the model-reference scheme operators, by the formula

$$\mathbb{K}(P)=P[I+Q(P-P_1)]^{-1}R \tag{8.6a}$$

which is obtained from the following sequence of equations. For any (u, v, y) in $\mathfrak{X}^3$ satisfying (8.3), we have

$$v=Ru-Q(P-P_1)v$$

$$[I+Q(P-P_1)]v=Ru$$

$$y=Pv=P[I+Q(P-P_1)]^{-1}Ru \tag{8.6b}$$

in which the inverse exists in $\mathbf{A}$ as $Q(P-P_1)$ is strictly causal. Since K_{13} maps u into y, K_{13} must coincide with the last operator of (8.6b), and (8.6a) is true.

The set of c.l. operators of the model reference scheme (8.3) is defined as in the preceding Section VII, except that K is augmented by the operator R.

For any P_1 in $\mathbf{A}_s$, $\mathbb{K}$ will be called (nominal) *plant invariant* iff $\mathbb{K}(P_1)=P_1$; and, for any R_1 in $\mathbf{A}$, *plant invariant* $(\times R_1)$ iff $\mathbb{K}(P_1)=P_1R_1$. From (8.4) it is clear that $\mathbb{K}$ has these properties whenever $R=I$ or $R=R_1$, respectively.

An operator $0_r\in\mathbb{B}$ is a *right zero*[19] of P_1 iff $P_10_r=0$.

Theorem 7:

a) Any feedback scheme (8.1) with (P, F, U) in $\mathbf{A}_s^2\times\mathbf{A}$ is equivalent to a model reference scheme (8.3) with (P, P_1, Q, R) in $\mathbf{A}_s^3\times\mathbf{A}$, in which (F, U) and (P_1, Q, R) are related by (8.4)–(8.5); and vice versa. If (8.4)–(5) hold, then we have the following.

b) If P assumes the (nominal) value $P_1\in\mathbf{A}_s$, then $\mathbb{K}(P_1)=P_1R$, and if the feedback scheme is closed-loop stable, then Q and R of the model reference scheme are stable (although F and U may be unstable).

c) The differences $\mathbb{K}(P)-\mathbb{K}(P_1)$ and $P-P_1$ are related by the formulas

$$\mathbb{K}(P)-\mathbb{K}(P_1)=(I-P_1Q)(P-P_1)\{I+Q(P-P_1)\}^{-1}R \tag{8.7a}$$

$$=\mathbb{E}(P, P_1, Q)(P-P_1)R. \tag{8.7b}$$

d) For any $R_1\in\mathbb{B}$, $\mathbb{K}$ is plant invariant $(\times R_1)$ iff $R=R_1+0_r$, where $0_r\in\mathbb{B}$ is any right zero of P_1.

Proof:

a) If (u, v, y) satisfies (8.1), then we have

$$(I-QP_1)v=(I-QP_1)(Uu-Fy) \quad \text{[by (8.1b)]}$$

$$=(I-QP_1)[(I+FP_1)Ru-Q(I-P_1Q)^{-1}Y] \quad \text{[by (8.5b) and (8.4b)]}$$

$$=(I-QP_1)[(I-QP_1)^{-1}Ru-(I-QP_1)^{-1}Qy]$$

$$=Ru-Qu$$

where the second last equation was obtained using the identity $(I+FP_1)=(I-QP_1)^{-1}$ [see (3.5)] and Property 2.1a. Hence, (u, v, y) satisfies (8.3). The reverse assertion is proved similarly. It follows that (8.1) and (8.3) are equivalent.

b) From (8.4), $\mathbb{K}(P_1)=P_1R$; if (8.1) is closed-loop stable, then the c.l. operators $K_{32}=F(I+P_1F)^{-1}$ and $K_{12}=(I+FP)^{-1}U$ are stable, and equal Q and R.

c) From (8.4) the following sequence of equations is obtained:

$$\mathbb{K}(P)-\mathbb{K}(P_1)=P[I+Q(P-P_1)]^{-1}R-P_1R$$

$$=\{P-P_1[I+Q(P-P_1)]\}[I+Q(P-P_1)]^{-1}R$$

from which (8.7a) follows; (8.7b) is obtained by Proposition 2.1a.

d) $\mathbb{K}$ is plant invariant $(\times R_1)$ iff $\mathbb{K}(P_1)-P_1R_1=P_1(R-R_1)=0$, i.e., $R-R_1=0_r$. Q.E.D.

Remark: Plant invariant schemes allow us to divide any control-law synthesis into two stages: 1) filtering of plant uncertainty $P-P_1$ and 2) design of a control law for a nominal plant P_1, with the assurance at least that the filtering stage will improve the design. In general, there is no "separation principle" to guarantee optimality of the division.

Either stage may come first. Therefore, in our theorems, P can be interpreted either as a plant without controller, for which a controller will be designed eventually; or, as a plant with controller attached, which requires additional filtering only to the extent that P differs from P_1.

B. Stabilizing Feedbacks

In the rest of Section VIII it will be assumed that: P and P_1 belong to a subspace $\mathbb{B}_r$ of the strongly causal operators $\mathbb{B}_s$, on which a right seminorm $\|\cdot\|_r$ is defined; P lies in a ball $b(P_1, \delta)$, defined as in Section VII-A, of what can be interpreted either as uncertainty or perturbations; and the equivalence conditions (8.4), (8.5) hold, so that the feedback and model reference schemes (8.1), (8.3) are equivalent.

Q in $\mathbf{A}_s$ and R in $\mathbf{A}$ are sought which give low sensitivity and maintain c.l. boundedness. In view of Theorem 7b), the assumption that Q is in $\mathbb{B}_s$ and R in $\mathbb{B}$, i.e., that both are stable, can be made without loss of generality.

[19] More simply, 0_r is an operator whose range is in the nullspace of P_1. The term "zero" is appropriate for a normed algebra.

For any $b(P_1, \delta)$, the definitions of a bounded operator, c.l. bounded scheme, and stabilizing Q were given in Section VII-A. The set $\mathbf{B}_s(P_1, \delta)$ of Q in $\mathbf{B}_s$ stabilizing for $b(P_1, \delta)$ was introduced.

Proposition 8.1: Under the present hypotheses, Propositions 7.1 and 7.2, are valid for the feedback scheme (8.1) and model reference scheme (8.3), and the set of Q in $\mathbf{B}_s$ stabilizing for $b(P_1, \delta)$ coincides with $\mathbf{B}_s(P_1, \delta)$.

Proof: As R is in $\mathbf{B}$, the set K augmented by R consists of operators bounded on $b(P_1, \delta)$ iff the unaugmented set K consists of operators bounded on $b(P_1, \delta)$. Therefore, $\{K_{ij}\}_{i,j=1}^{6}$ are bounded iff $\{K_{ij}\}_{i,j=1}^{5}$ are bounded. The conclusion follows. Q.E.D.

C. *Sensitivity to Plant Perturbations or Uncertainty*

Suppose that a pair of seminorms $(\|\cdot\|_W, \|\cdot\|_1)$ is defined on the spaces $(\mathbf{B}, \mathbf{B}_r)$, respectively. We assume $\|\cdot\|_W$ to be a left seminorm, but leave open the possibility that $\|\cdot\|_1$ is not a right seminorm, and assume instead that $\|\cdot\|_1 \leq \|\cdot\|_r$, i.e., $\|\cdot\|_1$ is dominated by the right seminorm $\|\cdot\|_r$. A weighted seminorm $\|\cdot\|_v$ is assumed to be defined on the set of products $\mathbf{B}\cdot\mathbf{B}_r$, and to have the multiplicative property, $\|CD\|_v \leq \|C\|_W \|D\|_1$ for all C in $\mathbf{B}$ and D in $\mathbf{B}_r$. The pair $(\|\cdot\|_W, \|\cdot\|_1)$ will be called *aligned* iff there is a $D \in \mathbf{B}_r$ for which the preceding inequality is an equality for all C in $\mathbf{B}$.

For the plant P in any ball of uncertainty $b(P_1, \delta)$, and any $(F \in \mathbf{A}_s,\ U \in \mathbf{A})$, or $(Q \in \mathbf{B}_s,\ R \in \mathbf{B})$, the *sensitivity to plant perturbations* (or uncertainty) of the equivalent schemes (8.1)–(8.3) is defined to be

$$\nu(P_1, Q, R; \delta) = \sup_{P \in b(P_1, \delta)} \{\|\mathbf{K}(P) - \mathbf{K}(P_1)\|_v \delta^{-1}\} \tag{8.8}$$

for Q stabilizing (i.e., in $\mathbf{B}_s(P_1, \delta)$), and $\nu(P_1, Q, R; \delta) = \infty$ otherwise. For any stabilizing Q, (8.7) gives the equation

$$\nu(P_1, Q, R; \delta) = \sup_{P \in b(P_1, \delta)} \{\|\mathbf{E}(P, P_1, Q)(P - P_1)R\|_v \delta^{-1}\}. \tag{8.9}$$

The following lemma relates the disturbance and plant perturbation sensitivities, η and ν, to each other, and to the $\|\cdot\|_W$-singularity measure $\mu(P_1)$, when $R = I$.

Lemma 8.2: $\nu(P_1, Q, I; \delta)$ is a monotone nondecreasing function of $\delta \geq 0$ satisfying the inequality

$$\nu(P_1, Q, I; \delta) \leq \eta(P_1, Q; \delta); \tag{8.10}$$

and if $\|\cdot\|_1 = \|\cdot\|_r$ and the pair $(\|\cdot\|_W, \|\cdot\|_r)$ is aligned then

$$\nu(P_1, Q, I; \delta) \geq \mu(P_1). \tag{8.11}$$

Remark: $\|\cdot\|_v$ can coincide with the principal norm $\|\cdot\|$. For example, if W is in $\mathbf{B}_r$, $\|W\| = 1$, $\|C\|_W \triangleq \|CW\|$, and $\|D\|_1 \triangleq \|W^{-1}D\|$, then the principal norm has the multiplicative property $\|CD\|_v \leq \|C\|_W \|D\|_1$ and may be used as the $\|\cdot\|_v$ norm. Even though $\|\cdot\|$ provides no weighting, (8.10) shows that sensitivities ν smaller than 1 can be achieved whenever $\nu(P_1) < 1$. In effect, the plant perturbations supply their own weighting.

This example has the alignment property, as $\|CW\|_v = \|C\|_W \|W\|_1$.

Proof of Lemma 8.2: $\nu(P_1, Q, I; \delta)$ is a $\uparrow$ function of δ, as it is a sup over sets $b(P_1, \delta)$ which are nested and nondecreasing with $\delta \in \Delta$, and is ∞ for $\delta \notin \Delta$. For any P in $b(P_1, \delta)$ and stabilizing Q, we have the inequalities

$$\begin{aligned}\|\mathbf{E}(P, P_1, Q)(P - P_1)\|_v \delta^{-1} &\leq \|\mathbf{E}(P, P_1, Q)\|_W \|P - P_1\|_1 \delta^{-1} \\ &\leq \|\mathbf{E}(P, P_1, Q)\|_W\end{aligned} \tag{8.12}$$

by the multiplicative property of $\|\cdot\|_v$, and as

$$\|P - P_1\|_1 \leq \|P - P_1\|_r \leq \delta$$

is obtained by taking the sup of both sides of (8.12) over all P in $b(P_1, \delta)$. If $\|\cdot\|_1 = \|\cdot\|_r$ and $(\|\cdot\|_W, \|\cdot\|_r)$ is aligned there is an operator $D \in \mathbf{B}_r$ for which $\|\mathbf{E}(P, P_1, Q)D\|_v = \|\mathbf{E}(P, P_1, Q)\|_W \|D\|_r$. The operator $P_0 \triangleq P_1 + \delta D \|D\|_r^{-1}$ is in $b(P_1, \delta)$ and $\|\mathbf{E}(P_0, P_1, \delta)(P - P_1)\|_v \delta^{-1} = \|\mathbf{E}(P_1, P_1, Q)\|_W \geq \mu(P_1)$. The sup in (8.9) must have the lower bound $\mu(P_1)$, and (8.11) follows. Q.E.D.

D. *Assumptions on R*

As $R = 0$ gives $\nu = 0$, the attainment of a small sensitivity is trivial unless R is constrained. In many problems a target value of the c.l. operator K_{13} is specified for some nominal value of the plant. By Theorem 7b), the target value can be attained iff it has the form $P_1 R$. We therefore make the following assumption.

Assumption 1: An $R_1 \in \mathbf{B}$ is given for which the equation $\mathbf{K}(P_1) = P_1 R_1$ must be satisfied, i.e., $\mathbf{K}$ is plant invariant $(\times R_1)$. By Theorem 7d), $R = R_1 + O_r$, where $O_r \in \mathbf{B}$ is any right zero of P_1. There are now two variables to be optimized, Q and O_r. Their simultaneous optimization can be difficult, and we make the following simplifying assumption.

Assumption 2: $O_r = 0$. Assumption 2 may constrain the class of allowable feedbacks and thereby give suboptimal sensitivities. However, it is clear from (8.7b) that there is no such constraint if the condition

$$(P - P_1)O_r = 0 \tag{8.13}$$

is fulfilled. Equation (8.13) is fulfilled whenever $\mathrm{N}(P_1)$ (nullspace of P_1) is contained in $\mathrm{N}(P)$ for all P in $b(P_1, \delta)$; for then, for any $u \in \mathcal{X}$, either $O_r u = 0$, or $O_r u \in \mathrm{N}(P_1)$ and so $O_r u \in \mathrm{N}(P)$. In either case, $(P - P_1)O_r u = 0$, i.e., (8.13) is true. If $\mathrm{N}(P_1)$ is trivial, then, of course, (8.13) is fulfilled. For example, $\mathrm{N}(P_1)$ is trivial for any nonzero P_1 in $\mathbf{H}_0^{\infty}$, because functions analytic and not identically zero in $\mathrm{Re}(s) \geq 0$ have at most a countable number of $\mathrm{Re}(s) \geq 0$ zeros.

As R in (8.6) is now fixed, cR can be absorbed into the weighting for some constant λ in $(0, 1]$ by replacing $\|\cdot\|_v$ with the new seminorm $\|C\|_v \triangleq \lambda \|CR\|_v$, and similarly for $\|\cdot\|_1$. Without loss of generality, we can therefore make the following assumption.

Assumption 3: $R=I$. We prefer to absorb R rather than to show it explicitly, as this approach reduces the number of variables and allows sensitivity to be related to weighted invertibility.

E. Minimal Sensitivity

For any P_1 in $\mathbf{B}_s$ and $\delta>0$, the *minimal sensitivity to plant perturbations* is defined to be

$$\nu(P_1,\delta)=\inf_{Q\in\mathbf{B}}\{\nu(P_1,Q,I;\delta)\}$$
$$=\inf_{Q\in\mathbf{B}}\sup_{P\in b(P_1,\delta)}\{\|\mathbf{E}(P,P_1,Q)(P-P_1)\|_v\delta^{-1}\}$$

provided the last sup is replaced by ∞ for nonstabilizing Q. For fixed P_1, $\nu(P_1,\delta)$ is a function of $\delta\geq 0$ which represents the dependence of perturbation sensitivity on plant uncertainty.

Theorem 8: For any nominal plant P_1 in $\mathbf{B}_s$, the minimal sensitivity to plant perturbations $\nu(P_1,\delta)$ is a monotone nondecreasing function of $\delta\geq 0$.

$\nu(P_1,\delta)$ satisfies the upper bound conditions $\nu(P_1,\delta)\leq\eta(P_1,\delta)$ and $\lim_{\delta\to 0}\nu(P_1,\delta)\leq\mu(P_1)$. If $\|\cdot\|_1=\|\cdot\|_r$, and $(\|\cdot\|_W,\|\cdot\|_1)$ are aligned, then $\lim_{\delta\to 0}\nu(P_1,\delta)=\mu(P_1)$ and, for $\delta=\|P_1\|_r$, $\|P_1\|_v\|P_1\|_r^{-1}\leq\nu(P_1,\delta)\leq\|I\|_W$.

Proof: $\nu(P_1,\delta)$ is a $\uparrow$ function of $\delta\geq 0$, as by definition it is a sup over Q in $\mathbf{B}_s$ of functions which are $\uparrow$ by Lemma 8.2. The inequality $\nu(P_1,\delta)\leq\eta(P_1,\delta)$ is obtained by taking $\sup_{Q\in\mathbf{B}}$ of both sides of (8.10). It follows by Theorem 5 that $\lim_{\delta\to 0}u(P_1,\delta)\leq\mu(P_1)$, and that $\nu(P_1,\delta)\leq\|I\|_W$ for $\delta=\|P_1\|_r$. If alignment and $\|\cdot\|_1=\|\cdot\|_r$ are assumed, then taking the $\inf_{Q\in\mathbf{B}}$ of both sides of (8.11) gives the inequality $\nu(P_1,\delta)\geq\mu(P_1)$, and the conclusion that $\lim_{\delta\to 0}\nu(P_1,\delta)=\mu(P_1)$ follows.

As the operator $P=0$ lies in $b(P_1,\|P_1\|_r)$ and, for each Q in $\mathbf{B}_s$, $\|E(0,P_1,Q)(-P_1)\|_v\|P_1\|_r^{-1}=\|P_1\|_v\|P_1\|_r^{-1}$; therefore $\|P_1\|_v\|P_1\|_r^{-1}\leq\nu(P_1,\|P_1\|_r)$. Q.E.D.

Remark: The ratio $\|P_1\|_v\|P_1\|_r^{-1}$ is a measure of the amount by which the $\|\cdot\|_v$ norm weighs down the nominal plant. Provided norms are used for which the weighting is not excessive, in fact whenever the ratio exceeds $\mu(P_1)$, there must be points in the interval $[0,\|P_1\|_r]$ at which any increase in plant uncertainty causes a worsening of minimal sensitivity.

Appendix I

Proof of Proposition 2.1:

a) If $(I+PQ)^{-1}$ exists in B, let $R\triangleq I-Q(I+PQ)^{-1}P$. Now, $(I+QP)R=I$, because

$$(I+QP)R=I-Q(I+PQ)^{-1}P+QP-QPQ(I+PQ)^{-1}P$$
$$=I-Q(I+PQ)^{-1}P+Q\left[I-PQ(I+PQ)^{-1}\right]P$$
$$=I-Q(I+PQ)^{-1}P+Q(I+PQ)^{-1}P=I.$$

Similarly, $R(I+QP)=I$, so R is the inverse of $(I+QP)$ in B. The required formula is obtained by multiplying both sides of the equation $(I+PQ)P=P(I+QP)$ by $(I+PQ)^{-1}$ on the left, and by $(I+QP)^{-1}$ on the right.

b) If P is in R and P^{-1} in B, there is a contradiction: any F in B belongs to R as $F=PP^{-1}F$, so R is not a proper subspace and not a radical.

c) Let $R\triangleq(I+P)^{-1}$ and observe that $R=I-PR$. By the triangle inequality, $\|R\|\leq 1+\|P\|\|R\|$ and, as $\|P\|<1$, $\|R\|\leq(1-\|P\|)^{-1}$. Q.E.D.

Proof of Proposition 3.5: Since $\|(I-PQ)\Pi\|=1$ when $Q=0$, the infimum α satisfies $\alpha\leq 1$. Suppose $\alpha\neq 1$; then there is a Q in $\mathbf{B}_s$ for which $\|(I-PQ)\Pi\|\triangleq\alpha_1<1$. It will now be shown that given any $\epsilon>0$, there is a Q_1 in $\mathbf{B}_s$ for which $\|(I-PQ_1)\Pi\|<\epsilon$, and so the proposition is true.

Let n be any integer for which $\alpha^n<\epsilon$. Now, $(I-PQ)^n=I+a_1PQ+\cdots+a_n(PQ)^n$ is a polynomial in PQ, in which every term except the first has P as a left factor, and Q in $\mathbf{B}_s$ as a factor. Therefore, there is a Q_1 in $\mathbf{B}_s$ for which the equation $(I-PQ)^n=(I-PQ_1)$ holds. Now as $\Pi\mathbf{B}$ is invariant under all operators in $\mathbf{B}$, we get $\|(I-PQ_1)\Pi\|=\|\{(I-PQ)\Pi\}^n\|\leq\|(I-PQ)\Pi\|^n\leq\alpha_1^n<\epsilon$, as claimed. Q.E.D.

Appendix II
Unstable Plants: Remarks

Consider the problem of input–output stabilizing an unstable plant P_0 in $\mathbf{H}^{\infty}_{e0}$ by identity feedback, given that the *a priori* information about P consists of $\hat{p}(j\omega)$ and the number N_+ of $\mathrm{Re}(s)\geq 0$ poles of P. By Nyquist's criterion, uncertainty as to N_+ translates into uncertainty[20] as to the number of unstable poles of $(I+P_0)^{-1}$. N_+ must not be underestimated if stability is to be assured. However, it is impossible to be sure that N_+ has not been underestimated from purely input–output measurements. For example, let s_i, $i=1,2,\cdots n$, be any finite set of frequencies, and $\hat{p}_m(s_i)$ a set of n measurements of $\hat{p}_0(s)$ of tolerance ϵ, i.e., $|\hat{p}_m(s_i)-\hat{p}_0(s)|<\epsilon$. It is impossible to deduce N_+ from the measurements, as it is always possible to find a dipole, $(s+\alpha)(s+\alpha+\delta)^{-1}$, with $\delta>0$ so small that multiplication of $\hat{p}_0(s)$ by the dipole changes $\hat{p}_0(s_i)$ by less than ϵ, but increases N_+ by 1. In our particular setting, it appears that input–output stabilization cannot be accomplished using solely input–output data. The implication is that some information concerning internal structure is essential, and that stabilization is not a legitimate problem for a purely input–output theory.

On the other hand, desensitization can be achieved using input–output data. If P is in H^{∞}_0, and $p(t)$ satisfies appropriate restrictions on smoothness and convergence to 0 as $t\to\infty$, then $\hat{p}(j\omega)$ has a finite modulus of continuity, and $\hat{p}$ can be located in a ϵ-ball of H^{∞} by a finite set of measurements [14]. This is sufficient for the purpose of desensitization, as sensitivity depends continuously on $\hat{p}$ in H^{∞} (see Theorems 5 and 7).

[20] A more complete discussion is in Zames and El-Sakkary [18].

Synopsis

The main results of this paper are the Theorems 1–7.

Section II: Spaces of frequency responses and algebras of input–output mappings are defined, and their relevant properties summarized.

Section III: A decomposition principle is derived. The disturbance attenuation problem is separated into two independent stages: stabilization, followed by desensitization of a stable system. The second stage is a model reference scheme.

Section III′: Some constraints on sensitivity norms are displayed. It is shown that the induced operator norm is useless as a measure of sensitivity.

Section IV: The concepts of a weighted seminorm, approximate inverse, and measure of singularity are introduced and illustrated by examples.

Section V: The plant is assumed to be stable and known precisely. Sensitivity is defined. Approximate invertibility is shown to be a necessary and sufficient condition for sensitivity reduction in Theorem 2, and optimal sensitivity is shown to be equal to the measure of singularity. An example of sensitivity minimization is solved in Section V-A. Unstable plants are discussed in Section V-B. Identity and inverting sequences are introduced in Section V-C, and a lemma on products of inverting sequences is proved.

Section VI: Devoted to multivariable systems. In Theorem 3, an optimal sequence of compensators is derived for a plant factorable into a product of nearly invertible and noninvertible factors. Corollary 6.2 specializes this result to a single-input single-output plant with a RHP zero. Corollary 6.2 gives conditions under which the sensitivity of multivariable systems without RHP zeros can be made arbitrarily small. Theorem 4 shows that sensitivity can never be made small if there are zeros in any heavily weighted part of the RHP.

Section VII: The plant is assumed to lie in a ball of uncertainty around a nominal value. Feedbacks stabilizing over a ball are defined. Optimal sensitivity is shown to be a monotone nondecreasing function of uncertainty in Theorem 5, and various bounds are obtained.

Section VIII: Problem 2 is formulated, again in terms of an equivalence between feedback and model reference schemes, this time subject to a plant invariance property, in Theorem 6. Sensitivity to plant perturbations is defined, bounded, and optimal sensitivity is shown to be a monotone nondecreasing function of plant uncertainty in Theorem 7.

Acknowledgment

The author thanks C. A. Desoer for many useful suggestions concerning the draft of this paper.

References

[1] H. W. Bode, *Network Analysis and Feedback Amplifier Design*. Princeton, NJ: Van Nostrand, 1945.

[2] M. Horowitz, *Synthesis of Feedback Systems*. New York: Academic, 1963.

[3] G. Zames, "Nonlinear operators—Cascading inversion, and feedback" M.I.T., *R.L.E., Quart. P.R.*, vol. 53, pp. 93–107, Apr. 1959.

[4] ——, "Nonlinear operators for system analysis," Res. Lab. Electron., M.I.T., Tech. Rep. 370, Sept. 1960.

[5] ——, "functional analysis applied to nonlinear feedback systems," *IEEE Trans. Circuit Theory*, vol. CT-10, pp. 392–404, Sept. 1963.

[6] C. A. Desoer, "Nonlinear distortion in feedback amplifiers," *IRE Trans. Circuit Theory*, vol. CT-9, pp. 2–6, Mar. 1962.

[7] ——, "Perturbation in the I/O map of a nonlinear feedback system caused by large plant perturbation," *J. Franklin Inst.*, vol. 306, pp. 225–235, Sept. 1978.

[8] J. B. Cruz, Jr. and W. R. Perkins, "A new approach to the sensitivity problem in multivariable feedback system design," *IEEE Trans. Automat. Contr.*, vol. AC-9, pp. 216–223, 1964.

[9] W. A. Porter and C. L. Zahm, "Basic concepts in system theory," System Eng. Lab., Univ. of Michigan, Tech. Rep. 33, 1969.

[10] R. M. Desantis and W. A. Porter, "A generalized Nyquist plot and its use in sensitivity analysis," *Int. J. Syst. Sci.*, vol. 5–12, pp. 1143–1153, 1974.

[11] J. C. Willems, *The Analysis of Feedback Systems*. Cambridge, MA: M.I.T., 1971.

[12] G. Zames, "Feedback hierarchies and complexity," in *Proc. IEEE Conf. Decision Contr.*, Dec. 1976, addenda.

[13] ——, "On the metric complexity of causal linear systems, ϵ-entropy and ϵ-dimension for continuous time," *IEEE Trans. Automat. Contr.*, vol. AC-12, pp. 222–230, Apr. 1979.

[14] ——, "Realizability conditions for nonlinear feedback systems," *IEEE Trans. Circuit Theory*, vol. CT-11, pp. 186–194, June 1964.

[15] R. G. Douglas, *Banach Algebra Techniques in Operator Theory*. New York: Academic, 1972.

[16] B. A. Francis and W. M. Wonham, "The internal model principle for linear multivariable regulators," *J. Appl. Math. Optimiz.*, vol. 2, pp. 170–194, 1975.

[17] D. M. Bensoussan, "High-gain feedback and sensitivity," Ph.D. dissertation, Dep. Elec. Eng., McGill Univ., to appear.

[18] A. K. El-Sakkary, "The gap metric for unstable systems," Ph.D. dissertation, Dep. Elec. Eng., McGill Univ.; also in G. Zames and A. K. El-Sakkary, *Proc. 18th Allerton Conf.*, Oct. 1980.

[19] M. A. Naimark, *Normed Rings*. Groningen: Noordhoff, 1964.

[20] A. Feintuch, "Strong Causality Conditions and Causal Invertibility," *SIAM J. Contr. Optimiz.*, vol. 18, pp. 317–324, May 1980.

[21] C. A. Desoer and W. S. Chan, "Feedback interconnection of lumped linear time invariant systems," *J. Franklin Inst.*, vol. 300, pp. 335–351, 1975.

[22] W. Rudin, *Functional Analysis*. New York: McGraw Hill, 1973.

[23] G. Zames, "Feedback and optimal sensitivity: Model reference transformations, weighted seminorms, and approximate inverses," in *Proc. 17th Allerton Conf.*, Oct. 1979, pp. 744–752.

Index

About the Editor

Tamer Başar received the B.S.E.E. from Robert College, Istanbul, Turkey in 1969 and the M.S., M.Phil., and Ph.D. degrees in engineering and applied science from Yale University during 1970–1972. After stints at Harvard University, Marmara Research Institute (Gebze, Turkey), and Boğaziçi University (Istanbul), he joined the University of Illinois at Urbana-Champaign (UIUC) in 1981, where he is currently the Fredric G. and Elizabeth H. Nearing Professor of Electrical and Computer Engineering and Director of the Decision and Control Laboratory.

Dr. Başar has authored or coauthored over 150 journal articles and book chapters, and also numerous conference publications in the general areas of optimal, robust, and adaptive control, large-scale and decentralized systems and control, dynamic games, stochastic control, estimation theory, stochastic processes, information theory, and mathematical economics. He is coauthor of *Dynamic Noncooperative Game Theory* (Academic Press, 1982; second edition 1995; the latest edition in the SIAM Series in Classics in Applied Mathematics, 1999) and *H^∞-Optimal Control and Related Minimax Design Problems* (Birkhäuser, 1991; second edition 1995); he has also edited or coedited several volumes on control and dynamic games. His current research interests are robust nonlinear and adaptive control, control of communication networks, risk-sensitive estimation and control, and robust identification.

Dr. Başar has played active roles in many scientific organizations and, in particular, IEEE and IFAC (International Federation of Automatic Control). Currently, he is the president of the IEEE Control Systems Society, the deputy editor-in-chief of the IFAC journal *Automatica*, the managing editor of the *Annals* of the International Society of Dynamic Games, a member of the IEEE Publication Board, and member of editorial boards of various journals. A member of the National Academy of Engineering and a Fellow of IEEE, some recent awards and recognitions Dr. Başar has received are the Medal of Science of Turkey (1993), Distinguished Member Award (1993) and Axelby Outstanding Paper Award (1995) of the IEEE Control Systems Society, and Nearing Distinguished Professorship at UIUC (1998).